W9-AQR-788

# EVOLUTIONARY BIOLOGY

# EVOLUTIONARY BIOLOGY SECOND EDITION

Douglas J. Futuyma
STATE UNIVERSITY OF NEW YORK AT STONY BROOK

SINAUER ASSOCIATES, INC. • PUBLISHERS
Sunderland, Massachusetts

EVOLUTIONARY BIOLOGY, Second Edition



For information address Sinauer Associates, Inc., Sunderland, Massachusetts 01375

Printed in U.S.A.

Sources of the scientists' photographs appearing in Chapter 1 are gratefully acknowledged:

Photographs of C. Darwin and A. R. Wallace courtesy of The American Philosophical Society Library

Photograph of R. A. Fisher courtesy of Joan Fisher Box

Photograph of J. B. S. Haldane courtesy of Dr. K. Patau

Photograph of S. Wright courtesy of Doris Marie Provine

Photograph of J. Huxley from the papers of Julian Sorrell Huxley, Woodson Research Center, Rice University Library

Photograph of E. Mayr courtesy of Harvard News Service and E. Mayr

Photograph of G. L. Stebbins, G. G. Simpson, and Th. Dobzhansky courtesy of G. L. Stebbins

**Library of Congress**
**Cataloging-in-Publication Data**

Futuyma, Douglas J., 1942–
Evolutionary biology.
Bibliography: p.
Includes index.
1. Evolution. I. Title.
QH366.2.F87 1986 575 86-15531
ISBN 0-87893-188-0
ISBN 0-87893-183-X (International student ed.)

6 5 4

*To those graduate students, present and past,*
*friends and colleagues, who give me length of days*

## ABOUT THE COVER

Phylogenetic (cladistic) relationships among the subfamilies and tribes of swallowtail butterflies, according to an analysis by James S. Miller. The study of the systematics of this group has figured in the analysis of coevolutionary relationships between herbivorous insects and their host plants. Photographs courtesy of J. S. Miller.

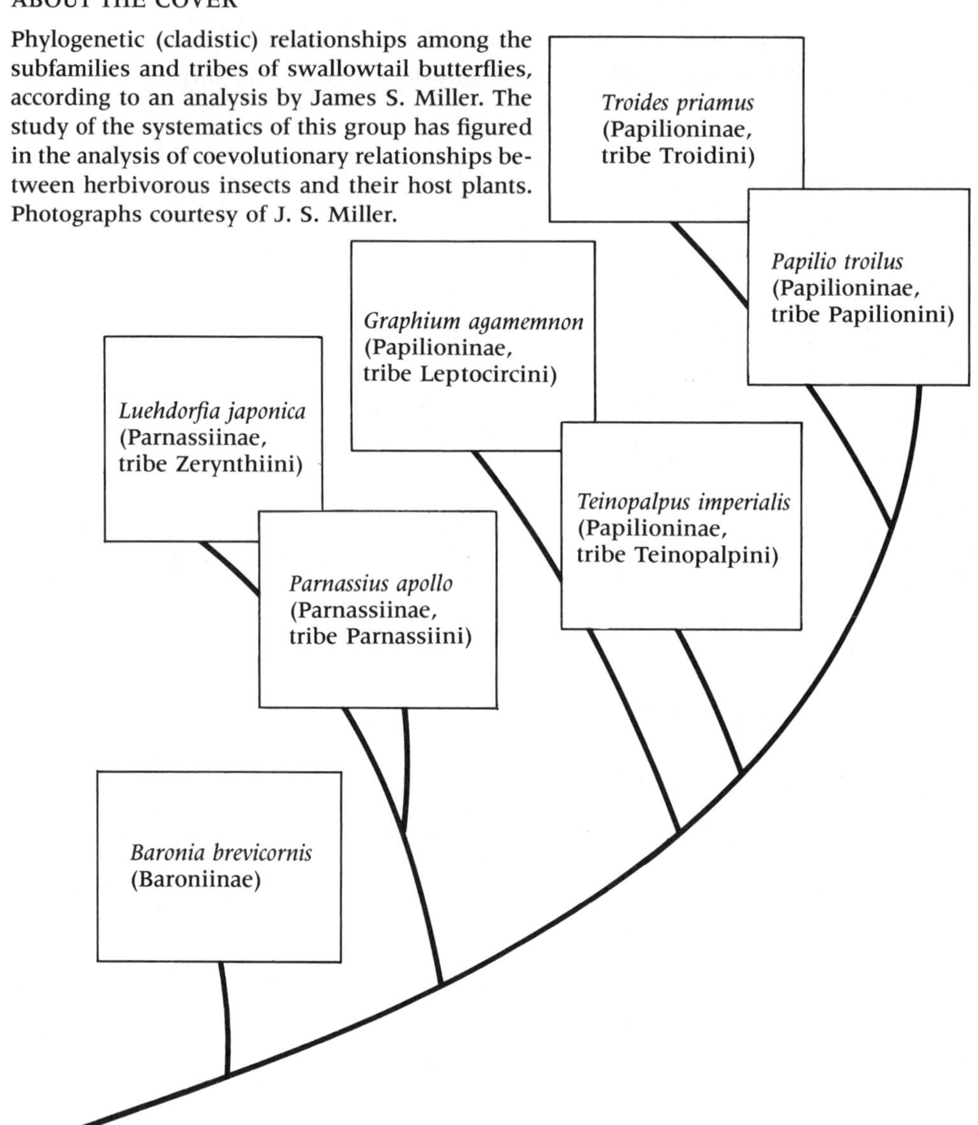

# CONTENTS


*Preface* xi

**Chapter One**
**The Origin and Impact of Evolutionary Thought** **1**

ORIGINS OF EVOLUTIONARY THOUGHT 2
*THE ORIGIN OF SPECIES* 6
CONCEPTIONS AND MISCONCEPTIONS OF EVOLUTION 7
EVOLUTION SINCE DARWIN 8
THE MODERN SYNTHESIS 10
EVOLUTION SINCE THE SYNTHESIS 13
HOW EVOLUTION IS STUDIED 13
EVOLUTION AS FACT AND THEORY 15

**Chapter Two**
**The Ecological Context of Evolutionary Change** **18**

ADAPTATION AND ENVIRONMENT 19
THE ECOLOGICAL NICHE 19
SPATIAL DISTRIBUTION 21
POPULATION GROWTH 21
THE EFFECT OF DENSITY ON POPULATION GROWTH 24
THE BIOTIC ENVIRONMENT: PREDATORS AND PREY 26
BENEFICIAL INTERACTIONS AMONG SPECIES 29
COMPETITION AMONG SPECIES 30
COMPLEX INTERACTIONS AMONG SPECIES 34
DIVERSITY AND STABILITY OF COMMUNITIES 35
ENVIRONMENTAL PATTERNS 37

**Chapter Three**
**Heredity: Fidelity and Mutability** **42**

TWO PRINCIPLES OF GENETICS 43
THE GENETIC MATERIAL 45
GENE STRUCTURE 47
REPEATED AND NONREPEATED DNA 48
REPLICATION, RECOMBINATION, AND SEGREGATION 50
GENOTYPE AND PHENOTYPE 53
THE CONTROL OF GENE EXPRESSION 56
DEVELOPMENT 58
MUTATION: THE ORIGIN OF GENETIC VARIATION 60
CHANGES IN THE KARYOTYPE 60
GENE MUTATIONS 65
RATES OF MUTATION 72
PHENOTYPIC EFFECTS OF MUTATIONS 75
THE RANDOMNESS OF MUTATIONS 76
RECOMBINATION: THE AMPLIFICATION OF VARIATION 76
EXTERNAL SOURCES OF VARIATION 78

**Chapter Four**
**Variation** **81**

THE HARDY-WEINBERG THEOREM 82
VARIATION IN QUANTITATIVE TRAITS 87
VARIATION WITHIN NATURAL POPULATIONS 90
VARIATION IN PROTEINS 96
THE ORGANIZATION OF GENETIC VARIATION 99
GENETIC VARIATION AMONG POPULATIONS 102
GEOGRAPHIC VARIATION 103
SPECIES 111
INTRASPECIFIC VARIATION AND HIGHER TAXONOMIC CATEGORIES 116

**Chapter Five**
**Population Structure and Genetic Drift** **119**

THE THEORY OF INBREEDING 120
THE GENETIC STRUCTURE OF INBRED POPULATIONS 127
POPULATION SIZE, INBREEDING, AND GENETIC DRIFT 129
EFFECTIVE POPULATION SIZE 131
MUTATIONS IN FINITE POPULATIONS 133
THE FOUNDER EFFECT 133
GENE FLOW 136
EFFECTIVE POPULATION SIZES AND GENE FLOW IN NATURAL POPULATIONS 139
GENETIC DRIFT IN NATURAL POPULATIONS 142
EVOLUTION BY GENETIC DRIFT 143
NONRANDOM MATING BASED ON PHENOTYPE 146

**Chapter Six**
**Effects of Natural Selection on Gene Frequencies** 149

DIFFERENTIAL SURVIVAL AND REPRODUCTION 150
INDIVIDUAL SELECTION 150
THE EFFECT OF ENVIRONMENT ON FITNESS 152
LEVELS OF SELECTION 153
MODES OF SELECTION 154
CONSTANT FITNESSES AND DIRECTIONAL SELECTION 155
ACCOUNTING FOR GENETIC VARIATION 159
INFERIOR HETEROZYGOTES 170
THE ADAPTIVE LANDSCAPE 172
INTERACTIONS OF EVOLUTIONARY FACTORS 172
POPULATION FITNESS AND GENETIC LOAD 175
THE NEUTRALIST–SELECTIONIST CONTROVERSY 177
ESTIMATES OF THE STRENGTH OF NATURAL SELECTION 181

**Chapter Seven**
**Selection on Polygenic Characters** 184

TWO LOCI 185
DIRECTIONAL SELECTION AT TWO LOCI 188
MULTIPLE EQUILIBRIA 188
POLYGENIC INHERITANCE 195
HERITABILITY AND THE RESPONSE TO SELECTION 200
GENETIC CORRELATIONS 205
RESPONSES TO ARTIFICIAL SELECTION 207
GENETIC AND DEVELOPMENTAL HOMEOSTASIS 210

**Chapter Eight**
**Speciation** 218

THE BIOLOGICAL SPECIES CONCEPT 219
THE GENETICS OF SPECIES DIFFERENCES 219
MODES OF SPECIATION 223
ALLOPATRIC SPECIATION 223
PARAPATRIC SPECIATION 227
SYMPATRIC SPECIATION 228
GENETIC THEORIES OF SPECIATION 231
THE FOUNDER EFFECT 238
SELECTION FOR REPRODUCTIVE ISOLATION 242
TIME REQUIRED FOR SPECIATION 244
THE SIGNIFICANCE OF SPECIES AND SPECIATION 246

**Chapter Nine**
**Adaptation** 250

PROBLEMS IN RECOGNIZING ADAPTATION 251
THE ADAPTATIONIST PROGRAM 254
LEVELS OF SELECTION 258
GROUP SELECTION 264
THEORETICAL APPROACHES TO MODELING ADAPTATION 266
ADAPTATION: SPECIAL TOPICS 272
THE EVOLUTION OF LIFE HISTORY CHARACTERISTICS 272
SEXUAL SELECTION 276
THE EVOLUTION OF RECOMBINATION AND SEX 279

**Chapter Ten**
**Determining the History of Evolution** 285

DEFINITIONS 286
CLASSIFICATION 287
CONTENDING SCHOOLS OF SYSTEMATICS 288
DIFFICULTIES OF PHYLOGENETIC INFERENCE 292
INFERRING PHYLOGENY FROM MORPHOLOGICAL DATA 299
PHYLOGENETIC INFERENCE FROM MACROMOLECULES 307

**Chapter Eleven**
**The Fossil Record** 317

DATING THE PAST 318
THE HISTORY OF THE EARTH 319
THE ORIGIN OF LIFE 322
PRECAMBRIAN LIFE 323
THE PALEOZOIC ERA 324
THE MESOZOIC ERA 334
THE CENOZOIC ERA 340

**Chapter Twelve**
**The History of Biological Diversity** 345

CHANGES IN DIVERSITY 346
IS DIVERSITY REGULATED? 351
PATTERNS OF ORIGINATION 354
PATTERNS OF EXTINCTION 359
THE DISTRIBUTION OF EXTINCTION RATES 360
MASS EXTINCTIONS 364
TRENDS IN EVOLUTION 366

**Chapter Thirteen**
**Biogeography** 373

THE IMPORTANCE OF PHYLOGENETIC ANALYSIS 374
GEOGRAPHIC PATTERNS 375

CAUSES OF GEOGRAPHIC DISTRIBUTIONS 378
EVIDENCE USED IN HISTORICAL BIOGEOGRAPHY: PALEONTOLOGY 381
EVIDENCE USED IN HISTORICAL BIOGEOGRAPHY: SYSTEMATICS 382
HISTORY AND THE COMPOSITION OF REGIONAL BIOTAS 386
ARE COMMUNITIES IN EQUILIBRIUM? 388
REGIONAL VARIATIONS IN SPECIES DIVERSITY 391
THE ORIGINS OF DOMINANT GROUPS 392

**Chapter Fourteen**
**The Origin of Evolutionary Novelties 396**

RATES OF EVOLUTION 397
PUNCTUATED EQUILIBRIUM 401
REGULARITIES OF PHENOTYPIC EVOLUTION 409
ALLOMETRY AND HETEROCHRONY 412
THE ORIGIN OF HIGHER TAXA 419
THE ADAPTIVE CONTEXT OF EVOLUTIONARY INNOVATIONS 423
GENETICS, DEVELOPMENT, AND EVOLUTION 425
THE GENETIC AND DEVELOPMENTAL BASIS OF MORPHOLOGICAL EVOLUTION 425
HOMEOTIC MUTATIONS IN DROSOPHILA 430
CONSERVATISM AND CHANGE IN DEVELOPMENTAL PROGRAMS 435
EVOLUTIONARY CONSTRAINTS AND PHENOTYPIC GAPS 436
DEVELOPMENTAL INTEGRATION AND MACROEVOLUTION 439
NEO-DARWINISM AND ITS CRITICS 440

**Chapter Fifteen**
**Evolution at the Molecular Level 443**

THE USES OF MOLECULAR INFORMATION IN EVOLUTIONARY STUDIES 444
TECHNIQUES 445
VARIATION IN SINGLE DNA SEQUENCES 447
RATES OF SEQUENCE EVOLUTION 447
EVOLUTIONARY CHANGES IN THE LOCATION AND NUMBERS OF GENES 451
UNEQUAL CROSSING OVER AND THE EVOLUTION OF DUPLICATE GENES 451
MOBILE GENETIC ELEMENTS 456
EFFECTS OF TRANSPOSABLE ELEMENTS 457
EVOLUTION OF THE SIZE OF THE GENOME 459
THE EVOLUTION OF GENE FAMILIES 464
ADAPTIVE EVOLUTION FROM A MOLECULAR PERSPECTIVE 472
EVOLUTION OF GENES AND PROTEINS 472
HORIZONTAL GENE TRANSFER 478
MOLECULAR BIOLOGY AND EVOLUTIONARY BIOLOGY 478

**Chapter Sixteen**
**The Evolution of Interactions Among Species 482**

COEVOLUTION 483
THE EVOLUTION OF RESOURCE UTILIZATION 484
COEVOLUTION OF COMPETING SPECIES 485
EVOLUTION OF PREDATOR–PREY RELATIONSHIPS 492
MUTUALISM 497
GENETIC STUDIES OF COEVOLUTION 500
EVOLUTION AND THE STRUCTURE OF COMMUNITIES 502

**Chapter Seventeen**
**Human Evolution and Social Issues 505**

THE PROBLEM OF OBJECTIVITY 506
THE PHYLOGENETIC POSITION OF THE HUMAN SPECIES 507
THE HOMINOID FOSSIL RECORD 511
CULTURAL EVOLUTION 516
THE PHYSICAL AND MENTAL EVOLUTION OF THE HUMAN SPECIES 518
GENETIC VARIATION WITHIN POPULATIONS 519
EVOLUTION AND HUMAN BEHAVIOR 528
TWO VIEWS OF HUMAN NATURE 529
VARIATION IN BEHAVIORAL TRAITS 532
VARIATION IN INTELLIGENCE 533
EVOLUTION AND SOCIETY 535

*Appendix I*
*Means, Variances, and Correlations* 541

*Appendix II*
*List of Symbols* 547

*Chapter-opening illustration credits* 549

*Glossary* 550

*Literature Cited* 557

*Index* 590

# PREFACE

In 1979, when the first edition of this book appeared, evolutionary biology was flourishing more actively than it had been for several decades; since then it has become even more exuberant, more commanding of the full sweep of biology. An evolutionary approach has become *de rigueur* in ecology and behavior; paleontology and systematics have grown in strength and have begun to reforge their bonds with population biology; evolutionary morphology, physiology, and development are entering a renaissance; a field of molecular evolution has grown into adolescence, if not beyond; the annual volume of information and theory on subjects old and new has grown so that new journals have emerged to fill the need. If seven years ago I felt presumptuous in attempting a textbook on evolutionary biology, today I feel the same with even greater force. Were it not for the forbearance of colleagues whom I have badgered shamelessly for information, I could not have hoped to face so strong a tide of information and ideas.

Every subject in biology, and in science generally, bears in it the excitement of discovery and of the birth and growth of ideas. Evolutionary biology has its own special rewards as well: the esthetic satisfaction, to those who seek it, of taking as its subject not only general principles but the study of the diversity of living things; the intellectual satisfaction of immersion in the most philosophical of the biological sciences and the most synthetic (comprehending and unifying, as it does, all of biology from the molecular to the ecological); the intellectual challenge of grappling with quesitons that, because they so often must be approached by logic or circumstantial evidence rather than by direct observation, may never be fully answered. With every year, new evidence or, just as often, new ways of viewing old evidence, forces us to reconsider entrenched ideas, to refine or abandon old hypotheses, to stay intellectually young.

I have tried to convey the dynamic state of evolutionary biology by treating controversial issues, by raising unresolved problems, by presenting plausible arguments only to finish them with a querulous "but," and by posing questions after each chapter that often admit no easy answers but are nonetheless an integral and important part of the book. As in the first edition, the text provides not a clean enumeration of facts and principles that pretend to eternal verity, but essays in reasoning, evidence, and uncertainty that some students may find difficult, but which I hope capture more faithfully the reality of this and every other science. We deal, after all, not with irrefutable facts but with hypotheses that may be cast aside by tomorrow's experiments; not with unquestionable principles but with-concepts formulated by the fallible human mind. If science teaches us anything, it should teach us to doubt, to question every statement, however authoritative the source.

This book is designed for use in advanced undergraduate and beginning graduate courses. I have tried also to serve graduate students and biologists in other fields by providing an entry into the literature on as many evolutionary topics as I felt capable of addressing. I have assumed that the reader has some knowledge of genetics (other than population genetics, which I develop from first principles). The mathematics in the text is kept to a minimum, and does not go beyond college algebra; a bit of calculus will be encountered in some of the boxed material, to which most mathematical derivations are relegated. The appendix on elementary statistics should be read by those to whom statistical concepts are unfamiliar. It will be useful to come to this book with some elementary knowledge of the essential principles of evolution, and of ecology, physiology, biochemistry, and development. It is hardly possible to appreciate the full scope of evolutionary biology without some familiarity with the taxonomy, anatomy, and natural history of plants and/or animals, but biology curricula these days unfortunately include less and less of this fundamentally important information, so I have done my best to define terms and to use familiar organisms as examples.

In this edition, I have tried to cover recent developments as fully as possible, and have expanded the material on some subjects (especially paleontology and molecular evolution) that the first edition treated inadequately. The first edition was structured to begin with what I thought was concrete and familiar—organisms—and to proceed through the abstract and unfamiliar—genetics—back to organisms, in a "commodious vicus of recirculation," as Joyce explained in *Finnegan's Wake.* I have regretfully concluded that this conceit is not the best for our age, in part because students are generally less familiar now with organisms than with DNA, and in part because almost every course on evolution begins with genetics and works up to historical evolution. Acceding to the times, I have reorganized the sequence of topics. The book begins, as before, with a history of the subject and with elementary ecology and genetics, but then proceeds through the genetics of evolutionary change to speciation and adaptation, on to historical evolution (systematics, paleontology, biogeography) and then to a historical, genetic, and developmental perspective on macroevolution. It ends with chapters on the special topics of molecular evolution, coevolution, and human evolution.

As before, my deepest debt is to my friends, colleagues, and students, too numerous to name, from whom I have learned so much, and who have given so generously of their knowledge and insights. I am very grateful to James Ajioka, Jody Hey, and Gabriel Moreno for reading several of the chapters in manuscript, and to the reviewers who provided immeasurably useful criticism, information, and advice: James Brown, Ted Case, David Jablonski, Malcolm Kottler, Russell Lande, Jack Sepkoski, Michael Wade, David Wake, Bruce Walsh, Kenneth Weiss, and David Wilson. John Leguyader helped greatly in compiling the illustrations, and Andy Sinauer, Carol Wigg, and Joe Vesely have provided guidance and material support at every stage. I am grateful to the Section of Ecology and Systematics, Cornell University, for its hospitality, and especially to Bob Bouma, Paul Feeny, Jim Liebherr, Amy McCune, Karl Niklas, and Deborah Rabinowitz for their friendship and support during the sabbatical leave on which I wrote much of the text. Above all, I give thanks to the faculty and graduate students of the Department of Ecology and Evolution at Stony Brook for their intellectual vitality and especially their friendship.

DOUGLAS J. FUTUYMA

# EVOLUTIONARY BIOLOGY

# The Origin and Impact of Evolutionary Thought

# Chapter One

"Old ideas give way slowly, for they are more than abstract logical forms and categories. They are habits, predispositions, deeply engrained attitudes of aversion and preference. Moreover, the conviction persists—though history shows it to be a hallucination—that all the questions that the human mind has asked are questions that can be answered in terms of the alternatives that the questions themselves present. But in fact intellectual progress usually occurs through sheer abandonment of questions together with both of the alternatives they assume—an abandonment that results from their decreasing vitality and a change of urgent interest. We do not solve them: we get over them. Old questions are solved by disappearing, evaporating, while new questions corresponding to the changed attitudes of endeavor and preference take their place. Doubtless the greatest dissolvent in contemporary thought of old questions, the greatest precipitant of new methods, new intentions, new problems, is the one effected by the scientific revolution that found its climax in the *Origin of Species*."

So concluded the philosopher John Dewey in his essay "The Influence of Darwin on Philosophy" (1910). A century after the publication of Darwin's book, philosophers could still affirm that "there are no living sciences, human attitudes, or institutional powers that remain unaffected by the ideas that were catalytically released by Darwin's work" (Collins 1959).

The theory of biological evolution is the mature expression of two revolutionary streams of thought antithetical to a world view that had long prevailed. First, the concept of a changing universe had been replacing the long-unquestioned view of a static world, identical in all essentials to the Creator's perfect creation. Darwin more than anyone else extended to living things, and to the human species itself, the conclusion that mutability, not stasis, is the natural order.

Second, people had long sought the causes of phenomena in purposes: the will of God, or Aristotelian final causes (the purposes for which events occur) rather than efficient causes (the mechanisms that cause events to occur). But Darwin showed that material causes are a sufficient explanation not only for physical phenomena, as Descartes and Newton had shown, but also for biological phenomena with all their seeming evidence of design and purpose. By coupling undirected, purposeless variation to the blind, uncaring process of natural selection, Darwin made theological or spiritual explanations of the life processes superfluous. Together with Marx's materialistic theory of history and society and Freud's attribution of human behavior to influences over which we have little control, Darwin's theory of evolution was a crucial plank in the platform of mechanism and materialism—of much of science, in short—that has since been the stage of most Western thought.

## ORIGINS OF EVOLUTIONARY THOUGHT

In every scientific discipline, the prevalent ideas and even the questions asked are the products of a historical development. Thus to understand the concerns of modern evolutionary biology, it is essential to known something of the history of the subject.

Although the notion of a dynamic world was not foreign to the ancient Greeks, the nonstatic, largely mythological explanation of the origins of living things offered by Empedocles and Anaximander gave way to the philosophy of Plato, which was incorporated into Christian theology and had a permanent and dominant effect on subsequent Western thought. Foremost in Platonic philosophy was the concept of ειδος, the "form" or "idea," a transcendent ideal form imperfectly imitated by its earthly representatives. The "idea" is an eternal, unchanging essence; thus despite variation among triangles, the sum of the angles of any triangle is 180°, and this "essential" property of triangles distinguishes them absolutely from rectangles. The triangles or horses we see in the material world, according to Plato, are but imperfect copies of the true, perfect Triangle and Horse that exist in the transcendent world of ideas. In this philosophy of essentialism, variation has no meaning; only essences matter.

Christian theology adopted an almost literal interpretation of the Bible, including special creation (the direct creation of all things in effectively their present form). But it also incorporated Platonic essentialism in the concept of plenitude (Lovejoy 1936). The eternal, unchanging essences of all things exist in the mind of God, but it would be an imperfection in God to deny material existence to something of which He conceived. Since God is perfect, He must have bestowed existence on everything that existed as His idea. Because to deny existence to anything at any time would introduce imperfection into His creation, all things must have been created in the beginning, and nothing God saw fit to create could have become extinct. Moreover, because order is clearly superior to disorder, God's creations must fit a pattern: the *Scala Naturae*, or Great Chain of Being. This "ladder of life," perceived in the gradation from inanimate material through plants, "lower animals," and humans to angels and other spiritual beings, must be perfect and have no gaps; it must be permanent and unchanging, and every being must have its fixed place according to God's plan. Since this natural order was created by a perfect God, that which is natural is good, and this must be the best of possible worlds. This natural hierarchy extended to higher and lower social classes in human societies. To aspire to change the social order must be immoral, and biological evolution is unthinkable.

The role of natural science in this view was to catalogue the links of the Great Chain of Being and to discover their order so that the wisdom of God could be revealed and appreciated. "Natural theology," as in John Ray's "The Wisdom of God Manifested in the Works of Creation" (1691), held up the adaptations of organisms as evidence of the Creator's beneficence. The profoundly influential work of Linnaeus in classification (*Systema Naturae* 1735; *Species Plantarum* 1753) was likewise undertaken *ad majorem Dei gloriam*, "for the greater glory of God."

These traditional views gave way under the development of empirical science. Hallowed concepts such as the central position of the earth in the universe were challenged. Newton, Descartes, and others developed strictly mechanistic theories of physical phenomena. By the end of the eighteenth century, the concept of a changing world was applied to astronomy by Kant and Laplace, who entertained notions of stellar evolution; to geology as evidence of changes in the earth's crust and of the extinction of species came to light; and to human affairs as the Enlightenment introduced the ideas of progress and human perfectibility.

Geologists came to recognize that sedimentary rocks had been laid down at different times, and began to realize that the earth might be very old; the great French naturalist Buffon suggested in 1779 that it might be as much as 168,000 years old. Fossils that characterized different strata were widely held to reflect a succession of catastrophes such as great floods; some held that there had been numerous successive creations as well. By 1788, however, James Hutton had developed the principle of UNIFORMITARIANISM, which held that the same processes are responsible for both past and present events. This implied that the earth was very old, with "no vestige of a beginning—no prospect of an end," as Hutton put it. Uniformitarianism was vigorously championed by the great geologist Charles Lyell, whose *Principles of Geology* (1830–1833) greatly influenced Darwin to adopt a uniformitarian view of geological and biological change, even though Lyell himself espoused a steady-state view of the earth that did not admit biological evolution (Mayr 1982a).

By the late eighteenth century, the possibility not only of successive special creations, but of the continuing origin of new species by more natural means was bruited about. Maupertuis, Diderot, and Goethe, among others, entertained speculations that new forms of life might originate either by spontaneous generation from inanimate matter, or by the unfolding (the literal meaning of "evolution") of immanent potentialities within existing species. "Evolution" in this sense meant not the modification of species, but the manifestation of essences that lay dormant in earlier species. Only Buffon, in 1766, articulated the possibility that different species had arisen by variation from common ancestors, but he immediately provided evidence against this position.

The earliest uncompromising advocate of evolution was Jean-Baptiste de Lamarck (1744–1829), who first presented an extended exposition of his theory in the *Philosophie Zoologique* (1809). Lamarck argued not that different living things have descended from common ancestors, but that lowly forms of life arise continually from inanimate matter by spontaneous generation, and progress inevitably toward greater complexity and perfection by "powers conferred by the supreme author of all things"—that is, by an inherent tendency toward complexity. Lamarck held that the particular path of progression taken is guided by the environment, and that a changing environment alters the needs of the organism, to which the organism responds by changing its behavior, and consequently uses some organs more than others. In other words, use and disuse alter morphology, which is transmitted to subsequent generations. This theory clearly would apply more to animals than to plants.

Lamarck is unfortunately and unjustly remembered mostly as someone who was wrong. But the inheritance of acquired characters on which his theory depended was not original with him; it was a widely held belief that Darwin himself incorporated into *The Origin of Species*. Lamarck merits respect as the first scientist who fearlessly advocated evolution and attempted to provide a mechanism to explain it. His ideas were almost universally rejected not because he embraced the inheritance of acquired characters, but because leading naturalists of the day found no evidence for evolution. In particular, Georges Cuvier (1769–1832), the founder of comparative anatomy and one of the most highly respected biologists and paleontologists of the nineteenth century, devastated Lamarck by arguing that the fossil record did not reveal gradual intermediate series of ancestors and descendants, and that organisms are so harmoniously constructed and perfectly adapted that any change would destroy the integrity of their organization. Lyell too, in his *Principles of Geology*, marshalled evidence against evolution in general and Lamarck in particular.

In part because of Lamarck, evolution was a common topic of discussion by

JEAN-BAPTISTE DE LAMARCK

the middle of the nineteenth century (Lovejoy 1959). Robert Chambers's anonymously published "Vestiges of the Natural History of Creation" (1844), for example, was a widely read, rather fanciful evolutionary tract that advanced Lamarckian ideas. But the evidence for evolution had not been marshalled in full, and since Lamarck had been discredited, no satisfactory mechanism of evolution was recognized. Curiously, the concept of the "struggle for existence" was used to account for extinction, but no one recognized that it could explain the modification of species except William Wells (1818) and Patrick Matthew (1831), who described the concept of natural selection almost as afterthoughts in publications that were devoted to other topics and were little read.

The career of Charles Robert Darwin (1809–1882) began with his voyage on H.M.S. *Beagle* (December 27, 1831–October 2, 1836) as ship's naturalist. Apparently an orthodox member of the Church of England during the voyage, he seems not to have accepted the notion of evolution until March 1837, when the ornithologist John Gould pointed out that Darwin's specimens of mockingbirds (not finches) from the Galápagos Islands were so distinct from one island to another as to represent different species (Sulloway 1979). This revelation seems to have led Darwin to doubt the fixity of species, and to set about gathering evidence on the "transmutation of species." He was concerned not only to amass evidence for evolution, but to conceive of a mechanism that could account for it. The theory of natural selection began to emerge on September 28, 1838 when, as Darwin recounts in his autobiography, "I happened to read for amusement Malthus on *Population,* and being well prepared to appreciate the struggle for existence which everywhere goes on from long-continued observation of the habits of animals and plants, it at once struck me that under these circumstances favourable variations would tend to be preserved and unfavourable ones to be destroyed." Darwin's reading of Malthus's *Essay on the Principle of Population* (1798), which argued that unchecked growth of the human population must lead to famine,

CHARLES ROBERT DARWIN

ALFRED RUSSEL WALLACE

may well have been an integral part of his search for a mechanism of evolution, rather than an "amusement" (M. Kottler, personal communication).

Twenty years passed between this memorable event and Darwin's first publication on the subject. Perhaps out of fear of the kind of hostility that Lamarck's and Chambers's speculations had met, Darwin occupied himself with acquiring evidence on evolution and worked on a four-volume taxonomic monograph on barnacles that occupied him for eight years. In 1844 he wrote but did not publish an essay on natural selection, and in 1856 began work on what was to be his "big book," *Natural Selection*. But this book was never completed, for in June 1858 he received a manuscript entitled "On the Tendency of Varieties to Depart Indefinitely from the Original Type" from a young naturalist, Alfred Russel Wallace (1823–1913). Wallace, who had been making a living by collecting biological specimens in South America and the Malay Archipelago, had independently conceived of natural selection. At the urging of his friends Charles Lyell and Joseph Hooker, Darwin had extracts from his 1844 essay presented along with Wallace's manuscript at the Linnaean Society of London on July 1, 1858; neither the presentation nor the publication of the essays evoked much response at all. Following his friends' advice, Darwin then published an "abstract" of his big book on November 24, 1859 under the title *The Origin of Species by Means of Natural Selection, or The Preservation of Favoured Races in the Struggle for Life*—a book that sold out its first printing in one day and set in motion a controversy that has still not entirely subsided.

### *THE ORIGIN OF SPECIES*

Wallace's theory of natural selection was as carefully reasoned as Darwin's, so he fully deserves credit as the co-discoverer of the chief mechanism of evolution. But although Wallace continued to work on evolutionary topics, especially biogeography, for much of his life, he neither presented a sweeping synthesis as Darwin did in *The Origin of Species* nor explored the ramifications of evolution as Darwin did in his numerous later works. It has been cogently argued (e.g., Ghiselin 1969) that all of Darwin's books, ranging from *The Various Contrivances by which Orchids are Fertilised by Insects* to *The Formation of Vegetable Mould, through the Action of Worms,* explore the ideas and principles that inhere in *The Origin of Species.*

*The Origin of Species* has two separate theses: that all organisms have descended with modification from common ancestors, and that the chief agent of modification is the action of natural selection on individual variation. Darwin was the first to marshal on so grand a scale the evidence for the first thesis, the historical reality of evolution, by drawing on all relevant sources of information: the fossil record, the geographic distribution of species, comparative anatomy and embryology, and the modification of domesticated organisms. Much of his argument consists of showing how naturally observations in these areas, such as the vestigial wings of flightless beetles, follow from the supposition of common ancestry, and how implausible they are under the hypothesis of special creation. In developing his arguments, Darwin was among the first to use what has come to be called the HYPOTHETICO-DEDUCTIVE METHOD, whereby a hypothesis is tested by determining whether the deductions drawn from it conform to observation. This was not a generally accepted methodology in Darwin's day, when science was supposed to be done by induction (drawing conclusions, as if they were self-evident, from an accumulation of individual observations), but it is now generally considered the most powerful method of science (see Ghiselin 1969 and Hull 1973 on the complex subject of the philosophical aspects of Darwin's methods).

As we have seen, the idea of evolution was not original with Darwin; but the evidence he provided was so strong that it elicited violent reaction from the opposition (especially on religious grounds) and converted all but a handful of prominent scientists to belief in evolution within twenty years. However, his truly original idea, the mechanism of natural selection, convinced few, and indeed fell deeper and deeper into disrepute until the late 1920s. The concept of the "struggle for existence" had been used to explain the extinction of species, but as long as species were viewed as Platonic essences or "types," selection could only eliminate the inferior, not give rise to novelty. The insight of Darwin and Wallace lay in recognizing that variation among individual organisms of a species was not mere imperfection, but the material from which selection could fashion better adapted forms of life. Darwin transformed the Malthusian principle of competition by applying it not only to competition among species, but to competition among individual organisms within a species. Darwin's replacement of essentialism by an emphasis on variation—what Mayr (1976, Chapters 3 and 19; 1982a) has called "population thinking"—was the foundation of his theory and his most revolutionary contribution to biology.

## CONCEPTIONS AND MISCONCEPTIONS OF EVOLUTION

Like all important concepts, evolution generates controversy; like many important concepts, it has been used as a foundation or just as a rationale for philosophical, ethical, or social views. In the broadest sense, evolution is merely change, and so is all-pervasive; galaxies, languages, and political systems all evolve. BIOLOGICAL EVOLUTION (or ORGANIC EVOLUTION) is change in the properties of populations of organisms that transcends the lifetime of a single individual. The ontogeny of an individual is not considered evolution; individual organisms do not evolve. The changes in populations that are considered evolutionary are those that are inheritable via the genetic material from one generation to the next. Biological evolution may be slight or substantial; it embraces everything from slight changes in the proportion of different alleles within a population (such as those determining blood types) to the successive alterations that led from the earliest protoorganism to snails, bees, giraffes, and dandelions.

The several mechanisms of evolution include natural selection, which accounts for the diverse adaptations of organisms to different environments. Neither natural selection nor any of the other mechanisms are providential; natural selection, for example, is merely the superior survival or reproduction of some genetic variants compared to others under whatever environmental conditions happen to prevail at the moment. Thus natural selection cannot equip a species to face novel future contingencies, and it has no purpose or goal—not even survival of the species. As environments vary, so do the agents of natural selection—so although trends may be discerned in the evolution of certain groups of organisms, there is no necessary reason to expect a consistent direction in the evolution of any one lineage, much less a direction that all life should follow. Natural selection, moreover, being as purely mechanical as gravity, is neither moral nor immoral.

The utterly impersonal, mechanistic nature of the mechanisms of evolution seems to be so difficult to grasp—and so repugnant to those who believe that all things exist for a purpose—that meanings have frequently been read into evolution that neither Darwin nor modern evolutionary biologists have intended. Some have equated evolution with "progress" from "lower" to "higher" forms of life, but it is impossible to define any nonarbitrary criteria by which progress can be measured. The very word "progress" implies direction, if not advance toward a goal, but neither direction nor goal are provided by the mechanisms of evolution. Least of all, despite popular conceptions, can evolution be conceived of as being directed toward the emergence of the human species. The misrepresentation of evolution as progress was so apparent to Darwin that he reminded himself in his notebook "never to say higher or lower" in reference to different forms of life, although he did not always follow his own admonition.

One of the legacies of the medieval scholastic principle of plenitude is the "naturalistic fallacy:" the supposition that what is "natural" is "good." Thus "natural laws" are taken not merely as regularities of nature, but become morally binding principles that, as Collins (1959) says, "offer a cosmic backing for the transition from *is* to *ought*." Evolution and natural selection, like hurricanes and friction, *are*, but whether they *ought* to be is a question that falls outside the realm of science. Nevertheless, despite the utter amorality (not immorality) of evolution, natural selection has been taken as a morally proper "law of nature" that ought to guide human behavior. The "law" of natural selection and the "progress" perceived in evolution were used by Marx to justify class struggle, while the so-called Social Darwinists of the late nineteenth and early twentieth centuries, finding in natural selection an ethical principle that Darwin denied, justified untrammeled economic competition, capitalism, and imperialism (Hofstadter 1955). With more charity but as little logic, Kropotkin (1902) and others pointed to the evolution of cooperative social behavior in animals as justification for more cooperative economic institutions, and the evolutionary biologist Julian Huxley attempted to develop an "evolutionary ethics" that would lead inexorably to higher consciousness and humanitarianism. Throughout all such thinking runs the supposition that what is natural is good, what is unnatural is bad—a philosophically indefensible posture that persists today when people justify their opposition to birth control or homosexuality by terming them "unnatural." This supposition long predates evolutionary theory, which has merely served as a rationalization for persistent biases.

## EVOLUTION SINCE DARWIN

The two major theses of *The Origin of Species*—that organisms are products of a history of descent with modification from common ancestors, and that the principal mechanism of evolution is the natural selection of hereditary variations—have their counterpart in the two major fields of study that constitute evolutionary biology: the study of evolutionary history and the elucidation of evolutionary mechanisms. The questions, methods, and training of investigators in these two fields tend to be quite different, and rather few biologists have contributed substantially to both fields. Quite commonly investigators in one of these major areas have been ill-informed on the questions and theories that constitute the other, and have been hampered by their incomplete understanding.

The most immediate impact of *The Origin of Species* was to provide a conceptual framework for the study of comparative morphology, descriptive embryology, paleontology, and biogeography, for "relationships" among organisms were now understood to mean common ancestry rather than affinity in the scheme of creation (such as the *Scala Naturae*). The data of these studies provided a basis for classification, which was widely adopted as a framework for describing evolutionary affinity. Biogeography and much of paleontology were devoted to inferring the history of spatial distribution and temporal change. The data of systematics provided, and continue to provide, an immense amount of information on evolutionary trends, patterns of adaptation, the kinds of evolutionary transformation that organisms' features undergo, the intermediate stages in evolutionary sequences, and patterns of variation within species. The concept of the species itself, partly because of studies in systematics, has been transformed. Many naive ideas, such as Ernst Haeckel's theory that ontogenetic change repeats phylogenetic transformation, have been revised. Although many systematists have devoted themselves chiefly to classification and only incidentally to discovering the evolutionary history of the organisms they study, their work has continued to provide data from which such historical inferences can be made. Systematics has experienced a renaissance in the last twenty years or so, chiefly through the development of more explicit methods for deducing phylogenetic history from such data. Very recently the data of systematics have been greatly enriched by molecular biology; we can now compare not only the phenotypes of organisms, but the sequences of their DNA.

The study of evolutionary mechanisms has had a somewhat more turbulent history. During Darwin's lifetime, the hypothesis of natural selection was understood by few, and accepted by still fewer (see, e.g., Hull 1973). Moreover, the nature of inheritance was not understood. The observation that offspring are generally intermediate between their parents (in characters such as size) was the basis of a widespread belief in BLENDING INHERITANCE, which may be likened to the mixing of two paints or dyes. If blending inheritance obtains, as the engineer Fleeming Jenkin pointed out in 1867, a population will quickly become homogeneous, so natural selection will have no effect; and newly arisen variations will similarly be lost by homogenization. At the same time, the belief that environmentally induced variations could be inherited was widespread, and provided an alternative to natural selection that Darwin himself gave more credence to in his later years. In 1883, August Weismann, an ardent advocate of natural selection, proposed that the germ plasm is entirely separate from and immune to any influences from the soma (the rest of the body), and vigorously rejected any influence of the environment on heredity. Weismann's ideas were strongly attacked by the "neo-Lamarckians" of the period, but were widely accepted after the recognition of Mendel's work. Nevertheless, Lamarckian ideas had their advocates well into the twentieth century, and they have not entirely disappeared today (see Fitch 1982 and Pollard 1984 for contrasting views).

The discovery in 1900 of Gregor Mendel's demonstration of particulate inheritance should have led to immediate acceptance of Darwin's theory of natural selection. Given Mendel's theory, the "problem" of blending inheritance held no force; recombination among loci can amplify variation, and new stable variants can arise by mutation. Instead, Mendelian genetics was initially interpreted as a death blow to Darwin's theory. To oversimplify a complex history (for a complete discussion see Provine 1971), Hugo deVries, William Bateson, and other early Mendelians dismissed continuous variation among individuals as inconsequential and largely nongenetic, and emphasized the role of discontinuous variants that displayed Mendelian ratios and clearly particulate inheritance. Because they took "species" to be forms that differ discretely in morphology, they believed that species arise in one or a few steps as discrete mutations. If species can arise purely by mutation, their origin does not require natural selection. Thus Darwin's key principles of natural selection and gradual change were dismissed. (A historically important extension of this view was Richard Goldschmidt's claim in his 1940 work, *The Material Basis of Evolution*, that new species or even higher taxa arise instantaneously by "systemic mutations" not of individual genes but of the entire chromosome complement.) In the early twentieth century, then, the Darwinian theory of evolutionary change was at a nadir; it was rejected not only by Mendelian geneticists, but by many paleontologists, who espoused "orthogenetic," or directional theories that relied either on intrinsic goal-directed or perfecting principles (as had Lamarck), or on the assumption that the genetic constitution of organisms constrains them to evolve only in certain directions.

## THE MODERN SYNTHESIS

Many of the emphases in evolutionary theory since about 1930—the great importance of natural selection, the nature of speciation as the gradual acquisition of reproductive isolation among populations, and the gradual evolution of higher taxa—cannot be understood except in light of the disrepute into which Darwinism had fallen. Modern evolutionary theory has its foundation in the EVOLUTIONARY SYNTHESIS or MODERN SYNTHESIS that from about 1936 to 1947 forged the contributions of genetics, systematics, and paleontology into a new NEO-DARWINIAN THEORY that reconciled Darwin's theory with the facts of genetics (Mayr and Provine 1980).

Among the information that contributed to this development were the decisive demonstrations by geneticists that acquired characters are not inherited, and that continuous variation has precisely the same Mendelian basis as discontinuous variation, entailing the segregation of numerous particulate genes, each with small phenotypic effect. Some naturalists and systematists provided evidence that the variation within and among geographic races had a genetic basis, and that some geographic variations are adaptive. In the late 1920s, Sergei Chetverikov and his coworkers in Russia began to reveal extensive hidden genetic variation in natural populations of *Drosophila*, a program of research that Theodosius Dobzhansky (1900–1975) extended greatly after he moved from Russia to the United States in 1927. Systematists developed the understanding that species are not morphological types, but variable populations that are reproductively isolated from other such populations.

The Evolutionary Synthesis emerged not from new information so much as from new concepts. The theory of population genetics, initiated in 1908 by G. Hardy and W. Weinberg's independent proofs of the "Hardy-Weinberg theorem," was adumbrated in 1926 by Chetverikov but developed in full by Ronald A. Fisher (1890–1962) and John B. S. Haldane (1892–1964) in England and Sewall Wright (1889–1988) in the United States. Fisher in *The Genetical Theory of Natural Selection* (1930) and Haldane in *The Causes of Evolution* (1932) developed fully the

RONALD A. FISHER

J. B. S. HALDANE

SEWALL WRIGHT

mathematical theory of gene frequency change under natural selection, and showed that even slight selective differences could bring about evolutionary change. From 1917 onward, Wright (notably Wright 1931, 1932) developed a comprehensive genetical theory that embraced not only selection but inbreeding, gene flow, and the effects of chance (random genetic drift). The content and power of these theories were not fully evident to most biologists, but their major conclusions had an impact.

The elements of theoretical population genetics and extensive data on genetic variation and the genetics of species differences were masterfully synthesized in 1937 in one of the most influential books of the period, Dobzhansky's *Genetics and the Origin of Species*. Ernst Mayr (b. 1904), in *Systematics and the Origin of Species* (1942), elucidated the nature of geographic variation and speciation, incorporating many of the genetical principles that Dobzhansky had articulated. George Gaylord Simpson (1902–1984), in *Tempo and Mode in Evolution* (1944) and its successor *The Major Features of Evolution* (1953), likewise drew on Dobzhansky

JULIAN HUXLEY

ERNST MAYR

and Wright in showing that paleontological data were fully consistent with neo-Darwinian theory. Julian Huxley's *Evolution: The Modern Synthesis* (1942), perhaps the most comprehensive synthesis of genetics and systematics, did much to establish neo-Darwinism in England. In Germany, the zoologist Bernhard Rensch (b. 1900) independently developed a neo-Darwinian interpretation of evolution in *Neuere Probleme der Abstammungslehre* (1947), the second edition of which appeared in English translation in 1959 as *Evolution Above the Species Level.* Botanical genetics and systematics were synthesized by G. Ledyard Stebbins (b. 1906) in *Variation and Evolution in Plants* (1950), which, like all the major works listed here, showed that neo-Darwinian principles of genetic change accounted for the origin not only of species but of other (higher) taxonomic levels as well.

The major tenets of the evolutionary synthesis, then, were that populations contain genetic variation that arises by random (i.e., not adaptively directed) mutation and recombination; that populations evolve by changes in gene frequency brought about by random genetic drift, gene flow, and especially natural selection; that most adaptive genetic variants have individually slight phenotypic effects so that phenotypic changes are gradual (although some alleles with discrete effects may be advantageous, as in certain color polymorphisms); that diversification comes about by speciation, which ordinarily entails the gradual evolution of reproductive isolation among populations; and that these processes, continued for sufficiently long, give rise to changes of such great magnitude as to warrant the designation of higher taxonomic levels (genera, families, and so forth).

G. LEDYARD STEBBINS GEORGE GAYLORD SIMPSON THEODOSIUS DOBZHANSKY

## EVOLUTION SINCE THE SYNTHESIS

Although the Modern Synthesis brought about a broad concensus on these tenets, there was still room for disagreement on certain issues. These were chiefly quantitative in nature: Do populations contain a great deal of genetic variation, or is the rate of evolution limited by the rate at which favorable variations arise? Is every feature shaped in fine detail by natural selection, or does genetic drift play a role? Moreover, some aspects of biology, such as developmental biology, were not yet fully integrated into the synthesis.

Since the Modern Synthesis, the study of evolutionary mechanisms has expanded to incorporate new information, new questions, and new controversies. The elucidation of the molecular basis of heredity since 1953, when Watson and Crick proposed the structure of DNA, has provided a deeper understanding of the nature of mutation and genetic variation, and has increasingly revealed new phenomena that have enriched and sometimes challenged neo-Darwinian theory. Molecular and other data, as well as an expansion of mathematical theory, have established random genetic drift as a major agent of evolutionary change along with natural selection. The concept of natural selection itself has expanded to include not only differential survival and reproduction of individual organisms, but also, at least in principle, of genes, groups of relatives, populations, and species.

Much of neo-Darwinian theory, including its formalization in the mathematical models of population genetics, is highly abstract; the allele frequencies and selection coefficients that appear in the models apply to traits in general, not to particular features of morphology, physiology, or behavior. The real features of organisms have increasingly been brought into evolutionary theory by the development of specific models of life history characteristics, particular aspects of behavior, modes of reproduction, and the like. Entire fields such as behavior and ecology have become incorporated into evolutionary biology. As this has happened, it has become increasingly clear that the real and present features of organisms cannot be understood solely in terms of existing genetic variations and selection pressures. Rather, present features are in part determined by the developmental processes that translate genotypes into phenotypes, and these in turn are the products of evolutionary history. The study of evolutionary mechanisms cannot be divorced from the study of developmental biology and of history, the subject matter of systematics and paleontology. The synthesis of these several areas of evolutionary biology has only begun.

## HOW EVOLUTION IS STUDIED

As indicated above, evolutionary biology consists of two principal endeavors: inferring the history of evolution and elucidating its mechanisms. In the study of evolutionary history, as in any historical study, inferences about past events are made from typically incomplete and often misleading data. Sometimes there are historical records (i.e., fossils); quite often there are not, and past events must be inferred from present patterns. Some past evolutionary events, such as the genealogical relationships among species or higher taxa, are determined by the methods of systematics (Chapter 10). More recent events, such as those determining patterns of genetic variation within and among populations, can often be inferred from the fit between data and the predictions of population genetics. Although the historical existence of an event can often be inferred with some confidence, its causes are often much more difficult to elucidate, and frequently must remain a matter for speculation. From the evidence of comparative morphology and paleontology, we are confident that flowering plants arose from gymnosperms, but we do not know whether the great diversification among flowering plants was caused by their insect pollination, their protected seeds, or their versatile vegetative architecture. That is, the causes of singular events such as the diversification of angiosperms or the origin of World War I are extremely difficult to discern, even in principle. Causes of repeated events, however, such as the frequent evolution of succulence in desert plants, may be inferred if they are correlated with particular conditions. The use of such correlations, the COMPARATIVE METHOD, has been a powerful tool in evolutionary biology.

All sciences progress by formulating and reformulating models and testing them by observation and experiment. A model may be expressed verbally, as when we postulate that isolation of two populations will permit them to diverge in genetic composition because they do not interbreed; or it may be mathematical, as when we formulate the degree of interbreeding that may permit divergence to occur. The mathematical regularity of Mendelian inheritance is the basis of a genetical theory of the mechanisms of evolution that is probably the most comprehensive and intricate body of mathematical theory in biology. All models, however, no matter how mathematically forbidding, are deliberately oversimplified descriptions of nature. They are not meant to describe the complexities of reality, but to abstract the most general and important features of a process.

Because the factors operating in any individual circumstance are far more numerous and complex than a general model can include, the models do not predict future events in detail; the equations of population genetics do not predict the detailed evolutionary path of a population any more than the equations of physics can predict the weather a year from now. Rather, the role of models in evolutionary biology is to specify the conditions under which conceivable events are likely or unlikely to occur, and so to provide a restricted range of possible explanations for observations (Lewontin 1985). Mathematically formulated models also stipulate the quantitative relationships among interacting forces that will permit one event or another to occur. For example, a simple model will show that a mutation that is deleterious when heterozygous cannot increase in frequency in a large population, even if the mutation is highly advantageous when homozygous. A more complex model shows that such a mutation may indeed increase in a small population, and the smaller the population, the more deleterious the mutation may be and still have an appreciable probability of increase. Thus if we observe differences among populations in alleles that fit this description, we will be inclined to infer that the populations have been small at some time in the past.

Models are extremely important in evolutionary theory, but empirical study is obviously important as well. First, observation and experiment are required to describe the phenomena that require explanation—to pose questions. Second, only empirical studies can tell us which of the conditions that a theoretician can conceive of actually occur, and how frequently: we can conceive of mutations that are deleterious when heterozygous and advantageous when homozygous, but only observation can tell us if such mutations exist and are common. Likewise, observation can tell us which of the innumerable complexities that might exist are important, and which can be ignored. Moreover, theory will often provide several competing explanations for an observation; empirical study is then required to decide which explanation is most likely.

Observation, finally, has another function besides providing support to our intellectual pursuit of understanding the laws of nature: it is the vehicle by which we achieve the esthetic satisfaction of knowing nature. Some biologists have purely intellectual motivations for studying orchids, butterflies, or bacteria; some also find beauty in the pollination of orchids or the structure of a cell. To describe the diversity and history of living things, to provide knowledge for its own sake, is to enrich humanity.

## EVOLUTION AS FACT AND THEORY

A few words must be said about "the theory of evolution," which most people take to mean the proposition that organisms have evolved from common ancestors. In everyday speech, "theory" often means a hypothesis or even a mere speculation. But in science, "theory" means "a statement of what are held to be the general laws, principles, or causes of something known or observed," as the *Oxford English Dictionary* defines it. The theory of evolution is a body of interconnected statements about natural selection and the other processes that are thought to cause evolution, just as the atomic theory of chemistry and the Newtonian theory of mechanics are bodies of statements that describe causes of chemical and physical phenomena. In contrast, the statement that organisms have descended with modifications from common ancestors—the historical reality of evolution—is not a theory. It is a fact, as fully as the fact of the earth's revolution about the sun. Like the heliocentric solar system, evolution began as a hypothesis, and achieved "facthood" as the evidence in its favor became so strong that no knowledgeable and unbiased person could deny its reality. No biologist today would think of submitting a paper entitled "New evidence for evolution;" it simply has not been an issue for a century.

Almost without exception, opponents of evolution today maintain their position not on grounds of logical argument, much less on grounds of evidence, but on the basis of emotions and religious beliefs. In recent years, creationism has re-emerged in the United States and elsewhere not as a scientific issue but as a social phenomenon, part of a larger reactionary ideology that poses a real threat to the integrity and quality of public education. The creationists' alternative to evolution is simply a literal interpretation of the first chapters of the biblical Book of Genesis, often clothed in the language of biology but lacking scientific substance.

In this book, I have not presented analyses of creationism or an explicit discussion of the evidence for evolution; I have treated these subjects at length elsewhere (Futuyma 1983a). Other books on the subject include those by Kitcher (1982), who treats the issues chiefly from a philosophical point of view; Newell (1982), who emphasizes the fossil record; and Godfrey (1983), who has compiled essays by many authors on various facets of the conflict.

## SUMMARY

Evolution, a fact rather than a hypothesis, is the central unifying concept of biology. By extension it affects almost all other fields of knowledge and must be considered one of the most influential concepts in Western thought. Its tenets have frequently been misinterpreted (for example, "evolution" is often equated with "progress") and the objective science of evolutionary biology has often been extended into the subjective realm of ethics and used illegitimately as justification for both pernicious and humanitarian economic, social, and scientific policies.

The recognition that evolution has happened was widespread before Darwin and, together with other advances in science, was a major change in the Western world view, which believed a static, orderly Great Chain of Being to be natural and therefore good. Darwin and Wallace, by providing the idea of natural selection, transformed speculation into scientific theory. Most of their ideas have been validated by more than a century of subsequent research in which evolutionary biology, especially through the growth of genetics, has become an ever more intricate, all-embracing, and sophisticated body of explanatory principles.

## FOR DISCUSSION AND THOUGHT

1. Analyze and evaluate Emerson's couplet,

   Striving to be man, the worm
   Mounts through all the spires of form.

2. Although the position has become increasingly untenable in the light of historical scholarship, some authors have claimed that although Darwin was a good observer, he was not very original or creative. Contrast and evaluate the antithetical views of this subject held by Himmelfarb (1959) and Ghiselin (1969).
3. One unsettling conclusion that some people draw from evolution is that humans are nothing more than animals. This is an example of what has been termed the *reductionist fallacy.* In what sense it is a fallacy? Does the fact that an organism consists of an evolved set of biochemical systems mean that life is nothing more than molecules in motion, and that all of biology can be explained in biochemical terms? Does our evolutionary relationship to other animals mean that all our characteristics and activities have the same causes as theirs?
4. Contrast the position of Social Darwinists with T.H. Huxley's 1893 statement, "Let us understand, once and for all, that the ethical progress of society depends not on initiating the cosmic process [of natural selection], still less in running away from it, but in combating it."
5. Biologists hold that evolution does not have a purpose. Thus it is meaningless to ask what purpose is served by the existence of tapeworms or aardvarks or humans. Many people equate the absence of a purpose with "chance." Analyze the concept of "chance" or "randomness." In what sense is chance responsible for evolution?
6. The idealization of progress is so much a part of our thinking that it may come as a revelation to realize that this is primarily a Western concept, and that even in the West it is a recent view, stemming from no earlier than the sixteenth century. Why is this concept so appealing? Is it possible to determine objectively whether the history of evolution, or of society for that matter, has been one of progress?
7. Discuss the proposition that it is not possible even in principle to determine the causes of a singular (unique) historical event.

8. What are the arguments for and against the creationists' demand that creationism be taught as an alternative theory to evolution? Do all alternative theories on a topic have equal claim to our attention?
9. The philosopher of science Karl Popper (1968) proposed that a theory is not scientific unless it could be refuted if it were wrong. Some people have claimed that evolutionary theory is so flexible that it could be stretched to explain any imaginable observation; thus it would not be refutable or testable, and so is not a scientific theory. Is it true that no conceivable evidence could disprove the reality of evolution? Is Popper's criterion a valid one for scientific theories? Can a distinction be made between scientific and nonscientific theories? (See Kitcher's 1982 discussion of this issue in relation to evolution.)

## MAJOR REFERENCES

The references at the end of each chapter usually include some that provide basic background on the subject, but most are major works that provide a comprehensive treatment and an entry into the technical literature.

Greene, J.C. 1959. *The death of Adam: Evolution and its impact on Western thought.* Iowa State University Press, Ames. 388 pages. A scholarly but easily readable description of the history of evolutionary thought up to *The Origin of Species* and its aftermath.

Mayr, E. 1982. *The growth of biological thought: Diversity, evolution, and inheritance.* Harvard University Press, Cambridge, MA. 974 pages. A detailed comprehensive history of systematics, evolutionary biology, and genetics that bears the personal stamp of one of the major figures in the Evolutionary Synthesis.

Ghiselin, M.T. 1969. *The triumph of the Darwinian method.* University of California Press, Berkeley. 287 pages. A study of Darwin's scientific methods and their application in his major works.

Bowler, P. 1984. *Evolution: The history of an idea.* University of California Press, Berkeley. 412 pages. A history that incorporates the results of recent studies.

Hofstadter, R. 1955. *Social Darwinism in American thought.* Beacon Press, Boston. 248 pages. A study of some social applications and abuses of evolution.

Futuyma, D.J. 1983. *Science on trial: The case for evolution.* Pantheon, New York. 251 pages. An introduction to evolution with a summary of the evidence and an analysis of the claims of so-called scientific creationists.

Kitcher, P. 1982. *Abusing science: The case against creationism.* MIT Press, Cambridge, MA. 213 pages. A devastating analysis of so-called scientific creationism by a philosopher of science.

# The Ecological Context of Evolutionary Change

# Chapter Two

Many features of organisms are adaptations to their environment. Indeed, much of biology, whether it be biochemistry, anatomy, physiology, or ecology, consists of the study of adaptations, those features that, having evolved by natural selection, enable organisms to survive and reproduce in the face of the innumerable contingencies that beset them. However, the environment of an organism, the totality of factors that influence its activities, achievements, and ultimate fate, can be difficult to characterize. My aim in this chapter is to review those aspects of ecology—the study of interactions between organisms and their environment—that most pertain to evolution; to characterize the concept of environment.

## ADAPTATION AND ENVIRONMENT

It is important to bear in mind that not all of evolution consists of the development of adaptations by natural selection; many other factors, including chance, influence evolution. But even if we restrict our attention to evolution by natural selection, there are numerous selective factors besides those imposed by the external ecological world. Chief among these are the internal relationships among biochemical and developmental pathways, and among different organs, that impose selection by requiring that new features be compatible with the rest of the organism's internal organization.

The ecological environment includes both abiotic and biotic features. Climate, salinity, soil type, availability of water, and other physical and chemical features are important. Also, other species—including prey, predators, pathogens, competitors, and mutualists—are a preeminent feature of every species' environment. Moreover, other members of an individual's species, with which the individual may mate, compete for resources or mates, or interact in various social contexts, are also important features of its environment. Thus properties of the population such as its density, sex ratio, or genetic composition may impinge importantly on each individual's prospects of survival and reproduction.

The features of the environment that are important vary from species to species because of their different evolutionary histories; it is not an exaggeration to say that by virtue of past evolution, species create their own environments (Lewontin 1983). To a predatory beetle, the chemical composition of the plants among which it forages is largely irrelevant; but if a beetle evolves the habit of herbivory, plant chemicals that act as toxins, repellents, or feeding stimulants become all-important. Beetle species whose larvae feed on the upper surfaces of leaves contend with a much hotter and drier environment than those that feed on lower surfaces; the beetle larvae that bore in plant stems inhabit a different environment entirely. Not only the past evolution of a species determines its environment; so do its present activities, for species deplete resources, release toxic metabolites, and alter their surroundings in numerous other ways. Thus species and environment alter each other reciprocally. It is an error to think of species simply as passive sufferers of harsh external fate; they are active participants in a dialectical interchange between organism and environment.

## THE ECOLOGICAL NICHE

Each of a great many factors affects the ability of a species to survive and reproduce, and thus in part determines whether it can persist in a particular locality. For example, a clam might tolerate a certain span of temperatures and feed on

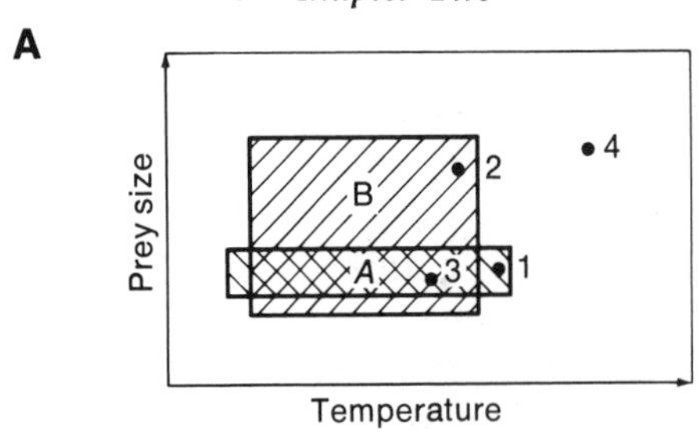


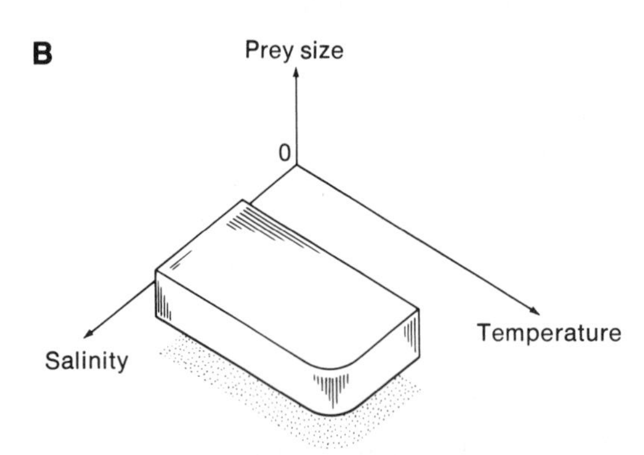


FIGURE 1

**(A) Two dimensions of the ecological niche of each of two species, A and B; perhaps they are estuarine bivalves, each restricted to a range of temperatures and of particle sizes they can ingest. Each point in the space represents a possible environment, a combination of particle size and temperature. If a locality presents only environments in the neighborhood of point 3, the species compete intensely. If there are some warm microhabitats (near point 1) to which A is adapted and some large prey items (near point 2) as well as smaller ones (points 1, 3), the species should be able to coexist. If the only microhabitats are near point 4, neither species will occur. (B) Three dimensions of the niche of a species are represented. If a locality contains combinations of the three variables that lie within the solid figure (the species' niche), the species may persist. Notice that the combination of high temperature and high salinity is inimical to the species.**

plankton in a specific size range. Following Hutchinson (1957), we may draw a two-dimensional graph in which a single point represents an environment with a particular temperature and a particular prey size. Part of this space then represents the range of possible environments in which the species can persist (Figure 1A). We may add a third axis corresponding to, say, salinity, thus defining a three-dimensional space (Figure 1B) containing a region that represents tolerable combinations of temperature, prey size, and salinity. To take into account other environmental factors we would have to draw many more axes (which we can do only in imagination). We conceive of an *n*-dimensional space, with one axis for each of *n* environmental factors. Within this space is a region consisting of a cloud of points, each representing a particular combination of temperature, prey size, salinity, copper concentration, starfish abundance, and so forth, that together constitute environments conducive to the survival and reproduction of the clam population. Hutchinson called this region, consisting of the set of possible environments in which the species can persist, the fundamental ECOLOGICAL NICHE of the species. Because different genotypes may have different tolerance limits for various axes of the niche, the ecological niche can evolve, and indeed will often vary among different geographic populations of a species.

With respect to any given environmental variable, a population may have a narrow niche or a broad niche; it may be relatively SPECIALIZED (or STENOTOPIC) or GENERALIZED (or EURYTOPIC). A species usually is specialized in some respects and generalized in others. For example, larvae of the fall cankerworm (*Alsophila pometaria*) feed on many species of trees but do so only in the spring; larvae of another geometrid moth, the lesser maple spanworm (*Itame pustularia*), feed only on maple foliage, but pass through several generations during spring and summer.

Rather than referring to a species as specialized, then, it is best to specify in what respect it is specialized.

In some contexts, it is useful to focus on only one or two axes of the niche. For example, when ecologists refer to the degree of "niche overlap" of two species of birds, they often are referring specifically to the species' common use of a single variable such as the size of the seeds they eat.

## SPATIAL DISTRIBUTION

The distribution of a species may be described in geographic terms (e.g., the bog turtle *Clemmys muhlenbergi* is restricted to eastern North America) and in ecological terms (the bog turtle is restricted to sphagnum bogs and similar habitats). Geographic distributions are often limited partly by history; the species may not yet have dispersed to potentially suitable areas. Some geographic limits are also set by ecological factors such as climate or competition with other species (Krebs 1978, Brown and Gibson 1983). Biogeography, the study of geographic distributions of species, is the subject of Chapter 13.

Within its geographic range, the distribution of every species is patchy to varying degrees because of spatial variation in physical features, the availability of resources, and other species that act as competitors, predators, or parasites. Many species, especially those that have very exacting requirements, consist of several or many small populations that are often widely separated. The recognition that a species is divided into local populations that often consist only of dozens or hundreds of individuals has had an important impact on our understanding of evolutionary processes (Chapter 5). Moreover, small populations are very susceptible to extinction, and the persistence of a species in a particular locality often depends on a continual influx of migrants from other populations. For example, what appeared to be a single population of checkerspot butterflies actually proved to be three subpopulations that fluctuated independently in size; over the course of ten years, subpopulations became extinct on several occasions, and were reestablished by new immigrants (Brussard et al. 1974). Suitable habitats for a species that has a low capacity for dispersal may be unoccupied because they have not been recolonized.

## POPULATION GROWTH

In any locality the density and persistence of the population of a species depends on its capacity to increase in numbers and the factors that limit its abundance. An understanding of these population dynamics (see Wilson and Bossert 1971, Ricklefs 1979) is essential to an appreciation of evolutionary theory.

In the most common formulation of population growth, we suppose that the rate of change ($dN/dt$) in the number of individuals in a population ($N$) is determined by birth ($b$) and immigration ($I$), and by death ($\delta$) and emigration ($E$). For simplicity, immigration and emigration are usually ignored by supposing them equal, but as we have just noted, immigration may be crucial to the persistence of the population.

If $b$ is the per capita birth rate and $\delta$ is the probability that an individual will die in the time interval $dt$, the per capita rate of change in population size is $r = b - \delta$ and the population as a whole grows at the rate

$$dN/dt = bN - \delta N = rN$$

Then $N_t$, the population size at time $t$, depends on $r$ and the initial size $N_0$:[1]

$$N_t = N_0e^{rt}$$

In populations such as annual insects that have discrete generations, the replacement rate $R$ is used instead of $r$, and the expression for population growth is $N_{t+1} = RN_t$. $R$ may be defined as the ratio of the number of individuals born in two successive generations.

As long as the birth rate exceeds the death rate ($r > 0$), the population will increase in an exponential fashion (Figure 2); if unchecked, as Darwin noted, the descendants of a single pair of any species would cover the earth in short order. The rate of increase $r$ actually depends on age structure, for a population made up mostly of individuals at the height of their reproductive powers grows more rapidly than if either juvenile or senile individuals predominate. However, if the environment is constant so that birth and death rates specific to each age class do not change, the population ultimately attains a stable age distribution: the entire population grows at the rate $r$, but the proportion made up of each age class remains constant. In such a population the rate of increase $r$ is affected by:

1. *Survival.* The higher the fraction $l_x$ of newborns that survive to each age $x$, the greater the growth rate (Figure 3A). The $l_x$ curve determines $\delta$.
2. *Fecundity.* The higher the number of offspring ($m_x$) produced by an average female of age $x$, the higher the birth rate $b$ (Figure 3B).
3. *The age at which reproduction begins* (Figure 4). Individuals that reproduce early in life are more likely to have grandchildren by some time $t$ in the future than those that reproduce at an advanced age. Thus populations of organisms with a short generation time have a more rapid potential rate of increase than those with a long generation time. In a growing population, a female who reproduces early has a greater REPRODUCTIVE VALUE than a female who reproduces later in life; she contributes more to the future population size (Fisher 1930, Slobodkin 1961).

[1] Most of the algebraic symbols used in this book are defined in Appendix II. In this equation, as elsewhere, $e$ is the base of natural logarithms, 2.718.

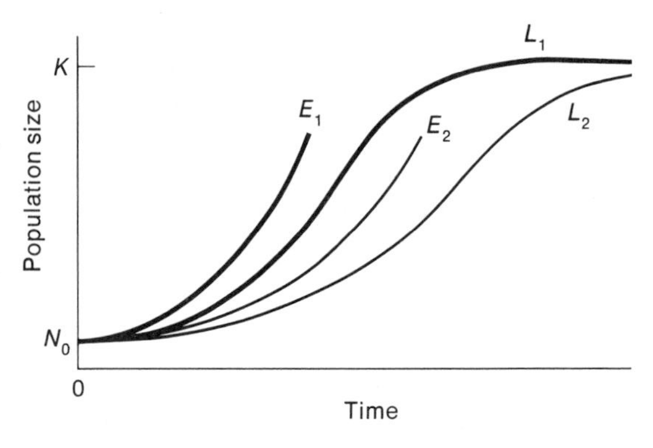


FIGURE 2

**Idealized growth of a population. Curves $E_1$ and $E_2$ describe exponential (density-independent) growth in different environments (e.g., higher and lower temperatures). They could equally well represent the growth of genetically different populations that differ in $r_m$. Curves $L_1$ and $L_2$, corresponding to $E_1$ and $E_2$ respectively, depict density-dependent growth according to the logistic equation. The equilibrium population size is $K$. In reality $K$ might differ between the environments that cause the differences in the rate of population growth. Real populations do not show such smooth curves or constant equilibrium densities, even under constant environmental conditions.**

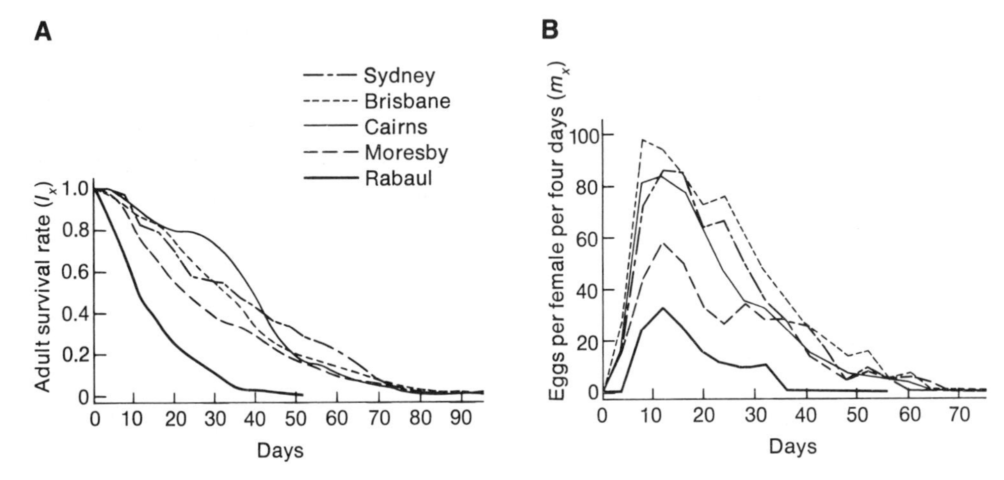


FIGURE 3

**Genetic variation in (A) survivorship ($l_x$) and (B) fecundity ($m_x$) schedules in *Drosophila serrata*. For strains taken from five Australian localities, the figures show the fraction of adult flies that survive to different ages and the fecundity per surviving female per four-day interval at 25°C. (From Birch et al. 1963)**

It is possible to calculate the per capita rate of increase $r$ that a population would have if it attained a stable age distribution by determining $l_x$ and $m_x$ for each age $x$. This may be done either by measuring these variables for individuals of each age in a mixed-age population, or by following a COHORT (a group of individuals of the same age) from birth to the death of the last individual, and measuring $l_x$ and $m_x$ at each age. The rate of increase $r$ is then found by balancing the equation

$$\sum_{x=1}^{L} l_x m_x e^{-rx} = 1$$

where $L$ is the maximum age attained.

A population's potential rate of increase is often very different from its actual

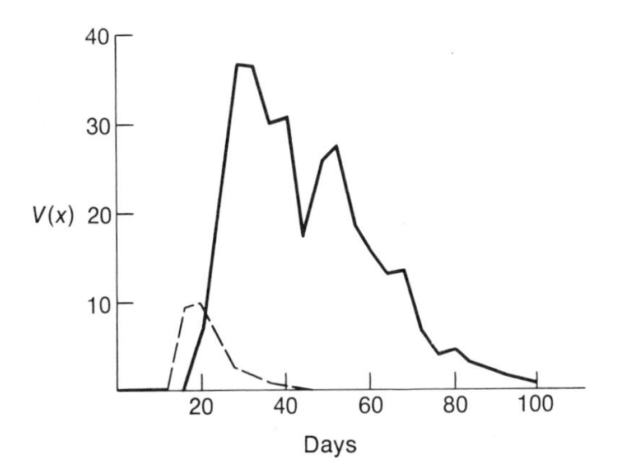


FIGURE 4

**The $V(x)$ function for *Drosophila serrata* at 20°C (solid line) and at 25°C (dashed line). $V(x)$ is the contribution of a female of age $x$ to the rate of population growth. Despite the tenfold difference in total production of offspring at these temperatures, both $V(x)$ functions give the same value of $r_m$ because of the importance of offspring produced early in life at 25°C. (After Lewontin 1965)**

rate of increase. The death rate $\delta$ may be great enough to equal the birth rate $b$, so that $r = 0$ and the population size is constant. The actual rate of increase, $r$, is different from the INTRINSIC RATE OF NATURAL INCREASE, $r_m$, which is the per capita rate of increase that a population with a stable age distribution would have in a given environment if it were utterly free of those factors, such as predation and scarcity of food, that reduce population growth. This $r_m$ differs from one environment to another; a bacterial culture given nutrients *ad libitum* grows more rapidly at higher temperatures than at lower. When a population is growing at its environment-specific intrinsic rate of natural increase $r_m$, its growth is said to be DENSITY-INDEPENDENT.

## THE EFFECT OF DENSITY ON POPULATION GROWTH

As a population grows, the age-specific birth and death rates change, so the actual rate of increase $r$ is not constant. At very low densities $r$ may be less than maximal because survival and reproduction may depend on favorable social interactions or the need to find mates. But as density increases, the population may deplete its resources, poison its environment with metabolic wastes, or engender the build up of predators or pathogens. Population growth then becomes DENSITY-DEPENDENT. In the most commonly used (and unrealistically simple) model of density-dependent growth, it is supposed that as density increases, the birth rate declines and/or the death rate increases linearly, so that the actual growth rate $r$ declines with slope $c$ from its intrinsic value $r_m$ according to the equation:

$$r = r_m - cN$$

A little algebra and the relation $r_m/c = K$ yields the LOGISTIC EQUATION of population growth in a limited environment:

$$dN/dt = r_m N(K - N)/K$$

According to this equation, population growth proceeds sigmoidally, slowing down and finally stopping when the equilibrium density $N = K$ is reached (Figure 2). Properly speaking, $K$ is merely the equilibrium density, which may well be set by, say, predation; but it is often referred to as the CARRYING CAPACITY of the environment. This phrase is suitable if density is limited by scarcity of resources such as food or space.

All species have the capacity to grow exponentially, but none do so indefinitely. In reality the approach to equilibrium is seldom a smooth sigmoid curve, for several reasons. One of these is that the effect of density on birth rates may not be felt until some time after the density has increased (i.e., there is a time lag). When this occurs, the population may oscillate in numbers as it approaches equilibrium and even oscillate violently and unpredictably around its theoretical equilibrium density rather than settling down to a constant size (May and Oster 1976).

In most natural populations there is considerable REPRODUCTIVE EXCESS: more individuals are born than survive to reproduce. This realization was critical in Darwin's formulation of the idea of natural selection; as he said in *The Origin of Species*, "... as more individuals are produced than can possible survive, there must in every case be a struggle for existence, either one individual with another of the same species, or with the individuals of distinct species, or with the physical conditions of life. . . . owing to this struggle, variations, however slight and from whatever cause proceeding, if they be in any way profitable to the individuals of a species, in their infinitely complex relations to other organic beings and to their physical conditions of life, will tend to the preservation of such individuals, and will generally be inherited by the offspring." Density-dependent reduction of population growth is not necessary for the operation of natural selection (Chapter 6), but it does offer considerable opportunity for selection to occur.

### The nature of the limits to density

The factors that limit population density in nature are the subject of a long-standing controversy (Cold Spring Harbor Symposium 1957, Den Boer and Gradwell 1970; see any of several ecology books for a summary, e.g., Krebs 1978, Ricklefs 1979). Andrewartha and Birch (1954), among others, argued that many populations are in a state of effectively continual exponential growth that is interrupted by environmental changes such as inclement weather before they become dense enough to cause a scarcity of resources or a buildup of natural enemies. Controlling factors such as weather are not responsive to population density, and so are termed density-independent. A contending school of thought (e.g., Lack 1954, Nicholson 1958, Slobodkin et al. 1967) denied the importance of density-independent limiting factors, and argued that most populations usually are kept near equilibrium by density-dependent factors such as a shortage of energy, nutrients, nest sites or other resources, predation and disease, or, in some animals, behavioral interactions such as territoriality.

Probably the most common current view is that most populations experience both density-independent and density-dependent growth at different times, with the relative proportions varying among populations and species. There is abundant experimental evidence that plants and sessile intertidal invertebrates are often limited by competition for space (and light and water in the case of plants); that many populations of predators are limited by scarcity of food; and that predators can act as density-dependent limiting factors of their prey, such as phytophagous insects and other herbivores. But it is also the case that weather can cause fluctuations in populations that may well be density-independent. Density-independent growth may be commonly experienced by FUGITIVE SPECIES that occupy very temporary resources and persist only by continual dispersal from one patch of resources to another. The fly larvae that develop in rotting fruits or mushrooms occupy a resource that may become uninhabitable before the density of larvae can become very great. But even in this case, there is evidence that mushroom-feeding flies are food-limited and compete for resources (Grimaldi and Jaenike 1984). It seems likely that most populations that persist for very long are at least occasionally limited by density-dependent factors.

### Responses to density

Organisms have many responses that mitigate the ill effects of high density. Most notable is the tendency to disperse. Virtually all plants, for example, have adaptations for dispersing their seeds; one of the worst places for a seed to germinate is next to the parent plant, which is not only a bigger competitor but is also a focus of activity for herbivores and pathogens (Janzen 1970). In some species of animals, high density triggers a complex of adaptive responses that include dispersal. Many species of aphids (plant lice) that are wingless under favorable conditions give birth to winged offspring if they are crowded or if the quality of their food plant deteriorates, as it does when it is heavily infested. As successive generations of the plague locust *Schistocerca gregaria* experience more and more crowding, these grasshoppers experience a hormonal change that affects their physiology, morphology, and behavior. They store fat, develop long wings and increased pigmentation, become more gregarious, and finally depart in enormous swarms, to land, with luck, in greener pastures.

Too sparse a population, however, may also be inimical to an individual's reproductive success. Thus fishes find protection from predators by schooling, birds may find food more effectively by watching other members of a flock than by foraging singly, and *Drosophila* larvae, by burrowing through their medium, promote the growth of the yeast on which they feed. There are many advantages of aggregation and social behavior (Wilson 1975) that can outweigh the disadvantages that aggregation necessarily entails, such as competition for food.

The limiting factors that prevent unlimited population growth may vary in time and space; for example, a population of herbivorous insects may commonly be controlled by natural enemies, but occasionally escape this control, reach higher density, and become food-limited. Moreover, several factors may operate simultaneously to govern population density. Birth and death rates may be sensitive to different factors as density increases; thus it is possible, for example, for scarcity of food to lower the birth rate and for increased predation to increase the death rate simultaneously.

It is extremely important to distinguish the effects an environmental factor has on a *population* from the effects it has on *individuals* within the population; it is the latter, as we shall see, that determines evolution by natural selection. Thus a population may be limited by food rather than predation, but some individuals nevertheless suffer predation, and there will be selection in favor of genotypes that can avoid predators. Predation may be "good" for a prey population by preventing it from becoming extinct by overexploiting its resources, but predation is not good for the individuals that suffer it.

## THE BIOTIC ENVIRONMENT: PREDATORS AND PREY

The subject of density-dependent population growth is integrally related to that of interactions between predators and their prey, or, more generally, between consumers and the species they consume. Because a consumer feeds on a dynamically growing population, its carrying capacity is not a constant, but a variable. As the prey population increases, that of the predator does likewise until it becomes so abundant that the prey population declines. The two populations, then, mutually limit each other's density: the prey population is limited by predation, and the predator population by food. In theory, either stable or unstable oscillations in the densities of both species may occur; if unstable, extinction of the prey, and thereafter the predator, may result. In many cases, predator–prey interactions are indeed unstable, and result in extinction. For example, a South American cichlid (*Cichla ocellaris*; Figure 5) introduced into Gatún Lake in Panama has locally extinguished several other species of fishes (Zaret and Paine 1973). The interaction may be more stable, however, if the prey species itself is partly

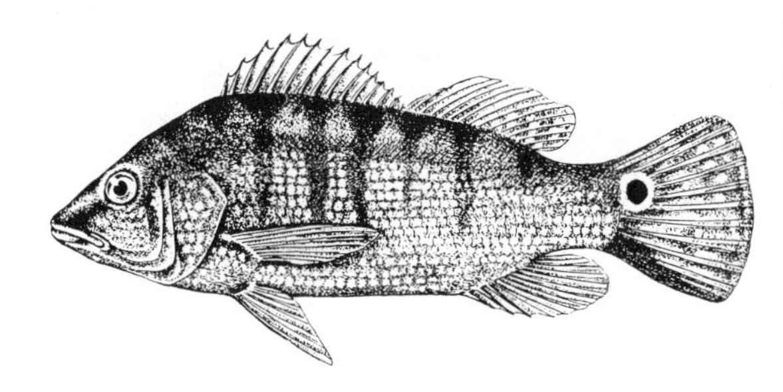

FIGURE 5
**_Cichla ocellaris_, a South American species introduced into Gatún Lake, Panama, where it is extinguishing native species. (From Sterba 1962)**

limited by the availability of its own resources, so that it does not grow as rapidly when predation is reduced.

Although some predator–prey combinations are unstable, other combinations persist for any of several reasons. Predators are often unable to find all the potential prey, some of which remain inaccessible. A prey species may survive through dispersal: the interaction may be locally unstable, so that local populations become extinct, but populations of prey, and consequently of predators, may build up elsewhere as vacant localities are colonized. In some instances, some of the age classes in a prey population are resistant to predators; for example, wolves attack chiefly defenseless young and infirm old moose rather than the reproductive age classes, which can effectively defend themselves. Moreover, many predators (and other consumers) consume several species. In some cases, predators focus their food-seeking efforts to a disproportionate degree on the most common kinds of prey. Whether by moving to places where a particular food type is concentrated, or by forming a SEARCH IMAGE that enables them to find food more effectively, they lessen their impact on uncommon species of prey (Figure 6).

Although no species is entirely free from predation, all have escaped some of their potential predators and parasites by evolving mechanisms of defense.

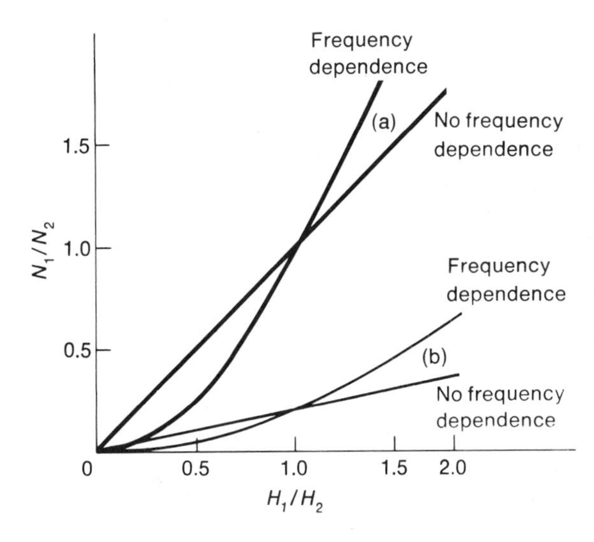


FIGURE 6
**Frequency-dependent predation, or "predator switching." $N_1/N_2$ is the ratio of two species (or genotypes) of prey consumed by a predator; $H_1/H_2$ is the ratio of their abundances in the environment. Curves _a_ describe a predator that does not prefer either species; curves _b_ describe a predator that has a fivefold preference for prey species 2. In each case, the line is straight if the predator shows no frequency-dependent behavior. The curved lines display switching: each prey species is taken disproportionately more frequently if it is more abundant relative to the other. (After Murdoch and Oaten 1975)**

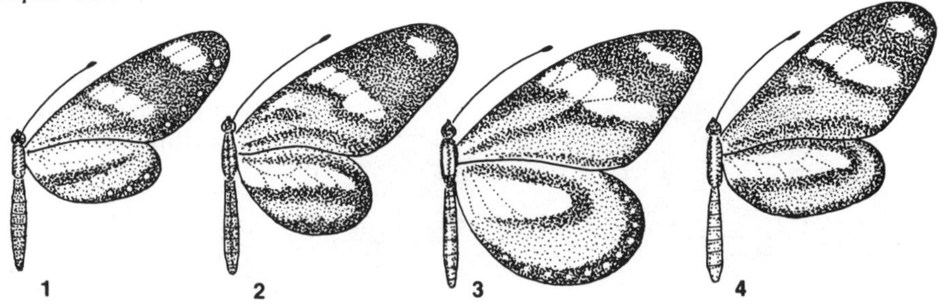


They may become inaccessible by hiding, fleeing, or by being too big (or too small) to eat. They may avoid notice by crypsis—imitating the background in form, color, and pattern. Many plant and animal species possess spines, stinging hairs, protective armor, or noxious chemicals that render them unpalatable to some predators. Distasteful animals frequently advertise their unpalatability by a warning (APOSEMATIC) coloration or pattern; after a few encounters with such distasteful organisms, naive predators quickly associate the color pattern with their unpleasant experience and refrain from attacking such prey for some time. Aposematically colored species are often the models in systems of mimicry for convergence to a common color pattern in otherwise dissimilar species. In some cases a palatable species masquerades as an unpalatable one (BATESIAN MIMICRY), while in other cases several unpalatable species converge in appearance, each species gaining protection from its similarity to the other ones (MÜLLERIAN MIMICRY; Figure 7).

A community of organisms displays a bewildering variety of defense mechanisms, which continue to evolve in diversity. If two species share the same defense system, the population of a predator adapted to counter this defense grows as a function of the combined abundance of the two species of prey. One prey species may therefore become extinct, because the other maintains the predator at high density. Thus the dynamics of two such prey populations, viewed without reference to the predator, has the appearance of competition between them, even if they are not actually competing for any limiting resources (Holt 1977, Bender et al. 1984). Moreover, because individuals of each prey species suffer from the presence of the other, the evolution of different defense systems is favored (Slobodkin 1974, Levin and Segel 1982). The complex issue of coevolutionary changes in predator–prey systems is one of the topics discussed in Chapter 16.

**Parasitism**

Most species are resources not only for predators, but for parasitic species of microbes, fungi, or animals. The ecology of parasites has been insufficiently studied, but they constitute a large fraction of the world's species, and have doubtless had a pronounced influence on the evolution of their hosts (Price 1980, Holmes 1983, May and Anderson 1983). Two of the most important aspects of the ecology of parasites are their reproductive capacity and dispersal. Whereas a single prey individual may be only one of many that a predator must eat before

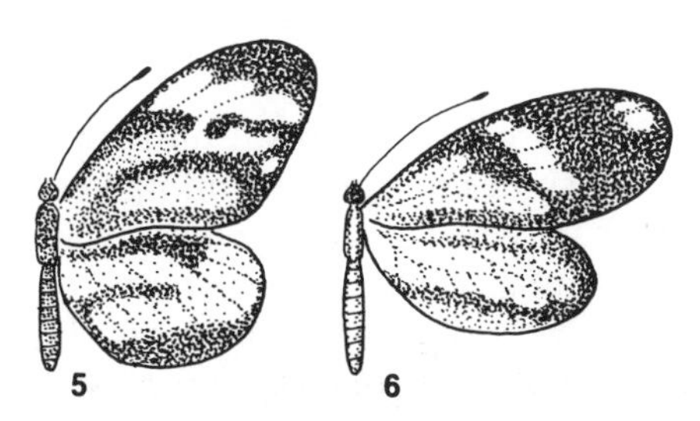


FIGURE 7
**Two forms of mimicry in tropical American butterflies. Species 1–4 are all distasteful and are similar to each other (Müllerian mimicry). Species 5 and 6 are palatable, but resemble the unpalatable species (Batesian mimicry). (Redrawn from Wickler 1968)**

it can reproduce, a single host individual may yield an enormous number of parasite offspring. Thus the intrinsic rate of natural increase of a parasite is often many times greater than that of its host.

For many parasites, dispersal to new hosts is fraught with difficulty, and may be accomplished by any of a vast number of remarkable adaptations, such as infestation of a secondary host that is consumed by the primary host (in which the parasite reproduces sexually). The population density of the host often affects the probability of encounter, and therefore the growth rate of the parasite population. The genetic structure of a parasite population is often different from its ecological structure: the parasite population may be reproductively a single unit, yet ecologically be subdivided into groups of individuals that interact with each other within their individual hosts (Wilson 1980). Parasites are commonly quite highly host-specific, and in this sense are more specialized than most predators, which usually prey on a considerable variety of species.

## BENEFICIAL INTERACTIONS AMONG SPECIES

COMMENSALISM is a relationship in which one species benefits from another species that is not affected by the interaction; examples might include the dispersal of seeds on the fur of mammals, or the use of woodpecker holes by owls and chickadees. In many cases the evolution of one species may be affected by the other, but not vice versa. MUTUALISM is an interaction in which individuals of each species exploit the other as a resource, and so profit from the interaction. Examples include plants and pollinating animals, legumes and nitrogen-fixing bacteria, and the association between many vascular plants and fungi, which form a mycorrhiza, a root–fungus combination in which the fungus uses carbohydrates synthesized by the plant and the plant profits from a more rapid uptake of water and nutrients.

Mutualisms may sometimes evolve from commensal relationships, but often seem to evolve from parasite–host relationships (Boucher et al. 1982; Chapter 16). Indeed, the line between antagonism and mutualism is often a fine one; the same fungus that enhances the growth of a plant in some environments may reduce it in others (Boucher et al. 1982). Some mutualisms involve suites of more or less interchangeable species; thus most bees gather pollen from several or many plant species, and most bee-pollinated plants can be pollinated by several or many species of bees. Other, often highly obligate, mutualisms are highly species-specific; for example, leaf-cutter ants (*Atta*) feed exclusively on a particular fungus that they grow in their nests. The fungus is not found elsewhere.

## COMPETITION AMONG SPECIES

Competition among species may take either of two general forms. Competition by EXPLOITATION occurs when one individual consumes a resource and makes it unavailable to others; the competing individuals may never actually meet. Competition by INTERFERENCE occurs when two individuals directly interact, and one loses the encounter. They may fight over food or territorial space; one individual may poison another (the phenomenon of allelopathy); or individuals may eat each other (which may be treated mathematically as a kind of competition). Competing individuals may be members of the same species (INTRASPECIFIC COMPETITION) or of different species (INTERSPECIFIC COMPETITION).

The most common model of interspecific competition is an extension of the logistic equation for population growth, in which the term $K - N$ is extended to $K_i - N_i - \alpha_{ij}N_j$:

$$dN_i/dt = r_{mi}N_i(K_i - N_i - \alpha_{ij}N_j)/K_i$$

This expresses the growth rate of species $i$ ($dN_i/dt$) as a function of the relation between the carrying capacity for that species ($K_i$) and the number of individuals of both species $i$ ($N_i$) and its competitor $j$ ($N_j$); as $N_i$ and $N_j$ increase, the term in parentheses decreases and $dN_i/dt$ declines. The COMPETITION COEFFICIENT $\alpha_{ij}$ expresses the impact of one individual of species $j$ on the growth rate of population $i$, relative to the impact of an additional individual of species $i$ on its own population; $\alpha_{ij}$ equals 1 if the two species are equivalent in their competitive interaction, and is less than 1 if an individual of species $j$ is a less serious competitor, from the viewpoint of population $i$, than a conspecific individual. Species $j$ might be a less serious competitor if, for example, it tends to use somewhat different resources, so that on average competition for resources is less intense between species than within species.

Oversimplifying the result of this model, it may be said that species can coexist stably if competition within each species is more intense than between species (i.e., $\alpha_{ij} < 1$ and $\alpha_{ji} < 1$). If one species is a stronger competitor than the other (e.g., $\alpha_{ij} > 1$, $\alpha_{ji} < 1$), the inferior competitor will become extinct. If the species are competitively identical, their relative proportions will fluctuate by chance until one or the other becomes extinct (the ecological equivalent of genetic drift; see Chapter 5). Thus this theory expresses "Gause's axiom," also known as the COMPETITIVE EXCLUSION PRINCIPLE, that competing species cannot indefinitely coexist if they are limited by precisely the same resources. If they use only some of the same resources, they may be able to coexist.

This principle is often invoked to explain the observation that coexisting species typically differ in the resources they utilize. For example, several species of anoline lizards that coexist on West Indian islands forage in somewhat different microhabitats (Schoener 1968); species of triclad flatworms coexist only where the diversity of prey animals is high enough to provide each species with a food supply that it can largely call its own (Reynoldson 1966). Competitive exclusion is also a plausible explanation of cases in which species have complementary distributions (e.g., Diamond 1975). For example, the altitudinal distribution of the salamander *Plethodon glutinosus* is broader in mountain ranges where it is the only species than in some ranges in which *P. jordani* occupies the higher altitudes (Hairston 1951). Competitive exclusion has been historically documented in some cases; for example, within a few years after the parasite wasp *Aphytis lingnanensis* was introduced from Asia to control olive scale in California, it replaced the species *A. chrysomphali* that had flourished after its introduction some years earlier (DeBach 1966). Numerous studies have shown that competition exists among species of plants and of many animals in nature; when the density of one species is experimentally altered, the density of others changes in a compensatory fashion (Connell 1983, Schoener 1983).

When two species compete, which one wins may depend on subtle environmental factors. Brown (1971b) found that the chipmunk *Eutamias dorsalis* excludes *E. umbrinus* from low altitudes by chasing it away from food. But at higher altitudes the more arboreal *E. umbrinus* can get more food than *E. dorsalis* by quickly returning after being chased. It can do this because the branches of trees at these altitudes form an interlocking network, providing an effective highway that is not available in the sparser forests lower on the mountain. In interference competition of this kind, one species usually responds behaviorally to the presence or absence of another. For example, the lizards *Anolis sagrei* and *A. cristatellus* use different microhabitats where they coexist, but if *A. cristatellus* is removed, *A. sagrei* quickly expands into the vacant microhabitat (Salzburg 1984). Some species, though, do not have this flexibility: the cricket *Allonemobius allardi* is replaced by *A. fasciatus* in moist habitats, but neither species expands its habitat range if the other is removed (Howard and Harrison 1984).

As Darwin first realized, closely related species are likely to compete most intensely, but sometimes unrelated species compete as well. Ants and rodents compete for seeds (Brown et al. 1979); terrestrial plant succession is a study in competitive exclusion of pioneer species by often unrelated late-successional forms. Nevertheless, most studies of competition have focused on related species, often with the purpose of showing that they differ sufficiently in resource use to coexist. According to Gause's axiom, each species can persist only by differing somewhat from all others, so that a community should contain more species if each is specialized for a different resource than if the species are more generalized, and so overlap more in resource use (MacArthur 1972). A considerable number of studies have taken morphological differences among species as an index of differences in feeding habits, and have tried to show that there is a lower limit to the similarity of coexisting species. These efforts have been vigorously criticized (e.g., Simberloff 1983) and defended (e.g., Case et al. 1983; see also Harvey et al. 1983). In at least some cases, species that coexist do seem to be more dissimilar than if species had been assembled into communities at random. The Galápagos Islands, for example, harbor several species of ground finches with similar beak sizes and feeding habits, but these species do not coexist on any one island (Schluter and Grant 1984).

One cannot attribute all differences in resource use to competition, nor will coexisting species necessarily differ in resource use. Even species that do not compete can differ in feeding habits; for example, it is probable that many species of herbivores are limited by predation rather than resources (Slobodkin et al. 1967), and there is little experimental evidence that they compete (Lawton and

Strong 1981, Schoener 1983); yet related herbivorous insects often are specialized for feeding on different hosts. Predation on a superior competitor may prevent it from eliminating inferior competitors, and competitively inferior fugitive species may persist by colonizing new areas as they are eliminated by competition in others (Horn and MacArthur 1972). Finally, more elaborate mathematical models than the one we considered show that it is sometimes theoretically possible for species to coexist stably even if they are limited by the same resource (Armstrong and McGehee 1980, Abrams 1983, Chesson and Case 1986).

### Competition and adaptive radiation

Beginning with Darwin, biologists have invoked competition as an important force affecting the diversification of species. On the one hand, competitors may prevent a species from evolving to use a particular resource; this is illustrated by its opposite, ECOLOGICAL RELEASE, whereby a species uses a broader range of resources in the absence of a competitor. This may be reflected by morphology. For example, the males of many woodpeckers have larger beaks than the females and feed in somewhat different parts of trees. This dimorphism is far more pronounced in *Centurus striatus*, the only species of woodpecker on Hispaniola, than in species that inhabit continents, where several species of woodpeckers usually coexist (Selander 1966).

ADAPTIVE RADIATION is a term used to describe diversification into different ecological niches by species derived from a common ancestor. Impressive examples include the Hawaiian honeycreepers (Figure 8) and the cichlid fishes of the great lakes of Africa (Chapter 8; see Figure 13 in Chapter 8); in both groups, closely related species differ greatly in feeding habits and in associated morphology. As in many examples of adaptive radiation, these two groups have diversified in regions in which the diversity of other birds or fishes is low. It is likely, therefore, that evolutionary changes in the use of resources are often constrained or prevented by competing species, but can be appreciable when the competition is relieved.

In theory (see Chapter 16), interspecific competition can favor the evolution of divergence so that species come to differ in the resources they use. Evolutionary responses of this kind may foster adaptive radiation, but they are not necessarily responsible for the adaptive radiations that have occurred. Evidence that species evolve in response to interspecific competition is provided by instances of CHARACTER DISPLACEMENT (Brown and Wilson 1956), defined as a greater difference between SYMPATRIC (in the same geographic locality) than ALLOPATRIC (in different localities) populations of two species. A few examples of this phenomenon are known (Figure 9), but not very many (Grant 1972; see Chapter 4).

FIGURE 8

**Adaptive radiation in the Hawaiian honeycreepers (Drepanididae). Only the major forms are shown; some similar species have been omitted. Arrows indicate possible evolutionary changes by which the beak shapes may have evolved through intermediate stages. This family, including thin-billed foliage gleaners, long-billed nectar feeders, woodpecker-like forms such as *Hemignathus wilsoni*, and thick-billed seed eaters, fills many of the ecological roles that on continents are filled by a variety of different families of birds. (After Bock 1970)**

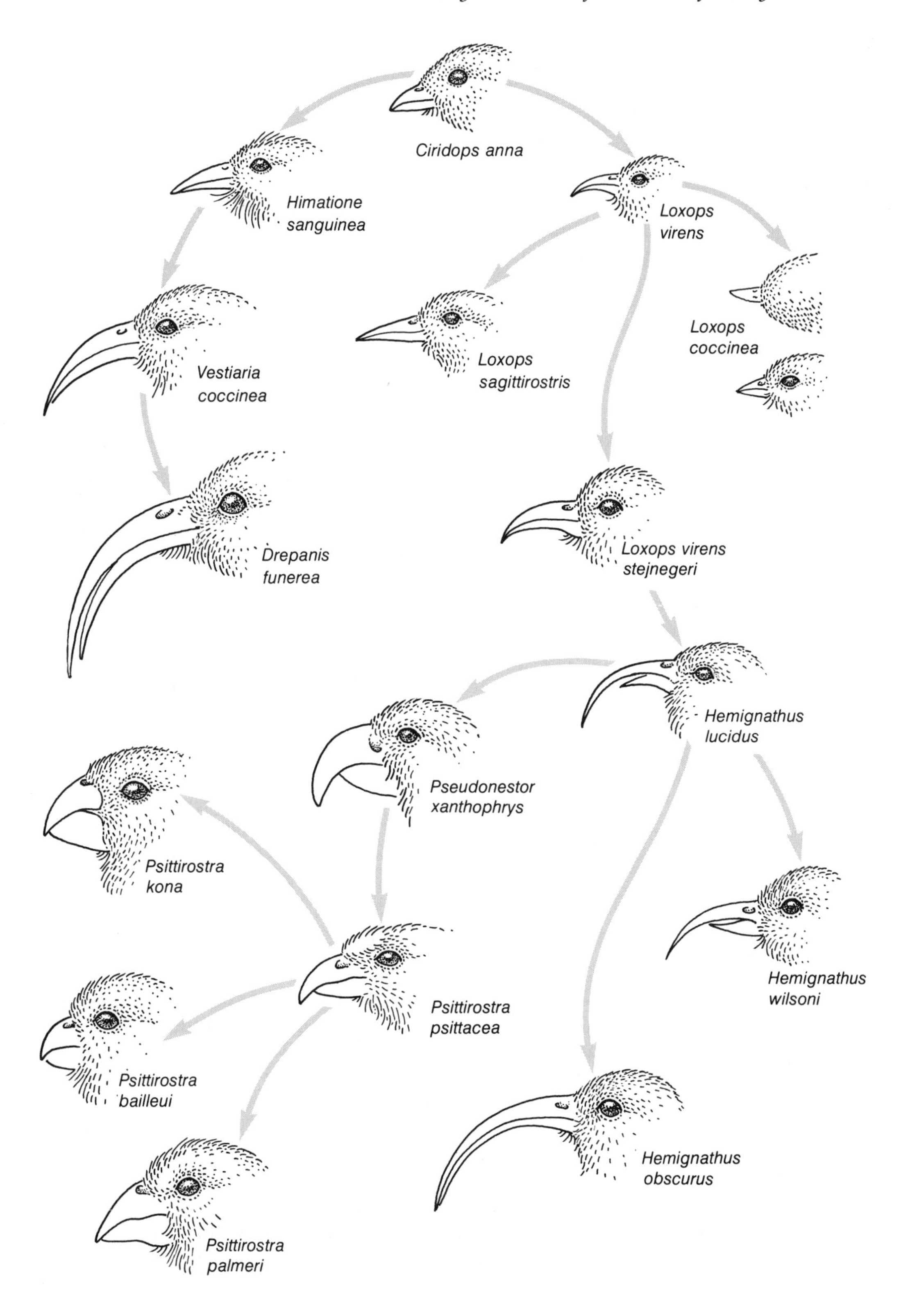
Ciridops anna
Himatione sanguinea
Loxops virens
Loxops coccinea
Loxops sagittirostris
Vestiaria coccinea
Drepanis funerea
Loxops virens stejnegeri
Hemignathus lucidus
Pseudonestor xanthophrys
Psittirostra kona
Psittirostra psittacea
Hemignathus wilsoni
Psittirostra bailleui
Psittirostra palmeri
Hemignathus obscurus

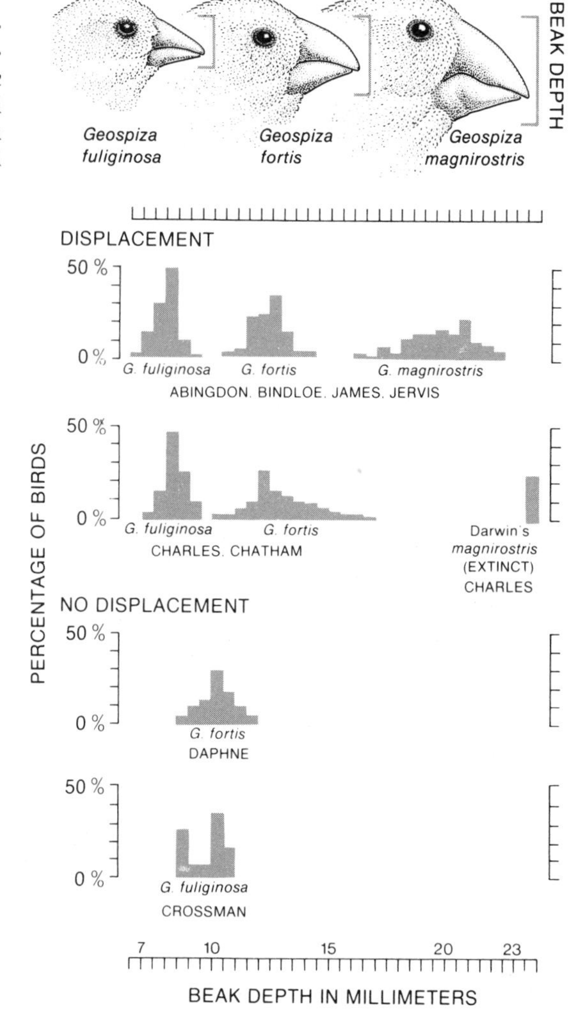


FIGURE 9

**One of the instances of character displacement among the ground finches of the Galápagos Islands. The beak depth of allopatric *Geospiza fortis* and *G. fuliginosa* (on Daphne and Crossman Islands) is similar, but differs on islands where they both occur. (Modified from Lack 1947)**

## COMPLEX INTERACTIONS AMONG SPECIES

We are accustomed to thinking in simplistic, linear terms: the more abundant mice are, the more abundant the weasels that feed on them. But interactions among species are seldom so simple (Price et al. 1980). Predation can sometimes prevent a superior competitor from excluding an inferior one. The intensity of competition between two species can be modified by the presence of a third competitor (Vandermeer 1969, Neill 1974). Two consumers may enhance each other's population growth by preventing competitive exclusion among the species they feed on (Vandermeer 1980). A species may have either a net beneficial or a net deleterious effect on another, depending on the presence or absence of a third species. For example, oropendolas (large tropical American orioles, *Zarhynchus wagleri*) are often aggressive toward giant cowbirds (*Scaphidura oryzivora*) which lay eggs in oropendola nests. This response is not surprising, because young cowbirds compete for food with young oropendolas. But young cowbirds can enhance the survival of oropendola nestlings by picking botfly larvae off them. However, oropendola nests in trees that harbor large wasp nests are rather free from botflies because wasps chase botflies away. Remarkably, the oropendolas are hostile to cowbirds when they nest in the vicinity of wasps, but tolerant of cowbirds in the absence of wasps. Even more remarkably, cowbirds that lay in oropendola nests associated with wasps lay mimetic eggs, while those that parasitize oropendolas that are not associated with wasps tend to lay non-mimetic eggs (Smith 1968).

Some species have a profound impact on the distribution and population dynamics of others simply through their physical effect on the environment. For example, the periwinkle (*Littorina littorea*), an herbivorous snail introduced to the American coast from Europe, acts like a little bulldozer when it grazes. When periwinkles are removed, the community of barnacles and encrusting algae is replaced by a community of filamentous algae and benthic worms and snails that live in the sediment that accumulates (Bertness 1984).

Finally, we should bear in mind that many species subsist on the *interaction* between other species. Numerous parasites, for example, depend on predator–prey relationships, as in some trematodes that develop sequentially, first in snails, then in snail-eating fishes, and finally in piscivorous birds. The development of species diversity in a community is self-reinforcing, as interactions among species create new ecological niches.

## DIVERSITY AND STABILITY OF COMMUNITIES

Ecologists have devoted a great deal of attention to the question of what determines the number of species (species diversity) in communities, and whether there exist forces that organize communities into a predictable structure (e.g., MacArthur 1972, Cody and Diamond 1975, Brown 1981). The number of species—of birds, for example—often is related to the area embraced, more or less according to the relation $S = cA^z$, where $S$ is the number of species, $A$ is the area, and $c$ and $z$ are constants; $z$ often has a value of about 1.2. The theory of island biogeography (MacArthur and Wilson 1967) explains the relationship by noting that the number of species in a region will be set by the balance between the rate of extinction and the rate of immigration of new species into the area (Figure 10). There is some evidence that extinction rates are higher in small areas, which support smaller populations, than in large areas (Diamond 1984).

It is unlikely that communities contain a random assortment of species that have chanced to invade them; rather, it appears that interactions among species permit some combinations to persist while excluding species that do not fit in. One reason for the belief that communities are organized is that faunas that have independently adapted to similar environments in different parts of the world sometimes appear convergently similar (Figure 11)—although this is not always true (Orians and Paine 1983). Moreover, as we have seen, related species in a community sometimes appear to be less similar in their use of resources than if the communities had been assembled by chance. Although there is considerable controversy about the interpretation of such data, few ecologists would altogether deny that some combinations of species are stable, whereas others are unstable, in the sense that one or more of the species will be unable to invade the com-

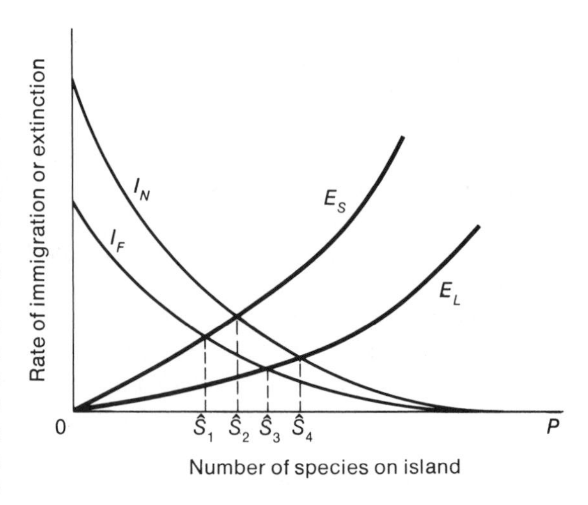


FIGURE 10

**MacArthur and Wilson's model of species diversity on islands or similar patches of habitat. The number of species $S$ on the island increases as new species immigrate and decreases as species already present become extinct. When the rates of immigration ($I$) and extinction ($E$) are equal (where the curves cross), the number of species is at equilibrium $\hat{S}$. The greater the number of species on the island, the smaller the number of immigrants that are new; hence the immigration curve declines. Even if the probability of extinction of each species is constant, the more species there are, the more extinctions there will be; hence the extinction curve rises. Immigration rates are likely to be higher for near ($I_N$) than far ($I_F$) islands; extinction rates are likely to be greater on small ($E_S$) than large ($E_L$) islands. Hence $\hat{S}$ should be lowest on distant small islands ($\hat{S}_1$) and greatest on near large islands ($\hat{S}_4$). (After MacArthur and Wilson 1967)**

munity, or will become extinct if another invades. There is some evidence (Elton 1958, Moulton and Pimm 1983) that complex communities consisting of many species may be more resistant to invasion by additional species than are more simple communities consisting of fewer species. The factors that enhance the stability of a community are complex and poorly understood: it is certainly not the case that increased species diversity automatically enhances stability (May 1973). Evolutionary adjustments of species to each other may sometimes enhance stability, but often will not (Chapter 16).

## ENVIRONMENTAL PATTERNS

Any physical or biotic feature of the environment that impinges on individuals of a species has several different properties which may affect the course of adaptation. For instance, there is its average value. We commonly presume that some average conditions are harsher, or more difficult to adapt to, than others. This may well be true, but it is hard to measure harshness, because it is relative to the capabilities of the particular species. A warm rainforest would presumably be as harsh for a caribou as the tundra would be for a spider monkey. Ephydrid fly larvae that inhabit pools of crude oil seem to do quite well, although this environment is uncongenial for the vast majority of species. Adaptation to a "harsh" environment is more likely for some species than others, and one of the major challenges to evolutionary biology lies in understanding why some adaptive changes are more likely than others.

Virtually all environmental factors vary. It is important to distinguish *spatial* variation from *temporal* variation. The factors that affect white-tailed deer (*Dama virginiana*) in Maine do not appreciably affect the survival of white-tailed deer in Florida. Local populations, exposed to different states of the environment, may take quite different evolutionary paths. However, the spatial scale over which the

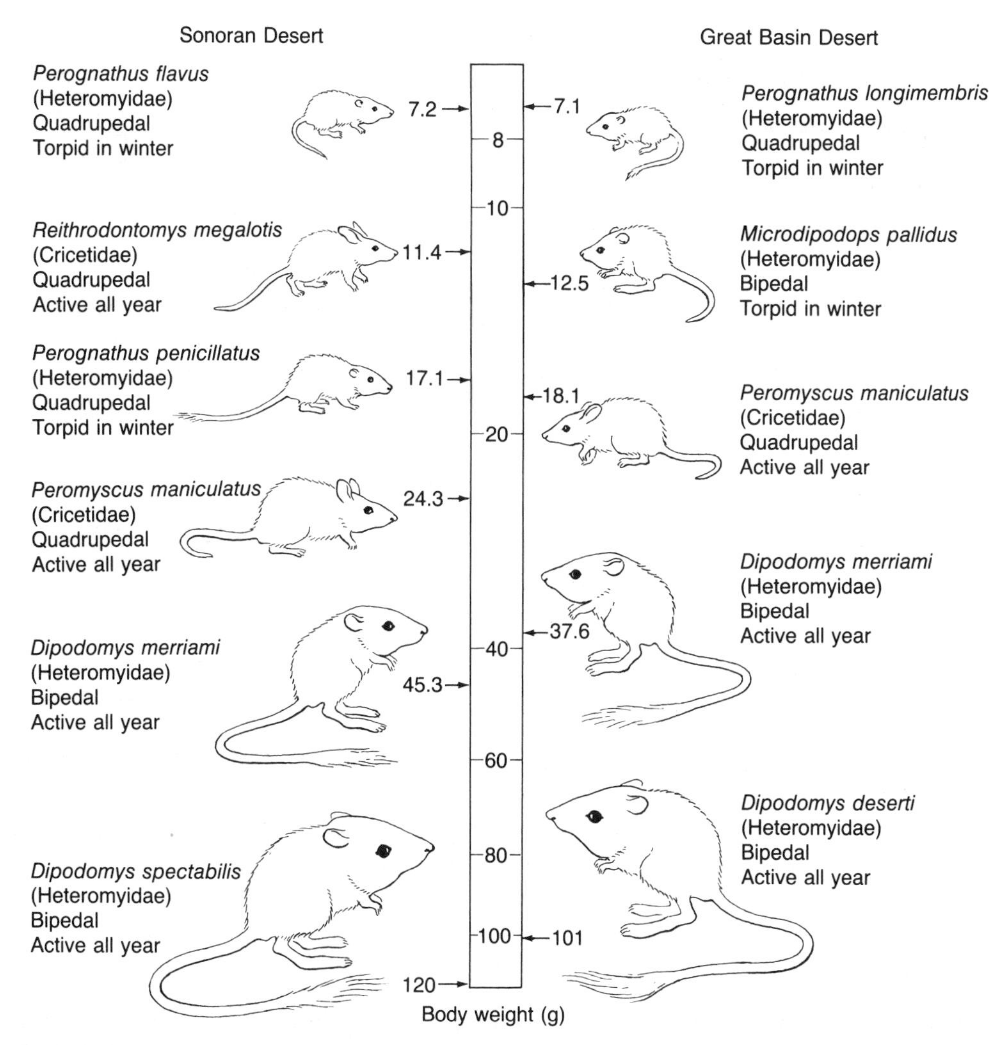


FIGURE 11

**Similarity of the communities of rodents in the Sonoran Desert of southern Arizona and the Great Basin Desert of Nevada. The distribution of body sizes is similar, although only two species occur in common. In each community, larger rodents eat larger seeds, so resources are similarly partitioned among species. The larger size of *Peromyscus maniculatus* and *Dipodomys merriami* in the Sonoran Desert has been interpreted as an evolutionary response to competition with the greater number and size of smaller species. (From Brown 1975)**

environment varies effectively depends on the distance over which members of the species typically move. Thus a single genetic population of the migratory monarch butterfly (*Danaus plexippus*) experiences the temperature regimes of both Canada and Mexico, whereas different populations of a nonmigratory species experience very different temperature regimes at different latitudes.

Temporal variation in the environment may be described in several ways.

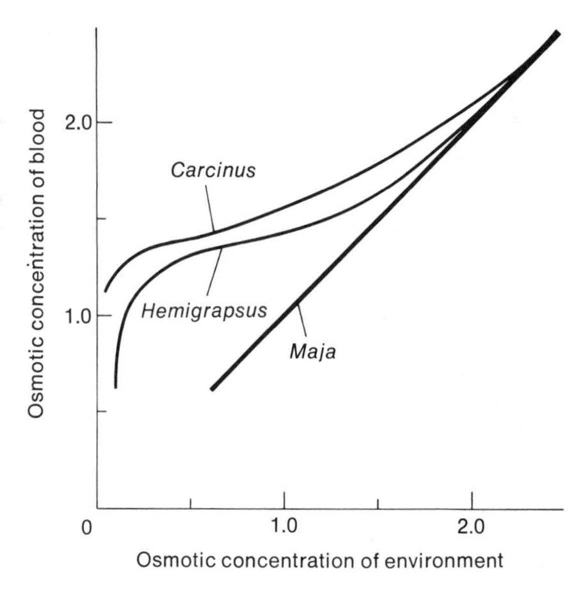


FIGURE 12

**Differences in physiological homeostasis among species of crabs. The osmotic concentration of the blood conforms to that of the environment in *Maja*, a marine crab. *Hemigrapsus* and *Carcinus*, shore crabs that experience greater fluctuations in salinity, regulate their blood osmoconcentration. (After Prosser and Brown 1961)**

The ABSOLUTE RANGE of the variable includes extremes that may be quite rare; a really cold winter once a century or an outbreak of a defoliating insect that ordinarily poses little threat to a tree species may have a dramatic impact on a population. Such unusual events must be part of the evolutionary experience of all species, but we can seldom assess their impact on evolution, even in theoretical terms (Lewontin 1966).

Organisms typically evolve adaptations to the more commonly experienced range of variation. (This may be statistically described by the *variance*; see Appendix I.) For example, estuarine species of crabs can acclimate (i.e., adjust physiologically) to a wider range of salinities than more exclusively marine species (Figure 12). Physiological acclimation, modification of behavior (including learning), and enzyme induction are among the ways in which the phenotype of an individual animal or plant may change during its life to meet the demands of a changing environment (Schmalhausen 1949, Thoday 1953, Slobodkin 1968). Not only the magnitude, but the predictability of environmental variation affects the course of evolution. For example, seasonal changes in the temperate zone are predictable, but the exact time at which they occur is not; so many birds have adapted to seasonal change by using a predictable event, the change in photoperiod (day length), as a cue for reproduction and migration. In contrast, rainfall in many deserts is so unpredictable from one year to the next that some desert birds come into reproductive condition not in response to photoperiod, but in direct response to the availability of water or green vegetation.

An important aspect of environmental variation is its frequency. The effect of the frequency of oscillation depends on the "response time" of the physiological function that is affected by the environmental change. For example, because of its large size a lion has a relatively low metabolic rate and can store considerable fat, so it requires a large meal only every few days; but if food is available that infrequently to a shrew, it will quickly die because of its far higher rate of

FIGURE 13
**Adaptive phenotype plasticity in response to ecological conditions: nongenetic differences in leaf form in *Sagittaria sagittifolia* under terrestrial (left) and aquatic (right) conditions.**

metabolism. Thus an environment that is effectively constant for a large homeotherm is lethally inconstant for a small homeotherm.

It is often useful to distinguish (MacArthur and Levins 1964) between environmental fluctuations with a period less than the life span of an individual organism (FINE-GRAINED VARIATION) and those with a period greater than an individual's life span (COARSE-GRAINED VARIATION). Fine-grained variation is often met with homeostatic changes in physiology or behavior that are reversible. Adaptation to coarse-grained variation often entails a developmental switch into one of several irreversible alternative phenotypes. For example, if the rotifer *Brachionus* is exposed early in life to the chemicals that exude from the predatory rotifer *Asplanchna*, it develops long protective spines (Gilbert and Waage 1967). In many semiaquatic plants the shape of the mature leaves depends on whether or not they are submerged during their development (Figure 13). The distinction between fine-grained and coarse-grained environmental variation holds for both spatial and temporal heterogeneity. The variety of plant species in a forest is a fine-grained mosaic to a leaf-eating monkey that in its lifetime can sample all the available plants. But to an individual gall wasp that undergoes its entire development in a single leaf, the forest is a coarse-grained environment.

## Time scales of environmental change

The consequences of environmental variation depend in part on the time scale over which a variable fluctuates. Some variation exists on a physiological time scale, within the lifetime of an individual organism, whether because the individual moves about from one environment to another or because it is buffeted by change even if it stays in place. A major consequence of such variation is to select for genotypes that have homeostatic abilities or that can avoid exposure to inimical changes (e.g., by dormancy). There exists variation also on an ecological time scale of greater than a generation—changes in physical or biotic variables that bring about changes in population density, including local extinction. Ecological succession is just one example of such change: early successional species such as dandelions become excluded locally by late-successional species, and persist only as other patches of favorable habitat become open elsewhere. As long as suitable habitats become available at a fairly constant rate, however, such *populations* may be said to experience a fairly constant environment, even if many of the *individuals* that compose these populations experience a deteriorating environment.

As we consider longer time spans, variation on the ecological time scale grades into variation on a geological time scale. A new environment that persists for several to many generations will often bring about changes in the genetic composition of the population by natural selection, as well as changes in population density. Some environmental changes do not threaten the population with extinction even if the population does not adapt by change in its genetic composition; others may bring about extinction unless the population adapts genetically; still others cannot be countered by genetic change, and result in extinction.

It is well to bear in mind that environmental changes are sometimes not as great as they seem for a particular species, because the species may have properties that filter out the changes—the ability to seek out appropriate microhabitats, the ability to acclimate, and so on. But it is also important to realize that environmental change is universal and pervasive. During the late Pleistocene (a mere 12,000 years ago), most modern ecological communities did not exist in their present locations, or consist of the particular mixture of species that coexist today. Fifty million years ago, the ancestor of the opossum looked much like its modern descendant, but it moved amidst creodonts, condylarths, titanotheres, and innumerable other creatures that have since become extinct.

## SUMMARY

Organisms must adapt not only to the average states of the environment, but to its pattern of variation as well. The amplitude, frequency, and predictability of environmental fluctuations affect the pattern of adaptation or even preclude adaptation altogether. Environmental variation is universal, especially because the other organisms with which a species interacts are such an important part of its environment. Among these organisms are other members of its own species, which act as mates, social consorts, or competitors. The rate of population growth commonly declines as the density of a population increases, because of the increasing scarcity of resources or the increasing prevalence of predators or disease. The "struggle for existence" so engendered has important effects on the evolution of reproductive rates, life history characteristics, dispersal mechanisms, feeding habits, and defense mechanisms. Indirectly or directly, it affects all the characteristics of organisms. Interactions with other species, including predation, competition and symbiosis, affect a species' abundance and distribution in time and space and have led to many of the adaptations that make up the diversity of the living world.

## FOR DISCUSSION AND THOUGHT

1. The deserts of southwestern North America are home to many species of bees. Each appears in years of high rainfall and feeds on only a few of the many species of plants that also appear in response to rain. Do these species experience a variable environment?
2. Can you find in the literature any cases in which the reasons why a species has *not* become adapted to an environment are really understood? Over a geographic transect the environment usually changes gradually, but a species' range stops at some point. Why should it be incapable of adapting enough to extend just a few more miles?
3. Hairston et al. (1960) argued that although physical factors such as weather may limit the geographic and local distribution of a species, such factors cannot limit population density, since the species is adapted to these factors. Discuss.
4. If organisms are supposed to be adapted to the environments they usually encounter, why can some tropical species like ostriches be kept outdoors all year in a temperate zone climate such as that of New York or London?
5. Haldane (1956) argued that any characteristic that makes an individual less subject to the factor that limits population density is advantageous. Does this mean that evolution necessarily leads to an increase in population size?
6. What factors are likely to determine the outcome of competition between a species with generalized feeding habits and one with specialized feeding habits, if the diet of the generalist includes the food of the specialist? Why is it that a community often contains both generalized and specialized species? What factors might influence species to become more or less specialized?
7. Discuss the proposition that there exist empty ecological niches awaiting the evolution of species to occupy them.
8. Suppose a species of herbivore evolves to be more efficient in digesting its food, so that it has a higher birth rate per amount of food consumed. Will its equilibrium population density increase? (See Roughgarden 1983a.)
9. In temperate zone forests, each tree on average loses about 10 percent of its photosynthetic tissue to herbivores during a growing season. Similarly, it is normal for most free-living mammals to carry parasites. Should we therefore expect such organisms to have evolved to compensate for this drain on their energy, so that they allocate some energy to herbivores and parasites? Does it follow that herbivores and parasites are not deleterious, since their hosts have evolved to take their toll into account?
10. Many species of weeds and pests, such as the gypsy moth in North America, are abundant pests where they are not native, but are much less abundant in their region of origin. Discuss possible reasons for this; are they necessarily evolutionary reasons?

## MAJOR REFERENCES

Wilson, E.O. and W.H. Bossert. 1971. *A primer of population biology.* Sinauer Associates, Sunderland, MA. 192 pp. An exceptionally lucid introduction to the simpler mathematical formulations of population biology.

Ricklefs, R.E. 1979. *Ecology.* Second Edition. Chiron Press, New York. 966 pp. A comprehensive introduction to ecology.

Pianka, E.R. 1983. *Evolutionary ecology.* Third Edition. Harper & Row, New York. 416 pp. An introduction to modern ecology from an evolutionary perspective.

Whittaker, R.H. 1975. *Communities and ecosystems.* Macmillan, New York. 383 pp. An introductory but comprehensive survey of community ecology.

# Heredity: Fidelity and Mutability

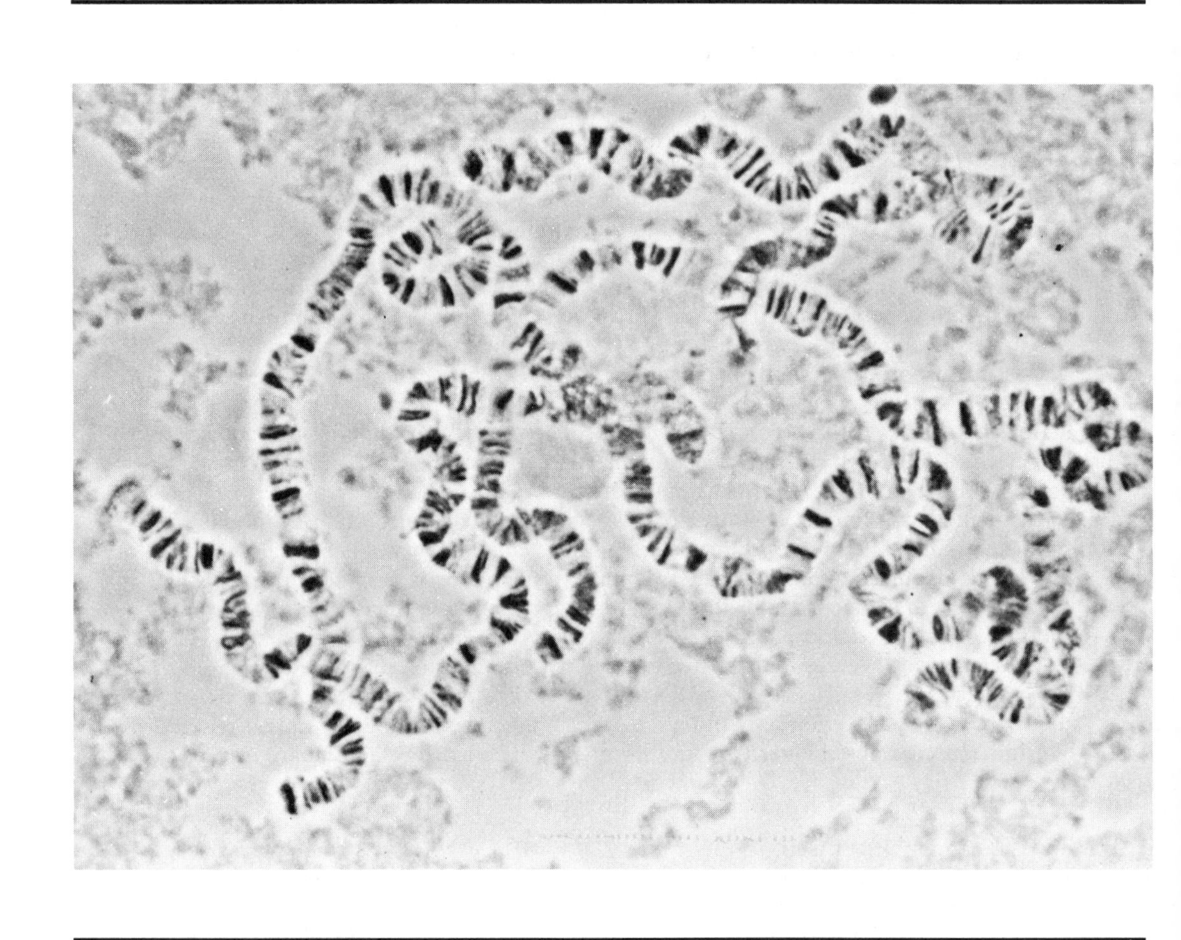

# Chapter Three

"The fertilised germ of one of the higher animals, subjected as it is to so vast a series of changes from the germinal cells to old age, . . . is perhaps the most wonderful object in nature. It is probable that hardly a change of any kind affects either parent, without some mark being left on the germ. But on the doctrine of reversion, . . . the germ becomes a far more marvellous object, for, besides the visible changes which it undergoes, we must believe that it is crowded with invisible characters, proper to both sexes, to both the right and left side of the body, and to a long line of male and female ancestors separated by hundreds or even thousands of generations from the present time: and these characters, like those written on paper with invisible ink, lie ready to be evolved whenever the organisation is disturbed by certain known or unknown conditions."

The study of evolution has been indissolubly bound to the study of heredity since its inception. In this passage from *The Variation of Animals and Plants Under Domestication* (Volume II, pp. 35–36), we see Darwin in 1868 struggling to develop a theory of inheritance, unaware that Mendel had published the solution two years earlier. We see Darwin falling into error—changes that affect parents *do not* leave a mark on the "germ"—but we see also that with his usual insight he perceived the fact of hidden variation and the crucial distinction between GENOTYPE and PHENOTYPE. The genotype is a blueprint for an organism, the set of instructions for development received from its parents. The phenotype is the manifestation, in a series of developmental stages, of the interaction of this information with the physical and chemical factors—the environment in a very broad sense—that enable the blueprint to be realized.

## TWO PRINCIPLES OF GENETICS

Among the most important principles of heredity are that the flow of information from genotype to phenotype is unidirectional, and that the units of heredity retain their identity from generation to generation. The prevailing view of heredity in Darwin's day was one of BLENDING INHERITANCE: the intermediate offspring of a cross between a large and a small animal was interpreted in the same terms as the intermediate color of a dye produced by mixing strong and weak solutions. But if hereditary factors lose their identity and become intermediate when combined, the variation (variance) among organisms in a population will be halved in every generation and will quickly vanish. Mutations would then have to arise at a phenomenal rate to account for variation, but inheritance is usually so faithful that mutation rates are evidently not that great. Darwin was so distressed by this problem that he ultimately conceded that the environment might induce hereditary variation, and so came to rank the inheritance of environmentally acquired characteristics as a potent evolutionary force.

With Mendel's discovery of genes, this problem disappeared. The pink-flowered plant produced by red- and white-flowered parents does not pass on pink-determining factors; instead, red- and white-determining alleles combine anew to yield red- and white-flowered plants. The same principles hold for POLYGENIC traits: continuously varying characters that are affected by the combined action of genes at several or many loci, each of which affects the trait only slightly (Figure 1; see also Figure 3 in Chapter 5).

The faithful replication of new mutations renders unnecessary the theory of the inheritance of acquired characteristics, which moreover is contradicted by

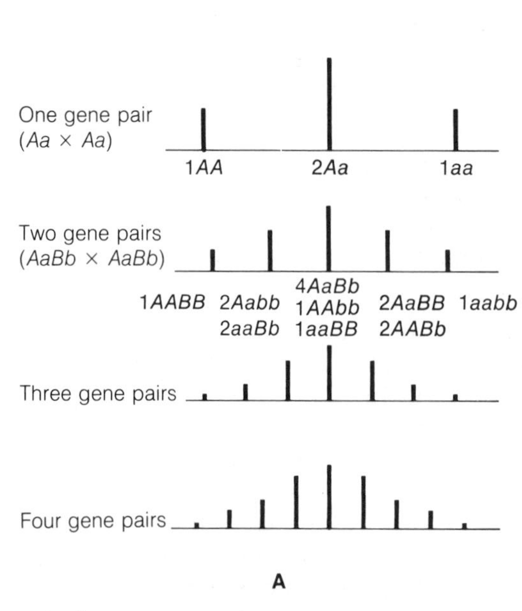


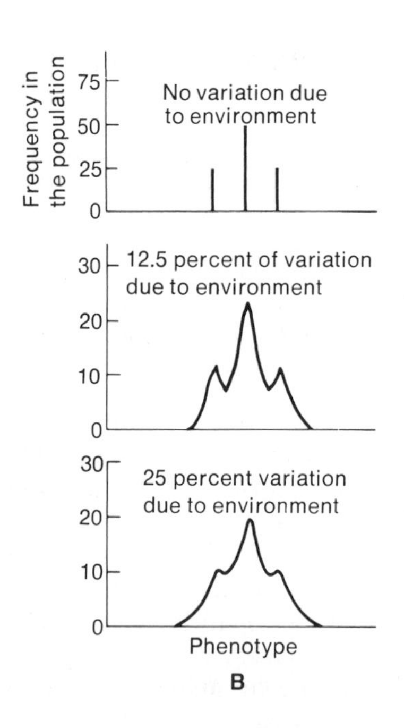


FIGURE 1

**Continuous variation. The abscissa represents a phenotypic character such as height, the ordinate the frequency of phenotypes in a population. (A) All individuals of a genotype have exactly the same phenotype. As we increase the number of loci (pairs of alleles, or genes), and decrease the contribution of each to the phenotype, the number of phenotypic classes increases. (B) For one pair of alleles, we superimpose the phenotypic variation that each genotype expresses because of variation in the environment. In a highly variable environment, the genotypes may overlap greatly in phenotype. Both (A) and (B) represent segregation in the $F_2$ generation of unlinked genes in a cross between heterozygotes. (A after Strickberger 1968; B after Allard 1960)**

fact. There is no evidence that an adaptive response of an organism's body to the environment can be translated backwards from, say, protein molecules into the DNA of germ cells.

However, the experiences of the parent can influence the characteristics of the offspring in some cases. Organisms inherit not only chromosomes but cytoplasm, especially in the egg. CYTOPLASMIC INHERITANCE (reviewed by Jinks 1964 and Grun 1976) is commonly based on the transmission of self-replicating bodies such as mitochondria, chloroplasts, and intracellular virus particles. In ciliate protozoans, the molecular organization of the cell surface acts as a template for the additional cell surface formed after cell division, so nongenetic alterations of the surface are transmitted from one generation to the next. Moreover, MATERNAL EFFECTS, usually persisting for only one generation, can arise if offspring are affected by the constituents of the egg cytoplasm. These sometimes are influenced by the mother's physiological state.

A few experiments have yielded puzzling results. For example, Durrant (1962), who studied flax, and Hill (1967), who studied tobacco, both found that variations in plant characteristics such as height that were induced by fertilizer treatments were transmitted to the plants' descendants for at least three generations. It has recently been discovered that some environmental shocks induce an increase in the number of copies of certain genes in flax, perhaps by interrupting DNA replication, which then starts again so that the earliest-replicating genes are replicated more than once (Cullis 1983). This might explain the peculiar observations in flax and tobacco. Several authors (e.g., Gorczynski and Steele 1981) have reported that phenotypic changes induced by drugs or antigens may be carried over into subsequent generations. The interpretation and reality of such effects is controversial (Campbell 1982). One possible interpretation is that the treatments selected for cytoplasmic particles of DNA (such as viruses) that conferred resistance and became incorporated into the genome. If this is indeed the mechanism, the phenomenon is compatible with neo-Darwinian principles (Fitch 1982), because the phenotypic alteration did not cause the genetic particles to change.

## THE GENETIC MATERIAL

Except in RNA viruses, the genetic material of all organisms is DNA. In prokaryotes this is organized into a single circular "chromosome;" in eukaryotes it is organized, together with histone proteins, into a set of linear chromosomes that reside in the nucleus. Eukaryotes also carry DNA, in circular form, in mitochondria and chloroplasts. The double-stranded helical molecule of DNA that runs uninterrupted through the length of a chromosome consists of a series of base pairs (BP), each consisting of a purine (adenine, A, or guanine, G) coupled to a specific pyrimidine (thymine, T, couples with A, while cytosine, C, couples with G). A single strand is so structured that polarity is evident from one end (the 5′ end, usually depicted on the left) to the other (the 3′ end). This polarity has important consequences during the formation of a DNA or RNA copy from a DNA template, as all known polymerases (copying enzymes) move only in the 3′ to 5′ direction on the template strand, creating a copy of opposite polarity (i.e., the copy grows from 5′ to 3′). In TRANSCRIPTION, one of the strands of the DNA duplex acts as a template for the synthesis of a single-stranded RNA molecule of opposite polarity in which the DNA positions held by G, C, T and A on the coding strand are occupied respectively by the complementary bases C, G, A and U (uracil). All or part of the RNA molecule may be TRANSLATED into a polypeptide or protein, with each successive triplet of bases in the RNA either coding for one of 20 different amino acids or acting as a "stop" signal. Each such triplet is called a CODON. Transfer RNA (tRNA) and ribosomal RNA (rRNA) are not translated into polypeptides.

The genetic code is often represented as an RNA code (Table I). There are numerous SYNONYMOUS CODONS: two or more codons with the same meaning. Much of the synonymy lies in the third base position; thus, for example, ACU, ACC, ACA and ACG all code for the amino acid threonine. An alteration of the DNA that does not alter the polypeptide product is referred to as a "silent substitution." The genetic code of mitochondria differs slightly from that of the nuclear genes, and in fact differs among major taxa such as mammals and yeasts; but with this and a few other minor exceptions, the genetic code is universal among both prokaryotes and eukaryotes as far as known. Jukes (1983) and others have speculated on the history by which the universal code has evolved.

TABLE I

**The genetic code. Amino acids specified by the 64 nucleotide triplets in mRNA**

| FIRST NUCLEOTIDE | SECOND NUCLEOTIDE: U | | C | | A | | G | | THIRD NUCLEOTIDE |
|---|---|---|---|---|---|---|---|---|---|
| U | UUU | Phe | UCU | Ser | UAU | Tyr | UGU | Cys | U |
| U | UUC | Phe | UCC | Ser | UAC | Tyr | UGC | Cys | C |
| U | UUA | Leu | UCA | Ser | UAA | Chain End | UGA | Chain End | A |
| U | UUG | Leu | UCG | Ser | UAG | Chain End | UGG | Trp | G |
| C | CUU | Leu | CCU | Pro | CAU | His | CGU | Arg | U |
| C | CUC | Leu | CCC | Pro | CAC | His | CGC | Arg | C |
| C | CUA | Leu | CCA | Pro | CAA | Gln | CGA | Arg | A |
| C | CUG | Leu | CCG | Pro | CAG | Gln | CGG | Arg | G |
| A | AUU | Ile | ACU | Thr | AAU | Asn | AGU | Ser | U |
| A | AUC | Ile | ACC | Thr | AAC | Asn | AGC | Ser | C |
| A | AUA | Ile | ACA | Thr | AAA | Lys | AGA | Arg | A |
| A | AUG | Met | ACG | Thr | AAG | Lys | AGG | Arg | G |
| G | GUU | Val | GCU | Ala | GAU | Asp | GGU | Gly | U |
| G | GUC | Val | GCC | Ala | GAC | Asp | GGC | Gly | C |
| G | GUA | Val | GCA | Ala | GAA | Glu | GGA | Gly | A |
| G | GUG | Val | GCG | Ala | GAG | Glu | GGG | Gly | G |

The names of the amino acids abbreviated in the table are: Ala, alanine; Arg, arginine; Asn, asparagine; Asp, aspartic acid; Cys, cysteine; Gly, glycine; Glu, glutamic acid; Gln, glutamine; His, histidine; Ile, isoleucine; Leu, leucine; Lys, lysine; Met, methionine; Phe, phenylalanine; Pro, proline; Ser, serine; Thr, threonine; Trp, tryptophan; Tyr, tyrosine; Val, valine.

## Some important methods

The methods, especially the molecular methods, of genetics are more appropriately treated in textbooks of genetics than of evolution. However, three methods that have been increasingly used in evolutionary studies bear mention at this point.

ELECTROPHORESIS separates macromolecules (e.g., proteins or nucleic acid fragments) on the basis of mobility differences owing to their size, conformation, and net charge. Extracts from one or more organisms are placed in a porous gel, often of starch, to which an electric field is applied. After this has induced movement of the charged molecules, their location is established by autoradiography (if they have been labeled with radioactive isotopes) or by reaction with a substrate; the product of this reaction is visualized by a reagent such as a dye.

In NUCLEIC ACID HYBRIDIZATION, fragments of DNA are denatured to form single strands, which by base pairing form duplexes with other single strands of either DNA or RNA. The more complementary the sequences of two paired strands, the more stable the duplex, and the higher the temperature required to dissociate them. The strength of association, then, is a measure of the degree of complementarity in base pair sequence of, for example, DNA from two species. The kinetics of the reaction enable one to estimate the number of copies of a DNA sequence; for example, if denatured DNA from a single organism has many copies of a sequence, the high encounter rate of like sequences will cause rapid duplex formation. Similarly, by mixing the messenger RNA (mRNA) from an organism with its DNA, the kinetics of the association between them can be used to estimate the diversity of different mRNAs.

A partial map of the base pair sequence of a region of DNA can be obtained by the use of RESTRICTION ENZYMES. These bacterial enzymes, which normally function to attack foreign DNA such as that of viruses, cut DNA at certain recognition sequences; for example, the enzyme *Eco*R1 (from *Escherichia coli* bacteria) recognizes the sequence GAATTC. The pattern of fragments of DNA produced by "digesting" it with this enzyme can be used to determine how many such recognition sequences occur, and the use of several such enzymes in concert enables one to determine the relative positions of and distances between these sequences.

The complete base pair sequence of a fragment of DNA can now be determined by breaking the molecule with reagents that are specific for each of the four bases. The methods by which the sequence is interpreted are too complex to be summarized here (see Lewin 1985, Chapter 3).

## GENE STRUCTURE

To a first approximation, one may think of a gene as a DNA sequence that is transcribed into an RNA transcript that codes for a single polypeptide. But genes are actually very complex (Figure 2) and exceedingly difficult to define.

For one thing, some sequences of DNA code for transfer RNAs and ribosomal RNAs that are not translated. For another, in bacteria a single RNA transcript frequently carries information for several proteins with different but coordinated functions. Moreover, in eukaryotes and some prokaryotes, the RNA transcript of a single "gene" often consists of coding regions (EXONS) separated by several or many noncoding "intervening sequences" (INTRONS). This primary transcript is processed by enzymes to form an mRNA molecule from which the introns are

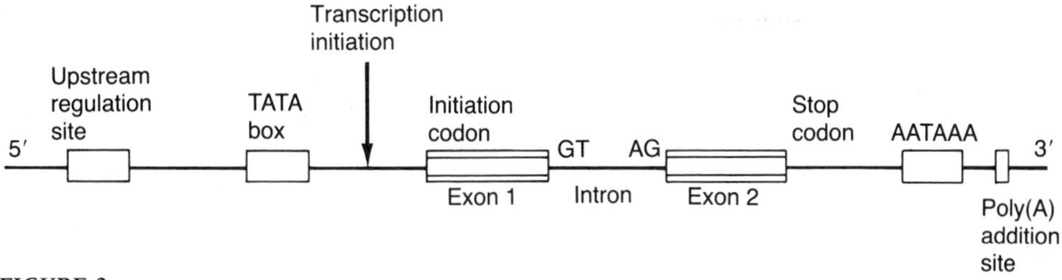


FIGURE 2

**Diagram of a eukaryotic gene. The upstream regulation site controls the initiation of transcription. The "TATA box," with a sequence similar to TATA, also helps to fix the site at which transcription is initiated. Each intron (nontranscribed region between translated regions called exons) begins with GT and ends with AG. Processing of the RNA is thought to require the AATAAA sequence and a tail of adenines [Poly(A)]. (From Li 1983)**

spliced out. A gene for collagen in the chicken is 40,000 bp long (40 kilobases, kb), but consists of more than 50 short exons that together code for an mRNA transcript of about 5000 bp (5 kb). The several exons of certain genes appear to correspond to different functional regions (domains) of the protein product for which the gene codes, but in other cases no such correspondence is evident. In many cases, an RNA transcript is variously spliced, yielding different messengers and ultimately different proteins. There exist several structurally distinguishable classes of introns that probably differ in evolutionary origin (Sharp 1985).

To complete our description of the gene's complexity, we note that transcription of a gene requires that the polymerase recognize a PROMOTER, typically a short nontranscribed sequence upstream (i.e., in the 5′ direction) from the starting point of transcription. Sequences that promote or enhance transcription may also extend downstream from the starting point, and sites upstream from the promoter often regulate transcription as well. Genes are often separated by "spacer" sequences that in at least some cases affect transcription.

## REPEATED AND NONREPEATED DNA

The size of the genome varies greatly among different organisms (Figure 3), from less than 400 bp in some virus-like particles to more than $10^{11}$ bp in the haploid (gametic) complement of some vascular plants. The high DNA content in some plants is the consequence of polyploidy (possession of multiple sets of chromosomes), but polyploidy does not explain why the DNA complement of some salamanders is more than ten times that of mammals, nor why it differs more than a hundredfold among salamander species. Generously estimating the average RNA transcript at 10,000 bp, the average mammalian genome has enough DNA to code for more than 300,000 genes. In *Drosophila,* the number of functional genes has been estimated at about 10,000, based on the close correspondence between sites of mutation and the bands that are visible in the salivary gland chromosomes; but this number of genes could account for less than 10 percent of the amount of DNA. The number of different RNA transcripts that can be discerned during development of the sea urchin *Strongylocentrotus purpureus* is less than 20,000; the coding regions in these transcripts would account for less than 3 percent of the haploid complement of DNA. The number of genes estimated to code for proteins in a eukaryote, then, appears to account for only a small fraction of the DNA.

This paradox was partly resolved by the discovery that much of the DNA in eukaryotes consists of repetitive sequences, some of which code for gene products and some of which do not. In contrast, the genome in viruses, mitochondria, and prokaryotes generally lacks a substantial amount of repetitive sequences. In eukaryotes, most of the sequences that code for polypeptides are represented by single copies, although as we will see, these often form clusters of similar sequences. Single-copy sequences constitute up to 90 percent of the DNA in "lower" eukaryotes such as fungi, but only about 20 percent of the DNA in some plants and amphibians.

Some protein-coding genes, such as those coding for histones, are represented by up to several hundred copies. These contribute to the MODERATELY REPETITIVE fraction of DNA, as do the genes coding for rRNA and tRNA. The frog *Xenopus laevis,* for example, has 450 copies of the genes coding for 18S and 28S

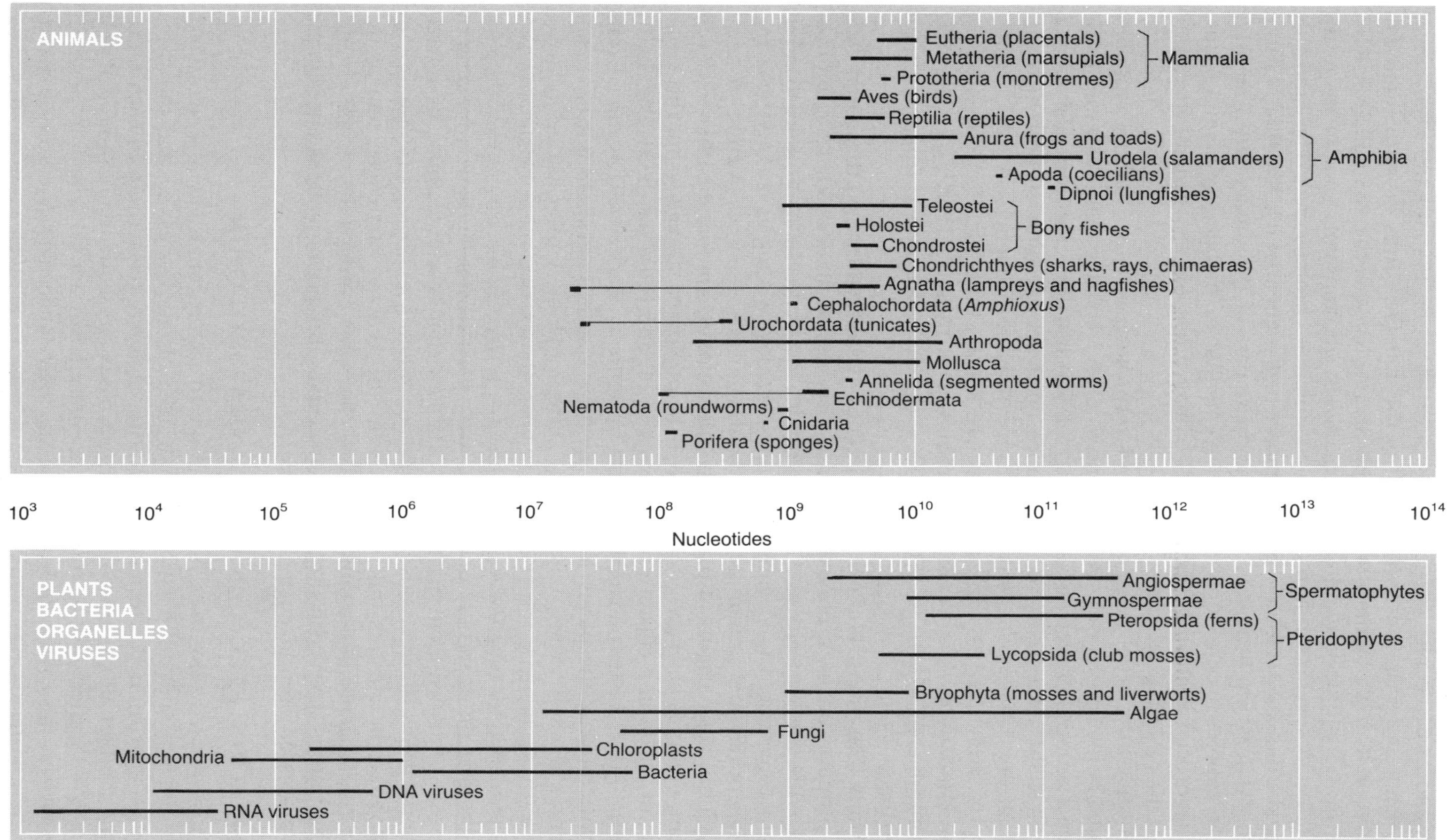


FIGURE 3

**The range of DNA contents, measured as the number of nucleotide base pairs per haploid cell, in each of a variety of groups of organisms. Note the logarithmic scale. (From Wagner et al. 1980)**

rRNA and 24,000 copies of the gene for 5S rRNA. The 18S and 28S rRNA genes are a single transcription unit, and all of the 450 such units lie sequentially, in tandem, on a single chromosome. Likewise, in *Drosophila* the 18S and 28S rRNA genes form a single tandem cluster on the sex chromosomes, but in mammals they fall into several tandem clusters on different chromosomes. The tandemly repeated units are separated by nontranscribed sequences of DNA that vary in length. At least in *Drosophila* and mammals, part of the moderately repetitive fraction appears to consist of genetic elements that are mobile (see below) or that may have been mobile in the past. This may prove to be a general feature of many eukaryotes.

HIGHLY REPETITIVE DNA consists of very short (usually 5–12 bp) tandemly repeated units. Several different sequences may be repeated in this fashion. These sequences, often referred to as SATELLITE DNA, usually constitute only about 5 percent of the total DNA, but sometimes much more; about 40 percent of the DNA of *Drosophila virilis* consists of three satellite sequences with $1.1 \times 10^7$, $3.6 \times 10^6$, and $3.6 \times 10^6$ copies respectively. Much of the satellite DNA occurs in HETEROCHROMATIC REGIONS of the chromosomes: those regions, especially near the centromeres, where the DNA is very tightly coiled and is apparently not transcribed.

Many if not most of the structural genes of eukaryotes (i.e., those that code for a product such as a polypeptide) are members of GENE FAMILIES: groups of genes with similar sequences. For example, the hemoglobin protein in adult humans consists of two $\alpha$ and two $\beta$ polypeptides (written as $\alpha_2\beta_2$). The $\alpha$ and $\beta$ polypeptides are similar in amino acid sequence. A small fraction of adult hemoglobin has $\delta$ chains in place of $\beta$. Other globins are found earlier in development: embryos have $\zeta_2\epsilon_2$, $\zeta_2\gamma_2$, and $\alpha_2\epsilon_2$ hemoglobins. The genes for the $\alpha$-like chains ($\alpha$, $\zeta$) lie in a cluster on chromosome 16, and those for the $\beta$-like chains ($\beta$, $\delta$, $\gamma$, $\epsilon$) in a cluster on chromosome 11 (Figure 4). Each gene has three exons and two introns; the genes in a cluster have closely corresponding base pair sequences in their exons, whereas the introns exhibit greater variation in sequence. By sequencing the DNA, it has been shown that the $\alpha$ cluster has two genes that code for identical $\alpha$ chains, and the $\beta$ cluster has two genes coding for $\gamma$ chains that differ in only one amino acid. Each cluster, moreover, has one or more PSEUDOGENES: nontranscribed sequences that resemble those of related functional genes. In the mouse *Mus musculus*, globin pseudogenes lie not only in the globin gene clusters, but elsewhere in the genome; moreover, one of the globin pseudogenes of the mouse, $\psi\alpha3$, has the same exons as the related $\alpha$ gene, but the introns have been precisely eliminated. This pseudogene is an example of a "processed pseudogene" (see below).

## REPLICATION, RECOMBINATION, AND SEGREGATION

DNA is replicated prior to each cell division. In each chromosome of a eukaryote, replication proceeds bidirectionally from each of many sites, so that the chromosome is partitioned into many replicating regions (REPLICONS). In each replicon, the double helix is unwound, pairs of bases are sequentially separated at a moving "replication fork," and complementary bases are brought into position along each of the two strands. Because the DNA molecule is helically twisted, separation of the strands would require the end of the molecule to rotate wildly

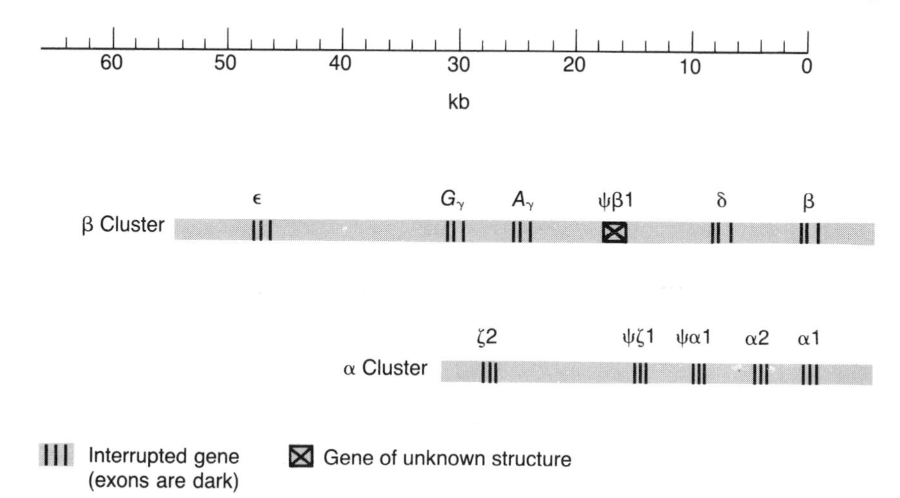


FIGURE 4

**The α and β globin families of human genes. Each gene is depicted as three dark lines representing three exons. Pseudogenes are denoted ψ. The genes are separated by regions of several kilobases in which no other functional genes have been identified. (From Lewin 1985)**

were it not for enzymes that cut ("nick") one or both strands, allowing short segments of one or both strands to unwind before they are rejoined.

During meiosis, the DNA duplex molecules of homologous chromosomes recognize each other by some unknown mechanism and become precisely aligned in synapsis. Crossing over may then occur. In one model of crossing over, nicks are made at corresponding points in a single strand of each of the two aligned molecules and each free end is joined with the broken strand of the other molecule (Figure 5). Base pairs then sequentially separate and reunite with those of the other DNA molecule, forming a heteroduplex sequence of greater or lesser length. According to this model, recombination is observed when the heteroduplex region is terminated by a second pair of nicks in the strands that did not participate in the exchange. The frequency of crossing over between any two sites is expressed by the RECOMBINATION FRACTION (often denoted *R*), which may range from effectively zero for very closely linked sites to 0.5, the frequency of recombination between genes on different chromosomes (INDEPENDENT ASSORTMENT), or between genes on the same chromosome that are so frequently separated by crossing over as to assort independently. Two loci on the same chromosome that are infrequently separated by crossing over are said to be more tightly linked than loci that are more frequently separated by crossing over.

Ordinarily, homologous chromosomes are drawn to opposite poles in the first meiotic division by movement of their centromeres. This results in a 1:1 segregation into the gametes of alleles on the two homologues.[1] Cases of SEGREGATION DISTORTION, also called MEIOTIC DRIVE, exist, whereby one allele is

[1] Throughout this book, I will describe the products of meiosis as gametes. This is not true of ferns and other forms with alternation of generations, in which the product of meiosis is a haploid gametophyte that forms haploid gametes.

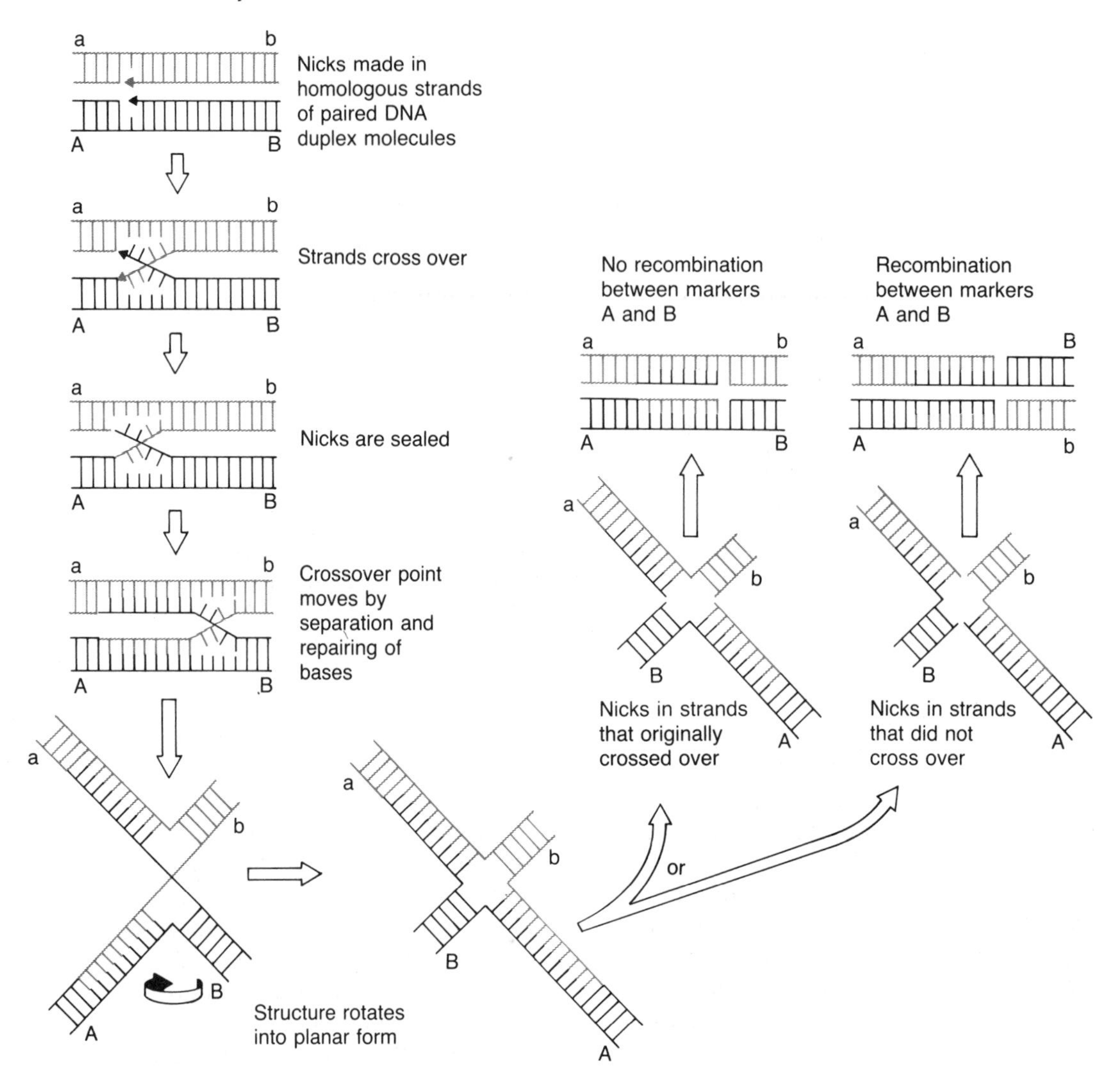


FIGURE 5

**A model of recombination between paired DNA molecules that carry different sequences at sites A and B (in the same gene or different genes). One strand of each DNA duplex is nicked, the free ends cross over and are joined, and the crossover point moves by separation of nucleotide pairs and repairing of nucleotides of the crossed strands, forming a heteroduplex region in each molecule. A second pair of nicks ends the movement. Crossing over between sites A and B is observed if the nicked strands are not those that were nicked at first. (From Lewin 1985)**

carried by more than one half of the gametes of a heterozygote. If the segregation ratio is known, the genotypic proportions among the offspring of any two parents can be probabilistically predicted if the gametes unite at random. Assuming, for example, the mating $AA' \times AA'$ and a 1:1 segregation ratio, the probability that two $A$ gametes will unite is $(\frac{1}{2})^2$, that of two $A'$ gametes is the same, and that of a union between *A* and *A′* is 2 × ½ × ½, since the union of *A* egg with *A′* sperm and of *A′* egg with *A* sperm are independent events. Thus the genotypes *AA*, *AA′*, and *A′A′* appear among the progeny in the familiar Mendelian ratio 1:2:1.

## GENOTYPE AND PHENOTYPE

Because genes exert their effects on the phenotype via biochemical reactions, they do not operate *in vacuo*. Their effects depend on the chemical and physical milieu in which these reactions take place. Quite often this milieu depends on other genes, so that the phenotypic effect at one locus depends on the genotype at one or more other loci: the phenomenon of EPISTASIS. The external environment, too, affects phenotypic expression, sometimes in the same way that alteration of the genes does. For example, altered phenotypes called PHENOCOPIES that are similar to the altered wing veins exhibited by such mutants as *crossveinless* and *Curly* can be produced in nonmutant *Drosophila* by a temperature shock during critical periods of development. An organism is not just its genes; it is constructed of materials derived from the environment, under environmental conditions that influence the rate of biochemical reactions. If we were to raise individuals of the same genotype under a wide range of environmental conditions, we would find that virtually every gene varies to some degree in phenotypic expression. Some characteristics, like body weight in animals or growth form in plants, would vary more than others, such as the number of vertebrae in a mammal or the structure of a cell membrane. The more invariant characteristics are said to be more highly CANALIZED, or DEVELOPMENTALLY BUFFERED, into a more restricted set of developmental channels.

Different genotypes are canalized to different extents, so that certain alleles are more variable in PENETRANCE (the fraction of individuals in which their phenotypic effect is manifest) and in EXPRESSIVITY (the magnitude of the phenotypic effect) than others. Each genotype has its own NORM OF REACTION (Figure 6), a variety of expressions under different environmental conditions. In some cases a wide reaction norm, the production of greatly different phenotypes under different environmental conditions, is disadvantageous or pathological. Our tendency to develop scurvy when deprived of vitamin C is not advantageous. But very often, especially if the environment fluctuates greatly, the most advantageous genotype is the one whose phenotypic expression varies with prevailing conditions. Examples include the ability of a weasel to develop pigmented fur in summer and white fur in winter, and the capacity of many plants to produce tough, heavily waxed sun leaves and more delicate shade leaves. Similarly, some behavioral traits are more readily altered by interaction with the environment than others, and more readily in some species than others—as in the phenomenon of learning.

Because each aspect of the phenotype is a product not of the genes alone or of the environment alone, but of the interaction between the two, it is fallacious to say that a characteristic is "genetic" or "environmental." We can only ask if *differences* among individuals are attributable more to genetic or to environmental factors. The answer is likely to depend on the particular group of individuals we examine. If they are genetically homogeneous, much of the variation is environmental; if they have all been exposed to homogeneous environmental conditions, more of the variation will be genetically based. These elementary principles have

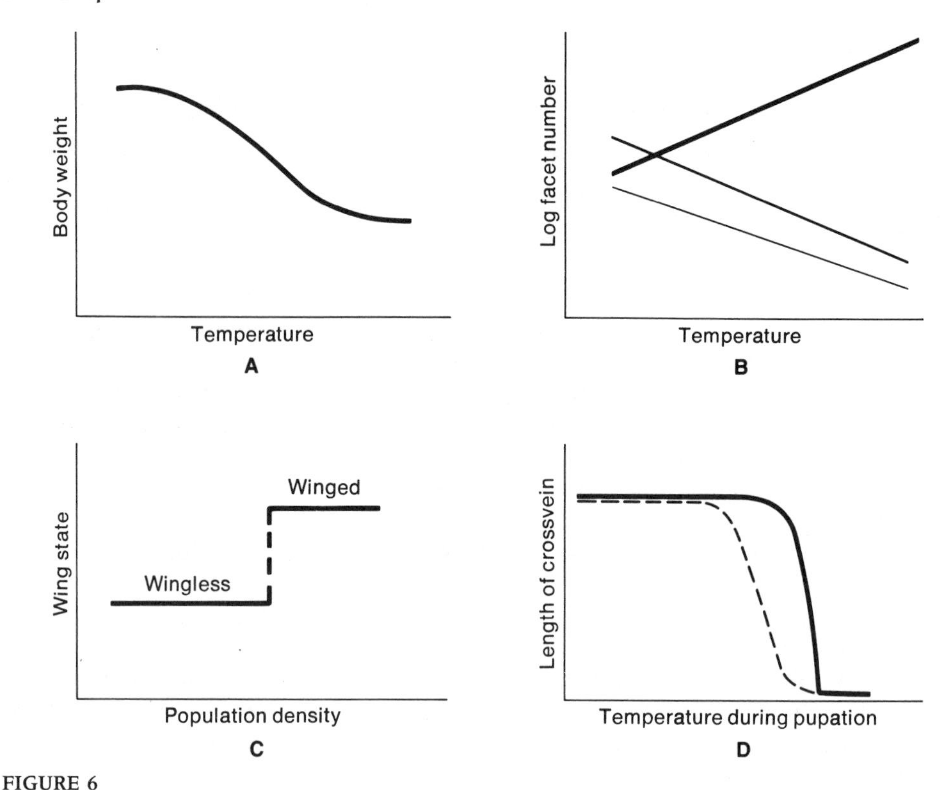


FIGURE 6

**Four schematic illustrations of real reaction norms. In each case the phenotype developed depends on environmental conditions; the norm of reaction of a genotype is the variety of different phenotypic expressions it can manifest. (A) The response of body weight to temperature in *Drosophila* and many other insects. (B) Number of eye facets, in relation to temperature, of different mutant genotypes at the *Bar* locus in *Drosophila*. (C) A developmental "switch," as found in some aphids which develop wings if sufficiently crowded at a critical period in development. (D) Differential sensitivity of two *Drosophila* genotypes to a heat shock that affects development of the crossvein in the wing. (B after Hersh 1930)**

sometimes been ignored, as in arguments over whether aggressive behavior or sex role stereotypy are or are not "innate" or "genetic" in humans. These behaviors are both genetic and environmental, in that both genes and environment are prerequisite to their existence. The question is, rather, whether the behavior varies more as a function of genetic or environmental variation, and to what degree.

The phenotypic effect of a structural gene that codes for an enzyme is likely to depend on which cells express the gene and at what point in development, on the amount of enzyme synthesized, and on the activity of the enzyme. If an enzyme acts by, say, converting precursor X into pigment Y, the amount of pigmentation might be proportional to the amount of enzyme (Figure 7). So if $A$ codes for a relatively active enzyme and $A'$ for a relatively inactive enzyme, the amount of pigment in $AA$ genotypes might then be twice that in $AA'$ genotypes, relative to the amount in $A'A'$. Such alleles are said to have ADDITIVE effects: the

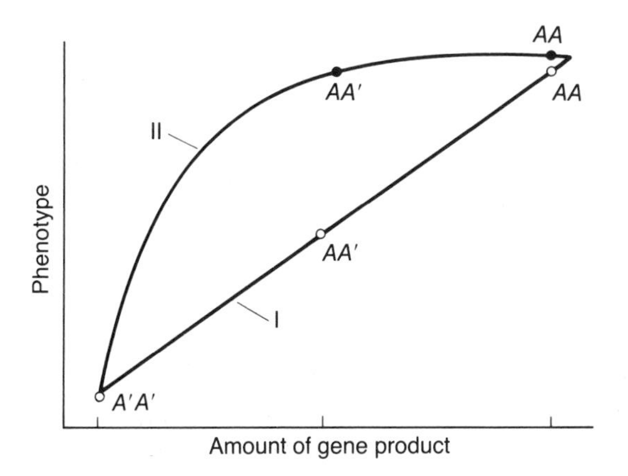


FIGURE 7

**Two possible relationships between amount of gene product (e.g., an enzyme) and a phenotypic character such as intensity of pigmentation. In both cases the amount of gene product in the heterozygote is intermediate between that of the two homozygotes. In curve I, the phenotype is proportional to the amount of gene product. The phenotype shows additive inheritance, and neither allele is dominant. In curve II, the amount of gene product of the heterozygote is sufficient for complete expression of the phenotype, and allele *A* is dominant.**

phenotype of the heterozygote is intermediate between that of the homozygotes. But if the enzyme produced by even a single *A* allele is active enough to convert all the substrate into pigment, *AA′* and *AA* would then have the same phenotype, and allele *A* is said to be DOMINANT over the RECESSIVE *A′*. (This is one of several possible explanations of dominance.) We can also imagine that *A* and *A′* code for functional enzymes that are most active under different conditions of temperature, pH, or other variables—as has been well documented for variant forms of many enzymes. It would then be conceivable that enzymatic activity in the heterozygote could exceed that of either homozygote, and give rise to a more extreme phenotype. A more extreme phenotype of the heterozygote is termed OVERDOMINANCE.

Almost every characteristic of an organism is POLYGENIC, i.e., affected by the action of several or many genes. Quite often each gene contributes independently to the character, so that the phenotype is the sum of the contributions of the individual loci. If there is epistasis between loci, the whole is not equal to the sum of its parts; for example, mutations at the *white* locus in *Drosophila* cause an incapacity to form the precursor to eye pigments formed subsequently by enzymes produced at other loci.

As each character is affected by several genes, so conversely each gene typically affects several characters: it is PLEIOTROPIC. Pleiotropy can arise at several levels. For example, mutation of a gene coding for a transfer RNA could affect the amino acid composition of every protein and is likely to be lethal. Mutation of a gene whose transcript is spliced alternatively into different mRNAs that are translated into different proteins could affect the activity of gene products with different functions. Mutational alteration of a polypeptide that has several biochemical functions will have correspondingly diverse effects; for example, a polypeptide that in monomeric form acts as an alanine dehydrogenase is known to act as a glutamic dehydrogenase when it is aggregated into a polymer. Alteration of a gene product that is incorporated into many tissues affects all of them. For example, the *achondroplasia* mutant in the rat causes an inability to suckle, faulty pulmonary circulation, occlusion of the incisors, and arrested development, all stemming from an organism-wide abnormal development of cartilage early in ontogeny. Pleiotropy in this case is "direct," meaning that the gene acts in all the affected tissues. "Relational" pleiotropy arises when developmental interactions engender a suite of diverse consequences from a single primary effect of gene action. For example, sickle-cell hemoglobin in humans, differing from normal hemoglobin by a single amino acid substitution in the β chain, causes the red blood cells to sickle; this causes destruction of the cells and interference with circulation, which in turn have numerous physical and physiological effects (Figure 8).

## THE CONTROL OF GENE EXPRESSION

With few exceptions, the complement and structure of genes is thought to be much the same in each of the somatic cells of a multicellular organism's body. Certain cells may become polyploid, as in the mammalian liver, and certain genes may become amplified; for example, in *Drosophila* ovarian follicle cells, genes that code for the chorion proteins of the eggshell are multiplied in number as much as sixtyfold. On the whole, however, the gene complement is stable, even though the expression of many genes varies among cell types and developmental stages.

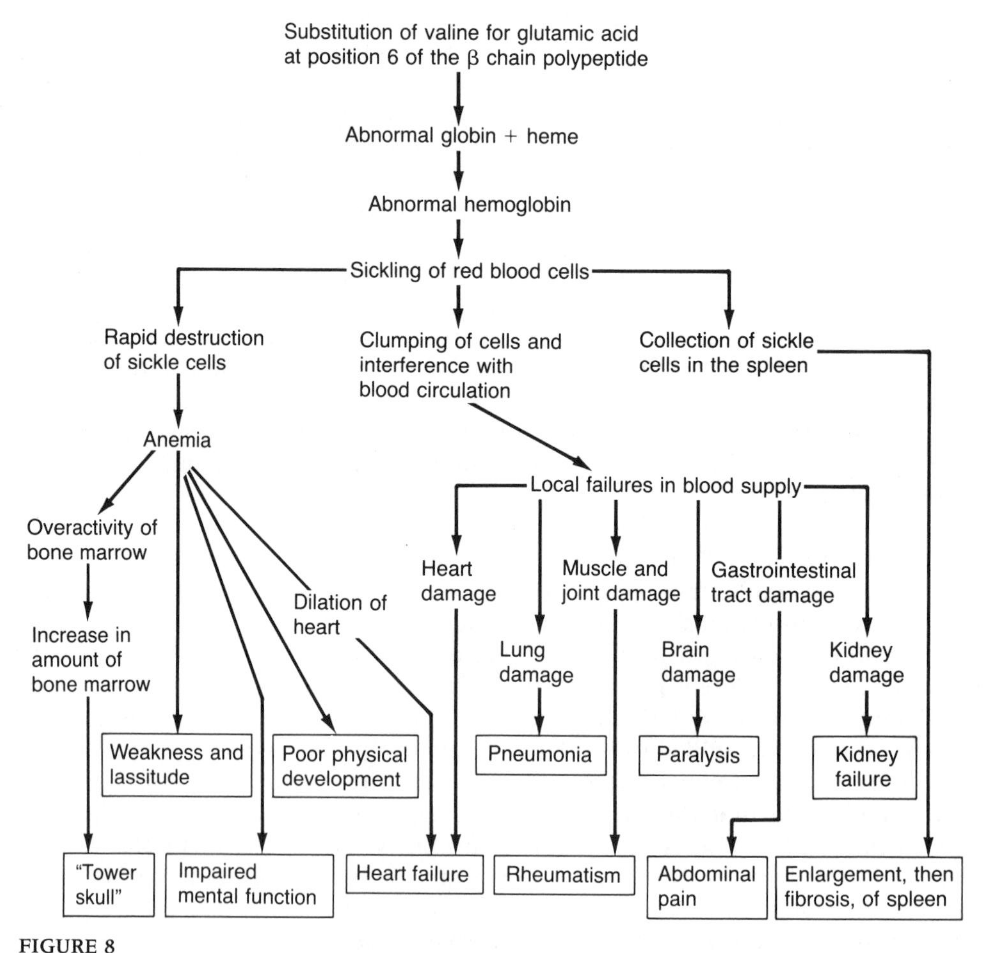


FIGURE 8

**Pleiotropic effects of the amino acid substitution in the β chain of human hemoglobin that results in sickle-cell anemia. (From Raff and Kaufman 1983)**

Several mechanisms might account for tissue-specific expression. These include regulation of the transcription of a DNA sequence into an RNA transcript; selective processing of primary RNA transcripts into mature mRNAs; regulation of the translation of mRNA into polypeptide products; posttranslational regulation of the activity of polypeptides; and regulation of the rate at which the gene products are degraded relative to the rate of synthesis. There is evidence that each of these operates in particular instances.

In bacteria, the expression of many genes is regulated at the level of transcription. In many cases, several genes with coordinated functions, together with a regulatory sequence called the operator, form an OPERON. One kind of operon is exemplified by the β-galactosidase system, in which a separate regulator gene produces a repressor protein that binds to the operator and prevents transcription. An inducing substance (lactose) that enters the cell binds to the repressor, dissociating it from the operator and leaving the operon free for transcription of its several structural genes (Figure 9).

Operons have not been identified as such in eukaryotes, although nontranscribed DNA sequences upstream from structural genes can have a regulatory effect, since mutations in these regions sometimes prevent gene expression. Britten and Davidson (1971; also Davidson and Britten 1973), extrapolating from bacterial operons, have offered a model of eukaryotic gene regulation that has attracted considerable attention (Figure 10). Each structural gene, they proposed, is governed by one or more adjacent receptor genes, each of which is activated by the product of an integrator gene. A "sensor" associated with the integrator gene responds to a stimulus (e.g., a hormone or an inducing factor released from another tissue), causing the integrator to activate the receptors. If the same receptor sequence were associated with a "battery" of different structural genes, they would be regulated coordinately, as one might expect to occur in the differentiation of a given cell type. However, there is little direct evidence in favor of this model.

Although the mechanism of control is poorly understood, much if not most of the regulation of gene expression in eukaryotes appears to occur by the differential transcription of genes in different tissues and at different times in

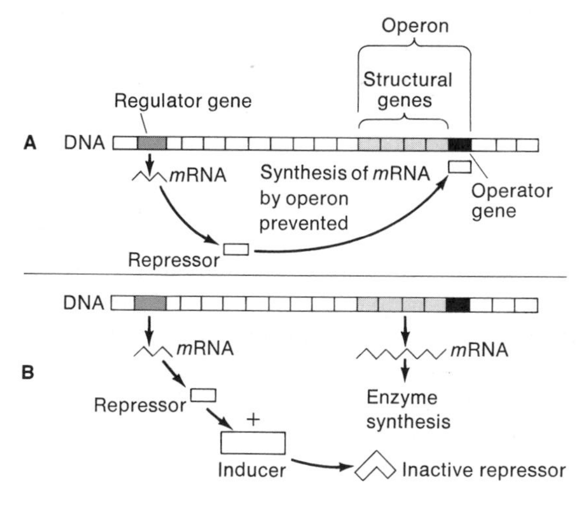


FIGURE 9

**Model of gene regulation in bacteria. (A) The structural genes in the operon are inactive, since the operator gene is bound by a repressor. (B) An inducer substance from outside the cell binds to the repressor, derepressing the operator. Transcription of mRNA at the structural genes associated with the operator then occurs. (After Strickberger 1968)**

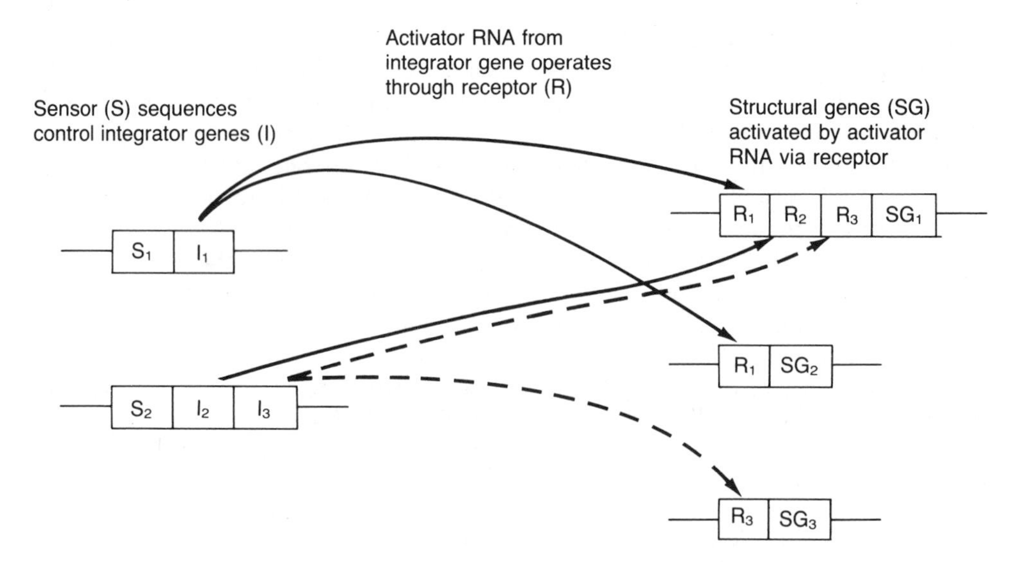


FIGURE 10

**The Britten-Davidson model for coordinated regulation of transcription in eukaryotes. A stimulus activates a sensor gene (S), which turns on an integrator gene (I). Its product, an activator RNA, activates one or more receptors (R) that turn on associated structural genes (SG). An integrator might activate different receptors, and a structural gene might be turned on by one of several receptors that are activated by different integrators. (From Gilbert 1985)**

development (Darnell 1982). At least some regulation, though, is posttranscriptional (Gilbert 1985). In sea urchins, for example, the kinds of primary RNA transcripts in the nucleus (heterogeneous nuclear RNA, or hnRNA) seem to be much the same throughout development; in each developmental stage a subset of these is processed into mRNA.

## DEVELOPMENT

The development of an organism from a fertilized egg includes processes of CYTODIFFERENTIATION, whereby cells acquire different biochemical and structural features, and MORPHOGENESIS, the acquisition of the three-dimensional form of tissues, organs, and structures. For the most part, the mechanisms of cytodifferentiation and morphogenesis are not understood in detail. In this gap in our knowledge—between primary gene action and the development of complex phenotypes—lies much of what we do not yet understand about evolution (Chapter 14).

The mechanisms of morphogenesis include cell movement, cell adhesion, the localized control of cell division, interactions among tissues, and cell death. Properties of cell surfaces govern cell-specific patterns of adhesion, whereby cells of like kind adhere more strongly to each other than to other kinds of cells—a phenomenon that is important in governing form. In response to stimuli that are generally not well understood, cells may move as groups—as in gastrulation—or as individuals, as in the case of vertebrate neural crest cells that ultimately differentiate into autonomic nerve ganglia, dermal pigment cells, and other tissues. Groups of cells often differentiate and acquire distinctive form in response to INDUCTION by other groups of cells; for example, the central nervous system of vertebrates differentiates in response to inducing factors emanating from the notochord. Inducing factors in some cases appear to form a concentration gradient as they diffuse from a source, with the concentration providing "positional information" that specifies the site of development of structures and the form they take. A gradient emanating from the posterior edge of the wing bud in a chick embryo, for example, appears to specify the position and form of the developing digits. Such a gradient in an inducing substance is an example of a PREPATTERN: a spatial arrangement of chemical factors that determine the spatial organization of features such as pigments, bones, or epidermal structures.

The three-dimensional form of a structure is influenced in part by factors that govern the rate of cell division, so that cells multiply more rapidly in certain sites or dimensions than in others. The causes of local variation in rates of cell division are not understood. Nor are the causes of localized cell death, which can also influence form. The digits of most vertebrates, for example, owe their form in part to the death of cells in the interdigital regions.

The mechanisms of development are known to be far more diverse and complex than I have discussed, and we will return to some of them in more detail in Chapter 14. Several major principles of development, however, bear emphasizing at this point. The chemical and physical properties of cells and tissues can be altered by mutation of the genes. But the organization of proteins and other molecules into cell constituents bestows on cells, tissues, and organs certain "self-organizing" properties that affect form. For example, dissociated cells in vitro will often reaggregate into groups of like cells, and these groups may spontaneously achieve the same spatial arrangement relative to each other that they have in the living embryo. Much of the "information" for making an embryo resides in the genes and in the content and organization of the cytoplasm of the fertilized egg—which itself is informed by the genotype of the mother. But this genetic information is translated into structures such as cell surfaces that have their own information, carrying out activities without continual instruction from the genes (Gerhart et al. 1982).

The interactions among the parts of an organism provide a certain capacity for correcting errors and coordinating the development of different structures. For example, the strength and form of a bone sometimes develop in response to, and are maintained by, the forces applied by the muscles that insert on it. But although interactions within the organism can provide integration and homeostasis, they can also amplify small deviations into large ones. For example, the double ovary of the nematode *Caenorhabditis elegans* develops from four cells, one of which gives rise to a "distal tip cell" early in differentiation. When this cell is ablated, one of the arms of the ovary fails to develop, even though the cell primordia from which it normally arises have not been altered (Raff and Kaufman 1983). Numerous other embryological experiments show that an alteration of an early embryonic structure can, by tissue induction and other direct or indirect influences, have diverse and drastic effects on structures that develop later.

## MUTATION: THE ORIGIN OF GENETIC VARIATION

Heredity is a conservative force conferring stability on biological systems (Dobzhansky 1970). Yet no mechanism composed of molecules and subject to the impact of the physical world can be perfect. Mistakes in the copying produce altered sequences of DNA—MUTATIONS—that are perpetuated.

"Mutation" is as difficult to define as "gene." Until recently, a gene was defined as a DNA sequence that codes for a polypeptide, but we may now argue whether or not introns count as part of genes, or whether the term gene should also embrace nontranslated regulatory regions and nontranscribed sequences. "Mutation" is similarly a vague term. It is often defined as a change in the base pair sequence of a gene, but sometimes the term is broadly used to include changes in the number and structure of chromosomes (the karyotype). Recombination may be said to differ from mutation in that it is usually a reciprocal exchange of DNA sequences ("genes") that are themselves not altered. However, crossing over is not always reciprocal, and it may occur within the confines of a gene and so alter the base pair sequence (see below). Thus some recombinational events are, in effect, mutations.

## CHANGES IN THE KARYOTYPE

The subject of mutational changes in chromosomes is very complex because of the immense variety of meiotic mechanisms and chromosomal behavior that animals and plants display. It is treated extensively by Stebbins (1950, 1971), White (1973), and Swanson et al. (1981). It is useful to bear in mind that the loss of a significant amount of genetic material, a whole chromosome or a major part of a chromosome, almost always reduces the viability of a gamete or an organism. Such deficiencies may occur when a chromosome arm loses its centromere, and thus is lost during nuclear division. (Exceptions exist, as in nematodes and hemipteran insects, in which even small fragments of chromosomes have centromeric activity and travel to the poles of the meiotic spindle.) A cell that lacks one or more of the normal complement of chromosomes is termed ANEUPLOID.

### Polyploidy

If the first meiotic division fails to occur, unreduced gametes are formed—for example, diploid gametes ($2N$) in a diploid organism. The union of such a gamete with a haploid ($N$) gamete yields a triploid ($3N$) zygote. A triploid is usually highly sterile, because each of its gametes receives an imbalanced complement of chromosomes: two from some triplets (trivalents) of homologues, one from others (Figure 11A). However, two unreduced ($2N$) gametes may unite to form a tetraploid ($4N$) zygote. In an AUTOTETRAPLOID, in which all four homologues come from one species and so have similar DNA sequences, all four chromosomes of each homologous set are aligned together on the meiotic spindle. Because these quadrivalents need not segregate in pairs, the gametes receive unbalanced chromosome complements, so autotetraploids are usually sterile (Figure 11B). Nevertheless, a few naturally occurring autotetraploid populations of plants have been described, such as certain fireweeds (*Epilobium*; Mosquin 1967). Autotetraploidy is most common in parthenogenetic forms, in which meiosis does not occur or is

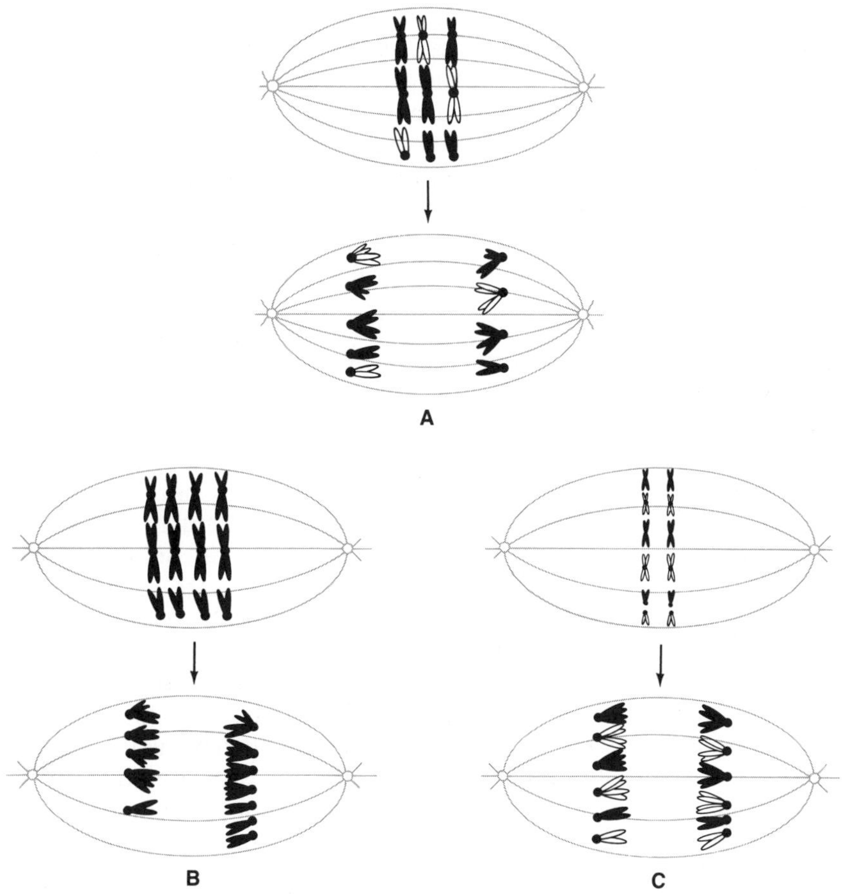


FIGURE 11

**Diagrams of meiotic metaphase and anaphase in polyploids. (A) A triploid typically produces gametes with unbalanced chromosome complements, as does an autotetraploid (B). An allotetraploid (C), if formed by polyploidization of a hybrid between two diploid species (with "white" and "black" chromosomes respectively), forms gametes with balanced chromosome complements.**

organized so that the parent's chromosome complement is reconstituted in the egg. The brine shrimp *Artemia salina*, for example, consists not only of a diploid sexual form but also of asexual diploids, triploids, tetraploids, pentaploids, octaploids, and decaploids.

Among natural populations, there is a full spectrum from autopolyploids to allopolyploids. ALLOPOLYPLOIDY, the formation of a polyploid by hybridization of genetically differentiated diploid populations (e.g., different species), is common among plants, although not among animals. If the genetic content of the chromosomes of the two parental forms is sufficiently different, the tetraploid hybrid is fertile because its chromosomes form $2N$ sets of bivalents (pairs) rather than $N$ sets of irregularly segregating quadrivalents (Figure 11C). Such tetraploids behave chromosomally as if they were diploid, although at many individual loci they are tetraploid (i.e., have four copies of the same gene). Perhaps because of the doubled gene dosage, polyploids are often larger and more robust than diploids, have higher enzyme and hormone levels, and differ from diploids in many physiological and ecological respects (Levin 1983).

Over evolutionary time, the identical (duplicate) genes that a tetraploid has derived from its diploid ancestors can experience different mutations, so that initially identical loci can diverge and the tetraploid becomes diploid not only in chromosomal behavior but in gene structure (Stebbins 1950). Divergence of duplicate loci, including loss of function, has been documented in several organisms.

### Chromosome rearrangements: Inversions

Several kinds of structural rearrangements, while leaving the number and variety of genes unchanged, alter the positions of genes relative to one another. Some of these rearrangements alter the number of chromosomes.

An INVERSION arises when two breaks occur in the same chromosome and the segment between them is rotated 180 degrees. This could occur if breaks occur at the point of overlap in a chromosome loop. PERICENTRIC INVERSIONS include the centromere; PARACENTRIC INVERSIONS do not. Because a mutation such as an inversion is rare after it first arises, it will be carried in heterozygous condition. During meiosis in a heterozygote (a HETEROKARYOTYPE) for a paracentric inversion, the chromosomes pair by forming a loop (Figure 12). If a single crossover occurs within the inverted region, one of the recombined strands lacks a centromere and so is lost; the other recombined strand lacks some loci, has others in duplicate, and possesses two centromeres, which break the chromosome as they move toward opposite poles during anaphase. Consequently neither daughter nucleus receives a full complement of genes, so the only viable gametes formed by a heterokaryotype come from cells in which no single crossovers occurred within the inversion. Hence the fertility of an inversion heterokaryotype is reduced, and since viable gametes contain only nonrecombinant chromosomes, inversions appear to suppress crossing over.

An inverted chromosome may itself experience further inversions that may overlap or lie within the limits of the first inversion. In *Drosophila,* individual regions of the chromosomes of the salivary glands can be identified by their distinctive banding patterns, so that it is possible to identify numerous such inversions and specify the steps by which each arose. For example, the banding sequence

$$\text{ABCFEDGH} \leftrightarrow \text{ABCDEFGH} \leftrightarrow \text{AFEDCBGH} \leftrightarrow \text{AFEDGBCH}$$

can be derived from each other by the single steps indicated by the arrows, although the direction of change cannot be specified from this information alone.

Although paracentric inversions generally cause partial sterility in heterozygotes, they do not do so in many of the true flies of the order Diptera, including *Drosophila.* In these flies, there is no crossing over in males, so the production of sperm does not suffer, and meiosis in females is peculiarly arranged so that dicentric chromosomes pass into the polar bodies and only the intact chromatid is incorporated into the egg nucleus.

If crossing over occurs within a heterozygous *pericentric* inversion, each recombinant strand has only a single centromere, but part of the gene sequence is

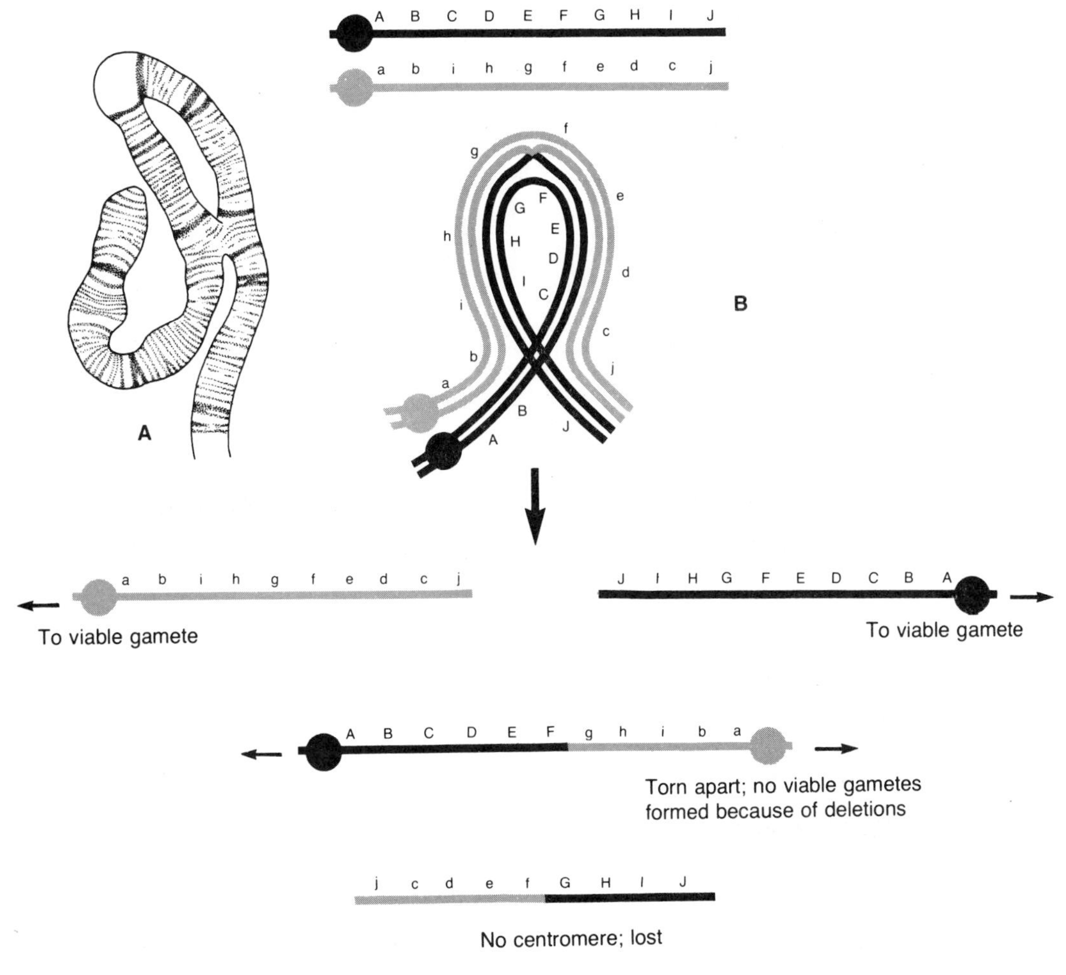


FIGURE 12

**Chromosomal inversions in *Drosophila*. (A) Synapsed chromosomes in a *Drosophila pseudoobscura* heterozygous for *Standard* and *Arrowhead* sequences. (B) The behavior of inversions in heterozygous (heterokaryotypic) form. Crossing over is suppressed because crossover products lack centromeres or substantial blocks of genes. (A redrawn from Strickberger 1968)**

repeated and part is deleted. Hence gametes that receive recombinant chromatids are inviable, and the fertility of the heterokaryotypic individual is reduced. This is true of Diptera as of other organisms.

## Translocations

Exchange of segments of the two nonhomologous chromosomes is a RECIPROCAL TRANSLOCATION (Figure 13). In the heterozygous condition in which translocations first occur, chromosomes achieve gene-for-gene synapsis by assuming rather contorted attitudes. Of the three obvious ways in which the four members of this aggregate can segregate two by two, only one yields viable gametes with complete

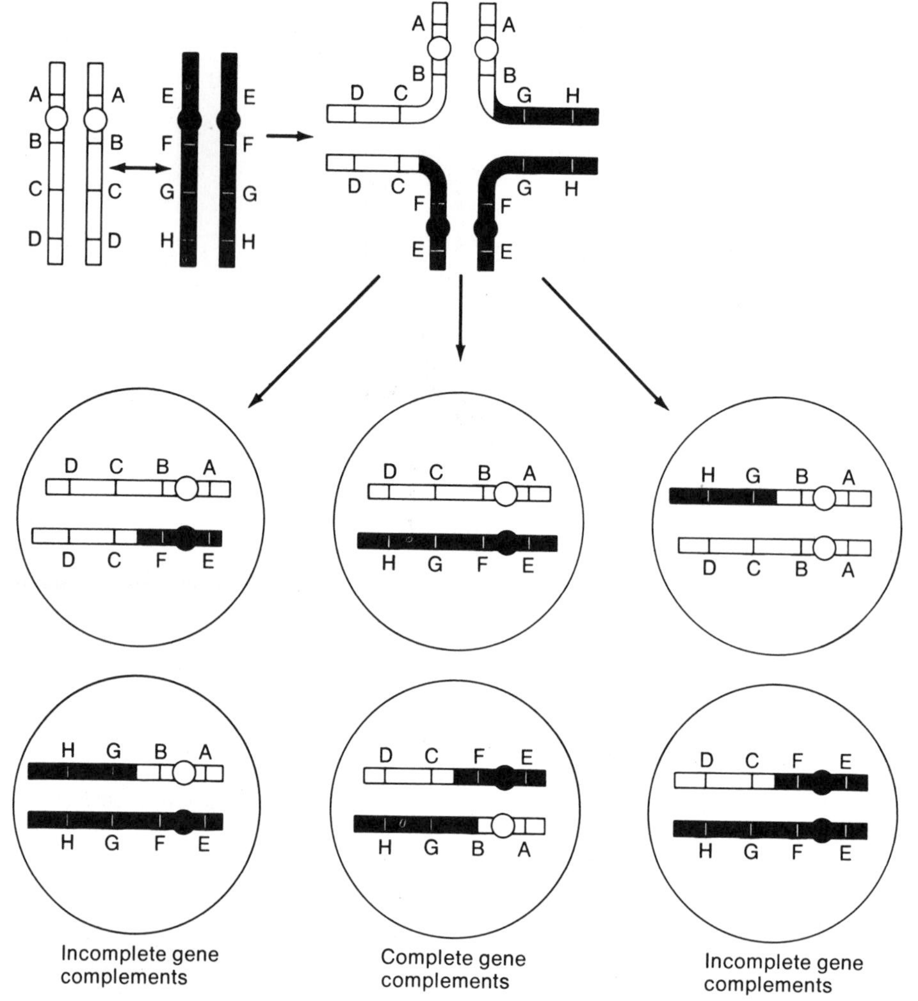


FIGURE 13

**A reciprocal translocation between chromosomes arises first in heterozygous form; synapsis during meiosis in the translocation heterozygote is as shown diagrammatically, and in anaphase meiosis yields one of the three outcomes shown.**

gene complements. Heterozygosity for translocations often reduces fertility by as much as 50 percent, and sometimes more. Nevertheless, the species-typical karyotypes of related species often differ by reciprocal translocations, so that homologous genes are carried in different linked combinations.

### Fusion and fission of chromosomes

Metacentric chromosomes are those in which the centromere lies somewhere in the middle; acrocentric chromosomes have the centromere very near the end. The two arms of a metacentric chromosome in one organism are often homologous to two different acrocentric chromosomes in another. A reciprocal translocation between the short arms of two acrocentrics is thought to be the mechanism by

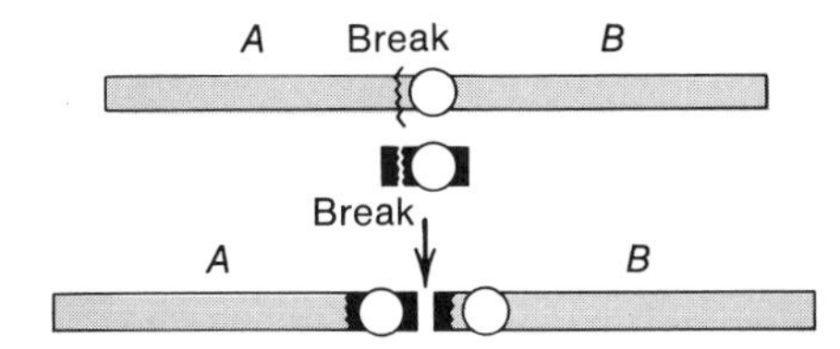


FIGURE 14

**A possible mechanism for increase in chromosome number: dissociation of a metacentric chromosome into two acrocentrics by translocation with a minute "donor" chromosome. (From White 1973)**

which a metacentric chromosome may arise (a "Robertsonian fusion"). Conversely, a metacentric may dissociate into two acrocentrics by a reciprocal translocation with a minute donor chromosome (Figure 14). In the heterozygous condition, a metacentric chromosome usually segregates normally from the homologous acrocentrics with which it pairs, so the gametes are usually euploid (i.e., have a complete haploid genome).

## GENE MUTATIONS

Most structural rearrangements of chromosomes, although they may lower fertility in the heterozygous condition, do not have noticeable effects on morphological or physiological characters (Lande 1979, John 1981). There are exceptions, though, that are usually referred to as POSITION EFFECTS. For example, transfer of genes from one sex chromosome to the other (from the X to the Y or vice versa) often affects their expression. The expression of some mutant genes in *Drosophila* is altered if their position is shifted, as by an inversion. Although it is possible that in such cases the gene is brought under the influence of different regulatory sequences, most position effects are thought to be caused by the inactivating effect of adjacent heterochromatin.

The vast majority of hereditary changes in phenotype, however, are the consequence of alterations of either the base pair sequence of genes or of the number of copies of a gene. Mutations are of many kinds, and their variety appears still to be growing as more molecular information comes to light.

### Point mutations

One of the several DNA polymerases that act to replicate DNA is capable of "proofreading" so that if the wrong base is entered into the strand that is being synthesized, the enzyme often removes it and inserts the right one. Several other enzymes, moreover, can repair errors in DNA replication. Nevertheless, errors of replication do occur, giving rise to altered DNA sequences. Different states of a gene that are recognized by the segregation of different phenotypes are called ALLELES. With modern methods, variations in the base pair sequence can often be identified. Each such variant sequence is called a HAPLOTYPE.

In classical genetics, a mutation that maps to a single gene locus is referred to as a POINT MUTATION; in our molecular era, this term is often restricted to describe a substitution of one base pair for another (Figure 15). A TRANSITION is a substitution of a purine for a purine or a pyrimidine for a pyrimidine (A ↔ G and C ↔ T substitutions). In a TRANSVERSION, a purine is replaced by a pyrimidine or vice versa (A or G ↔ C or T). Transitions appear to be more frequent than transversions (Gojobori et al. 1982). Some base pair substitutions in sequences that code for polypeptides alter amino acid sequences; for example, a transversion

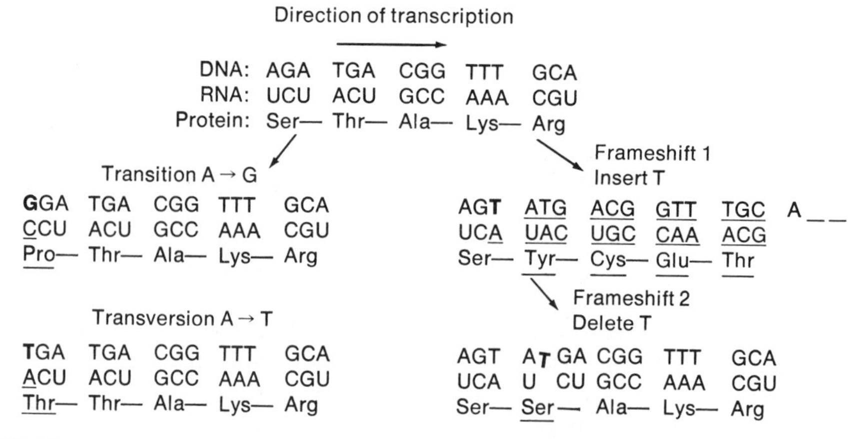


FIGURE 15

**Examples of kinds of point mutations. Notice the magnitude of change caused by a frameshift mutation and the reconstitution of much of the original message by a second frameshift mutation.**

from the RNA triplet GAA (or GAG) to GUA (or GUG) substitutes valine for glutamic acid. This is the mutational event that caused the abnormal β chain in sickle-cell hemoglobin ($Hb^S$). Because of the redundancy of the RNA code, many (about 24 percent, assuming random substitution) codon substitutions are SILENT: they do not change amino acid sequences. Even base pair changes that do alter amino acid sequences may not have discernible effects on the organism's phenotype because an amino acid substitution does not always affect the protein's function. Moreover, a change in one of a number of repeated genes may be masked by the activity of the other gene copies. Mutations in the DNA of a mitochondrion or a chloroplast may have little immediate phenotypic effect because each cell has many of these organelles (Brown 1983).

Single base pair alterations can have other effects. A mutation from an amino acid-coding triplet to a termination codon (e.g., UAU to UAA in the RNA code) will terminate translation so that no functional product is formed. The deletion or insertion of a single base changes the "reading frame" so that the sequence is read as a new series of codons, thereby altering the amino acid sequence downstream from the lesion. This is a FRAMESHIFT MUTATION. A second insertion or deletion may reestablish the original reading frame so that only part of the nucleotide sequence is read in altered triplets. Frameshift mutations appear usually to produce nonfunctional products.

At the phenotypic level, BACK MUTATION, or reversion of a mutant allele to the original form, is observed. True reversion at a particular base pair, however, is unlikely; if all base pair substitutions were equally likely, the probability of a true reversion in a sequence of 423 base pairs (the length of the coding sequence for human hemoglobin α) would be 1/1269 (1/423, the probability of mutation at that site, × 1/3, the probability of mutation to the original base pair).

Microbial studies have shown that most back mutations result from a second amino acid substitution that restores the function altered by a substitution elsewhere in the protein (Allen and Yanofsky 1963). Thus in a population, each allele recognized by its phenotypic effect is likely to be a group of phenotypically indistinguishable isoalleles, some with base pair substitutions that do not alter protein function, and others with complementary substitutions that together maintain protein function. Because many more mutations can abolish gene function than can restore it, the rate of back mutation observed at the phenotypic level is much lower than the rate of "forward" mutation.

### Recombinational change at the gene level

Adjacent genes on a chromosome are different nucleotide sequences along a single long molecule. Thus the mechanisms of recombination need not act only between genes: INTRAGENIC RECOMBINATION also occurs. In a heterozygote for alleles that code for, say, the amino acid sequences Val–Thr–Arg–Leu and Glu–Thr–Arg–Gly, recombination could give rise to the new polypeptide sequence Val–Thr–Arg–Gly. A polymorphism of the enzyme 6–phosphoglycerate dehydrogenase in Japanese quail (*Coturnix coturnix*) appears to have arisen in this manner (Ohno et al. 1969). By intragenic recombination, variation begets variation: the more alleles (haplotypes) there are, the more new alleles can come into being. This mechanism could generate new alleles at fairly high rates (Watt 1972, Golding and Strobeck 1983). New base pair sequences arising by intragenic recombination could well code for different amino acids than either parent sequence does.

The mechanism of recombination appears to be closely related to the mechanism of GENE CONVERSION, a phenomenon that has been most extensively studied in fungi. The four products (gametes) of meiosis in a heterozygote should carry the two alleles ($A_1$ and $A_2$) in a 1:1 ratio. Occasionally, however, they occur in different ratios, such as 1:3. The $A_1$ allele has been replaced specifically by an $A_2$ rather than any of the other alleles ($A_3$, $A_4$, etc.) to which it might have mutated; it seems to have been converted into $A_2$ (or vice versa). Several models have been suggested for gene conversion. In one model (Figure 16), a break in both strands of the DNA duplex on one chromosome is corrected by repair enzymes that excise a length of DNA on either side of the break, and replace the missing sequence by copying from the homologous strands on the other chromosome. This changes a strand of the DNA of one allele into the sequence of the other allele. The gene conversion may extend for up to several thousand bases. In some cases gene conversion is unbiased (conversion of $A_1$ to $A_2$ is as likely as the converse), but cases of biased gene conversion have been described whereby one allele is preferentially converted to the other.

During synapsis, genes on homologous chromosomes are not always perfectly aligned. UNEQUAL CROSSING OVER between misaligned chromosomes (Figure 17) gives rise to a tandemly duplicated region on one chromosome and a complementary deletion on the other, the length of which depend on the magnitude of misalignment. A duplication may be advantageous, for the duplicated genes may produce greater amounts of gene product, as Hansche (1975) found in yeast. A duplication of the acid monophosphatase locus increased in frequency in a laboratory population, because the doubled amount of enzyme enabled the cells to exploit more efficiently the low concentrations of phosphate in the medium. Deletions of coding regions, however, are usually deleterious.

Gene duplication has been very important in amplifying the size and infor-

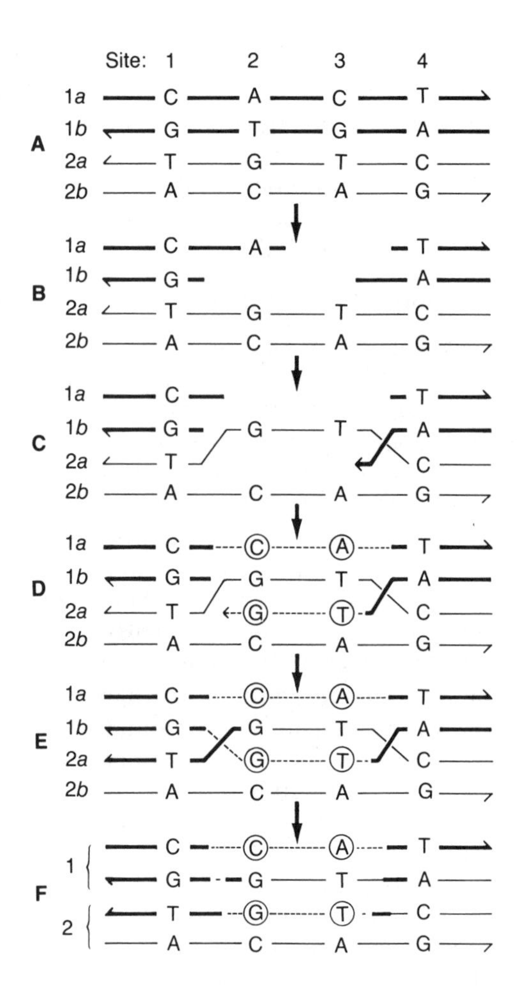


FIGURE 16

**A model of gene conversion. (A) Two DNA duplex molecules, on homologous chromosomes 1 and 2, pair. Each DNA duplex consists of two antiparallel strands *a* and *b*, with the 3′ ends marked by arrows. The genes are shown to differ for four base pairs. (B) A break in gene 1 is enlarged by endonuclease enzymes; the enlargement is extended for strand 1*b* in step (C). (C) The free 3′ end of strand 1*b* invades duplex 2, displacing strand 2*a* into a loop that may extend indefinitely to the left as strand 1*b* is repaired. (D) Strand 1*b* is resynthesized at the free 3′ end by addition of nucleotide bases complementary to strand 2*b*. The gap in strand 1*a* is repaired by addition of bases complementary to the loop in strand 2*a*. (E) The 3′ end of 1*b* is ligated to the 5′ end of the intact part of 1*b*. There are now two crossover points. (F) At each crossover point the crossed strands are broken and rejoined as shown. Sites 2 and 3 of gene 1 have now been converted to the sequence of gene 2. (After Szostak et al. 1983)**

mation content of the genome (Ohno 1970), and in generating families of genes that may diverge in nucleotide sequence (e.g., the globin family). A duplication event may make unequal crossing over subsequently more likely, because the gene in the first position on one chromosome may recognize and pair with the gene in the second position on the other. Unequal crossing over would then generate a chromosome with three tandem copies and another chromosome with one.

## Mobile genetic elements

In addition to DNA sequences that faithfully occupy particular sites on nuclear chromosomes and in the genomes of the mitochondria and chloroplasts, cells carry numerous nucleic acid sequences that carry on lives of their own; that is, their dynamics are not strictly tied to the replication of the nuclear DNA during the cell cycle. Bacteria and some eukaryotes often carry in their cytoplasm independently replicating circular DNA molecules known as PLASMIDS, some of which can affect the phenotype of the cell by, for example, conferring resistance to antibiotics. Some plasmid-like particles (episomes) have the capacity to become

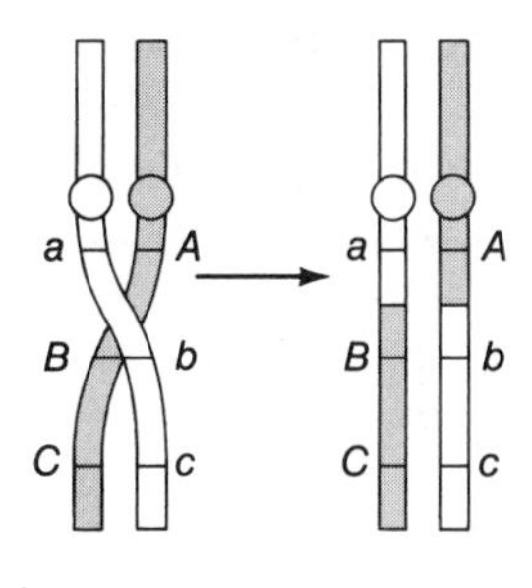


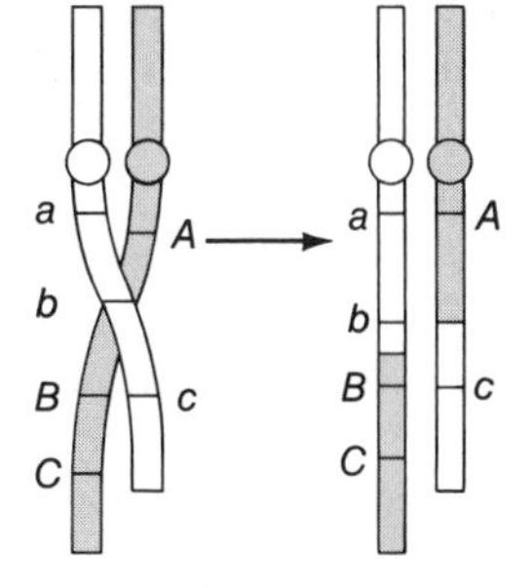


Equal crossing over

Unequal crossing over

FIGURE 17

**Ordinarily, as at left, crossing over is reciprocal. But in unequal crossing over, shown at right, a segment of one chromosome, marked by locus *B*, is transferred to the other chromosome. Thus one chromosome suffers a deficiency, which if large causes inviability. The other chromosome bears a tandem duplication for one or more loci such as *B*. Such a duplication may persist and is sometimes advantageous.**

integrated into the bacterial chromosome. In this respect they resemble viruses, including those viruses of bacteria known as phages. DNA viruses become integrated into the host genome, and may either be replicated with it, or use the biochemical machinery of the host to make free copies of themselves and a protein coat. The mature virus particles, consisting of a nucleic acid surrounded by protein, burst free from the cell (often destroying it in the process) and infect other cells or organisms. In this process, they sometimes carry with them part of the host's genome. Viruses, then, can transfer genetic material among individual organisms, and even among organisms of different species (Reanney 1976).

The traditional view that information flow proceeds only from DNA to RNA to protein has been changed by the discovery in the early 1970s of the phenomenon of REVERSE TRANSCRIPTION (Baltimore 1985, Temin 1985). Viruses that exhibit reverse transcription (retroviruses) have as their genetic material a single-stranded RNA molecule that includes a gene for the enzyme reverse transcriptase. This enzyme uses the RNA sequence as a template for copying a duplex DNA molecule (cDNA), which becomes integrated into the host genome, apparently at random sites, where it is transcribed into more virus RNA. The transcription process sometimes extends to neighboring host genes, and can thus alter gene expression (as in some cases of tumor formation). At least occasionally, the reverse transcriptase can also reverse transcribe other mRNA sequences (a process that is now routinely used by molecular geneticists to make cDNA copies of mRNAs). Some DNA viruses, such as the hepatitis B virus, are also known to be reverse transcribed from RNA into DNA. Processed pseudogenes, such as the $\psi\alpha 3$ globin pseudogene of the mouse, seem to have arisen by reverse transcription from mRNA to cDNA. Numerous processed pseudogenes have been described, and it appears that as much as 20 percent of a mammalian genome may consist of reverse-transcribed sequences (see Rogers 1985, Temin 1985, Walsh 1985a for summaries). This fraction includes the *Alu* family of moderately repetitive DNA, consisting of about 300,000 copies of 300-bp segments that vary only slightly in sequence.

Short DIRECT REPEATS (i.e., repeated sequences with the same 5′ to 3′ polarity) of about 3–30 bp flank most mobile elements, as well as many sequences that are thought to have been inserted by reverse transcription. Such repeats are thought to be created by enzymes that make a staggered cut in the DNA duplex so that each strand has an overlapping end (Figure 18). Ligation of the mobile element onto the overhanging end, followed by replication of the single-stranded region, generates the flanking repeats. In addition to the direct flanking repeats, the mobile element itself often contains repeats that may be either direct or inverted (i.e., having the same sequence read from opposite directions).

TRANSPOSABLE ELEMENTS (also called transposons or, in the popular press, "jumping genes") generally insert copies of themselves (sometimes many copies) into recipient sites elsewhere in the genome, while the "donor" copy remains in place; less commonly, the donor copy may be excised during the process. The mechanisms of insertion are partially understood for some bacterial transposons, but otherwise are not well understood. Some transposable elements in eukaryotes, such as the *Ty* family of transposons in yeast and the *copia* family in *Drosophila* (Figure 19), appear to transpose via reverse transcription of their RNA transcript; their structure resembles that of retrovirus DNA, and they appear to encode reverse transcriptase (Baltimore 1985).

Transposable elements very frequently affect the function of genes at or near the site of insertion, and so have a mutational effect. In fact, transposons appear to account for many of the morphological mutations of *Drosophila,* such as those of the *white* locus that affect eye color (Spradling and Rubin 1981, Zachar and Bingham 1982, Engels 1983, Shapiro 1983). The *copia* and *P* transposable elements in *Drosophila melanogaster* have been especially studied in this respect. *P* elements characterize some strains of *Drosophila* but not others (*M* strains). When *P*-carrying chromosomes are introduced into a strain with *M*-type cytoplasm, a syndrome called HYBRID DYSGENESIS ensues in the germ cells of the hybrid flies (Kidwell et al. 1977, Engels 1983): crossing over occurs in males, the offspring are highly sterile, and both chromosome rearrangements and gene mutations appear at high frequency in the offspring of hybrids. Hybrid dysgenesis in *Drosophila melanogaster*

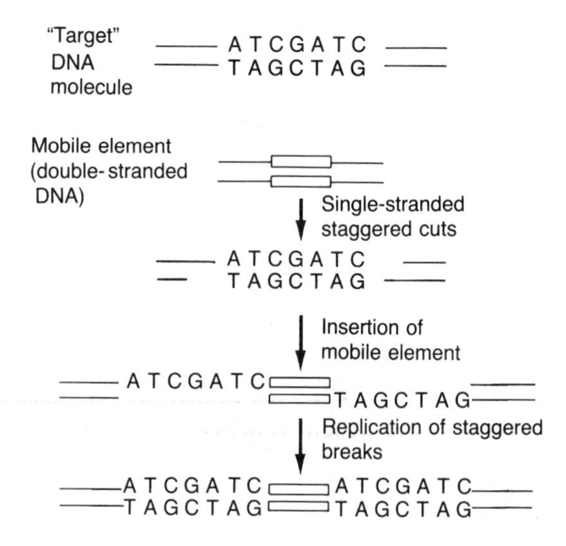


FIGURE 18

**A model for the origin of direct flanking repeats in a DNA molecule when a transposable element becomes inserted.**

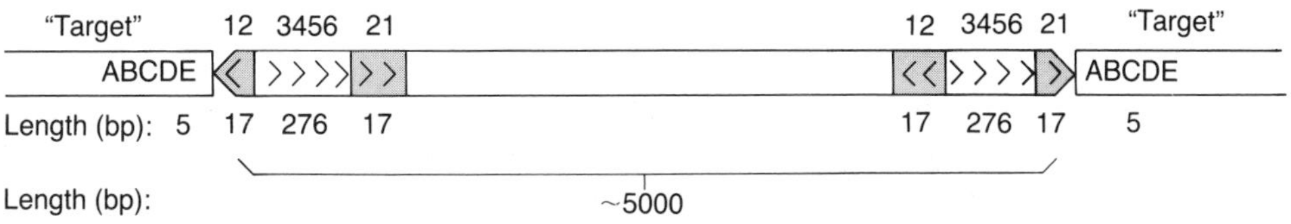

"Target" 12 3456 21 12 3456 21 "Target"
ABCDE ABCDE
Length (bp): 5 17 276 17 17 276 17 5
Length (bp): ~5000

FIGURE 19

**Schematic illustration of a *copia* sequence inserted into a target DNA sequence; it is flanked by 5-bp repeats (A–E) generated by the insertion process. The *copia* element begins and ends with 276-bp direct repeats (symbolized 3456), each flanked by 17-bp inverted repeats (symbolized 12,21). (After Lewin 1985)**

has been shown to substantially increase genetic variation in the number of abdominal bristles, a polygenic character (Mackay 1984).

Transposable elements can exercise their mutagenic effect by interrupting structural or regulatory regions of the genes into which they become integrated, thereby disrupting function. Also, many mobile elements include stop and start signals for transcription, and so can alter gene expression even if they are not inserted into a regulatory region. Moreover, when two or more copies of a transposable element exist on a chromosome, recombination between them can cause a deletion or inversion of the region between them (Figure 20). A sequence deleted by this mechanism may be inserted, along with the transposons that bear it, into another site in the genome. The *FB* transposable elements of *Drosophila* are known to move sequences of hundreds of kilobases.

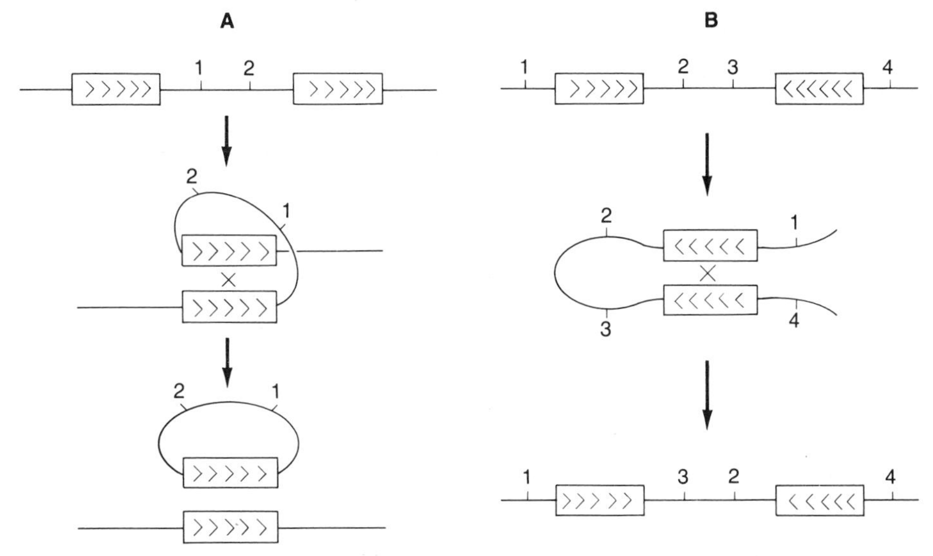


FIGURE 20

**Recombination between repeated sequences can result in deletions and inversions. The boxes represent repeats, with the polarity of base pair sequence indicated by the arrows within. (A) Recombination between two direct repeats excises one repeat and deletes the sequence between the two copies. (B) Recombination between two inverted repeats inverts the sequence between them. (After Lewin 1985)**

## RATES OF MUTATION

Mutation rates are typically scored by the frequency with which a new mutant arises among the progeny in a population (often a laboratory population) of nonmutant individuals. Hence they are expressed as the number of mutants per gamete per generation. A given mutation may have arisen anywhere in the lineage of cells leading to that gamete, not necessarily during meiosis. Possibly for this reason, the overall mutation rates for multicellular eukaryotic organisms, in which mutations are counted per organismal generation, appear to be higher than in unicellular organisms such as bacteria (Table II).

The spontaneous rate of origin of a given class of chromosomal mutation (e.g., reciprocal translocation) is about $10^{-4}$ to $10^{-3}$ per gamete per generation (Lande 1979), although any particular rearrangement rarely arises and may be considered effectively unique if it can be described with sufficient precision. Such precision is usually lacking, especially when mutation rates at a particular locus are scored. When mutations are scored by their effects on morphology (e.g., mutation from red to white eye color at the *white* locus of *Drosophila*) or by their effect on the mobility of an enzyme in electrophoresis (see Chapter 4), numerous alterations of the DNA will have the same phenotypic effect, and many will not be detected at all. By a variety of indirect methods, the average mutation rate per base pair of the DNA has been estimated in bacteria at about $10^{-10}$ to $10^{-9}$ per cell division (Drake 1974, Lewin 1985), and in *Drosophila* at about $10^{-8}$ per generation (Neel 1983).

Based on their morphological or physiological effects, mutations at individual loci generally appear to arise at a rate of $10^{-6}$ to $10^{-5}$ per generation (Table III), with similar estimates arising from the study of individual proteins (Mukai and Cockerham 1977, Neel 1983). Estimates vary greatly among loci, however, and phenomena such as hybrid dysgenesis make it likely that mutation rates may be elevated sporadically within natural populations. Transposable elements, and for

TABLE II
**Comparative spontaneous mutation rates**

| Species | Base pairs per genome | Mutation rate per base pair replication | Mutation rate per genome per generation |
|---|---|---|---|
| Bacteriophage lambda | $4.7 \times 10^4$ | $2.4 \times 10^{-8}$ | 0.001 |
| Bacteriophage T4 | $1.8 \times 10^5$ | $1.1 \times 10^{-8}$ | 0.002 |
| *Salmonella typhimurium* | $3.8 \times 10^6$ | $2.0 \times 10^{-10}$ | 0.001 |
| *Escherichia coli* | $3.8 \times 10^6$ | $4.0 \times 10^{-10}$ | 0.002 |
| *Neurospora crassa* | $4.5 \times 10^7$ | $5.8 \times 10^{-11}$ | 0.003 |
| *Drosophila melanogaster*[a] | $4.0 \times 10^8$ | $8.4 \times 10^{-11}$ | 0.93 |

(After Drake 1974)
[a]The *Drosophila* values are per diploid genome per generation of flies, not per generation of cells as in the other species.

TABLE III
**Spontaneous mutation rates of specific genes**

| Species and locus | Mutations per 100,000 cells or gametes | |
|---|---|---|
| | Forward | Back |
| *Escherichia coli* | | |
| streptomycin resistance | 0.00004 | |
| resistance to T1 phage | 0.003 | |
| arginine independence | 0.0004 | |
| *Salmonella typhimurium* | | |
| tryptophan independence | 0.005 | |
| *Neurospora crassa* | | |
| adenine independence | 0.0008–0.029 | |
| *Drosophila melanogaster* | | |
| yellow body | 12 | |
| brown eyes | 3 | |
| eyeless | 6 | |
| *Zea mays* (corn) | | |
| sugary seed | 0.24 | |
| *I* to *i* | 10.60 | |
| *Homo sapiens* | | |
| retinoblastinoma | 1.2–2.3 | |
| achondroplasia | 4.2–14.3 | |
| Huntington's chorea | 0.5 | |
| *Mus musculus* (house mouse) | | |
| *a* (coat color) | 7.1 | 0.047 |
| *c* (coat color) | 0.97 | 0 |
| *d* (coat color) | 1.92 | 0.04 |
| *ln* (coat color) | 1.51 | 0 |

(After Dobzhansky 1970)

that matter mutation of genes that affect replication, can elevate mutation rates and so may be called "mutator genes" (e.g., Ives 1950, von Borstel et al. 1973). For example, the allele *mutT* in *Escherichia coli* increases the frequency of A–T to C–G transversions throughout the genome. Because a mutator allele will generally be associated with the mutations it causes, and because most mutations are deleterious, mutator alleles will often be removed with them from the population by natural selection; hence, mutator alleles appear to be rare in natural populations (Chapter 9).

It is not possible to estimate the mutation rate at individual loci that contribute to polygenically inherited traits, but it is possible to estimate the rate at which genetic variation (variance) in such a trait is increased by mutation (reviewed by Lande 1976b). Estimates of the mutational contribution to variance in morphological traits are usually made in a population that has been inbred (Chapter 5) so that it initially lacks genetic variation (e.g., Clayton and Robertson 1955). The rate of mutation of genes that affect viability in *Drosophila melanogaster* can be estimated by allowing recessive mutations to accumulate in chromosomes that are held in heterozygous condition with a "balancer" chromosome that masks the effect of recessive mutants. Periodically some of the sheltered chromosomes are brought into homozygous condition so that their effect on viability can be measured (see Figure 6 in Chapter 4). Over the course of generations, the mean viability of such homozygotes declines and the variation among chromosomes increases (Mukai et al. 1972; Figure 21). From this and other work, it appears that per fly per generation, at least one new mutation affecting viability occurs somewhere in the genome.

The contribution to the genetic variance of a character that arises per generation by mutation is frequently expressed by the ratio $V_M/V_E$, where $V_E$ is the variance in the phenotype caused by nongenetic factors and $V_M$ is the increase in genetic variance caused by mutation that would be observed if $V_E$ were zero. This ratio enables comparison of the rate of increase of variation in different characters that vary in $V_E$. $V_M/V_E$ generally increases steadily with the passage of generations, and for many characters is about 0.001 to as much as 0.030 (or more) per generation. A value of 0.001 implies that a completely invariant population would achieve in 1000 generations a genetic variance equal to the magnitude of nongenetic variance. This is a rather low value, in the sense that a genetically invariant population would not display the rapid changes in response to selection that are actually observed, if the response to selection depended entirely on new mutations. Rather, observed responses to selection are based on the larger store of variation in the population that has originated by mutation in the past (Chapter 7). But the mutation rate is high in the sense that it can rather quickly replenish variation that is lost, and may indeed account for much of the genetic variation in natural populations (Lande 1976b; but see Turelli 1984 for a contrary view).

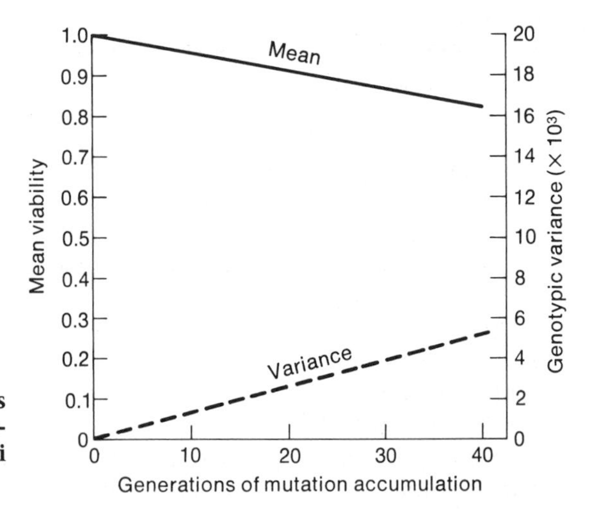


FIGURE 21

**Effects of the accumulation of spontaneous mutations on viability. Mean viability decreases and variation increases. (After Mukai et al. 1972)**

## PHENOTYPIC EFFECTS OF MUTATIONS

Effects of mutation on the phenotype vary enormously. Silent substitutions, as well as many others, may have no discernible effect, although synonymous codons sometimes have different effects on the rate of translation of mRNA into protein. Mutations in polygenic characters frequently have such slight effects that they can be measured in aggregate, but cannot be isolated for individual study. Other mutations have drastic effects: in *Drosophila,* a single mutation such as *singed* changes the shape of all the bristles, and *Curly* changes the shape of the wing. HOMEOTIC MUTANTS, by acting on organs with related developmental patterns, convert one into the other: the antenna into a leg-like structure (*antennapedia*), or the metathorax with balancer organs into a mesothorax with wings (*bithorax*).

The components of complex new adaptations can also be shown to arise by mutation. For example, an amidase enzyme in the bacterium *Pseudomonas aeruginosa* can provide sources of carbon and nitrogen by metabolizing acetamide and propionamide. It metabolizes butyramide with very low efficiency, but synthesis of the enzyme is repressed by butyramide, as well as by valeramide and phenylacetamide. Clarke (1974) was able to obtain a mutant that was not repressed by butyramide, as well as mutants with structurally different amidases that metabolized butyramide more efficiently. From these strains, moreover, she selected further mutants with different enzyme structures that were able to metabolize valeramide and phenylacetamide. Thus enzymes with entirely new metabolic capacities, as well as changes in regulation of the synthesis of these enzymes, arose by mutation. Clarke notes that a related species, *P. cepacia,* has two different amidases, one that metabolizes acetamide and another that acts on valeramide and phenylacetamide. It is possible that a single amidase locus was duplicated and that the duplicate loci diverged in structure and function. Similar observations on the mutational origin of complex adaptations have been reported by other authors (e.g., Hall 1983) and will be considered further in Chapter 15.

Many mutations have deleterious effects on the organism's likelihood of survival and/or reproduction (i.e., fitness). This is the case with many of the classical mutations in *Drosophila,* such as *white* eye and *cut* wing, which in caged populations are rapidly replaced by nonmutant alleles (e.g., Ludwin 1951, Rendel 1951). Advantageous mutations frequently, although not inevitably, have more subtle effects on the phenotype, as is the case for many of the alleles that contribute to polygenic variation. Some mutations appear to be NEUTRAL, neither increasing nor decreasing fitness. In both *Drosophila pseudoobscura* (e.g., Yamazaki 1971) and in *Escherichia coli* (e.g., Dykhuizen and Hartl 1980), sensitive tests showed that alleles at an enzyme-coding locus did not change in frequency in experimental populations over the course of many generations. Whatever difference in fitness among genotypes existed must have been exceedingly small.

Whether an allele is detrimental, neutral, or salubrious, however, frequently depends on environmental circumstances. Even seemingly harmful mutations can be advantageous in certain environments. For example, the ability to synthesize essential amino acids is an important adaptation, yet Zamenhof and Eichhorn (1967), Dykhuizen (1978) and others have shown that *E. coli* mutants incapable of such syntheses frequently are competitively superior to wild-type bacteria if these amino acids are supplied to them (reviewed by Dykhuizen and Hartl 1983a). For human populations in a malaria-free environment, hemoglobin *S*, which in heterozygous form causes sickle-cell anemia and in homozygous form causes sickling disease, is unquestionably harmful. Who would have supposed that such a gene could confer protection against malaria?

The genetic background in which a mutation arises often determines its fate. The simplest instance is that of a genetic priority effect: a new, possibly beneficial mutation is not advantageous if its function is already served by an established gene. For example, several different hemoglobin mutants reduce human susceptibility to malaria, and certain of these have complementary distributions among the human populations of the Old World tropics; none has increased within the range of the others (Livingstone 1964). Frequently, moreover, the genetic background at other loci determines whether a mutation is advantageous, disadvantageous, or neutral (Chapters 6 and 7).

## THE RANDOMNESS OF MUTATIONS

Mutations occur at random. This does not mean that all loci mutate at the same rate, nor that all imaginable mutations are equally likely. Nor does it mean that mutations are independent of the effect of the environment; environmental mutagens increase mutation rates. Mutation is random in that the chance that a specific mutation will occur is not affected by how useful that mutation would be. As Dobzhansky (1970) said, "It may seem a deplorable imperfection of nature that mutability is not restricted to changes that enhance the adaptedness of their carriers. However, only a vitalist Pangloss could imagine that the genes know how and when it is good for them to mutate."

Resistance to a source of mortality, such as a toxin, is based on alleles that are already in the population or come about through mutation, irrespective of the presence of the source. Two elegant experiments illustrate this. Lederberg and Lederberg (1952) used the technique of replica plating in bacteria to show that mutations for resistance to an antibiotic occur independently of resistance to the drug and are not induced by it (Figure 22). In a conceptually similar experiment Bennett (1960) showed that the evolution of DDT resistance in *Drosophila* was based on mutations already present in the population. He demonstrated this by breeding from the siblings of the most resistant flies in each generation. These breeding flies were not exposed to DDT, yet after 15 generations the stock was highly resistant to DDT even though none of its ancestors had been exposed to the insecticide.

As far as we know, the environment does not evoke the appearance of favorable mutations, nor can a population accumulate mutations in anticipation of a change in its environment, as we shall see in Chapter 6.

## RECOMBINATION: THE AMPLIFICATION OF VARIATION

Genetic variation arises not only from mutation, but from recombination as well. In eukaryotes this results from two processes that are often, but not always, associated with each other. These are sexual reproduction (the union of potentially different gametes) and the formation of gametes genetically different from those that united to form the individual that produces them. The latter also entails two

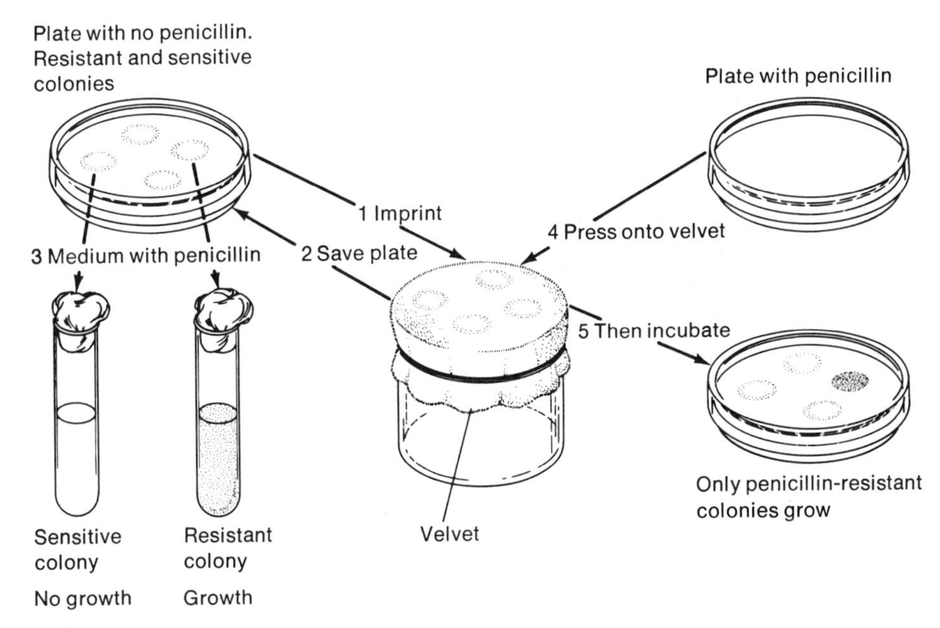


FIGURE 22

**The method of replica plating, showing that mutations for penicillin resistance arise spontaneously and are not induced by penicillin. In step 3 the number of resistant colonies from a plate without penicillin is scored; in step 5 the number of colonies capable of growth when exposed to penicillin is measured. The two numbers are the same; thus penicillin (step 5) does not induce resistant mutations. (Modified from Srb et al. 1965)**

processes that are not invariably associated: independent segregation of nonhomologous chromosomes, and crossing over between homologous chromosomes.

These processes vary greatly among different organisms. Sexual reproduction can involve two individuals or one, as in self-fertilizing plants. In ameiotic parthenogenesis, meiosis does not occur and the offspring develops from an unreduced egg. Some uniparental organisms have meiosis, independent segregation, and crossing over. For example, the diploid chromosome number is restored in a parthenogenetic form of the moth *Solenobia triquetrella* by fusion of two of the four products of meiosis. Some organisms, such as the males of *Drosophila* and many other Diptera, have normal meiosis but no crossing over. In males of the fungus gnat *Sciara*, only the chromosomes inherited from the mother are transmitted to the sperm cells, so segregation of chromosomes is anything but independent.

These variations in the meiotic machinery illustrate that recombination rates evolve. Even the frequency of crossing over can be subject to precise genetic control. For example, Chinnici (1971) and others have shown that the crossover frequency between a pair of loci is under genetic control and can be altered by artificial selection (Figure 10 in Chapter 9) without affecting the frequency of crossing over elsewhere on the chromosome.

The new combinations of genes that arise by recombination may confer properties on offspring that transcend those of either parent. If, for instance, + and − represent alleles at each of five loci that equally affect body size, two parents with genotypes + − − + −/− + + − + and − − + − +/+ + − + − are much the same size; but their offspring could include the genotypes + + + + +/+ + + + + and − − − − −/− − − − −, spanning the range of possible sizes. Thus variation can arise at a far higher rate than by mutation alone. However, extreme homozygous genotypes arise with low frequency. If the five loci we have postulated segregate independently, the probability is only $1/1024$ that an offspring from this mating will have the genotype + + + + +/+ + + + +. Most individuals in the population will have a mix of + and − alleles and will be intermediate in size. Hence the population contains latent variation for extreme genotypes, but these genotypes are rare. Moreover, an unusual genotype that is generated by recombination will be broken apart by recombination in the next generation; thus recombination tends to stabilize the range of variation (Eshel and Feldman 1970, Felsenstein 1974, Williams 1975). However, a newly arisen combination can be perpetuated if recombination is restricted. Chromosome inversions that suppress crossing over have this effect, as does parthenogenesis.

## EXTERNAL SOURCES OF VARIATION

In addition to the genetic variation that arises within a population by mutation and recombination, a population often has genetic variations that have entered by GENE FLOW from other populations (Chapter 5). Usually these are other populations of the same species; in some cases, however, hybridization occurs between different species that are typically reproductively isolated. The grass *Bothriochloa intermedia* seems to have incorporated genes from numerous other grasses: from *B. ischaemum* in Pakistan, from *B. insculpta* in eastern Africa, from *Dichanthium annulatum* in Pakistan and India, and from *Capillipedium parviflorum* in northern Australia (Harlan and deWet 1963). The variation arising from hybridization often transcends that within either parent species (Figure 23).

Hybridization may sometimes provide enough genetic variation for adaptation to environments that would otherwise be closed to a species. Laboratory populations of hybrids between the fruit flies *Dacus tryoni* and *D. neohumeralis* were better able to adapt to high temperatures than were populations of either "pure" species, leading Lewontin and Birch (1966) to postulate that the hybridization known to occur naturally between these species may have been responsible for their range expansion over the last century. The progeny of hybrids inherit some alleles that typify each parent species, but sometimes they also bear distinctive alleles that may have arisen in the hybrids by intragenic recombination (Sage and Selander 1979, Woodruff 1981).

Allopolyploid plants are most prevalent in stringent habitats (Stebbins 1970) and abundantly demonstrate the evolutionary success that hybrid genotypes may enjoy. Allopolyploidy does not in itself create new morphological features of the kind that distinguish genera or other higher categories of classification, but allopolyploid stocks have often evolved into phyletic lines that achieve taxonomic distinction. Most of the families of plants have high chromosome numbers that are thought to have originated by polyploidy.

As noted earlier in this chapter, genes are occasionally transferred between unrelated species by viruses or other parasitic agents (Chapter 15). Such transfers

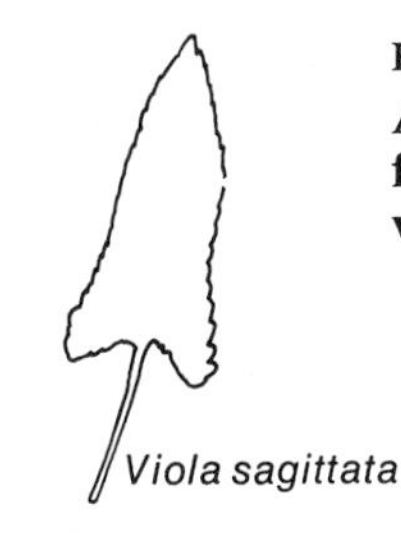


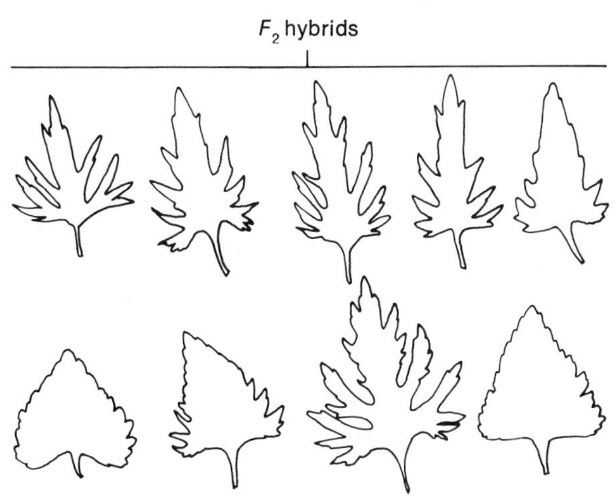


FIGURE 23

**An example of the variation that may arise from hybridization between two species of violets. (After Stebbins 1950)**

appear to occur rather frequently among different genera of bacteria, and there is evidence from protein structure that transfers may occur between prokaryotes and eukaryotes, as well as among eukaryotes. There is little evidence so far that transfers of this kind have occurred frequently in evolution, but it is possible that they have occasionally had important consequences.

## SUMMARY

The phenotype of an organism is the consequence, in a succession of developmental events from conception to death, of the interplay between its genotype and its environment. Because the units of heredity (DNA molecules) replicate themselves in their passage from generation to generation, genetic variations can persist once they have arisen. Variations in the genetic program arise by errors in replication and by more or less orderly processes of recombination; changes acquired by an organism during its lifetime through interacting with its environment are not inherited. The mechanisms of development, whereby genotype and environment interact to form the phenotype, are important to evolutionary studies, but in general they are not well understood.

Knowledge of the molecular structure of the genome and of the molecular bases of replication, recombination, and gene function has grown at a progressively greater rate. Phenomena such as interrupted genes, pseudogenes, and reverse transcription have been revealed that were unknown or only suspected a decade ago. As we will see in subsequent chapters, much of this information has been incorporated into evolutionary theory, but the evolutionary implications of some of these phenomena are only beginning to become clear. In many contexts, however, evolutionary theory is not greatly affected by the new molecular biology; it is often adequate to develop an evolutionary model in the same genetic terms that sufficed before the details of molecular structure and function were understood. As in all sciences, models and descriptions can be validly phrased at different levels of organization. For some purposes it is necessary to use descriptions at the molecular level; for others, nonmolecular descriptions of segregating genes, of phenotypic characters, and of whole organisms are more appropriate.

## FOR DISCUSSION AND THOUGHT

1. The near universality of the genetic code is often used as evidence that all living things are monophyletic (have a single origin). How do variations in the code in mitochondrial DNA affect this claim? What evidence besides the genetic code argues for a monophyletic origin?
2. What kind of evidence would be necessary to demonstrate that acquired characteristics actually can be inherited? What pitfalls would an experimenter have to avoid?
3. Discuss the functions that nontranscribed DNA could perform for an organism. Does nontranscribed DNA necessarily have a function?
4. Discuss the problems that arise if one attempts to envision how the mechanisms of meiosis or of DNA replication evolved.
5. Is the DNA code imposed on organisms as their necessary mechanism of heredity, or has the genetic code itself evolved? Might there have been other molecular mechanisms of heredity or other meanings for nucleotide triplets? Why is the genetic code nearly universal?
6. In this chapter I have referred to mutations as errors, but I have not referred to crossing over in these terms. What evidence or reasoning leads to this distinction? Is it valid? Have the processes of mutation evolved as adaptations (see Chapter 9)?
7. Transposable elements might sometimes cause adaptive mutations and in this sense have importance in evolution, or they might invariably cause maladaptive mutations and in this sense be less important. What kinds of observations or experiments might cast light on whether or not they cause adaptive mutations?
8. How can one account for the homogeneity of repetitive sequences of DNA?
9. The proximate causes of many developmental phenomena can be found in the properties of self-assembling proteins, the structure of cell surfaces, the mutual influences of developing tissues, and so forth. In what sense, then, can we say that development is genetically programmed—or is it?
10. Why would an evolutionary biologist wish to understand the mechanisms of development? To what kinds of evolutionary questions might such understanding provide answers?

## MAJOR REFERENCES

Klug, W.S. and M.R. Cummings. 1983. *Concepts of genetics*. Charles E. Merrill, Columbus, OH. 614 pages. This and other recent textbooks of genetics provide background for much of the material in this chapter.

Lewin, B. 1985. *Genes II*. Wiley, New York. 716 pages. A comprehensive, detailed treatment of the molecular aspects of genetics. Most of the otherwise unattributed statements about molecular genetics in this chapter are based on this book.

Gilbert, S.F. 1985. *Developmental biology*. Sinauer Associates, Sunderland, MA. 726 pages. A clear and comprehensive treatment of molecular and organismal aspects of development.

# Variation

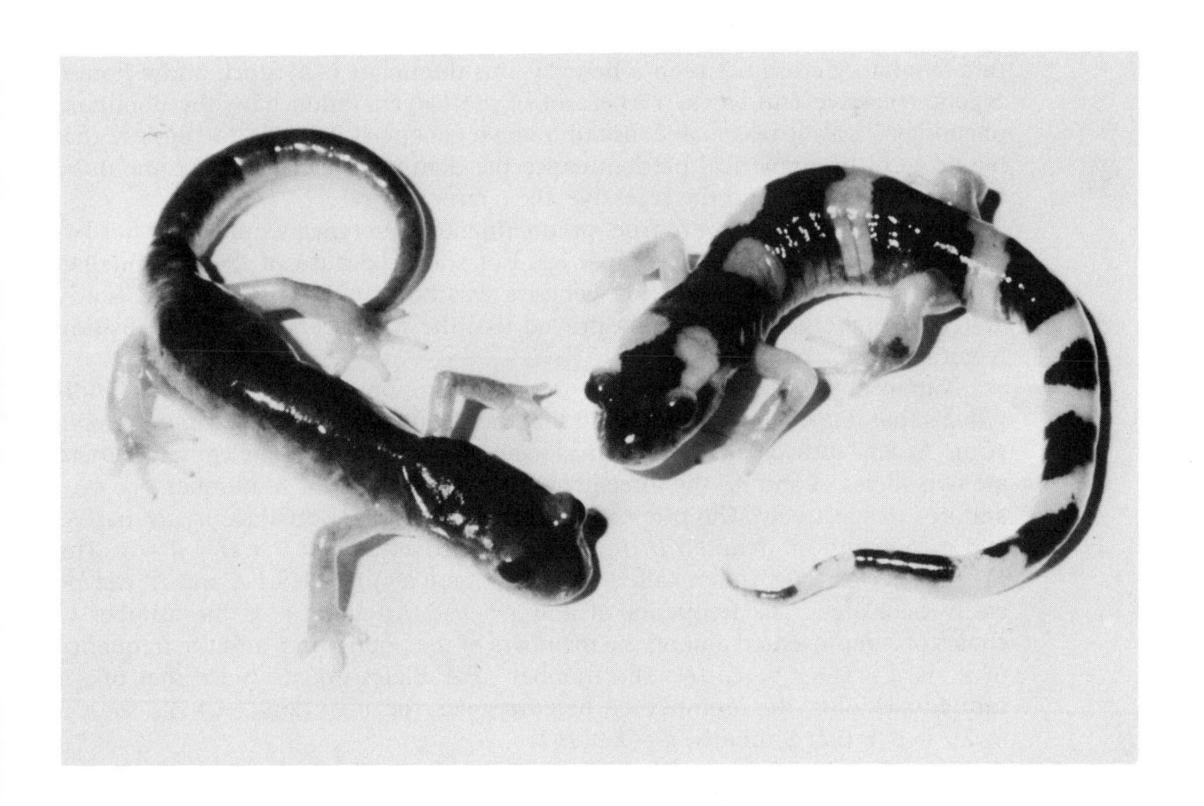

# Chapter Four

Evolutionary biology, like any science, seeks to understand the complexity of nature by formulating generalizations: we attempt to simplify for the sake of greater comprehension. In doing so, we run the risk of missing much of the beauty and excitement of biology inherent in the wondrous variety of living beings. But more than esthetics is at stake, for variation is at the heart of the scientific study of the living world. As long as essentialism, the outlook that ignores variation in its focus on fixed essences, held sway, the possibility of evolutionary change could hardly be conceived, for variation is both the product and the foundation of evolution. Few other sciences take variation as a primary focus of study, as does evolutionary biology.

## THE HARDY-WEINBERG THEOREM

In a laboratory cross between a homozygous dominant (*AA*) stock and a homozygous recessive (*aa*) stock, 75 percent of the $F_2$ generation have the dominant phenotype. Isn't it reasonable, asked a naive geneticist in 1908, to suppose that by virtue of its numerical predominance the dominant allele will become more and more common, and the recessive allele rarer and rarer?

If this supposition were true, predominant phenotypes would exist not because of their advantageous properties, but solely because of their Mendelian ratios. The science of population genetics was born in 1908 when G.H. Hardy and W. Weinberg independently proved that the naive geneticist's supposition is not true.

Suppose that a diploid, sexually reproducing population consists of $N$ individuals that interbreed at random, so that their genes combine in a common GENE POOL. At any autosomal locus, such as locus $A$, there are $2N$ gene copies. If there are two alleles, $A$ and $A'$, the three genotypes $AA$, $AA'$, and $A'A'$ number $n_{AA}$, $n_{AA'}$, and $n_{A'A'}$ respectively. The proportions, or GENOTYPE FREQUENCIES, $n_{AA}/N$, $n_{AA'}/N$, and $n_{A'A'}/N$ will be denoted $D$, $H$, and $R$ respectively. Thus $D + H + R = 1$. The ALLELE FREQUENCIES (often called GENE FREQUENCIES) of alleles $A$ and $A'$ can be easily calculated. The frequency of $A$ is $p = n_A/2N$, where $n_A$ is the number of copies of $A$ represented among the members of the population, and the frequency of $A'$ is $q = 1 - p = n_{A'}/2N$. The number of $A$ alleles equals twice that of $AA$ individuals plus the number of heterozygotes, or $n_A = 2n_{AA} + n_{AA'}$, so $p = n_A/2N = D + H/2$. Similarly, $q = R + H/2$.

If mating is random within the population, the frequency of each type of mating can be calculated; for example, the frequency of $AA \times AA$ matings is $D^2$ and that of matings between $AA$ and $AA'$ is $2DH$ (since there are two such crosses, $AA \times AA'$ and $AA' \times AA$, each with probability $DH$). By calculating the frequency of each cross and the proportion of each genotype among the progeny of that cross, it is shown (Box A) that after one generation of random mating the frequencies of the genotypes $AA$, $AA'$ and $A'A'$ will be $p^2$, $2pq$, and $q^2$ respectively, no matter what the initial genotype frequencies $D$, $H$ and $R$ were. Among these progeny, moreover, the frequency of $A$ is $p^2 + \frac{1}{2}(2pq) = p(p + q), = p$—the same allele frequency as in the previous generation.

The relative abundances of the alleles $A$ and $A'$ do not change from one generation to the next; the only change in the genetic complexion of the population is a redistribution of the genotypes into frequencies that will be retained in all subsequent generations. Whether either allele is dominant over the other

## A *Derivation of the Hardy-Weinberg theorem*

| MATING | FREQUENCY OF MATING | PROGENY AA | PROGENY AA′ | PROGENY A′A′ |
|---|---|---|---|---|
| $AA \times AA$ | $D^2$ | $D^2$ | | |
| $AA \times AA'$ | $2DH$ | $DH$ | $DH$ | |
| $AA \times A'A'$ | $2DR$ | | $2DR$ | |
| $AA' \times AA'$ | $H^2$ | $H^2/4$ | $H^2/2$ | $H^2/4$ |
| $AA' \times A'A'$ | $2HR$ | | $HR$ | $HR$ |
| $A'A' \times A'A'$ | $R^2$ | | | $R^2$ |
| Total | $(D + H + R)^2 = 1$ | $(D + ½H)^2 = p^2$ | $(D + ½H)(½H + R) = 2pq$ | $(½H + R)^2 = q^2$ |

Note: $D$ = initial frequency of $AA$, $H$ = that of $AA'$, $R$ = that of $A'A'$; $D + H + R = 1$. The progeny totals are calculated by recognizing that $p = D + ½H$, $q = ½H + R$.

is irrelevant; dominance describes the phenotypic effect, not the abundance, of an allele. This then is the HARDY-WEINBERG THEOREM, the foundation of the entire genetical theory of evolution: under the conditions we have implicitly assumed, a single generation of random mating establishes binomial genotype frequencies, and neither these frequencies nor the allele frequencies $p$ and $q$ will change in subsequent generations.

In both natural and laboratory populations genotype frequencies are often very close to these theoretical expectations. For example, of 1612 specimens of the scarlet tiger moth *Panaxia dominula* that Ford and his associates collected (Ford 1971, p. 136), 1469 had white spotting ($AA$), 5 had little spotting ($A'A'$), and 138 were intermediate ($AA'$). The frequency $p$ of the allele $A$ was thus $D + H/2$, or $1469/1612 + ½(138/1612) = 0.954$. The frequency of $A'$ was 0.046. The genotype frequencies, according to the Hardy-Weinberg theorem, should be $(0.954)^2 = 0.9101$ $AA$, $2(0.954)(0.046) = 0.0878$ $AA'$, and $(0.046)^2 = 0.0021$ $A'A'$. The numbers of the three genotypes that we should expect in a sample of 1612 moths are, then, 1467 $AA$ (approximately $0.9101 \times 1612$), 142 $AA'$, and 3 $A'A'$. Because the expected and observed numbers are very similar, we can conclude that this locus was in Hardy-Weinberg equilibrium.

### Extensions of the Hardy-Weinberg theorem

The Hardy-Weinberg theorem can be extended in several ways. Let the three alleles $A$, $A'$, and $A''$ have gene frequencies $p$, $q$, and $r$, where $p + q + r = 1$. Then the genotype frequencies after one generation of random mating are given by the multinomial expansion $(p + q + r)^2$:

| $AA$ | $AA'$ | $AA''$ | $A'A'$ | $A'A''$ | $A''A''$ |
|---|---|---|---|---|---|
| $p^2$ | $2pq$ | $2pr$ | $q^2$ | $2qr$ | $r^2$ |

If there are many, say $k$, alleles with frequencies $p_1, p_2 \ldots p_k$, the genotype frequencies are given by $(p_1 + p_2 + \ldots + p_k)^2$.

The Hardy-Weinberg theorem can be extended to cases of polyploidy also. In a tetraploid plant two alleles $A$ and $A'$ with frequencies $p$ and $q$ assort into five genotypes ($AAAA, AAAA', \ldots, A'A'A'A'$). If the chromosomes assort randomly in meiosis to form diploid gametes, these genotypes are formed in accord with the expansion $(p + q)^4$. In general, for an $n$-ploid organism, the genotype frequencies are given by $(p + q)^n$. In some cases, though, meiosis in polyploids is not that simple, and the theorem must be modified accordingly.

If a locus is sex-linked rather than autosomal, two-thirds of the genes[1] in a population with a 1:1 sex ratio are carried by the homogametic sex (e.g., the human female) and one-third by the heterogametic sex (e.g., the human male). Because human males receive their sex-linked genes only from their mothers, the gene frequency among the males must equal the frequency among the females in the previous generation; and the gene frequency in females must be the average of that of their mothers and fathers. As a result the allele frequency of the population as a whole remains constant, but within each sex it oscillates toward an equilibrium that is the same for both sexes (Figure 1).

The extension of the Hardy-Weinberg theorem to two (or more) loci is complicated, but it is very important. Let $p_1$ and $q_1$ be the frequencies of alleles $A$ and $A'$ at one locus, and $p_2$ and $q_2$ the frequencies of alleles $B$ and $B'$ at a second locus. The loci recombine and segregate in the double heterozygote $AA'BB'$ at a rate $R$ which ranges from 0 (for loci that do not recombine) to ½ (for independently segregating loci). If the loci are on the same chromosome, then $R$ is the crossover distance between the loci.

The critical concept in multiple-locus theory is GAMETE FREQUENCY. There are nine possible zygotic genotypes ($AABB, AABB', \ldots, A'A'B'B'$), but only four possible gametes ($AB, AB', A'B, A'B'$) that unite at random to form zygotes. Denote the frequencies of these gametes by the respective symbols $g_{00}, g_{01}, g_{10}$, and $g_{11}$; note that $g_{00} + g_{01} + g_{10} + g_{11} = 1$. The frequency $p_1$ of the allele $A$ is $g_{00} + g_{01}$, since these are the only gametes that carry $A$. The frequency $p_2$ of $B$ is $g_{00} + g_{10}$, and similarly for the other two alleles.

If the probability that a gamete contains allele $A$ is independent of its carrying allele $B$, the frequency ($g_{00}$) of the $AB$ gamete is then the product of the allele frequencies, $p_1p_2$. Similarly, in this case $g_{01} = p_1q_2$, $g_{10} = q_1p_2$, and $g_{11} = q_1q_2$. The gametes $AB$ and $A'B'$ are COUPLING GAMETES, and $AB'$ and $A'B$ are REPULSION GAMETES. The term $D$, defined as $D = g_{00}g_{11} - g_{01}g_{10}$, expresses the degree to which coupling and repulsion gametes differ in frequency. If the alleles at the two loci are randomly combined into gametes, coupling and repulsion gametes are equally common, and $D = (p_1p_2 \times q_1q_2) - (p_1q_2 \times q_1p_2) = 0$.

[1] The word "gene" is sometimes used rather loosely in the evolutionary literature. Generally it means a gene locus, as when we refer to the *white* gene of *Drosophila*, which has several allelic states (wild type, *white*, etc.). Sometimes, as in this sentence, it refers to a particular copy of the gene in a population; thus its usage here is shorthand for "two-thirds of the copies at any particular locus." In a population of 100 females and 100 males, 200 genes copies ("genes") of a sex-linked gene are carried by the females. This usage does not specify what fraction are of one allele or another. To make matters even more confusing, allele frequencies are often referred to as "gene frequencies," an imprecise usage that I will attempt to avoid.

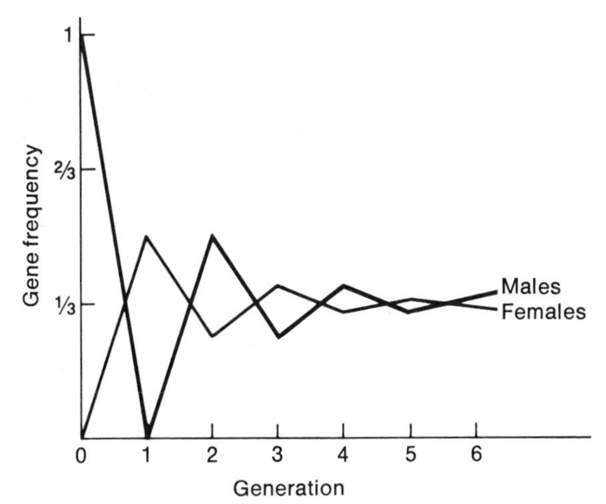


FIGURE 1

**Approach to equilibrium at a sex-linked locus in a species in which females have two X chromosomes, males one. The allele has an initial frequency of zero in females, one in males. Its frequency in the population is ⅓ throughout. (From Crow and Kimura 1970)**

However, the alleles may not be combined at random. For example, if a population has just been formed by mixing two homozygous stocks *AABB* and *A′A′B′B′* in equal numbers, the allele frequencies $p_1$, $p_2$, $q_1$ and $q_2$ all equal ½, but only two kinds of gametes are formed, *AB* and *A′B′*, each with a frequency of ½. Thus $D = (½ \times ½) - (0 \times 0) = ¼$. The population is said to be in a state of GAMETIC EXCESS, or LINKAGE DISEQUILIBRIUM. Often termed the COEFFICIENT OF LINKAGE DISEQUILIBRIUM, $D$ ranges from $+¼$ when only coupling gametes exist, to $-¼$ when all gametes are in the repulsion phase, assuming that $p_1 = p_2 = ½$.

Several points of interest are proven in Box B. For each of the two loci viewed separately, a single generation of random mating establishes the genotype frequencies $p^2$, $2pq$, and $q^2$ just as if the other locus did not exist; the Hardy-Weinberg theorem holds true. However, the degree of association between alleles *A* and *B* and between *A′* and *B′* persists for a while. There is a correlation between alleles at the two loci that is visible among the zygotes as an excess of some genotypes (those carrying *AB* and *A′B′* combinations) and a deficiency of others (those with *AB′* and *A′B* combinations in their genetic makeup). The strength of the correlation is measured by $D$, which only gradually declines toward zero as recombination assorts the alleles into random combinations. The rate at which $D$ approaches zero (a state of LINKAGE EQUILIBRIUM) depends on how tightly the loci are linked; the value of $D$ in the $n$th generation is $D_n = D_0(1 - \mathrm{R})^n$, where $D_0$ is the initial value of $D$ (Figure 2). Thus even if two loci are linked, the characteristics they determine are not necessarily correlated with one another in the population. Conversely, two characteristics that are correlated with one another may be determined by unlinked genes, for even independently segregating loci can show some degree of gametic excess for a number of generations.

## Assumptions of the Hardy-Weinberg theorem

The Hardy-Weinberg theorem and its extensions are based on certain assumptions; violations of these assumptions cause changes in the frequencies of alleles, of genotypes, or of both. The discrepancies between an "ideal" Hardy-Weinberg

## B Two Loci

Let $p_1$ and $q_1$ be the frequencies of alleles $A$ and $A'$, and $p_2$ and $q_2$ be the frequencies of $B$ and $B'$. The initial frequencies of the gametes $AB$, $AB'$, $A'B$, and $A'B'$ are $g_{00}$, $g_{01}$, $g_{10}$, and $g_{11}$ respectively. The frequency of an allele is the sum of the frequencies of the gametes that carry it, e.g., $p_1 = g_{00} + g_{01}$.

The gametes combine at random into zygotes; thus the frequency of AABB is $g_{00}^2$, that of $AABB'$ is $2g_{00}g_{01}$, that of $AA'BB'$ is $2g_{00}g_{11} + 2g_{10}g_{01}$, and so on. The frequency of each gamete type produced by this population of zygotes is the sum of the frequency of the zygotes that produce the gamete, weighted by the proportion of the zygote's gametes that are of the appropriate type. Note that the double heterozygote undergoes recombination at rate $R$. A proportion $R/2$ of the gametes of the repulsion heterozygote ($AB'/A'B$) are of type $AB$, and a proportion $(1 - R)/2$ of the gametes of the coupling heterozygotes ($AB/A'B'$) are of type $AB$; i.e., those produced without recombination. The gamete $AB$ is thus produced by genotypes $AB/AB$, $AB/AB'$, $AB/A'B$, $AB/A'B'$, and $AB'/A'B$. Its new frequency ($g'_{00}$) is accordingly

$$\begin{aligned} g_{00}' &= g_{00}^2 + (\tfrac{1}{2})2g_{00}g_{01} + (\tfrac{1}{2})2g_{00}g_{10} + (\tfrac{1}{2})(1 - R)2g_{00}g_{11} + (\tfrac{1}{2})(R)2g_{10}g_{01} \\ &= g_{00}^2 + g_{00}g_{01} + g_{00}g_{10} + g_{00}g_{11} - Rg_{00}g_{11} + Rg_{10}g_{01} \\ &= g_{00}(g_{00} + g_{01} + g_{10} + g_{11}) - R(g_{00}g_{11} - g_{10}g_{01}) \end{aligned}$$

Because the first term in parentheses equals 1 and the second term in parentheses equals the coefficient of linkage disequilibrium $D$, this simplifies to

$$g_{00}' = g_{00} - RD$$

Thus in one generation the excess of coupling gametes $g_{00}$ is reduced by a fraction proportional to the recombination rate and the amount of gametic excess. It can similarly be shown that $g_{01}' = g_{01} + RD$, $g_{10}' = g_{10} + RD$, and $g_{11}' = g_{11} - RD$. These expressions describe the approach to linkage disequilibrium.

Note that the allele frequencies do not change; for example, if $p_1'$ is the frequency of $A$ in the next generation,

$$p_1' = g_{00}' + g_{01} = (g_{00} - RD) + (g_{01} + RD) = g_{00} + g_{01} = p_1$$

Moreover, each locus viewed separately is in Hardy-Weinberg equilibrium. For example, the frequency of $AA$ after one generation is

$$(g_{00}')^2 + 2g_{00}'g_{01}' + (g_{01}')^2 = (g_{00}' + g_{01}')^2 = p_1^2.$$

population and real populations are the ingredients of evolution. The assumptions underlying this theorem are:

1. The size of the population is infinite, or effectively infinite. But in a finite (real) population the frequency of an allele may fluctuate from generation to generation due to chance events (genetic drift).
2. Individuals mate with one another at random. But the pattern of breeding in a population is often not random, and this has important consequences.
3. All alleles are equally competent in making copies of themselves, which enter the gene pool in the gametes. If alleles differ in their replacement rates, their

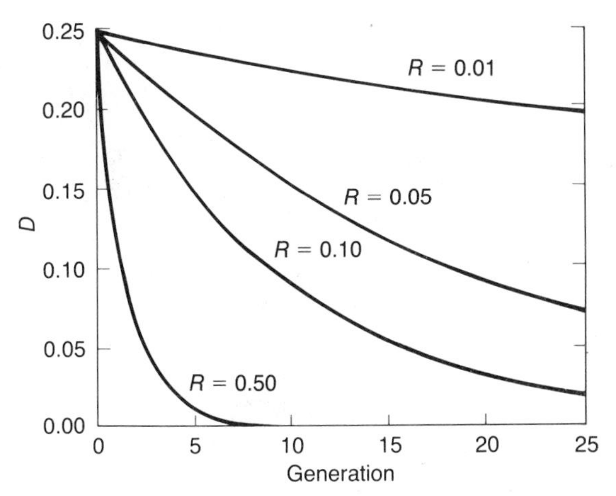


FIGURE 2

**The decay of linkage disequilibrium between two loci, each with two equally frequent alleles, from an initial state of complete linkage disequilibrium. Linkage disequilibrium declines less rapidly for more tightly linked loci (those with a lower rate of recombination, *R*). (From Hedrick 1983)**

frequencies may change. This phenomenon is called selection. A corollary assumption is that alleles segregate in a one-to-one ratio into the gametes produced by heterozygotes.

4. There is no input of new copies of any allele from any extraneous source. If there were, and if one allele entered the population at a greater rate than another, the allele frequencies would change. There are two possible sources of new copies: migration of genes (alleles) from another population (gene flow), and the transformation of one allele into another (mutation).

## VARIATION IN QUANTITATIVE TRAITS

We shall shortly consider examples of genetic variation in characteristics that segregate as discrete phenotypes; but first we must consider, from a theoretical point of view, the kind of variation that is most evident: the slight differences among individuals in characteristics such as hair color, body size or number of eggs. Most characteristics show such continuous (or "quantitative" or "metric") variation (Figure 3). Some characteristics, such as the number of scutellar bristles in *Drosophila*, vary discretely, but only because a fly can have either four or five bristles, but not four and a half. It is possible to show that this is merely a discrete expression of an underlying continuum of bristle-making materials of some kind (Rendel 1967, Wright 1968). Such characteristics are referred to as MERISTIC (countable) or THRESHOLD (discontinuous) traits.

Consider for a moment a single locus at which the average value of some characteristic such as tail length is $3a$, $2a$, and $a$ for genotypes $AA$, $AA'$ and $A'A'$ respectively. The value of $a$ might be, say, 2 cm. The relations among the phenotypes remain the same if we subtract $2a$ from each value, so that $AA$, $AA'$, and $A'A'$ have the standardized scores $a$, 0 and $-a$ respectively. These are the "phenotypic values" of the genotypes, and express how much each genotype deviates from the midpoint between the phenotypes of the homozygotes. In this example, in which the heterozygote is precisely intermediate between the homozygotes, the alleles are said to have ADDITIVE EFFECTS, because the phenotypic consequence of replacing one $A'$ allele with $A$ is precisely half that of replacing two $A'$ alleles with $A$ alleles. Neither allele is dominant.

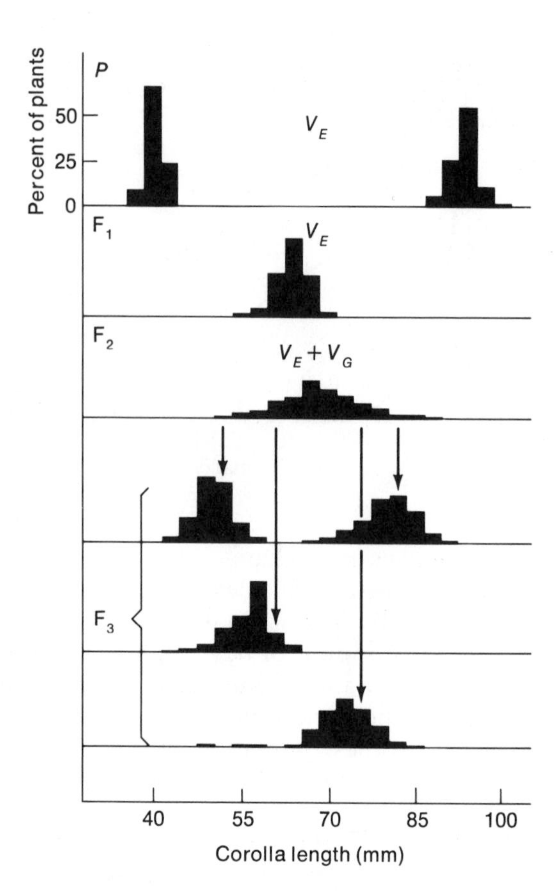


FIGURE 3

**Inheritance of a metric, or continuously varying, trait: corolla length in the tobacco plant, *Nicotiana longiflora*. The two parental strains (*P*) are inbred; all the phenotypic variation in these and in the $F_1$ offspring is environmental ($V_E$). In the $F_2$ and $F_3$ generations, segregation occurs and the variance has both genetic ($V_G$) and environmental components. Four $F_3$ families are shown, from crosses between parents whose means are indicated by arrows. The mean of the offspring is nearly that of the parents, indicating that most of the genetic variance is additive. (From Cavalli-Sforza and Bodmer 1971, after Mather 1949)**

The MEAN PHENOTYPE in the population is the average of these three values weighted by the frequencies of the genotypes, and equals $2ap - a$, as indicated in Box C. Thus the mean depends on the allele frequencies, and is greatest when allele *A* is fixed in the population (i.e., when $p = 1$). The most useful measure of the amount of variation is the VARIANCE (see Appendix I), which is the average squared deviation of observations from the population mean. A general expression for the variance is $V = \Sigma_i f_i(x_i - \bar{x})^2$, that is, the sum over all values that the trait may take ($x_i$), of the squared deviation of the value from the population mean ($\bar{x}$), weighted by the frequency of that value among the members of the population ($f_i$).

In Box C, the variance of a trait that is additively determined by two alleles at one locus is shown to be $V = 2pqa^2$ in a randomly mating population. Thus the magnitude of the variance is directly proportional to the frequency of heterozygotes ($2pq$) in the population. Note, moreover, that the frequency of heterozygotes is greatest when the two alleles are equally common (i.e., when $p = q = 0.5$). In this example, the genetic variation in the trait is called the ADDITIVE GENETIC VARIANCE ($V_A$) because the alleles contribute to the phenotype in an additive fashion.

The variance is a useful measure of variation because the total variance is the sum of individual components of variance. For example, each of several

## C *The Mean and Variance of a Quantitative Character*

Let the genotypes $AA$, $AA'$, and $A'A'$ have mean phenotypic values, $x_i$, equal to $a$, 0, and $-a$ respectively. The alleles have additive effects on the phenotype. If $p$ and $q$ are the frequencies of alleles $A$ and $A'$ respectively, then if the population is in Hardy-Weinberg equilibrium the population mean is $\bar{x} = p^2(a) + 2pq(0) + q^2(-a) = a(p^2 - q^2) = a(p + q)(p - q) = a(p - q)$ since $p + q = 1$. This may also be written as $\bar{x} = 2ap - a$.

The variance, $V = \sum_i f_i(x_i - \bar{x})^2$, is

$$\begin{aligned} V &= p^2[a - (2ap - a)]^2 + 2pq[0 - (2ap - a)]^2 + q^2[-a - (2ap - a)]^2 \\ &= 2a^2\{pq + [2p^2(1 - 2p + p^2) + 2pq(2p^2 - 2p + pq)]\} \end{aligned}$$

Replacing the first term in parentheses with $q^2$, and substituting $1 - p$ for $q$ in the second parenthetical term,

$$\begin{aligned} V &= 2a^2[pq + 2p^2q^2 + 2pq(p^2 - p)] \\ &= 2pqa^2[1 + 2p(1 - p) + 2(p^2 - p)] \end{aligned}$$

The terms within the brackets sum to one, leaving

$$V = 2pqa^2.$$

variable loci may contribute to a trait. If the phenotype is a simple sum of the contributions of all the loci, the total additive genetic variance, $V_A$, is $\sum_k V_{Ak}$, where $V_{Ak}$ is the additive genetic variance at the $k$th locus. This formula would generally apply to a case such as the following, in which each allele $A$ contributes one unit and each allele $B$ two units to the phenotype:

| | | Locus *B* | | |
|---|---|---|---|---|
| | | *BB* | *BB'* | *B'B'* |
| Locus *A* | *AA* | 6 | 4 | 2 |
| | *AA'* | 5 | 3 | 1 |
| | *A'A'* | 4 | 2 | 0 |

Almost every characteristic is affected by several or many loci, each of which contributes such a slight amount to the total variation that it is extremely difficult to perform genetic crosses that might determine the genotype of an individual, much less to determine allele frequencies at individual loci. Thus the trait will often vary in very small increments from one individual to another. Moreover, each genotype may be phenotypically variable to some extent because its development is directly affected by the environment and by random events during ontogeny ("developmental noise"). For example, genetically identical *Drosophila* reared in a homogeneous environment vary slightly in wing length, and a single fly is often bilaterally asymmetrical, because of uncontrollable variations in the molecular mechanisms of development. The magnitude of all of the nongenetic variation (usually termed the ENVIRONMENTAL VARIANCE, $V_E$) may be measured in several ways (Chapter 7); for example, it is sometimes possible by inbreeding or cloning to obtain a number of individuals with the same genotype. Then whatever phenotype variation they display is a consequence of the environment or of developmental noise. By this or other means, it is possible to describe the phenotypic variance ($V_P$) as the sum of the genetic variance ($V_G$) and the environmental variance ($V_E$), assuming that genetic and environmental effects on the phenotype are independent and additive:

$$V_P = V_G + V_E$$

We will see in Chapter 7 that the additive genetic variance $V_A$ is sometimes just a portion of the total genetic variance $V_G$. We will also see how it is possible to measure the HERITABILITY ($h^2$) of a trait, which is defined in the narrow sense as

$$h^2 = V_A/V_P$$

the proportion of the phenotype variance that is attributable to the additive effects of alleles on the trait. For the moment, suffice it to say that the higher the heritability is, the more closely offspring resemble their parents, and indeed the magnitude of the phenotypic resemblance between relatives offers one way of measuring the heritability of a character. Although the genetic component of variation is the material of evolutionary change, many characteristics have a considerable nongenetic component to their variation (i.e., they have low heritability). For example, *Drosophila* flies are larger if reared at cool temperatures, or under uncrowded conditions, than if they are warm or crowded; the growth form of most plants is very plastic and depends on light, water, and soil conditions; the rotifer *Brachionus* develops spines or not, depending on the presence or absence of chemicals produced by a predatory rotifer species (Chapter 2).

## VARIATION WITHIN NATURAL POPULATIONS

Most characteristics vary to at least some extent; commonly the phenotypic standard deviation (the square root of the phenotypic variance) is about 5–10 percent of the mean. That a considerable fraction of this variation is often (but not always) genetically based can be shown in several ways. One method is to show that there is a correlation between relatives. For example, Boag (1983) calculated the heritability of several traits in populations of Darwin's finches in the Galápagos Islands by measuring banded adults and their offspring, and found that even in a variable natural environment, most of the phenotypic variation had a genetic basis (Table I). This method, however, runs the risk of confounding genetic effects with non-genetic correlations between relatives, such as maternal effects (Chapter 3) or environmental effects shared by members of a family. One way to exclude such effects is to transfer offspring at random among nests, so that they are reared by unrelated individuals. For example, Smith and Dhondt (1980) were able to confirm by this technique that morphological traits of song sparrows (*Melospiza melodia*) were correlated with those of their true parents rather than their foster parents.

Another way of revealing genetic variation is to select artificially for change in the trait by breeding only from individuals that are either higher or lower than the population mean, for selection can alter a population from one generation to another only if some of the variation is hereditary. In *Drosophila*, artificial selection

TABLE I
**Heritability of morphological features of the medium ground finch *Geospiza fortis*[a]**

| Trait | Mean | Coefficient of variation | Phenotypic variance | Heritability |
|---|---|---|---|---|
| Body weight (gm) | 15.86 | 8.05 | 1.630 | .91 ± .09 |
| Tarsus length (mm) | 18.75 | 3.42 | 0.411 | .71 ± .10 |
| Bill length (mm) | 10.74 | 6.64 | 0.509 | .65 ± .15 |
| Bill width (mm) | 8.74 | 6.66 | 0.339 | .90 ± .10 |

(Data from Boag 1983)
[a]First three columns are from 44 adult birds measured in 1976; final column from pooled 1976 and 1978 data. Phenotypic variances were calculated from Boag's Table 4. Coefficient of variation= 100 × (standard deviation/mean). Estimates of heritability are based on regression of offspring values on mean of their two parents, and are followed by their standard errors.

has accomplished changes in numerous morphological traits, as well as geotactic behavior, sexual behavior, developmental rate, fecundity, dispersal ability, feeding preferences, resistance to insecticides and other toxins, crossover rates between linked genes, and many other characteristics (summarized by Lewontin 1974a). Indeed, the history of breeding improved strains of domestic plants and animals testifies to the ubiquity of genetic variation in almost every characteristic. The response to selection is attributable primarily to preexisting genetic variation, not to the occurrence of new mutations as selection proceeds, for highly inbred homozygous populations hardly respond to selection at all (Clayton and Robertson 1955).

To be sure, some characteristics do not respond immediately to selection. For example, *Drosophila* normally has one median anterior ocellus (simple eye) and two bilaterally symmetrical posterior ocelli. It is possible to select strains that lack the median ocellus or that have fluctuating asymmetry, i.e., random absence of one of the posterior ocelli. But it was not possible to select a strain of flies that consistently lacked a specific (i.e., right or left) posterior ocellus; there appear to be developmental constraints on the expression of genetic variation (Maynard Smith and Sondhi 1960). But even some traits that ordinarily show no phenotypic variation are genetically variable and can sometimes respond to selection. For example, *Drosophila melanogaster* almost always has four bristles on the scutellum; yet by introducing into a *Drosophila* population a mutant allele that disrupts the usual developmental pathway, latent variation in bristle number caused by genes other than the mutant allele becomes visible and provides a basis for an evolutionary response to natural selection (Rendel 1967; see Chapter 7). Thus there are genetic variants that are prevented by homeostatic developmental pathways from exhibiting their effects.

### Visible polymorphisms

In classical genetics it is impossible to know whether a gene exists, much less to be able to study its properties, unless alternative alleles at that locus can be identified. So the early studies of genes in populations focused on discrete variants such as the ABO blood types or white eyes in *Drosophila*. But in natural populations of most species we seldom find a characteristic with two or more discrete phenotypes that would tempt us to perform a Mendelian cross. Populations of eastern gray squirrels (*Sciurus carolinensis*), for example, contain occasional black, melanistic individuals, but such "mutants" are rare. For this reason, classical geneticists distinguished the prevalent wild type from mutants and assumed that populations were genetically homogeneous. In this classical view of population structure, a wild type allele (often denoted as +) is prevalent at most loci; the only variation is due to occasional mutations which, since they seem rare, must be deleterious. In this view evolution is a very slow process, since it must await rare advantageous mutations to provide adaptation to environmental change.

Occasionally, however, two or more discretely different phenotypes are fairly common within a population. When the rarest of them exceeds some arbitrarily high frequency—say one percent or so—this condition is referred to as POLYMORPHISM. The color patterns of the snow goose (*Chen caerulescens*) and the California king snake (*Lampropeltis getulus*) (Figure 4), for example, are so dimorphic that in

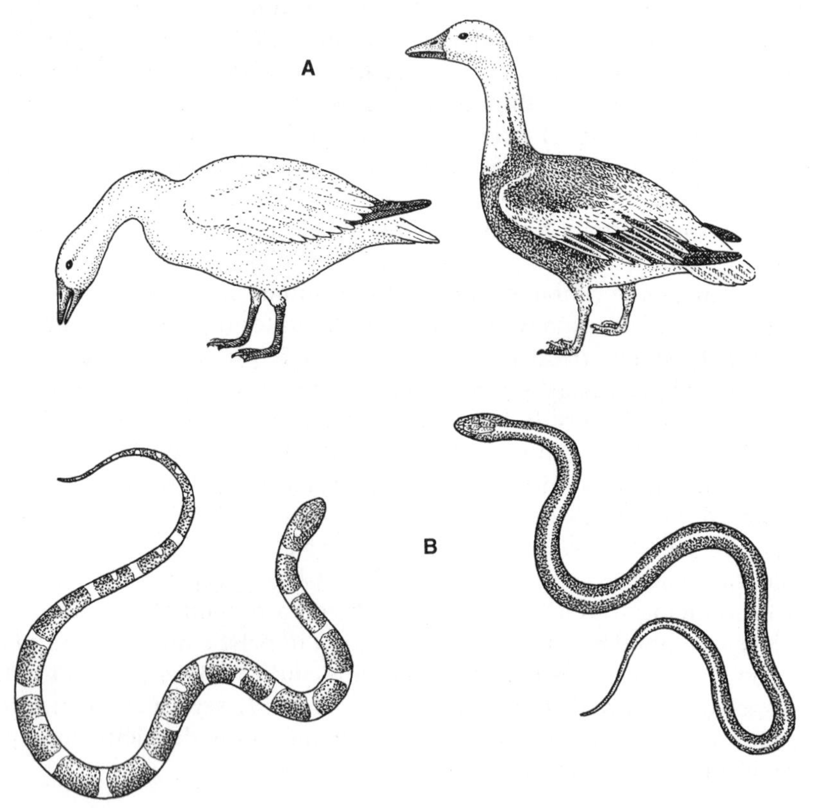


FIGURE 4

**Polymorphic variation. (A) Different color forms ("snow goose" and "blue goose") of *Chen caerulescens*. (B) The two patterns of the king snake *Lampropeltis getulus* in California. In both cases the two forms are often found in the same litter. (A redrawn from Pough 1951, B redrawn from Stebbins 1954)**

both cases the different forms were originally described as separate species. However, the color forms interbreed freely, and in both cases the difference in color pattern appears to be caused primarily by a single locus.

Some polymorphisms are TRANSITIONAL; we observe the population while one allele is replacing another. The most famous case of such a polymorphism is the melanistic form of the moth *Biston betularia*, a form that was almost unknown before the Industrial Revolution, but by 1895 constituted about 98 percent within some populations in England (Kettlewell 1973, Bishop and Cook 1980). During this period the melanistic form increased, but it did not completely replace the nonmelanistic form. After the phase of transitional polymorphism, the alleles arrived at a stable intermediate frequency, forming a BALANCED POLYMORPHISM.

A complex balanced polymorphism has been studied in the land snail *Cepaea nemoralis*, in which the background color of the shell is brown, pink, or yellow, and the shell has zero to five bands, the number of which is controlled by several loci (Chapter 7). The relative numbers of the various forms differ greatly from one place to another, even between localities less than a mile apart. Fossils show that the polymorphism has persisted at least since the Pleistocene (Diver 1929). That the gene frequencies are affected by natural selection is suggested by their correlation with habitat (Figure 5). Open fields and hedgerows are predominantly populated by yellow, banded snails, while brown, unbanded individuals are more

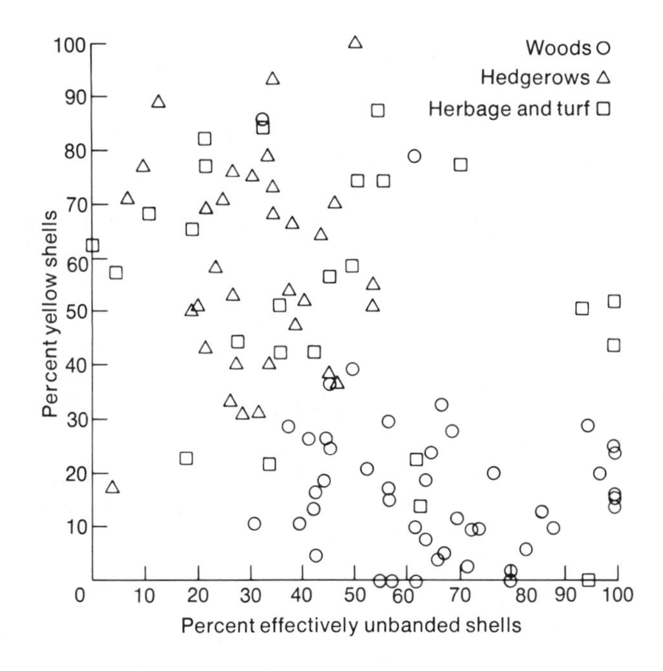


FIGURE 5

**Microgeographic variation in the frequencies of shell color and banding variants in the land snail *Cepaea nemoralis*. Each symbol represents a sample from the habitat indicated. The samples were taken in the neighborhood of Oxford, England. In wooded habitats, there is a high proportion of unbanded, brown or pink shells; in more open habitats, more of the snails have yellow, banded shells. (From Cain and Sheppard 1954)**

common in woods. Cain and Sheppard (1954) and Murray (1962) showed that this correlation is at least partly caused by predation by thrushes, which are less likely to find yellow snails than other forms in grassy areas where the yellow coloration is more cryptic. Moreover, the genotypes differ in their susceptibility to extreme temperatures (Lamotte 1959, Jones et al. 1977).

The marine copepod *Tisbe reticulata* has a color polymorphism (Battaglia 1958). In laboratory crosses the offspring consistently show an excess of heterozygotes, which have higher survival than either homozygote. The magnitude of the difference in viability depends on culture conditions; the excess of heterozygotes is greater in highly crowded than in less dense cultures and is also influenced by salinity. In Chapter 6 we will see that if the fitness of the heterozygote is greater than that of either homozygote, both alleles remain in a balanced polymorphic state. The polymorphism in *Tisbe* is one of the better documented cases of such heterozygote superiority. It also illustrates that the relative viabilities of genotypes vary with environmental conditions and that the effect of a gene on survival may not be due to the gene's obvious morphological manifestation. It seems implausible that the coloration of these copepods was itself responsible for their ability to survive in the laboratory; rather, the coloration seems to be a pleiotropic effect, perhaps not adaptive in itself, of a gene whose unknown physiological effects are responsible for the existence of the polymorphism.

### Cryptic variation

Although a fair number of such obvious polymorphisms are known, they are more the exception than the rule. Early geneticists were rather surprised to find that many populations actually harbor a great many mutants. Most of these are rare recessives and are brought to light by inbreeding stocks of, say, *Drosophila*, thus making them homozygous. From an extensive study of *D. mulleri*, Spencer (1957) estimated that every wild fly carries somewhere in its genome an average of one mutant allele that would cause some morphological abnormality in homozygous condition. At any particular such locus, the mutant allele has a low frequency, about $10^{-3}$ (Lewontin 1974a).

Such morphological mutations are known at only a small fraction of the loci that *Drosophila* is thought to have; moreover, it may be easy to miss seeing many of them by simply looking at flies. A better estimate of the true prevalence of mutations comes from the study of more unambiguously recognizable genes—modifiers of viability. The study of such genes requires special techniques that can be applied only in some species of *Drosophila*. Through a series of crosses between a wild fly and a laboratory stock that carries a dominant marker gene and an inversion that prevents crossing over (Chapter 3), it is possible to bring a single chromosome from a wild population into homozygous condition at all of its loci (Figure 6). The crosses ultimately produce a family of flies of which one-fourth are expected to be homozygous for the wild chromosome. The degree of deviation from the expected proportion is a measure of the inviability of the flies that are homozygous for the wild chromosome; if no wild-type flies appear, the wild chromosome must carry a recessive lethal allele. Performing such crosses with many different wild flies makes it easy to determine what fraction of wild chromosomes are lethal, semi-lethal, subvital, quasi-normal, or supervital in homozygous condition.

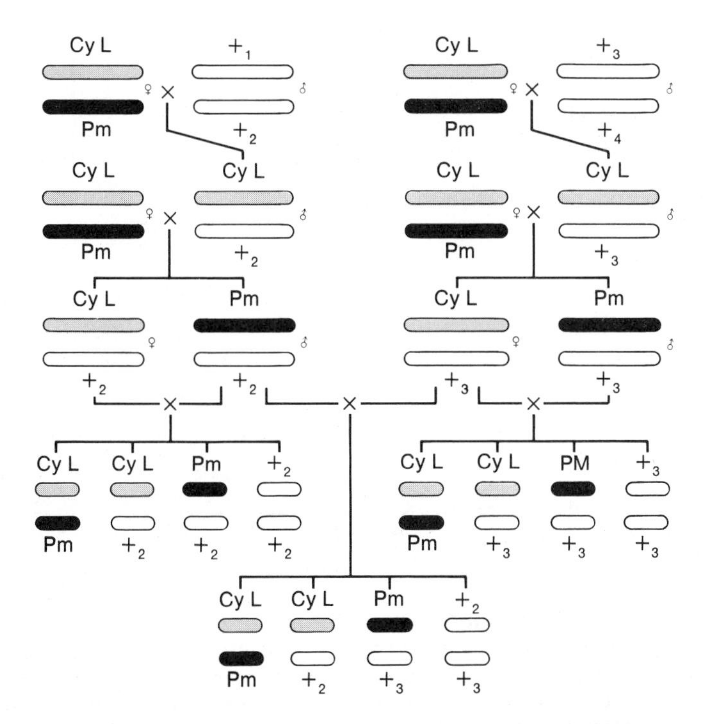


FIGURE 6

**Crossing technique for "extracting" a chromosome from a male *Drosophila melanogaster* and making it homozygous to detect recessive alleles. The process is shown for two wild-type males, each of which is crossed to a laboratory stock that has dominant mutant alleles *Cy* (curly wing) and *L* (lobed wing) on one chromosome and *Pm* (plum eye color) on the other. These chromosomes also carry inversions that prevent crossing over. The crosses shown produce wild-type flies (i.e., those without the *Cy, L,* or *Pm* markers) that are homozygous for either of two chromosomes ($+_2$ and $+_3$), as well as flies heterozygous for these same two chromosomes. The viability of these homozygotes and heterozygotes is measured by their proportion in the family relative to the expected 1:1:1:1 ratio. (From Dobzhansky 1970)**

Studies of this kind (reviewed by Dobzhansky 1970, Lewontin 1974a, and Simmons and Crow 1977) have been performed on many *Drosophila* populations; some typical results are illustrated in Figure 7. Most heterozygotes for randomly chosen wild chromosomes have quasi-normal viability, but homozygotes are in general less viable. In fact, about 10 percent of the chromosomes are virtually lethal in homozygous condition. Almost every individual carries abnormal alleles. This also appears to be the case in corn (*Zea mays;* Crumpacker 1967), Douglas fir (*Pseudotsuga menziesii;* Sorensen 1969) and probably in human populations as well (Morton et al. 1956).

Moreover, any two quasi-normal, seemingly identical chromosomes from different flies are genetically different. When they are allowed to recombine, and the recombined chromosomes are then made homozygous, the newly synthesized chromosomes vary in their viability and are sometimes even lethal. So the original chromosomes must have had different alleles, which in new combinations interact to affect viability.

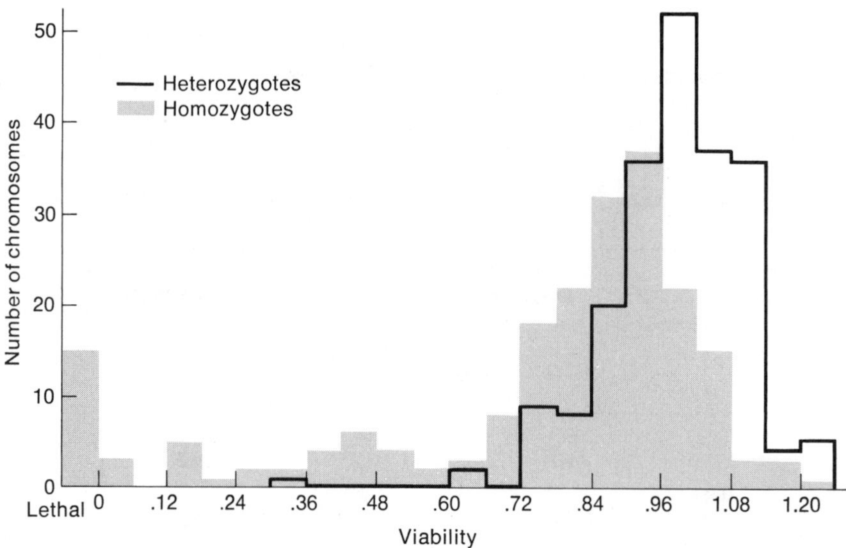


FIGURE 7

**The distribution of relative viabilities of chromosomes extracted from a wild population of *Drosophila pseudoobscura* by the method illustrated in Figure 6. The chromosomes vary greatly in the viability they confer, especially when homozygous. The average viability of heterozygotes is higher than that of homozygotes, indicating that many chromosomes carry deleterious recessive alleles. (From Lewontin 1974a)**

The revelation of all this variation was something of a shock in the 1920s and 1930s, when genetic uniformity was taken for granted. It resulted in a swell of new opinion among many geneticists, led by Theodosius Dobzhansky: a population is an immensely diverse assortment of genotypes, and there is no such thing as a wild-type, or normal, genotype. Rather the norm *is* diversity. The words "normal" and "abnormal" begin to lose their meaning.

A fly with two lethal chromosomes is usually healthy as long as the chromosomes were derived from different wild individuals. Therefore most of the lethal genes in a wild population are not allelic to one another (Wright et al. 1942), so lethal alleles must occur at many different loci. And the lethal allele at any one locus is actually very rare; its frequency is generally about $10^{-3}$ to $10^{-2}$. Dobzhansky and his associates held the view that some lethal alleles are slightly beneficial in heterozygous condition and are kept in the population by heterozygous advantage (Dobzhansky et al. 1960, Wallace and Dobzhansky 1962). The weight of the evidence (see Simmons and Crow 1977) indicates that some deleterious recessives are indeed slightly beneficial in heterozygous condition, but that the majority of them are partially dominant, i.e., they reduce fitness both in the heterozygous and homozygous state.

## VARIATION IN PROTEINS

It was not possible to tell what fraction of loci were polymorphic or what the heterozygosity of an average locus was as long as it was impossible to count the number of loci that exhibit no genetic variation. But it may be possible to do so if most loci code for proteins (especially enzymes). Different forms of, say, alcohol dehydrogenase are coded for by different alleles; but two individuals with the same form of the enzyme are presumably genetically the same. So it is possible to identify an invariant gene locus by finding an invariant enzyme. By taking a random sample of all kinds of enzymes and determining what fraction of them is genetically variable, one can estimate the fraction of polymorphic loci. Knowing which individuals are homozygous and which are heterozygous for each enzyme, one can calculate the proportion of heterozygotes (the heterozygosity, $H$) at each locus.

The most common technique for distinguishing different genetic forms (ALLOZYMES) of the same enzyme is gel electrophoresis. Extracts from several individuals are placed in a porous gel, often of starch, across which an electric potential is applied. If the amino acid composition of a given enzyme varies and if the amino acid substitutions carry different charges, the enzymes differ in net charge and move at different rates, so they are separated in the gel. Their positions are then found by flooding the gel with a substrate on which the enzyme acts, together with a stain that reacts with the product of the enzyme–substrate reaction to yield a colored band that indicates the position of the enzymes. Thus different genotypes are identified by differences in the position of the bands on the gel (Figure 8). These differences usually prove to be genetic (i.e., inherited in a Mendelian fashion in crosses), so an experienced worker can recognize different genotypes and be quite sure that they are genetically different, even in species that cannot be crossed in the laboratory.

The first assays of genetic variation by protein electrophoresis were published in 1966, when Harris reported on variation at 10 enzyme loci in humans and Lewontin and Hubby investigated 18 loci in each of several populations of *Drosophila pseudoobscura*. Lewontin and Hubby's data (Table II) were a surprise to the classical school of geneticists, for they led to the conclusion that an average *Drosophila* population is polymorphic at no less than 30 percent of its loci and

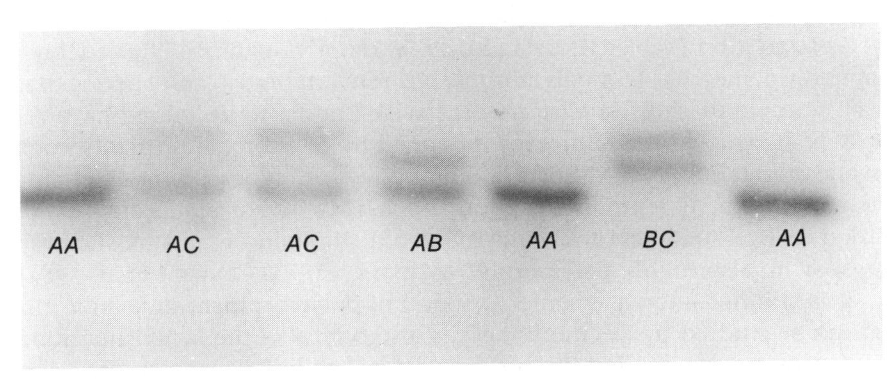


FIGURE 8

**Polymorphism in an enzyme, leucine aminopeptidase, in the mussel *Mytilus edulis*. A starch gel with the enzyme profile of seven individual mussels representing four genotypes. The enzyme is a monomer; hence homozygotes have one band and heterozygotes have two. (Gel courtesy of R. K. Koehn)**

TABLE II
**Polymorphism and heterozygosity at 18 enzyme loci in *Drosophila pseudoobscura***

| Population | Number of loci polymorphic | Proportion of loci polymorphic | Proportion of genome heterozygous per individual |
|---|---|---|---|
| Strawberry Canyon | 6 | 0.33 | 0.148 |
| Wildrose | 5 | 0.28 | 0.106 |
| Cimarron | 5 | 0.28 | 0.099 |
| Mather | 6 | 0.33 | 0.143 |
| Flagstaff | 5 | 0.28 | 0.081 |
| Average | — | 0.30 | 0.115 |

(After Lewontin and Hubby 1966)

that the polymorphic loci have so many alleles (2–6) at such high frequencies that an average fly is likely to be heterozygous at about 12 percent of its loci. If these loci (mostly coding for soluble enzymes) are truly a random sample of the roughly 10,000 loci in *Drosophila,* there must be about 2000–3000 polymorphic loci in a local population, and an average fly has 700–1200 heterozygous loci. If so, any two flies from a single population differ at about 25 percent of their loci. This amount of variation is truly staggering. Moreover, Harris's data indicated that humans, with 30 percent polymorphic loci and 10 percent average heterozygosity, were similarly highly variable.

At least at the loci that can be studied by electrophoresis, these early estimates of genetic variation were, if anything, underestimates. At some highly variable loci, the enzyme products of several different alleles have similar electrophoretic mobility (Coyne 1976, Coyne et al. 1978). The term ELECTROMORPH refers to an apparent protein variant observed on a gel that may actually be a class of variants with the same electrophoretic mobility.

Many species have been studied by electrophoresis, and almost all have high levels of variation (Table III; Selander 1976, Hamrick et al. 1979). Vertebrates are commonly somewhat less polymorphic than invertebrates, and species that form small local populations or that are otherwise known to be inbred have reduced levels of heterozygosity. But even populations of inbreeding plants consist of considerable numbers of different homozygous genotypes (Hamrick et al. 1979), and many asexually reproducing populations likewise are genetically very diverse (Parker 1979, Harshman and Futuyma 1985). Haploid organisms such as *Escherichia coli* are also highly polymorphic (Milkman 1973, Caugant et al. 1981).

It is still not known whether the level of polymorphism in soluble enzymes that can be studied by electrophoresis is also typical of the far greater number of loci that cannot be studied by this technique. The loci usually studied may well not be a representative sample of the genome, for some classes of enzymes seem consistently more polymorphic than others (Johnson 1976, Selander 1976). However, specific DNA sequences, such as mitochondrial DNA, have been shown by analysis of restriction sites and by direct sequencing to vary abundantly within

TABLE III
**Genetic variation at allozyme loci in animals and plants**

| | Number of species examined | Average number of loci per species | Average proportion of loci | |
|---|---|---|---|---|
| | | | Polymorphic per population | Heterozygous per individual |
| Insects | | | | |
| *Drosophila* | 28 | 24 | 0.529 | 0.150 |
| Others | 4 | 18 | 0.531 | 0.151 |
| Haplodiploid wasps[a] | 6 | 15 | 0.243 | 0.062 |
| Marine invertebrates | 9 | 26 | 0.587 | 0.147 |
| Marine snails | 5 | 17 | 0.175 | 0.083 |
| Land snails | 5 | 18 | 0.437 | 0.150 |
| Fish | 14 | 21 | 0.306 | 0.078 |
| Amphibians | 11 | 22 | 0.336 | 0.082 |
| Reptiles | 9 | 21 | 0.231 | 0.047 |
| Birds | 4 | 19 | 0.145 | 0.042 |
| Rodents | 26 | 26 | 0.202 | 0.054 |
| Large mammals[b] | 4 | 40 | 0.233 | 0.037 |
| Plants[c] | 8 | 8 | 0.464 | 0.170 |

(After Selander 1976)
[a]Females are diploid, males haploid
[b]Human, chimpanzee, pigtailed macaque, and southern elephant seal
[c]Predominantly outcrossing species

natural populations (e.g., Avise et al. 1979, Brown 1980). Thus variation is not limited to loci that code for the enzymes that are studied by electrophoresis. Variation in DNA sequences will be treated more fully in Chapter 15.

## THE ORGANIZATION OF GENETIC VARIATION

When allele frequencies at one or more loci have been determined it is possible, as we have seen, to predict the frequency of each genotype by assuming that each locus is in Hardy-Weinberg equilibrium and that the pairs of loci are in linkage equilibrium. Deviations from these predicted values indicate that organizing factors may be operating to produce nonrandom associations of alleles within or among loci.

At a single locus, deviations from Hardy-Weinberg frequencies can entail either an excess or a deficiency of heterozygotes. For example, Avise and Smith (1974) found that heterozygotes for the enzyme glutamate oxalate transaminase were present in excess in a population of bluegill sunfish (*Lepomis macrochirus*). Such an excess could mean that the survival of heterozygotes is higher than that of homozygotes. Substantial heterozygote excess is rather uncommon, but considerable deficiencies of heterozygotes are very commonly observed. These are usually caused by some kind of inbreeding effect (Chapter 5); for example, they are typically observed in self-fertilizing plant species (Table IV). Populations of such plants are genetically variable, but consist mostly of a variety of homozygous genotypes.

Nonrandom associations among alleles at different loci constitute linkage disequilibrium. For example (Ford 1971), the flowers of the primrose *Primula vulgaris* are heterostylous: there are two forms (Figure 9). The "pin" form, with long style and low-set anthers, is inherited as if it were recessive to the "thrum" form, which has a short style and elevated anthers. The reciprocal situation of style and anthers in pin and thrum plants enhances the likelihood that pollen will be deposited by insects on the stigma of a plant of the opposite type; thus cross-fertilization is facilitated. Occasionally plants with both a long (pin) style and elevated (thrum) anthers are found. These prove to be recombinant genotypes: pin and thrum plants differ not at one locus, but at two very closely linked loci, one of which (*G, g*) controls the style and the other (*A, a*) the anthers (but see Chapter 7 for further complications). *G* is almost always associated with *A*, and *g* with *a*; thrum plants are *GA/ga* and pin plants *ga/ga*. The exceptional plants with the combination of pin and thrum characteristics are *gA/ga*. Thus *G* and *A* are inherited together as if they were one locus—a SUPERGENE that is only rarely broken up by crossing over. The prevalent combinations, *GA* and *ga*, are advantageous compared to the rare combinations *Ga* and *gA*, and are maintained in high frequency by natural selection.

Quite a few similar cases have been described in which favorable combinations of alleles at closely linked loci are held in linkage disequilibrium; for

TABLE IV

**Genotype frequencies and inbreeding coefficients for three loci in the self-fertilizing wild oat *Avena fatua***

| Genotype | Frequency | Frequency of recessive allele | Expected frequency[a] | Inbreeding coefficient ($F$)[b] |
|---|---|---|---|---|
| *BB* | 0.712 | | 0.610 | |
| *Bb* | 0.138 | 0.219 | 0.342 | 0.597 |
| *bb* | 0.150 | | 0.048 | |
| *HH* | 0.583 | | 0.444 | |
| *Hh* | 0.167 | 0.334 | 0.445 | 0.625 |
| *hh* | 0.250 | | 0.112 | |
| *LsLs* | 0.775 | | 0.702 | |
| *Lsls* | 0.125 | 0.162 | 0.272 | 0.539 |
| *lsls* | 0.100 | | 0.026 | |

(After Jain and Marshall 1967)

[a]Genotype frequencies expected under Hardy-Weinberg equilibrium.

[b]$F$, defined in Chapter 5, measures the deficiency of heterozygotes relative to frequency expected under Hardy-Weinberg equilibrium.

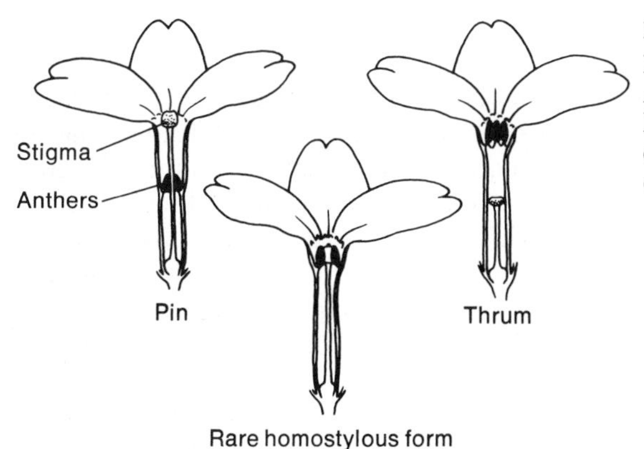


FIGURE 9
**Heterostyly in the primrose *Primula vulgaris*. Above, the "pin" and "thrum" phenotypes, at left and right respectively. Below, the homostyled phenotype formed by occasional crossing over. (From Ford 1971)**

example, this is the case for the several genes that control the polymorphism in shell color and banding pattern in the snail *Cepaea nemoralis*. However, if we may judge from electrophoretic data, most pairs of loci depart only slightly, if at all, from linkage equilibrium (Langley 1977; Hedrick et al. 1978). The most detailed analyses of linkage disequilibrium have been made in *Drosophila melanogaster*, because allele combinations can be determined for individual chromosomes that are extracted from wild flies (Figure 6) and the loci can be mapped precisely. In this species, most loci are in linkage equilibrium, but there is a tendency for closely linked loci to show slight gametic excess (Langley 1977).

Linkage disequilibrium is commonly pronounced when there exist mechanisms that restrict recombination. The most extreme such mechanism is asexual reproduction, and it is common for asexual populations to be dominated by a few clones that differ at several or many loci, so that these display strong linkage disequilibrium in the population as a whole. Populations of self-fertilizing plants likewise can manifest linkage disequilibrium if they are dominated by a few homozygous genotypes. For example, Hamrick and Allard (1972) found two alleles at each of six loci in the self-fertilizing wild oat *Avena barbata*. Of the $2^6$, or 64, possible homozygous genotypes, only two were common, in moist and dry microhabitats respectively. The adaptive differences between these genotypes may not be caused by the specific loci studied, but by other loci that are in linkage disequilibrium with those that can be seen.

Many populations of *Drosophila* are polymorphic for paracentric chromosome inversions that generally suppress crossing over. Dobzhansky and Pavlovsky (1953) postulated that each inversion type carried a COADAPTED combination of alleles at various loci, i.e., a combination that functions well together. Data corresponding to this hypothesis emerged when Prakash and Lewontin (1968) found that different inversions tend to carry different alleles at two electrophoretic loci. Similar cases of association between inversions and alleles at loci within the inversions have since been found by other workers. However, there is no direct evidence that these associations are caused by coadaptation; they might exist simply because different mutations occurred within the several inversions (Nei 1975).

## GENETIC VARIATION AMONG POPULATIONS

At many loci, allele frequencies differ from one population to another, so that the variation that arises within populations becomes transformed into variation among populations. This has long been obvious from the study of morphological variation, but enzyme electrophoresis has provided a powerful tool for assessing the degree to which populations actually differ in allele frequencies. When allele frequencies have been determined for each of a number of such loci, the genetic similarity or difference (sometimes called GENETIC DISTANCE) between pairs of populations can be expressed by any of several indices (Table V). To cite just one example, Schaal and Smith (1980) determined allele frequencies at 13 enzyme loci in five populations of a leguminous herb, the tick trefoil (*Desmodium nudiflorum*), in Ohio and Michigan. Six of the loci were polymorphic in at least one population; at each such locus, the same allele was most common or fixed in all the populations. At the locus that showed the greatest geographic variation in allele frequencies, the most common allele ranged in frequency from $p = 0.54$ to $p = 1.00$. The genetic distance (Nei's $D$; see Table V) between populations ranged from 0.0011 to 0.0232. Overall, 53 percent of the genetic diversity revealed was within populations, and 47 percent was among populations.

Very often, especially in species with limited mobility, even very closely situated populations differ not only in allozyme frequencies, but in one or more physiological or morphological traits. In plants, morphologically or physiologically different forms, called ECOTYPES, often are found in a mosaic pattern in association with different microhabitats. For example, in the hawkweed *Hieracium umbellatum* four ecotypes, associated respectively with sand dunes, sandy fields, seaside cliff faces, and forests, are morphologically distinct (Turesson 1922). Metal-impregnated soils near mines are occupied by metal-tolerant populations of grasses that are genetically distinct from conspecific plants on nontoxic soils a few feet away (Antonovics et al. 1971). Quite often, the adaptation of different populations of a species to similar environmental conditions is indistinguishable

TABLE V
**Measures of genetic similarity and difference between populations $X$ and $Y$**

| | |
|---|---|
| Nei's index of genetic similarity (Nei 1972) | $I_N = \dfrac{\sum_{i=1}^{m} (p_{ix}p_{iy})}{\left[\left(\sum_{i=1}^{m} p_{ix}^2\right)\left(\sum_{i=1}^{m} p_{iy}^2\right)\right]^{1/2}}$ |
| Nei's index of genetic distance (Nei 1972) | $D_N = -\log_e I_N$ |
| Rogers' index of genetic similarity (Rogers 1972)[a] | $S_R = 1 - \left[1/2 \sum_{i=1}^{m} (p_{ix} - p_{iy})^2\right]^{1/2}$ |

$p_{ix}$= frequency of allele $i$ in population (or species) $X$
$p_{iy}$= frequency of allele $i$ in population (or species) $Y$
$m$= number of alleles at the locus

[a] $S_R$ usually gives values similar to but lower than $I_N$.

phenotypically, but has a different genetic basis from one population to another (Cohan 1984). Thus convergent evolution can occur among different populations of the same species.

## GEOGRAPHIC VARIATION

Variation among populations exists on a larger geographic scale as well. Commonly, the farther apart populations are, the more different they are in allele frequencies and in genetically based phenotypic characteristics, although there is often not a strict correlation. The degree of divergence ranges from very slight to very great.

The study of geographic variation, beginning with Darwin and Wallace, has been one of the most important approaches to the study of evolution. Because so many historical evolutionary events are inaccessible to direct observation, many evolutionary hypotheses can be tested only by examining extant organisms. Much as an ecologist can trace the historical course of succession by piecing together the individual stages that exist in various places at the present time, an evolutionary biologist can use the varying levels of differentiation among populations and species to infer the course of evolutionary change. Such observations have indicated that evolution is generally a gradual process, for differences among populations range from the immeasurably small, through varying degrees of differentiation, to levels of behavioral, chromosomal, and developmental distinctiveness that are characteristic of different species. New species must therefore be formed by the same processes that engender genetic differences among conspecific populations.

### Patterns of geographic variation

Geographically differentiated forms of a species may be PARAPATRIC, meeting along a narrow border along which they interbreed, or they may be totally ALLOPATRIC (separated). Because it is often difficult to tell whether allopatric forms would interbreed or not if given the opportunity, it is sometimes rather arbitrary whether they are considered different species or not. For example, island populations of drongos (*Dicrurus*; Figure 10)—crow-like birds of the Old World tropics—differ in the form of their crests, which may be used in courtship displays, but whether or not they would interbreed is unknown. Mayr and Vaurie (1948), in fact, treated them as geographic races, or SUBSPECIES, of a single species. In zoological taxonomy the term subspecies means a recognizably different geographic population, or set of populations, that is given a formal Latin name. (Botanists sometimes apply subspecies names to sympatric, interbreeding genotypes.) A species that is divided into subspecies (geographic races) is often called a POLYTYPIC SPECIES, or *Rassenkreis*. In contrast, a group of similar species that have parapatric distributions but do not interbreed where they meet is called a SUPERSPECIES, or *Artenkreis*. For example, most of the species of chickadees (*Parus*) in North America meet only along narrow geographic or altitudinal borders, where competition apparently excludes each from the other's range (Lack 1969). In cases like the drongos, the difference between an *Artenkreis* and a *Rassenkreis*, between allopatric species and allopatric subspecies, is arbitrary.

In many cases, subspecies differ in a number of characters that show similar patterns of geographic distribution. For example, eastern populations of the northern oriole (*Icterus galbula*) differ from western populations in the color pattern of the head, back, tail, and wing, and in allele frequencies at several enzyme loci (Corbin et al. 1979). Despite their numerous differences, these populations interbreed in a HYBRID ZONE in the Great Plains. Such patterns of concordant geographic variation in many characters are usually interpreted to mean that there has been SECONDARY CONTACT between formerly isolated populations that diverged from a common ancestor.

However, geographic variation can take many forms (Figure 11). For example, different characters sometimes show independent patterns of geographic variation, as in the rat snake *Elaphe obsoleta* (Figure 12). In this instance, body color (brown, yellow, orange, or black), the blotched pattern, and the possession of stripes are independently distributed; if subspecies were described only on the basis of body color, they would have a different distribution than if they were defined on the basis of stripes. Patterns of this kind suggest that the characteristics have differentiated independently, perhaps in response to different environmental factors. A PRIMARY ZONE OF INTERGRADATION along which a character changes abruptly is one in which the character is thought to have differentiated *in situ*, without prior allopatry of the populations. However, it is often difficult to tell if a zone of intergradation between genetically different populations is primary or secondary (Endler 1977).

A gradual change in a character along a geographic transect is referred to as a CLINE. This term is also used for a gradual change in the frequency of an allele along a transect. Clines may extend over the whole geographic range of a species;

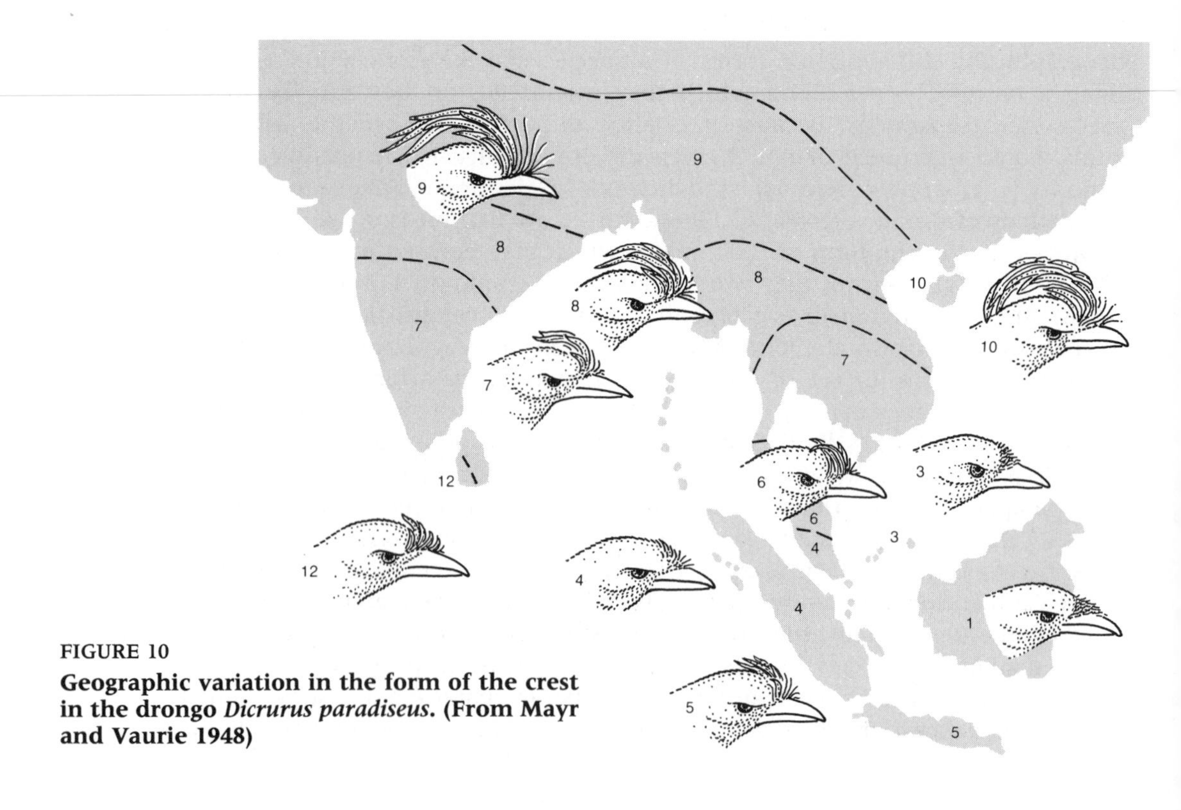


**FIGURE 10**

**Geographic variation in the form of the crest in the drongo *Dicrurus paradiseus*. (From Mayr and Vaurie 1948)**

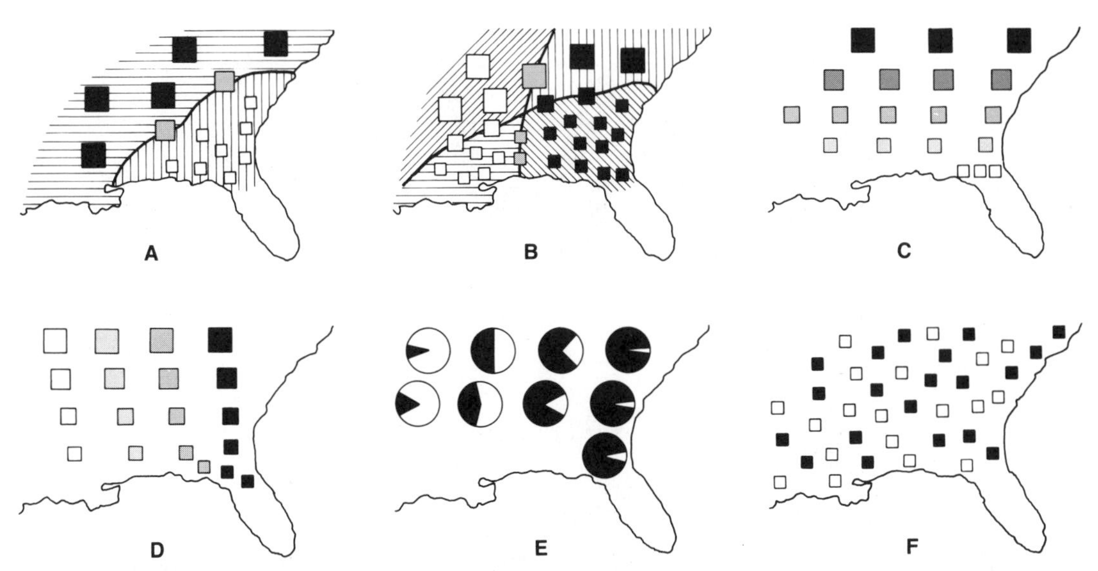


FIGURE 11

**Highly diagrammatic representations of some common patterns of geographic variation. (A) Two classical subspecies that interbreed along a narrow border. Size and color are correlated. (B) Abrupt transition in each of two characters that have discordant distributions. (C) Concordant clines in each of two characters. (D) Discordant clines in each of two characters. (E) An east-west cline in the frequency of black and white individuals; each "pie" diagram represents proportions in a sample from a single locality. (F) A mosaic distribution of two phenotypes, as might be observed if one (black) were a wetland ecotype and the other (white) an upland ecotype.**

for example, body size in the white-tailed deer (*Odocoileus virginianus*) increases gradually with increasing latitude over most of North America. (This relationship between body size and latitude is so common in mammals and birds that it has been dubbed "Bergmann's rule.") Alternatively, as in the contact zones between subspecies of the rat snake, there may be a short, steep cline between two widespread forms; this is sometimes called a "step cline." A cline may be established for several reasons. Among these are interbreeding between formerly isolated populations, and geographic variation in selection pressures that affect the character. For example, in the clover *Trifolium repens*, the proportion of plants that produce cyanide increases from north to south (Figure 13). The frequency of cyanogenic plants is apparently determined by a balance between the advantage they derive from being unpalatable to herbivores and the disadvantage they suffer when frost disrupts the cell membranes, releasing cyanide within the plants' tissues (Jones 1973).

Clines in different characteristics may be concordant (parallel) or discordant (independent). But to determine whether characteristics vary concordantly, it is important first to eliminate the correlation *within* populations caused by developmental patterns or pleiotropy (Gould and Johnston 1971). The lengths of the upper arm and the forearm, for example, may vary concordantly simply because they are two aspects of one characteristic—arm length. An initially long list of

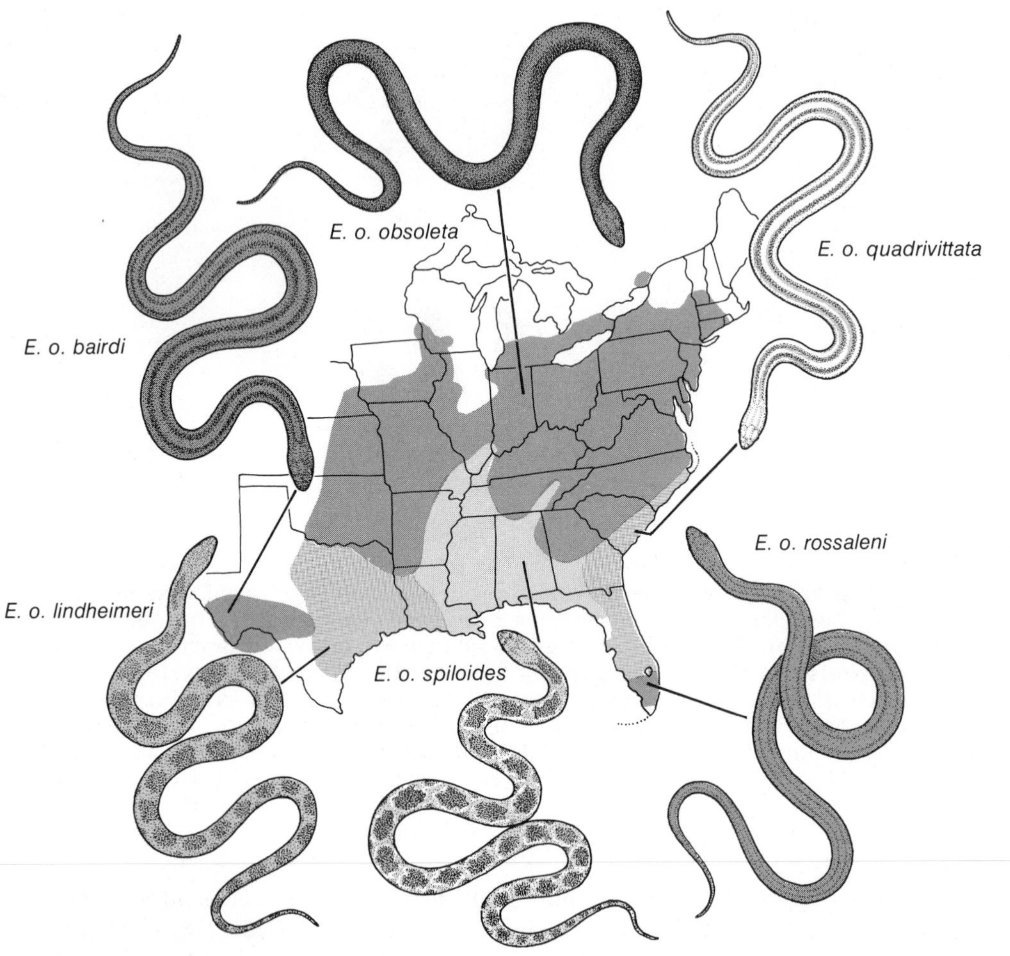


FIGURE 12

**Classical subspecies in the rat snake *Elaphe obsoleta*. These allopatric geographic races interbreed where their ranges meet. The subspecies are defined by several characters, such as color and pattern. Some of the color differences are pronounced: for example, *E. o. obsoleta* is deep black, *E. o. quadrivittata* is bright yellow, *E. o. rossalleni* is orange. On the other hand, *E. o. lindheimeri* and *E. o. spiloides* differ only slightly. (Redrawn from Conant 1958)**

characteristics may be reduced by the statistical technique of factor analysis to a shorter list of truly independent characters, or factors (Figure 14). For example, Thomas (1968) found that 16 measurements of the mouthparts, legs, and other features of larval rabbit ticks (*Haemaphysalis leporispalustris*) were so intercorrelated that they could be reduced to three independent characters: a "body size factor" that described overall variation in size, an "appendage factor" that accounted for variation in most of the leg segments, and a "capitulum factor" that expressed variation in the mouthparts and some of the leg segments. The body size factor varied clinally with latitude; ticks are larger in the north. The capitulum factor

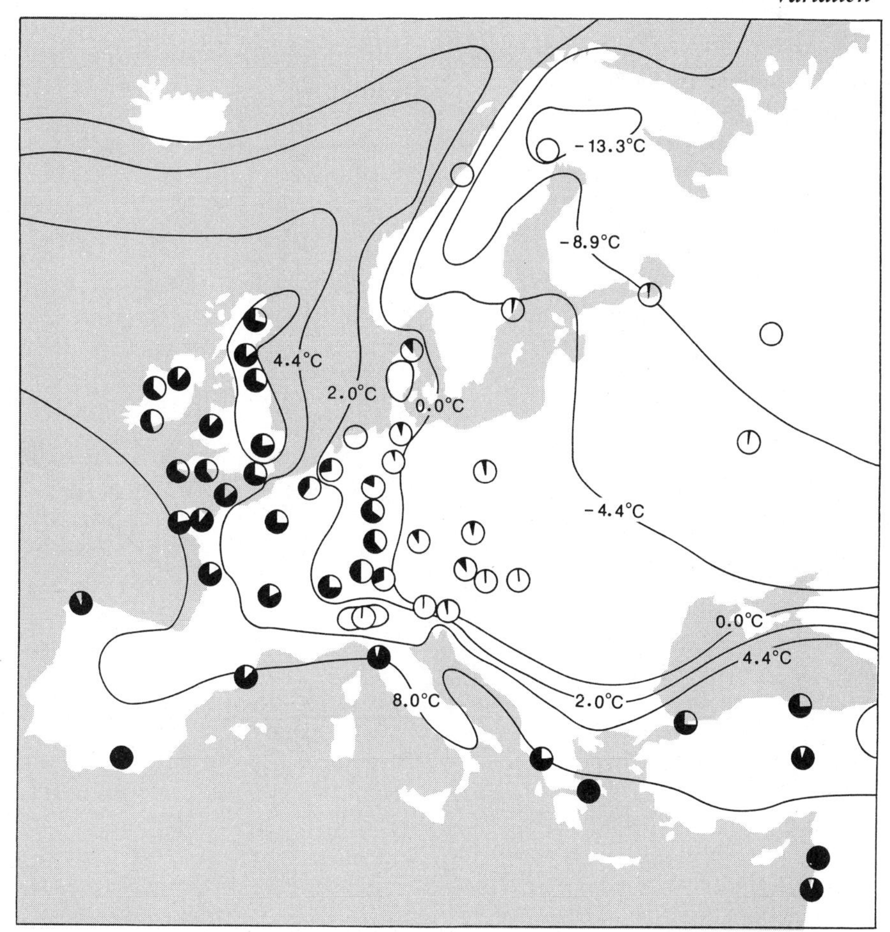


FIGURE 13

**Frequency of the cyanide-producing form in populations of white clover (*Trifolium repens*), represented by the black section of each circle. The cyanogenic form is more common in warmer regions. Thin lines are January isotherms. (Modified from Jones 1973, after Daday 1954)**

varied from east to west. Thomas suggested that large size is advantageous in cold northern areas where ticks need to store fat during hibernation, and that short appendages many be advantageous in the arid west where a low surface-to-volume ratio reduces desiccation.

## Subspecies, races, and typology

To Linnaeus and other early biologists, species were immutable units created in the beginning by God. Variation within species represented mere imperfections in creatures that, but for the faults of the material world, would conform to the *type*, the Platonic "idea" or ειδος in the mind of God. This conception of variation as an unimportant epiphenomenon carried over into the thinking of early taxonomists, who established a system of taxonomic practice in which specimens

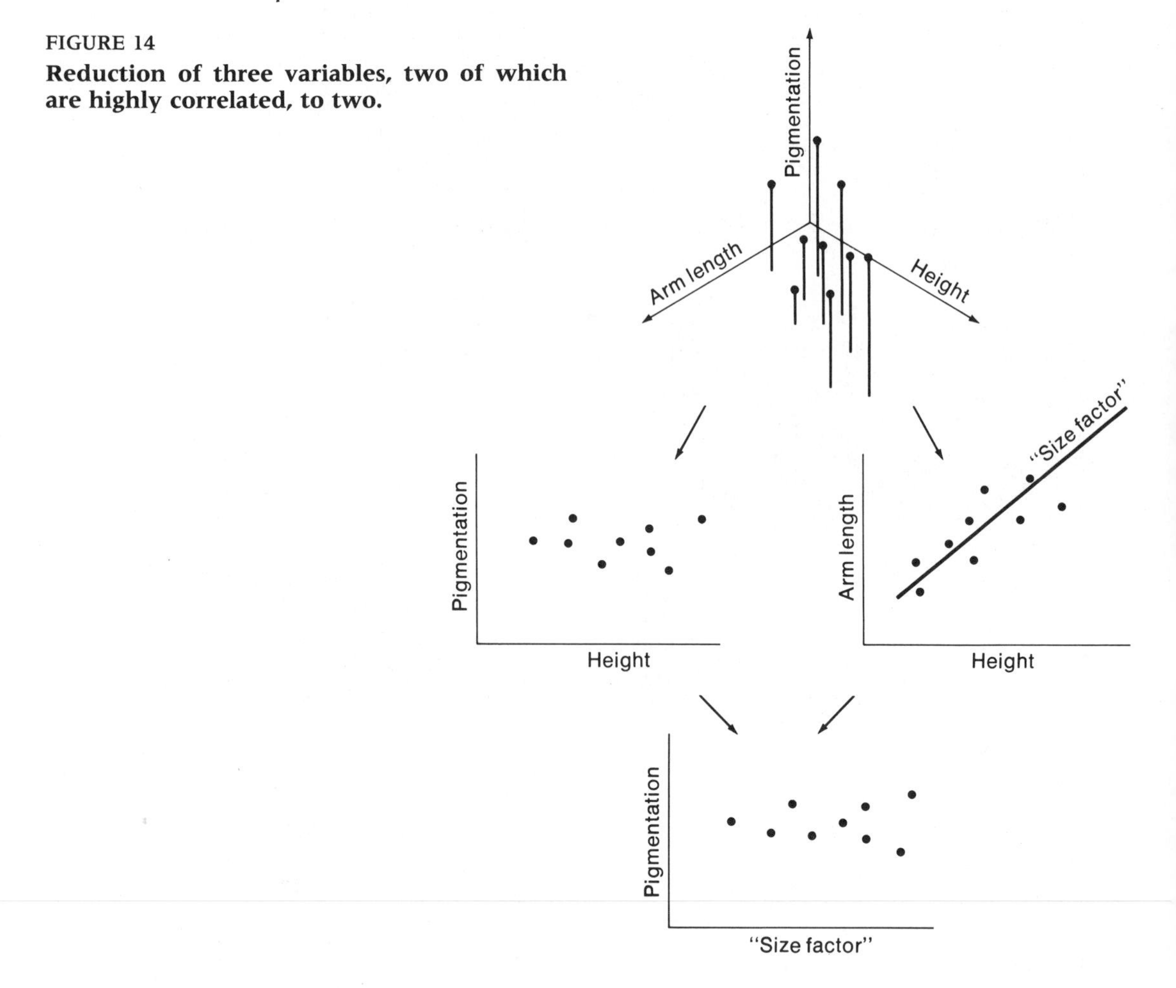


FIGURE 14
**Reduction of three variables, two of which are highly correlated, to two.**

were assigned to a species if they conformed to the type specimen, or holotype, on the basis of which the species was originally described. By the mid-nineteeth century, this practice was extended to subspecies, or geographic races, which were named and classified as if they too were discrete unvarying entities.

To dismiss variation as unimportant and to classify specimens into discrete categories is a manifestation of essentialism, a world view that Mayr (1963) has called "typological thinking." It is the kind of thinking that dichotomizes: either/or, black/white, good/evil, normal/abnormal. Mayr has shown that the replacement of typological thinking by the recognition of variation was pivotal in the development of the modern view of evolution.

Taxonomists previously argued at length about how many subspecies, or races, a species consisted of; in some quarters there is still argument about how many human races (subspecies) should be recognized. But because so many characters show independent patterns of geographic variation, and because so many loci are polymorphic and vary in allele frequency from one population to another, some combination of characters will distinguish every population from all others, so there is no clear limit to the number of subspecies that could be recognized. Many workers therefore accept Wilson and Brown's (1953) argument that the subspecies is so arbitrary a concept that it should be abandoned. For example, almost every human population differs from every other in allele frequencies at some loci, and the genetic diversity within a single "race" is much greater that the genetic differences among "races;" in fact, the differences among races account for only about 15 percent of the genetic diversity in the entire human species (Lewontin 1972, Nei and Roychoudhury 1972). The concept of race, masking the overwhelming genetic similarity of all peoples and the mosaic patterns of variation that do not correspond to racial divisions, is not only socially dysfunctional but is biologically indefensible as well.

### Geographic variation in ecological and reproductive characteristics

Geographic variation is displayed by almost all characteristics, including those most closely associated with a species' ecological role and those which by further differentiation may result in speciation. Among ecological characteristics, for example, the capacity for physiological acclimation varies ecotypically in the goldenrod *Solidago virgaurea*, in which the range of photosynthesis is greater in plants from exposed habitats than in those from shaded habitats (Björkman and Holmgren 1963). Similarly, there is considerable genetic variation both within and among populations of the fungus-feeding fly *Drosophila tripunctata* in preference for different kinds of food (Jaenike and Grimaldi 1983).

Very often the pattern of geographic variation is what one would readily predict if populations have become adapted to their local environments, but sometimes the nature of the adaptive variation is surprising. For example, in nature, larval development in montane populations of the green frog (*Rana clamitans*) proceeds more slowly than in lowland populations; this is a direct consequence of lower temperatures at high altitudes. When reared at the same low temperature in the laboratory, however, larvae from montane populations develop faster than lowland larvae. Thus the genetic differences among populations run counter to the phenotypic differences observed in the field, as if the montane frogs have compensated genetically for the slow development that their environment imposes (Berven et al. 1979). This pattern has been termed COUNTERGRADIENT VARIATION.

Species vary in characteristics that affect their interactions with other species. This sometimes takes the form of character displacement (Chapter 2), in which species differ more where they are sympatric than where they are allopatric. Some instances of character displacement appear to reflect evolution to reduce interspecific competition, as in the Galápagos finches (Figure 9 in Chapter 2). In other cases, it is manifested as an accentuation of the premating isolating mechanisms (see below) that prevent different species from mating with each other. For example, males and females of two partially differentiated species in the *Drosophila paulistorum* complex interbreed more readily in the laboratory if they are from allopatric than from sympatric populations; those with a history of exposure to one another have evolved greater mating discrimination (Ehrman 1965). Wing coloration, a cue for eliciting courtship, differs to a greater extent between sympatric than allopatric populations of two species of damselflies (*Calopteryx maculata* and *C. aequabilis*; Waage 1979). Character displacement has also been described for the vocalizations that prevent interbreeding between the Australian tree frogs *Litoria verreauxi* and *L. ewingi* (Littlejohn and Loftus-Hills 1968), and for flower color in *Phlox.* Allopatric populations of two species of this plant have pink flowers, but one species frequently has white flowers where it is sympatric with the other. Insects transfer pollen less frequently between differently colored *Phlox* flowers than among pink ones (Levin and Kerster 1967).

Postmating isolating mechanisms such as hybrid sterility or inviability (see below) also vary geographically. For example, the form and number of chromosomes frequently differs among populations, as in the plant *Clarkia unguiculata,* in which populations differ by three to seven reciprocal translocations and form sterile hybrids (Vasek 1964). Hybrid sterility can also be caused by genic differences; for example, Californian populations of the plant *Streptanthus glandulosus* do not appear to differ in chromosome structure, but vary greatly in their ability to form fertile $F_1$ offspring in the laboratory. The more geographically distant two populations are from each other, the lower the fertility of their hybrid (Kruckeberg 1957). In eastern Asia the gypsy moth (*Porthetria dispar*) consists of "weak" and "strong" geographic populations that Goldschmidt (1940) termed "sex races." A cross between a weak female and a strong male produces daughters that are phenotypically male; if the female is from a "half-weak" race, the daughters are sterile intersexes. There is a general north–south cline from strong to weak (Figure 15). Certain local populations of gypsy moths would be incapable of exchanging genes if they were to meet. Thus barriers to gene exchange, the very

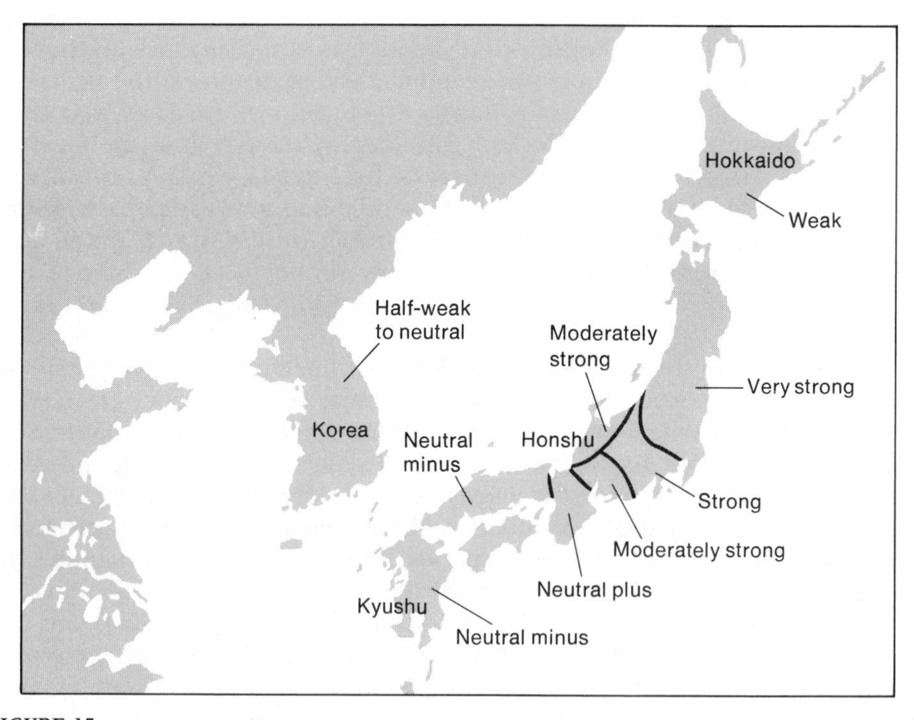


FIGURE 15

**The distribution of the "sex races" of the gypsy moth *Porthetria* (= *Lymantria*) *dispar* in eastern Asia. (After Goldschmidt 1940)**

properties that define species, vary geographically. This is evidence that speciation is often a gradual process in which geographic populations diverge into different species.

## SPECIES

As we have seen, subspecies are quite arbitrarily defined; and as we shall see, the same is true of genera, families, and other higher categories of classification. But one taxonomic category, the species, has been thought by many to be real and nonarbitrary, and to play a critical role in evolution.

Species were originally defined typologically by their morphological differences, but this criterion was never applied absolutely. Some very different-looking forms, such as different sexes or developmental stages, are obviously conspecific, as are different genotypes that interbreed within a population, such as the polymorphic geese or king snakes illustrated in Figure 4. Conversely, some forms—SIBLING SPECIES—are morphologically virtually indistinguishable, yet differ in other biological characteristics and do not interbreed. *Drosophila pseudoobscura* and *D. persimilis,* for example, were first recognized as different species by differences in their chromosomes and in ecological characteristics; only later were slight differences in the male genitalia discovered. Thus the generally accepted criterion for species is not morphological degree of difference, but evidence that two forms do not interbreed in nature. This criterion has led to widespread acceptance of the BIOLOGICAL SPECIES CONCEPT, enunciated by Mayr (1942) as follows: "Species are groups of actually or potentially interbreeding populations, which are reproductively isolated from other such groups."

The species as a biological concept is sometimes different from the species as a taxonomic category. For example, the criterion of interbreeding does not apply to asexually reproducing organisms, yet predominantly asexual forms are given names by taxonomists (e.g., *Escherichia coli*) based on phenotypic distinctiveness, just as sexually reproducing forms are. In paleontology, ancestral and descendant members of a single evolving lineage are often given different names (e.g., *Homo erectus* and *H. sapiens*), but these merely express degree of morphological difference, not reproductive isolation.

Populations are not assigned to different biological species if they are merely isolated by topographic barriers such as bodies of water. If they resemble one another closely, they are presumed to be potentially capable of interbreeding; yet it must be admitted that they are rather seldom put to the test, so perhaps it is not very important whether allopatric populations are called different species or not. Hybrid sterility is not the sole criterion of species; in both ducks and orchids, for example, species hybridize in captivity and produce fertile offspring, yet they do not exchange genes in nature. Conceptually, individuals are members of the same species if their genes could descend through the generations to unite in the same individual under natural conditions. By their passage from generation to generation, genes from white-tailed deer (*Odocoileus virginianus*) in Colorado might ultimately unite with those from deer in Florida, so these populations are considered conspecific. Yet white-tailed deer and mule deer (*O. hemionus*) are sympatric in Colorado and do not interbreed; their evolutionary paths through time are distinct.

The evidence that two populations are reproductively isolated sometimes comes from direct observations or experiments on their mating propensities or the sterility of hybrids produced in the laboratory. More often, however, it is inferred from the absence of individuals intermediate in characteristics that are known or presumed to be inherited. A discrete difference in only one trait may represent only a polymorphism in a single species, but when two or more alleles or traits distinguish two forms, the absence of recombinants is likely to indicate that the forms do not interbreed.

### Isolating mechanisms

The biological characteristics that cause sympatric species to exist—i.e., to maintain distinct gene pools—are usually called ISOLATING MECHANISMS, although this may be an unfortunate term because it suggests that these patterns evolved specifically to prevent interbreeding, whereas this appears seldom to be the case (Chapter 8). PREMATING isolating mechanisms are those that prevent gametes from uniting to form hybrid zygotes; POSTMATING isolating mechanisms operate thereafter (Table VI). Premating isolation sometimes has an ecological basis, as in species of spadefoot toads (*Scaphiopus*) that seldom meet because they occupy different soil types (Wasserman 1957), and in parasites that mate on different species of hosts. Species may be temporally isolated, as are plants that have different flowering seasons (Grant and Grant 1964) or insects that mate at different times of night (e.g., fireflies; Lloyd 1966).

Even if ecological or temporal isolation between sympatric species is incomplete, they usually do not interbreed because of physiological or behavioral features (Levin 1978). Animal-pollinated plants that differ in the form or coloration of their flowers tend to attract different pollinators. ETHOLOGICAL, or sexual, isolating mechanisms in animals include differences in the courtship behavior of males, in the vocalizations or chemical signals (pheromones) by which one sex attracts the other, or in the color patterns or other morphological features by which an individual recognizes a potential mate. Female fireflies, for example, respond to the light pattern emitted by males of their own species (Figure 16). It is sometimes possible to induce hybridization by altering such characters, showing that they indeed prevent interbreeding. Smith (1966), for example,

TABLE VI
**A classification of isolating mechanisms in animals**

1. Mechanisms that prevent interspecific crosses (premating mechanisms)
   a. Potential mates do not meet (seasonal and habitat isolation)
   b. Potential mates meet but do not mate (ethological isolation)
   c. Copulation attempted but no transfer of sperm takes place (mechanical isolation)
2. Mechanisms that reduce full success of interspecific crosses (postmating mechanisms)
   a. Sperm transfer takes place but egg is not fertilized (gametic mortality)
   b. Egg is fertilized but zygote dies (zygote mortality)
   c. Zygote produces an $F_1$ hybrid of reduced viability(hybrid inviability)
   d. $F_1$ hybrid zygote is fully viable but partially or completely sterile, or produces deficient $F_2$ (hybrid sterility)

(From Mayr 1963)

FIGURE 16

**The flight paths and flash patterns of male fireflies (Lampyridae) of nine species. Females respond to the patterns of their own species and not to others. (From Lloyd 1966)**

induced species of gulls (*Larus*) to interbreed by changing the contrast between the eye and the feathers of the face.

In angiosperms, even if pollen is transferred to the stigma of another species, it may not germinate; or if it does, the pollen tubes may fail to reach the ovules because they grow slowly; there is physiological incompatibility to a greater or lesser extent between pollen and pistil.

Under laboratory conditions, and sometimes in nature as well, some species that ordinarily do not interbreed will do so, thereby revealing any of several postmating barriers to gene exchange. The $F_1$ hybrid may not develop; for instance, hybrids between the frogs *Rana pipiens* and *R. sylvatica* do not develop beyond the early gastrula stage (Moore 1961). If hybrids develop, they may fail to survive in nature. For example, hybrids between the buttercups *Ranunculus millanii* and *R. dissectifolius*, which occupy wet and dry habitats respectively, cannot compete successfully with either parent in the parent's habitat and so in nature occur only in intermediate, disturbed habitats (Briggs 1962). Hybrids that do survive often fail to reproduce, either because they are sterile or because they have inappropriate mating behavior. Rao and DeBach (1969) found that male hybrids between species of parasitic wasps (*Aphytis*) were accepted by hybrid females but not by females of either parent form.

Hybrid sterility can have a chromosomal or a genic basis, although the distinction sometimes breaks down. Gross differences in chromosome structure, such as multiple translocations, commonly cause sterility because many or most of the gametes are deficient for whole chromosomes (ANEUPLOIDY) or for portions of chromosomes (Stebbins 1950, Grant 1981). Mispairing of chromosomes in meiosis sometimes occurs even if the chromosomes do not differ visibly under the microscope, but it appears likely that multiple submicroscopic differences in structure may be responsible (Stebbins 1982).

The diploid hybrid between the primroses *Primula verticillata* and *P. floribunda* is sterile, but tetraploid hybrids called *P. kewensis* have been produced by artificial means, and these are fully fertile. Whereas chromosomes of the diploid hybrid do not pair properly, each chromosome in the tetraploid can pair with its newly made partner, so that meiosis proceeds normally (Stebbins 1950). This example (and others like it) shows that the sterility of the diploid hybrid is caused by incompatibility in chromosome structure rather than in genes. In contrast, the chromosomes of the hybrid between *Drosophila pseudoobscura* and *D. persimilis* pair normally, yet the males have atrophied testes and are sterile. By a series of elaborate crosses Dobzhansky (1936) showed that sterility was due to differences between the species in genes that are scattered over all the chromosomes. Genic sterility appears to be prevalent in *Drosophila* (Ehrman 1962) and probably in many other animal groups as well.

If two sympatric populations maintain distinct gene pools, there must exist efficacious premating barriers, postmating barriers, or both. In many groups that appear to undergo very rapid speciation, isolation appears to exist primarily at the premating level; fertile interspecific hybrids can be obtained rather readily among the orchids, ducks, and cichlid fishes, for example. Similarly, many of the hundreds of species of hummingbirds (Figure 7 in Chapter 8) are virtually indistinguishable as females, but the males differ in the elaborately developed plumes and bright colors that are displayed during courtship. It appears likely that in these groups, rapid divergence in premating reproductive characters and little else is responsible for speciation.

### Limitations of the biological species concept, and hybridization between species

The biological species concept is not universally accepted (e.g., Sokal and Crovello 1970, Levin 1979, Raven 1980). As we have noted, it simply does not apply in some cases, as in asexual organisms or temporally separated fossil populations.

More importantly, however, it is often not very useful to apply the concept to spatially separated populations, because gene flow can be limited to such short distances that widely separated populations evolve virtually independently (Levin 1979).

There are numerous cases, moreover, in which parapatric or sympatric populations are in an intermediate stage of speciation, and are only partially reproductively isolated. Such partially isolated populations, or SEMISPECIES, usually interbreed along a HYBRID ZONE that can persist for long periods of time. In northeastern South America a hybrid zone between two forms of the mimetic butterfly *Heliconius* has existed for at least 200 years (2000 generations), but is only about 50 kilometers wide (Turner 1971). Quite often some alleles display a very steep cline in frequency at the hybrid zone, while others infiltrate farther into the range of each semispecies; this is known as INTROGRESSIVE HYBRIDIZATION (Anderson 1949). For example, a form of the Australian grasshopper *Caledia captiva* with acrocentric chromosomes hybridizes along a zone about 200 meters wide with another form that has metacentric chromosomes. The contact zone is thought to be about 8000 years old, and to have formed when the populations established contact after being isolated during a dry Pleistocene period. Whereas the cline in the X chromosome is very abrupt, some of the autosomes change in frequency over a considerably longer distance (Shaw 1981). Different cline widths for different loci suggest that natural selection is acting on some loci more strongly than on others (Chapter 8).

Enzyme electrophoresis has revealed that hybrid zones sometimes contain rare alleles that are not found in either parental semispecies (Sage and Selander 1979, Golding and Strobeck 1983). It is possible that these are consequences of higher mutation rates in hybrid genomes, or that they have been formed by intragenic recombination between different alleles carried by the parent populations. Whatever their origin, they appear to represent one way in which hybridization can introduce genetic variations into a population.

Occasionally in animals, and frequently in plants, considerable localized interbreeding occurs between sympatric populations of what are generally considered different species. Such "hybrid swarms" often occur in disturbed habitats where the ecological component of reproductive isolation has broken down. For example, *Iris hexagona*, which grows in exposed tidal marshes of the Mississippi delta, and *I. fulva*, which occupies the shady margins of streams, produce an $F_1$ hybrid that is only partially fertile; yet populations of hybrids with various combinations of characteristics occur in lumbered, drained swampy areas (Riley 1938). Even without disturbance of the habitat, some degree of hybridization may occur, as in oaks, in which hybridization is quite common (Palmer 1948). Ecological isolation appears often to be the strongest barrier to gene exchange among plant populations, so that hybridization is common at the interface between different habitats. Even when the $F_1$ hybrids have low fertility because of aneuploidy, hybrid genotypes frequently can persist and multiply by asexual reproduction or polyploidy (Grant 1981). Thus in many genera of plants, biological species are ill-defined (Raven 1980); fully reproductively isolated populations are clearly different species, but in other cases the species concept is not readily applied.

## INTRASPECIFIC VARIATION AND HIGHER TAXONOMIC CATEGORIES

If species are hard to define, genera and other higher taxonomic categories are even more so; in fact, they are clearly arbitrary constructs. A higher taxon is typically defined by one or more characters that distinguish its species from those of other, higher taxa. For example, the taxon *Quercus* (the oak genus) is distinguished by having an acorn as its fruit. Such differences, having evolved over long periods, are typically much greater than those within a species or between very closely related species. There are exceptions to this generalization, however. For example, the tribe Helenieae in the sunflower family (Asteraceae) was traditionally distinguished from the tribe Heliantheae by the lack of ray florets and bracts. Clausen et al. (1947) discovered a small population of plants that lacked these features, but was unlike any known member of the Helenieae. They erected a new genus for it, naming it *Roxira serpentina*. However, it proved to be an aberrant ecotype of *Layia glandulosa* in the Heliantheae; crosses with this species showed that two loci control the presence or absence of ray florets and bracts (Figure 17).

The kinds of characteristics that define certain higher taxonomic categories often vary within species in other taxa in which these characters are not diagnostic. The number and form of the teeth are highly distinctive of mammalian taxa, but vary within species of many other vertebrates. Even as fundamental a character as the number of cotyledons, which distinguishes the two great classes of angiosperms, varies intraspecifically in the shrub *Pittosporum*, and it has been possible to select tricotyledonous strains of snapdragons (Stebbins 1974).

FIGURE 17

**A plant described as a new genus, *Roxira* ($P_2$, at upper right), is actually conspecific with *Layia glandulosa* ($P_1$). The distinguishing characters are determined by only a few genes, as illustrated by segregation in the $F_2$ generation. (From Clausen et al. 1947)**

## SUMMARY

Genetic variation within a population is described by the number and frequency of alleles at each locus, and the degree to which alleles are organized into nonrandom combinations. There is considerable genetic variation within most populations, but nonrandom organization is generally not strongly pronounced. Differences in allele frequencies among populations of a species are the basis of geographic variation, which can take on many different patterns. Substantial genetic differences are frequently evident even among closely situated populations. The pattern of geographic variation in a trait is sometimes correlated with that of other traits, and sometimes not. There is a spectrum of genetic differentiation among populations, to the point at which characters that affect interbreeding may so differ that the populations may be defined as distinct species.

## FOR DISCUSSION AND THOUGHT

1. Answer the following misguided query: "The Hardy-Weinberg theorem holds only when there is no mutation, gene flow, or selection, and only in a randomly mating population. But there are no gene loci that really meet these assumptions, so what good is the theorem?"
2. Evaluate the pros and cons of constructing a mathematical model of genetic changes in populations rather than a simple verbal description. Apply your evaluation to the model of linkage disequilibrium.
3. Suppose you find that in a wild population of beetles, black-winged specimens usually (but not always) have yellow tarsi, while brown-winged specimens usually have brown tarsi. What can you say about the genetic basis for these traits?
4. Evolution by natural selection is commonly supposed to entail the "survival of the fittest," whereby the best genotype prevails over inferior ones. How does our recognition of extensive genetic variation modify this view?
5. Early geneticists were familiar with continuous variation in morphological characters and with the effects of artificial selection. Yet they were surprised at and reluctant to accept the notion of extensive genetic variation, as revealed by lethal genes and enzyme polymorphisms. Why?
6. What differences between chromosomal variation (e.g., inversion polymorphism) and genic variation are important for evolution? Can a species have one without the other?
7. How can morphological or electrophoretic data be used to determine whether two sympatric forms are different species? How about allopatric forms?
8. Different geographic populations of a species usually differ in some characteristics but are quite uniform in many other characters—otherwise they would not be referred to the same species. What might account for such uniformity?
9. In some polymorphic species, such as the snow goose, the different morphs mate more frequently among themselves than with each other (e.g., Cooke and Cooch 1968). If such assortative mating is less than 100 percent perfect, should the forms be considered different species?
10. The question has frequently been raised whether new species arise by the same processes that cause genetic change within populations, or if qualitatively different processes give rise to new species. How does the material described in this chapter bear on this question?

## MAJOR REFERENCES

Hartl, D.L. 1981. *A primer of population genetics.* Sinauer Associates, Sunderland, MA. 191 pages. A useful introduction to elementary population genetic theory of the kind presented in this and succeeding chapters.

Lewontin, R.C. 1974. *The genetic basis of evolutionary change.* Columbia University Press, New York. 346 pages. A description of genetic variation as revealed by classical methods and electrophoresis, together with a review of the history of the subject and a thorough analysis of the applications of population genetic theory to data.

Mayr, E. 1963. *Animal species and evolution.* Harvard University Press, Cambridge, MA. 797 pages. This classic work on evolution contains a wealth of information on geographic variation and the nature of species in animals, with important interpretations of speciation and other evolutionary phenomena.

Stebbins, G.L. 1950. *Variation and evolution in plants.* Columbia University Press, New York. 643 pages. A seminally important work that summarizes extensive early information on variation and the nature of species in plants.

Dobzhansky, Th. 1970. *Genetics of the evolutionary process.* Columbia University Press, New York. 505 pages. This and its predecessors, the several editions of *Genetics and the origin of species,* are among the most influential books on evolution from the viewpoint of population genetics. The material on genetic variation is out of date but has enormous historical importance.

# Population Structure and Genetic Drift

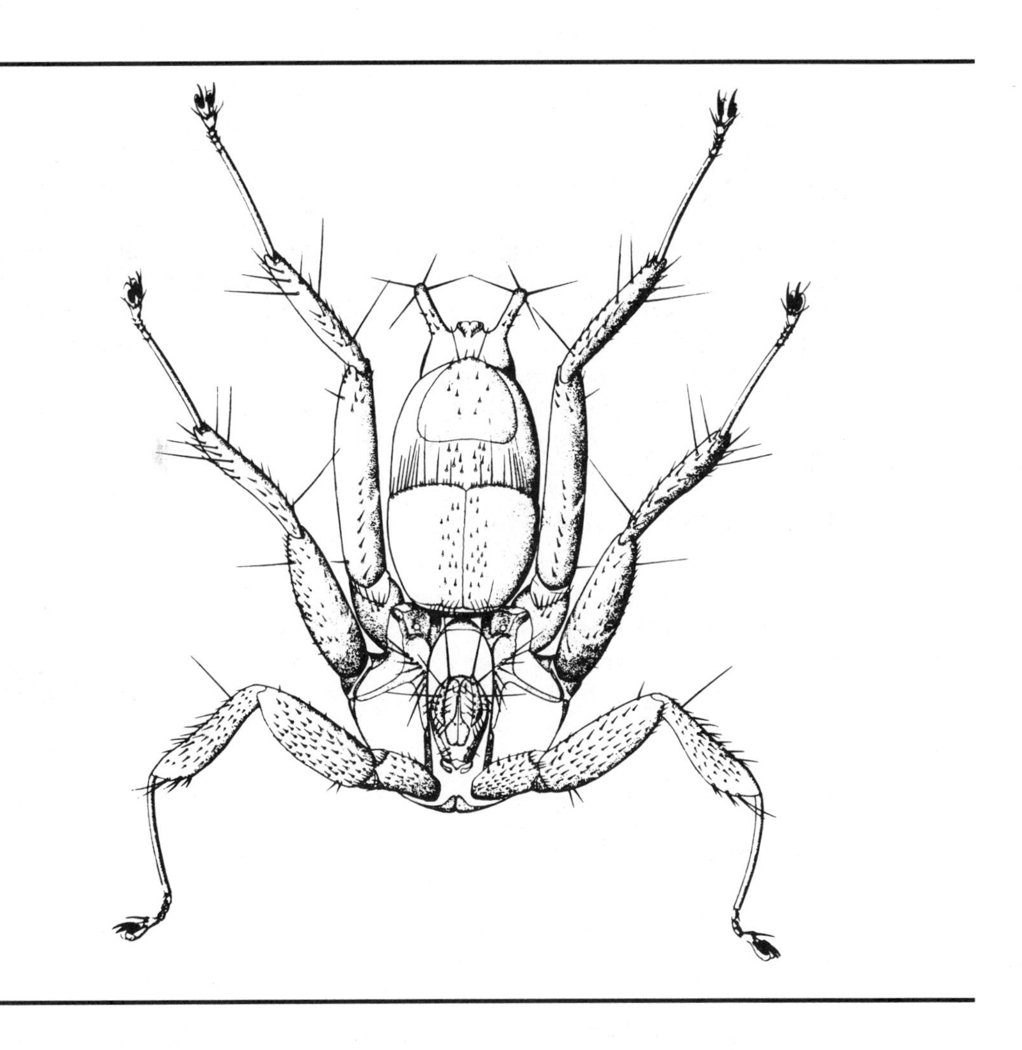

# Chapter Five

The Hardy-Weinberg theorem, the cornerstone of the genetical theory of evolution, assumes that a population is effectively infinite in size and that all individuals in the population mate at random. In reality, however, populations are finite in size, and mating may be structured nonrandomly in several ways. For example, the likelihood that two individuals mate may depend on their phenotypic characteristics. This has consequences that are discussed at the end of this chapter and again in Chapter 8, which treats the origin of new species.

The most common deviation from random mating occurs when mating is more likely among related than among unrelated individuals. One major reason for this is that the capacity for dispersal is limited in most species, so that mates are likely to have been born near each other and to be related to some extent. A truly PANMICTIC species, one in which all individuals of the species over its entire geographic range form a single, randomly mating population, is a rarity. One of the few possible exceptions is the common eel (*Anguilla rostrata*), in which individuals from freshwater drainages along the entire Atlantic coast of North America (and perhaps Europe as well) are thought to migrate to a single area near Bermuda to breed. But the vast majority of species are fragmented into local breeding populations, which may or may not be discrete (Figure 1), and which may be large or small in size. Among such populations there may be much or little GENE FLOW: interbreeding among resident individuals and immigrants from other populations. Finally, a local population may be panmictic, meaning that all its members interbreed at random, or it may be INBRED because individuals tend to mate with their relatives.

## THE THEORY OF INBREEDING

The mathematical theory of inbreeding was pioneered by Wright (1921, 1931, 1969); Crow and Kimura (1970) provide clear derivations.

Imagine that it is possible to trace in a population the copies of a gene that are descended from a particular gene carried by some individual. Such copies are called IDENTICAL BY DESCENT (Figure 2). Because they are identical, they are the

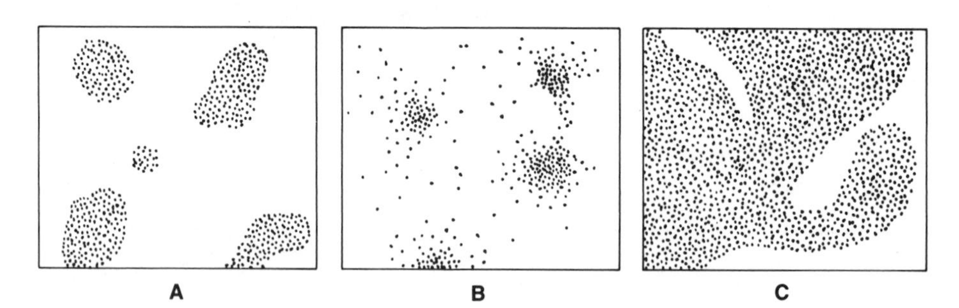


FIGURE 1

**Some patterns of spatial distribution; each dot represents an individual. (A) Discrete populations, corresponding to the "island model" if mating is more frequent within than between the populations. (B) Perhaps the most common pattern in nature, ill-defined populations between which density is low. (C) A more or less uniform distribution, corresponding to the "isolation by distance" model except that regions of unfavorable habitat are acknowledged.**

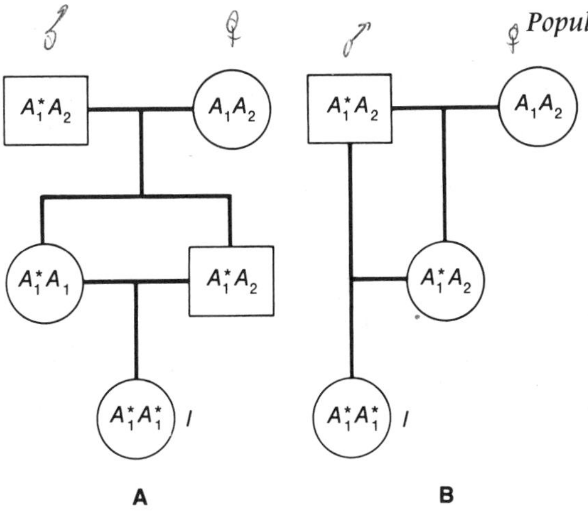

FIGURE 2

**Two pedigrees showing inbreeding due to (A) sib mating and (B) parent–offspring mating. Squares represent males, circles females. Copies of an $A_1$ allele, $A_1^*$, are traced through three generations. Individual I possesses alleles that are identical by descent at this locus. The average inbreeding coefficient of an individual produced by either of these consanguineous matings is ¼, since this is the probability that I will have alleles identical by descent. Note in (A) that I's mother is also homozygous for allele $A$, but the two copies are not identical by descent.**

same allele. There may be other representations of the same allele in the population, but we focus attention on those that are descended from one particular segment of DNA. Clearly, individuals that inherit genes that are identical by descent must be related. A population is said to be inbred if the probability that offspring inherit two gene copies that are identical by descent is greater than would be expected under purely random mating. This probability, referred to as the INBREEDING COEFFICIENT ($F$) of the population, increases the more closely related mating individuals tend to be. The most extreme form of inbreeding, for example, is self-fertilization ("selfing") such as occurs in many plants. The probability that an offspring of a self-fertilizing plant will inherit two copies of one of its parent's genes is ½. The average inbreeding coefficient of a population, $F$, ranges from 0 in the case of purely random mating, to 1 when all individuals are homozygous for alleles that are identical by descent.

Inbreeding in a large population does not change the frequencies of alleles. Rather, it increases the proportion of homozygotes of all kinds, and decreases the proportion of heterozygotes (Figure 3). For example, if a locus has two alleles $A$ and $A'$, with frequencies $p$ and $q$ (= $1 - p$), the frequencies of the genotypes $AA$,

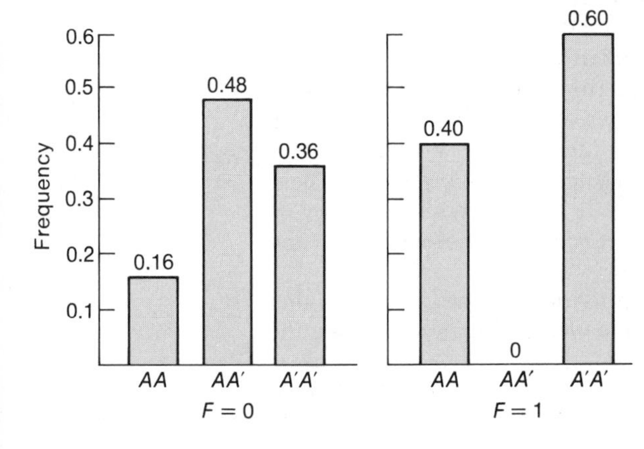


FIGURE 3

**Genotype frequencies in a population with allele frequencies $p = 0.4$ and $q = 0.6$, when mating is random ($F = 0$) and when the population is completely inbred ($F = 1$).**

$AA'$, and $A'A'$ are not $p^2$, $2pq$, and $q^2$ as in a randomly mating (Hardy-Weinberg) population, but rather (see Box A)

$$\begin{array}{ccc} D & H & R \\ p^2 + Fpq & 2pq(1 - F) & q^2 + Fpq \end{array}$$

The frequency of the allele $A$ is

$$p^2 + Fpq + (1/2)(2pq)(1 - F) = p$$

just as in a randomly mating population. Note that in a fully inbred population ($F = 1$), the frequency of $AA$ homozygotes is $p$, that of $A'A'$ is $q$, and that of heterozygotes is 0.

As long as nonrandom mating among relatives continues, $F$ increases to-

## A *Genotype Frequencies in Inbred Populations*

In Figure 2A, a brother and a sister, the offspring of unrelated parents, mate and have a daughter $I$, who inherits a gene, $A_1^*$, from both her parents. This gene is IDENTICAL BY DESCENT, because both of $I$'s parents inherited it from their father. The INBREEDING COEFFICIENT $f$ of an individual $I$ is defined as the probability that $I$ is AUTOZYGOUS—that is, that $I$ has two copies of an allele that are identical by descent. In this example, the probability that $I$ is homozygous for any particular one of her grandparents' genes is $(1/2)^4 = \frac{1}{16}$. For example, the probability that $A_1^*$ was transmitted to both $I$'s parents is $(1/2)^2 = 1/4$ and the probability that both her parents transmitted $A_1^*$ is likewise $(1/2)^2 = 1/4$; thus $1/4 \times 1/4 = \frac{1}{16}$. But since $I$ could have been homozygous for any of the four genes possessed by her grandparents, the probability that she is autozygous is $f = 4 \times \frac{1}{16} = 1/4$.

The average inbreeding coefficient of a population, $F_{IS}$ (or $F$), is the probability that the average individual is autozygous; thus $1 - F$ is the probability that an individual is ALLOZYGOUS (i.e., not autozygous). Note that an individual can be homozygous, but not autozygous, if its two gene copies are not identical by descent (e.g., the mother of $I$ in Figure 2A). If $p$ and $q$ are the frequencies of alleles $A_1$ and $A_2$ in a population with an average inbreeding coefficient $F$, then the frequency ($D$) of $A_1A_1$ homozygotes is $p^2(1 - F) + pF$. Of the fraction $1 - F$ of the population that is allozygous, $p^2$ have the genotype $A_1A_1$, according to the Hardy-Weinberg theorem. Of the fraction $F$ that is autozygous, and hence necessarily homozygous, the probability that an individual carries $A_1$ is $p$. Hence $D = p^2(1 - F) + pF = p^2 + Fpq$ (since $1 - p = q$). Similarly, the frequency ($R$) of $A_2A_2$ homozygotes will be $q^2(1 - F) + qF = q^2 + Fpq$. The frequency ($H$) of heterozygotes is given by the probability of being allozygous $(1 - F)$ times the proportion of allozygous individuals that are heterozygous ($2pq$). Hence in a population inbred to the extent $F$, the proportion of heterozygotes is $H_F = 2pq(1 - F)$. These results may be extended to multiple alleles: for any allele $A_i$ with frequency $q_i$, the proportion of $A_iA_i$ homozygotes is $q_i^2(1 - F) + Fq_i$, and the proportion of any particular heterozygote $A_iA_j$ is $2q_iq_j(1 - F)$.

The COEFFICIENT OF CONSANGUINITY of two individuals is the probability that their offspring will be autozygous. As the example we have analyzed shows, the coefficient of consanguinity of two siblings is $1/4$.

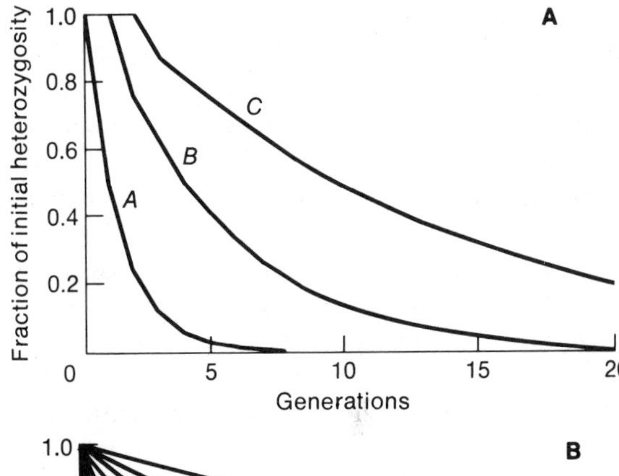


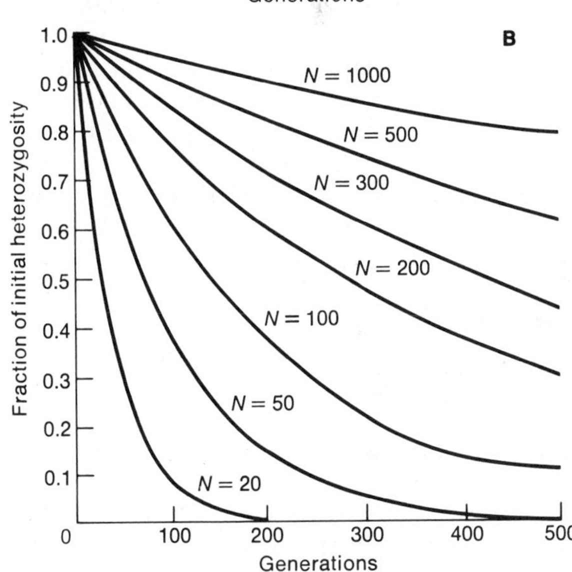


FIGURE 4

**Decrease in heterozygosity due to (A) inbreeding systems of mating and (B) finite population size. In (A) systems of mating are exclusive self-fertilization (curve A), sib mating (curve B), and double first cousin mating (curve C). In (B), *N* equals population size. (A from Crow and Kimura 1970, B from Strickberger 1968)**

ward 1. This is most clearly seen by considering the most extreme form of inbreeding, exclusive self-fertilization. Only half of a heterozygote's offspring are heterozygous, so in each generation the frequency of heterozygotes is halved. If mating entails other kinds of consanguineous unions, then $H$ approaches 0 and $F$ approaches 1 at a lesser rate (Figure 4). If, however, some matings are among unrelated individuals, $F$ may reach an equilibrium value below 1.

Because all loci in the genome are affected simultaneously by the pattern of mating, these results apply to all loci. If, when inbreeding begins, there are $k$ loci, each with two alleles, and there is no linkage disequilibrium among any of the loci, a large population, when it becomes fully inbred, will consist of a large number ($2^k$) of different homozygous genotypes. But any initial linkage disequilibrium among loci will decay at a slower rate in an inbred than in an outbred population because the frequency of heterozygotes, in which recombination occurs, is reduced under inbreeding. Thus if only gametes $AB$ and $A'B'$ are initially present, and loci $A$ and $B$ are tightly linked, the population will consist mostly of the genotypes $AABB$ and $A'A'B'B'$ when it becomes fully inbred.

Inbreeding, in itself, actually increases phenotypic variation, as measured by the variance of a trait (Box B), as long as the phenotype of a heterozygote does not lie outside the range spanned by the homozygotes. This is merely because homozygous genotypes at either end of the phenotypic spectrum increase in frequency, whereas heterozygotes decrease. Inbreeding causes a decline in the mean phenotype if individuals with the greatest value of the trait are dominant homozygotes or are heterozygotes. (If heterozygotes have the greatest value, the locus is said to be OVERDOMINANT). Such a decline, called INBREEDING DEPRESSION, is generally attributed to increased frequency of homozygotes for recessive alleles. Conversely, outbreeding, by increasing the frequency of heterozygotes and so concealing more of the recessive alleles, can increase the mean—a phenomenon called HETEROSIS. When recessive alleles lower survival or reproduction (i.e., when they cause lower fitness), they will be eliminated more rapidly from an inbred than from an outbred population because inbreeding exposes such alleles in homozygous form. Thus the frequency of deleterious alleles should be lower in inbred populations. This also means that if a randomly mating population that harbors deleterious recessive alleles becomes inbred, there will initially be considerable inbreeding depression as the alleles are exposed, but the population will eventually regain high average fitness as natural selection purges the population of the deleterious alleles. Such a population will have less genetic variation than a randomly mating population, not because of inbreeding in itself, but because of the combined action of inbreeding and natural selection.

**Inbreeding and outbreeding in animals and plants**

The inbreeding coefficient of a population, $F$, can be estimated directly if individuals are marked so that pedigrees can be obtained from information on matings over several generations (see Box A; also Wright 1969, Crow and Kimura 1970, Hedrick 1983). Alternatively, $F$ can be estimated indirectly if allele frequencies and genotype frequencies are known at one or more loci, from the relation $H_F = 2pq(1 - F)$. Thus $F$ is given by the frequency of heterozygotes relative to that expected under random mating ($H_0$): $F = (2pq - H_F)/2pq$, or $F = (H_0 - H_F)/H_0$, where $H_F$ is the observed proportion of heterozygotes.

The incidence of inbreeding in natural populations varies greatly among species. Self-fertilization is unusual in animals, even among hermaphroditic species, such as those snails in which courtship commonly leads to mutual insemination. Inbreeding in animals depends on mating among close relatives; the likelihood that this will occur depends on how far individuals have dispersed from their birthplace when they reproduce. The inbreeding coefficient of a population of great tits (*Parus major*) was estimated from pedigree information at only $F = 0.0036$ (Bulmer 1973), but inbreeding can be more intense, as in certain parasitic Hymenoptera, in which males hardly disperse at all and mate with their sisters almost as soon as they emerge from the host (Askew 1968). Selander (1970) found that among the house mice (*Mus musculus*) in a barn in Texas, the genotypes $AA$, $AA'$, and $A'A'$ at an esterase locus had the frequencies 0.226, 0.400, and 0.374 respectively. Since $p = 0.226 + 0.200 = 0.426$, the Hardy-Weinberg proportions should be 0.181, 0.489, and 0.329. The heterozygotes were less prevalent than expected, and $F = (0.489 - 0.400)/0.489 = 0.182$. This implies that the population is reproductively subdivided into extended family groups. We may

## B *The Effect of Inbreeding on a Metric (Continuously Variable) Character*

Assume that genotype $AA$ increases some character such as height by an amount $a$ over the background level; that $A'A'$ decreases height by an amount $-a$; and that $AA'$ increases height by an amount $d$. If $d = a$, the allele $A$ is dominant; if $d > a$, there is overdominance; if $d = 0$, there is no dominance, since the heterozygote is exactly intermediate between the homozygotes. Thus we have

| | $AA$ | $AA'$ | $A'A'$ |
|---|---|---|---|
| Phenotype | $a$ | $d$ | $-a$ |
| Frequency | $p^2(1-F) + pF$ | $2pq(1-F)$ | $q^2(1-F) + qF$ |

If $F = 0$, the mean $\bar{x}_0$ equals $p^2a + 2pqd - q^2a$. In a population inbred to the degree $F$, the mean $\bar{x}_F$ is

$$\begin{aligned}\bar{x}_F &= a[p^2(1-F) + pF] + d[2pq(1-F)] - a[q^2(1-F) + qF] \\ &= (p^2a + 2pqd - q^2a) + apqF - apqF - 2pqdF \\ &= \bar{x}_0 - 2pqdF\end{aligned}$$

Thus if there is no dominance ($d = 0$), inbreeding does not change the mean. If $A$ is dominant or overdominant ($d > 0$), the mean declines linearly with the degree of inbreeding (this is called inbreeding depression).

Now consider the effect of inbreeding on the variance (*Appendix* I), making the simplifying assumption that $d = 0$.

The variance $V_F$ is the sum of the squared deviations of the phenotypes from the mean, weighted by the frequency of each phenotype, or

$$V_F = [p^2(1-F) + pF]\,(a - \bar{x}_F)^2 + [2pq(1-F)]\,(0 - \bar{x}_F)^2 + [q^2(1-F) + qF]\,(-a - \bar{x}_F)^2$$

When $F = 0$, the variance $V_0$ is

$$V_0 = p^2(a - \bar{x}_0)^2 + 2pq(0 - \bar{x}_0)^2 + q^2(-a - \bar{x}_0)^2$$

and in a totally inbred population ($F = 1$) the variance $V_1$ is

$$V_1 = p(a - \bar{x}_1)^2 + q(-a - \bar{x}_1)^2$$

Since $d = 0$, we have $\bar{x}_0 = \bar{x}_1 = \bar{x}_F$, and then the equation for $V_F$ can be written

$$p^2(a - \bar{x})^2(1-F) + 2pq\bar{x}^2(1-F) + q^2(-a - \bar{x})^2(1-F) + pF(a - \bar{x})^2 + qF(-a - \bar{x})^2$$

Thus by substitution $V_F = (1-F)V_0 + FV_1$, so the total variance is decreased within $[(1-F)V_0]$, and increased among $(FV_1)$, lines. We may simplify this by noting from Appendix I that the variance $V = \Sigma_i f_i(x_i - \bar{x})^2$ can also be written $V = \Sigma_i f_i x_i^2 - \bar{x}^2$. Then $V_0$ becomes $p^2a^2 + q^2(-a)^2 - \bar{x}^2$, and $V_1$ becomes $pa^2 + q(-a)^2 - \bar{x}^2$. Now note that $\bar{x}_0$ (hence also $\bar{x}_1$) is $p^2a + 2pq(0) - q^2a = a(p^2 - q^2) = a\,(p - q)$. Hence $\bar{x}^2 = a^2(p^2 - 2pq + q^2)$. Substituting this into the expressions for $V_0$ and $V_1$, we obtain $V_0 = 2pqa^2$ and $V_1 = 4pqa^2$. Hence $V_1 = 2V_0$: the variance in the completely inbred population is twice that in the randomly mating population. By substituting $V_1 = 2V_0$ in the expression $V_F = (1-F)V_0 + FV_1$, we obtain $V_F = V_0\,(1 + F)$.

Thus, for additively acting genes, the phenotypic variance of the whole population increases linearly with the degree of inbreeding.

deduce that mice tend not to wander far from their birthplace before reproducing, or are prevented from mating except with relatives. Indeed, behavioral studies indicate that house mice form small groups that repel outsiders, especially males.

In higher plants the distance to which pollen and seeds are dispersed affects the frequency of consanguineous matings. In addition, plants are frequently hermaphroditic and often capable of self-pollination. A single plant may have separate male and female flowers (e.g., in ragweed, *Ambrosia*, or corn, *Zea*), but transferring pollen from one flower to another on the same plant is genetically the same as pollination within a single flower. But even in such plants there are many ways of reducing the likelihood of selfing. Stigmas commonly are receptive to pollen only after the plant has shed its own pollen. In other cases the flower is so constructed that a pollinator is unlikely to transfer pollen from the anthers to the stigma of the same flower (Figure 5). Heterostyly (Chapter 4) is one such mechanism. Many plants, moreover, are self-incompatible because of several or many alleles at a locus that affects the stigma's receptivity to pollen. In such species, genotype $S_iS_j$ is receptive only to pollen with a haploid genotype other than $S_i$ or $S_j$. Because $S_i$ pollen will not grow on an $S_iS_j$ stigma, plants are necessarily heterozygous at this locus, so self-fertilization is avoided. The number of such alleles can be very great. The entire evening primrose species *Oenothera organensis* consisted of only about 500 individuals when Emerson (1939) studied it, yet it had 45 self-incompatibility alleles.

However, many plants are self-compatible and engage in a great deal of self-fertilization (Jain 1976). In some species, such as some violets and grasses, some or all of the flowers are CLEISTOGAMOUS: self-fertilization occurs within the flower, which never opens. The proportion of cleistogamous versus CHASMOGAMOUS flowers (those that are open to cross-pollination) on a plant is a partially heritable trait that varies among individuals and among populations of the grass *Danthonia*

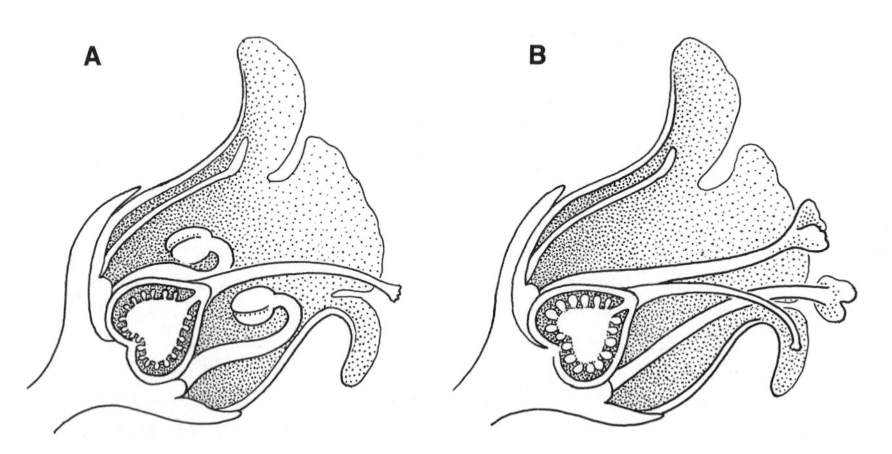


FIGURE 5

**One of many mechanisms that promote outcrossing in plants. (A) In the flower of *Salvia nodosa* the pistil is at first exposed to pollinators and the stamens are reflexed. (B) Later the pistil becomes less accessible and the stamens are exposed. (After Faegri and van der Pijl 1971)**

*spicata* (Clay 1982). The frequency of self-fertilization in self-compatible species is also affected by the degree of PROTANDRY, a term used by botanists for the delay between the dehiscence of the anthers and the time at which the stigma becomes receptive. The degree of protandry is highly heritable in the plant *Gilia achilleifolia*, a relative of the phlox, and is correlated with the amount of outcrossing, which varies greatly from one population to another (Schoen 1982).

## THE GENETIC STRUCTURE OF INBRED POPULATIONS

Numerous investigators have used genotype frequencies at enzyme loci to estimate the inbreeding coefficient in plant populations. As theory predicts, the frequency of heterozygotes is reduced in self-fertilizing species compared to strictly outcrossing species (Brown 1979; see also Table IV in Chapter 4). It is often possible to measure the degree of outcrossing independently by comparing the genotypes of progeny and their parents. When this is done, the frequency of heterozygotes is often higher than the incidence of self-fertilization would predict, although heterozygotes are less common than they would be if there were random crossing. This suggests that the frequencies of these genotypes are affected in part by factors other than the mating system, such as natural selection.

Linkage disequilibrium among alleles at different loci is also common in populations of inbreeding plants (Brown 1979). We have already noted (Chapter 4) the case of the self-fertilizing wild oat *Avena barbata*, in which alleles at five loci were found to be organized into only two common homozygous combinations (Hamrick and Allard 1972).

Inbreeding in itself should not change the number or frequency of alleles at a locus, yet the "virtual heterozygosity" of selfing plant populations—the proportion of heterozygotes that would occur if the population suddenly underwent purely random mating—is considerably lower than in outcrossing species (Hamrick et al. 1979). This difference may be in part because outcrossers have greater levels of gene flow as pollen moves from one local population to another. The difference in genetic variation, however, may also be caused by linkage disequilibrium. Suppose alleles $A$ and $A'$ code for two allozymes revealed by electrophoresis, and that neither allele confers higher fitness than the other. At a closely linked locus that is not revealed by electrophoresis, allele $B'$ reduces fitness relative to $B$. If by chance the population begins in a state of linkage disequilibrium, so that $A$ is associated with $B$ and $A'$ with $B'$, inbreeding will increase the frequencies of $AABB$ and $A'A'B'B'$ but since $B'B'$ is deleterious, it is eliminated from the population—and with it goes the allozyme allele $A'$.

The phenomenon of inbreeding depression has long been known; for example, fecundity (or yield in corn and other crops) and viability decline drastically as a population becomes more inbred in the laboratory (Figure 6; Table I). In human populations, the deleterious consequences of inbreeding include a higher incidence of mortality, mental retardation, albinism, and other physical abnormalities (Schull and Neel 1965, Cavalli-Sforza and Bodmer 1971, Stern 1973). From pedigree studies of inbreeding, Morton et al. (1956) estimated that each of us, on average, carries the equivalent of three to five lethal recessive genes in heterozygous condition. Inbreeding depression can be pronounced in natural populations of animals and plants as well: within populations of the great tit

FIGURE 6

**Heterosis (hybrid vigor) and its converse, inbreeding depression. The two corn plants at left are two inbred strains; their $F_1$ is to their right, followed by successive self-fertilized generations from the hybrid. (From Jones 1924)**

(*Parus major*), nestling mortality is up to 70 percent greater among the offspring of related than of unrelated birds (Greenwood et al. 1978, van Noordwijk and Scharloo 1981). In species of plants that engage in both selfing and outcrossing, it is common to find that the fitness of offspring produced by self-fertilization is less than half that of offspring produced by outcrossing to an unrelated individual (Schemske 1983a).

TABLE I

**Inbreeding depression in rats**

| Year[a] | Nonproductive matings (percent) | Average litter size | Mortality from birth to 4 weeks (percent) |
|---|---|---|---|
| 1887 | 0 | 7.50 | 3.9 |
| 1888 | 2.6 | 7.14 | 4.4 |
| 1889 | 5.6 | 7.71 | 5.0 |
| 1890 | 17.4 | 6.58 | 8.7 |
| 1891 | 50.0 | 4.58 | 36.4 |
| 1892 | 41.2 | 3.20 | 45.5 |

(After Lerner 1954. Data from Ritzema Bos)

[a]The years 1887–1892 span about 30 generations of parent × offspring and sib matings.

## POPULATION SIZE, INBREEDING, AND GENETIC DRIFT

Suppose that a population consists of $N$ diploid breeding individuals in each generation, so that there are $2N$ gene copies at a locus. If all matings, including self-fertilization, are equally probable, the likelihood that in generation $t$, a given gene will unite with a gene from the same parent that is identical by descent is $1/2N$, and the likelihood of uniting with a different gene is $1 - (1/2N)$. But this "different gene" may well be identical by descent if the parents (generation $t - 1$) were already inbred, with an average inbreeding coefficient of, say, $F_{t-1}$. In generation $t$ the average inbreeding coefficient is therefore

$$F_t = 1/2N + (1 - 1/2N)F_{t-1}$$

Thus any population of finite size becomes more inbred in time. In the process, the proportion of heterozygotes, $H$, declines. By substituting $(H_0 - H_t)/H_0$ for $F_t$ and $(H_0 - H_{t-1})/H_0$ for $F_{t-1}$, it is easily shown that

$$H_t = (1 - 1/2N)H_{t-1}$$

or

$$H_t = H_0(1 - 1/2N)^t$$

That is, the proportion of heterozygotes declines by a fraction $1/2N$ in each generation (Figure 4B). The smaller the population ($N$), the faster heterozygosity declines.

Because this conclusion is extremely important in evolutionary theory, it is worth pursuing by a less formal, more intuitive approach. The gene copies possessed by the population today are only a sample of the genes present in the previous generation, since some of their genes, by chance, were not transmitted to any of the surviving offspring. Similarly, the genes in the parental generation were only a sample of those present one generation earlier, and these in turn must have been sampled from an even larger number of the genes that were present in the distant past. Therefore, if we now go forward rather than backward in time, we see that successively fewer of the original genes are represented by copies in successive generations. Thus the probability of identity by descent must increase, so the population must become more and more homozygous. But the loss of heterozygosity in this case differs from that in a large population in which there is mating among relatives. With inbreeding, the gene frequencies remain the same, but various homozygous genotypes increase in frequency. In contrast, when the loss of heterozygosity is a consequence merely of small population size, the population remains approximately in Hardy-Weinberg equilibrium. But the frequency of heterozygotes decreases because one or another allele is increasing in frequency—namely, that allele represented by the original gene copy that by chance has left the greatest number of descendants in subsequent generations. Since it is only chance that dictates which gene leaves the greatest number of descendants, the allele that comes to predominate in a different (but initially identical) population, also of size $N$, could well be a different allele.

Yet another way of viewing the process is to imagine a very large number of populations, often called DEMES, each with $N$ breeding individuals ($2N$ genes). Initially, the frequencies of alleles $A$ and $A'$ are $p$ and $1 - p$ in each population.

If we sample $2N$ gene copies at random to make the next generation of surviving offspring, the probability that exactly $i$ of them will be $A$ is given by the binomial distribution

$$\frac{2N(2N-1)\ldots(2N-i+1)}{i!}p^i(1-p)^{2N-i}$$

The mean of this distribution is $\bar{x} = p$, and the variance is $V = p(1-p)/2N$ (see Appendix I). That is, after one generation, the average frequency of $A$ among all the populations is still $p$, but the populations vary in gene frequency to the extent $p(1-p)/2N$; and the smaller the demes, the greater the variation.

This process of random change in allele frequencies is termed RANDOM GENETIC DRIFT (or just genetic drift). So far, we have considered the effects of genetic drift on the frequencies of different alleles that are NEUTRAL with respect to fitness—that is, we suppose that they do not differ at all in their effect on survival or reproduction.

As genetic drift proceeds, the variation among populations increases from generation to generation. Suppose, for example, that some colonies have drifted from $p = 0.50$ to $p = 0.55$ (approximately the same number will have drifted from 0.50 to 0.45). Among these, some will increase and others decrease in gene frequency, with a variance of $(0.55)(0.45)/2N$, so that the span of gene frequencies is extended. After $t$ generations, the variance in gene frequency among the whole ensemble of populations will be

$$V_t = p(1-p)[1-(1-1/2N)^t]$$

After some time, the distribution of gene frequencies will be so spread out that all possible gene frequencies between 0 and 1 are equally represented among the populations, and some populations will reach the frequencies 0 and 1 (Figure 7). When $p = 1$, the $A$ allele is said to be FIXED, and the population is monomorphic $AA$. When $p = 0$, the $A'$ allele has been fixed. But if one allele has been fixed, the other cannot reappear, except by a new mutation or gene flow from other populations. The remaining polymorphic populations continue to drift toward 0 or 1, and in each generation $1/2N$ of these populations will become fixed for either $A$ or $A'$. Eventually, all populations will be fixed for one allele or the other, and the variance in gene frequency becomes $p(1-p)$.[1] Clearly, the allele that was initially most common has the greater chance of being fixed; in fact, the probability that a selectively neutral allele is fixed is precisely its initial frequency, $p$ (assuming that it is not in linkage disequilibrium with an allele at another locus that is affected by natural selection). Thus in our hypothetical example, a fraction $p$ of the populations will become monomorphic $AA$, and a fraction $q$ $(= 1 - p)$ will be monomorphic $A'A'$. As in the case of complete inbreeding, heterozygotes are absent when this process has gone to completion.

The process of random genetic drift may be viewed as a random fluctuation in allele frequency, leading eventually to either fixation or loss of the allele. The process occurs in every population that is finite (i.e., in all populations), but is more rapid the smaller the population. Genetic drift has numerous evolutionary consequences, two of which merit special emphasis: genetic drift results in the loss of genetic variation within populations, and in the genetic divergence of populations, entirely by chance.

[1] The above expression for $V_t$ becomes $p(1-p)$ when $t$ becomes very large.

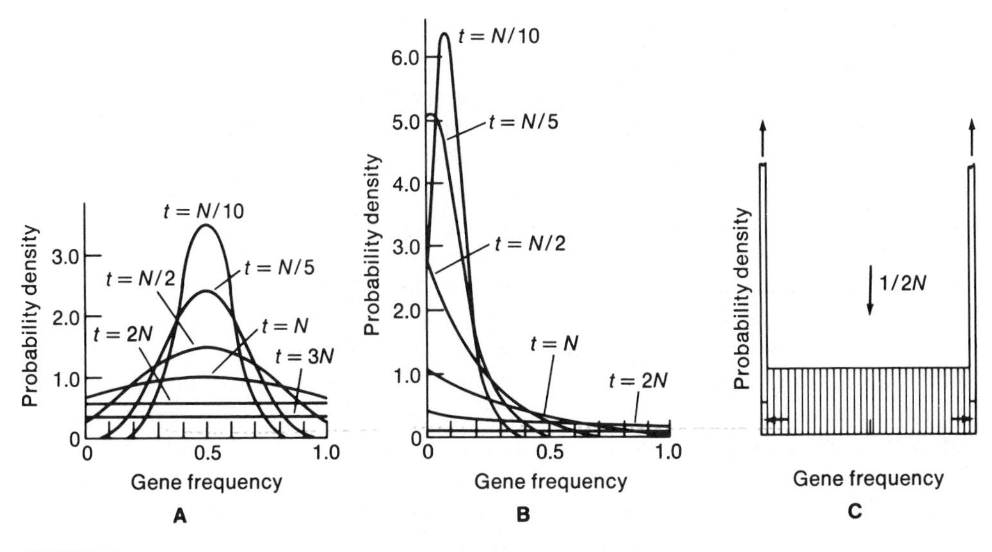


FIGURE 7

**(A) Probability distribution of allele frequencies between 0 and 1 under genetic drift when the initial allele frequency is 0.5. (B) Same, with initial allele frequency 0.1. *N* is the effective population size. After $t = N/10$ generations, the probability distribution is given by the uppermost curve. This may be thought of as the distribution of allele frequencies among populations, each of size *N*, that began with the same allele frequency. With the passage of generations, the curve becomes lower and broader, until at $t = 2N$ generations, all allele frequencies between 0 and 1 are equally likely. These panels do not show the proportion of populations fixed at 0 or 1. (C) After $t = 2N$ generations, the proportion of populations in which the allele is lost ($q = 0$) or fixed ($q = 1$) increases at the rate of $1/4N$ per generation, and each allele frequency class between 0 and 1 is decreasing at rate $1/2N$ per generation. (A and B after Kimura 1955; C after Wright 1931)**

## EFFECTIVE POPULATION SIZE

It might be thought that genetic drift is important only in rare species. But because most species consist of rather small, localized breeding populations, these populations can diverge substantially by genetic drift, giving rise to geographic variation in the species. Even in species like the ponderosa pine (*Pinus ponderosa*) that are distributed rather evenly over vast expanses of terrain, individuals mate within a local NEIGHBORHOOD. The number of individuals in the neighborhood within which an individual is likely to mate approximately at random is $4\pi s^2 D$, where $D$ is the density of the population (the number of individuals per unit area) and $s$ is the standard deviation of the distances between the birth sites of individuals and the birth sites of their offspring. These distances are often small, as we will see.

But there are further complications. An ecologist may determine that the size of a population is $N$, but if some of those $N$ individuals do not reproduce, the population is actually smaller from a genetic point of view. If only half the individuals in a population of $N = 50$ breed, $1/50$ rather $1/100$ ($1/2N$) of the heterozygosity is lost per generation. The population has an EFFECTIVE SIZE $N_e$ of 25; it is $N_e$ that determines the rate of genetic drift. One factor that can lower the effective population size is an unbalanced sex ratio. If, for example, males guard harems of females against other males, the few males that reproduce contribute disproportionately to subsequent generations, so the rate of genetic drift is inflated. If the breeding population consists of $N_m$ males and $N_f$ females, the effective population size is $N_e = 4N_mN_f/(N_m + N_f)$. Thus 100 tribes of mice, each with one male and four females, constitute an effective population of 320 rather than 500.

This is just a special instance of the more general case. To the extent that some individuals leave more offspring than others, the future population will trace its ancestry to relatively few individuals. If $\bar{k}$ and $\sigma_k^2$ are, respectively, the mean and the variance in the number of offspring per parent that survive to reproduce, and if $\sigma_k^2 > \bar{k}$, the effective population size is lower than the number of individuals in the population. Anything that causes such variation reduces effective population size, including selection, which by definition constitutes inequality of reproductive success. For example, Eisen (1975) found that genetic variation for growth rate in mice was reduced virtually to zero after 14 generations of strong artificial selection that reduced population size to 4 or 8 pairs; but there was still variation in selected lines that had been kept at 16 pairs.

The effective population size is further lowered if generations overlap so that offspring can mate with their parents (Felsenstein 1971, Giesel 1971), or if the population fluctuates in size from generation to generation. Fluctuations in population size put the population through BOTTLENECKS, during which genetic variation is reduced. The effective size $N_e$, in this instance, is approximately found by the harmonic mean population size,

$$\frac{1}{N_e} = \frac{1}{t}\sum_{i=1}^{t}\frac{1}{N_i}$$

For example, if in five successive generations a population consists of 100, 150, 25, 150, and 125 individuals, $N_e$ is approximately 70 rather than the arithmetic mean 110; it is more strongly affected by the lower than by the higher population sizes.

It is usually extremely difficult to obtain enough information about a natural population—dispersal distances, effective sex ratio, mating frequency across generations, variance in reproductive success—to measure the effective population size. Even when only one or a few of these factors are measured, such as variance in reproductive success, $N_e$ in human populations and in laboratory populations of *Drosophila* commonly turns out to be less than 75 percent of the actual number of individuals (Crow and Kimura 1970). Merrell (1968) counted the number of egg masses laid by leopard frogs (*Rana pipiens*) in ponds in which the total number of frogs had been estimated; assuming that each egg mass came from a different pair of parents, he found that only 1 to 67 percent of the frogs had bred successfully.

We will return to estimates of effective population size in considering gene flow. For the present, it suffices to say that for a great many species, it is likely that the effective size of local populations is in the hundreds or thousands, and often much less. Such estimates, moreover, do not take into account bottlenecks which many, perhaps most, populations have experienced, since we do not know their past histories of population fluctuation.

## MUTATIONS IN FINITE POPULATIONS

In a very large population, the rate of gene frequency change due to mutation alone is likely to be low, because the mutation rate at a locus is likely to be about $u = 10^{-5}$ or $10^{-6}$ per gamete per generation (see Chapter 3). If we assume that allele $A$ (with frequency $p$) mutates to $A'$ (with frequency $q = 1 - p$) at the rate $u$, and that the rate of mutation from $A'$ to $A$ is $v$, the change in the frequency of $A'$ is

$$\Delta q = u(1 - q) - vq$$

per generation. Thus if $q = 0.5$, $u = 10^{-5}$, and $v = 10^{-7}$, $q$ will have changed to only 0.50000495 by the next generation. At this rate, it takes about 70,000 generations to get halfway to the equilibrium gene frequency (Crow and Kimura 1970). The equilibrium gene frequency, found by setting $\Delta q = 0$ and solving for $q$, is

$$\hat{q} = u/(u + v).$$

Thus if $u = 10^{-5}$ and $v = 10^{-7}$, the stable equilibrium frequency is $\hat{q} = 0.99$, but it will take a very long time to get there; mutation *by itself* is a very weak force.

The derivation above assumes that there are only two distinguishable alleles, as is sometimes the case for gross phenotypic characteristics. If we recognize as alleles any difference in the nucleotide sequence of a large gene, however, a given nucleotide sequence can mutate to any of many thousands of other sequences, and the rate at which such sequences will back-mutate to a specific nucleotide sequence is negligible. Thus under mutation alone, any one allele will become extremely rare, and the population will contain a vast number of different nucleotide sequences.

In a finite population of effective size $N_e$, alleles are lost by genetic drift, and mutation is therefore important because it replenishes the genetic variation. If mutation rates and population size remain constant long enough, the rate of loss of alleles and the rate of gain of alleles by mutation will eventually achieve a balance (Box C). The level of genetic variation as measured by the average proportion of heterozygotes at a locus will then be

$$\hat{H} \approx \frac{4N_e u}{4N_e u + 1}$$

Thus $\hat{H}$ depends on the flux of new mutations, which is, again, proportional to the population size $N_e$ and the mutation rate $u$ (Figure 8). For example, if $u = 10^{-6}$ and $N_e = 30{,}000$, $\hat{H}$ will be about 0.11. Bear in mind throughout this discussion that $N_e$ is *effective* population size, and that we are only considering hypothetical alleles that are precisely equivalent to one another in their effects on survival and reproduction. After a bit more theory, we must consider whether any genes exist that meet these assumptions.

## THE FOUNDER EFFECT

The genetic accidents inherent in small population size may be important during colonization. If an island or a patch of newly available habitat previously unoccupied by a species is colonized by one or a few individuals, all the genes in the

## C *Mutations in Finite Populations*

Under genetic drift alone, the frequency of heterozygotes in a population of effective size $N_e$ declines as the inbreeding coefficient rises, according to the relation

$$F_t = 1/2N_e + (1 - 1/2N_e)F_{t-1}$$

This is the probability that two uniting genes are identical by descent. But they will not be identical by descent if either of them has just mutated. If $u$ is the rate at which an allele mutates to any of the other possible alleles, the probability that a gene has not mutated is $1 - u$, so the probability that neither has mutated is $(1 - u)^2$. Thus the probability that two uniting genes are identical by descent is

$$F_t = [1/2N_e + (1 - 1/2N_e)F_{t-1}]\,(1 - u)^2$$

When the rate of loss of alleles by genetic drift is balanced by the gain of new alleles by mutation, $F_t = F_{t-1} = \hat{F}$, and $\hat{F}$ is found to be

$$\hat{F} = \frac{(1 - u)^2}{2N_e - (2N_e - 1)\,(1 - u)^2}$$

Ignoring terms with $u^2$ because this is very small, this relation becomes approximately

$$\hat{F} \approx \frac{1}{4N_e u + 1}$$

and since heterozygosity $H = 1 - F$, the heterozygosity at equilibrium ($\hat{H}$) is

$$\hat{H} \approx \frac{4N_e u}{4N_e u + 1}$$

Thus if $N_e u \gg 1$, heterozygosity will be quite high; if $N_e u \ll 1$, genetic drift predominates and heterozygosity is low.

This, however, is the average heterozygosity we expect at an average locus. There is variance around this value: at some loci, heterozygosity is higher, and at some it is lower as one allele becomes fixed by genetic drift and others are lost. How often do new mutations become substituted (fixed) by genetic drift?

Recall that if the present frequency of an allele is $q$, the probability that it will eventually be fixed (if genetic drift is the only factor affecting allele frequencies) is $q$. If a population of size $N$ has $2N$ gene copies and one of them is a new mutation, its frequency is $q = 1/2N$. Its most probable fate is to be lost by genetic drift [with probability $1 - (1/2N)$]. But the chance that it eventually becomes fixed is its present frequency, $1/2N$. In any generation, the number of new mutations at this locus is the number of genes ($2N$) times the mutation rate $u$, or $2Nu$. If the fixation rate per generation is at a steady state, the number of mutations fixed will be the number that arose in some particular generation in the past, $2Nu$, multiplied by the proportion that are fixed, ($1/2N$). Thus

$$2Nu \times 1/2N = u$$

is the substitution rate of mutations per generation. It is striking that this number does not depend on the population size; the effect of a larger population in generating more mutations is nullified by the slower rate of drift toward fixation as population size increases.

It can be shown that if an allele is fixed by genetic drift, the average number of generations it takes from its time of origin to its time of fixation is $4N_e$.

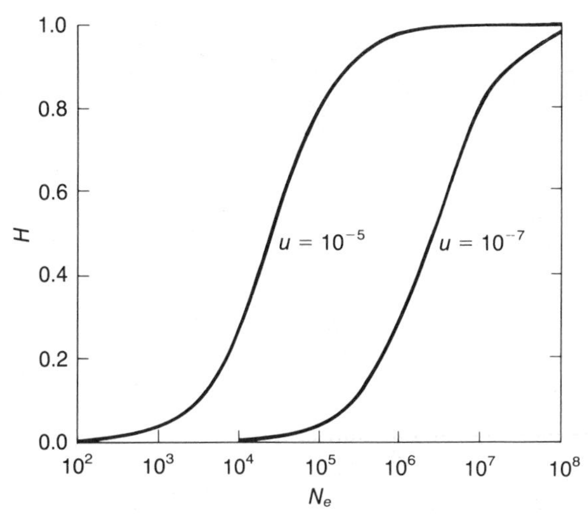


FIGURE 8
**The relation between heterozygosity, *H*, and effective population size, $N_e$, at equilibrium when the rate of loss of genetic variation by genetic drift is balanced by the rate of gain through mutation. The relation is shown for two mutation rates (*u*) for the model that assumes an unlimited number of possible alleles (the infinite-allele model). (From Hedrick 1983)**

population to which they give rise stem from the few carried by the founders and from subsequent mutations and immigrants.

A colony founded by a pair of diploid individuals can have at most four alleles at a locus, although there may have been many more alleles in the population from which the colonists came. But even though the number of alleles is reduced, the degree of heterozygosity, and hence the genetic variance, is almost as high as in the source population. [On average it is $(1 - 1/2N)H_0$, where $H_0$ is the proportion of heterozygotes in the source population and $N$ is the number of colonists.] This is simply because rare alleles contribute little to the source population's level of heterozygosity, and these are the very alleles that are likely to be missing from the colony. However, the heterozygosity of the colony declines rapidly by genetic drift if the colony remains small, and it only slowly builds back up by mutation and drift unless natural selection increases the frequency of rare alleles (Nei et al. 1975; Figure 9). If the colony grows rapidly, however, the amount of genetic variation will not be greatly reduced.

The same principle applies to any population that undergoes a severe bottleneck in population size. Mayr (1954, 1963) argued that the genetic constitution of a population may be so affected by this FOUNDER EFFECT that it accelerates the population's development into a new species. This hypothesis is discussed in Chapter 8.

An extreme example of the bottleneck effect is provided by the northern elephant seal (*Mirounga angustirostris*). This species was reduced by hunting to about 20 animals in the 1890s; since then the population has grown to more than 30,000. The effective population size must have been even smaller than 20 at the population's nadir, for the species is polygynous; less than 20 percent of the males service all the females. Among species that have been examined for electrophoretic variation, this seal is almost unique; Bonnell and Selander (1974) found no genetic variation in a sample of 24 electrophoretic loci. Genetic variation has been observed in the southern elephant seal (*M. leonina*), which was never as severely reduced in numbers.

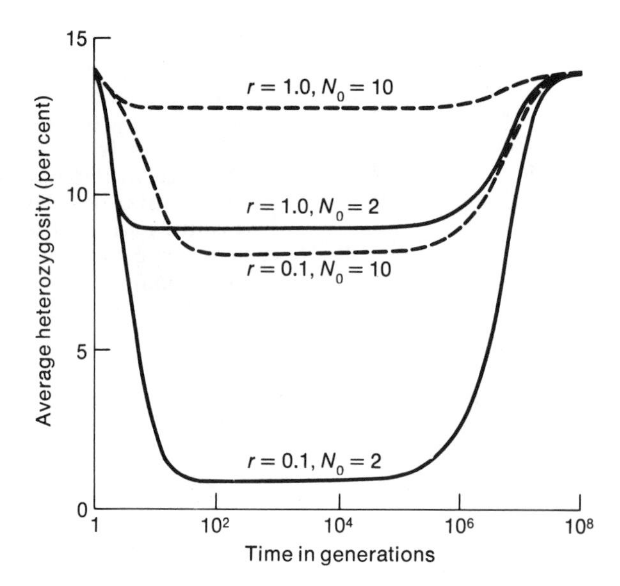


FIGURE 9

**Effects on genetic variation of a bottleneck in population size, as in a newly founded population beginning with two individuals (solid lines) or ten individuals (dashed lines). Heterozygosity declines more substantially because of inbreeding after colonization if the growth rate of the population is low ($r = 0.1$) than if high ($r = 1.0$). (After Nei et al. 1975)**

## GENE FLOW

The degree to which a population can be delimited from other populations depends on the level of GENE FLOW between them. The rate of gene flow $m_{ij}$ from population $j$ to population $i$ is the proportion of individuals breeding in population $i$ that have immigrated from population $j$ during that generation. If $m_{ij}$ is very high, near 0.5, the two "populations" are in effect one panmictic population, the total size of which is the sum of the two. Thus the rate of gene flow influences the effective size of the population.

Several models of gene flow exist that correspond to differences in population structure. These include:

1. The "continent-island" model, in which there is effectively one-way movement from a large ("continental") population to a smaller, isolated population.
2. The "island" model, in which migration occurs at random among a group of small populations.
3. The "stepping-stone" model, in which each population receives migrants only from neighboring populations.
4. The "isolation-by-distance" model in which gene flow occurs among local neighborhoods in a continuously distributed population.

Most of the models consider gene flow which occurs at an approximately constant rate in each generation. Genes may be carried either by the movement of gametes (e.g., pollen) or of individual organisms (which in the case of plants are likely to be seeds). It is important to recognize that the amount of genetically effective movement (measured by the gene flow rate $m$) is often much less than the movement of organisms, many of which do not succeed in breeding after settling into another population. Social interactions (e.g., territoriality), as well as physical and biological vicissitudes, reduce the likelihood of breeding. Indeed, some insect species such as certain leafhoppers invade northern areas in large numbers every summer, but are incapable of surviving the winter except in the southern areas where they originate (Ross et al. 1964); in such cases, ecological dispersal occurs at a high rate, but there is no effective genetic dispersal.

Whatever model of population structure is used, gene flow has the effect of homogenizing the genetic composition (Box D), so that if gene flow is the only factor operating (i.e., if population sizes are so large that genetic drift may be ignored and if the alleles are selectively neutral), all the populations will converge to the same allele frequency (generally an average of the initial gene frequencies). If natural selection favors different genotypes in different populations, the gene frequency in each will arrive at an equilibrium determined by the relative strength of selection and gene flow (Chapter 6). If alleles are selectively neutral, but population sizes are so small that genetic drift is important, convergence of gene frequencies caused by gene flow is counteracted by random divergence caused by genetic drift. A useful measure of variation in allele frequencies among different populations is $F_{ST}$, the variance in allele frequency ($V_q$) standardized by the mean ($\bar{q}$):

$$F_{ST} = V_q/[\bar{q}(1 - \bar{q})]$$

As Box D shows, $F_{ST}$ is approximately equal to

$$F_{ST} \approx 1/[4N_e m + 1]$$

in the case of the island model, and decreases in proportion to the product ($N_e m$) of the effective population size and the rate of gene flow. If migrants arrive at random from all the subpopulations that make up the larger population (or METAPOPULATION), even a low amount of gene flow greatly reduces the divergence among populations caused by genetic drift. For example, if each of a set of idealized "island" populations has the same effective size $N_e$ and if only one individual in each population in each generation is an immigrant, $m = 1/N_e$ and $F_{ST} = 0.2$. All these principles hold qualitatively not only for discretely separated populations, but for the isolation-by-distance model. In this model, as the average distance of gene dispersal increases, neighborhood size increases, and variations in genetic composition from one locality to another become less pronounced (Figure 10).

The average rate of gene flow among established populations of a species is often quite low, and is especially likely to be low if immigrating individuals must compete with residents to survive and reproduce. However, the effective rate of gene flow among populations of a species can be considerably higher than the average rate might suggest if local populations frequently become extinct and the sites are recolonized by individuals drawn from several other populations (Slatkin 1977, 1981, Maruyama and Kimura 1980). As a newly founded population grows in size, its gene frequencies are a mixture of those in the populations from which the colonists came, and these in turn will contribute to the genetic composition

## D *Gene Flow of Neutral Alleles*

Suppose, following the continent–island model, that in each generation the proportion of breeding individuals in a local population that are derived from a "source" population is $m$, and that the frequency of allele $A'$ is $q$ in the local population and $q_m$ among the immigrants. Let $q_0$ be the allele frequency in generation 0, and $q_1$ the allele frequency a generation later. Then within the local population, $q_1 = (1 - m)q_0 + mq_m$, and the change in gene frequency in a single generation is $\Delta q = q_1 - q_0 = -m(q_0 - q_m)$. This reaches equilibrium ($\Delta q = 0$) when $q_0 = q_m$, so the local allele frequency comes to equal that in the source population. Similarly, in the island model of gene flow among a multitude of populations the equilibrium allele frequency in population $i$ will approach $\bar{q}$, the mean allele frequency.

If the populations vary in allele frequency, with $\bar{q}$ the mean allele frequency and $V_q$ the variance in allele frequency among populations, the total frequency of heterozygotes, summed over all populations, will be less than if they formed a panmictic population, and equals $H = 2\bar{q}(1 - \bar{q}) - 2V_q$. This reduction of the frequency of heterozygotes, termed the WAHLUND EFFECT, measures the degree of subdivision of the population as a whole. Another such measure is the probability that two genes taken at random from two different subpopulations are identical by descent. Sewall Wright terms this probability $F_{ST}$, and contrasts it with $F_{IS}$ (the usual inbreeding coefficient within a population, i.e., the probability that two genes taken at random from within a population are identical by descent) and with $F_{IT}$ (the probability that two genes taken at random from the entire set of populations are identical by descent). The relationship among these inbreeding coefficients is

$$F_{ST} = \frac{(F_{IT} - F_{IS})}{(1 - F_{IS})}$$

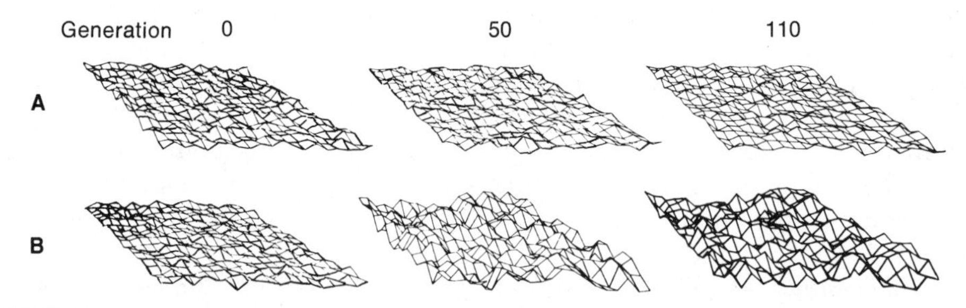


FIGURE 10

**Effects of neighborhood size on spatial variation in allele frequencies in the isolation by distance model. We are looking down from an angle at an area populated by a 100 × 100 array of equally spaced individuals. Local allele frequency differences from place to place within the region are shown by different heights above the horizontal plane after 0, 50, and 110 generations of mating. Random changes in allele frequency were simulated by a computer. In series A each individual had an equal probability of mating with any other; in series B each mated equiprobably within a neighborhood of nine individuals. Genetic differentiation is more pronounced when neighborhood size is small. (From Rohlf and Schnell 1971)**

$F_{ST}$ is closely related to the Wahlund measure of subdivision:

$$F_{ST} = \frac{V_q}{\bar{q}(1 - \bar{q})}$$

If the populations are finite in size, $F_{ST}$ tends to increase as the populations diverge by genetic drift, but to converge because of gene flow. By genetic drift, $F_{ST}$ — the probability of autozygosity — increases from generation $t - 1$ to $t$ by the relation

$$F_t = (1/2N_e) + [1 - (1/2N_e)]F_{t-1}$$

but this is true only if neither of the uniting gametes has just entered the population from another population. The probability that neither gamete is an immigrant is $(1 - m)^2$, so

$$F_t = \{1/2N_e + [1 - (1/2N_e)F_{t-1}]\}(1 - m)^2$$

At equilibrium, $F_t = F_{t-1} = \hat{F}$:

$$\hat{F} = \frac{(1 - m)^2}{2N_e - (2N_e - 1)(1 - m)^2}$$

If $m$ is small, this becomes approximately

$$\hat{F} \approx \frac{1}{4N_em + 1}$$

Since $F_{ST} = V_q/\bar{q}(1 - \bar{q})$, the equilibrium variance in allele frequency among populations is approximately

$$V_q = \frac{\bar{q}(1 - \bar{q})}{4N_em + 1}$$

Thus variation among populations declines as $N_em$ increases.

of other populations as the process is repeated. The higher the rate of extinction and recolonization, the higher the rate of gene flow will be, and the lower the variance in allele frequencies among populations. This may have the effect of lowering the effective population size and level of genetic diversity in the species as a whole, because gene flow in this case counteracts the tendency of genetic drift to augment the total genetic diversity of a species by fixing different alleles in different populations.

## EFFECTIVE POPULATION SIZES AND GENE FLOW IN NATURAL POPULATIONS

Gene flow and effective population size are among the most difficult parameters to estimate in natural populations, but a variety of direct and indirect approaches have been used; frequently it is possible to estimate only the product $N_em$, rather than each component separately.

By marking individual deer mice (*Peromyscus maniculatus*), Howard (1949) showed that at least 70 percent of males and 85 percent of females breed within 150 meters of their birthplace, and consanguineous matings are rather common.

In the rusty lizard (*Sceloporus olivaceus*), juveniles disperse an average of 78 meters from their hatching site to their final home, and females wander about 29 meters from their home range to lay eggs; the neighborhood size is about 10 hectares (Kerster 1964). In these species, the effective sizes of local populations are quite small. Even migratory birds commonly have lower gene flow and smaller neighborhood sizes than their power of flight might suggest, because they usually return to the vicinity of their birthplace (Greenwood and Harvey 1982). From banding data, Barrowclough (1980) estimated that the local effective sizes of several species of noncolonial birds such as wrens and finches ranged from about 175 to 7700 individuals.

Dobzhansky and Wright (1943) estimated neighborhood size in *Drosophila pseudoobscura* by releasing genetically marked flies at a single point and recapturing them at baits placed at various distances from the site of release. The flies moved an average of 133 meters the first day and about 90 meters per day thereafter. Assuming that females mate within two days after emerging, the neighborhood may be calculated from these data to encompass 24,500 square meters; multiplying this by the average density of the wild population during summer gives a neighborhood size of about 25,700 flies. However, this is likely to be an overestimate because the population is greatly reduced in density during winter and because the abnormally high density at the point of release may have induced an abnormally high rate of dispersal (Wallace 1966).

Gene flow in higher plants is accomplished by dispersal of seeds and pollen. By planting recessive homozygotes at various distances from a strain marked with a dominant gene and then examining the distribution of heterozygous progeny, Bateman (1947a,b) was able to measure pollen dispersal in both wind-pollinated (maize) and insect-pollinated (radish) crops. The proportion of maize seeds carrying the dominant allele decreased exponentially with distance and was reduced to one percent at only 40–50 feet from the pollen source (Figure 11). Similarly, in insect-pollinated plants most pollen is carried only very short distances, although a small proportion is carried to much greater distances and may contribute importantly to gene flow (Levin and Kerster 1974). The level of gene flow in plants is generally much less than the actual amount of seed and pollen dispersal, and it appears that the effective size of local populations is often in the dozens or hundreds (Levin and Kerster 1974, Levin 1981).

Differences among species in the relative values of $N_em$ have frequently been estimated from $F_{ST}$, a measure of the among-population variance in the frequency of alleles that are presumed to be only weakly, if at all, affected by natural selection. Allozymes have been most frequently used for this purpose. As we shall see, there is considerable controversy about whether or not allozyme frequencies are strongly affected by selection. Hence it may be more accurate to estimate $N_em$ from the distribution of rare alleles, which may be disadvantageous everywhere, than from alleles that are common in some populations but not others, a pattern that could be caused by natural selection. Slatkin (1981) took a sporadic distribution of individual rare alleles to indicate low $N_em$, and concluded that gene flow is indeed low in species thought to have low powers of dispersal, such as salamanders.

In general, $F_{ST}$ values calculated from allozyme frequencies indicate that the biological differences among species predict fairly well differences in gene flow,

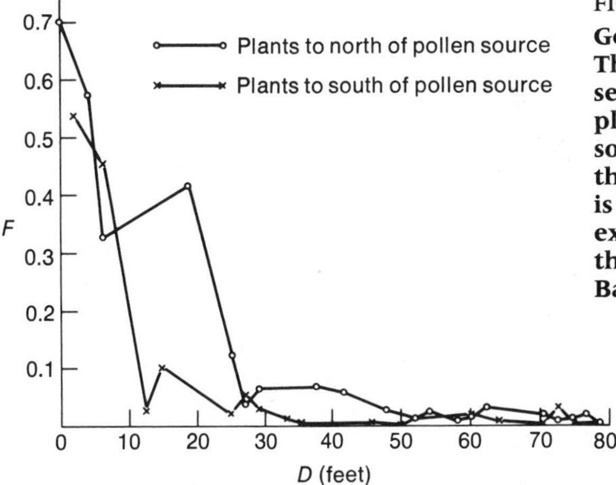


FIGURE 11

**Gene flow in corn, a wind-pollinated plant. The vertical axis *F* gives the proportion of seeds fertilized by a different genetic strain in plants set at distances *D* to the north and south of that strain. Prevailing winds affect the amount of gene flow, but in both cases it is quite restricted. Gene flow may be more extensive in other wind-pollinated plants that have lighter pollen than corn. (After Bateman 1947b)**

effective population size, and genetic differentiation among populations. For example, populations of a wingless species of water strider are genetically more differentiated, and have lower within-population heterozygosity, than those of a species that often has wings (Zera 1981). Similarly, there is much less geographic variation in allele frequencies in the migratory monarch butterfly *Danaus plexippus* than in nonmigratory species of butterflies (Eanes and Koehn 1978).

Extinction and recolonization of populations have been studied very little, so their effects on gene flow in nature are unknown. Little is known even about causes of population extinction other than habitat destruction by human activities. One of the few exceptions is Jaeger's (1980) study of *Plethodon shenandoah*, a salamander that is confined to dry talus slopes by competition with a related species. During a severe drought, a few individuals of *P. shenandoah* survived in a talus slope that had moist patches of soil, but a population that occupied a talus slope without moist patches became extinct.

Two studies exemplify the possible impact of extinction and colonization. The gut bacterium *Escherichia coli* has a high level of allozyme polymorphism, but much less than might be expected of a species that numbers in the many millions. Electrophoresis of several proteins revealed relatively few genotypes, and some of these came from unrelated human and nonhuman hosts (Selander and Levin 1980). Within a single human host, there was very rapid turnover in genotypes (Caugant et al. 1981). These observations together suggest that there is a very high rate of genotype loss and recolonization of hosts, so that the effective population size might be orders of magnitude less than it seems (Kimura 1982).

An extreme case of the impact of extinction and colonization is that of the red-backed salamander *Plethodon cinereus* (Highton and Webster 1976). Populations in unglaciated regions of North America vary considerably in allozyme frequencies (the average identity index of Nei was only $\bar{I} = 0.87$), whereas all populations surveyed over a wide region in the north, from which the most recent glacier retreated less than 10,000 years ago, were genetically very similar ($\bar{I} = 0.99$). Northern populations have almost no unique alleles, whereas southern populations have many local, rare alleles that are not found in the north. It is likely that most populations in the glaciated area were colonized from a rather restricted region in the south.

## GENETIC DRIFT IN NATURAL POPULATIONS

Several authors have attempted to determine whether the theory of drift is sufficient to explain observations of the rate of genetic change in populations. Some (e.g., Kerr and Wright 1954, Buri 1956) have found that frequencies of mutant alleles in small laboratory populations of *Drosophila* change in rather close accord with theoretical predictions (Figure 12) if both population size and the independently measured intensity of natural selection against the mutants are taken into account. Some observations of natural populations are at least qualitatively consonant with the predictions of the theory of drift. For example, Selander (1970) found that the mean allele frequency at two protein loci (esterase, *Es*-$3^b$, and hemoglobin, *Hbb*) was much the same in small and large populations of house mice, but the variation in gene frequency was greater among small than among large populations (Table II).

In some cases gene frequency differences among populations correspond quantitatively to what one would expect if genetic drift were the only important factor. For example, the cattle of Iceland are descended from those brought from Norway by Vikings about one thousand years ago. For each of several blood group loci, Kidd and Cavalli-Sforza (1974) calculated $F_{ST}$ between Icelandic populations and Norwegian populations, and compared it to the value expected from

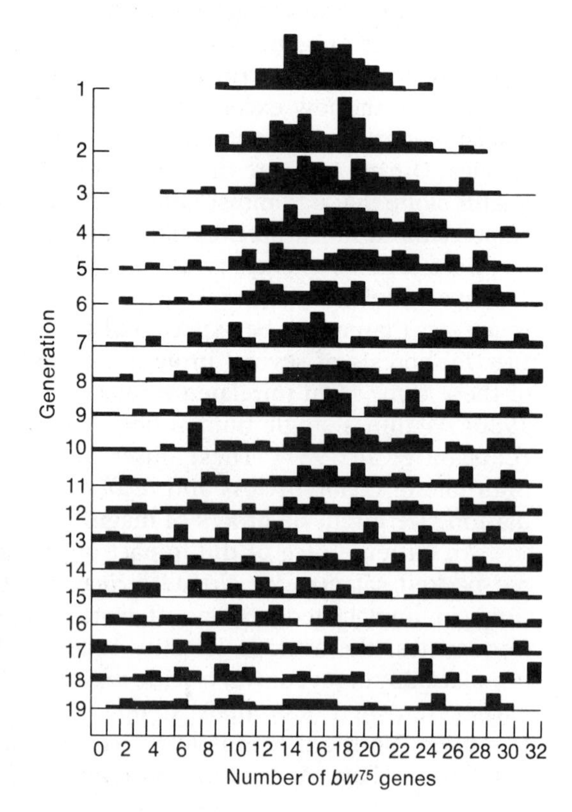


FIGURE 12

**Experimental demonstration of genetic drift. The number of copies of an allele, $bw^{75}$, in each of many replicate populations of *Drosophila melanogaster* maintained in the laboratory at 16 flies for 19 generations. In each population, the frequency of the allele fluctuated, so the variation in gene frequency increased. After about 12 generations all gene frequency classes have become about equally frequent. (From Buri 1956)**

TABLE II
**Frequency of alleles at two loci relative to population size of house mice**

| Estimated population size | Number of populations sampled | Mean allele frequency | | Variance of allele frequency | |
|---|---|---|---|---|---|
| | | *Es - 3*[b] | *Hbb* | *Es - 3*[b] | *Hbb* |
| Small (median size 10) | 29 | 0.418 | 0.849 | 0.0506 | 0.1883 |
| Large (median size 200) | 13 | 0.372 | 0.843 | 0.0125 | 0.0083 |

(After Selander 1970)

the equation $F_{ST} = 1 - (1 - 1/2N)^t$. From historical records it was possible to estimate how long the Icelandic cattle had been isolated, the sex ratio, and the fluctuations in population size. The theoretical value of $F_{ST}$ calculated from these estimates was quite close to the observed value for each of seven loci. Thus these genetic differences can be explained purely by genetic drift. This analysis does not prove that these differences among populations are due to drift rather than to natural selection, but it suggests that there is no compelling reason for believing that the differences are due to anything but chance.

## EVOLUTION BY GENETIC DRIFT

This chapter has been concerned only with the effects of inbreeding, genetic drift, and gene flow on alleles that are selectively neutral, i.e., those that do not differ in their effects on survival or reproduction. Whether or not many alleles conform to this assumption is the subject of considerable controversy (Chapter 6). For alleles that are indeed approximately neutral, an extremely important theoretical conclusion (Box C) is that new alleles at a locus should be substituted (fixed) at a constant rate $u$ per generation, where $u$ is the rate at which new alleles arise by mutation. Consequently different populations (or species) will diverge at a constant rate, and at any given time a population should have a considerable number of alleles drifting toward either loss or fixation (Figure 13), the number per locus depending on the mutation rate and the effective population size.

Proponents of the neutral theory, such as Kimura (1983a,b) and Nei (1983) interpret much allozyme variation in this way, holding that it reflects the transient passage of neutral alleles. In their view, much of the variation observed at the molecular level (in allozymes and variation in protein and DNA sequences) is neutral, and much of the divergence among species at the molecular level has been caused by genetic drift. One prediction of the neutral model is that the rate of evolution should be higher in molecules (proteins, DNA sequences) that are not subject to strong functional constraints than in those that are; in a weakly constrained molecule, fewer mutations will be selected against, and more will have a neutral effect, because they are less likely to disrupt the molecule's function. The average rate of evolution of a molecular sequence may be estimated by counting the proportion of sites that differ between two species, and dividing by the time since their most recent common ancestor (judged from the fossil record).

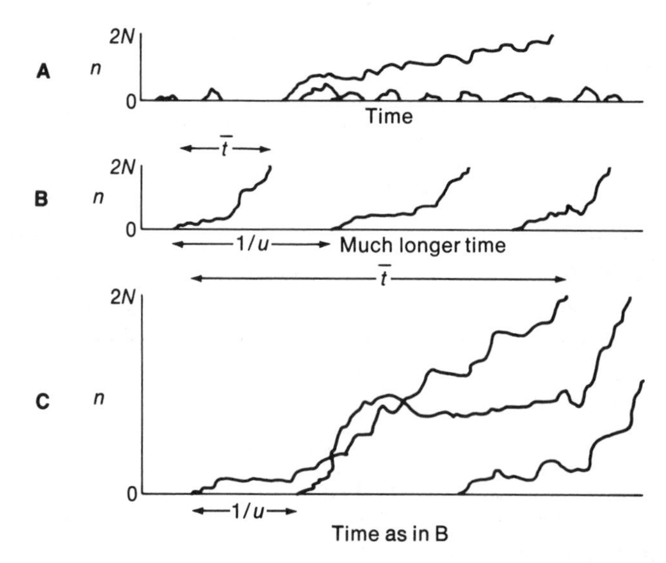


FIGURE 13

**Gene substitution by genetic drift. The number of copies *n* of a mutant fluctuates at random; most mutations are lost, but an occasional mutant drifts to fixation (A). For those that reach fixation in (B) and (C), the interval between mutation events (1/*u*) is shorter in a larger population (C) than in a smaller population (B), and it takes longer ($\bar{t}$) for the mutation to be fixed. Hence at any time more selectively neutral alleles will be in the larger population. (From Crow and Kimura 1970)**

Thus humans and carp diverged about 400 million years ago (in the Devonian), when the teleost line diverged from the fishes that later gave rise to the amniotes. The number of differences in amino acid sequence of two species is often converted into the number of nucleotide differences in the corresponding genes, using the genetic code.

As the neutral theory predicts, the rate of evolution is greater in functionally less constrained molecules (Kimura 1982, 1983a). For example, histone H4 is a strongly constrained protein that binds to DNA and is essential to the structure of chromosomes. Its average rate of evolution has been $0.008 \times 10^{-9}$ amino acid substitutions per year per site, the term "substitution" meaning the fixation of a different amino acid within a species. In contrast, fibrinopeptides have evolved at an average rate of $8.3 \times 10^{-9}$ substitutions per year. These peptides have little function because they are discarded after being cleaved from fibrinogen during blood clotting. Similarly, the proinsulin molecule consists of three peptides, two of which are joined to form insulin while the third is cut off and discarded. The rate of evolution of the "active" peptides has been $0.4 \times 10^{-9}$ substitutions per year, while that of the discarded peptide has been $2.4 \times 10^{-9}$.

Because it is now fairly easy to obtain direct nucleotide sequences of specific genes, these comparisons can be made at the level of the DNA as well. Comparisons of homologous genes among species have shown that the second position in the triplet (codon) that codes for an amino acid evolves most slowly, the first position somewhat more rapidly, and the third position most rapidly—at a rate up to one hundredfold that of the second position (see Table II in Chapter 15). At least half the nucleotide substitutions at the third position are synonymous: they do not result in an amino acid substitution, and are unlikely to affect the phenotype of the organism or its fitness. Moreover, the rate of substitution of synonymous nucleotides is sometimes the same in two genes, even though the proteins for which they code differ greatly in the rate at which their amino acid composition changes (Kimura 1983b). Finally, the rate of nucleotide substitution has been much higher in introns than in exons of the same gene (e.g., in the genes that code for globins of various mammals), and the α-3 gene of mice, a nontranscribed pseudogene derived from the functional α-1 globin gene by gene duplication (see Chapters 3 and 15) has diverged much faster from the α-1 gene of the mouse than the α-1 genes of the mouse and rabbit have diverged from each other. It is exceedingly unlikely that such rapid evolution in nonfunctional molecules and in synonymous codons can be ascribed to natural selection, whereas it is predicted by the neutral theory.

Proponents of the neutral theory claim that protein and DNA sequences have diverged among species at a constant rate, as the theory of genetic drift predicts (Kimura 1983a; Wilson et al. 1977). If this is so, macromolecules offer a MOLECULAR CLOCK by which the time since common ancestry can be estimated. Molecular information for taxa that have a good enough fossil record is used to calibrate the "clock," which can then be used to estimate the time since divergence for pairs of taxa that have an inadequate fossil record (see Chapter 10). Some workers claim, in contrast, that proteins evolve at highly inconstant rates. For example, Goodman et al. (1982) argue that globins evolved rapidly when the loci were duplicated early in vertebrate evolution and diverged to take on different functions (e.g., myoglobin and the α and β hemoglobin chains). Goodman and his colleagues also present evidence that the rate of amino acid substitution in globins has slowed down considerably during the divergence of humans and anthropoid apes.

Some data certainly suggest that at least over long periods of evolutionary time, the average rate of amino acid substitutions has been the same for independently evolving molecules. For example, both teleost fishes and amniotes have α and β hemoglobins, whereas agnathans (e.g., lampreys) have only a single globin locus. Thus the gene duplication that produced the α and β hemoglobins occurred before the divergence of teleost fishes from the amniotes; the α and β chains have been diverging within these lineages for 400 million years. The number of nucleotide substitutions that correspond to amino acid differences between the α and β chains in humans is almost the same as the number that distinguish the carp α chain from the human β chain (Kimura 1982). Langley and Fitch (1974) inferred the number of nucleotide substitutions entailed in the divergence of several proteins among a number of species of vertebrates, and plotted them against the divergence time (determined from the fossil record) of pairs of these species (Figure 14). Their analysis indicates that the rate of divergence has been rather constant, differing among lineages with a variance only about twice that expected of a random process with a constant mean rate, such as radioactive decay (see Hudon 1983). Although the constant rate of divergence predicted by the neutral theory has not yet been confirmed in detail, it seems to hold as a rough approximation.

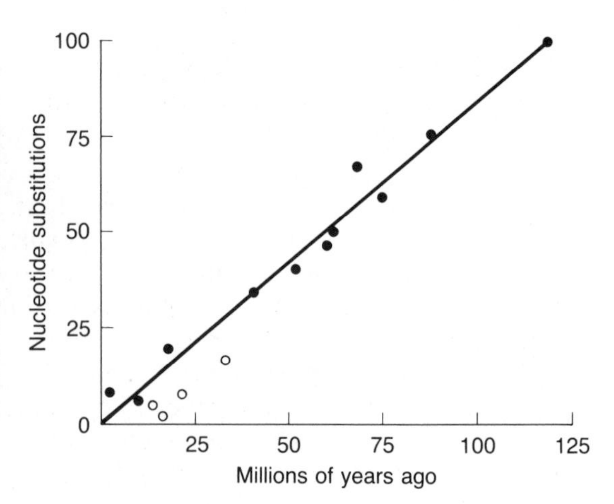


FIGURE 14
**Nucleotide substitutions versus time since divergence, illustrating approximate constancy of the rate of molecular evolution. Each point represents a pair of living species of mammals whose most recent common ancestor, based on fossil evidence, is indicated on the abscissa. The ordinate is the number of nucleotide substitutions inferred from the difference in amino acid sequence of seven proteins, data from which were pooled for this analysis. The four open circles represent pairs of primate species, which appear to have diverged more slowly than other groups of mammals. (From Selander 1982, after Langley and Fitch 1974)**

## NONRANDOM MATING BASED ON PHENOTYPE

This chapter opened with a discussion of deviations from random mating that arise owing to mating among relatives. As we have seen, inbreeding affects genotype frequencies at all loci in the genome in the same way. We briefly note here another form of deviation from nonrandom mating that will be explored further in Chapters 7 and 8.

In some cases, individuals mate nonrandomly on the basis of one or more phenotypic traits. For example, in sexual selection, mating success in one sex (usually males) depends on phenotypic traits, as in an African finch, the widowbird *Euplectes progne*, females of which choose long-tailed over shorter-tailed males (Andersson 1982a). Mate selection is sometimes inversely frequency-dependent: in some *Drosophila* species, for example, males are more successful in mating if their genotype is rare than if it is common (Ehrman 1967). Finally, in ASSORTATIVE MATING, males and females of like phenotype tend to mate with each other; for example, in the United States, tall women tend to marry tall men and short women, short men (Spuhler 1968). Reproductive isolation between sympatric species may often be viewed as a form of assortative mating.

All these forms of nonrandom mating based on phenotypic traits affect the frequencies of alleles and genotypes, but *only* at those loci that govern the particular phenotypic character in question (and at loci that are closely linked to them). Suppose, for example, that two loci contribute additively to variation in height so that $AABB$ is the tallest genotype and $A'A'B'B'$ the shortest. Under continued perfect assortative mating, heterozygotes will decrease in frequency and the population at equilibrium will consist only of the genotypes $AABB$ and $A'A'B'B'$. In contrast, not only these genotypes but also $A'A'BB$ and $AAB'B'$ would be present in this population if it were completely inbred. An unlinked polymorphic locus will not be affected by assortative mating if the locus does not influence the phenotypic character (e.g., height); thus the allele frequencies of this locus will remain the same in each of the two populations ($AABB$, $A'A'B'B'$) that develop from the assortative mating process. Note that if assortative mating is complete, these two populations constitute different species.

## SUMMARY

Inbreeding distributes genes from the heterozygous to the homozygous state. A geographically subdivided population in which inbreeding occurs becomes increasingly homozygous within subpopulations, but the genetic variation among subpopulations more than compensates for the homogeneity within each, so that the variation of the population as a whole increases. In each subpopulation gene frequencies can fluctuate by chance, until one allele or another becomes fixed and genetic variation at that locus is lost, unless it is restored by mutation or immigration. This process, genetic drift, happens in any finite population, but is more rapid in small than in large populations. Because almost every species is distributed in semi-isolated populations that are often effectively smaller than they seem, some of the genetic changes in most species are likely to be random. Gene flow among populations counteracts the effects of genetic drift. Although the level of gene flow is high in some species, it appears to be very low in others.

If many of the mutations that arise at a locus are selectively neutral, populations and species may be expected to diverge in genetic composition at a constant rate. Some data on protein and DNA sequences conform to this theory, but whether or not the divergence is so constant that macromolecules may be used as evolutionary "clocks" is disputed.

## FOR DISCUSSION AND THOUGHT

1. Many characteristics may reduce the incidence of inbreeding in plants: self-sterility, heterostyly, dioecy, progressive flowering such that few flowers on a plant are open at any one time, protogyny or protandry, dispersal of seeds or pollen. What evidence is required to state confidently that the advantage of outbreeding is the *raison d'être* of any one of these characteristics? In what other ways might each of these be advantageous?
2. If individuals from a normally outcrossing species are inbred, a decrease in fitness is usually observed among their progeny. How, then, can inbreeding mechanisms such as cleistogamy evolve? What advantages are inherent in inbreeding?
3. Suppose a population was founded a few thousand generations ago by a pair of individuals but has had a large constant size since then. Does it still have a small effective population size because of the initial bottleneck? What determines when the effect of the bottleneck is so diminished that we can consider it a large population?
4. Given the impact that inbreeding has on phenotypic variation, what effect is it likely to have on the rate at which a population's genetic composition is changed by natural selection?
5. What effects do monogamy versus polygamy (polygyny or polyandry) have on effective population size? How might these mating systems evolve in birds or mammals?
6. How small does a population have to be before genetic drift has an important effect on the gene frequencies at a locus? Why is this not a very meaningful question?
7. What evidence would be sufficient to determine whether the geographic variation in some characteristic of a species was primarily adaptive (caused by natural selection) or random (caused by genetic drift)?
8. Ehrlich and Raven (1969) argue that gene flow among populations of a species usually proceeds at a very low rate. Why then, they ask, don't species show far more geographic variation than they do? How is it possible that we can recognize white-tailed deer from the Atlantic to the Pacific coast of North America as members of one species?
9. As long as widely separated populations remain the same species, the only evolutionary change in the species as a whole must be due to mutations that are advantageous

throughout the species, in all the environments it inhabits. Since there must be few such mutations, evolutionary change must necessarily be slower in widespread species than in isolated local populations. Discuss.

10. The neutral theory proposes that species should diverge at the molecular level at a rate that is constant per *generation*, but the data seem to suggest a constant rate of divergence per unit *time*. Discuss how this paradox might be resolved (see Chapter 10).

## MAJOR REFERENCES

Crow, J.F., and M. Kimura. 1970. *An introduction to population genetics theory.* Harper & Row, New York. 591 pages. A major exposition of mathematical population genetics, containing extensive treatment of inbreeding and stochastic effects.

Kimura, M., and T. Ohta. 1971 *Theoretical aspects of population genetics.* Princeton University Press, Princeton, N.J. 219 pages. Emphasizes the role of genetic drift in evolution, from a largely theoretical standpoint.

Wright, S. 1968-1978. *Evolution and the genetics of populations.* Vol. 1: *Genetic and biometric foundations,* 469 pages; Vol. 2: *The theory of gene frequencies,* 511 pages; Vol. 3: *Experimental results and evolutionary deductions,* 613 pages; Vol. 4: *Variability within and among natural populations,* 580 pages. University of Chicago Press, Chicago. A major, highly technical, treatise on population genetics by one of the founders of the field.

Hedrick, P.W. 1983. *Genetics of populations.* Science Books International, Boston. 629 pages. An introduction to the theory of population genetics, with empirical examples, that treats the subjects of this chapter clearly.

Kimura, M. 1983. *The neutral theory of molecular evolution.* Cambridge University Press, Cambridge, England. A comprehensive discussion of the neutral theory by its foremost architect.

Slatkin, M. 1985. Gene flow in natural populations. *Annu. Rev. Ecol. Syst.* 16: 393–430. A comprehensive review of the theory and empirical studies of gene flow and its interaction with natural selection.

# Effects of Natural Selection on Gene Frequencies

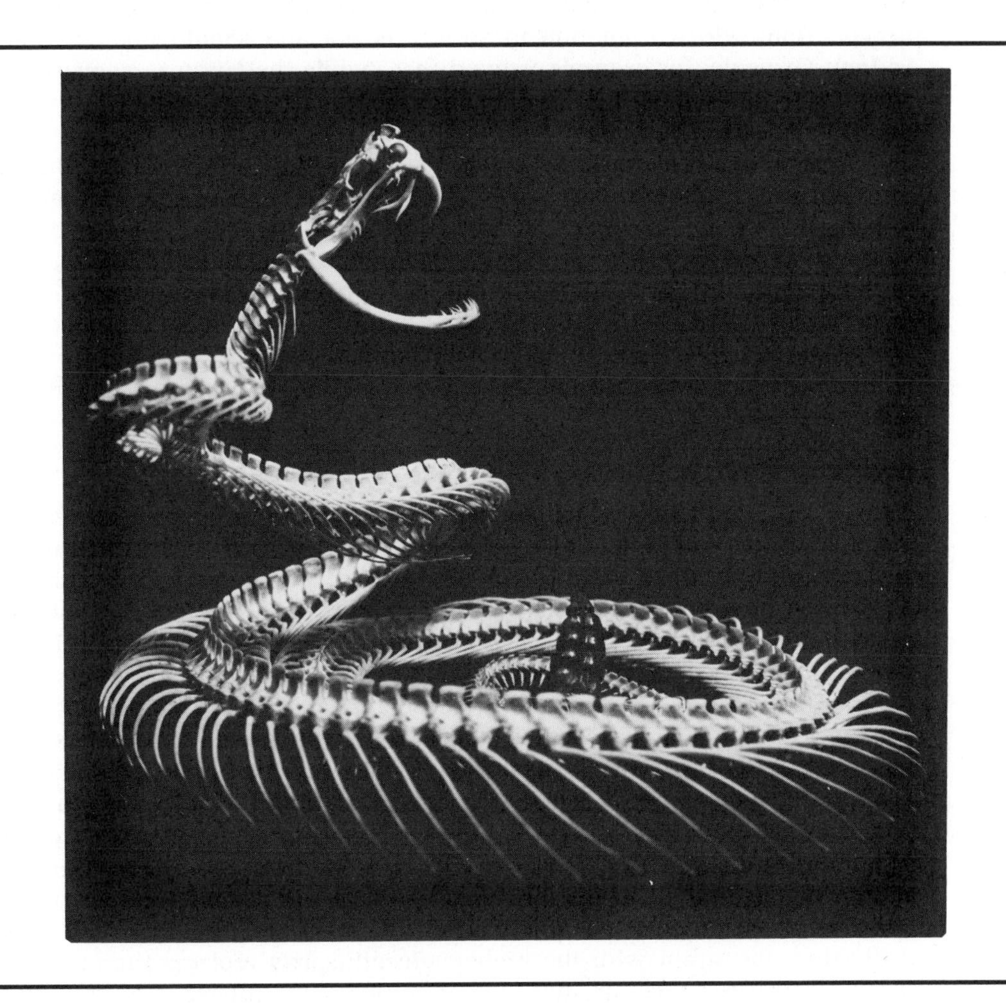

# Chapter Six

Natural selection was the central principle of Darwin's theory of evolutionary change, and remains the preeminent concept in evolutionary biology. Genetic drift clearly plays an important role in evolution, but we must turn to selection to find the explanation of many of the complex characteristics of organisms, and the innumerable features that fit them to their ways of life.

Although exceedingly simple in principle, natural selection can be diverse in its operations. Although it is an elementary concept, it has been commonly misunderstood, and a great deal of nonsense has been uttered in its name. It is viewed by some as a dark force, as the inexorable action of an unfeeling, uncaring universe, and by others as the creative agent of nature, instilling progress into history. It has been invoked as a natural law, with the moral force of an ethical precept, from which even human society is not and should not be exempt. Andrew Carnegie, for example, argued that "while the law may sometimes be hard for the individual, it is best for the race, because it insures the survival of the fittest in every department."

These misconceptions stem largely from the metaphorical ways in which the concept has been expressed, even by Darwin himself: "Nature, if I may be allowed to personify the natural preservation or survival of the fittest, cares nothing for appearances, except insofar as they are useful to any being. . . . Man selects only for his own good: Nature only for that of the being which she tends." But as Darwin noted, such poetical expressions can lead us to view natural selection as "an active power or Deity," omniscient, omnipotent, and, depending on one's point of view, either beneficent, shaping species into perfect form, or malevolent, red in tooth and claw.

## DIFFERENTIAL SURVIVAL AND REPRODUCTION

Natural selection, however, has none of these qualities. It is not providential, it is neither moral nor immoral, it carries no ethical precepts—"it" is not an active agent with physical properties, much less a mind. It is no more than a statistical measure of the difference in survival or reproduction among entities that differ in one or more characteristics. Selection is not caused by differential survival and reproduction; it *is* differential survival and reproduction, and no more.

Natural selection in the broadest sense applies to all of nature, not just to genotypes and phenotypes. Some isotopes decay more rapidly than others, so the longest-lived are the most prevalent. Of the possible orbits celestial bodies may have taken, some are more stable than others, so a form of selection shapes the universe. Selection of a kind determines the species composition of ecological communities, for species that cannot coexist with others have been eliminated. Within populations, selection may entail struggle and competition among individuals, or it may not: if two strains of bacteria in a chemostat differ in their rate of division, the strain with the higher rate ultimately replaces the other completely, even if resources are not in short supply. The difference in growth rate is natural selection, but there is no violent struggle, there is no goal, and there is no question of morality.

## INDIVIDUAL SELECTION

At least theoretically, selection can operate whenever different kinds of self-reproducing entities that beget descendants like themselves differ in their rate of survival or reproduction. In biological evolution, such entities can be individual genes or larger portions of the genome, individual organisms that differ in genotype, populations of organisms, or species. Most of the theory of natural selection centers on differential survival and reproduction of individual organisms that differ in phenotype because of genetic differences at one or more loci.

Natural selection operates whenever genotypes differ in FITNESS. In an evolutionary context, fitness is measured only by a genotype's rate of increase relative to other genotypes, not by attributes that enter into our everyday uses of the term (such as "physical fitness"). Consider first a population of asexually reproducing genotypes, in which the rate of increase of a genotype is not complicated by recombination. If generations overlap, the rate of increase of genotype $i$ is $r_i$, the instantaneous per capita growth rate that we encountered in the expression for population growth, $dN/dt = rN$ (Chapter 2). The growth rate $r$ depends on the values of $l_x$ (the probability of survival to age $x$) and $m_x$ (the average fecundity at age $x$). If $l_x$ and $m_x$ are measured for ages $x = 1, 2, \ldots L$, where $L$ is the maximum life span, $r$ is found from the equation

$$\sum_{x=1}^{L} l_x m_x e^{-rx} = 1$$

If generations do not overlap, the replacement rate $R$ ($\approx e^r$) measures a genotype's fitness (Figure 1). $R$ is calculated as the product of the genotype's mean fecundity times the probability of survival to reproductive age.

For sexually reproducing genotypes, fitnesses are considerably more complex to calculate (see Crow and Kimura 1970), because the frequency of a genotype in each generation depends on the survival and fecundity of all those genotypes which may give rise to it by interbreeding. The fitness of a sexual genotype may be approximated by measuring its values of $l_x$ and $m_x$ and calculating the rate of increase $r_i$ or the replacement rate $R_i$ that a hypothetical pure population of the genotype would have.

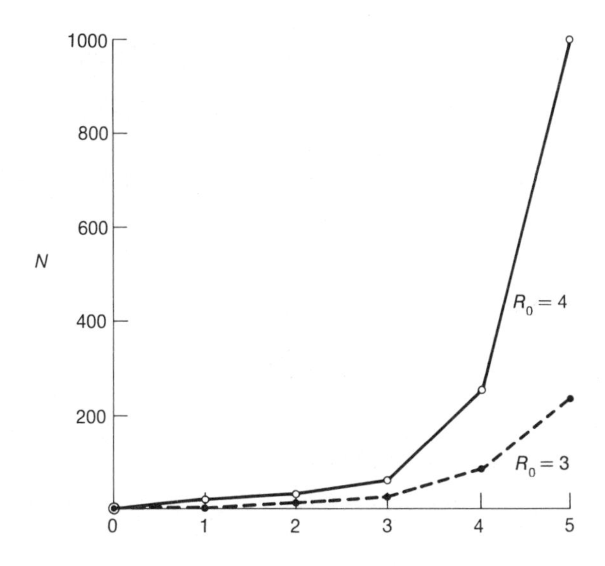


FIGURE 1

**The growth of two asexually reproducing genotypes in a population with discrete generations. The proportion of the more prolific genotype ($R_0$ = 4) rapidly becomes far larger than that of the less fecund genotype ($R_0$ = 3). The differential growth rate of the genotypes is an instance of natural selection.**

Clearly many differences between genotypes may contribute to differences in fitness. A difference in any value of $l_x$ up to the age of last reproduction may constitute a difference in fitness; but after reproduction ceases, the individuals do not contribute to subsequent generations, so any differences in survival after this age usually have no impact on their genetic fitness. Likewise, differences in fitness may arise from differences in $m_x$, the average fecundity of a female of age $x$. In addition, males have an analogue of $m_x$, namely the number of offspring they father at age $x$; thus differences in fitness may arise from genetic differences in the ability of males to obtain mates, and, indeed, in whether they succeed in mating with highly fecund or less fecund females.

The course of selection may be most easily illustrated by considering an asexual population with discrete generations, in which two genotypes $A$ and $B$, with replacement rates $R_A$ and $R_B$, number $N_A$ and $N_B$ respectively. Let the population size be $N_A + N_B = N$, and let $p = N_A/N$ and $q = N_B/N$ be the frequencies of $A$ and $B$ respectively. The rate of increase in the population as a whole is then $pR_A + qR_B = \bar{R}$.

After a single generation of population growth, the number of $A$ individuals is $N_A' = N_A R_A$, the total population size is $N' = N\bar{R}$, the new proportion of genotype $A$ is

$$p' = \frac{pNR_A}{N(pR_A + qR_B)} = \frac{pR_A}{pR_A + qR_B}$$

and the change in the proportion of $A$ is

$$\Delta p = \frac{pR_A}{pR_A + qR_B} - p = \frac{pq(R_A - R_B)}{\bar{R}}$$

Thus as long as $R_A > R_B$, $p$ increases ($\Delta p > 0$) until $A$ has replaced $B$ entirely. The rate of change depends on the relative values of $R_A$ and $R_B$, and is the same whether $R_A$ and $R_B$ have the values 2 and 1 or the values 2000 and 1000. Thus, although the absolute fitnesses are $R_A$ and $R_B$, it is often convenient to describe selection by the use of RELATIVE FITNESSES, for example by assigning a fitness value of 1 to the genotype with the highest fitness. The difference between the relative fitness of the most fit genotype and that of a less fit genotype is often denoted $s$, the SELECTION COEFFICIENT. Thus if $R_A = 200$ and $R_B = 150$, the relative fitnesses of genotypes $A$ and $B$ are 1.0 and 0.75 respectively, and $s = 0.25$. The relative fitness of a genotype $i$ is often denoted by $w_i$, and the arithmetic mean relative fitness of the individuals in a population by $\bar{w}$. (Thus $\bar{w} = \sum_i f_i w_i$, where $f_i$ is the frequency of the $i$th genotype.)

## THE EFFECT OF ENVIRONMENT ON FITNESS

Some alleles, such as those that cause embryonic death or severe developmental abnormalities, are always deleterious. But many alleles and genotypes vary in fitness, relative to other alleles or genotypes, depending on the environment. Even such a seemingly harmful allele as that coding for sickle-cell hemoglobin can be advantageous and rise in frequency if malaria is prevalent. In *Drosophila pseudoobscura*, different chromosome inversions, with names such as *Standard* (*ST*) and *Chiricahua* (*CH*), carry different alleles at some of the loci they embrace. The genotypes *ST*/*ST*, *ST*/*CH*, and *CH*/*CH* appear not to differ in fitness in the laboratory

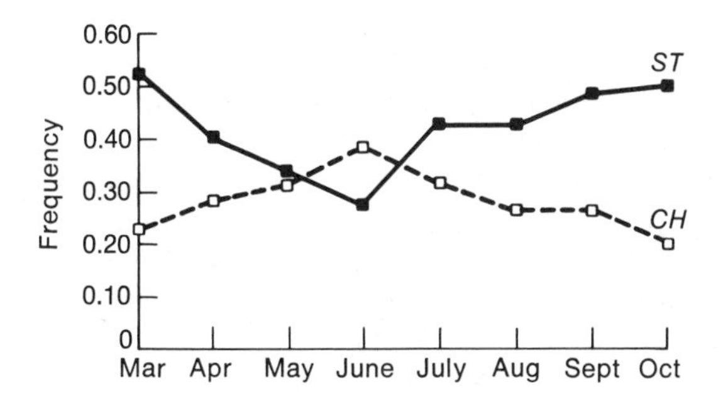


FIGURE 2
**Seasonal changes in the frequencies of two chromosome inversions in *Drosophila pseudoobscura*. (After Dobzhansky 1970)**

at 16°C, but at 25°C their relative fitnesses were estimated to be $ST/CH = 1$, $ST/ST = 0.89$, and $CH/CH = 0.41$ (Dobzhansky 1970). Consequently the frequency of an allele may increase in some localities but not others, or it may vary as environmental conditions fluctuate. The frequencies of *ST* and *CH* chromosomes, for example, change seasonally in natural populations (Figure 2).

Changes in allele frequencies are determined by immediate conditions and are not influenced by future environments. Hence there can be neither goal nor predestined direction in evolution. Natural selection cannot even shape the genetic properties of a population so as to reduce the risk of future extinction, for the genetic changes impelled by current circumstances may well prove disadvantageous in the future.

## LEVELS OF SELECTION

Under most circumstances, selection acts simply by the differential reproduction and survival of genetically different individuals within a population. In some cases, however, selection may act at the level of individual genes, which may reproduce at a rate different from that of the organisms that carry them. For example, in the phenomenon known as SEGREGATION DISTORTION or MEIOTIC DRIVE, one allele is transmitted to more than 50 percent of the gametes of a heterozygous individual; all other things being equal, such an allele will become fixed in a population. This is GENIC SELECTION. On the other hand, it is possible that populations that differ in allele frequency may become extinct at different rates, or give rise to new populations at different rates. This is GROUP SELECTION, or INTERDEMIC SELECTION. It is at least conceivable that an allele that is relatively deleterious to individual organisms may be beneficial to the population as a whole, by reducing the risk of extinction or increasing the chance of forming new populations, and so could become fixed in the species. Only in this way might a characteristic evolve that is good for the species but not for the individual. Finally, we can imagine SPECIES SELECTION, whereby speciation rates or extinction rates of whole species are affected by whether or not they possess a particular characteristic; consequently, the proportion of species in a higher taxon that bear one trait or another might change. Whether or not selection at the gene, population, or species level is important in evolution is a matter of some debate; we will treat the issue in Chapter 9. For the moment we will focus on individual selection.

## MODES OF SELECTION

The opportunity for selection exists when differences in some phenotypic characteristic result in consistent differences, on average, in rates of survival or reproduction. "On average" is a key phrase; selection may favor tall plants compared to short plants in a particular species, but not every tall plant will survive and reproduce better than every short plant. When such phenotypic differences are correlated with survival or reproduction, the opportunity for selection exists, but there is no evolutionary response to selection unless the differences in phenotype are at least partly due to genetic differences. If the phenotypic variation is caused solely by the direct impact of the environment, the distribution of phenotypes in the next generation will be the same as previously, even if some phenotypes reproduce more than others (assuming the environment is the same in both generations).

There are three major modes of selection, depending on the relationship between phenotype and fitness (Figure 3). If the relationship is monotonic, so that an extreme phenotype is most fit, selection is DIRECTIONAL. If intermediate

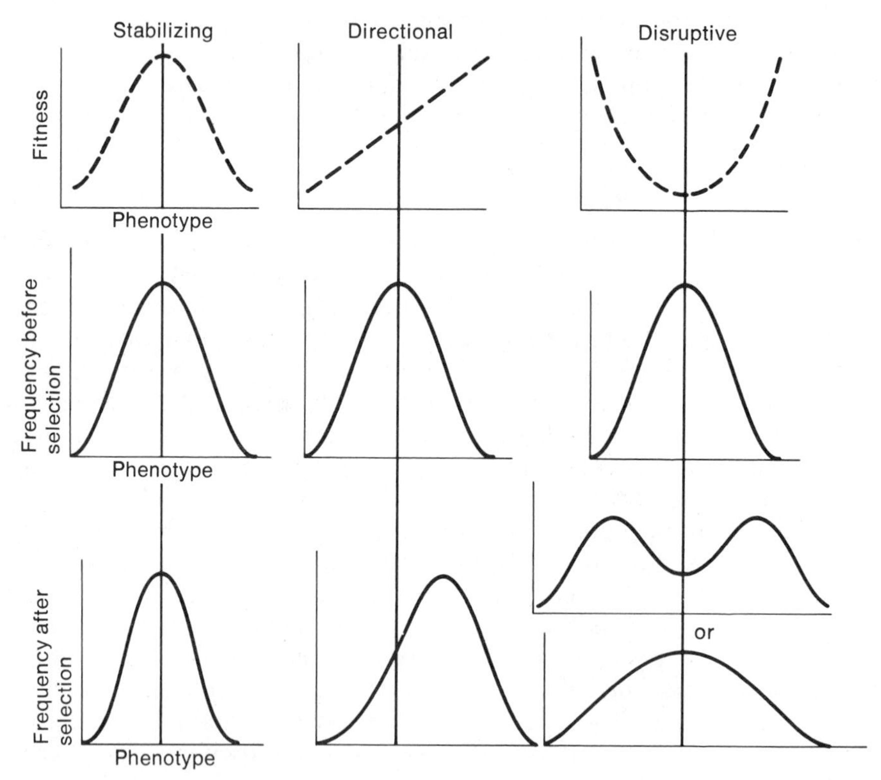


FIGURE 3

**Effects on the mean and variation of a quantitative character of three modes of selection. The relation between fitness and phenotype may be due to natural or to artificial selection, or to both acting together. (From Cavalli-Sforza and Bodmer 1971)**

phenotypes are most fit, we speak of STABILIZING or BALANCING selection. If two or more phenotypes have high fitness, but intermediates between them have low fitness, selection is DIVERSIFYING or DISRUPTIVE. In addition, sometimes the fitness of a phenotype depends on its frequency in the population (FREQUENCY-DEPENDENT SELECTION), so that there is no simple, constant relation between phenotype and fitness. It is convenient, in fact, to distinguish models of selection in which fitnesses are constant within a given environment, from models, such as frequency-dependent selection, in which fitness is variable. But even for models of constant fitness, the outcome of any particular mode of selection depends on the relationship, the "mapping," between phenotype and genotype. The remainder of this chapter will treat the simplest models, in which phenotypic differences are caused by only a single locus.

## CONSTANT FITNESSES AND DIRECTIONAL SELECTION

If there are two alleles at a locus, and one of the homozygous genotypes has the highest fitness, selection is directional, and the advantageous allele will be fixed in the population if no other factors intervene. For example, suppose allele $A$ is advantageous and is fully dominant with respect to fitness, meaning that the fitnesses of $AA$ and $AA'$ are equal and higher than that of $A'A'$. If genotypes $AA$, $AA'$, and $A'A'$ have frequencies $p^2$, $2pq$, and $q^2$ respectively, and if we denote their fitnesses as 1, 1, and $1 - s$ respectively, the change in the frequency of $A$ after one generation of selection will be

$$\Delta p = spq^2/(1 - sq^2)$$

(Note that this is the complement of the decrease in frequency of $A'$, i.e., $\Delta q$, given in Box A.)

This expression shows that the advantageous allele increases in frequency ($\Delta p$ is positive), at a rate that is proportional to the strength of selection ($s$) and to the frequency of both the advantageous allele ($p$) and that of the deleterious allele that is being replaced ($q$). In this case, in fact, the rate of genetic change will be greatest when $p = 1/3$. This is a particular example of an important general principle: the rate of genetic change is great only if both alleles are common in the population (Box A). An advantageous allele will increase much more slowly if it has just arisen by mutation (so that $p$ is very small) than if it is common. If the environment changes so that the genotype $AA$, formerly not advantageous, becomes the fittest genotype, the population will adapt to the change more rapidly if $A$ is at intermediate frequency than if it is rare. This shows the crucial importance of genetic variability, as measured by heterozygosity, and indicates why evolutionary biologists have been so concerned with discovering the amount of genetic variation and explaining why it exists.

The equilibrium allele frequency is found by setting $\Delta p = 0$ and solving for $p$; it is evident that in the case of an advantageous dominant allele, the equilibrium frequency is $\hat{p} = 1$: the allele is fixed, and the deleterious recessive allele is eliminated. However, when a deleterious recessive becomes rare, $\Delta q$ is very small, so elimination proceeds slowly; in fact, because rare recessive alleles occur primarily in heterozygous condition, in which they are masked by advantageous dominant alleles, they are seldom eliminated altogether by selection. Natural populations should therefore have many deleterious recessive alleles in very low

## A *Allele Frequency Change by Natural Selection*

Suppose two alleles are in Hardy-Weinberg equilibrium and have the following frequencies and fitnesses in a sexual population with discrete generations:

| Genotype | $AA$ | $AA'$ | $A'A'$ |
|---|---|---|---|
| Frequency | $p^2$ | $2pq$ | $q^2$ |
| Fitness ($w$) | 1 | $1 - s_1$ | $1 - s_2$ |

The *average fitness* of individuals in the population is

$$\bar{w} = p^2 + 2pq(1 - s_1) + q^2(1 - s_2)$$
$$= 1 - 2s_1q + 2s_1q^2 - s_2q^2$$

Note that

$$d\bar{w}/dq = -2s_1 + 4s_1q - 2s_2q$$
$$= 2(-s_1 + 2s_1q - s_2q)$$

Assume that selection occurs by prereproductive mortality; after selection has operated, the frequency of $A'$ will be the proportion of $A'$ genes that survive, divided by the proportion of all genes that survive ($\bar{w}$), or

$$q' = \frac{\frac{1}{2}(2pq)(1 - s_1) + q^2(1 - s_2)}{1 - 2pqs_1 - q^2s_2}$$

The change in allele frequency is

$$\Delta q = q' - q = \frac{pq(2s_1q - s_1 - s_2q)}{1 - 2pqs_1 - q^2s_2}$$

The term in parentheses is $\frac{1}{2}d\bar{w}/dq$, and the denominator is $\bar{w}$. Making these substitutions,

$$\Delta q = \frac{pq}{2\bar{w}} \cdot \frac{d\bar{w}}{dq}$$

This is a general expression for genetic change in a single generation by natural selection when fitnesses are constant, and may be used to find expressions for particular cases. Note that the magnitude of allele frequency change ($\Delta q$) is proportional to the intermediacy of allele frequencies ($pq$). This expression, although adequate when selection acts by viability differences, is not accurate if selection operates by fertility differences, which can create deviations from Hardy-Weinberg proportions (Bodmer 1965).

1. ADVANTAGEOUS ALLELE DOMINANT, DELETERIOUS ALLELE RECESSIVE.

Supposing that allele $A'$, with frequency $q$, is recessive and deleterious, the fitnesses of genotypes $AA$, $AA'$, and $A'A'$ may be written as 1, 1, and $1 - s$, respectively. Then $\bar{w} = p^2(1) + 2pq(1) + q^2(1 - s) = 1 - sq^2$, and $d\bar{w}/dq = -2sq$. Substituting this into the general equation, we find

$$\Delta q = \frac{-spq^2}{1 - sq^2}$$

2. ADVANTAGEOUS ALLELE PARTIALLY DOMINANT, DELETERIOUS ALLELE PARTIALLY RECESSIVE.

Let $h$ describe the degree of dominance with respect to fitness, so that the fitness of $A'A'$ is $1 - s$ and that of $AA'$ is $1 - hs$. If $h = 0$, allele $A$ is dominant; if $h = 1$, $A'$ is dominant; if $h = \frac{1}{2}$, neither allele is dominant. Letting the fitnesses of $AA$, $AA'$, and $A'A'$ be 1, $1 - hs$, and $1 - s$, respectively, $d\bar{w}/dq = -2hs + 4qhs - 2sq$; substituting into the expression for $\Delta q$ and omitting the tedious algebra, we have

$$\Delta q = \frac{-spq[q + h(p - q)]}{1 - sq(2hp + q)}$$

The main point to grasp from this expression is that the magnitude of the numerator, and hence of $\Delta q$, increases with $h$, so the degree of dominance affects the rapidity of allele frequency change. A rare deleterious allele ($q < \frac{1}{2}$) will decrease in frequency more rapidly, the more dominant it is.

frequency, as is in fact the case (Chapter 4). A deleterious dominant allele has no such protection, however, and such alleles are seldom found in natural populations, although they have been observed to come into existence by mutation in the laboratory. Just as a rare deleterious allele is eliminated more rapidly if it is dominant than if it is recessive, rare advantageous alleles increase more rapidly if they are dominant (Box A; Figure 4). Thus if an advantageous phenotype arises independently by dominant or by recessive mutations, a dominant mutation is more likely to be fixed.

Finally, it should be pointed out that as the advantageous genotype increases in frequency in the population, the average fitness of individuals in the population (denoted $\bar{w}$) increases. Under a wide range of conditions, the rate of increase of mean fitness is approximately equal to the additive genetic variance (see Chapter 4) in fitness, which in turn is directly proportional to the frequency of heterozygotes (Box B). This relationship was determined by R.A. Fisher (1930), who called it the "fundamental theorem of natural selection."

## Examples of directional selection

Several investigators have measured the viability and fecundity of different genotypes in the laboratory, in organisms such as *Drosophila* and the flour beetle *Tribolium*. It is possible to allow a polymorphic population to grow in a container, termed a "population cage," to which food is added periodically, and to trace changes in allele and genotype frequencies over the course of generations. In some cases, at least, the allele frequencies change at the rate predicted from the estimated fitness values; an example is the decline in frequency of a recessive

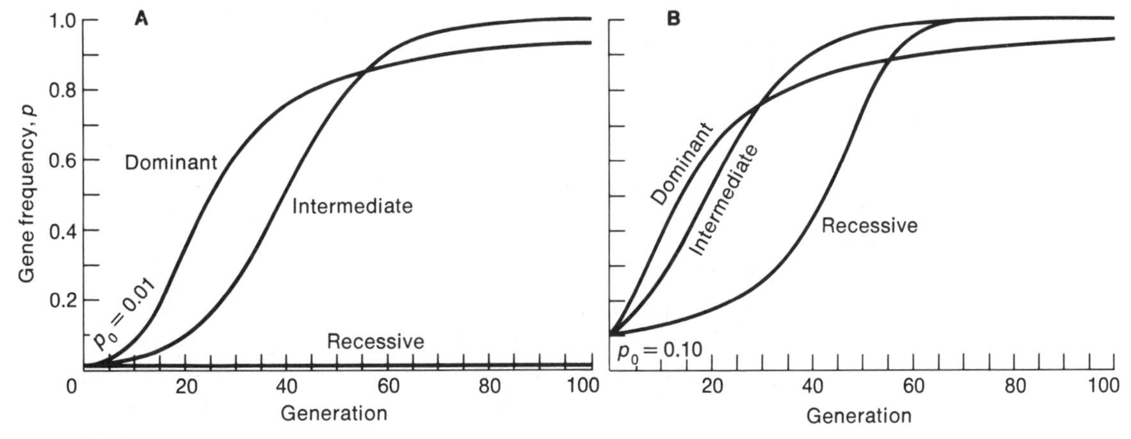


FIGURE 4

**Increase of an advantageous allele from an initial frequency of (A) $p_0 = 0.01$, as if recently arisen by mutation, and (B) $p_0 = 0.10$, as if already present as a polymorphism. Change in allele frequency is shown when the advantageous allele *A* is dominant with respect to fitness; neither dominant nor recessive ("intermediate") to its allele *A′*; and recessive. Fitness values of *AA*, *AA′*, and *A′A′* are respectively 1, 1, 0.8 for the dominant case; 1, 0.9, 0.8 for the intermediate case; and 1, 0.8, 0.8 for the recessive case. Note that a newly arisen recessive mutation hardly increases at first because it is so rarely exposed in homozygous form. The intermediate allele reaches fixation before either the dominant or the recessive because the dominant reaches fixation only as the deleterious recessive *A′* is eliminated—a very slow process. The final approach to fixation of the recessive is very rapid because a deleterious dominant *A′* is being eliminated, a rapid process. Notice that a new advantageous dominant mutation is fixed more rapidly than an equally advantageous recessive, a possible reason for the prevalence of dominant alleles in natural populations. (Computer simulation courtesy of J.P.W. Young)**

## B *Increase of Fitness Under Natural Selection*

Suppose the fitness of the heterozygote is precisely intermediate between that of the homozygotes, so that fitness is additively inherited. Denote the fitnesses of $AA$, $AA'$, and $A'A'$, the frequencies of which are $p^2$, $2pq$, and $q^2$, by $1 + s$, 1, and $1 - s$ respectively. Then the variance in fitness is $V = 2pqs^2$ (Box C in Chapter 5). Also, the change in allele frequency $\Delta q$, which if selection is weak may be written as $dq/dt$ (the continuous analogue of $\Delta q$), is

$$\frac{dq}{dt} = \frac{-spq}{1 + s(1 - 2q)}$$

Since $\bar{w} = 1 + s(1 - 2q)$, $d\bar{w}/dq = -2s$.
Now the rate of change of the average fitness, $\bar{w}$, is

$$\frac{d\bar{w}}{dt} = \frac{dq}{dt} \cdot \frac{d\bar{w}}{dq}$$

By substituting the expressions for $dq/dt$ and $d\bar{w}/dq$, we find that

$$\frac{d\bar{w}}{dt} = \left(\frac{-spq}{1 + s(1 - 2q)}\right)\left(-2s\right) = \frac{2pqs^2}{1 + s(1 - 2q)}$$

Thus the rate of change of average fitness is directly proportional to the additive genetic variance in fitness, $2pqs^2$. R. A. Fisher (1930), using a different derivation, concluded that in general the rate of change in fitness is actually equal to the additive genetic variance in fitness, a statement that he termed the "fundamental theorem of natural selection."

lethal allele in experimental populations of *Tribolium castaneum* (Dawson 1970; Figure 5).

The best known and most carefully studied case of directional selection at a single locus in natural populations is the increase in frequency of the black form of the moth *Biston betularia* in England. Museum collections provided evidence that it increased in frequency from about 1 percent to more than 90 percent in some areas within less than a century after the onset of the Industrial Revolution. The allele for black coloration is not quite completely dominant over that for gray coloration. From the rate of change of allele frequency, it may be calculated that the black form had at least a 50 percent selective advantage in some areas, and Kettlewell (1955) has shown that the forms differ greatly in susceptibility to predation by birds. The black form has not become fixed in most localities, however, and it is clear that its frequency is affected by other factors than predation alone; for example, viability differs among the genotypes even in the absence of predation (Lees 1981).

The evolution of "industrial melanism" in *Biston betularia* and a number of other species of moths is one of many examples of rapid evolutionary change in response to human alteration of the environment (Bishop and Cook 1981). Populations of hundreds of species of insects have evolved resistance to various insecticides (Wood 1981), some plants have become resistant to toxic heavy

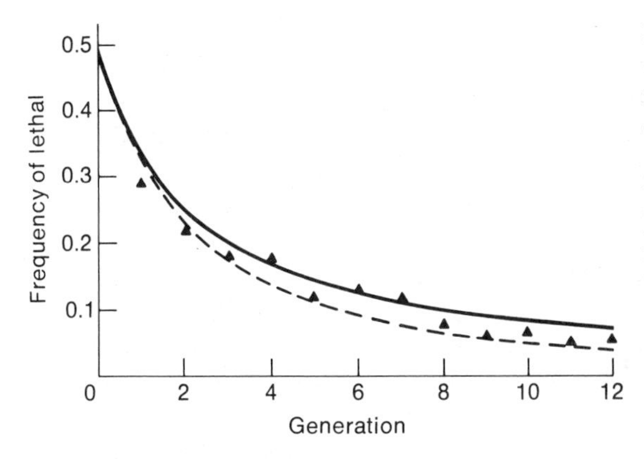


FIGURE 5

**Observed decrease, represented by triangles, in the frequency of a recessive lethal allele in a laboratory population of the flour beetle *Tribolium castaneum*, compared to two theoretical curves. The solid line is the expected change in frequency of a completely recessive lethal; the broken line is that of a lethal that lowers the fitness of heterozygotes by 10 percent. (From Dawson 1970)**

metals in the vicinity of mines (Antonovics et al. 1971), and many species of weeds have developed resistance to herbicides. In some cases, resistance is attributable to the action of several genes (Chapter 7). A single gene substitution, however, confers resistance to triazine in the pigweed, *Amaranthus hybridus*, one of about 28 species that have evolved resistance to this herbicide (Hirschberg and McIntosh 1983). The herbicide inhibits photosynthesis by binding to a polypeptide, consisting of 317 amino acids, in the chloroplast membrane. The gene that encodes this polypeptide has been sequenced, and proves to differ at only three nucleotide positions between resistant and susceptible strains. Two of the nucleotide substitutions are synonymous; resistance appears to be due to a single nucleotide substitution that changes serine to glycine in the polypeptide of the resistant form. It is interesting that the same amino acid substitution appears to confer triazine resistance in an unrelated plant, the nightshade *Solanum nigrum*.

## ACCOUNTING FOR GENETIC VARIATION

If selection acts at a locus, and a homozygous genotype is most fit, the less advantageous alleles should be eliminated, and the population will be monomorphic. However, many loci are variable in virtually every population that has been examined (Chapter 4). Much of the effort of population geneticists has been devoted to accounting for the existence of this variation. There are five classes of explanations for the genetic variation we observe:

1. The genotypes do not actually differ appreciably in fitness, and allele frequencies are accounted for by genetic drift (Chapter 5).
2. The locus is not at equilibrium, and is in a state of TRANSIENT POLYMORPHISM either because selection is driving one allele toward fixation or because the locus is closely linked to other polymorphic genes (Chapter 7).
3. Fixation by selection is counteracted by mutation.
4. Fixation by selection is counteracted by gene flow.
5. Selection acts on the locus in such a way as to maintain a stable polymorphism. This is the "balanced view" of genetic variation, which Dobzhansky (1955) espoused and contrasted with the "classical view", which assumed that one genotype, homozygous for the "wild type" allele, is most fit.

### Selection and recurrent mutation

If a deleterious allele arises repeatedly by mutation at a rate $u$, its frequency will ultimately reach an equilibrium between input from mutation and loss by selection. For example, the equilibrium frequency of a deleterious recessive allele is $\hat{q} = (u/s)^{1/2}$, where $s$ is the coefficient of selection against $A'A'$ (Box C). If selection is weak, the allele will be rather common; if strong, rare. For example, if $u = 10^{-5}$ and $s$ is only $10^{-4}$, then $\hat{q} \approx 0.32$ at equilibrium; if $s = 10^{-2}$, then $\hat{q} \approx 0.1$. A recessive lethal ($s = 1$) should have an equilibrium frequency of $\sqrt{u}$, or about 0.003, in large populations. Recurrent mutation may explain the existence of rare variants, such as albinos in many species, or the deleterious alleles in *Drosophila* populations. As we noted in Chapter 4, there is evidence that some of these alleles may actually confer superiority when heterozygous (e.g., Dobzhansky et al. 1960), but that most are deleterious both in homozygous and heterozygous condition (Simmons and Crow 1977).

### Selection and gene flow

In much the same way that a locally deleterious allele can be reintroduced into a population by mutation, it may be repeatedly introduced by gene flow from other populations in which it may be advantageous. If within a population, selection against the allele is weak relative to the rate at which it enters by gene flow, the deleterious allele can persist at an appreciable frequency. Indeed, a population that inhabits a very small patch of a distinctive environment will not be able to differentiate genetically in response to local selection pressures if it is flooded with immigrants from surrounding populations in which different alleles are favored (Nagylaki 1975).

Similarly, if a species is distributed over an environmental gradient that favors one allele at, say high altitude and another at low altitude, a cline in allele frequency will be established by selection. If gene flow along the gradient is appreciable, the cline will be less steep than if there were no gene flow (Slatkin

## C *Balance Between Selection and Recurrent Mutation*

If allele $A$ mutates to a deleterious recessive allele $A'$ at a rate $u$, the frequency ($q$) of $A'$ will increase by $\Delta q = up$ per generation. Selection reduces the frequency of $A'$ by

$$\Delta q = \frac{-spq^2}{(1 - sq^2)}$$

per generation (Box A). If $q$ is near 0 because $A'$ is highly deleterious, $1 - sq^2 \approx 1$, so the factors acting together give

$$\Delta q \cong up - spq^2 = p(u - sq^2)$$

At equilibrium, $\Delta q = 0$, so $0 = u - sq^2$ and the equilibrium frequency is

$$\hat{q} = (u/s)^{1/2}$$

## D *Selection and Gene Flow*

Assume that in an "island" population, $i$, allele $A'$ has frequency $q_i$ and that the genotypes $AA$, $AA'$, and $A'A'$ have fitnesses 1, $1 - s$, and $1 - 2s$ respectively in this environment. In each generation, a fraction $m$ of the breeding population in population $i$ are immigrants from a "continental" population where $A'$ has a selective advantage and has a frequency $q_m$. From Chapter 5, gene flow changes $q_i$ in each generation by the amount $\Delta q_i = m(q_m - q_i)$. Within population $i$, the change in allele frequency due to selection (from Box A) is

$$\Delta q_i = \frac{-sq_i\,(1 - q_i)}{1 - 2sq_i}$$

If $s$ is small, this is approximately $\Delta q_i = -sq_i\,(1 - q_i)$. Adding the two factors,

$$\Delta q_i = m(q_m - q_i) - sq_i\,(1 - q_i)$$

At equilibrium, $\Delta q_i = 0$, and the equilibrium frequency is given by the quadratic equation

$$\hat{q}_i = \frac{(m + s) \pm [(m + s)^2 - 4smq_m]^{1/2}}{2s}$$

This may be simplified by various approximations (Li 1976) to give the following conclusions: if $m \approx s$, $\hat{q}_i \approx (q_m)^{1/2}$; if $m >> s$, $\hat{q}_i$ approaches $q_m$; if $m << s$, $q_i \approx mq_m/s$.

When different alleles are favored in two populations that exchange migrants reciprocally, the mathematical treatment is much more complex (see Roughgarden 1979 for a brief treatment).

When a species is distributed continuously or in a linear series of populations over an environmental gradient such that $AA$ has a selective advantage $s$ on one side of a boundary and a selective disadvantage $-s$ on the other side, a cline in allele frequency will be established. The length of the cline, say the distance from the point where $p = 0.2$ to where $p = 0.8$, will be greater, the more gene flow there is; it equals $(V/s)^{1/2}$, where $V$ is the variance in the distance traveled by migrants (Slatkin 1973). Thus the lower the amount of gene flow, the steeper the cline will be.

1973, Endler 1977), and the allele frequency at many points along the gradient will not be what selection alone would dictate (Box D; Figure 6). An example of a localized cline has been described in the wind-pollinated grass *Agrostis tenuis*, in which copper-tolerant genotypes have a selective advantage in copper-impregnated soils near a mine, but are competitively inferior to nontolerant genotypes on non-copper-laden soils nearby. The frequency of copper-tolerant individuals declines rapidly away from copper-laden soils, but is still moderately high, especially downwind from the copper-tolerant population, as a consequence of the movement of pollen from tolerant plants (McNeilly 1968; Figure 7).

In many species, the rate of gene flow among populations is so much higher than the mutation rate that it must be a potent force contributing to polymorphism. For example, the blue mussel *Mytilus edulis* is polymorphic for aminopep-

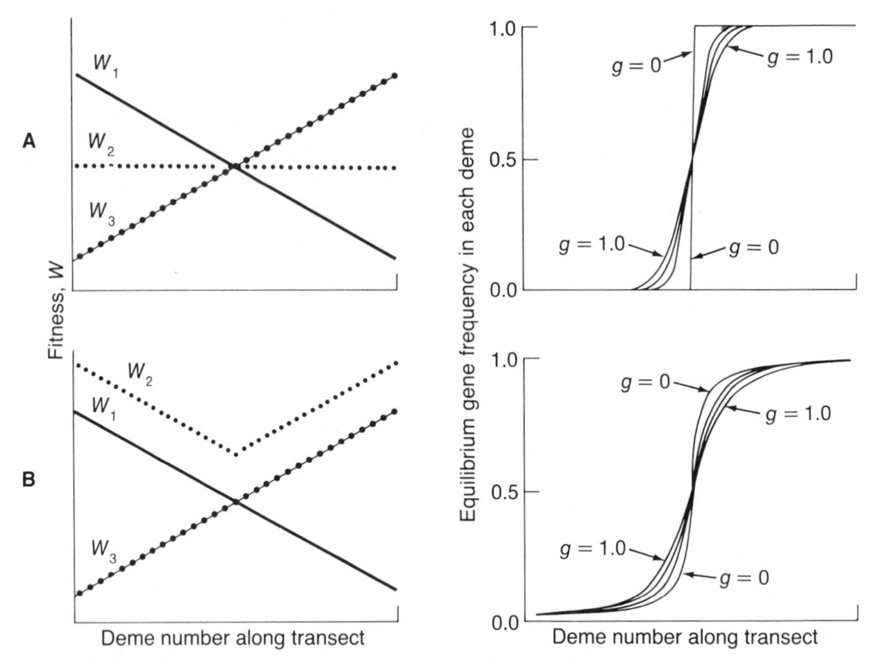


FIGURE 6

**Allele frequency clines determined by the balance between selection and gene flow. The horizontal axis in each case represents a transect along an environmental gradient that causes different relative fitnesses $w_1$, $w_2$, $w_3$ of the genotypes *AA*, *AA′*, and *A′A′*. The fitness relationships are shown on the left. In model (A), heterozygotes have intermediate fitness in each population (deme); in (B) heterozygotes are most fit. At right are the allele frequencies in each deme at equilibrium. Gene flow ($g$) consists of an equal exchange of immigrants between each deme and the neighboring demes on either side; it ranges from 0 ($g$ = 0) to 100 percent ($g$ = 1.0), when the breeding population of each deme is derived entirely from the neighboring demes. Note that in (A), when $g$ = 0 there is an abrupt change in the frequency of allele *A* from 0 to 1, even though the change in the environment is gradual. In these models the form of the cline is determined primarily by selection, even in the face of massive gene flow. (After Endler 1973)**

tidase I, coded for by the locus *Lap*. Under osmotic stress, this enzyme helps to maintain cell volume by metabolizing proteins into free amino acids. The specific activity of the allozyme coded by the allele $Lap^{94}$ is higher than that of other allozymes, and this allele has a higher frequency in populations exposed to oceanic salinity than in low-salinity estuaries, where $Lap^{94}$ has a selective disadvantage because it imposes too great a drain on the animal's nitrogen economy (Koehn et al. 1983). In each generation, the frequency of $Lap^{94}$ in estuarine populations is reduced by mortality, but it is increased each spring by the incursion of vast numbers of larvae from the oceanic population (Koehn et al. 1980). It is likely that immigration maintains the allele at higher frequencies in the estuarine population than if selection alone were acting.

## Heterozygous advantage

Several forms of natural selection can, in themselves, maintain stable polymorphism. The simplest such form of selection is HETEROZYGOUS ADVANTAGE, also

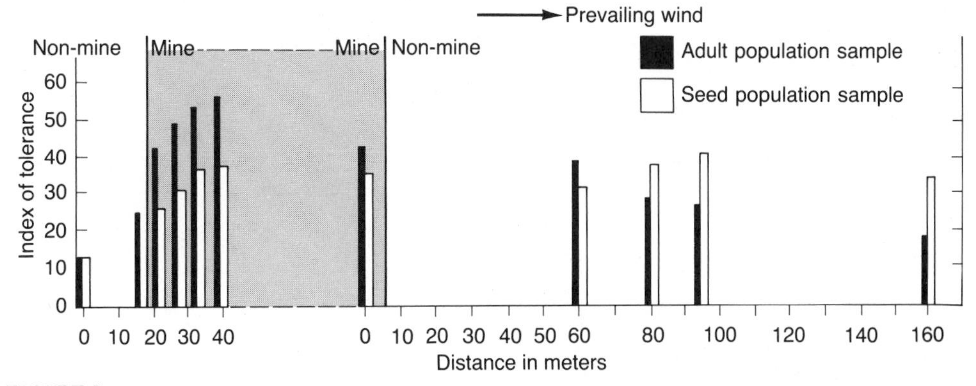


FIGURE 7

**Genetic variation in average copper tolerance of samples of the grass *Agrostis tenuis* along a transect through the edge of a copper mine. The copper-impregnated part of the transect is shaded. Note that although the average copper tolerance is higher in the copper-rich area, it is also high in the grasses immediately downwind due to gene flow from the copper area. The white and black bars represent individuals grown from seed and individuals taken as adults from the site. Adults are more tolerant in the mine area than if they had grown from a random sample of seed; this indicates there is selection against nontolerant genotypes arising from gene flow. In the nonmine area, copper tolerance is selected against: adults are less tolerant than if they had grown from a random sample of seed. (From Macnair 1981)**

called overdominance for fitness or heterosis for fitness. If the heterozygote is fitter than either homozygote, both alleles are maintained in the population by the superior survival or reproduction of the heterozygote, and random mating will regenerate both homozygous genotypes even as they are being eliminated by natural selection. If the relative fitnesses of genotypes $AA$, $AA'$, and $A'A'$ are $1 - s$, 1, and $1 - t$ respectively, the gene frequencies converge from any initial values to the stable equilibrium $\hat{p} = t/(s + t)$, $\hat{q} = s/(s + t)$ (Box E; Figure 8). The position of the equilibrium depends on the balance of fitness of the two homozygotes.

One of the few well-documented cases of heterozygous superiority at a single locus is that of sickle-cell hemoglobin. Persons homozygous for $Hb^S$ suffer severe anemia and usually die before reproducing, so $t \approx 1$. Heterozygotes have slight anemia (sickle-cell trait), but in parts of Africa where falciparum malaria is prevalent, their survival rate is higher than that of homozygotes for normal hemoglobin $Hb^A$, who suffer a disadvantage of $s \approx 0.15$. The expected frequency of $Hb^S$ is therefore about $\hat{q} = 0.13$, which is about the frequency actually observed in parts of West Africa (Allison 1961, Cavalli-Sforza and Bodmer 1971). Several other hemoglobin variants, such as $Hb^C$ and thalassemia, also appear to have heterozygous advantage under malarial conditions in human populations.

Apparent heterozygous advantage has been described for a number of other polymorphic genes, such as a color polymorphism in the copepod *Tisbe reticulata* (see Chapter 4). *Drosophila pseudoobscura* that are heterozygous for chromosome inversions sometimes display higher viability, fecundity, or mating ability than chromosome homozygotes (Dobzhansky 1970). It is often observed that relatively heterozygous individuals are fitter than more homozygous individuals. For ex-

## E *Selection in Favor of Heterozygotes*

Let the genotypic frequencies and fitnesses at a locus be

| Genotype | $AA$ | $AA'$ | $A'A'$ |
|---|---|---|---|
| Frequency | $p^2$ | $2pq$ | $q^2$ |
| Fitness | $1 - s$ | 1 | $1 - t$ |

The change in allele frequency per generation is

$$\Delta q = q' - q = \frac{pq + (1 - t)q^2}{(1 - s)p^2 + 2pq + (1 - t)q^2} - q$$

which after some algebra becomes

$$\Delta q = \frac{pq(sp - tq)}{1 - sp^2 - tq^2}$$

$\Delta q$ is either positive or negative, depending on the sign of $(sq - tq)$, so $q$ can increase or decrease toward equilibrium. At equilibrium, $\Delta q = 0 = sp - tq$. Solving for $q$ (recalling that $p = 1 - q$), the equilibrium frequencies are

$$\hat{q} = \frac{s}{s + t} \text{ and } \hat{p} = \frac{t}{s + t}$$

Note, further, that $\overline{w} = 1 - sp^2 - tq^2 = 1 - s + 2sq - sq^2 - tq^2$. Then $d\overline{w}/dq = 2(s - sq - tq)$. The value of $q$ that maximizes $\overline{w}$ is found by setting $d\overline{w}/dq = 0$ and solving for $q$. This value is $q = s/(s + t)$. Thus, as is generally the case when genotypic frequencies are constant, average fitness is maximized when the allele frequency is at its stable equilibrium. In this case there is only one stable equilibrium, but in more complex cases there are multiple stable equilibria, some of which may correspond to lower average fitnesses than others (Box F).

ample, the mean growth rate of oysters (*Crassostrea virginica*) is positively correlated, and the variance is negatively correlated, with the number of allozyme loci for which they are heterozygous (Zouros et al. 1980).

Despite such observations, it is very difficult to show that specific loci confer higher fitness when they are heterozygous, because closely linked loci that are not directly observed may cause the difference in fitness. Suppose that at the observed locus, $A$, the fitnesses of genotypes $AA$, $AA'$ and $A'A'$ are 1, 1, and $1 - s_1$ respectively, and that at a closely linked locus the genotypes $BB$, $BB'$, and $B'B'$ have fitnesses 1, 1, and $1 - s_2$ respectively. If there is linkage disequilibrium, so that allele $A'$ is generally associated with $B$, and $A$ with $B'$, the homozygotes $AA$ and $A'A'$ will often actually represent genotypes $AAB'B'$ and $A'A'BB$. These genotypes may have fitnesses $1 - s_2$ and $1 - s_1$ respectively, whereas the fitness of the heterozygote ($AB'/A'B$) will be 1. Thus there will be "pseudo-overdominance" at the $A$ locus, caused by the superiority of the two-locus heterozygote, even though neither locus individually is overdominant. Close linkage of a selectively neutral (or even deleterious) allele to a superior allele at a neighboring locus can temporarily maintain a neutral allele in a population, and can increase its frequency by "hitchhiking" with the superior allele as the latter increases in frequency (Maynard Smith and Haigh 1974).

In principle, it is possible for a single locus to be overdominant for fitness if the heterozygote has average superiority over a variety of environmental conditions, even though it is not superior in any one environment (MARGINAL OVERDOMINANCE). Dobzhansky and Levene (1955) tested the viability at several temperatures and on several food media of *Drosophila* strains that were heterozygous versus homozygous for whole chromosomes, and found that although some of the homozygotes were superior under certain conditions, survival of the heterozygotes was more uniform, and on average higher, over the range of environments. The environmentally-caused variance ($V_E$) in morphological traits, growth rates, and other characters is frequently higher in homozygotes than in heterozygous genotypes, which appear to be better buffered against environmental perturbations of development (Lerner 1954, Zouros et al. 1980).

It is possible to devise models whereby heterogeneity of temperature or of internal cellular conditions could cause heterozygotes for an enzyme to be most fit, even if the enzyme activity of the heterozygote is intermediate between that of the homozygotes under any one condition (Gillespie and Langley 1974, Berger 1976). Some authors have postulated that in enzymes that form dimers, the activity of a heterodimerous enzyme, a combination of polypeptides produced by different alleles, might transcend that of homodimers produced by homozygotes. For example, in alcohol dehydrogenase of maize, the homodimer of $C^mC^m$ homozygotes has low activity but high stability, the homodimer of *FF* homozygotes has high activity and low stability, but the heterodimer in $C^mF$ heterozygotes has both high activity and high stability (Schwartz and Laughner 1969). In general, however, multimeric enzymes are less polymorphic than monomeric enzymes (Kimura 1983a), which suggests that the assembly of subunits coded by different alleles seldom provides a basis for overdominance.

Although heterozygous advantage is theoretically a potent force for maintaining polymorphism, and has been sought by many workers, there are few unequivocal examples of single loci that display this effect. Moreover, it seems unlikely that this mode of selection could maintain all the allozyme variability in natural populations even in principle. Haploid populations of *Escherichia coli*, for example, are genetically variable (Chapter 5). Moreover, the theory of heterozy-

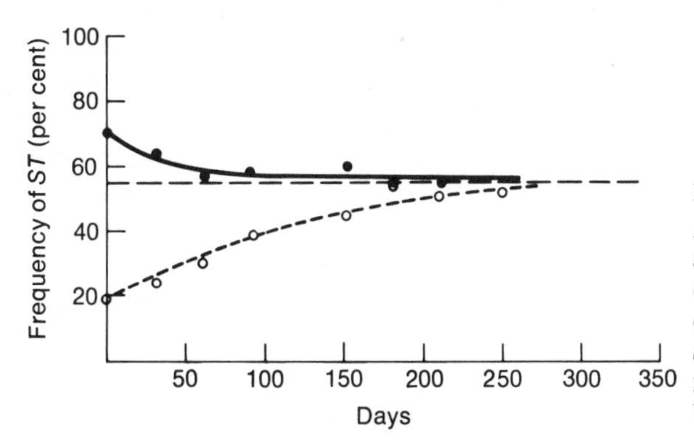


FIGURE 8

**Approach to equilibrium when heterozygotes are most fit. Points are frequencies of *Standard* chromosomes in two laboratory populations of *Drosophila pseudoobscura* carrying both *Standard* and *Arrowhead* chromosomes, which are most fit in heterozygous condition. (From Wallace 1968a, after Dobzhansky 1948)**

gous advantage is quite complex when there are more than two alleles, as is the case for many allozyme loci. Seldom will more than two or three alleles be maintained as a stable polymorphism, unless all the heterozygotes have much the same fitness and are far superior to all the homozygotes (Wright 1969; Lewontin et al. 1978).

### Frequency-dependent selection

In the models we have considered so far, the fitness of each genotype is constant within a given environment. There are numerous cases, however, in which fitness varies. One of the most important of these is FREQUENCY-DEPENDENT SELECTION (Ayala and Campbell 1974), in which the fitness of a genotype (or of an allele) is affected by its frequency within the population. One way of viewing this is to consider that the allele frequency itself alters the selective environment within which the genotypes function. The simplest model is one in which allele $A$ is dominant and the fitness of each genotype depends on the phenotype frequencies. Thus we might write the fitness of both $AA$ and $AA'$ as $1 + s_1(1 - q^2)$, so that fitness is a function of the frequency, $1 - q^2$, of the dominant phenotype. Similarly, the fitness of the recessive homozygote $A'A'$ might be $1 + s_2(q^2)$. If $s_1$ and $s_2$ are positive, there is positive frequency dependence, so that a genotype enjoys an increased advantage if it conforms to the majority phenotype. Either allele $A$ or $A'$ will then have a selective advantage if it is sufficiently frequent, and will then become fixed. It is easy to imagine such situations: for example, predators that learn to avoid distasteful prey on the basis of their aposematic (warning) coloration are more likely to encounter common than rare phenotypes, so prey with a common aposematic phenotype are more likely to have a selective advantage than those with rare aposematic phenotypes.

Conversely, if in this model $s_1$ and $s_2$ are negative, each phenotype is more greatly favored when rare than when common, and this will clearly lead to a stable polymorphism (at which $\hat{q} = [s_1/(s_1 + s_2)]^{1/2}$). Many alleles may persist in polymorphic condition if such INVERSE FREQUENCY DEPENDENCE applies to many different genotypes. It is interesting that when selection is frequency dependent, the average fitness ($\bar{w}$) does not necessarily increase to its theoretical maximum, and in fact may decline.

So far, there is no clear evidence that any allozyme polymorphisms are maintained by frequency-dependent selection (Kimura 1983a), but numerous other polymorphisms are governed by this mechanism. Indeed, it is likely that there is a frequency-dependent component in virtually all selection that operates in natural populations, for interactions among members of a population affect the selective advantage of almost all traits, and such interactions usually give rise to frequency-dependent effects. For example, if different genotypes are differentially adapted to two different limiting resources, competition favors rare over common genotypes. Suppose that plants of genotypes $A$ and $B$ are competitively superior in moist and dry patches of soil respectively, but that genotype $A$ is initially rare, so that moist patches are occupied primarily by plants of genotype $B$. Within the moist patches, individuals of the $A$ genotype are primarily competing with inferior competitors as long as $A$ is rare; but as $A$ becomes more common, it loses its advantage, because each individual of genotype $A$ is competing primarily against equally competent plants, of its own genotype. The same would occur in dry patches if $B$ is rare, as the dry-adapted genotype $B$ increases in frequency.

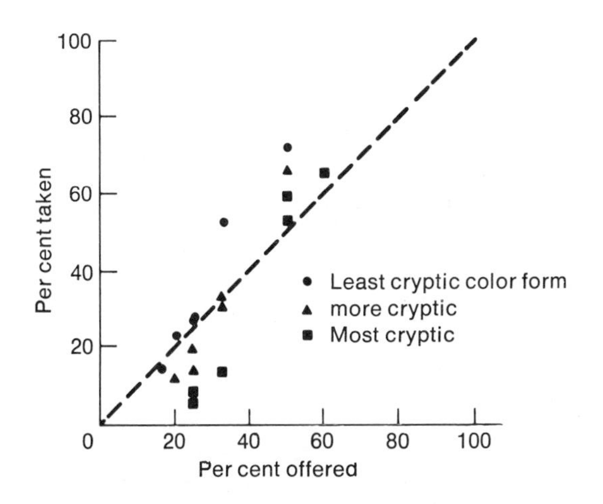


FIGURE 9

**An example of frequency-dependent selection: predation by a fish on three color forms of the corixid bug *Sigara distincta*. Each suffers disproportionately higher predation (percentage taken) when it is the more common form. Compare with Figure 6 in Chapter 2. (After Clarke 1962, based on data of Popham 1942)**

Frequency-dependent selection has been documented for a great variety of phenomena. Self-compatibility alleles in plant populations are powerfully governed in this manner (Chapter 5), because each pollen genotype can germinate on fewer stigmas as the frequency of its allele rises in the population. Because many predators form a "searching image" for the most common type of prey and tend to ignore rarer phenotypes, polymorphism in prey can be maintained by predation (Figure 9). It is likely that polymorphisms in both parasites (including pathogenic bacteria and fungi) and their hosts are maintained in this fashion, if each genotype of the host is resistant to certain genotypes of the parasite, and each genotype of the parasite can attack only certain genotypes of the host (Clarke 1976, May and Anderson 1983). Although some such "gene for gene" systems in parasites and their hosts have been described (see Chapter 16), this is a largely unexplored topic. Mating preferences, too, may have a frequency-dependent effect, as in some species of *Drosophila* in which females mate preferentially with less common genotypes of males (Ehrman 1967). Most of the models of the evolution of phenomena such as sex ratio and aggression and other behavioral traits are models of frequency-dependent selection (Chapter 9).

### Heterogeneous environments

Although variation in the environment can sometimes maintain genetic variation in a population, this is by no means inevitable. Suppose that in an insect that has a spring and a fall generation each year, the genotypes at a locus have the following fitnesses so that different homozygotes are favored in spring and fall:

| | $AA$ | $AA'$ | $A'A'$ |
|---|---|---|---|
| Fitness in spring | 1.0 | 0.8 | 0.6 |
| Fitness in fall | 0.7 | 0.9 | 1.0 |

It is evident that because the disadvantage of the $A'$ allele in spring is not precisely balanced by its advantage in fall, $A'$ will experience a net decline in frequency over the entire year. Since there is no reason to expect a fluctuating environment to precisely balance the rises and falls in allele frequencies, fluctuating selection pressures will generally not maintain polymorphism indefinitely (Haldane and Jayakar 1963; Figure 10), although they may slow the rate at which an allele is

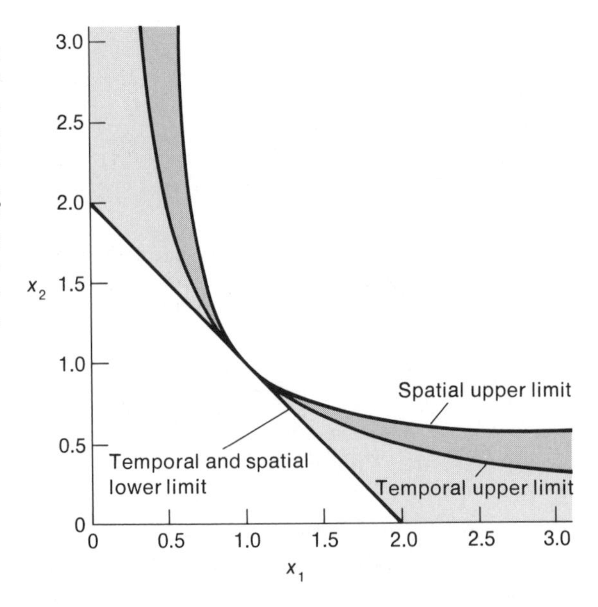


FIGURE 10

**The conditions for stable polymorphism maintained by environmental heterogeneity (multiple-niche polymorphism and polymorphism maintained by environmental fluctuations). The relative fitnesses of *AA*, *AA′*, and *A′A′* are 1, 1, and $x_1$ in environment 1; and 1, 1, and $x_2$ in environment 2. Combinations of $x_1$ and $x_2$ lying within the gray regions give polymorphism, but the set of such combinations is more restricted when the environment varies temporally than spatially. (Adapted from Hedrick et al. 1976)**

lost from a population. Thus, although the frequencies of chromosome inversions in *Drosophila pseudoobscura* fluctuate seasonally (Figure 2), the seasonal variation does not in itself explain the persistence of the polymorphism.

Let us now consider spatial variation in the environment, imagining that each genotype *i* has a fixed fitness, $w_{ij}$, in an environment of type *j*. Suppose, for example, that the fitnesses are as follows:

| | *AA* | *AA′* | *A′A′* |
|---|---|---|---|
| Fitness in environment 1 | 1.0 | 0.9 | 0.8 |
| Fitness in environment 2 | 0.8 | 0.9 | 1.0 |

If the environments are spatially segregated so that they support separate populations, one population will become fixed for allele *A* and the other for *A′*. As the rate of gene exchange between the populations increases, the fate of an allele is determined by the genotypes' fitnesses in both areas rather than in just one. If the two habitat-associated populations interbreed fully at random, and then are reconstituted by the random settling of offspring into the two habitats, allele frequencies are determined by the fitness of each genotype averaged over the two environments.

If the environment is "fine grained" (Chapter 2), so that each individual experiences both environmental states, there will be a stable polymorphism if the arithmetic mean fitness of the heterozygote is greater than that of either homozygote (i.e., if there is marginal overdominance). This is not true in the hypothetical example given above. But if each individual experiences only one environmental state, as do plants that inhabit a mosaic of soil types or insects that feed on a single plant throughout their feeding stages, the environment is "coarse-grained," and polymorphism persists only if the geometric mean fitness of the heterozygote is greatest. The geometric mean fitness of genotype *i* is

$$1/\sum_j (c_j/w_{ij})$$

where $c_j$ is the proportion of the population emerging from the *j*th environment into the mating pool (Levene 1953). In our hypothetical case, the geometric mean fitness of the heterozygote (0.900) is greater than that of the homozygotes (0.889) if the two environments are equally frequent. A MULTIPLE-NICHE POLYMORPHISM, maintained by environmental variation, is more likely if the environmental variation is coarse-grained than if fine-grained, if the environments are spatially segregated than if intermingled, and if the variation is spatial than if temporal (Figure 10). Even so, polymorphism will persist only if the selection coefficients are rather large and bear a precise relationship to the frequency of the different environmental conditions (Felsenstein 1976, Hedrick et al. 1976, Maynard Smith and Hoekstra 1980).

Among the possible examples of multiple-niche polymorphisms (Hedrick et al. 1976, Futuyma and Peterson 1985) few have been documented in detail. A likely case is in the swallowtail butterfly *Papilio demodocus* (Clarke et al. 1963; Figure 11), in which larvae that feed on umbelliferous plants generally have a different color pattern from those that feed on citrus; there is some evidence that each pattern provides better cryptic protection against predators on one or the other plant. In several species of African mosquitoes, apparently conspecific strains tend to lay eggs either indoors (in drinking cisterns) or outdoors, and differ in color pattern and in allozyme frequencies (Tabachnik et al. 1979). Whether or not they form a panmictic breeding population is not known. There

FIGURE 11

**The umbellifer-associated (left) and citrus-associated (right) forms of the larva of the butterfly *Papilio demodocus*. Each appears to suffer less predation in nature when on the plant with which it is usually associated. (Courtesy of Sir Cyril Clarke)**

is little evidence that environmental variation explains much allozyme polymorphism, or that genetic variation in morphological traits is enhanced within populations that occupy a heterogeneous habitat or use a great variety of resources (Patterson 1983, Futuyma and Peterson 1985).

Several factors can make polymorphism more likely than in the simple models of environmental variation summarized above. As we have noted previously in this chapter, if there is competition for different resources to which different genotypes are differentially adapted, selection has a frequency-dependent component that helps to stabilize the polymorphism. Polymorphism is also more likely if the genotypes actually seek the environments to which they are best adapted. It might seem obvious that animals should do this, but whereas it is clear that different species select habitats to which their morphological and physiological features suit them, there is much less evidence on whether or not different genotypes within a species can do so. If habitat choice is governed by genes different from those that control morphological or physiological traits, recombination will break down the correlation that is necessary for adaptive habitat choice (unless the loci are very tightly linked). Alternatively, each genotype might learn what kind of resource it can handle most efficiently, and develop fidelity to the patch type to which it is best adapted. There is very little evidence on this subject. Kettlewell and Conn (1977) found that industrial melanistic moths (*Biston betularia*) prefer to sit on dark surfaces, while nonmelanistic genotypes prefer light surfaces; but no such correlation was discernible in another polymorphic moth (*Phigalia titea*; Sargent 1969), nor in a color-polymorphic grasshopper (*Circotettix rabula*) that is commonly found on matching backgrounds (Cox and Cox 1974). Although some authors have suggested that female herbivorous insects become "conditioned" to lay eggs on the same kind of host they fed on as larvae (e.g., Thorpe 1956), there is little evidence for this effect, although adult insects can learn to prefer certain hosts for oviposition (Jaenike 1982).

## INFERIOR HETEROZYGOTES

Sometimes two homozygotes both have high fitness, but the fitness of the heterozygote is lower. This could be the case if two homozygotes were adapted to different resources, but the heterozygote were adapted to neither. The model applies especially to the dynamics of chromosomal mutations such as translocations, which frequently reduce fertility greatly in heterozygous condition (Chapter 3), but which do not reduce fertility in homozygous condition.

If the fitnesses of genotypes $AA$, $AA'$, and $A'A'$ are $1 + s_1$, 1, and $1 + s_2$ respectively, there is an equilibrium at $\hat{q} = s_1/(s_1 + s_2)$, but it is an unstable equilibrium: if $q$ is any less than $\hat{q}$, $\Delta q$ is negative and A′ decreases in frequency; if greater, $\Delta q$ is positive and $A'$ increases and is fixed (Box F). The ultimate genetic composition of a population in this case depends on its initial allele frequency. As is usual when fitnesses are constant, allele frequency change accomplishes an increase in the average fitness of the population, but the population moves toward either "peak" of the fitness "surface" (Figure 12), depending on the "slope" on which it initially lies. It may well arrive at the lower of two peaks; but since natural selection cannot decrease average fitness in this model, the population is stuck on the lower peak and cannot be moved by natural selection to the higher peak, which presumably represents a population monomorphic for a better geno-

## F Selection Against Heterozygotes

Suppose genotypes $AA$, $AA'$, and $A'A'$, with frequencies $p^2$, $2pq$, and $q^2$, have fixed fitnesses $1 + s$, 1, and $1 + s_2$, respectively. Then $\overline{w} = p^2(1 + s_1) + 2pq + q^2(1 + s_2) = 1 + s_1p^2 + s_2q^2$, and (substituting $1 - q$ for $p$) $d\bar{w}/dq = 2s_1q + 2s_2q - 2s_1$. Using (from Box A) the relation

$$\Delta q = \frac{pq}{2\overline{w}} \cdot \frac{d\overline{w}}{dq}$$

we find

$$\Delta q = \frac{pq(s_2q - s_1 + s_1q)}{\overline{w}}$$

or, by reentering $p$ $(= 1 - q)$,

$$\Delta q = \frac{pq(s_2q - s_1p)}{\overline{w}}$$

Note that if $s_2 = s_1 = s$, the numerator is $pqs(q - p)$. $\Delta q$ is positive (so $A'$ becomes fixed) if $s_2q > s_1p$ (or if $q > p$ when $s_2 = s_1$), and is negative (so $A'$ is eliminated) if $s_2q < s_1p$. We observe, by setting $\Delta q = 0$ and solving for $q$, that there is an equilibrium at $q = s_1/(s_1 + s_2)$, but this is an unstable equilibrium; if $q$ exceeds this equilibrium value, $s_2q > s_1p$ so $\Delta q > 0$, whereas if $q$ is less than the equilibrium value, $\Delta q < 0$. The stable equilibria are at $q = 0$ (where $\overline{w} = 1 + s_1$) and at $q = 1$ (where $\bar{w} = 1 + s_2$). Average fitness is greater at either of these monomorphic equilibria than at intermediate allele frequencies. However, the average fitness will be higher at one monomorphic equilibrium than the other if $s_1 \neq s_2$; yet the allele frequency need not arrive at the higher of the two equilibria.

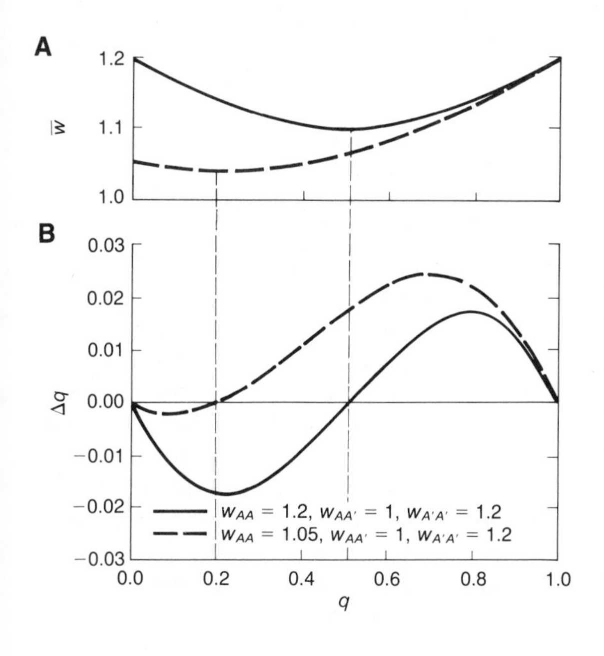


FIGURE 12

**Selection when the heterozygote is less fit than either homozygote. The solid curves represent a case in which the two homozygotes are equally fit, the broken curves a case in which the homozygotes differ in fitness. (A) The relation between mean fitness ($\bar{w}$) and the frequency $q$ of allele $A'$. Because the fitness values are constant, allele frequencies change under selection so that $\bar{w}$ increases. (B) The change in allele frequency in one generation, $\Delta q$. $\Delta q = 0$ at an intermediate allele frequency, but this equilibrium is unstable. $\Delta q$ is positive if $q$ is greater than the unstable equilibrium value, negative if less, which results in movement toward frequencies 1 and 0 respectively. The frequencies 0 and 1 are alternative stable equilibria. Note that the unstable equilibrium gene frequency is 0.5 when the homozygotes are equally fit. (After Hedrick 1983)**

type. This is the first instance we have encountered of a very important principle: a population is not necessarily driven by natural selection to the most adaptive possible genetic composition.

When heterozygotes are inferior to both homozygotes, there is often a pronounced transition (a steep cline in allele frequency) between populations dominated by the different homozygotes, because when the allele from the other population enters by gene flow, it is in the minority. Hence after one generation of mating it is primarily in heterozygous form and so is selected against. For example, in grasshoppers and in many other organisms, there are local "chromosome races" that often give way to one another over a kilometer or less (White 1978, Barton and Hewitt 1981). The *Vandiemenella viatica* group of flightless grasshoppers in Australia consists of numerous populations that are monomorphic for various pericentric inversions and chromosome fusions, heterozygotes for which have numerous aneuploid gametes. No two "chromosome races" are sympatric, but they meet in hybrid zones that are typically only 200–300 meters wide (White 1978).

## THE ADAPTIVE LANDSCAPE

In perhaps the most popular metaphor for evolutionary change, Sewall Wright (1932) envisioned a population as occupying a position on an ADAPTIVE LANDSCAPE of allele frequencies, in which peaks represent possible genetic compositions of a population for which $\bar{w}$ (average fitness) is high, and troughs represent possible compositions for which $\bar{w}$ is low (Figure 13). As gene frequencies change and $\bar{w}$ increases, the population moves uphill and comes to rest on an adaptive peak. There may be multiple peaks, some higher than others, but because selection can only increase fitness, the population cannot get from a lower peak to a higher one by natural selection alone because to do so would require moving downhill. Of course, if the environment changes, the fitnesses of the genotypes may change, as does the adaptive landscape. Genetic compositions that previously were troughs may become peaks, and natural selection then brings the population to a new peak; but the genetic composition will generally move to a local peak rather than the highest one in the landscape.

## INTERACTIONS OF EVOLUTIONARY FACTORS

We have now considered all the factors—mutation, gene flow, genetic drift, and selection—that affect the frequency of an allele in a population. Three of these, mutation, gene flow, and selection, are DETERMINISTIC: if the fitnesses, mutation rates, and rate of gene flow are the same for a number of populations that begin with the same allele frequency, all will attain the same equilibrium composition. Genetic drift, in contrast, is a STOCHASTIC factor that brings about random increase or decrease in an allele's frequency. In all real populations, genetic drift acts together with whatever selection, mutation, or gene flow may be occurring. Although selection and other deterministic factors may theoretically specify a particular equilibrium allele frequency, this or any other allele frequency $q$ will be attained with some probability, $\phi(q)$. A set of populations of the same size, with the same selective pressures, mutation rates, and rates of gene flow, would thus have a steady state distribution of allele frequencies centered around the theoretical deterministic equilibrium. Wright (1931) developed the equations for

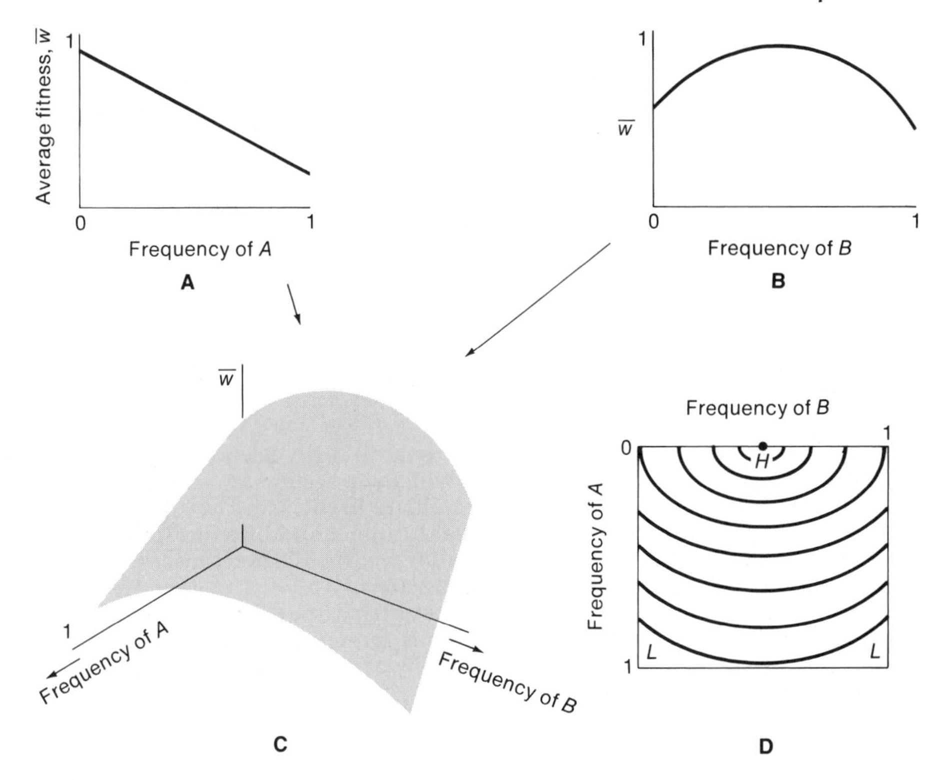


FIGURE 13

**A simple adaptive landscape. (A) Mean fitness, $\bar{w}$, declines as the frequency of allele *A* increases, at a locus at which the fitnesses are $1 - s$, $1 - s/2$, and 1 for genotypes *AA*, *AA'*, and *A'A'* respectively. (B) $\bar{w}$ is maximized when the allele frequency is ½ at the heterotic locus *B* at which genotypes *BB*, *BB'*, and *B'B'* have fitnesses $1 - t$, 1, and $1 - t$ respectively. In each of these single-locus cases, the mean fitness of the population is the average of the fitnesses of three genotypes, weighted by their frequencies. (C) These two loci considered together yield a fitness surface as shown if they do not interact epistatically. Each point on the floor of the three-dimensional space represents a population with particular allele frequencies at loci *A* and *B*; the height of the fitness surface above that point is the mean fitness of nine genotypes (*AABB . . . A'A'B'B'*) weighted by their frequencies. (D) The fitness surface in (C) is represented here as a "topographic map," or adaptive landscape, in which all points on a curve represent populations with different allele frequencies but equal mean fitness ($\bar{w}$). The population moves by natural selection toward the peak, *H*; minimal values of $\bar{w}$ are at the points *L*.**

such probability distributions, which needless to say are more complex than this introduction to the theory can bear. Figure 14 shows some examples for a deleterious allele. Notice that the probability of an allele frequency that departs substantially from the deterministic equilibrium becomes greater, as selection becomes weaker or the effective population size becomes smaller. As a general rule, it may be said that selection has the stronger influence on allele frequency if $4N_e s >> 1$, whereas genetic drift has the upper hand if $4N_e s << 1$. A weakly selected allele ($s$ small) is primarily affected by selection if the population is large,

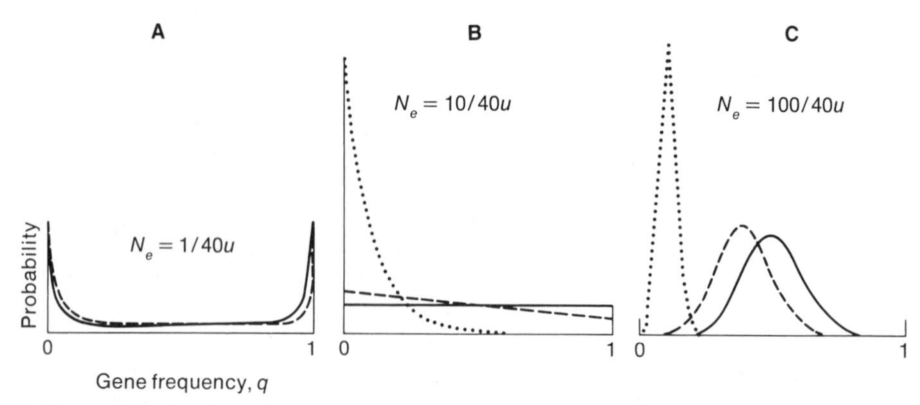


FIGURE 14

**The interaction of selection, mutation, and genetic drift. Each curve is the distribution of probabilities that a population will have frequency $q$ of a deleterious allele. Population size is small in (A), intermediate in (B), large in (C). Forward and back mutation rates are assumed to be equal ($u = v$). Solid lines indicate the selection coefficient against the allele is very small ($s = u/10$), dashed lines that it is quite small ($s = u$), and dotted lines that it is moderate ($s = 10u$). The gene frequency is is likely to be near its deterministic equilibrium (that specified by selection alone) of 0 if selection is strong and the population is large. Gene frequencies are much more variable in smaller populations, especially if selection is weak. (From Crow and Kimura 1970, after Wright 1937)**

but changes in frequency almost as if it were neutral if the population is small.

This conclusion has several important consequences. First, a mutation that has a slightly advantageous effect is less likely to be fixed in a small than in a large population. Second, the frequency of an allele may wander around the equilibrium that might be set by, say, heterozygous advantage, and may drift to fixation rather than being kept in a polymorphic state (see Robertson 1962). Third, the frequency of a deleterious allele may actually increase, if the population is small enough. These facts together mean that if the allele frequencies are at one equilibrium state in a small population, so that the population is on one of the adaptive peaks in the fitness landscape, genetic drift may carry the population "downhill" into an adaptive valley. Selection, then acting to increase fitness, may bring the population back uphill, along the slope of a different adaptive peak, perhaps a higher one (Figure 12). Thus genetic drift and selection may act in concert to accomplish what selection alone cannot, moving the genetic composition of a population from peak to peak until the highest peak—the optimal genetic composition—is attained (Wright 1977). This is part of Sewall Wright's SHIFTING BALANCE THEORY of evolution.

This theory accounts for why populations and species differ in chromosomal rearrangements such as translocations that reduce fertility when heterozygous (Wright 1941, Lande 1979, Hedrick 1981). Selection alone cannot cause such a rearrangement to increase from low initial frequency, but in a sufficiently small population, a rearrangement can increase by genetic drift to the critical frequency (about 0.5; see Box F) above which selection will cause its increase to fixation. From the proportion of related species that differ in various chromosomal rearrangements and the likely time since these species diverged from their common ancestors, Lande (1979) has estimated the rate of fixation of chromosome variants in several animal groups. Given this rate, as well as estimates of the strength of selection against heterozygotes and the rate of mutation to new chromosome variants, Lande calculated that the effective size of local populations of many mammals and insects has been in the range of a few tens to a few hundreds of individuals. In view of ecological estimates of effective population sizes (Chapter 5), these figures seem reasonable.

## POPULATION FITNESS AND GENETIC LOAD

When allele frequencies change by natural selection and the average fitness $\bar{w}$ rises, it seems reasonable to say that less well adapted genotypes are replaced by better adapted genotypes. What consequences does this process of adaptation have for the population as a whole?

The answer depends on how the trait being selected is related to the factors that regulate population growth. Recall that the equations describing the action of selection use *relative* rather than *absolute* fitness values. If fitness ($w$) is measured by a genotype's per capita rate of increase $r$, an increase in $\bar{w}$ would be manifested by an increase in the growth rate of the population if $w$ referred to absolute fitness. But since $w$ is so standardized as to measure relative fitness, one genotype may replace another, yet the growth rate of the population as a whole may not increase. It is useful to distinguish, as Wallace (1968b) has, between HARD SELECTION and SOFT SELECTION. Under hard selection, the mortality rate or fecundity of a genotype is fixed, so that an increase in $\bar{w}$ is reflected in an absolute increase in the growth rate of the population. For example, if an insect population is exposed to an insecticide that kills irrespective of population density, the size and growth rate of the population will at first be reduced, but will increase as resistant genotypes increase in frequency.

In contrast, soft selection operates when a population is regulated around an equilibrium density by density-dependent factors. Then on average the growth rate $\bar{r}$ of the population is zero, and most of the progeny in each generation die before reproducing. Nevertheless, genotypes that have higher fecundity or greater resistance to mortality factors will be disproportionately represented among the survivors. Mathematically, $\bar{w}$ increases under soft selection just as under hard selection, yet the population growth rate remains at zero, and population density does not increase as evolution proceeds. Thus in a population regulated by predation, greater fecundity or superior adaptation to temperature may evolve and the average individual in the population is better adapted, yet the population as a whole does not increase.

The distinction between hard and soft selection is important to understanding the literature on GENETIC LOAD (Muller 1950), which is defined as the difference between the average fitness ($\bar{w}$) and the fitness of the fittest genotype (whose fitness is set equal to 1). Thus the load $L = 1 - \bar{w}$. For example, a deleterious recessive allele that recurs by mutation reaches an equilibrium frequency of $\hat{q} = (u/s)^{1/2}$; $\bar{w}$ is $1 - sq^2$, so $\bar{w} = 1 - u$, and $L = u$, the mutation rate. If there were $n$ such loci, and the probability of death associated with each locus were independent of that of other loci, $\bar{w} = (1 - u)^n$, and $L = 1 - (1 - u)^n$, which could be a fairly large fraction, approaching 1 if $nu >> 1$. If these alleles were unconditionally deleterious, so that recessive homozygotes died whether the population were sparse or crowded, $\bar{w}$ would reflect absolute fitness, and a large load $L$ would imply that the growth rate, and very likely the size, of the population would be reduced. The genetic load attributed to recurrent mutations in natural populations of *Drosophila* has been estimated to be about $L = 0.01$ (Simmons and Crow 1977).

Every individual with a less-than-maximal fitness is part of the genetic load. So in addition to the mutational load we have calculated above, there are genetic loads associated with gene flow, recombination, genetic drift, and segregation, all of which increase the frequency of inferior genotypes that selection then tends to eliminate. For example, in a balanced polymorphism maintained by heterozygous advantage, $\bar{w}$ at equilibrium is $1 - [st/(s + t)]$, so the segregational load is $st/(s + t)$. For the case of the sickle-cell hemoglobin polymorphism, if $s = 0.15$ and $t = 1$, $L = 0.13$. If this represents hard selection, as it does in the sickle-cell case, 13 percent of the offspring born in each generation die before reproductive age.

If loci are in linkage equilibrium so that their effect on fitness is independent, the total load caused by polymorphisms that are maintained by hard selection is bounded by the reproductive excess in each generation, $R - 1$, where $R$ is the maximal replacement rate per generation (counting females only; see Chapter 2). With very many such loci, the mortality becomes so great that the population cannot be maintained. In the same vein, the replacement of one allele by another during directional selection entails a substitutional load, or COST OF SELECTION. For example, a breeder selecting for high milk yield in cattle must prevent the lower-yielding individuals from breeding, perhaps by slaughtering them. If the fraction killed in each generation is small, the herd remains large but genetic change is slow. If the fraction killed is large, genetic change is fast but the size or the growth rate of the herd is reduced. The cost of selection, the number of GENETIC DEATHS required to change the allele frequency from 0 to near 1, is about the same whichever alternative is chosen. To bring a favorable allele from a low frequency ($10^{-2}$) to a high frequency requires a number of deaths about 30 times the size of the population, if the population is kept at the same size (Haldane 1957). Since the number of genetic deaths per generation is limited by the species' fecundity, there can be rapid genetic change at just a few loci simultaneously or slow change at many loci. Haldane calculated that it might take about 300,000 generations for two populations to diverge completely in allele frequency at 1000 loci, assuming that the loci affect fitness independently and are always in linkage equilibrium.

If selection is soft, neither the substitutional load nor the segregational load associated with balanced polymorphisms will reduce population size, but the number of alleles substituted, or the number of polymorphisms maintained, may nevertheless be limited. Suppose we postulate that each of 300 loci is maintained polymorphic by overdominance. Then for every 100 offspring born, some must die because they are homozygous at locus $A$, some because they are homozygous at locus $B$, and so on. If selection is strong, the mortality must be great; if the mortality is less, selection is weaker and the polymorphisms are more subject to loss by genetic drift. If the population size of adults is regulated at $K$ in each generation and the fecundity per female is $R$, there are only $KR - K$ offspring per generation available to die. $R$ therefore sets the upper limit on the number of polymorphic loci at which there can simultaneously be strong balancing selection.

When Lewontin and Hubby (1966) reported that about 30 percent of the enzyme loci examined in *Drosophila pseudoobscura* are polymorphic, they pointed out that if these are a random sample of the genes, *Drosophila* must have about 3000 polymorphic loci. If each of these were maintained by heterozygous advantage, the selection coefficients $s$ and $t$ associated with each locus must be minuscule, because *Drosophila* simply isn't fecund enough to afford enough selective deaths in each generation to constitute strong selection at each of 3000 independently selected loci. However, several authors (King 1967, Milkman 1967, Sved et al. 1967, Wills 1978) argued that the genetic load argument does not apply with full force if the loci do not act independently. A whole individual lives or dies, so the death of one highly homozygous individual may account for selection at many loci simultaneously. Thus if only $K$ of the $KR$ offspring born each generation survive, but these are the most heterozygous individuals, selection may be strong enough at each locus to maintain it in a polymorphic state. But Lewontin (1974a) responded that this model predicts that inbreeding depression should be far more pronounced than it actually is. Hence if *Drosophila* is really polymorphic at thousands of loci, it is unlikely that the majority of the polymorphisms are maintained by balancing selection.

## THE NEUTRALIST–SELECTIONIST CONTROVERSY

Until the 1960s, a chief issue in population genetics was whether populations contained much or little genetic variation. That issue is now resolved, and the question has become whether most of the variation is selectively neutral and hence largely irrelevant to a population's capacity to respond to new forces of selection, or whether the genetic variants differ in fitness and so constitute the raw material for adaptation to new selective regimes. There is not a complete dichotomy in viewpoint. For example, genetic variations in a phenotypic trait such as body size could be neutral at some times, yet become significant if selection should subsequently favor a different mean size; likewise, alleles at a locus might be neutral on prevailing genetic backgrounds, yet affect fitness on different genetic backgrounds. Nevertheless, the question of whether or not much of the genetic variation we observe is maintained by selection is a major issue in evolutionary genetics (see reviews by Lewontin 1974, Nei 1975, Kimura 1983a, Koehn et al. 1983). Most of this literature has concerned the interpretation of allozymes, because it is virtually impossible to determine the frequencies of alleles that contribute to variation in most morphological and physiological traits, for which numerous loci, as well as the environment, generally contribute to the variance.

Not even the most confirmed neutralist would argue that all evolutionary changes in enzymes have occurred by genetic drift. The properties of an enzyme frequently differ among species and populations that inhabit different environments (Somero 1978); for example, a study of the kinetics of muscle lactate dehydrogenase of congeneric pairs of fishes from either side of Isthmus of Panama showed that the Pacific forms in all cases are better adapted to the cooler temperatures at which the fish live (Graves et al. 1983). It is nevertheless possible that only a few of the many enzyme variants that electrophoresis reveals within or among species differ enough to be fixed by selection, and that the rest of the variation is neutral. The evidence (Chapter 5) that evolutionary rates are highest for nontranscribed DNA and for relatively functionless polypeptides argues strongly for considerable genetic drift at the molecular level; but how much molecular variation and evolution does it explain? There are several ways of approaching this question.

### Amount and pattern of variation

The observation that enzymes are as variable in the haploid bacterium *Escherichia coli* as in diploid organisms (Caugant et al. 1981) indicates that heterozygote advantage is not a general explanation for allozyme polymorphism, although this leaves open the possibility that other modes of selection might be responsible. In contrast, the nature of polymorphism in alcohol dehydrogenase (*Adh*) in *Drosophila melanogaster* strongly suggests the operation of selection. Only two electrophoretically distinguishable alleles (*F* and *S*) are at all common; their frequencies vary clinally with latitude in several continents, but the polymorphism is ubiquitous (van Delden 1982). Kreitman (1983) determined the nucleotide sequence of a number of individual genes of both electromorphs (see Chapter 15). The *F* and *S* alleles differ at one nucleotide site that determines the single amino acid difference between the two allozymes. Both the *F* and *S* alleles vary considerably at other nucleotide positions, but all these other variations are either synonymous or are in noncoding regions of the gene. It appears that selection has been strong enough to eliminate other nucleotide substitutions that would alter amino acid sequence, yet has permitted one polymorphic site to remain. Either this is the only amino acid substitution that does not alter the function of the molecule and so is neutral—which seems unlikely—or selection maintains both alleles. In many other enzymes, however, there are numerous variants; in fact, the degree of polymorphism is correlated with the size of the molecule (Koehn and Eanes 1978). This can be interpreted to mean that a large molecule has more sites that are free to vary without affecting function.

### Distributions of allele frequencies

Several authors (e.g., Ewens 1977, Fuerst et al. 1977, Watterson 1978) have used the theory of genetic drift to predict the frequency distribution of alleles at a locus or set of loci, i.e., the number of alleles in each frequency category from nearly 0 to 1. The observed distribution of allele frequencies, when compared statistically to the theoretical distribution, has been claimed to fit the neutral model in some cases (Fuerst et al. 1977; Figure 15). However, discrepancies are fairly frequent, so some defenders of the neutral theory have felt obliged to modify the models by assuming that most mutations are slightly deleterious rather than strictly neutral (Ohta 1974, Nei 1983). Because either a selectionist or neutralist model can be made to fit the data by introducing various assumptions about factors such as selection, population size, or gene flow that are hard to measure accurately, this statistical approach is not a rigorous solution to the problem of polymorphism.

### Correlations with ecological factors

The most serious problem in assessing whether or not selection acts at an enzyme locus of interest is that it is extremely difficult to rule out the possibility that the locus is in linkage disequilibrium with closely linked loci that are the real "target"

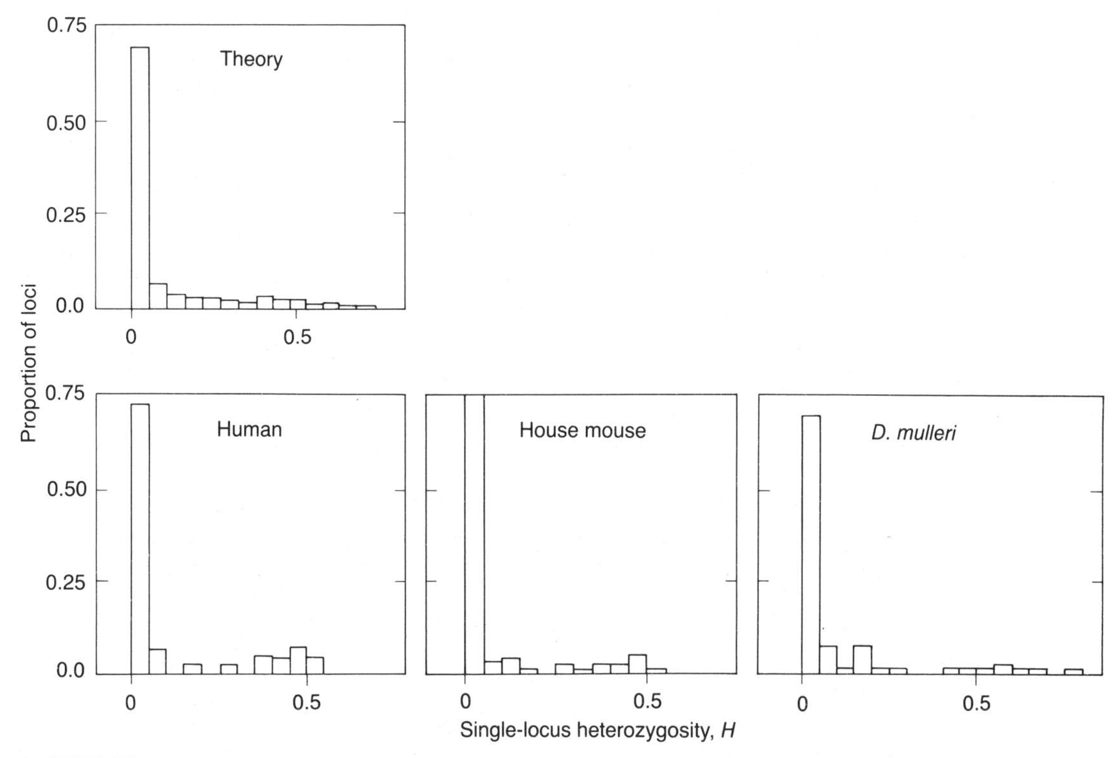


FIGURE 15

**Given the mean heterozygosity $\bar{H}$ over a set of loci, the variation in heterozygosity among loci can be predicted by the neutral theory of genetic drift. The theoretical distribution in the upper left panel is compared here to the heterozygosity at each of a number of allozyme loci analyzed by electrophoresis in populations of humans, house mice, and *Drosophila mulleri*. The ordinate is the fraction of loci displaying a particular value of heterozygosity. (After Nei et al. 1976)**

of selection. For example, Powell and Wistrand (1978) and several other authors have observed that allozyme heterozygosity is lost more slowly in experimental populations maintained in heterogenous than homogeneous environments. Some genes are clearly affected by the variation in the experimental environment, but not necessarily the enzyme loci under scrutiny. Species that use a wide variety of resources or occupy spatially heterogenous environments are not generally more polymorphic than ecologically more specialized species (Gooch and Schopf 1973, Mitter and Futuyma 1979, Futuyma and Peterson 1985), but in some species heterozygosity varies from population to population, and is correlated with meteorological variability (Bryant 1974).

The possibility that the frequency of allozyme alleles is affected by "hitchhiking" with other loci makes it difficult to be sure that clines in allele frequency are caused by selection. For example, the frequency of an allele at the lactate dehydrogenase locus varies latitudinally in a minnow (*Pimephales promelas;* Merritt 1972) and in an unrelated killifish (*Fundulus heteroclitus;* Place and Powers 1979). Although this suggests a response to an environmental variable such as temperature, the cline in itself does not provide the necessary evidence. In both these examples, however, biochemical studies showed that the allozyme that is more frequent in the north has higher enzyme activity at low temperatures than the allozyme that is more frequent in the south.

Another approach that indirectly circumvents the problem of linkage disequilibrium is to study closely related species. For example, the frequency of an allele at the phosphoglucoisomerase locus varies from one locality to another in parallel in two sympatric species of crickets (*Gryllus veletis* and *G. pennsylvanicus*; Harrison 1977). It is not very plausible to suppose that linkage disequilibrium with an unseen locus should have persisted since the species first diverged.

Clines can also be caused by gene flow between two formerly isolated populations that diverged in allele frequency by genetic drift. For example, several enzyme loci vary in allele frequency from east to west among populations of the land snail *Cepaea nemoralis* in the Pyrenees Mountains. Even though the environment differs much more between the northern and southern face of the mountain range than it does from east to west, the populations on northern and southern slopes do not differ in allele frequency at these loci—although they do differ in the frequencies of shell color morphs (see Chapter 4). Thus, although selection seems responsible for the shell color variation, the most likely explanation of the allozyme clines is that the mountains were invaded by eastern and western populations, perhaps following the retreat of the Pleistocene glaciers (Selander and Whittam 1983).

Patterns that might suggest the importance of genetic drift likewise may not be conclusive. For example, allozyme heterozygosity is usually lower in small than in large populations (e.g., Avise and Selander 1972, Lewontin 1974), but even though this pattern is consistent with genetic drift, it does not rule out the simultaneous action of natural selection.

### Experimental populations

One experimental approach to determining whether or not selection acts on allozyme variants has been to initiate replicate experimental populations of *Drosophila* with different allele frequencies to see if, over the course of generations, the replicates converge toward an equilibrium allele frequency determined by selection. In one of the first such experiments, Yamazaki (1971) found no evidence of convergence toward an equilibrium frequency for an esterase polymorphism in *Drosophila*. Several subsequent authors did find such convergence for other enzyme loci, but the rate of allele frequency change was so high that Mukai and Yamazaki (1980), in reviewing the data, concluded that selection was too strong to be attributed to a single locus, and that groups of linked loci were responsible for the changes.

Another experimental approach is to directly measure the viability, or other components of fitness, of different genotypes at an enzyme locus. The problem, again, is to rule out the possibility that neighboring loci are responsible for fitness differences. One way of minimizing this problem is to test the genotypes in an environment that should exert selection specifically on the locus of interest. For example, alcohol dehydrogenase metabolizes ethanol, and several workers have found that one of the two prevalent forms of this enzyme in *Drosophila melanogaster* increases survival in larvae that are reared in a medium that, like the rotting fruit in which this species often develops, contains this toxic substance (reviewed by van Delden 1982). Similar experiments with other polymorphic enzymes have provided strongly suggestive evidence that allozyme genotypes differ in fitness when the appropriate substrate for the enzyme is provided (Koehn et al. 1983).

This technique is especially useful if the organism can be genetically manipulated so as to make the genome uniform except at, or very near, the locus of interest. In *Drosophila melanogaster*, there are visible mutant alleles (flanking markers) that embrace and are closely linked to the loci that code for the enzymes glucose-6-phosphate dehydrogenase (G6PD) and 6-phosphogluconate dehydrogenase (6-PGD). The latter enzyme follows G6PD in the pentose shunt, one of the two pathways leading from glucose to pyruvate and thence to the Krebs cycle. By using flanking markers, Eanes (1984) was able to backcross alleles at both these loci into a genetic background that was homozygous except for about 3 percent of the chromosome in the immediate vicinity of the enzyme loci. He found that the common alleles at the G6PD locus did not differentially affect viability if the normal allele was present at the 6-PGD locus, but did affect viability if the flies were homozygous for a rare 6-PGD allele that has low enzyme activity. It is possible that when 6-PGD activity is low, the product of the G6PD-mediated step builds up, inhibiting further glycolysis, and that one of the G6PD allozymes produces this product at a higher rate than the other. However, even though the G6PD allozymes clearly differ in reaction rate, this difference seems not to affect fitness on the normal 6-PGD background genotype.

Because the bacterium *Escherichia coli* can be grown in huge numbers and has a short generation time, it is possible to devise very sensitive tests for selection in this species. Dykhuizen and Hartl (1980) used genetic techniques to backcross different alleles at the 6-PGD locus into a homogeneous genetic background, and compared the rate of population growth of the genotypes in a chemostat with gluconate, the substrate for this enzyme, as the sole source of carbon. Even though the experiment was sensitive enough to detect a selection coefficient (i.e., a difference in $r$) as low as 0.005, they found no evidence of fitness differences among the alleles, as long as the normal, functional allele was present at another locus, *edd*, that provides an alternative pathway for the metabolism of gluconate. However, some 6-PGD alleles did differ in fitness when placed in conjunction with a nonfunctional *edd* allele. Thus the study of the 6-PGD enzyme in both *Drosophila* and *E. coli* reveals that although the alleles are effectively neutral in populations with a normal genetic constitution, they provide potential variation for response to selection if the genetic background should change. In similar experiments with another variable enzyme (phosphoglucoisomerase) in *E. coli*, Dykhuizen and Hartl (1983b) found that the naturally occurring variants were selectively neutral when the populations were grown in a glucose medium, but differed in fitness when grown in a fructose medium. Thus both the external environment and the genetic constitution at other loci can determine whether the alleles at a locus are selectively significant or selectively neutral.

## ESTIMATES OF THE STRENGTH OF NATURAL SELECTION

When Darwin and Wallace proposed the theory of natural selection, they imagined that hereditary variations might differ only slightly in their effects on fitness. We now know that genotypes in natural populations often differ in fitness by

TABLE I

**Fitness of insecticide-resistant homozygotes in two species of *Anopheles* mosquitoes relative to that of susceptible homozygote (fitness = 1)** [a]

| *Anopheles* species | Insecticide | Fitness in environment with insecticide | Fitness in environment without insecticide |
|---|---|---|---|
| *culifacies* | DDT | 1.3–1.5 | 0.62–0.97 |
| | Dieldrin | 2.9–6.1 | 0.44–0.79 |
| *stephensi* | Malathion | 1.3–1.6 | — |
| | Dieldrin | 1.7–2.7 | — |
| | DDT | — | 0.96 |

(From Wood and Bishop 1981)

[a]Fitness estimated from the change in frequency of resistant genotypes in natural populations during insecticidal control and after termination of control. It was assumed that there was no immigration of susceptible mosquitoes. *A. culifacies* was studied in villages in India, *A. stephensi* in villages in India and Iran.

more than 10 or even 50 percent (Ford 1975), especially when the genotypes differ markedly in physiological or morphological traits that have clear adaptive significance. An allele for resistance to the rat poison warfarin has increased in many populations of Norway rats (*Rattus norvegicus*), in which it confers a strong selective advantage. In the absence of warfarin, however, the resistant homozygote suffers as much as a 54 percent disadvantage compared to the susceptible homozygote, because the resistant allele lowers the animal's ability to synthesize vitamin K, a necessary factor in blood clotting in wounds (Bishop 1981). Likewise, insect genotypes that differ in resistance to insecticides often differ greatly in fitness, both in the presence of insecticides and in their absence (Wood and Bishop 1981; Table I). In contrast, selection at many allozyme loci appears to be very weak or absent under prevailing conditions. But the direction, intensity, and even the existence of selection at a locus depends on the environment and on the genetic milieu in which the locus functions. Thus no generalizations can be made about the strength of selection; the concept of an average selection coefficient is meaningless. The neutralist-selectionist controversy over allozyme variation, for example, will almost surely not result in a clear "victory" for either side. It seems almost certain that some allozyme variants, such as the common variants of alcohol dehydrogenase in *Drosophila*, differ substantially in fitness, whereas others, such as some of the variations that have been studied in *E. coli*, are effectively neutral. Almost every type of enzyme polymorphism is now the subject of investigation in one or more species. It is through the analysis of many such loci that the relative importance of genetic drift and selection will gradually be resolved.

## SUMMARY

When changes in allele frequencies are caused by natural selection, the rate of change depends on the magnitude of the differences in fitness among genotypes, on the frequencies of the alleles, and on the degree of dominance. Selection may act to eliminate all but the most fit allele from a population or it may maintain stable genetic polymorphism by acting in various balancing modes, or by its conjunction with mutation or gene flow. The genetic variation maintained by selection, as well as some of the genetic variation that is selectively neutral, is important in providing the basis for adaptation to changes in the environment. It appears that some of the considerable genetic variation in natural populations is adaptively neutral, and that some is maintained by natural selection.

## FOR DISCUSSION AND THOUGHT

1. If allele frequencies change only in response to current, not future, selective pressures, how can we explain cases of apparent foresight, such as the migration of birds before the onset of cold weather?
2. How might one test Fisher's fundamental theorem of natural selection, that average fitness increases under natural selection? How can this theorem be reconciled with the fact that extinction is common?
3. *Drosophila* with the gene "grandchildless" have a normal number of offspring but have no grandchildren; it takes two generations for their low fitness to become evident. How many generations do we have to follow before we can decide which of two genotypes is more fit? How can we measure fitness?
4. Although we express the operation of natural selection in terms of the fitness of genes and genotypes, it is often said that selection acts on the phenotype. Integrate these two points of view.
5. If natural selection acts on several loci in a population, how will linkage disequilibrium among the loci affect the magnitude of the genetic load?
6. Darwin noted that "with all beings there must be much fortuitous destruction, which can have little or no influence on the course of natural selection" (*Origin*, p. 80). For example, the genotype of a small planktonic crustacean may have little influence on whether or not it is eaten by a baleen whale; it is either lucky or unlucky. What are the implications of Darwin's statement? Does it imply that such organisms cannot evolve by natural selection?
7. Suppose a population is limited by density-dependent factors, so that the individuals are engaged in a "struggle for existence." These factors are now alleviated, so that the population grows exponentially. What will be the effect on allele frequencies, i.e., on genetic variation? Will any natural selection occur during exponential growth?
8. Hebert (1974) found that in parthenogenetic populations of the water flea *Daphnia magna* the proportion of heterozygotes at two enzyme loci increased in successive generations and that heterozygotes had higher fecundity. Do these data show that heterozygotes at these loci have higher fitness than homozygotes?
9. Is the resolution of the neutralist–selectionist controversy about allozyme variation relevant to the question of whether or not genetic variation in morphological traits is maintained by natural selection? Is it relevant to our understanding of the causes of morphological evolution?
10. Does the theory of genetic load predict that species with high fecundity should evolve more rapidly than species with low fecundity? Is there any evidence that this has been the case?

## MAJOR REFERENCES

Spiess, E.B. 1977. *Genes in populations.* Wiley, New York. 780 pages. Discusses ways in which selection and fitness may be analyzed in practice.

Johnson, C. 1976. *Introduction to natural selection.* University Park Press, Baltimore. 213 pages. A good introduction to selection theory and its applications.

Lewontin, R.C. 1974. *The genetic basis of evolutionary change.* Columbia University Press, New York. 346 pages. Insightful analysis of early data on genetic variation and its interpretation from selectionist and neutralist viewpoints.

# Selection on Polygenic Characters

# Chapter Seven

Very early in the history of genetics, it became evident that there is not a one-to-one correspondence between genes and phenotypic traits; there are not separate genes for the femur and the tibia, for example. Rather, each gene has pleiotropic effects on several or many characters, and the development of each character is influenced by several or many genes. We have already seen (Chapter 6) that the selective advantage of a gene can depend on other genes, and that the analysis of selection at a particular locus is often complicated by its linkage to other loci. Thus it is necesary to develop a theory of allele frequencies in the context of other genes. This theory is mathematically formidable[1], and often difficult to test, but it demonstrates that evolution can take some surprising turns that are not predicted by the simple theory we have considered so far.

## TWO LOCI

In Chapter 4, we briefly considered the case of two loci, denoting the frequencies of alleles $A$ and $A'$ as $p_1$ and $q_1$ ($p_1 + q_1 = 1$) respectively, and of alleles $B$ and $B'$ as $p_2$ and $q_2$ ($p_2 + q_2 = 1$) respectively. The gametes $AB$, $AB'$, $A'B$, and $A'B'$ have frequencies $g_{00}$, $g_{01}$, $g_{10}$, and $g_{11}$ respectively. We defined the COEFFICIENT OF LINKAGE DISEQUILIBRIUM as $D = (g_{00} \times g_{11}) - (g_{01} \times g_{10})$—that is, the difference between the product of the frequencies of the coupling gametes $AB$, $A'B'$ and the frequencies of the repulsion gametes $AB'$, $A'B$. The loci are in complete linkage disequilibrium if only coupling gametes are present ($D = 1/4$ if $p_1$ and $p_2$ both are 0.5) or if only repulsion gametes are present ($D = -1/4$). If coupling gametes are present in excess, recombination in the double heterozygote ($AB/A'B'$) gives rise to repulsion gametes, so that $D$ declines from generation to generation at the rate $D_t = D_0(1 - R)^t$ where $R$ is the recombination rate (the map distance) between the loci and $t$ is the number of generations. This theory applies as long as the genotypes do not differ in fitness.

If natural selection is acting on these loci, however, there are some circumstances under which a state of permanent linkage disequilibrium may be maintained. For this to occur, it is at least necessary that the loci show EPISTASIS FOR FITNESS. Recall from Chapter 4 that the genetic variance in a trait is entirely additive when heterozygotes at each locus are intermediate between the homozygotes (no dominance) and the phenotypic value of two loci considered together is the simple sum of the phenotypic effects of the individual loci, as in this example:

| | $BB$ | $BB'$ | $B'B'$ | |
|---|---|---|---|---|
| $AA$ | 6 | 4 | 2 | (case 1) |
| $AA'$ | 5 | 3 | 1 | |
| $A'A'$ | 4 | 2 | 0 | |

In this example, $a_A$, the effect of replacing an $A'$ allele by an $A$ allele, is 1, while $a_B$, the corresponding effect at the $B$ locus, is 2. The phenotype $AABB$ (6) is found by adding the effects of each substitution of $A$ or $B$ for $A'$ or $B'$. If there is dominance at each locus but no epistatic interaction between the loci, the phenotypic values might be as follows:

| | $BB$ | $BB'$ | $B'B'$ | |
|---|---|---|---|---|
| $AA$ | 6 | 6 | 2 | (case 2) |
| $AA'$ | 6 | 6 | 2 | |
| $A'A'$ | 4 | 4 | 0 | |

[1] This chapter, however, will not delve into the mathematical intricacies.

In these cases, the loci do not interact, i.e., they do not modify each other's effect on the phenotype. In contrast, consider this hypothetical example:

| | $BB$ | $BB'$ | $B'B'$ | |
|---|---|---|---|---|
| $AA$ | 2 | 4 | 6 | (case 3) |
| $AA'$ | 3 | 4 | 3 | |
| $A'A'$ | 4 | 2 | 0 | |

In this case, the effect of replacing $A'$ by $A$ depends on the organism's genotype at the $B$ locus, and of replacing $B'$ by $B$ depends on the genotype at the $A$ locus: the loci interact. Any such interaction among loci is referred to as EPISTASIS in the literature of population genetics. The phenotype in question could be a morphological or physiological trait, or a component of fitness, or fitness itself.

Lewontin and Kojima (1960) analyzed the case in which the genotypes have these fitnesses:

| | $BB$ | $BB'$ | $B'B'$ |
|---|---|---|---|
| $AA$ | $a$ | $c$ | $a$ |
| $AA'$ | $b$ | $d$ | $b$ |
| $A'A'$ | $a$ | $c$ | $a$ |

They defined for this case a coefficient of epistasis $\epsilon = a + d - b - c$, which is zero if the fitness of each genotype is a simple sum of the fitness conferred by each locus separately (i.e., if fitness is additively determined). But if fitness is measured by the probability of survival, and if the effect of each locus is independent, the probability of survival of $AABB$ is the product, not the sum, of the probability of survival of $AA$ and the probability of survival of $BB$. For example, if these probabilities for locus $A$ are 0.8, 1, and 0.8 for genotypes $AA$, $AA'$, and $A'A'$, and similarly at locus $B$ they are 0.9, 1, and 0.9, then the fitness of $AABB$ is $W(AABB) = W(AA) \times W(BB) = 0.8 \times 0.9 = 0.72$, and similarly for the eight other genotypes. Then $a = 0.72$, $b = 0.9$, $c = 0.8$, $d = 1.0$, and $\epsilon = 0.02$, so there is epistasis for fitness.

The behavior of this system is found by analyzing the change in frequency of each of the four gametes. We can conceptually define the fitness of a gamete type, such as $AB$, as the average fitness of the genotypes in which this allele combination occurs, weighted by their frequencies in the population of reproducing individuals. Thus $w_{00}$, the average fitness of gamete type $AB$, is $\Sigma f_{ij} W_{ij}$, where $f_{ij}$ and $W_{ij}$ are the frequency and the fitness of these four zygote genotypes: $AB/AB$, $AB/AB'$, $AB/A'B$, and $AB/A'B'$. After one generation of selection, the frequency ($g_{00}$) of gamete $AB$ changes by the amount $\Delta g_{00} = [g_{00}(W_{00} - \bar{w}) - RDW_H]/\bar{w}$, where $\bar{w}$ is the mean fitness of the genotypes in the population. The term $-RDW_H$ takes into account the fact that in the heterozygotes $AB/A'B'$ (the fitness of which is $W_H$), recombination reduces the frequency of the $AB$ combination in proportion to its excess ($D$).

Thus the rate of increase of the combination $AB$ depends on its present frequency $g_{00}$, its average selective advantage $W_{00} - \bar{w}$, the degree $D$ to which the $AB$ combination is present in excess, and the rate $R$ at which recombination breaks down the $AB$ combination.

Lewontin and Kojima (1960) used this equation in the case in which the heterozygotes are most fit and the homozygotes have symmetrical fitnesses, so that the equilibrium allele frequency is 0.5 at each locus. One equilibrium has

$g_{00} = g_{10} = g_{01} = g_{11} = 1/4$, and $\hat{D} = 0$: that is, there is no linkage disequilibrium ($\hat{D}$ denotes the equilibrium value of $D$). Two other possible equilibria, with $\hat{D} \neq 0$, exist if $R < \epsilon/4d$. One of these has a permanent excess of coupling gametes: $g_{00} = g_{11} = 1/4 + |\hat{D}|$, where $|\hat{D}| = (1/4)[1 - 4Rd/(a + d - b - c)]^{1/2}$: the other possible equilibrium has an excess of repulsion gametes, so that $g_{00} = g_{11} = 1/4 - |\hat{D}|$, where $\hat{D}$ has the same value. Lewontin and Kojima showed that the condition of no linkage disequilibrium will be unstable, so that the system will move to a state of permanent linkage disequilibrium, if $R < \epsilon/4d$; if $R$ is greater than this value, the loci will come into linkage equilibrium ($\hat{D} = 0$). Thus there will be a permanent association between alleles at the two loci if they are closely linked ($R$ small) compared to the degree of epistatic interaction ($\epsilon$) between the loci in their effect on fitness. The association between $A$ and $B$ and between $A'$ and $B'$ (or vice versa) is built up in each generation by natural selection arising from the favorable interactions between the alleles, even as it is broken down in each generation by recombination. The magnitude of permanent linkage disequilibrium similarly depends on the balance of these factors; the expression for $\hat{D}$ decreases with an increase in $R$ or a decrease in the term $a + d - b - c$ (i.e., $\epsilon$). Even if two loci are closely linked, their alleles ultimately will not be associated unless epistasis for fitness exists; conversely, even alleles at loci on different chromosomes ($R = 1/2$) could show some degree of permanent association if they interact strongly enough.

These conclusions arise from a special case in which fitnesses are symmetrical; other cases (see reviews by Bodmer and Parsons 1962, Karlin 1975, Hedrick et al. 1978) are mathematically much more complex. In general, however, permanent linkage disequilibrium will be established only if epistasis for fitness is strong relative to the recombination rate. Epistasis usually will not in itself maintain a locus in a polymorphic condition if selection at that locus alone would not suffice to do so. And it is usually (although not always) the case that linkage disequilibrium increases the average fitness ($\bar{w}$) in the population, simply because it reduces the frequency of those genotypes that consist of unfavorable combinations of alleles.

Genetic drift can also give rise to linkage disequilibrium (Hill and Robertson 1968). Consider two loci that are so closely linked that each of the combinations $AB$, $AB'$ $A'B$, and $A'B'$ persists with little recombination, and so is transmitted as if it were one of four alleles. The frequencies of these combinations will then fluctuate by chance just as those of four alleles at a single locus would, so one or another combination will drift to excess. Because of this effect of chance, populations founded by a few individuals are likely to have considerable linkage disequilibrium among many of their loci. Powell and Richmond (1974) invoked this effect to explain the erratic changes in the frequency of a tetrazolium oxidase allele in experimental populations of *Drosophila paulistorum* subsequent to the founding of each population by three pairs of flies; it is likely that the allele was in linkage disequilibrium with different alleles at various other loci in each of the populations, so that blocks of genes that included the tetrazolium oxidase locus differed in fitness from one population to another. In populations that were founded by 50 pairs of flies, the allele frequency did not change substantially, presumably because there was little linkage disequilibrium among the loci.

## DIRECTIONAL SELECTION AT TWO LOCI

As the previous example implies, selection at one locus can change the frequencies of alleles at a linked locus, if there is initial linkage disequilibrium between them. This effect, often termed HITCHHIKING (Maynard Smith and Haigh 1974), is clearly stronger, the closer the linkage is between the loci. If two neutral loci *B* and *C* are both in linkage disequilibrium with a selected locus *A*, the association with *A* can give rise to linkage disequilibrium between *B* and *C* and changes in allele frequency at locus *A* can cause changes in allele frequencies at loci *B* and *C* (Thomson 1977).

Linkage has extremely important consequences when there is directional selection at both of two loci that interact epistatically in their effect on fitness (Felsenstein 1965). Suppose selection favors both *AA* and *BB* over other genotypes at these loci, either because these loci govern different favorable characters or because they contribute to the same characteristic. For example, the genotype *AABB* might have the greatest body size in a population that is undergoing selection for increased body size. (As we will soon see, fitness may be epistatically determined in this case even if the loci have additive effects on body size.) Then because the gamete *AB*, once incorporated into a zygote, increases body size more than any other combination does, selection will generate an excess of *AB* chromosomes, i.e., linkage disequilibrium. Moreover, if the population is already in linkage disequilibrium so that *AB* is present in excess ($D > 0$) when selection is initiated, the trait will respond more rapidly to selection than if the alleles were not associated ($D = 0$). Conversely, the response to selection will be slower if there is an excess of repulsion chromosomes $AB'$ and $A'B$ ($D < 0$), because the favored *AB* combination will become available to selection only as it is formed by recombination. If $D < 0$ before directional selection is imposed, but the new selection regime favors a composition with $D > 0$, tight linkage between the loci at first slows the response to selection, but then, when *D* passes from negative to positive, it quickly accelerates the response. A population exposed to a novel but constant selection pressure may show little or no response for a number of generations, and then respond quickly in a burst of evolutionary change. Such a lag in the response to selection is sometimes observed in artificial selection experiments (e.g., Mather and Harrison 1949).

## MULTIPLE EQUILIBRIA

We have already seen from Lewontin and Kojima's (1960) analysis that a given regime of selection may give rise to a polymorphic equilibrium with $D > 0$ or to one with $D < 0$ (if epistasis is strong enough compared to recombination); which of these two different genetic configurations is reached depends on whether linkage disequilibrium is initially positive or negative. If directional selection were then imposed on such a population, the response would be either very rapid or delayed, depending on which state the population was in initially. This is another example of the principle we encountered in Chapter 6, that the response to selection often depends on the past history of a population.

Epistatic interactions among loci can often be quite complex. Consider, as a heuristic example, the Australian grasshopper *Keyacris* (or *Moraba*) *scurra*, which is polymorphic for an inversion *A* on one chromosome and an inversion *B* on another. The nine genotypes do not occur in the frequencies expected under

Hardy-Weinberg equilibrium. From the disparity between observed and expected genotype frequencies, Lewontin and White (1960) estimated that in one population the fitnesses were

| | $AA$ | $AA'$ | $A'A'$ |
|---|---|---|---|
| $BB$ | 0.79 | 0.67 | 0.66 |
| $BB'$ | 1.000 | 1.006 | 0.66 |
| $B'B'$ | 0.83 | 0.90 | 1.07 |

From these fitnesses, Lewontin and White calculated an adaptive landscape. That is, the genotype frequencies ($f_i$) are calculated for various possible values of $q_A$ and $q_B$, and these frequencies, multiplied by the fitness values ($W_i$), give values of $\bar{w}$ ($= \Sigma f_i W_i$) that are plotted as a "contour map" as in Figure 1. Recall that if fitness values are constant, allele frequencies (in this case chromosome frequencies) should change until an adaptive peak is reached. This topography illustrates that the population should move either toward the equilibrium $q_A = 1$, $q_B = 1$ (the lower right corner of the topography) or the equilibrium $q_A = 0$, $q_B = 0.55$ (along the upper edge). The frequencies of the chromosomes did not in fact correspond to one of these stable equilibria, but rather to an unstable "saddle point" (Figure 1). Further analysis of this case has shown that this population

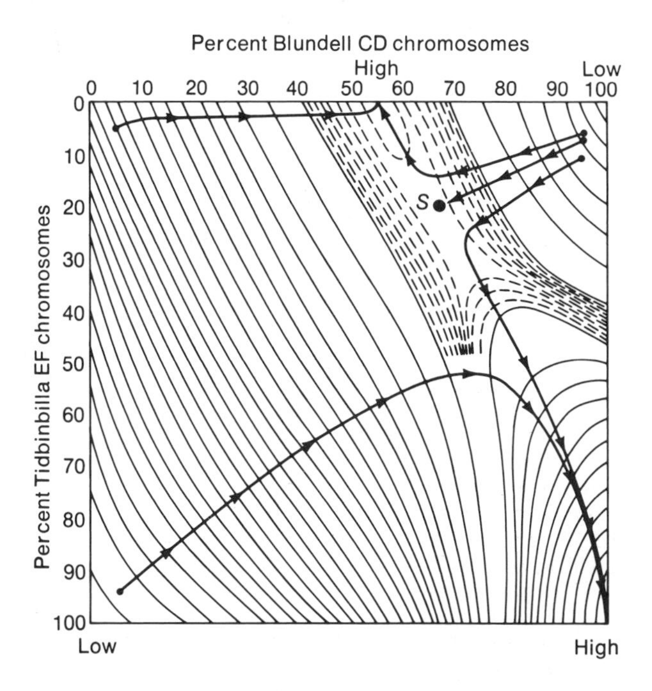


FIGURE 1

**The adaptive landscape for a population of the Australian grasshopper *Moraba scurra* that is polymorphic for both the *EF* and *CD* chromosomes. Based on genotype frequencies in the field, viabilities were calculated for each genotype, and from these the theoretical fitness $\bar{w}$ of populations of each possible chromosomal constitution was calculated. Compositions of equal $\bar{w}$ are indicated by contour lines; the dashed lines indicate finer distinctions of $\bar{w}$ than the solid lines. There are two peaks (high) and a saddle point (*S*). The trajectories are theoretical changes in genetic composition a population would follow from five initial states. (After Lewontin and White 1960)**

may actually be at a stable equilibrium if it is inbred (Allard and Wehrhahn 1964); moreover, it appears that the frequencies of the chromosomes change from year to year as the environment fluctuates, so there may not be fixed adaptive topography (Colgan and Cheney 1980). Lewontin and White's analysis shows how a multi-peaked landscape might be determined in principle, even if it does not adequately describe the dynamics of this population.

Cavener and Clegg (1981) monitored the change in frequency of alleles at loci for alcohol dehydrogenase (*Adh*) and α-glycerol phosphate dehydrogenase (α-*Gpdh*) in experimental populations of *Drosophila melanogaster*. The allele frequencies at both loci changed considerably in the populations exposed to ethanol, but not in control populations (Figure 2). By comparing the actual genotype frequencies to those expected under random mating and no selection, Cavener and Clegg estimated the viability of each genotype, much as Lewontin and White (1960) had. These estimates suggest considerable epistasis for fitness (Table I). Using these estimates, Cavener and Clegg simulated on a computer the expected changes in allele frequencies (Figure 2), and found that viability estimates from early generations predicted rather well the general course of allele frequency

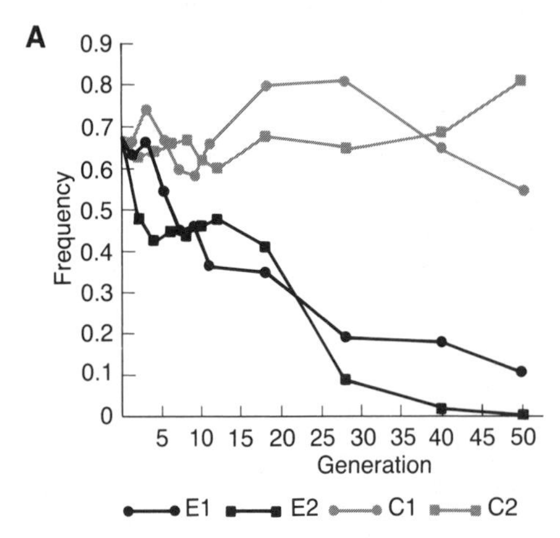


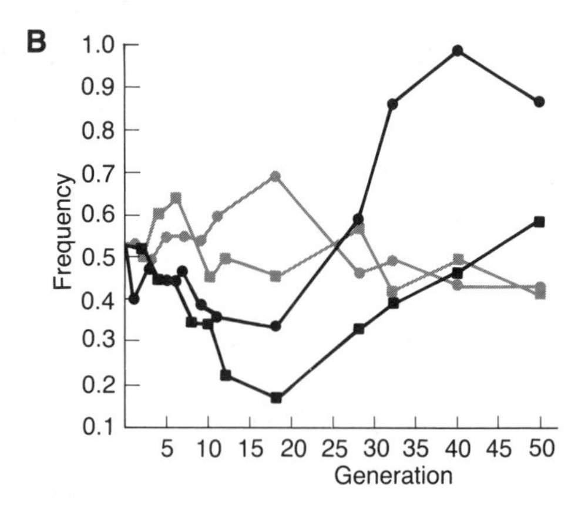


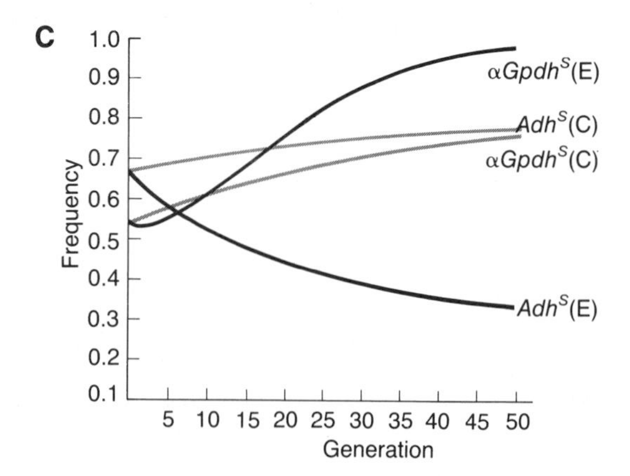


FIGURE 2

**(A) The frequency of the *Adh*$^S$ allele (for alcohol dehydrogenase) in four experimental populations of *Drosophila melanogaster* in the laboratory, over the course of 50 generations. C1 and C2 are control populations; E1 and E2 had ethanol in the medium. (B) The frequency of the α*Gpdh*$^S$ allele (for α-glycerol-phosphate dehydrogenase) in the same four populations. (C) Computer simulations of the changes in allele frequencies of the alleles at each locus in each of the two environments (C = control, without ethanol; E = with ethanol). The expected allele frequency changes were based on estimates of the fitness of each of the 9 genotypes in each of the environments, as described in the text. (From Cavener and Clegg 1981)**

TABLE I

**Viability of two-locus genotypes of *Drosophila melanogaster* in two experimental environments**[a]

(A) Ethanol Present

| | | *Adh* Genotype | | |
|---|---|---|---|---|
| | | *SS* | *SF* | *FF* |
| α-*Gpdh* Genotype | *SS* | 0.596 | 1.288 | 0.932 |
| | *SF* | 0.964 | 1.000 | 0.836 |
| | *FF* | 0.909 | 0.968 | 0.864 |

(B) Ethanol Absent

| | | *Adh* Genotype | | |
|---|---|---|---|---|
| | | *SS* | *SF* | *FF* |
| α-*Gpdh* Genotype | *SS* | 0.992 | 1.059 | 0.863 |
| | *SF* | 1.080 | 1.000 | 0.935 |
| | *FF* | 0.765 | 1.164 | 0.750 |

(Modified from Cavener and Clegg 1981)

[a]The viability of the double heterozygote, taken as a standard, has been set at 1.000 in each environment.

change over 56 generations of the experiment. Note that when $Adh^S$ has a high frequency, as it did initially, the *S* allele at the α-*Gpdh* locus declined, because on an *Adh-SS* background, α-*Gpdh-SS* is much inferior to *SF* or *FF*. But as the deleterious $Adh^S$ allele declined in frequency, the α-*Gpdh-S* allele reversed its course and increased, because the genotype *SS* at this locus has highest viability if it is associated with the *Adh* genotypes *SF* or *FF*. Thus epistatic interactions between loci, even if they are in linkage equilibrium, can cause reversals in allele frequency change: allele frequencies do not respond merely to the external environment, but to the "genetic environment" as well.

The importance of multiple peaks in the adaptive landscape cannot be overestimated. Frequently populations exposed to the same selective "problem" arrive at very different genetic "solutions." For example, when copper-tolerant populations of the plant *Mimulus guttatus* from different copper mines are crossed, the variation in copper tolerance is greater in the $F_2$ generation than within either parental population, indicating that the populations differ in the loci that confer tolerance. Different natural populations of houseflies (*Musca domestica*) have achieved resistance to DDT by any of several physiological or biochemical mechanisms; similarly, some populations of spider mites (*Tetranychus urticae*) have developed resistance to organophosphate pesticides by detoxifying the poison, while others have decreased the sensitivity of their nervous systems. These and numerous other examples are summarized by Cohan (1984).

To see further how uniform selection can cause populations to diverge in genetic composition, suppose two loci have equal and additive effects on a trait such as size:

| | $BB$ | $BB'$ | $B'B'$ |
|---|---|---|---|
| $AA$ | 4 | 3 | 2 |
| $AA'$ | 3 | 2 | 1 |
| $A'A'$ | 2 | 1 | 0 |

Suppose that selection has heretofore favored small size, so that $A'A'B'B'$ is the prevalent genotype, but that the environment changes, and there is now stabilizing selection for an optimum size of 2. We might describe the fitnesses of the genotypes by supposing, as many models do, that the fitness of a genotype declines as the square of the deviation of its phenotype from the optimum (Wright 1969). Thus we distinguish the *phenotypic* scale along which genotypes are measured from the *fitness* scale, which might in this case be the following:

| | $BB$ | $BB'$ | $B'B'$ |
|---|---|---|---|
| $AA$ | 0.6 | 0.9 | 1.0 |
| $AA'$ | 0.9 | 1.0 | 0.9 |
| $A'A'$ | 1.0 | 0.9 | 0.6 |

In this example, then, there is additivity on the phenotypic scale but epistasis on the fitness scale. Initially $A$ and $B$, both of which increase size, will each increase in frequency. Selection will establish the repulsion gametes ($AB'$ and $A'B$) in linkage disequilibrium, because both of the homozygotes ($AB'/AB'$ and $A'B/A'B$), as well as the double heterozygotes ($A'B/AB'$ and $AB/A'B'$) have the optimal body size. But a polymorphic population has a lower $\bar{w}$ (because recombination generates suboptimal genotypes) than a population that is monomorphic for either the $AB'$ or the $A'B$ combination. Either the alleles $A$ and $B'$ or $A'$ and $B$ will become fixed, the outcome being dependent on their initial frequencies. Thus stabilizing selection for an intermediate value of an additively determined polygenic trait does not maintain genetic variation at multiple loci (Lewontin 1964); moreover, the same regime of stabilizing selection in populations that differ only slightly in genetic composition can cause them to diverge (Cohan 1984).

### Epistasis and linkage disequilibrium in natural populations

Genes or blocks of genetic material with large phenotypic effect often show considerable epistatic interaction for fitness. For example, the viability of *Drosophila pseudoobscura* flies that are homozygous for two nonhomologous chromosomes is often lower than would be predicted from the homozygous effect of the individual chromosomes (Spassky et al. 1965). Components of fitness like fecundity and egg hatchability likewise show considerable epistasis (Table II). In contrast, when the variance in polygenic morphological traits like wing length and bristle number is analyzed in *Drosophila*, the genes are usually found to act additively rather than epistatically (Kearsey and Kojima 1967). It is possible that epistatic interactions among genes are expressed more strongly under natural conditions than in the laboratory. In tobacco (*Nicotiana rustica*), for example, the epistatic component of variation in several traits was more pronounced in both "good" and "bad" environments than in "average" environments (Jinks et al. 1973).

Closely linked genes that contribute to a single character or to functionally related characters often show strong linkage disequilibrium. Such clusters of

TABLE II
**Heritability and components of variance for some traits of animals**[a]

| Trait | Heritability ($h^2_N = V_A/V_P$) | $V_D + V_I$ | $V_E$ |
|---|---|---|---|
| *Cattle* | | | |
| Amount of white spotting in Friesian breed | 0.95 | | |
| Milk yield | 0.3 | | |
| Conception rate | 0.01 | | |
| *Pigs* | | | |
| Body length | 0.5 | | |
| Litter size | 0.15 | | |
| *Sheep* | | | |
| Length of wool | 0.55 | | |
| Body weight | 0.35 | | |
| *Chickens (White Leghorn)* | | | |
| Egg weight | 0.6 | | |
| Age at first laying | 0.5 | | |
| Egg production | 0.2 | | |
| Viability | 0.1 | | |
| *Mice* | | | |
| Tail length | 0.6 | | |
| Litter size | 0.15 | | |
| *Drosophila melanogaster* | | | |
| Abdominal bristle number | 0.52 | 0.09 | 0.39 |
| Length of thorax | 0.43 | 0.06 | 0.51 |
| Ovary size | 0.30 | 0.40 | 0.30 |
| Egg production in 4 days | 0.18 | 0.44 | 0.38 |

(After Falconer 1960, from various sources)
[a]Total phenotypic variance $V_P$ set equal to 1 for all traits.

genes are sometimes called SUPERGENES. For example, outcrossing in *Primula vulgaris* and many other plants is facilitated by heterostyly (Chapter 4), whereby some plants (thrum) have short styles and elevated anthers, and others (pin) have the reverse arrangement. In Chapter 4 we noted that the allele for short style ($G$) is in linkage disequilibrium with that for elevated anthers ($A$), and that the alleles $g$ and $a$ are likewise associated. The two forms of the plant differ, moreover, in the size of pollen grains, the length of stigmatic papillae, and in the compatibility between stigma and pollen. These characteristics, in aggregate, are controlled by at least seven tightly linked loci, inherited as a supergene in a state of extreme linkage disequilibrium (Ford 1975). Likewise, in the land snail *Cepaea nemoralis*, shell color and the type of bands on the shell are determined by five loci that form a single supergene, while another supergene consisting of six loci governs the presence or absence of certain bands, as well as the color of the body of the snail (Jones et al. 1977). In both *Primula* and *Cepaea*, the linkage disequilibrium is probably maintained by epistatic interactions for fitness, since the outcrossing mechanism in *Primula* depends on the right combination of features, and the cryptic appearance of *Cepaea* depends on both shell color and banding pattern.

Several experiments with allozymes, such as the experiment by Cavener and Clegg (1981) described above, suggest that different enzyme loci interact epistatically in their effect on fitness. The possibility exists that a chromosome is thickly studded with polymorphic loci, some of which are visualized as allozyme variants, that interact epistatically among themselves in determining fitness. If this were the case, the loci would become assembled into a few common sequences, so that there would be extensive linkage disequilibrium even among loosely linked loci (Lewontin 1974a). However, extensive studies have revealed that in populations of *Drosophila*, even tightly linked allozyme loci are very nearly in linkage equilibrium (Mukai and Voelker 1977, Charlesworth et al. 1979). This observation implies that if chromosomes are densely packed with polymorphic loci, many of them do not interact strongly, or that the polymorphisms are not maintained by natural selection.

Mechanisms that restrict recombination often give rise to associations among loci. The paracentric chromosome inversions for which many *Drosophila* populations are polymorphic tend to suppress recombination. As expected, certain alleles at enzyme loci located within the polymorphic sequence are associated with particular inversions (Prakash and Lewontin 1968, Mukai and Voelker 1977). Inbreeding, in concert with selection, gives rise to associations among loci; if only certain inbred genotypes pass the gauntlet of selection, they are likely to differ at numerous loci, as is the case in some predominantly self-fertilizing plants (Chapter 4). The extreme in suppression of recombination is asexual reproduction. Populations of asexually reproducing organisms typically consist of rather few clones that differ at numerous loci. For example, populations of the parthenogenetic earthworm *Octolasion tyrtaeum* that were polymorphic for ten allozyme loci consisted of only eight different genotypes; thus there was extreme linkage disequilibrium (Jaenike et al. 1980).

### Coadapted gene pools

Because of epistatic interactions, different populations of a species may be located on different adaptive peaks. Dobzhansky (1955) used the term COADAPTED GENE POOL to refer to the system of genes that interact harmoniously within a local population, but which interact disharmoniously when combined with genes from other populations. For example, the chromosome inversions *ST* and *CH* within Californian populations of *Drosophila pseudoobscura* confer high fitness when heterozygous, and form a stable polymorphism. But when Californian *ST/ST* flies are crossed with Mexican *CH/CH*, the heterozygous offspring do not have high fitness (Dobzhansky and Pavlovsky 1953). Dobzhansky inferred that the alleles within the inversions differed between California and Mexico, and did not form favorable combinations. A breakdown of coadaptation is also shown by the reduced viability and fecundity of the $F_2$ offspring from crosses between different geographic populations of *Drosophila pseudoobscura*, even when the $F_1$ generation displays heterosis (Wallace and Vetukhiv 1955). This phenomenon, termed "$F_2$ breakdown," implies that recombination between the two sets of genes has given rise to unfavorable combinations. The idea of coadapted gene pools is important in some theories of speciation (Chapter 8).

## POLYGENIC INHERITANCE

The variation in most traits is continuous, or quantitative, in kind, so that individuals vary in slight increments. Some of the phenotypic variation is directly caused by the environment, and some by genetic differences at several or many loci, each of which has a slight effect. A trait that is governed by many loci shows POLYGENIC inheritance.

Virtually every trait shows some degree of polygenic variation. Even the activity per gram of tissue of a specific enzyme such as alcohol dehydrogenase, for which variants that differ in kinetic properties can be identified, is modulated by other genes on several of the chromosomes (Laurie-Ahlberg 1985). In some cases the variation in enzyme activity stems from variation among individuals in the number of gene copies in a multigene family (Nielsen 1977); in other cases it seems likely that alleles at several loci modulate the transcription or activity of the gene product.

If the effect of each locus is small compared to the total amount of variation, it is very difficult to estimate the number of loci that contribute to the variation. Analysis of the variance of the progeny of crosses between phenotypically different populations or artificially selected lines often provides an estimate that 5–20 loci contribute to the variation in a morphological trait (Wright 1968, Lande 1981a). The methods used underestimate the number of loci to an immeasurable extent; in some cases there is reason to believe that 100 or more loci are involved. The prolonged response to artificial selection that is often seen in selection experiments (see below) likewise suggests that at least 15–20 loci are involved, although perhaps 5–8 loci may contribute most of the effect (Mather 1979). In those species of *Drosophila* in which stocks with different chromosome constitution can be constructed, it is almost invariably found that loci on each arm of each chromosome contribute to the variation in a trait. The number of loci that affect the development of a character may be considerably greater than the number that are polymorphic and contribute to the observed variation. More than 70 loci are known to affect eye color in *Drosophila melanogaster*, but few of them vary in natural populations.

### The analysis of phenotypic variance

Because individual loci that contribute a small amount to the variation in a polygenic trait usually cannot be identified and mapped to a specific location in the genome, quantitative genetics rests on the statistical analysis of variation, a topic introduced in Chapter 4. In the simplest case, the PHENOTYPIC VARIANCE $V_P$ — the variance that is actually measured in a population — is just the sum of the GENETIC VARIANCE $V_G$ and the ENVIRONMENTAL VARIANCE $V_E$:

$$V_P = V_G + V_E$$

The genetic variance is the variance among the mean phenotypic values of the various genotypes in the population. The environmental variance is the nonge-

TABLE III
**Relative asymmetry of sternopleural bristle number in *Drosophila melanogaster***

| Genotype[a] | Environment[b] | Total asymmetry |
|---|---|---|
| Homozygotes | Home | 363 |
| | Foreign | 451 |
| Intrapopulation heterozygotes | Home | 351 |
| | Foreign | 342 |
| Extrapopulation heterozygotes | Home | 466 |
| | Foreign | 442 |

(After Thoday 1955)
[a]Intrapopulation heterozygotes were heterozygous for two chromosomes taken from the same laboratory population; extrapopulation heterozygotes had chromosomes from two laboratory populations adapted to the same environment.
[b]"Home" is the environment to which the populations from which the chromosomes were taken were adapted; "foreign" indicates tests in another laboratory environment.

netic variation among individuals of a genotype, due to developmental noise and the direct effects of environmental differences on phenotypes. This environmental variance, $V_E$, is usually assumed to be the same for each genotype, but it may not be. The environmental variance of heterozygous genotypes is often lower than that of homozygotes (Lerner 1954, Thoday 1955). It appears that heterozygotes are often less sensitive than homozygotes to disruptions of development; for example, they tend to be more bilaterally symmetrical (Table III).

The expression $V_P = V_G + V_E$ will hold true only if the phenotypic difference among genotypes is the same in each environment to which they are exposed

FIGURE 3
**Genotype × environment interaction. (A) An idealized, hypothetical case in which there is no interaction between genotype and environment. At higher temperatures, each genotype has a lower phenotypic score (e.g., number of bristles), but the two genotypes, although different in phenotype, respond similarly to the environment. (B) The mean number of bristles on sternites 4 and 5 of the abdomen of male *Drosophila pseudoobscura* of 10 genotypes, reared at three temperatures. The genotypes respond differently to temperature, so there is an interaction between genotype and environment. (C) The curves from genotypes 5 and 10 in (B) are shown to illustrate the distribution of phenotypes that might be observed in a population of these two genotypes if the flies developed under a range of temperatures from about 17°–20°. Most of the phenotypic variation would be genetic, because at these temperatures the genotypes differ greatly in phenotype. (D) As for (C), but the environment varies from about 23°–25°. The population is genetically the same as in (C), and the amount of variation in the environment is similar, but most of the phenotypic variation would be environmental because the genotypes have similar responses at these temperatures. (From Gupta and Lewontin 1982)**

(Figure 3A); a change in environment would then simply increase or decrease each genotype's phenotype to the same extent. Very often, however, the reaction to a difference in environment differs among genotypes (Figure 3B). There is then a GENOTYPE × ENVIRONMENT INTERACTION ($V_{G \times E}$) that contributes to the phenotypic variance, so that $V_P = V_G + V_E + V_{G \times E}$. In this case the genetic variance is not readily distinguishable from the environmental variance (Figure 3C,D), because each depends on the other (Lewontin 1974b, Gupta and Lewontin 1982). We will proceed, as most workers in the field do, by ignoring the genotype × environment interaction, which in practice is often included in the term $V_E$.

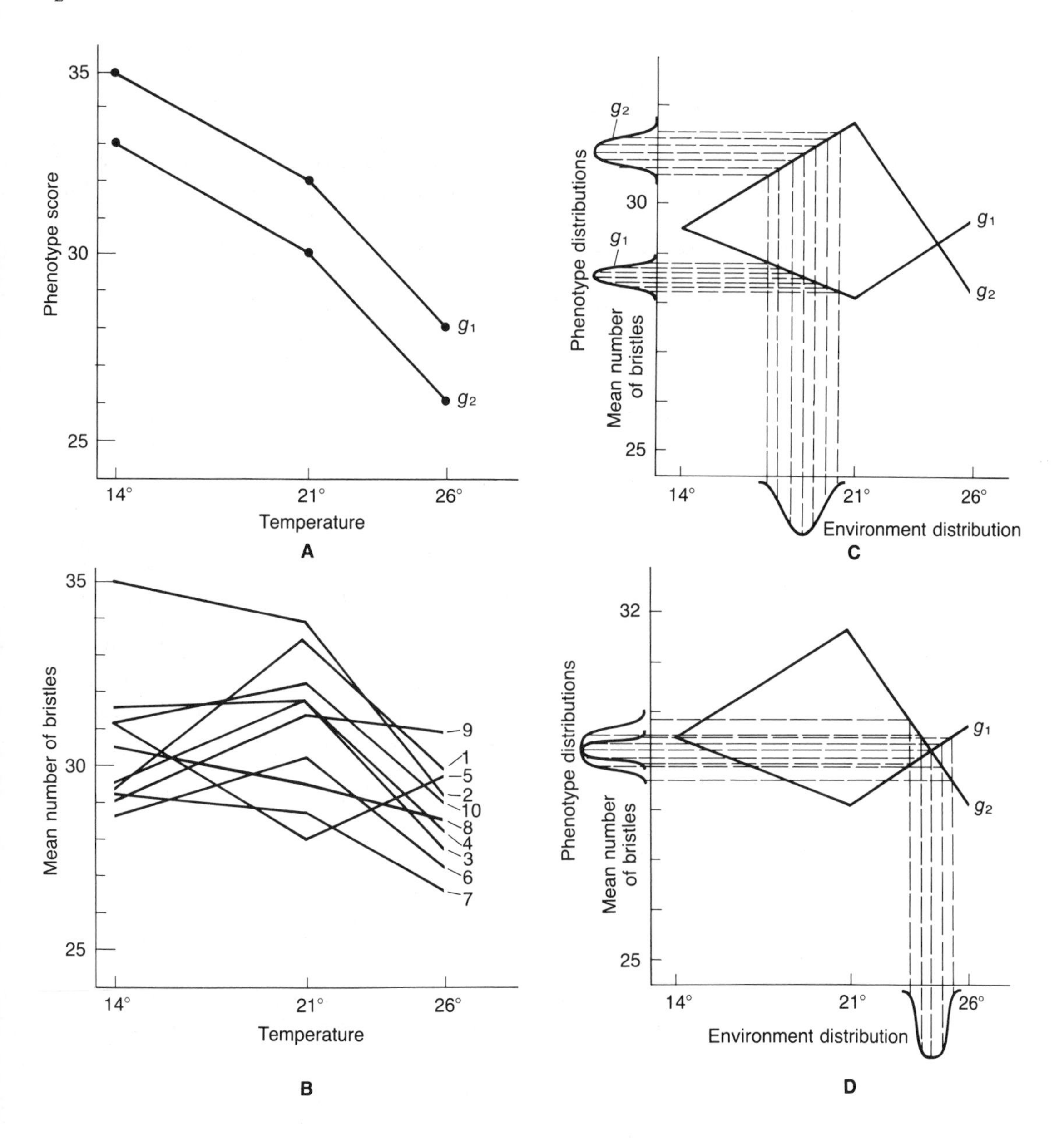

We will also assume that there is no COVARIANCE between genotype and environment. Such a covariance, or correlation, occurs if one genotype tends to occur in one environment and another occurs in a different environment. For example, genotypes that differ at the alcohol dehydrogenase locus in *Drosophila* differ in their tolerance to ethanol, giving rise to variance in viability. As larvae, these genotypes also differ in their tendency to avoid ethanol (Cavener 1979). Hence if there were fine-scaled variation in ethanol at the larval feeding sites, the variance in viability would be reduced by the ability of the genotypes to seek favorable environments.

## Heritability

The HERITABILITY of a trait is the fraction of the phenotypic variance that is attributable to genetic variation; it is denoted $h^2$. If $V_P = V_G + V_E$, the *heritability in the broad sense* is $h^2_B = V_G/V_P$. This is the heritability that would be measured if one cloned each of a number of genotypes, grew them together, estimated the variance among the genotype means, and divided this value by the sum of the variance among clones and the average variance within clones. Note that $h^2_B$ could not be estimated from a single individual in each clone, because it would then be impossible to determine whether the differences among individuals were caused by differences in their genotypes or in their environments.

It is important to recognize that any estimate of heritability is strictly valid only for the particular population analyzed, in its particular environment, because the amount of genetic variation may vary among populations, and the heterogeneity of the environment may vary in space and time.

The genetic variance $V_G$ can consist of several components. One, which was introduced in Chapter 4, is the ADDITIVE GENETIC VARIANCE, $V_A$, that is attributable to the additive effects of alleles within and among loci. In the hypothetical Case 1 on page 185, all the genetic variance is additive: the phenotypic effects of the two individual loci sum to the phenotypic effect of the loci taken together, and at each locus the heterozygote is precisely intermediate between the two homozygotes. If there is dominance at a locus, however, there is an extra component to the genetic variance, the VARIANCE IN DOMINANCE DEVIATIONS $V_D$. Moreover, if the loci interact epistatically in their influence on the phenotype (as in Case 3, page 186), the genetic variance includes yet another component, the EPISTATIC or INTERACTION VARIANCE $V_I$. The terms $V_D$ and $V_I$ themselves consist of a number of terms, the expressions for which are quite complicated.

Thus the genetic variance is a sum of terms

$$V_G = V_A + V_D + V_I$$

so that the phenotypic variance is

$$V_P = V_A + V_D + V_I + V_E$$

if we ignore the effects of genotype × environment interaction and covariance between genotype and environment.

The additive genetic variance is the term of particular interest; in fact, the centrally important concept in quantitative genetics is the *heritability in the narrow sense*,

$$h^2_N = V_A/V_P$$

The reason for its importance is that *the additive effects of genes are almost entirely responsible for the resemblance between parents and their offspring*. For example, suppose genotypes *AA*, *AA'*, and *A'A'* have phenotypes 6, 5, and 4 respectively. Then the mean of the parents (the MIDPARENT VALUE) in the cross *AA* × *A'A'* is (6 + 4)/2 = 5. This is the mean of the $F_1$ progeny (*AA'*), and the $F_2$ progeny as well [(¼ × 6) + (½ × 5) + (¼ × 4) = 5]. In contrast, if there is dominance, the average correlation between parents and offspring is lower. If both *AA* and *AA'* have a phenotype of 6 and *A'A'* of 4, the midparent value in the cross *AA* × *A'A'* is 5, the $F_1$ phenotype is 6, and the mean of the $F_2$ progeny is 5.5. It should be evident that not only dominance, but epistatic interactions among loci and direct effects of the environment on the phenotype will tend to lower the correlation between the phenotypes of parents and offspring.

Because of this relationship, it is possible to estimate heritability in the narrow sense (henceforth referred to simply as "heritability" unless otherwise indicated) by plotting the mean phenotype of pairs of parents (i.e., the midparent value) against the mean phenotype of their progeny, as in Figure 4. The higher the slope of this relationship (the slope being the least squares estimate of the regression coefficient), the higher the heritability; in fact, in principle the heritability $h^2_N$ equals this slope. Similar relationships among other relatives (e.g., sibs, half-sibs, or offspring regressed against one parent) may be used to estimate not only $h^2_N$, but $V_D$ and $V_I$ as well (Falconer 1981). In practice, regression of

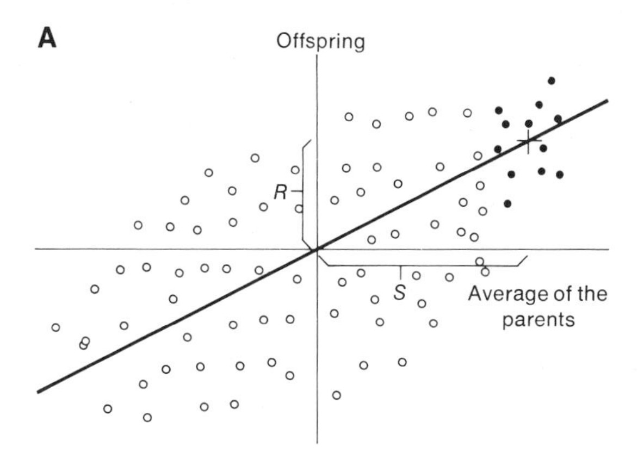


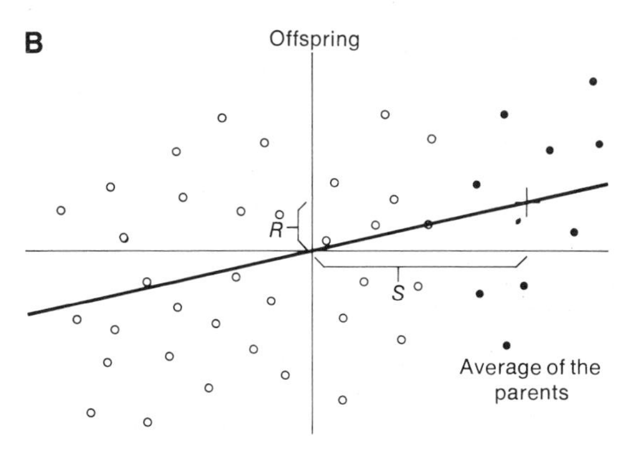


FIGURE 4

**Plot of phenotype of offspring against that of their parents. Each point represents the mean of a brood versus the mean of the brood's two parents (the "midparent" value). The sloping line is the regression of offspring on parent values. It is steeper if the relationship is strong (A) than if it is weak (B). The slope of this line is the "heritability in the narrow sense,"** $h^2_N$**. Solid points are parents selected for breeding and their progeny. The difference between the mean phenotype of selected parents and that of all potential parents is the selection differential** *S***; the difference between the mean phenotype of the offspring of the selected parents and that of all potential offspring is the response to selection** *R***. (A after Falconer 1981)**

offspring against midparent values may be less trustworthy than correlations among other relatives such as half-sibs, because nongenetic maternal effects may contribute to the similarity between mothers and their offspring.

Estimates of $V_A$, and especially $V_D$ and $V_I$, are often imprecise unless large samples and a careful breeding scheme are employed. Nevertheless, some estimates of variance components (Table II) show an interesting regularity in that the proportion of the genetic variance that is nonadditive ($V_D + V_I$) is relatively greater in traits such as egg production that ought to be strongly correlated with fitness, whereas the additive genetic variance is relatively greater for traits such as bristle number, whose effect on fitness does not seem critical, or at least consistent (Falconer 1981). This is to be expected from Fisher's fundamental theorem of natural selection (Chapter 6), which states that the rate of increase in mean fitness equals the additive genetic variance in fitness. If the population is at allele frequency equilibrium, mean fitness is not increasing, so the additive genetic variance in fitness itself must be zero, and the stronger the correlation between a trait and fitness, the closer to zero $V_A$ will be for that trait (ignoring the input of variation by mutation and gene flow). At equilibrium, the trait may display genetic variance, but it is nonadditive in nature. For example, if the heterozygote at one locus has highest fitness because its intermediate nose length is favored by selection, the correlation between phenotype and fitness is low (i.e., longer noses do not confer higher fitness). At allele frequency equilibrium [$\hat{q} = s/(s + t)$ from Chapter 6], the genetic variance in nose length will be additive, but the genetic variance in fitness will be entirely nonadditive ($V_D$). Recently it has become evident that natural populations harbor considerable additive genetic variance for characters such as clutch size (Findlay and Cooke 1983) and the timing of reproduction (van Noordwijk et al. 1980) that clearly affect fitness directly (see Istock 1981). If such traits are negatively correlated with each other, there may be little additive genetic variance in fitness itself, even though there is additive genetic variance in each component of fitness (Lande 1982; see below).

## HERITABILITY AND THE RESPONSE TO SELECTION

If the only source of phenotypic variation is the segregation of additively acting alleles, then $h^2_N = 1$, and the average phenotype of offspring in a large family equals that of their parent. If a group of parents is a selected sample whose mean differs from that of the whole population by an amount $S$ (the SELECTION DIFFERENTIAL), the population mean will change by just that amount in the next generation. The amount of change in the mean is the RESPONSE $R$, and in this case would equal the selection differential. If $h^2 < 1$, the slope of the offspring-midparent relation is lower (Figure 4), so the response to a given selection differential will be reduced. In fact, $R = h^2_N S$, and any factor that lowers the similarity between offspring and parents—whether it be dominance, epistasis, or environmental effects—reduces the response to selection. That is, it reduces the rate of evolutionary change.

The evolution of a polygenic trait thus depends on the heritability and on the way in which selection acts on the trait, i.e., the relation of the trait to fitness. The higher the correlation (strictly, the additive genetic covariance; Crow and Nagylaki 1976) between the phenotype and fitness, the more rapidly the trait will evolve. If $\bar{z}$ is the mean value of a phenotypic trait $z$, $V_A$ is its additive genetic variance, and $\bar{w}$ is mean fitness, the mean will change by an amount

$$\Delta\bar{z} = \frac{V_A}{\bar{w}} \cdot \frac{d\bar{w}}{d\bar{z}}$$

per generation (Lande 1976a). (This expression, in which $d\bar{w}/d\bar{z}$ is the slope of the relation between $\bar{w}$ and $\bar{z}$, assumes that $h^2$ and $V_P$ are constant.)

Under directional selection, the mean increases or decreases according to the equations given above. The change in the mean can be expressed either in simple units (e.g., 2 mm per generation) or in standard deviations. The latter is often more meaningful. For example, if we are told that in one generation the mean wing length in a bird population increased from 67.7 to 69.2 mm, we do not know whether we should be impressed or not. But if the phenotypic variance is 0.25 $mm^2$, the standard deviation is $\sqrt{0.25} = 0.5$ mm, and a change of 1.5 mm in the mean is a change of three standard deviations. Now, since 99.7 percent of a population lies within three standard deviations on either side of the mean (Appendix I), a shift of three standard deviations indicates that the mean has shifted to a value expressed by considerably less than one percent of the population before selection occurred. This is, then, a very large change. In fact, the values I have used for this example are those described by Boag and Grant (1981) for a population of the Galápagos finch *Geospiza fortis* that suffered a severe decline during an intense drought. When the food supply was reduced, the survivors turned to seeds larger than the species usually feeds on. Apparently large birds had greater access to the seeds, or a greater ability to feed on them. Since the heritability of size in this population is 0.76 (a rather high value), a considerable evolutionary increase in size would occur in this population if the stringent environmental conditions were to persist (although they did not, in this case). This example illustrates that directional selection can be stringent in natural populations, and that rapid evolution can sometimes be expected.

Directional selection seldom moves a character toward an ever more extreme mean; at some point, countervailing forces of selection come into play, so that stabilizing selection keeps the mean at some intermediate value. For example, Hecht (1952) found that in lizards of the genus *Aristelliger*, the mean body size was determined by the balance between the advantage that larger lizards have in defending territories and the disadvantage they suffer in being more susceptible to predation by owls. The reasons for the superiority of intermediate forms are often more obscure. For example, for four out of six characteristics (e.g., head width) of the milkweed beetle *Tetraopes tetraophthalmus*, males found mating were less variable and were closer to the population mean than those that were not found *in copula* (Mason 1964). Eggs of intermediate weight have the highest hatching success in ducks (*Anas platyrhynchos*; Rendel 1953) and chickens (*Gallus gallus*; Lerner and Gunns 1952), and human infants with intermediate weight at birth have the highest survival rate (Karn and Penrose 1951; see Figure 5). In a classic study, Bumpus (1899) found that intermediate-sized house sparrows (*Passer domesticus*) had higher survivorship during a storm than did those of either extreme size (but see Johnston et al. 1972).

Diversifying selection has been less extensively studied in natural populations than directional and stabilizing selection, although it may be quite common if different phenotypes in a resource-limited population are specialized for dif-

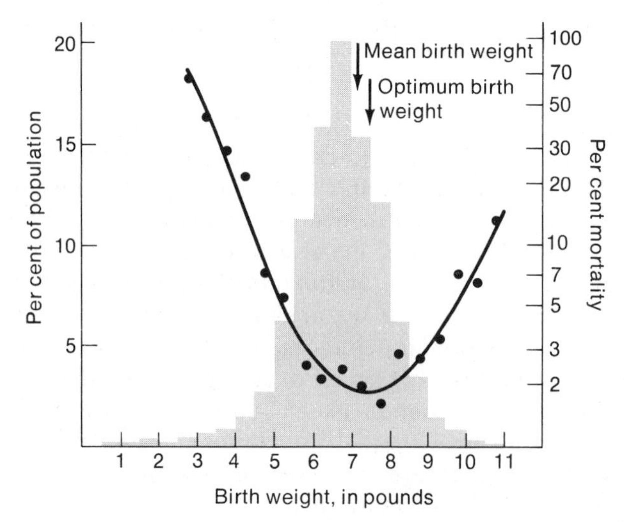


FIGURE 5

**Stabilizing selection for birth weight in humans. Early mortality is lowest at birth weight (optimum birth weight) near the mean for the population. The histogram represents the distribution of birth weights in the population; points represent percent mortality. (From Cavalli-Sforza and Bodmer 1971)**

ferent resources; in this case, selection on the character would be frequency-dependent (Slatkin 1979), as in the one-locus case (Chapter 6). Selection favors multiple discrete phenotypes in the swallowtail butterfly *Papilio dardanus* (Clarke and Sheppard 1960, Ford 1975). In southern Africa, some females of this palatable butterfly have black wings with white patches, some have the white reduced to small spots, and others have extensive reddish-brown areas on the wings. Each of these discretely different patterns closely resembles that of one or more unrelated, distasteful species, of which the *P. dardanus* females are Batesian mimics. The color patterns are inherited as if they were determined by multiple alleles at a single locus, but the locus appears to be a supergene. Moreover, the color pattern is stabilized by polygenic modifier loci, the effects of which become evident when these females are crossed with a nonmimetic race from Madagascar. Various intermediates among the mimetic forms then appear among the progeny. These results indicate that in the South African population, selection has favored modifier genes that restrict the expression of the major genes to one of several discrete alternative phenotypes.

## The effect of selection on genetic variation

The most immediate effect of stabilizing selection on the genetic architecture of a polygenic trait is to increase linkage disequilibrium. Suppose, for example, that a + allele at each of several loci on a chromosome increases the number of bristles, that a − allele decreases the number, and that inheritance of the trait is strictly additive. Given various chromosome types such as + + + −, + − − +, and so on, genotypes such as $\frac{++--}{--++}$ and $\frac{+-+-}{-+-+}$ will be favored because of their intermediate bristle number. If the loci are tightly linked, selection will initially increase the frequency of chromosomes that are in repulsion phase, such as $+ - + -$ and $- + - +$ (Mather 1941). When such chromosomes are frequent in a population, a great deal of variation is produced each generation by recombination. Extreme variants, however, will be rare. If the allele frequency is $p = 0.5$ at each of four loci, the genotype $\frac{++++}{++++}$ would arise with a frequency of only $(p^2)^4 = 0.0039$ if the loci segregated independently, and even less frequently if they were in linkage disequilibrium. Thus the latent genetic variation includes phenotypes far more extreme than would actually be observed in any but the largest population.

If the only effect of the alleles is on the character that is subject to stabilizing selection, and the alleles contribute additively to the phenotype, the superiority of intermediate phenotypes does not maintain genetic variation indefinitely (Wright 1935, Lewontin 1964, Felsenstein 1979), for two reasons. First, because some homozygous genotypes (e.g., $\frac{+-+-}{+-+-}$) have the same intermediate phenotype as heterozygotes (e.g., $\frac{+-+-}{-+-+}$) selection will generally fix homozygous combinations (see page 192). Second, because a very large number of genotypes have the same intermediate phenotype, the coefficient of selection at any particular locus is so small that the allelic variation at that locus is nearly neutral, and the allele frequencies fluctuate by genetic drift (Wright 1935, Kimura 1981). As the frequency of a + allele at one locus drifts toward fixation, the frequency of − alleles at one or more other loci is increased by selection so that the mean phenotype stays near the optimum. Eventually genetic drift should erode genetic variation altogether.

Directional selection increases the frequency of alleles that contribute to the advantageous phenotype. In principle, the additive genetic variance should change under selection, but how it does so depends on allele frequencies. If, for example, + alleles are rare at most of the loci and a new regime of selection favors an increase in the mean, the variance should increase as these alleles increase to intermediate frequencies, and then decrease as they approach fixation. The maximum possible degree of divergence, based on the existing additive genetic variance in a population, would be the range between a subpopulation selected to fixation for − alleles and one selected to fixation for + alleles at all the contributing loci. The limit to evolution, then, is determined by the number of loci that contribute to the variation and by the frequencies and phenotypic effects of alleles at these loci. But since none of these quantities can be readily measured, it is usually not possible to predict how far the mean will shift under a long-term regime of directional selection.

Under both stabilizing and directional selection, then, the genetic variance should be eliminated, yet as we have seen (Chapter 4), the additive genetic variance for most traits is quite considerable in magnitude. Moreover, even under prolonged selection in the laboratory, $V_A$ (and $h^2{}_N$) usually is not greatly diminished if the populations are kept large enough that genetic variation is not diminished by genetic drift (Figure 6). For example, Kaufman et al. (1977) imposed stabilizing selection on pupal weight in the flour beetle *Tribolium castaneum* by breeding from individuals with intermediate weight for 95 generations. The phenotypic variance $V_P$ decreased, but $V_A$ and $h^2$ hardly decreased at all. The decrease of $V_P$ came about by a reduction of $V_E$; selection had apparently favored genotypes that were better buffered against the effects of the environment. Enfield (1980) applied directional selection for increased pupal weight in *T. castaneum*

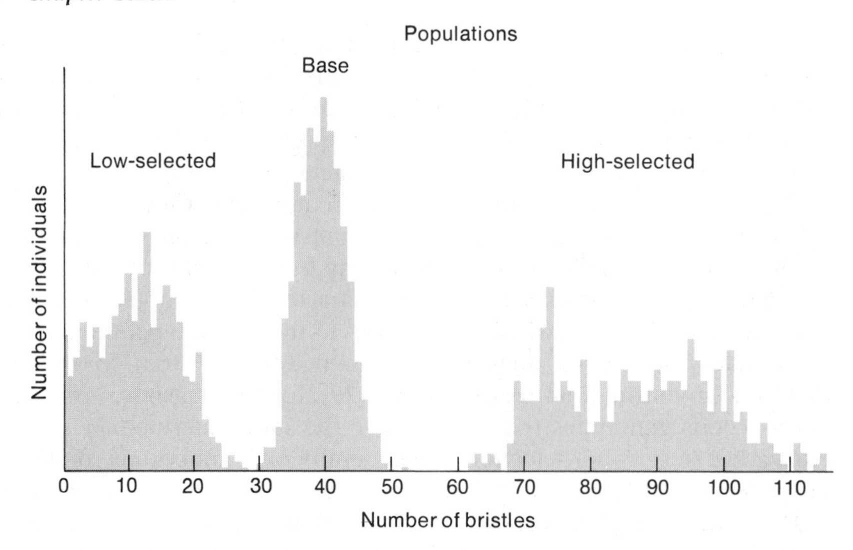


FIGURE 6

**Demonstration that directional selection need not diminish variation: the frequency distribution of abdominal bristle number in a stock of *Drosophila melanogaster* (center) from which lines were artificially selected for low (left) and high (right) bristle number for 34 and 35 generations respectively. (After Clayton and Robertson 1957 and Falconer 1981)**

and observed a continued increase in mean weight for 130 generations, by which time the mean had increased from 2450 to 5980 mg. The heritability of the trait was initially 0.28, but even after 130 generations of selection it still had the appreciable value of 0.18. Thus although heritability is expected to decline, it generally does not do so to any great degree. This observation justifies the assumption in many models of the evolution of quantitative traits that $h^2$ remains approximately constant in large populations, at least if selection is not very intense. But it raises the puzzling question of why the variation persists.

## Mutation in polygenic characters

Although the mutation rate at individual polygenic loci cannot easily be measured, the increase in genetic variance in a character that arises by new mutations can be estimated in homozygous lines (Chapter 3). This value, per generation, is generally about 0.001 times the environmental variance $V_E$ for bristle counts in *Drosophila* and for various traits of mice, *Mus musculus*, and corn, *Zea mays* (Lande 1976b, Hill 1982). In about 1000 generations, a large initially homozygous population would achieve a heritability ($V_A/V_P$) of about 0.5 in the absence of selection. The mutational variance appears generally to be symmetrical about and independent of the mean of the character, so that even if the mean has been selected to an extreme value, mutation gives rise to still more extreme phenotypes. The mutational variance is considerably greater than expected if the mutation rate per locus is only $10^{-7}$–$10^{-5}$ as is usually estimated, and if there are only 10–20 loci affecting the character. Either the number of loci is far greater than this, or the mutation rates of genes affecting quantitative traits are higher than the rate of mutation at allozyme loci or lethal loci (Turelli 1984). It is possible that quantitative variation arises from transposable elements or unequal recombination between tandemly duplicated genes at higher rates than single-locus mutation rates would predict.

Because the genetic variance of a character increases at an appreciable rate by mutation, it can balance the erosion of the genetic variance by stabilizing selection (Lande 1976b). However, it is not clear that stabilizing selection is weak enough, or the number of loci large enough, for recurrent mutation to explain the high levels of additive genetic variance that are actually observed (Turelli 1984). It seems likely that mutation and gene flow account for much of the variation, but more data and theory will be needed to resolve the problem fully.

## GENETIC CORRELATIONS

The correlation coefficient between traits $x$ and $y$, $r_{xy}$, ranges from $+1$ for perfectly positively correlated characters to $-1$ for negatively correlated characters (see Appendix I). The PHENOTYPIC CORRELATION $r_P$ between, say, body size and fecundity is what we measure if we simply take a random sample from a population. But just as the phenotypic variance has additive genetic, nonadditive genetic, and environmental components, so does the phenotypic correlation. An individual may deviate from the mean in both size and fecundity, for example, because her genotype affects both characters, because her environment affects both, or because of both effects. The correlation $r$ is related to the COVARIANCE ($cov$) between the traits (Appendix I) by the formulation $cov_{xy} = r_{xy}\sigma_x\sigma_y$, where $\sigma$ is the standard deviation in the trait.

The phenotypic covariance $cov_P$ between traits $x$ and $y$ is the sum of the covariance due to genotype and that due to environmental causes: $cov_P = cov_G + cov_E$. A similar but more complicated expression describes the relationship between the GENETIC ($r_G$) and ENVIRONMENTAL ($r_E$) CORRELATIONS that are components of the phenotypic correlation. The genetic and environmental correlations can differ substantially in magnitude and even in sign. For example, $r_G$, the genetic correlation (caused by additive effects of genes) between weight gain and thickness of back fat in pigs (*Sus scrofa*) was 0.13 in one experiment, whereas the environment-induced correlation $r_E$ was $-0.18$, so that the phenotypic correlation $r_P$ was approximately zero. Body weight and tail length in mice (*Mus musculus*), however, are positively correlated both genetically and environmentally: $r_G = 0.29$, $r_E = 0.56$, and $r_P = 0.45$ (Falconer 1981).

Such correlations cause CORRELATED RESPONSES to selection: as selection alters the mean of one character, that of an unselected character increases or decreases, depending on the sign of the genetic correlation (and at a rate determined by its magnitude, as well as the heritability of each trait). Thus selection on body weight in mice would bring about a correlated change in tail length. It is less obvious, but nonetheless true, that although back fat is not phenotypically correlated with weight gain in pigs, a genetic change in the mean of one character would carry with it a change in the other, because they are genetically correlated. The topic of genetic correlations will become important when we consider topics such as adaptation and the evolution of morphology.

Genetic correlations arise from two causes: pleiotropy and linkage disequilibrium. Two characters are genetically the same character to the extent that they are pleiotropically affected by the same genes. Unless the pleiotropic effects become modified or dissociated by the action of yet other genes, genetic correlations due to pleiotropy are essentially permanent. In contrast, genetic correlations due to linkage disequilibrium, such as the correlation between style length and anther height in *Primula vulgaris* (page 193), break down if recombination leads to linkage equilibrium. If two polygenic traits are each subject to stabilizing selection and they are governed by different sets of genes, linkage disequilibrium between the genes governing the two traits will not persist unless the loci are very tightly linked or selection strongly favors a correlation between the traits (Lande 1980). In a population that is at equilibrium under stabilizing selection, therefore, genetic correlations between traits are primarily caused by pleiotropy. Since virtually all genes appear to have numerous pleiotropic effects (Wright 1968), most characters are likely to be correlated with some other characters.

### Pleiotropy, linkage, and genetic variation

Pleiotropy may help to explain the persistence of the considerable genetic variance that is observed even in polygenic characters that strongly affect fitness. Several possibilities exist (Robertson 1955, Falconer 1981). One possibility is that variation in a quantitative trait such as bristle number does not in itself affect fitness, but that the trait in question is the pleiotropic by-product of genes that do affect fitness. For example, survival of *Drosophila melanogaster* in the larval stage is greater for genotypes that have intermediate bristle numbers as adults than those that have extreme bristle numbers (Kearsey and Barnes 1970). The bristle number itself cannot be responsible for the variation in viability because the larvae lack bristles. It is conceivable that heterozygotes at the individual loci have higher fitness than homozygotes because of their influence on whatever characters other than bristle number these loci affect, but there is no evidence that this is the case.

Another possibility is that the trait in question is subject to directional selection for increase in the mean, but that it is negatively genetically correlated with another such trait. The outcome could then give the appearance of stabilizing selection. For example, high fecundity should be advantageous, and directionally selected for increase, both earlier and later in an organism's life. However, Rose and Charlesworth (1981) found that female *Drosophila* that are fecund early in life lay fewer eggs later in life, and die earlier. Consequently, when a population was selected for high late fecundity by breeding only from the progeny of old females, the fecundity early in life declined (Rose and Charlesworth 1981). "Antagonistic pleiotropy," as the negative correlation between fitness-related traits has been termed, can stabilize the mean for each character at an intermediate value that gives the highest overall fitness.

Negative genetic correlations arising from pleiotropy may also explain why heritability remains high in a population subjected to prolonged directional selection; if the alleles for, say, high pupal weight have deleterious pleiotropic effects, these effects will counteract whatever selection (whether natural or artificial) might promote their increase in frequency, so that they may be arrested at intermediate frequencies and genetic variation persists. When directional selection for a particular trait is strong, negative genetic correlations also arise from linkage disequilibrium. Suppose that a small section of a chromosome carries several loci that affect the trait, among which are interspersed other loci that affect fitness. Among these chromosomes are some such as (1) + − + + *v* + and (2) − + − − *V* −, where + and − are alleles affecting the selected trait and *V* and *v* are, respectively, dominant advantageous and recessive deleterious alleles at a locus that affects viability. Selection for an increase in the trait rapidly increases the frequency of chromosome (1), and the allele *v* likewise increases in frequency because of linkage. There is then an association between *v* and the + alleles at several loci, and between *V* and the − alleles. If there are alleles like *v* at several or many loci, their accumulation by linkage with various + alleles will eventually so lower viability that the chromosomes on which they reside will cease to increase in frequency. Linkage disequilibrium, like pleiotropy, will tend to lower viability and other components of fitness if a trait is subjected to strong directional selection.

## RESPONSES TO ARTIFICIAL SELECTION

Ever since *The Origin of Species*, which Darwin opened with an analysis of variation and selection in domesticated organisms, evolutionary biologists have drawn useful inferences about the genetic basis of evolution from the study of artificial selection. Artificial directional selection generally entails breeding from individuals that exceed some cut-off point in the distribution of the trait of interest. Usually a fixed proportion of the population is retained for breeding, so the cut-off point changes as the character responds to selection (Figure 7). Although such selection is "artificial" in the sense that selection focuses on one trait rather than

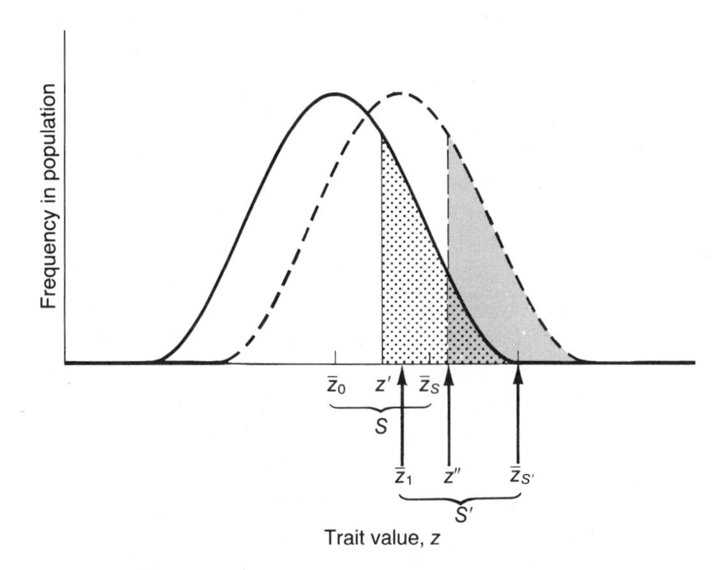


FIGURE 7

**Truncation selection in a program of artificial selection for increase in a character $z$. In generation 0, the population mean is $\bar{z}_0$. Individuals in the stippled region above the truncation point $z'$ are saved for breeding. Their mean is $\bar{z}_s$, so the selection differential is $S = \bar{z}_s - \bar{z}_0$. In the next generation the mean has increased to $\bar{z}_1$, and the same selection differential is maintained by breeding from the shaded portion above the truncation point $z''$. The selection differential is then $S' = \bar{z}_{s'} - \bar{z}_1$. The smaller the proportion saved for breeding, the larger the selection differential.**

overall fitness, natural selection is often very similar in its operation, especially hard selection (Chapter 6), under which only these individuals survive that possess a specific state of some characteristic such as toxin resistance. Artificial selection studies should not be dismissed as "unnatural." It is true, however, that laboratory populations differ from natural populations in several important ways: they are smaller, they usually do not receive gene flow from other populations, and their members cannot escape the stringent selection regime by dispersing to more favorable habitats.

The most common features of the response to artificial selection may be illustrated by the pioneering studies of Mather and Harrison (1949) and the more recent studies of Yoo (1980) on the number of abdominal bristles in *Drosophila melanogaster*. Mather and Harrison scored bristles on several segments of the abdomen on each of 40 flies (20 of each sex) in each generation, and bred from the top two of each sex. Yoo scored the bristles on a single segment and bred from 50 pairs out of 250 pairs scored in each generation. In each experiment, several replicate selection lines were derived from a single laboratory population. Because Yoo's selected populations were maintained at a larger size, we will examine this experiment in more detail. The general patterns of response were the same in both experiments, as they have been in many other such experiments on various organisms.

In the base population from which Yoo's selected lines were drawn, the mean bristle number was 9.35 in females and 6.95 in males. The phenotypic variances $V_P$ were 2.99 and 2.07 for males and females respectively, and the heritability for both sexes combined was 0.2. Selection for increased bristle number was imposed for 86 generations. By this time, mean bristle number in the various replicates had increased by 20–30 bristles over the original value—an increase of 12–19 phenotypic standard deviations, or 28–43 additive genetic standard deviations (the square root of the additive genetic variance $V_A$). On average, the mean had increasd by 316 percent, a truly impressive response.

However, the progress was by no means uniform (Figure 8). The replicate populations increased at different rates, and some (e.g., line *CRb*) showed temporary plateaus during which there was little increase, followed by brief periods of extremely rapid increase. Several of the populations (e.g., *Ua*) stopped responding to selection; they reached a SELECTION PLATEAU. The cessation of response was not caused by a loss of genetic variation, because variation in bristle number was high throughout the experiment and actually increased in later generations. This variation was partly genetic, because when selection was terminated (generations 86–122), mean bristle number declined abruptly, and then stabilized above the original level. The decline shows that alleles for low bristle number were still present, so that the mean could have been increased still further if these had been eliminated by further selection.

The decline when selection was relaxed shows that bristle number was negatively correlated with fitness; presumably the selection plateau was similarly caused by natural selection acting in opposition to artificial selection. In fact, some of Mather and Harrison's (1949) lines actually died out because of the reduction in fitness. Other characters also showed correlated responses: Mather and Harrison observed changes in the number of coxal and sternopleural bristles and in the number and form of spermathecae. It is likely that linkage disequili-

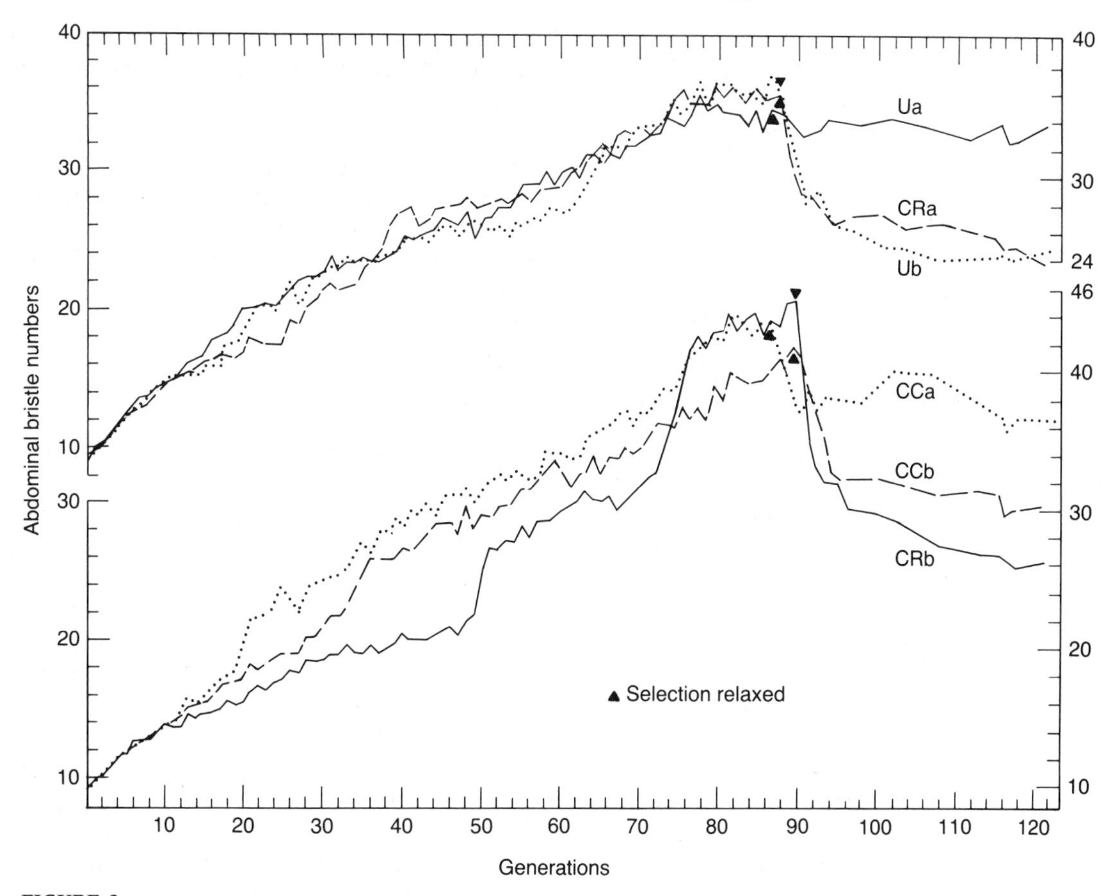


FIGURE 8

**Responses to artificial selection for increased number of abdominal bristles in six laboratory populations of *Drosophila melanogaster*. Mean bristle number in females is shown for two sets of populations separately for ease of visualization; both began at the same mean. The mean declined after artificial selection was terminated (relaxed) in each population, indicated by the black triangles. (From Yoo 1980)**

brium played some role in lowering fitness, because when Mather and Harrison resumed selection on lines in which they relaxed selection for a few generations, bristle number increased without as great a diminution of viability and fecundity. This suggests that during the period of relaxation, recombination dissociated deleterious alleles from the alleles for high bristle number, and that chromosomes free of deleterious alleles then increased in frequency when selection was resumed.

These experiments and others like them confirm the important role of linkage and pleiotropy in several respects. The selected character does not change at a steady rate, probably because new mutations occur and because recombination sporadically rearranges closely linked loci into new combinations that are favorable material for selection. Replicate populations vary in response, probably because patterns of linkage disequilibrium differ among the samples from which the populations are initiated; the frequency of polygenic combinations may also vary because of genetic drift as selection proceeds. Change in the character lowers some components of fitness because of pleiotropy and/or linkage disequilibrium, and the decrement in fitness eventually becomes great enough to reduce the response to selection, even while the population retains genetic variation. Finally, it is possible to select phenotypes that differ extremely from any that are present within the original population, at rates that are, as we shall see in Chapter 14, thousands of times faster than have been usual in the history of natural evolution. Although some of the overall response may well be based on new mutations (Hill 1982), much of it is based on preexisting genetic variation, which is typically considerably greater than the input of genetic variance by mutation.

## GENETIC AND DEVELOPMENTAL HOMEOSTASIS

The plateau that is often observed in artificial selection experiments is the consequence of the opposition between natural and artificial selection. Determining whether pleiotropy or linkage disequilibrium is the more important cause of the opposition is not easy. If pleiotropy is responsible, the same traits might be expected to show a correlated response in different populations selected for the same character, and this is sometimes observed (Clayton et al. 1957). If linkage disequilibrium is responsible, correlated responses should be more pronounced under strong than weak artificial selection, because stronger selection maintains more extensive linkage disequilibrium. This prediction too has been borne out in some experiments (Eisen et al. 1973). Although sternopleural bristles change as a correlated response to selection on abdominal bristles, Davies and Workman (1971) showed that the correlation was due more to linkage than to pleiotropy by successfully selecting these characteristics in opposite directions in the same population of flies. Moreover, careful mapping of many of the loci that affect these characters showed that they are on different sites on the chromosomes (Davies 1971).

Lerner (1954) referred to the tendency of a population to resist change, and to return toward its original state when selection is relaxed, as GENETIC HOMEOSTASIS. The basis of genetic homeostasis, he argued, is the superior fitness of heterozygotes, of which the common phenomenon of inbreeding depression gives evidence. Strong artificial selection increases homozygosity and so reduces fitness. One reason for the low fitness of homozygotes, Lerner argued, is their reduced capacity for developmental homeostasis.

DEVELOPMENTAL HOMEOSTASIS is the capacity of an individual's genotype to produce a proper, well-formed, adaptive phenotype in the face of the perturbations that can occur during the course of development. The development of a normal phenotype is canalized along proper channels and resists diversion into other paths. One example of the superior developmental homeostasis of heterozygotes in *Drosophila* is the environmental variance $V_E$ for characters such as bristle number. This increases directly in proportion to the number of chromosomes for which the flies are homozygous (Robertson and Reeve 1952). Moreover, bilateral asymmetry in bristle number—surely an example of uncanalized development—is greater in homozygous than in heterozygous flies (Thoday 1955). Among island populations of the lizard *Uta stansburiana* asymmetry is greater in small than in large populations, perhaps because of inbreeding, and is especially pronounced in populations that are most distinct from their continental ancestors, perhaps because of directional selection (Soulé 1967).

Developmental homeostasis does not automatically arise from heterozygosity, however; newly arisen heterozygous mutations do not reduce $V_E$ (Mukai et al. 1982). Developmental homeostasis is the consequence of alleles that have been retained in the population by natural selection, perhaps because of their effect on canalization. Moreover, developmental homeostasis depends on the joint action of genes that have been selected to interact harmoniously to produce a well-organized pattern of development, and so are coadapted (Dobzhansky 1955). For example, in *Drosophila* heterozygotes are more bilaterally symmetrical than homozygotes only if the chromosomes come from the same population. Heterozygotes formed by crossing flies from populations adapted to different environments were highly asymmetrical (Thoday 1955; see Table III).

Coadaptation of alleles at different loci implies that loci interact epistatically. So far we have given rather little consideration to epistasis, because the additive component of genetic variance is the basis of the response to selection. But we know that phenotypes emerge via complex, interacting developmental pathways. Thus we should examine in more detail the extent to which gene interaction has important effects on the phenotype. One context in which gene interaction might be important is the phenomenon of dominance.

## The problem of dominance

Early in the history of genetics, it became evident that among the spontaneous mutations that noticeably affect the phenotype of *Drosophila* and other organisms, a few are dominant over the wild-type allele, but the majority are recessive. This poses a problem, because there is no a priori reason why the recessive phenotype should not be favored by natural selection, in which case the recessive allele would be the prevalent wild-type, and the dominant allele would be the mutant. R. A. Fisher (1928) postulated that the wild-type allele is not intrinsically dominant over the mutations that arise from it, but that its dominance evolves. His reasoning was that natural selection would favor modifier alleles at other loci that would suppress the deleterious phenotypic effect of recurrent mutations at the locus in question. Thus in the presence of *mm* at the modifier locus, the phenotype of *AA′* would be intermediate between the phenotypes of *AA* and *A′A′*, but in the presence of the favored modifier allele *M*, *AA′* would have the same phenotype as *AA*, so the deleterious mutation *A′* would be recessive. This theory, then, implies strong epistasis between loci, and strong coadaptation: evolution at locus *M* is governed by its interaction with locus *A*.

In one of the most famous disputes in the history of evolutionary biology, Sewall Wright (1929) took strong exception to Fisher's theory. He admitted that in principle dominance could be altered by selection of modifiers, because earlier work by geneticists, including himself, had already demonstrated this. He argued, however, that because the deleterious mutation *A′* is eliminated rapidly by selection, its frequency in the population will be about equal to the recurrent mutation rate *u*, which is generally about $10^{-5}$. Since a modifier *M* can only be advantageous in conjunction with *A′*, the selection coefficient *s* in its favor cannot be much greater than *u*. An allele with such a small selective advantage is highly subject to drift; but Wright's principal argument was that the locus *M*, which surely does not exist solely for the sake of modifying dominance at the *A* locus, has other pleiotropic functions that will be much more important, and these will predominate in determining its allele frequency.

It is quite plausible, on the other hand, that modifiers should be selected to affect dominance at a locus that has a stable polymorphism, for then the heterozygote has a high frequency. This seems to have occurred in the evolution of the polymorphic mimetic forms of the butterfly *Papilio dardanus* (page 202), in which several alleles form a dominance series so that intermediate, nonmimetic phenotypes do not arise. The dominance breaks down when these forms are crossed with other populations that have different modifier alleles.

The explanation of dominance is often physiological, as Wright (1929) had suggested: recessive alleles are often those that do not produce a functional gene product. But this raises the question of why the amount of gene product produced by a heterozygote should suffice to yield the same phenotype as the dominant homozygote. One possibility is that the amount of a particular enzyme has little influence on the total flux through a biochemical pathway, because the flux is regulated by interactions among the several enzymes along the pathway (Kacser and Burns 1981). Another possibility (Sved and Mayo 1970) may be that selection has favored canalization of the phenotype. Thus we should examine canalization in more detail.

## Genetic assimilation and canalization

There is no clear distinction between genetic and environmental disruptions of development. For example, the crossveinless condition in the wing of *Drosophila* can be produced by any of many mutant alleles or simply by exposing pupae to heat shock (Waddington 1953). Such environmentally induced mimics of mutations, or PHENOCOPIES, show that there are numerous causes for the deflection of development from one pathway into another.

By selecting flies that developed the crossveinless condition in response to heat shock, Waddington developed a population in which most of the flies were crossveinless when treated with heat. But after further selection, a considerable portion of the population was crossveinless even without a heat shock, and the crossveinless condition was heritable. An initially acquired character had become genetically determined, a phenomenon that Waddington called GENETIC ASSIMILATION. This response was not Lamarckian inheritance (Chapter 3). Rather, genotypes differ in their degree of canalization, so that some are more easily deflected than others into an aberrant developmental pattern. When the aberrant pattern appears, selection for this character favors alleles that canalize development into the newly favored pathway (Figure 9). As such alleles increase in frequency, less environmental stimulus is required to produce the new phenotype. Genetic assimilation may sometimes be important in evolution. For example, high temperature during development reduces body size in *Drosophila*. This ontogenetic response is paralleled by genetic differences among geographic populations that experience different temperature regimes. Moreover, experimental populations of *Drosophila pseudoobscura* that were maintained at different temperatures for six years diverged so that even when reared at the same temperature, the cold-adapted flies were larger than those that were warm-adapted (Anderson 1966).

The relationship of canalization to dominance is illustrated by Rendel's

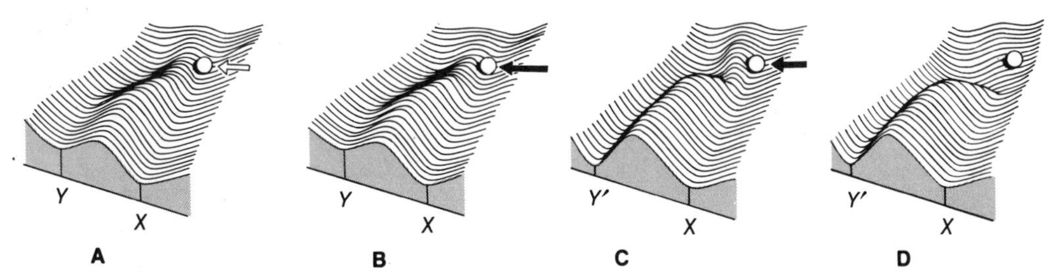


FIGURE 9

**Genetic assimilation visualized in terms of an "epigenetic landscape." The development of an organ is conceived as a ball moving along a developmental pathway. Each point on the surface is a conceivable morphology, but movement along some paths (valleys) is more likely than others (hills). In (A), development is generally along path X, but environmental stimuli (hollow arrow) may push it into path Y. In (B), mutations (solid arrow) may push development into path Y. If genetic changes (C, D) successively lower the barrier to path Y and deepen this channel, this may become the new canalized norm. (From Waddington 1956a)**

(1967) work on the scutellar bristles of *Drosophila melanogaster*. Wild-type (++) flies almost invariably have four bristles, as do heterozygotes for the recessive mutant *scute* (*sc*). The *scute* homozygote (*scsc*) has a lower, more variable number of bristles. Selection in a homozygous *scsc* population can change the mean bristle number, so there is polygenic variation for the trait that is ordinarily not expressed; scutellar bristle number in ++ flies is highly canalized.

Rendel postulated that bristle number is affected by an underlying character, perhaps a substance used in bristle formation, that has a continuous distribution in the population much like any other quantitative trait such as pupal weight. He postulated that natural selection has favored genes that regulate the relation between the amount of the substance (which he called "Make") and the number of bristles, so that ++ flies develop four bristles even if the amount of "Make" varies considerably (Figure 10). Thus the phenotype is canalized. If, then, the +*sc* heterozygote has less "Make" than the ++ heterozygote usually does, it will still fall within the region of canalization: the decrease in "Make" does not affect bristle number. Thus the + allele is dominant. Rendel's evidence for this hypothesis came from an experiment in which he selected for increased bristle number in an *scsc* population (in which bristle number is variable), and then backcrossed the + allele into this selected genetic background. Both the ++ and the +*sc* genotypes then had a higher mean bristle number than is normal for them, presumably because the polygenic loci so increased the amount of "Make" that these genotypes were now outside the region of canalization.

Although Rendel's model may not be correct in all details (Sheldon and Milton 1972), it suggests that in this case dominance is just one manifestation of canalization; that heterozygotes may be developmentally more stable than homozygotes if they lie closer to the center of the region of canalization; and that strong epistatic interactions between major genes such as the *scute* locus and genes such as those that govern canalization influence the phenotype. Why should such canalization (dominance) have evolved? Rendel noted that an increase in scutellar bristle number is often correlated with a decrease in abdominal bristles.

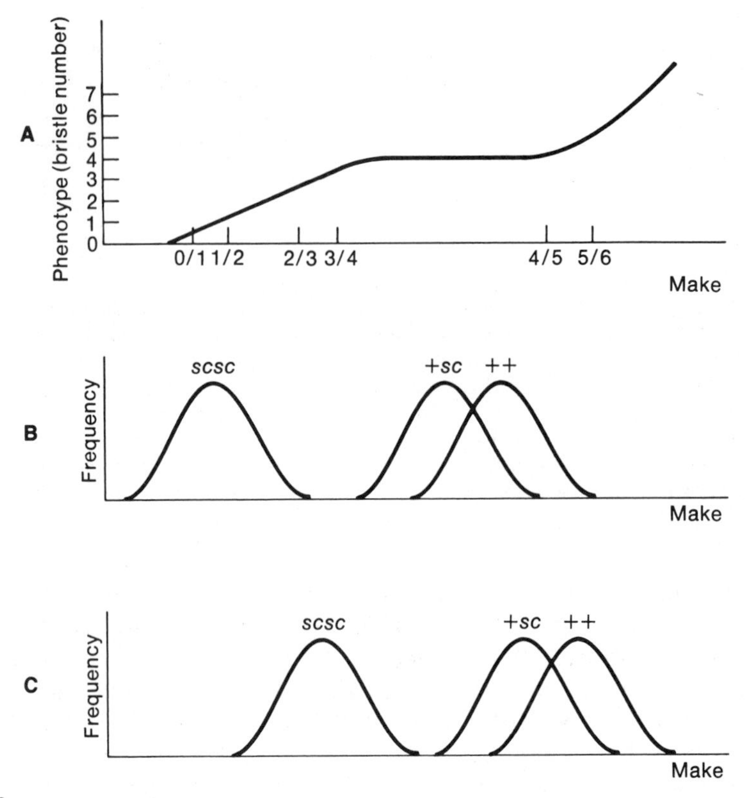


FIGURE 10

**Canalization and the response to selection. (A) The relationship between the number of scutellar bristles of *Drosophila melanogaster* and the value of an underlying variable called "Make." The notations 0/1 and so forth mark the threshold values of "Make" at which the bristle number is changed by 1. Development is canalized at 4 bristles; i.e., there is no change in bristle number despite a large change in "Make." (B) The distribution of "Make" in female flies of three genotypes at the *scute* locus (*sc*). Both ++ and +*sc* lie within the canalization plateau and have 4 bristles. However, *scsc* flies vary in bristle number from 0 to 3. (C) After modifier alleles at other loci have been selected to increase mean bristle number, all genotypes at the *scute* locus have more "Make." + + and +*sc* have been shifted to the edge of the canalization plateau, so they have a higher mean, and a more variable number of bristles than previously.**

He suggested that the two regions of the fly compete for a common pool of some limited substance ("Make"), and that canalization of a trait such as scutellar bristle number is adaptive because it achieves a developmental balance that prevents any one part of the phenotype from monopolizing energy and materials.

## Thresholds

Scutellar bristle number in *Drosophila* is an example of a THRESHOLD CHARACTER: a trait that is affected by a continuous distribution of some underlying trait, but which is expressed discontinously at the phenotypic level. Canalization prevents the expression of variation except when the underlying character (possibly some chemical substance) passes a threshold. Such characters are probably very important in evolution, and may be expected to evolve in spurts of rapid change, interspersed with periods of stasis. For example, the polygenic variation that affects the number of digits in mammals, although continuous for some underlying character, is expressed as a discrete number of digits. The range of variation in the underlying characters may be divided into canalized regions for either four or five digits (for example), separated by narrow intervals (thresholds) around which digit number is variable (Figure 11). Lande (1978) modified his model (page 201) of evolution of a polygenic character to take such thresholds into account. If $h^2$ is the heritability of the underlying character and $s$ is the selection coefficient in favor of a phenotype with one less digit, the mean ($\bar{z}$) of the continuous variable determining digit number will decrease at a rate of $\Delta\bar{z} = -0.066h^2s$ per generation when it lies within a zone of canalization, and at a rate $\Delta\bar{z} = -0.395h^2s$ when it lies at the threshold. So the rate of evolutionary loss of digits will be about six times more rapid at some times than others, even if selection is constant. Even if $h^2 = 0.1$ and selection is very weak ($s = 0.001$), it would take only a million years to go from four digits to zero, which is quite a rapid evolutionary change. In several groups of lizards, some species have fully formed limbs and closely related species have fewer digits or none at all. As Lande points out, these differences could have evolved very rapidly.

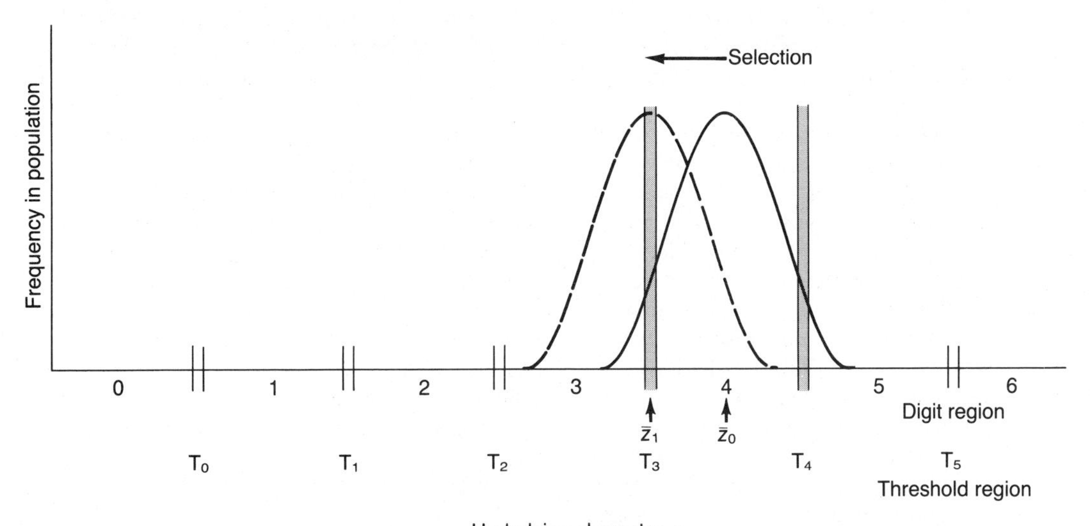


FIGURE 11

**Model for the evolution of a threshold character such as number of digits. An underlying character *z*—perhaps a substance that induces digit development (cf. "Make" in Figure 10)—is polygenically determined and continuously distributed in the population. Individuals whose value of *z* lies within a digit region have a stable (canalized) number of digits; if *z* lies in a threshold region, say $T_3$, digit number may be either 3 or 4. If $\bar{z}_0$ is the initial population mean and lies in digit region 4 (solid curve), most of the population has 4 digits if $h^2$ (expressed by the width of the curve) is low. With selection for reduced digit number, the mean is lowered into threshold region $T_3$ (broken curve, with mean $\bar{z}_1$) and digit number is more variable, so the rate of evolution increases. (After Lande 1978)**

## SUMMARY

The dynamics of allele frequency change at a locus are greatly affected by linkage and interaction with other loci. Selection for favorable combinations of genes can create strong associations (linkage disequilibrium) among alleles at different loci if they are tightly enough linked. Different gene combinations may confer high fitness, so that a population can evolve toward any of several or many stable genetic compositions. The rate of response to selection can be increased or decreased by linkage disequilibrium. Populations harbor enough additive genetic variance for most characters to change rapidly and extensively in response to selection, but pleiotropy and linkage disequilibrium, giving rise to negative genetic correlations between the selected character and other traits including fitness, reduce the response to selection, and cause the character to return toward its original state if selection is relaxed. These factors may help to account for the persistence of polygenic variation even in the face of stabilizing or directional selection, but recurrent mutation also contributes to the variation. Interactions among loci influence some traits, and help to account for developmental homeostasis and canalization. The rate of evolution may be irregular because of linkage disequilibrium and canalization.

## FOR DISCUSSION AND THOUGHT

1. Some characteristics, such as the growth forms of many plants, show a great deal of nongenetic plasticity compared to other traits. Explain why such traits should respond less rapidly to natural or artificial selection than less plastic traits. Under what conditions should developmental plasticity evolve?
2. Would you expect the genetic correlation between two traits to remain the same in a lineage over long periods of evolutionary time? Why or why not? If the genetic variation for two traits that strongly affect fitness is at an equilibrium set by the balance between mutation and selection, and if the traits are genetically correlated, is the genetic correlation likely to be positive or negative?
3. Some genotypes of corn manifest "general combining ability," meaning that they contribute a predictable increment to seed yield in the $F_1$ hybrid, no matter what strains they are crossed with. Other genotypes have "specific combining ability;" they contribute more to yield when crossed with some strains than others. How would you characterize such genotypes in terms of additive and nonadditive gene action? How would combining ability affect the response to natural or artificial selection?
4. Discuss in detail the evolutionary importance of genotype × environment interactions of the kind described by Lewontin (1974b). Can you see any ways in which this phenomenon might be important in human genetics?
5. Some authors have assumed that if species differ in some trait, the difference reflects a specific adaptation of each species to its special environment or way or life. Discuss how the phenomena described in this chapter bear on this assumption.
6. More than one author has suggested that genetic homeostasis is a mechanism whereby populations can retain the genetic variation necessary for adaptation to subsequent environmental changes. Criticize this view.
7. If modifier genes can influence dominance at a major locus, could they also act on the various effects of a pleiotropic gene and eliminate pleiotropy? How quickly could this happen? What would be the evolutionary consequences?
8. Distinguish between *genetic variation* and *selectable genetic variation*. Why is some genetic variation not selectable? Some experiments (e.g., Scossiroli 1954, Ayala 1966) have

shown that the response to selection can be enhanced by irradiation. Does this imply that the population had been genetically invariant?

9. Do terms such as canalization and developmental homeostasis *explain* the observations to which they refer? Speculate on the biochemical or developmental mechanisms that might be responsible for these phenomena.
10. Explain why a population might not adapt as quickly to each of several independent environmental factors simultaneously as to a single factor. (See Pimentel and Bellotti 1976 for an example.)

## MAJOR REFERENCES

Falconer, D.S. 1981. *Introduction to quantitative genetics.* Second edition. Longman, London and New York. viii + 340 pages. An unmatched introduction to quantitative genetics. Requires some knowledge of statistics, but not of advanced mathematics.

Lerner, I.M. 1958. *The genetic basis of selection.* Wiley, New York. 298 pages. A somewhat outdated but still excellent coverage of many of the same topics as Falconer.

# Speciation

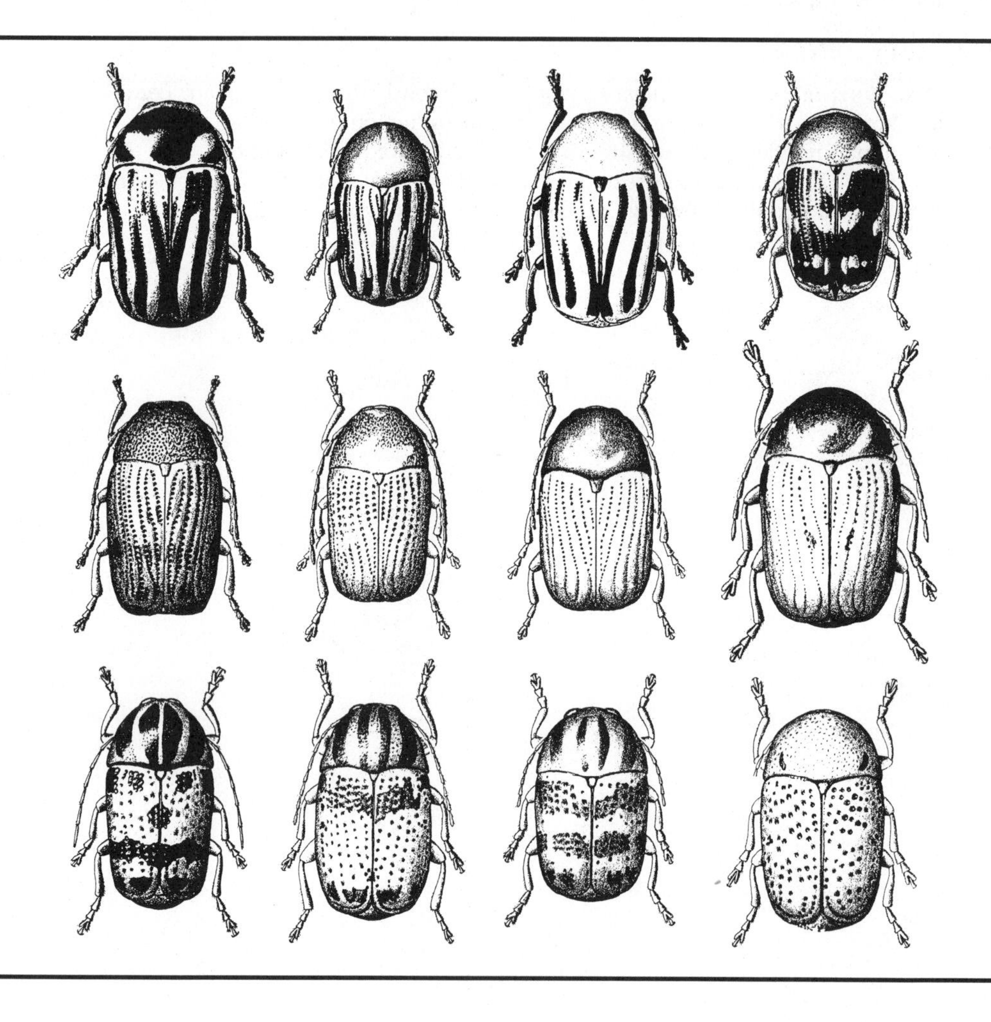

# Chapter Eight

The preceding chapters have explored the genetic mechanisms of evolution within populations, and have demonstrated the numerous factors that cause populations of a species to diverge. If evolution consisted only of the mechanisms we have described so far, there would be very little diversity. Diversity arises through CLADOGENESIS, the divergence of different genetic lines from common ancestors. The critical step in cladogenesis is speciation, the formation of two or more species from a single stock.

## THE BIOLOGICAL SPECIES CONCEPT

The term "species" refers both to a taxonomic category and to a biological concept (Chapter 4). Very often, but not always, taxonomic species are equivalent to biological species. But in reading the taxonomic literature on many groups, it is important to bear in mind that some taxonomists do not use the biological concept to define their taxonomic species, but rely on morphological differences to define species, without explicitly considering whether or not the morphological differences provide evidence for reproductive isolation. Moreover, as we saw in Chapter 4, populations are often only partially reproductively isolated; in these cases the biological species concept cannot be applied. Such borderline cases, which are especially prevalent in plants, do not invalidate the concept where it does apply; still less do they vitiate the concept of speciation as a process. Rather, they forcefully illustrate that in many instances species arise by the gradual evolution of barriers to gene exchange.

The evolution of new species is equivalent to the evolution of genetic barriers to gene flow (i.e., "isolating mechanisms") between populations (Table VI in Chapter 4). As indicated in Chapter 4, isolating mechanisms are numerous in kind (Levin 1978), and differ from group to group. Often the sterility or inviability of hybrids is a potential isolating mechanism, but usually, at least in animals, such postmating isolation does not come into play because premating isolation is sufficient to prevent interbreeding.

It is sometimes difficult to tell whether the variation among individuals from a locality represents one variable species or more than a single species. If specimens differ discretely in only one character, the variation may be only a single-locus polymorphism (e.g. the king snakes in Figure 4 in Chapter 4). If it is possible to recognize heterozygous genotypes but they are nevertheless absent, there is prima facie evidence for reproductive isolation; for example, the existence of two homozygotes but no heterozygotes at an enzyme locus sometimes provides such evidence. Except in self-fertilizing or parthenogenetic forms, a strong correlation among several variable characters provides evidence that there exist two or more species, if it may be assumed that the characters are under separate genetic control. Morphological criteria of this kind are the usual basis on which taxonomists describe species; sometimes it is also possible to identify premating isolation by direct observation.

## THE GENETICS OF SPECIES DIFFERENCES

Species often differ in numerous characteristics, not all of which are instrumental in their reproductive isolation. Genetic differences between species include alleles that already differed in frequency among populations before they achieved reproductive isolation, those that diverged in frequency during the process of speciation, and those that came to differ after reproductive isolation was established. To some extent, the process of divergence is continuous, proceeding both before and after speciation, although genetic differentiation can become more pronounced after reproductive isolation is attained. Among populations and species in the *Drosophila willistoni* complex, for example, there is progressively greater differentiation in allele frequencies at a number of enzyme loci as one passes from a comparison of local populations to semispecies, but fully isolated sibling species differ more appreciably (Figure 1). Electrophoretic analyses have shown that the "genetic distance" (Table V in Chapter 4) between closely related species is sometimes considerable, but in some groups reproductively isolated species differ only slightly in this respect (see reviews by Ayala 1975, Avise and Aquadro 1982). At least so far, there is little reason to believe that differentiation at allozyme loci affects the evolution of reproductive isolation.

The genetic basis of morphological differences between species can be de-

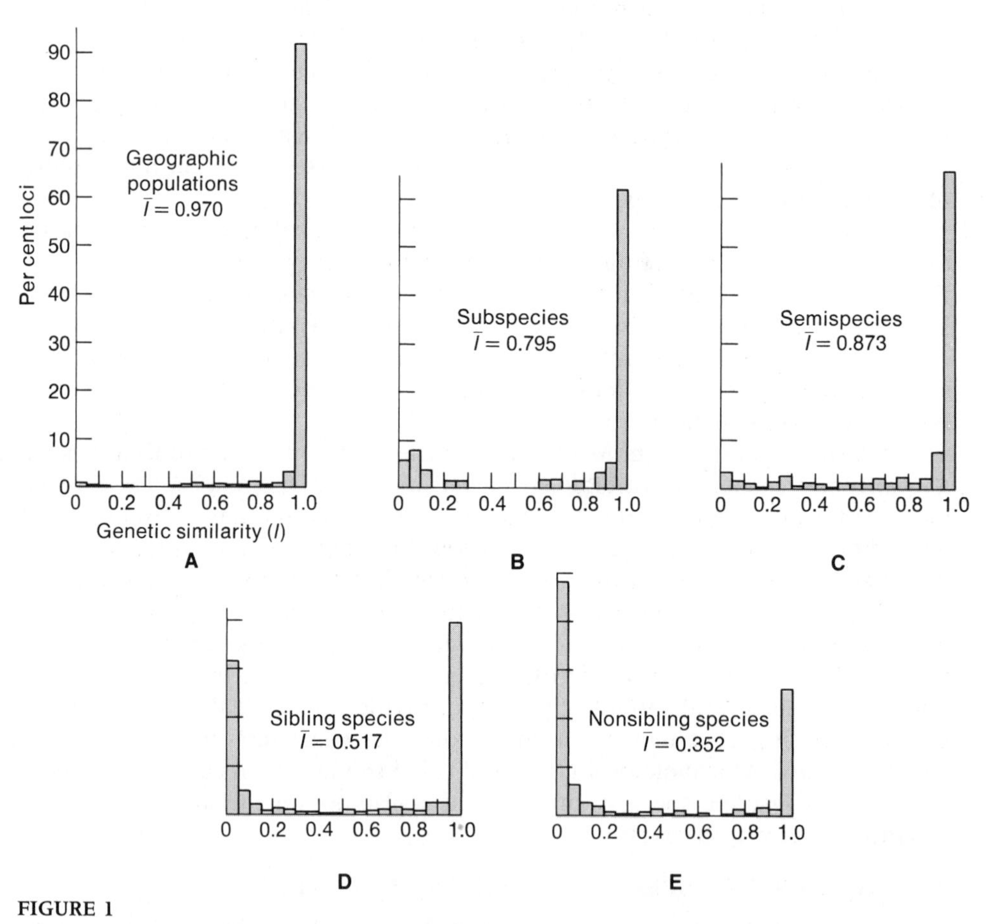


FIGURE 1

**The percentage of electrophoretic loci that exhibit various levels of gene frequency similarity among (A) geographic populations, (B) named subspecies, (C) semispecies, (D) sibling species, and (E) nonsibling species in the *Drosophila willistoni* complex. (From Ayala et al. 1974)**

FIGURE 2

**Males of (A) Lady Amherst's pheasant, *Chrysolophus amherstiae,* and (B) golden pheasant, *Chrysolophus pictus.* [Redrawn from lithographs by John Gould (1804-1881), in A. Rutgers, *Birds of Asia,* Methuen, London]**

termined only when hybrids are fertile enough to yield $F_2$ or backcross progeny. Especially in animals, segregation of different phenotypes among the progeny commonly shows that the difference in each morphological character is determined by several or many genes, that often have a more or less additive effect on the phenotype (Lande 1981a, Templeton 1981). For example, the male of the Hawaiian *Drosophila heteroneura* differs from the closely related *D. silvestris* in that its head is greatly extended on the sides, and probably functions in territorial or sexual display. The difference in head shape is controlled by a major gene on the X chromosome and about 8 or 10 additively acting modifier genes on the autosomes (Val 1977). Males of related species of pheasants usually differ in a striking array of brilliantly colored plumes, crests, and exaggerated wing and tail feathers that are prominently displayed in courtship and are doubtless the basis for reproductive isolation. From crosses between the pheasants *Chrysolophus pictus* and *C. amherstiae* (Figure 2), Danforth (1950) concluded that at least three major loci affect the color of the crest, numerous loci affect the form of the crest feathers, and at least two loci affect the color of the cape feathers. By analyzing backcross hybrids with mutant-marked chromosomes, Coyne (1983a) showed that genes on every chromosome affect the difference among related *Drosophila* species in the shape of a structure in the male genitalia, the only morphological feature by which these species can be distinguished. Although it was thought at one time that genital morphology is an isolating mechanism in insects, it seems not to be. The causes of divergence in genitalic morphology are unknown, but might include sexual selection (Eberhard 1986; see below).

Although many species differences are polygenically determined, there are numerous cases, especially in plants (Hilu 1983, Gottlieb 1984), in which a qualitative difference between species is due to one or two genes. In some of these cases, it is likely that a single gene mutation causes the loss of a character that originally evolved by changes in several genes.

The genetics of ethological (sexual) isolation are known in greatest detail for several groups of *Drosophila*. Almost invariably, genes on several of the chromosomes contribute to sexual isolation, although it is usually not known how many genes on each chromosome are involved. For example, the courtship signals of males of *Drosophila arizonensis* and *D. mojavensis* appear to differ because of genes on the Y chromosome and one autosome (Zouros 1981). In closely related species of crickets (*Teleogryllus*; Hoy et al. 1977) and of sulfur butterflies (*Colias*; Grula and Taylor 1980), genes controlling both the male courtship signal and the female response appear to reside on the sex chromosome.

When interspecific hybrids are sterile, the sterility may vary in degree, and can have several causes. Related species often have differences in chromosome structure such as reciprocal translocations or pericentric inversions that can lower fertility (Stebbins 1950, White 1973). Not all species differ in chromosome structure; for example, among Hawaiian *Drosophila*, in which even small rearrangements can be discerned in the banding pattern of the salivary chromosomes, related species sometimes have identical structure (Carson 1970). However, chromosomes of different species sometimes fail to pair properly in hybrids even when no structural differences can be discerned, perhaps because of submicroscopic structural differences (Stebbins 1982). Mispairing of chromosomes in hybrids between species of wheat (*Aegilops*) can be caused by a single gene mutation (Riley 1982).

Sterility can arise from an interaction between chromosomal genes and cytoplasmic factors transmitted through the egg, as in certain mosquitoes (*Culex*; Laven 1958); in some cases, the cytoplasmic agent is an intracellular microorganism (Ehrman and Williamson 1969). Hybrid sterility can also be caused by incompatibility between chromosomal genes. For example, although male hybrids between *Drosophila pseudoobscura* and *D. persimilis* are sterile, the females are fertile, so it is possible, by backcrossing, to construct progeny with any desired mixture of chromosomes. It has been determined that the sterility is due to an interaction between at least two sex-linked loci and autosomal loci located on each arm of each of the three autosomes (Dobzhansky 1970). The autosomal loci have an additive effect on the degree of sterility. Although the data from *Drosophila* generally point to polygenic effects on sterility, cases have been described in plants in which sterility is attributable to one or two loci (reviewed by Templeton 1981).

It seems clear, then, that the characters that isolate species differ genetically in the same way as characters that vary within species. Frequently they are polygenically controlled; sometimes they are caused by a few loci with strong epistatic interactions, modified to varying degrees by other genes (Templeton 1981). Thus there is nothing mystical or intangible in the difference between species; their differences are amenable to the same analyses as variations within species, and prove to have the same kinds of genetic foundations.

Although sterility and other isolating mechanisms have been genetically analyzed, the molecular basis of sterility, like that of most developmental phenomena, has not been determined. It is possible that differences in gene regulation are involved, but this in itself says little. It is known that in hybrid fishes, enzymes coded by paternal and maternal genes are expressed later or earlier in development than in non-hybrids, suggesting that the regulatory systems governing transcription or translation differ (Whitt et al. 1977). The morphology of hybrids frequently is developmentally unstable (Levin 1970); for example, even though the bristle pattern of *Drosophila simulans* and *D. melanogaster* is identical and almost invariant, hybrids between these species vary greatly in this respect (Sturtevant 1920–1921).

To the question, how great a genetic difference makes a species? there is no simple answer; it may be a meaningless question. As Mayr (1963, p. 544) put it, species differences cannot "be expressed in terms of the genetic bits of information, the nucleotide pairs of the DNA. This would be quite as absurd as trying to express the difference between the Bible and Dante's *Divina Commedia* in terms of the difference in the frequency of the letters of the alphabet used in the two works." But there is no evidence that closely related species differ, as do the Bible and the *Divina Commedia*, in their entirety. Rather, species owe their existence to specific characters governed by specific genes: sexual behavior, the timing of reproduction, and specific developmental pathways that become disrupted and result in sterility or inviability.

## MODES OF SPECIATION

Genetic barriers to interbreeding, or isolating mechanisms, can conceivably arise in many ways, and speciation can accordingly be classified into different modes. Two such classifications (Table I) emphasize, respectively, the geographic scale on which speciation may occur, and the genetic events entailed in the origin of reproductive isolation. These are in part related to each other, because certain genetic modes of speciation require geographic isolation, and others do not. Suppose, for example, that reproductive isolation by hybrid sterility or by a difference in mating behavior is based on the cumulative, perhaps additive, effect of several loci, so that genotypes *AABBCC* and *aabbcc* are reproductively isolated from each other, but not from other genotypes such as *AaBbCc*. It is generally unlikely that *aabbcc* will arise and form a reproductively isolated entity within the confines of a parental population *AABBCC*, because the numerous other "intermediate" genotypes will form a reproductive bridge between them. Moreover, the alleles *a, b,* and *c* will be selected against when rare, if they contribute to hybrid sterility and hence are disadvantageous in heterozygous condition. The most readily visualized means by which reproductive isolation with a polygenic basis can arise is by the interposition of an external barrier between two populations so that, in the absence of gene flow, different alleles can be fixed at each of the loci. Ernst Mayr (1942, 1963), whose writings have had a strong influence on speciation theory, has argued forcefully that geographic isolation is almost invariably necessary for speciation to occur.

## ALLOPATRIC SPECIATION

The evidence for allopatric speciation is extensive (Mayr 1942, 1963), coming most abundantly from studies of geographic variation. Species vary geographically in the very characteristics that can bar gene exchange between sympatric

TABLE I

**Two classifications of potential modes of speciation in sexual organisms**

BY GEOGRAPHY AND LEVEL (AFTER MAYR 1963)

1. Hybridization (maintenance of hybrids between two species)
2. Instantaneous speciation (through individuals)
   - A. Genetically: Macrogenesis (single mutation conferring reproductive isolation)
   - B. Cytologically
     - a. Chromosomal mutation (e.g., translocation)
     - b. Polyploidy
3. Gradual speciation (through populations)
   - A. Sympatric speciation
   - B. Parapatric (semigeographic) speciation
   - C. Allopatric (geographic) speciation
     - a. By isolation of a colony
     - b. Division of range by extrinsic barrier or extinction of intermediate populations

BY POPULATION GENETIC MODE (AFTER TEMPLETON 1982)

1. Transilience
   - A. Hybrid maintenance (selection for hybrid)
   - B. Hybrid recombination (selection for recombinants following hybridization)
   - C. Chromosomal (fixation of chromosomal mutation by drift and selection)
   - D. Genetic (founder event in a colony)
2. Divergence
   - A. Habitat (divergent selection without isolation by distance)
   - B. Clinal (selection on a cline with isolation by distance)
   - C. Adaptive (erection of extrinsic barrier followed by divergent microevolution)

The categories in Mayr's classification correspond approximately to those in Templeton's as follows: Templeton's 1A, 1B ≈ Mayr's 1; 1C = 2Ba, 2A = 3A, 2B = 3B, 1D = 3Ca, 2C = 3Cb. Templeton's classification does not include speciation by nonadaptive gradual divergence, polyploidy or macrogenesis, the latter included by Mayr only for historical reasons. I have reorganized these authors' classifications to facilitate comparison. Speciation by chromosomal mutation (Mayr's 2Ba) would better be placed in category 3 (gradual) than in category 2 (instantaneous).

species (Chapter 4); quite often, geographically more remote populations are more strongly isolated by sterility or ethological differences (when tested experimentally) than neighboring populations (e.g., Kruckeberg 1957, Oliver 1972). That remote populations might well not interbreed if they came into contact is illustrated by cases of CIRCULAR OVERLAP, in which a chain of races that are believed to interbreed curves back on itself, so that the highly divergent forms at the termini come into sympatry and do not interbreed. A possible example (Fox 1951) is a group of garter snakes (*Thamnophis*) in western North America (Figure 3).

Evidence for allopatric differentiation is provided by the frequent correspondence between biological and topographical discontinuities. For example, freshwater animals show the greatest regional diversity in mountainous regions where there are many isolated river systems. On islands, a species that is homogeneous over much of its continental range may diverge spectacularly in appearance, ecology, and behavior; the drongo (*Dicrurus*) described in Chapter 4 is an example. Cases of "double invasion," many of which are described by Mayr (1942), are especially interesting. For example, the bird *Acanthiza pusilla* is widespread on the Australian continent and has a slightly differentiated population on Tasmania, where a more markedly differentiated species, *A. ewingi*, also occurs. It is plausible that during a Pleistocene glaciation, when the sea level was lower, *Acanthiza* invaded Tasmania and differentiated into *A. ewingi* when it was isolated by a subsequent rise in sea level. A second invasion, perhaps during a later glacial period, may have established *A. pusilla* on the island.

Such multiple invasions explain why each island of an archipelago often has several related species, while a remote island of similar size may have just one. Most of the islands in the Galápagos Archipelago have several species of Darwin's finches, but there is only one species on Cocos Island, 600 miles away. The Cocos Island finch has been there long enough to have become morphologically very distinct, but it has not had the opportunity to proliferate into more than one species. Within the archipelago, in contrast, populations can diverge on different islands and recolonize the islands from which they came, having become reproductively isolated during their period of allopatry.

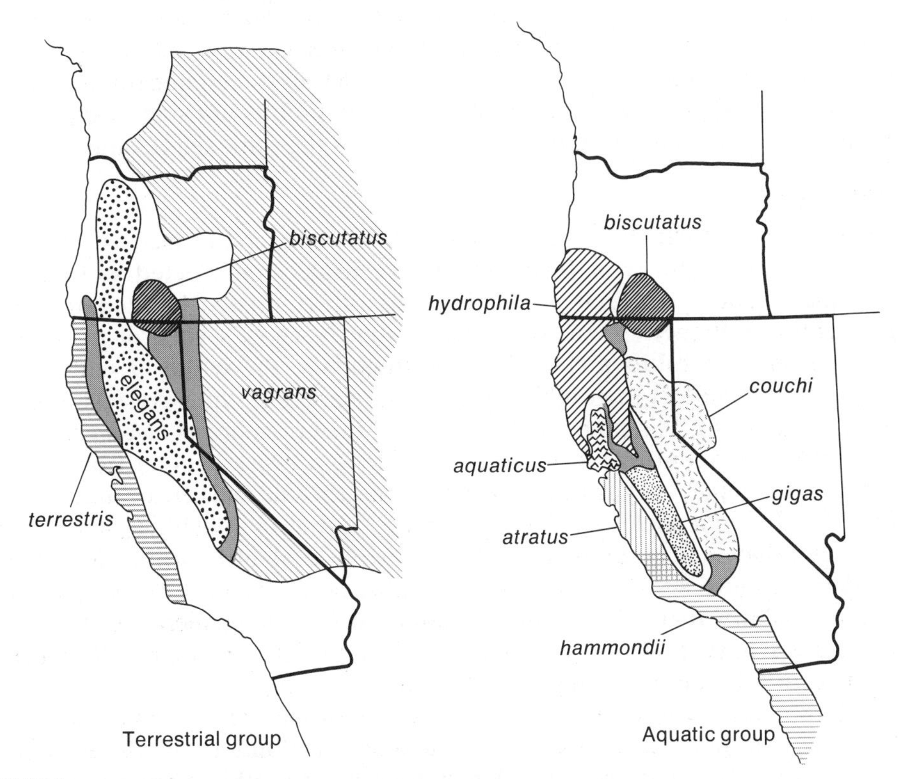


FIGURE 3

**Complex relationships among races of garter snake (*Thamnophis*). In the aquatic group, *hammondii, gigas, couchi, hydrophila, aquaticus,* and *atratus* form a sequence of allopatric subspecies that interbreed where they meet (gray areas); but *atratus* coexists with *hammondii* without interbreeding. Moreover, *hydrophila* interbreeds with *biscutatus* where they meet, but *biscutatus* also interbreeds with members of the terrestrial group, which is otherwise broadly sympatric with the aquatic group and does not interbreed with it. (After Fox 1951)**

The fossil record is usually far too incomplete to shed light on whether speciation is allopatric or not. Commonly, however, a new form, recognized by paleontologists as a distinct species, appears quite suddenly in the fossil record, suggesting that it invaded from some other region where it came into being. For example, a widespread form of the Devonian trilobite *Phacops rana*, with 18 rows of eye lenses, was quickly replaced by a form with 15 rows that seems to have originated in shallow epicontinental seas along the margin of the distribution of *P. rana* (Eldredge 1971). The contribution of geology to the study of speciation is usually more indirect, however, in the form of evidence on the past distribution of communities and climates. During the Pleistocene, for example, most of Australia was quite dry, with mesic vegetation persisting in the southwestern and southeastern portions of the continent. These regions are the present home of numerous pairs of closely related species, some of which meet in central Australia.

Finally, laboratory experiments have often shown that incipient reproductive isolation can develop between isolated populations of *Drosophila* and houseflies (*Musca domestica*) that are exposed to divergent artificial selection for responses to light, gravity, or chemical features of the food medium (e.g., del Solar 1966, Soans et al. 1974, de Oliveira and Cordeiro 1980). Ehrman (1964) tested for sexual isolation among six populations of *Drosophila pseudoobscura*, all derived from the same stock, of which two had been kept at each of three temperatures for more than four years. Some pairs of populations were slightly but significantly sexually isolated, even among those that had adapted to the same temperature.

### Hybrid zones

Hybrid zones, at which populations that differ in several or many characteristics interbreed to a greater or lesser extent, are usually interpreted as instances of secondary contact between populations that differentiated in allopatry, but did not achieve full species status. At such a zone, each of several or many loci (or chromosomes) exhibits a cline in allele frequency, the width of which may be greater for some loci than others (the phenomenon of "introgressive hybridization," Chapter 4). For example, two races of the grasshopper *Caledia captiva* form a hybrid zone about one kilometer wide along a 200-kilometer front in southeastern Australia. They differ in at least nine chromosomal rearrangements, are fixed for different alleles at four enzyme loci, and form inviable $F_2$ and backcross hybrids (Shaw 1981). Similarly, a long, narrow zone of hybridization along the border of France and Italy marks the contact between races of the grasshopper *Podisma pedestris* that differ in one chromosomal rearrangement. In this case the $F_1$ hybrids have low fitness not because of the chromosomal difference, but because of differences at numerous loci (Barton 1980).

Steep clines will persist for loci (or chromosome rearrangements) that have low fitness in heterozygous condition, and so contribute to hybrid inferiority. If the heterozygote at locus $A$ has lower fitness than either homozygote, neither allele will be able to increase in the population into which it is introduced by hybridization (see Chapter 6). But unless $AA'$ is entirely sterile or inviable, backcrossing between $AA'$ and $A'A'$ (or $AA$) will introduce alleles at other loci from one population into the other. Thus partial sterility or inviability of hybrids is only a moderately effective barrier to gene flow, except at loci that are very closely linked to those loci that reduce hybrid fitness (Barton 1979, 1983). Therefore two hybridizing populations will remain differentiated throughout the genome only if the chromosomes are thickly studded with loci that, like our hypothetical locus *A*, reduce hybrid fitness, for it is only then that most genes will be closely linked to a locus that restricts gene exchange. The low fitness of hybrids in some narrow hybrid zones seems, indeed, to be caused by many loci (Barton and Charlesworth 1984).

## PARAPATRIC SPECIATION

If selection favors different alleles in two adjacent, or PARAPATRIC, populations a cline in allele frequency is established, the width ($l$) of which is proportional to $\sigma/\sqrt{s}$, where $\sigma$ is the standard deviation in the distance to which individuals disperse, and $s$ is the strength of selection against the "wrong" allele (Slatkin 1973, Endler 1977). With sufficiently strong selection on loci that contribute to reproductive isolation, the populations can differentiate into reproductively isolated species (Endler 1977, Lande 1982b, Barton and Charlesworth 1984). Endler (1977) has argued that many of the hybrid zones that are usually attributed to secondary contact may actually have arisen in situ by the differentiation of parapatric populations, and that species may often arise parapatrically. Numerous cases of localized geographic variation show that considerable genetic divergence can arise despite gene flow (Chapter 4).

A major argument against parapatric speciation has been that the spatial pattern of selection should differ for different loci, so that their clines should be situated at different locations, rather than at the same location as is often observed in hybrid zones. However, the reduction of gene flow caused by selection against heterozygotes at one locus can establish concordant clines in allele frequency at closely linked loci if they are only weakly selected in their own right. Moreover, if the fitness of genotypes at one locus depends on the genotype at a locus that is subject to clinal selection (i.e., if there is epistasis), the geographic variation at one locus will engender parallel variation in the other (Clark 1966, Slatkin 1975, Barton 1979). It is difficult to tell if multilocus hybrid zones are evidence of incipient parapatric speciation. However, populations of certain grasses on toxic mine wastes have differentiated not only in tolerance to heavy metals, but in flowering time and in their degree of self-compatibility, and so are partially reproductively isolated from surrounding populations on non-toxic soils (Macnair 1981).

Another model of parapatric speciation is the STASIPATRIC speciation model of White (1968, 1978). White observed that in sedentary wingless grasshoppers, populations within the broad range of the species differ in chromosome configuration. White proposed that a chromosomal aberration—a partial isolating mechanism—arose within a population and expanded its range, forming an ever-expanding narrow hybrid zone. But a chromosomal mutation that lowers fertility enough to confer reproductive isolation cannot increase in frequency except by genetic drift in a very small population (Key 1968, Bengtsson and Bodmer 1976)—that is, one that is effectively isolated from the larger surrounding population and so is more or less allopatric to it. If speciation comes about by chromosomal divergence, it is most likely to be by the sequential fixation of numerous rearrangements, each of which lowers fertility only slightly (Walsh 1982). But these multiple rearrangements are likely to arise in and spread from different locations, and so have an overlapping pattern rather than conjointly distinguishing two species, unless they have arisen in allopatric populations. The stasipatric model is therefore not widely accepted.

## SYMPATRIC SPECIATION

Speciation would be sympatric if a biological barrier to interbreeding arose within the confines of a panmictic population, without any spatial segregation of the incipient species. Both *instantaneous* and *gradual* models of sympatric speciation have been proposed. Most models of sympatric speciation are highly controversial; the only exception is one mode of instantaneous speciation, speciation by polyploidy, that occurs in plants. If the hybrid between two diploid species becomes tetraploid, it will be largely reproductively isolated from its diploid parents because the triploid backcross progeny have a high proportion of aneuploid, inviable gametes. Limited interbreeding among the diploid and tetraploid forms, or between different tetraploids, may give rise to yet other polyploids (Figure 4). Such polyploid complexes have been described for many plant genera (Grant 1981).

If a single mutation or chromosomal change (such as polyploidy) confers complete reproductive isolation in one step, its bearer will not successfully reproduce unless there is close inbreeding (i.e., self-fertilization or mating with sibs that may also carry the new mutation). Among animals, such close inbreeding is unusual, but it does exist in groups such as the Chalcidoidea, parasitic Hymenoptera in which mating between brothers and sisters that emerge from the same host is common. Askew (1968) has suggested that the high species diversity in this group may have been facilitated by this mating system. However, reproductive isolation between closely related species is usually attributable to differences not at a single gene locus, but at several or many. Thus most speciation must be gradual, as initially incomplete barriers to gene flow become progressively more effective. Whether or not this can happen within the confines of a single breeding population is strongly debated (Mayr 1963, Maynard Smith 1966, Bush 1975a, Futuyma and Mayer 1980, Felsenstein 1981).

Many models of sympatric speciation (e.g., Maynard Smith 1966, Dickinson and Antonovics 1973, Felsenstein 1981) are based on disruptive selection, as when two homozygotes at one or more loci are adapted to different resources, and there is a multiple-niche polymorphism (Chapter 6). Suppose, for example, that in an herbivorous insect genotypes $AA$ and $A'A'$ are adapted to host plant species 1 and 2 respectively, and that $AA'$ is not well adapted to either. Each homozygote would have higher fitness if it mated assortatively (Chapter 5) with like genotypes, and did not produce unfit heterozygous progeny. Assortative mating might be conferred by locus $B$, which could influence either mating behavior or impel the insect to choose a specific host species on which to find a mate and lay eggs. Thus if $BB$ and $Bb$ mate only on host 1 and $bb$ mates only on host 2, the difference in host preference could constitute reproductive isolation. In many groups of herbivorous insects (e.g., treehoppers, *Enchenopa*; Wood and Guttman 1983) closely related species are restricted to different host plants for both feeding and mating, and Bush (1975a,b) has argued that sympatric speciation has been prevalent in such groups.

If the population is initially polymorphic at locus $A$ that provides adaptation

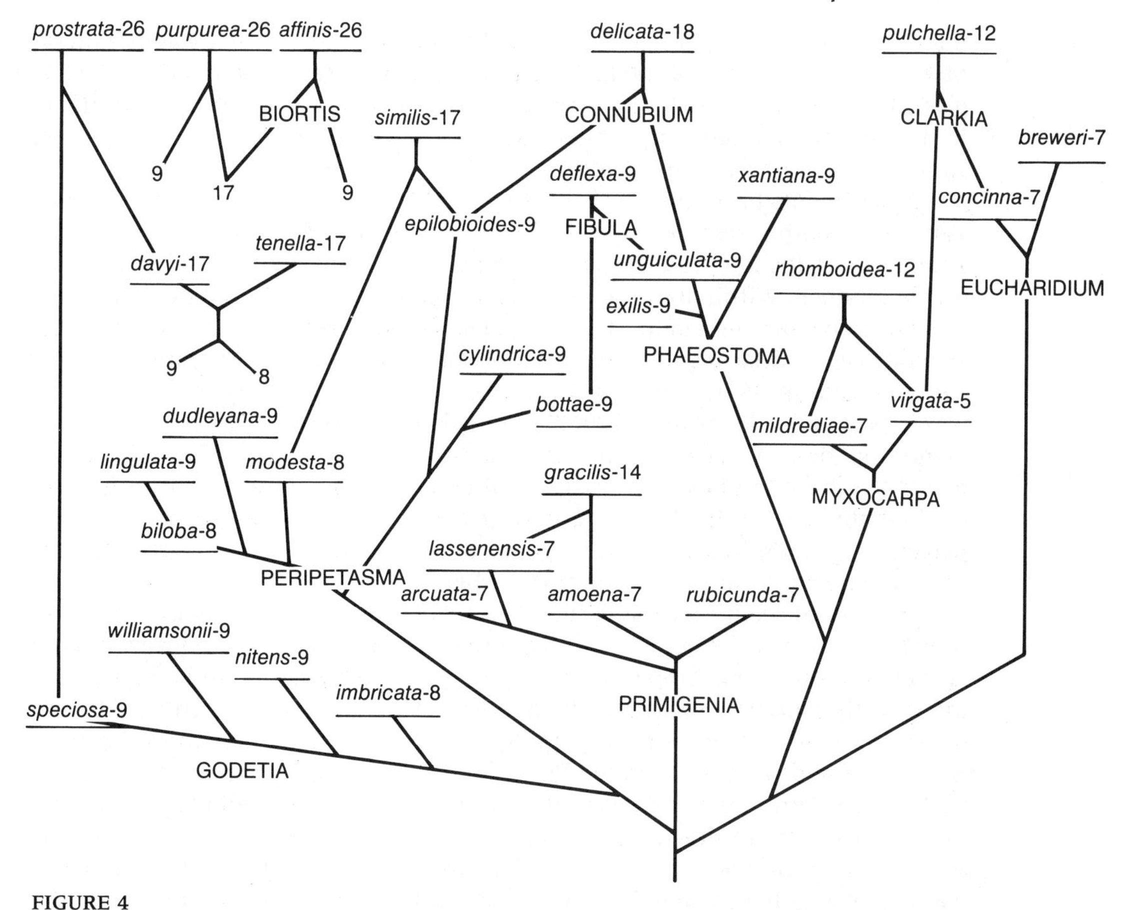


FIGURE 4

**Inferred phylogenetic relationships among polyploid forms in the plant genus *Clarkia*. The names in upper case letters are sections of the genus; those in lower case are species, with their usual gametic chromosome numbers indicated. Chromosome numbers without associated names represent hypothetical ancestors. Note that polyploid forms have arisen from hybridization between species. (After Lewis and Lewis 1955)**

to one host or the other, but is monomorphic (e.g., *BB*) at the locus that would provide assortative mating, the sympatric evolution of assortative mating requires that allele *b* increase in frequency, and that strong linkage disequilibrium be established between the loci, so that the prevalent genotypes are *AABB* (adapted to and mating only on host 1) and *aabb* (adapted to and mating only on host 2). Allele *b* will indeed increase in frequency if it is in linkage disequilibrium with *a*; the *ab* combination will be favored because *aabb* insects mate and lay eggs on the host on which their offspring are most fit. For this to occur, however, the loci must be tightly linked, selection based on adaptation to different hosts must be quite strong, and there must be little gene flow caused by factors such as environmentally induced variation in host preference or mate preference (Felsenstein 1981). If host preference or mate preference is controlled by several loci rather than one, recombination makes it very difficult for selection to establish linkage disequilibrium among all the loci, and so to establish two reproductively isolated groups (Felsenstein 1981). So sympatric speciation is theoretically possible, but it is likely to happen only under exceptional conditions such as single-locus control of host or mate preference.

Experiments have shown that with strong enough selection, some degree of assortative mating (reproductive isolation) can arise sympatrically. For example, Thoday and Gibson (1962) selected disruptively for *Drosophila* with high and low bristle numbers within a single laboratory population, and found that a tendency for assortative mating among high- and low-bristle flies developed within a few generations. Similar results have been obtained in other experiments (e.g., Thoday and Gibson 1970, Hurd and Eisenberg 1975, Rice 1985). A dramatic case was reported by Paterniani (1969) who planted a mixture of two genetically marked strains of maize (corn), and in each generation sowed seed from the plants that bore the lowest proportion of heterozygous kernels. Within five generations the level of intercrossing declined from about 40 percent to less than 5 percent (Figure 5) because of divergence in flowering time and a reduced receptivity of one strain to the pollen of the other.

It does not seem very likely that disruptive selection is commonly this severe in nature, nor that the genetic architecture that would favor sympatric speciation is common. Incipient sympatric speciation has been suggested to be occurring in the fruit fly *Rhagoletis pomonella* (Bush 1969). The native host of this fly is hawthorn, but it was found infesting apples in the 1860s and cherries in the 1960s. However, the degree of reproductive isolation and genetic divergence is not well understood (Futuyma and Mayer 1980). In two species of lacewing flies (*Chrysopa*) that are associated with different vegetation and emerge at somewhat different seasons, the difference in coloration that makes them cryptic in their respective habitats is largely controlled by one locus, and the difference in emergence time

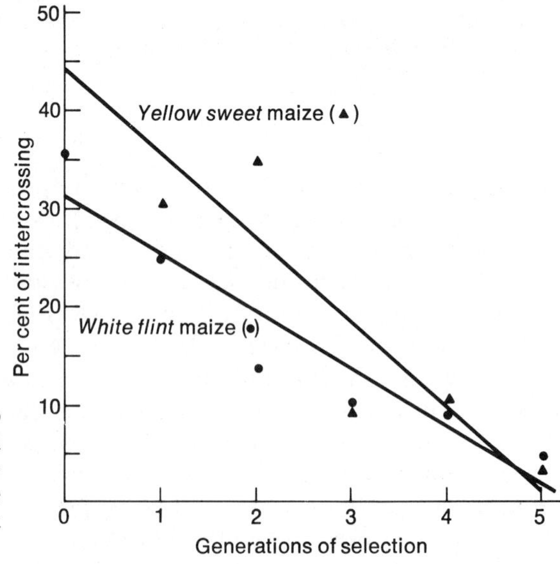


FIGURE 5

**Reduction in interbreeding between two strains of corn due to artificial selection against intercrossing. Initially *Yellow sweet* received more pollen from *White flint* than vice versa, but both strains declined in the amount of intercrossing. (After Paterniani 1969)**

by two loci. Because the genetic control of these characters is simple, as required by models of sympatric speciation, Tauber and Tauber (1982) have proposed that these species arose sympatrically. This interpretation has been contested (Henry 1985) because the species differ in their mating call, so that the seasonal and ecological differences may not be the only isolating factors. In the few cases that have been analyzed, host preference in insects appears to be controlled by several or many genes, which would not be favorable to sympatric speciation (Futuyma and Peterson 1985).

## GENETIC THEORIES OF SPECIATION

During speciation, populations diverge toward different genetic equilibria that are incompatible: either a character such as mating behavior diverges so that interbreeding does not occur, or the fitness of hybrids is so low as to bar substantial gene flow between them. One can visualize species, then, as occupying different peaks on an adaptive landscape of the kind described by Sewall Wright (Figure 13 in Chapter 6). The genetic theories of how populations come to occupy different, incompatible adaptive peaks are of two major kinds (Figure 6). One class of theories, which may be called speciation by GRADUAL DIVERGENCE (Templeton 1981), supposes that the forces of selection experienced by two spatially isolated populations differ, so they gradually move toward different adaptive peaks. The other class of theories, which might be termed speciation by PEAK SHIFT or by GENETIC REVOLUTION, supposes that the environments of two populations are the same, but that either of two (or more) genetic constitutions is favored by selection. For one of the populations to shift from one adaptive peak to the other, it must cross an adaptive valley (so mean fitness, $\bar{w}$, must decrease before it again increases). This can occur only by sufficient genetic drift to bring the allele frequencies into the domain of attraction of the other adaptive peak. Thus genetic drift destabilizes one adaptive gene complex enough for selection to mold a different gene complex that is incompatible with the ancestral one. Templeton (1981) has used the term "transilience" for such a peak shift.

### Speciation by divergence

When two populations diverge in genetic composition by adapting to different environments, some of the genetic differences between them may incidentally confer reproductive isolation when the populations later meet. It is easy to see, for example, how affiliation with a different habitat or host plant (e.g., for herbivorous insects), or a seasonal difference in the time of flowering or mating, could arise in this fashion. Ecological or seasonal isolation can readily break down, and this may be a reason for the prevalence of hybridization among plants that are hard to classify either as species or ecotypes. Gradual genetic divergence may also pleiotropically cause genetic incompatibility that results in sterility or inviability of hybrids; but since so little is known of the biochemical or physiological basis of postzygotic isolation, we do not know which characteristics may diverge by selection and effect reproductive isolation. It is not clear why selection in different environments should alter pheromones, courtship signals, or other such bases for premating isolation; it is possible that interactions with other species favor changes in these traits, but few examples can be cited (see below).

One explanation for the development of ethological isolation is SEXUAL SE-

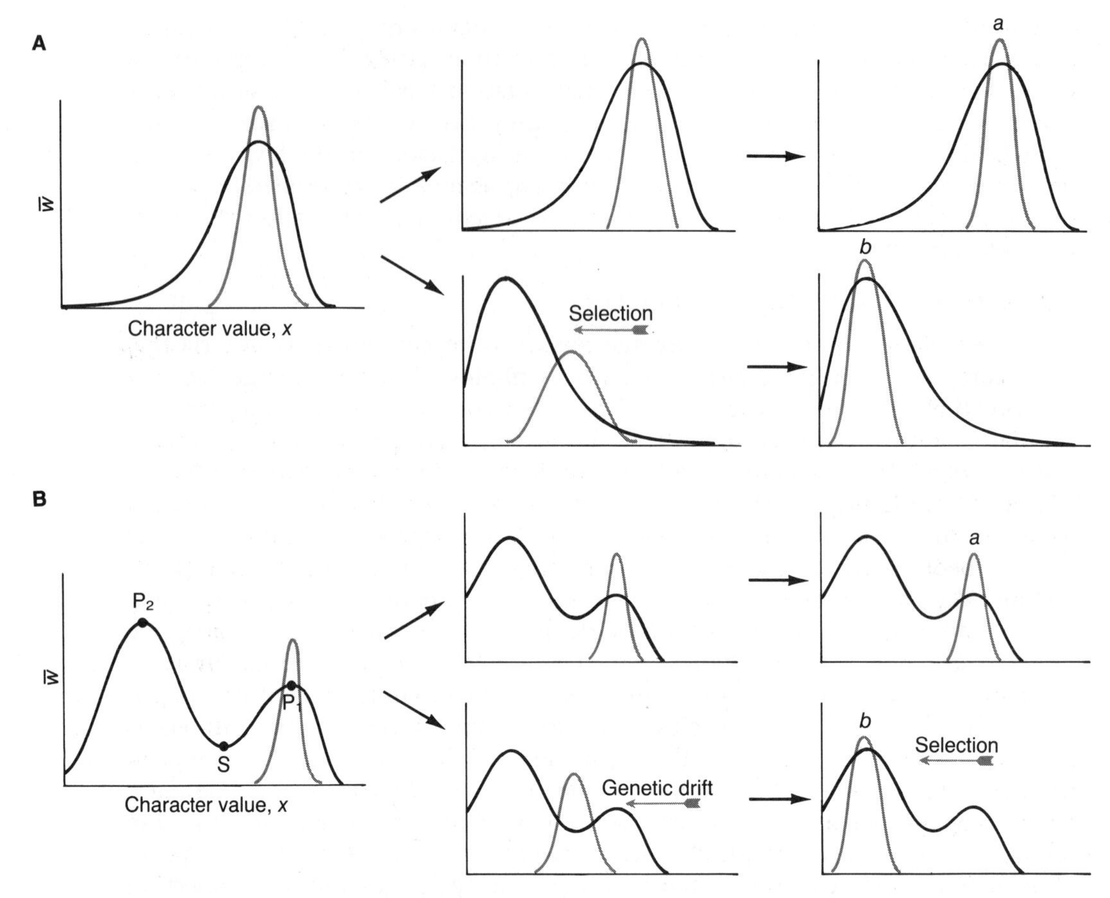


FIGURE 6

**Two theories of speciation. The solid line represents an adaptive landscape, i.e., mean fitness as a function of a character value *x* (*x* may also be interpreted as an allele frequency). The gray line represents the frequency distribution of a character in a population. (A) Speciation by divergence. One of two isolated populations inhabits a region where selection favors a lower value of the character. The populations diverge so that they come to occupy different equilibria *a* and *b*. (B) Speciation by peak shift. The ancestral population occupies one of two adaptive peaks ($P_1$) separated by an adaptive valley or "saddle" (*S*). In a small population derived from this ancestor, selection dictates the same adaptive landscape, but the character moves past the saddle by genetic drift, and then is moved by selection to peak $P_2$, so the populations arrive at different equilibria *a* and *b*. In either scenario, the populations may later become sympatric and retain their differentiation if the character difference confers reproductive isolation.**

LECTION, the term Darwin used for "the advantage which certain individuals have over others of the same sex and species solely in respect of reproduction." In *The Descent of Man, and Selection in Relation to Sex* (1874), Darwin explained many of the secondary sexual characteristics of males—features such as the antlers of deer, the bright nuptial colors of minnows and sticklebacks, the train of the peacock—as features that enable their bearers to father more offspring, by conferring superiority in inter-male combat or display or by rendering the male more attractive to females. We will treat the theory of sexual selection at greater length in Chapter 9. For the present, suffice it to say that under some theoretical conditions, if female preference for, say, red versus orange plumes diverges by genetic drift, both the male plumage color and the females' preference can rapidly diverge even more, in a self-accelerating "runaway" process (Fisher 1930, Lande 1981b, Kirkpatrick 1982, West-Eberhard 1983). A similar phenomenon can result in divergence of plants that have very specialized pollinators; a slight change in the average phenotype of a flower in a local population can select for a change in the preference of its pollinator, which in turn selects for further change in flower form (Kiester et al. 1984).

Speciation has been prolific in several groups that seem to fit these models: figs and their pollinating wasps, each of which is faithful to only one species of fig; orchids, the most diverse family of plants; and among animals, groups such as the Hawaiian *Drosophila*, ducks, birds of paradise, pheasants, and hummingbirds (Figure 7). These animals are all characterized by elaborate courtship displays, extreme sexual dimorphism, and generally a high degree of interfertility when they are experimentally hybridized. They appear to have diverged more in features associated with courtship than in any other respect. West-Eberhard

FIGURE 7

**Secondary sexual characteristics vary greatly among the males of hummingbirds, although the females are rather similar to each other. Differences in plumage among the males facilitate reproductive isolation. Left to right, above, are *Sappho sparganura, Ocreatus underwoodii, Lophornis ornata*; below, *Stephanoxis lalandi, Popelairia popelairii, Topaza pella*. (Redrawn from illustrations by A.B. Singer in Skutch 1973)**

(1983) has described the role of sexual selection in speciation in these and numerous other groups.

Another model of speciation by gradual divergence in isolated populations relies almost entirely on genetic drift (Nei et al. 1983). In the simplest form of the model, suppose a series of multiple alleles that affect a character such as flowering time, fertility, or courtship behavior can be arrayed in a linear fashion as in Figure 8, and that each allele can arise by mutation only from an adjacent allele in the array. Assume that fully fertile offspring arise from a mating either between like homozygotes (e.g., $A_0A_0 \times A_0A_0$) or between homozygotes that are one step away ($A_0A_0 \times A_1A_1$). Homozygotes that are two or more steps away ($A_0A_0$ and $A_2A_2$) are incompatible, either because they do not interbreed or because their offspring are sterile. This model may easily be extended to two or more loci that control, say, male courtship signal and female response.

Assume, now, that two isolated populations are initially monomorphic $A_0A_0$, and that mutant $A_1$ arises in one population and $A_{-1}$ arises in the other. Since in each population heterozygotes are fertile, genetic drift may fix $A_1$ in one population and $A_{-1}$ in the other, yielding populations that are reproductively incompatible. If speciation occurs in this way, it will proceed more rapidly in small than in large populations; moreover, the more loci that contribute to incompatibility in this fashion, the faster the process. Because even a little gene flow prevents populations from differentiating substantially by genetic drift (Chapter 5), speciation in this model requires that there be virtually no gene flow at all between the populations. The simplicity of this model makes it attractive, but since we do not know the relative importance of selection and drift at loci that contribute to reproductive isolation, the model's applicability to natural populations is not yet known.

### Molecular evolution and speciation

With the discovery of transposable sequences of DNA and of the amplification of DNA sequences (Chapters 3 and 15), there has ensued considerable speculation about the role of molecular changes in speciation, focusing especially on changes that may lead to postzygotic barriers to gene exchange (Dover and Flavell 1982, Rose and Doolittle 1983). One theory is that the amplification of different repeated DNA sequences in different populations might cause misalignment and failure of synapsis during meiosis in hybrids, resulting in sterility. However, it appears

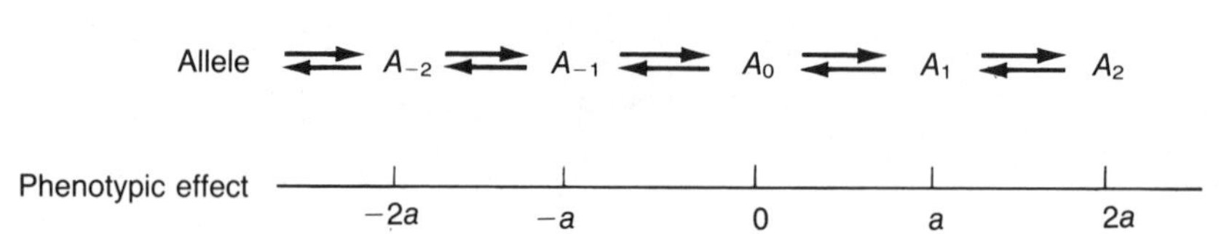


FIGURE 8

**Stepwise mutation model for alleles that contribute to hybrid sterility, hybrid inviability, or premating isolation. If the phenotypic difference caused by neighboring alleles in the series is slight, heterozygotes and homozygotes for the alleles may have virtually the same fitness, so allele frequencies may change by genetic drift. If alleles at either end of the series differ enough in their phenotypic effect, homozygotes are reproductively isolated. (From Nei et al. 1983)**

that chromosomes in grasses and in *Drosophila* can pair normally even if they differ considerably in the amount and distribution of repeated sequences (Rees et al. 1982, Rose and Doolittle 1983).

Another possible role of molecular phenomena is illustrated by the phenomenon of HYBRID DYSGENESIS in *Drosophila melanogaster* (Kidwell et al. 1977, Engels 1983). This phenomenon arises from an incompatibility between cytoplasmic factors and any of several "families" of transposable DNA sequences, of which the *P* element is best known. The *P* element is a sequence of about 3000 nucleotide base pairs that is transposed at fairly high frequency into many sites throughout the chromosomes; consequently, it acts like a "genomic parasite", multiplying at a faster rate than the genome as a whole. In a compatible (*P*) cytoplasmic background, it has no untoward effects, but when a sperm bearing the *P* element fertilizes an egg with an incompatible (*M*) cytoplasm, the resulting offspring displays (under some environmental conditions) a syndrome of "dysgenic" effects that include considerable sterility, recombination in males (which ordinarily does not occur in *Drosophila*), and a high incidence of chromosomal aberrations and other mutations. Curiously, almost all wild flies of this species appear to carry *P* factors (and have compatible, *P*-type, cytoplasm) at present, whereas laboratory stocks derived from flies captured before the 1950s lack *P* elements, have M-type cytoplasm, and display hybrid dysgenesis when crossed with recently captured flies. It appears likely that the *P* element arose recently in nature, and has swept rapidly throughout populations of this species, world wide, as if it were an epidemic virus (see Chapter 15).

At present, it seems improbable that such a sterility-causing element could in itself create a new species, because if dysgenic flies are entirely sterile, the element cannot spread, whereas if they can reproduce at all, there is nothing to prevent the element from becoming uniformly distributed throughout the species. A population into which *P* elements are introduced is not reproductively isolated from the *P*-bearing source population; it merely suffers lower fertility as susceptible flies are infected by the *P* element. Fertility will be restored if the cytoplasmic properties evolve by natural selection to become compatible with the transposable element.

### Speciation by peak shifts

The common feature of models of speciation by peak shifts (Wright 1931, 1932, 1977) is that for one or more loci, there are two or more different equilibrium allele frequencies dictated by selection. If the allele frequency is perturbed only slightly from one equilibrium, it returns to it, but if the frequency is changed sufficiently by genetic drift, it moves by natural selection to another equilibrium. Loci at which heterozygotes have a selective disadvantage provide the simplest illustration of this principle (Chapter 6).

The difficulty that such models must deal with is that for a population to move from a high frequency of one allele ($A$) to a high frequency of another ($A'$), the frequency of heterozygotes must increase. Genetic drift can most effectively increase the frequency of $A'$ if the fitness of heterozygotes is not severely reduced; but in that case, reproductive isolation is not appreciable. If the heterozygotes' fitness is very low, reproductive isolation is appreciable, but then selection strongly opposes genetic drift, which can increase the frequency of $A'$ only in very small populations. If $\bar{w}_P$ and $\bar{w}_S$ are, respectively, the average fitness at an adaptive peak and at the saddle point between two peaks (Figure 6), the probability of a peak shift as a result of genetic drift is proportional to $(\bar{w}_S/\bar{w}_P)^{2N_e}$, where $N_e$ is the effective population size (Barton and Charlesworth 1984). The lower the valley between two peaks ($\bar{w}_S/\bar{w}_P$), the more difficult it is to cross.

## Chromosomal differentiation

For this reason, speciation by the fixation of chromosomal aberrations that severely reduce fertility in heterozygous condition can occur only in a very small population (Lande 1979); if the population is of moderate size, reproductive isolation is more likely to occur by the successive fixation of numerous chromosomal aberrations that individually reduce the fertility of homozygotes only

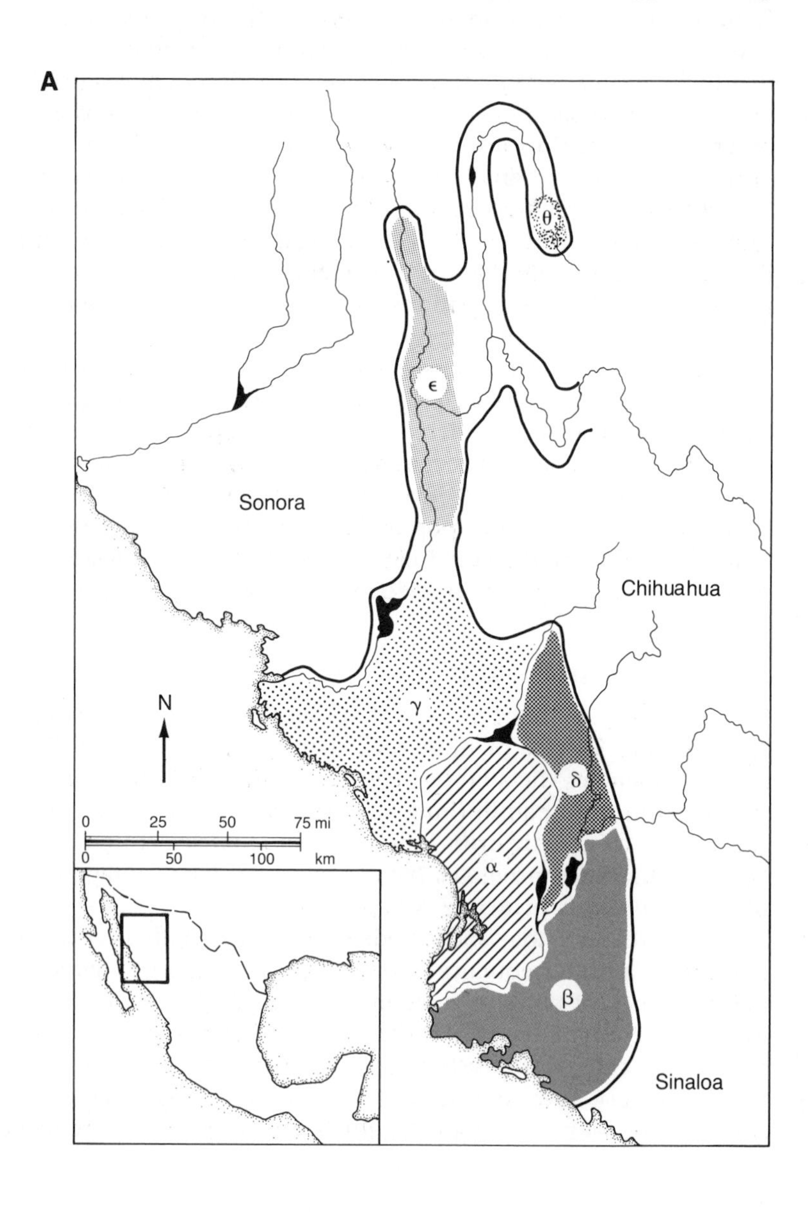

slightly (Walsh 1982). Closely related species of animals frequently differ by rearrangements such as fusions and fissions that only slightly reduce heterozygote fertility, but rarely differ by reciprocal translocations, which have a much greater heterozygous disadvantage (Lande 1979).

Substantial chromosomal differences among populations are frequently found in organisms such as wingless grasshoppers (White 1978) and burrowing rodents (e.g., Patton 1972, Nevo and Bar-El 1976) in which populations are small and experience little gene flow. A typical example of chromosomal variation is that of the pocket mouse *Perognathus goldmani,* in which different Mexican populations have different predominant chromosome numbers ranging from $2N = 50$ to $2N = 56$ (Figure 9; Patton 1969). Most of the differences are the consequence

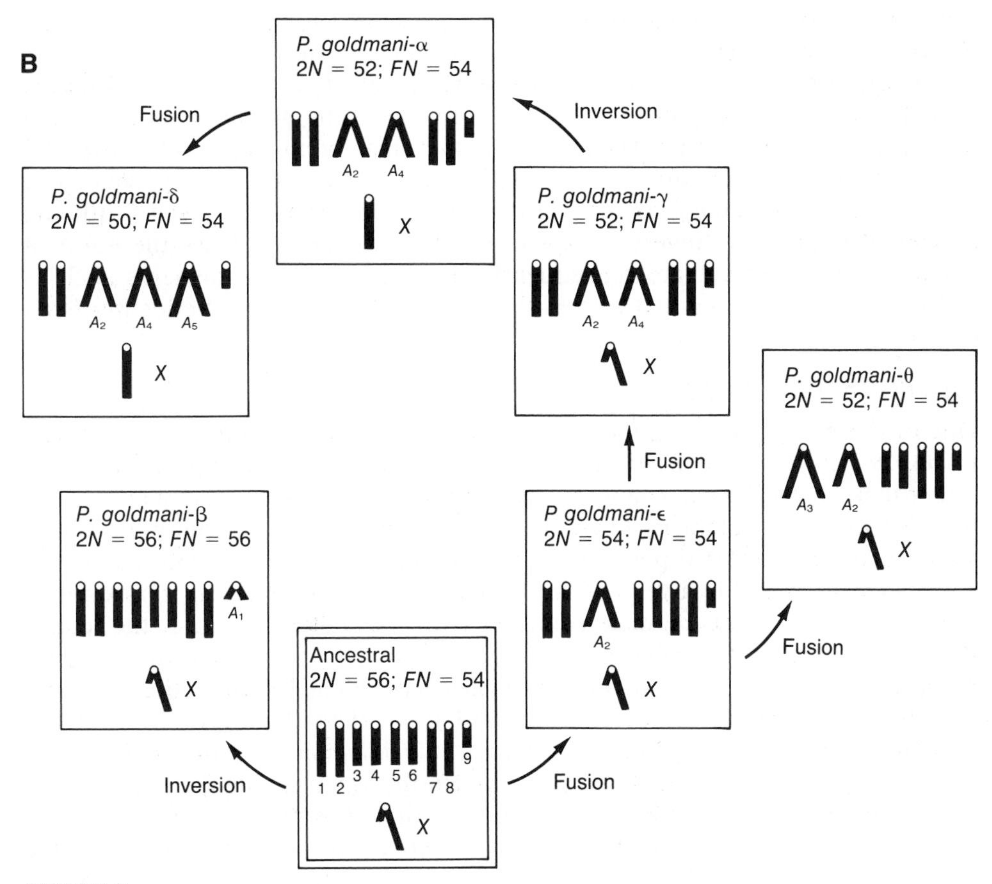


FIGURE 9

**Geographic variation in the karyotype of the pocket mouse *Perognathus goldmani* in northwestern Mexico. (A) The distribution of the chromosome "races," marked by Greek letters. Several meet along narrow borders. (B) The proposed phylogeny of changes in those chromosomes that differ among populations. $2N$ is the diploid number of chromosomes, *FN* the number of chromosome arms ("fundamental number"). Karyotypes β and ϵ, found at opposite ends of the range, are believed to be most similar to the hypothetical ancestral pattern, which has been deduced by comparisons with related species of *Perognathus.* (After Patton 1969)**

of fusion of acrocentric chromosomes to form metacentric chromosomes. The chromosome patterns at the northern and southern edges of the range are thought to resemble the ancestral karyotype, and those in the center of the range to represent derived conditions. Patton (1969) has postulated that the distribution of this arid-adapted species was divided by forested habitat during the Pleistocene. When the region became more arid and the forest retreated, the species spread from north and south, establishing colonies that became chromosomally differentiated.

Chromosomal differences appear to contribute importantly to reproductive isolation in plants, and again seem to arise in small, isolated populations. One of the most fully studied examples is *Clarkia lingulata*, which consists only of two colonies at the very edge of the range of its parent species, *C. biloba*, in California (Lewis 1973). The species differ by one translocation, at least two pericentric inversions, and by a fission of one chromosome in *C. biloba* into two in *C. lingulata*. Hybrids between them are sterile because of the chromosomal differences, but the species are morphologically almost identical. Lewis proposed that *C. lingulata* arose by "catastrophic selection," when a very small population became adapted to an unusually arid habitat. The possibility that small population size contributed to speciation is further supported by the fact that *C. lingulata* is less polymorphic than *C. biloba* at several allozyme loci (Gottlieb 1974). In partially self-fertilizing species of plants, the lowered frequency of heterozygotes reduces the effect of heterozygous disadvantage; consequently chromosome rearrangements such as reciprocal translocations that severely reduce fertility are more frequently fixed than in animals.

## THE FOUNDER EFFECT

One of the most influential theories of speciation by peak shift is Ernst Mayr's (1954, 1963, 1982b) theory of PERIPATRIC SPECIATION,which focuses on the genic, rather than chromosomal, changes that Mayr believes occur in small populations founded by a few colonists. Mayr was led to propose this theory by his observation that local, isolated populations that are peripheral to the main range of a species are often highly divergent. In the flycatcher *Petroica multicolor*, for example, the bright plumage of the male contrasts with the duller pattern of the female throughout eastern Australia, yet among various of the South Sea islands, geographic variation is so great that the males have "female" coloration in some islands, and the females have "male" coloration on others (Figure 10; Mayr 1942). Similarly, the small insectivorous lizard *Uta stansburiana* exhibits only subtle geographic variation throughout western North America, but populations on islands in the Gulf of California vary so greatly in body size, scalation, coloration, and ecological properties that some have been called separate species (Soulé 1966).

To explain the divergence of peripheral populations, Mayr (1954) proposed that genetic change could be extremely rapid and pervasive throughout the genome in small, localized populations that are founded by a few individuals and are cut off from gene exchange with the main body of the species. The selective pressures acting on the colony are likely to be different, because the environment of a small area is often more homogeneous than that of a larger area and hence the conflicting pressures that act on a widespread population may be less nu-

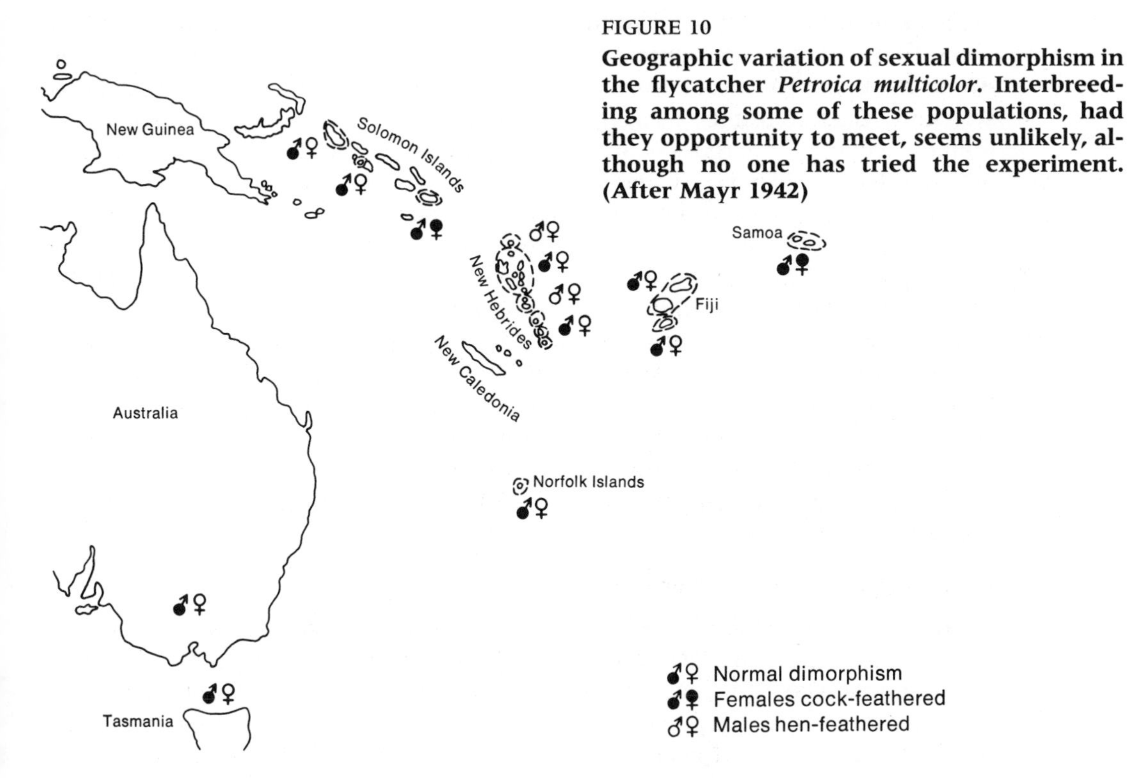


FIGURE 10

**Geographic variation of sexual dimorphism in the flycatcher *Petroica multicolor*. Interbreeding among some of these populations, had they opportunity to meet, seems unlikely, although no one has tried the experiment. (After Mayr 1942)**

merous. More importantly, Mayr argued, allele frequencies at some loci will differ from those in the parent population because of accidents of sampling (genetic drift). The colony will be less genetically variable, he said, because the few colonists carry only a fraction of the genetic variability in the population from which they came. Thus the "genetic environment" is altered, and because of strong epistatic interactions that affect fitness, selection will alter allele frequencies at many loci, bringing about a massive genetic change that Mayr termed a "genetic revolution."

Mayr (1954, 1963) used the term FOUNDER EFFECT to describe the initial alteration in allele frequencies by genetic drift, which sets off a cascade of genetic change at other loci. He argued that evolution in a widespread species is likely to be a slow process because its coadapted gene pool resists change (genetic homeostasis) and because gene flow among its populations opposes divergence, whereas evolution of newly founded colonies is likely to be more rapid, and may entail changes of such magnitude as to mark the origin not only of new species, but of new genera. Substantial evolution is likely to occur so rapidly and on such a localized geographic scale that it will seldom be documented in detail in the fossil record. Mayr's theory, in which speciation by the founder principle is the context in which most of evolution occurs, is the theoretical foundation of the idea of PUNCTUATED EQUILIBRIUM (Chapter 14) advanced by some paleontologists (Eldredge and Gould 1972).

A somewhat similar theory was advanced by Carson (1975), who argued that whereas many loci respond readily to selection (comprising the part of the genome that he terms the "open variability system"), certain loci have strong epistatic relationships and form a "closed variability system" that is resistant to selection. Carson thinks these blocks of loci are destabilized when a newly founded colony undergoes a flush of exponential population growth, during which selection is relaxed and recombinants that ordinarily have low fitness become prevalent. When the population then crashes to a low level, genetic drift and selection together determine which of the novel recombinants persist, thus bringing the population to the vicinity of a different adaptive peak. Carson (1982) suggests that genes controlling mate recognition and courtship behavior may be an example of the closed variability system. In a variant of this model, Templeton (1980) emphasizes not growth and decline of the colony, but a change in the frequency of alleles at a few loci of major effect, by genetic drift when the colony is founded. Polygenic modifier loci are then altered by selection to bring about a new coadapted state of a character, such as courtship behavior, that is controlled by the major loci. In Carson's and Templeton's scenarios, genetic drift initiates changes in only one or a few critical characters.

**Evaluation of the founder effect**

From a theoretical point of view, speciation by founder effects may not be very likely (Barton and Charlesworth 1984). For one thing, the depletion of genetic variation that Mayr postulated to occur during the foundation of a population by a few colonists will not be very great unless the population remains small for a considerable number of generations (Nei et al. 1975; see Chapter 5). Hence the new population will retain much the same capacity for immediate response to selection that the ancestral population has. Moreover, alleles that have major effects on fitness, and so are most likely to contribute to reproductive isolation and to engender selection at epistatically interacting loci if their frequency is changed, are least likely to undergo a peak shift by genetic drift, simply because this will be powerfully opposed by selection. If a population goes through a severe bottleneck in size for only a single generation (i.e., a founding event), genetic drift is unlikely to change allele frequencies sufficiently to move the genetic composition to the region of a different adaptive peak, especially if the alleles greatly lower the fitness of heterozygotes. It appears likely, then, that genetic drift will initiate a peak shift only if the population remains small for a considerable number of generations, and that the founding event per se may not be the impetus to speciation. Barton and Charlesworth (1984) argue that speciation more often occurs by adaptive divergence (Figure 6A) in response to different ecological selection pressures, than by stochastic peak shifts.

The principal arguments in favor of rapid, stochastically induced peak shifts leading to speciation are based on empirical evidence (Carson and Templeton 1984). For example, Powell (1978) found that among laboratory populations of *D. pseudoobscura*, taken from the same stock formed by mixing samples from several localities, incipient reproductive isolation developed if the populations went through several cycles of population expansion followed by severe reduction in size. Populations that were maintained at constant population sizes did not develop reproductive isolation. The most widely quoted data bearing on the founder effect are those of Dobzhansky and Pavlovsky (1957), who initiated replicate populations of *Drosophila pseudoobscura* with equal frequencies of *PP* and *AR* chromosome inversions. The *PP/AR* heterozygote is heterotic for fitness, so all the populations were expected to reach the same equilibrium frequency. However, after about 19 generations the variation in frequency was greater among populations founded by few (20) than among those founded by many (4000) flies (Figure 11). Dobzhansky and Pavlovsky interpreted this to mean that the relative fitnesses of the genotypes *PP/PP*, *PP/AR*, and *AR/AR* depend on the rest of the genotype in which these chromosomes operate, and that this "genetic environment" varied among the populations that were initiated with few founders.

Among the Hawaiian *Drosophila*, it is often possible to show from chromosomal evidence that one species has been derived from a "parent" species on a different, older island (see Chapter 10). In some cases, females of the ancestral species discriminate against males of the derived species, whereas females of the derived species are less discriminating. Kaneshiro (1983) has suggested that females in a sparse newly founded colony, in which it may be difficult to find mates, have been selected to accept males even if their courtship is aberrant because of stochastic changes in their genetic makeup. Ancestral and derived populations do not always show asymmetry in mating discrimination, however (Markow 1981), for females of derived species sometimes prefer conspecific males.

The biogeographic patterns that led Mayr to formulate his theory may be the best evidence that speciation is favored by small population size, although these patterns do not provide evidence that the founding event itself is crucial. Perhaps the best evidence comes from studies of the Drosophilidae of Hawaii, where this family has undergone an extraordinary adaptive radiation (Carson and Kaneshiro

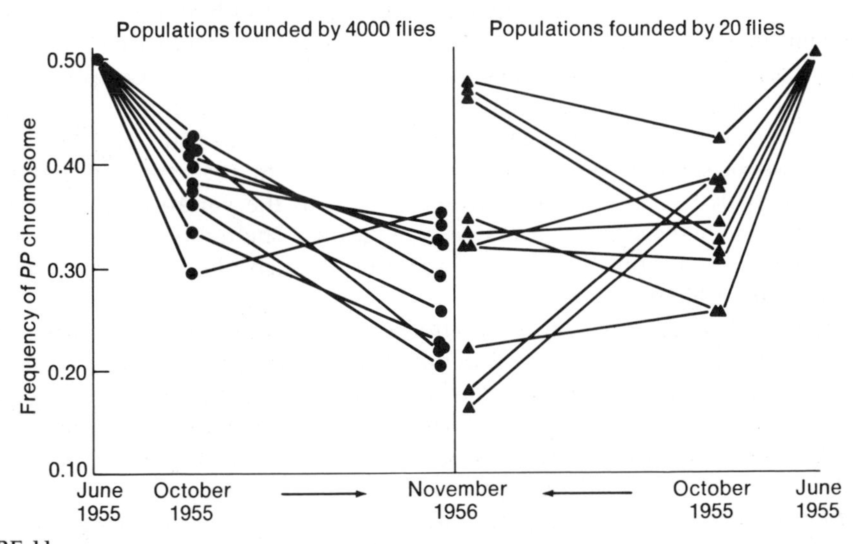


FIGURE 11

**Experimental demonstration of the founder effect. In experimental cultures of *Drosophila pseudoobscura*, genetic composition varies more among populations founded by few colonists than those founded by many. (After Dobzhansky and Pavlovsky 1957)**

1976, Carson and Templeton 1984). Almost every one of the more than 700 species of drosophilids in this archipelago is restricted to a single island. In experimental tests, pairs of ancestral and descendant species almost invariably are sexually isolated; intermediate stages of speciation are seldom found. Thus speciation by isolation on different islands appears to be very rapid—the island of Hawaii is less than 400,000 years old, but it has distinct species. Whereas ancestral species are often polymorphic for inversions, their immediate descendants are often monomorphic, which Carson and colleagues interpreted to mean that speciation is facilitated when free recombination occurs in blocks of epistatically interacting genes that are ordinarily held in linkage disequilibrium by inversions. However, descendant species are highly polymorphic at allozyme loci, which suggests they did not go through a prolonged bottleneck in population size.

Many, although not all, of the data on the genetic basis of reproductive isolation indicate that isolating mechanisms have a polygenic basis. If this proves to be generally the case, it will favor the view (Barton and Charlesworth 1984) that speciation proceeds by gradual, sequential allele substitutions rather than by stochastic shifts at a few major loci. On the other hand, these data indicate that related species differ genetically in a few critical characters, rather than by a massive reorganization of the genome as implied by the phrase "genetic revolution."

## SELECTION FOR REPRODUCTIVE ISOLATION

When two genetically differentiated populations establish contact, gene flow between them will reverse any progress they made toward becoming different species unless some combination of premating and postmating isolation factors reduces gene flow to a very low level (Figure 12). As backcrossing between hybrids and parental populations occurs, allele frequency differences diminish rapidly except at the particular loci that reduce fitness in heterozygous condition. As recombination occurs among loci that contribute to reproductive isolation, the genetic basis for the isolating mechanism breaks down. For example, Stebbins and Daly (1961) describe a population of sunflowers in which the species *Helianthus annuus* and *H. bolanderi* began hybridizing along a road in 1941. Morphological analysis of samples taken every year from 1951 to 1958 showed that a great variety of hybrid genotypes was present and that the variation in each of five characteristics increased in successive generations. The hybrids had highly inviable pollen in 1946, but by 1955 the majority of such plants had more than 80 percent viable pollen.

These considerations bear on the question of whether or not isolating mechanisms evolve specifically for the function of preventing hybridization. Dobzhansky (1937, 1970) urged the view that premating isolating mechanisms develop by natural selection to prevent the formation of unfit hybrids, the theory being that an allele for mating discrimination would increase in frequency if its bearer had fit, non-hybrid offspring. Mayr (1963) argued against this view, from his observation that some hybrid zones have persisted for long periods without any apparent development of sexual isolation. For example, a narrow hybrid zone between two European crows (*Corvus corone* and *C. cornix*) seems to have persisted since the Pleistocene. There are also numerous cases of hybrid zones between populations that differ by chromosome rearrangements that reduce the fertility

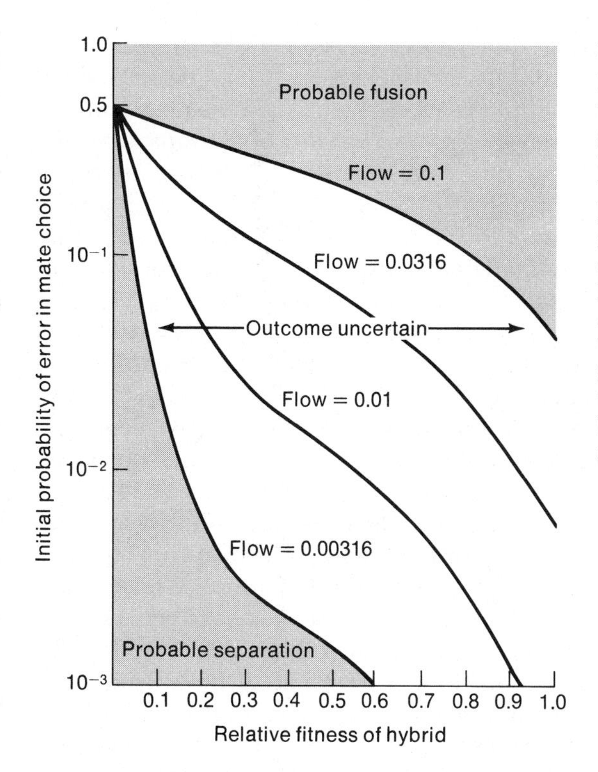


FIGURE 12

**Likelihood of speciation versus fusion when two previously isolated populations meet. A gene that increases discrimination in mating will increase and accomplish cessation of interbreeding (separation) if the fitness of the hybrid is low and the initial frequency of errors in mating is also low. "Flow" is the level of gene flow between the populations, at the levels of mating error and hybrid fitness indicated. The higher the initial degree of interbreeding (the weaker the initial prezygotic isolating mechanism), the lower the fitness of the hybrid (the more intense the postzygotic barrier) must be if speciation is to be consummated. (From Wilson 1965, after Bossert)**

of hybrids, but which nevertheless appear not to be sexually isolated (Barton and Hewitt 1981)

It is possible to enhance the degree of assortative mating by artificial selection (e.g., Koopman 1950, Paterniani 1969, Rice 1985), and in theory natural selection can enhance sexual isolation, especially if the fitness of hybrids is low and the loci that control assortative mating are closely linked to those that lower hybrid fitness (Caisse and Antonovics 1978, Felsenstein 1981, Sved 1981). However, if, as Stebbins and Daly (1961) found, the fitness of hybrids increases over time as gene flow reduces the degree of genetic difference between the populations, there may be little time for selection to develop highly effective sexual isolation de novo. Moreover, if selection favors assortative mating only in a narrow hybrid zone, gene flow from outside the zone will tend to prevent alleles for sexual isolation from increasing in frequency (Moore 1957, Barton and Hewitt 1981).

The evidence on this subject comes from the search for cases of character displacement for premating isolation, whereby sympatric populations of two species display greater premating isolation than allopatric populations (reviews by Levin 1968 and symposium papers in *American Zoologist* 14 [1974]). For example, Waage (1979) has shown that in two species of damselflies (*Calopteryx*), wing color, which is known to elicit courtship, differs strikingly where the species are sympatric, but not where they are allopatric. It is generally agreed, however, that rather few cases of reproductive character displacement—or "reinforcement of isolating mechanisms," as it is often called—have been convincingly demonstrated. The bulk of the evidence supports Mayr's view, that sexual isolation arises during the period of allopatric divergence. In some cases, it may arise through sexual selection or by genetic drift; in other cases, it may be caused by ecological sources of selection. For example, male sticklebacks (*Gasterosteus aculeatus*) have bright red nuptial coloration except in certain populations that are sympatric with an unrelated fish that preys on young sticklebacks. In these populations, male sticklebacks, which guard the young, are black, perhaps thereby attracting fewer of the predators to the nest. Females from these populations mate readily with black males, whereas females from other populations discriminate strongly against black males in favor of red ones (McPhail 1969).

Each species uses a "specific mate-recognition system" (Paterson 1982) of cues to find, recognize, and stimulate mates. This system evolves by environmental selection, sexual selection, and genetic drift whether the population is "threatened" by hybridization or not. Moreover, postmating barriers to gene flow will seldom be accentuated by natural selection, for this would entail the increase in frequency of alleles that reduce viability or fecundity. The only exception is the possibility that in a species with parental care (such as provision of yolk or endosperm), selection might favor the ability to abort hybrid offspring, so that the parent could allocate yolk or endosperm to nonhybrid offspring whose prospects for high fitness are better (Grant 1966, Coyne 1974). In general, then, neither premating nor postmating isolating mechanisms appear often to evolve for the sake of providing reproductive isolation. They appear not to be adaptations, selected to promote speciation or to maintain species identity, but byproducts of genetic change that transpires for other reasons. Because the word "mechanism" implies a function, or *raison d'être*, for a characteristic, the term "isolating mechanism" is often a misnomer.

## TIME REQUIRED FOR SPECIATION

How rapidly does speciation occur? The answer appears to differ depending on the kind of organism and on its population structure. At one extreme, some populations that have been isolated for many millions of years have diverged hardly at all in morphology and remain reproductively compatible. For example, some plant populations that have been isolated for at least 20 million years, such as American and Eurasian sycamores (*Platanus*; Stebbins 1950) and American and Mediterranean plantains (*Plantago*; Stebbins and Day 1967), form fertile hybrids. Likewise, European and American forms of certain birds, such as tits (*Parus atricapillus*), creepers (*Certhia familiaris*) and ravens (*Corvus corax*) are so similar that taxonomists classify them as the same species. Paleontologists have documented numerous cases in which organisms that are morphologically indistinguishable from modern species extend back into the fossil record for five to ten million years or more (Stanley 1979), although in such instances there is no evidence on the evolution of characters that might affect reproductive isolation. It seems likely that when a barrier separates two widely distributed forms, especially species that are ecologically generalized, the average forces of selection on either side of the barrier are similar, and so do not promote divergence. Especially when there is considerable gene flow among the populations on either side of the barrier, locally varying selection pressures become averaged. Moreover, there will be little opportunity for genetic drift to contribute to peak shifts.

On the other hand, speciation can sometimes be quite rapid. Although many pairs of populations that were isolated in separate regions by Pleistocene glaciation have not become different species, some populations have developed at least incipient reproductive isolation after being separated for at most 1.8 million years, and several genera of mammals, such as *Thalarctos* (polar bear) and *Microtus* (vole), appear to have originated during the Pleistocene (Stanley 1979). Species of pupfish (*Cyprinodon*) have developed in the Death Valley region since the extensive lakes that existed there in the Pleistocene became reduced to isolated springs 20,000–30,000 years ago. Several species of cichlid fishes are endemic to Lake Nabugabo, which is separated from Lake Victoria in Africa by a strip of low-lying land dated by radiocarbon analysis at 4,000 years (Fryer and Iles 1972). Lake Victoria itself is thought to be only 500,000–750,000 years old, but harbors about 170 species of cichlids, almost all in the genus *Haplochromis*. The extraordinary proliferation of cichlids in Lake Victoria and in the other great lakes of Africa may be a consequence of sexual selection, for these fishes have complex courtship behavior, and related species often differ in their nuptial coloration (Fryer and Iles 1972, Dominey 1984). These fishes provide a spectacular example of adaptive radiation, that has no doubt been facilitated by the paucity of other kinds of fishes in these lakes. In Lake Malawi, for example, there exist cichlids that specialize in feeding on benthic algae, insects, plankton, fishes, and molluscs. Some have extraordinary feeding habits: two genera feed on the scales of other fishes, and one species has the gruesome habit of plucking out other fishes' eyes. The form of the teeth in many of these species is so remarkably modified for specialized feeding (Figure 13) that were the species not obviously related, they

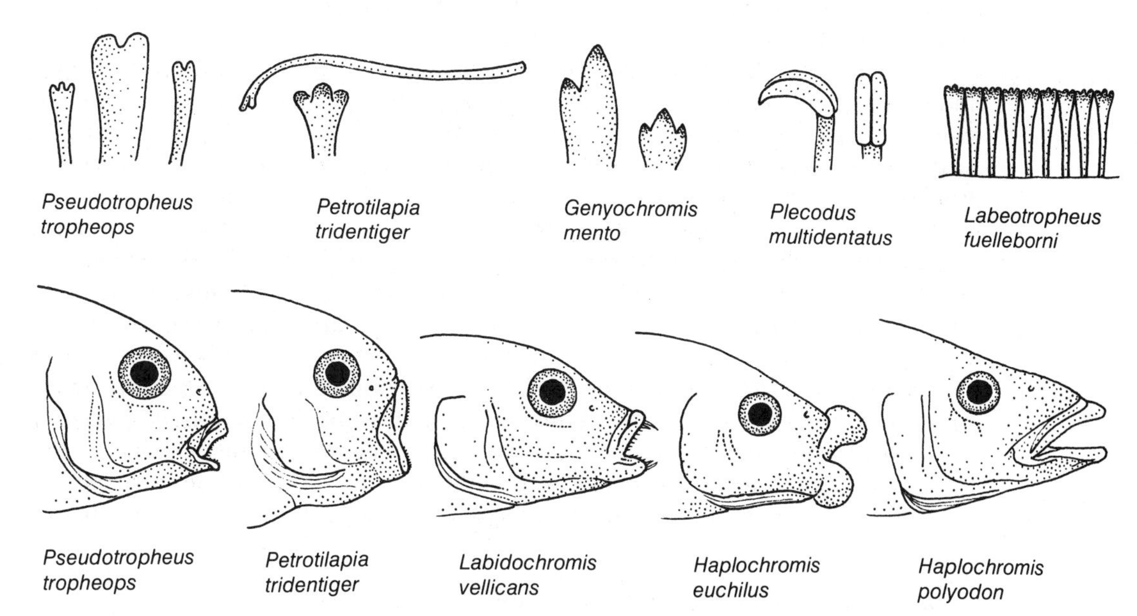


FIGURE 13

**A sample of the diverse tooth forms and head shapes among the Cichlidae of the African Great Lakes. The differences in morphology are associated with differences in diet and mode of feeding. (Redrawn from Fryer 1959 and Fryer and Iles 1972, after various sources. Courtesy of the Zoological Society of London)**

would probably be assigned to different families. These cichlids are the most spectacular of the "species flocks" that have evolved in many groups of fishes (Echelle and Kornfield 1984).

In one case, finally, a new biological species has arisen spontaneously in a laboratory. A strain of *Drosophila paulistorum* when first collected was interfertile with other strains but developed hybrid sterility after being isolated in a separate culture for just a few years (Dobzhansky and Pavlovsky 1971). The mechanism of speciation in this case is perhaps aberrant, since it was probably caused by adaptation to genetic change in an intracellular symbiont that renders hybrids sterile.

That some groups consist of many species and others of few does not in itself mean that they differ in the rate at which isolating mechanisms evolve, because the number of species is influenced both by the rate of speciation and the rate of extinction; competition may prevent many newly generated species from persisting. Nevertheless, in very species-rich groups, speciation does appear to proceed rapidly. In many such groups (e.g., Hawaiian *Drosophila*, African cichlids, orchids), numerous species have arisen very recently, for they are often genetically very similar (e.g., in their allozymes) and can be hybridized readily. Rapid speciation and the development of species richness appear to be facilitated by several factors. Low levels of gene flow and the isolation of small populations seem to be important. The founder effect in itself may not be critical in speciation, but the evidence that related species often differ in chromosome rearrangements that can only be fixed by genetic drift in small populations suggests that small population size plays a role (Coyne 1983b, Lande 1984). But whether or not the rate of chromosomal evolution is correlated with the rate of speciation is a matter of dispute (Bush et al. 1977, Lande 1979, Imai 1983).

Both the breeding system and ecology of species also seem to affect speciation rates. Speciation seems to be especially rapid in groups that have elaborate courtship display, and in which different populations may diverge rapidly by sexual selection. The high species diversity of groups such as many herbivorous insects and the cichlid fishes, in which many species are trophically highly specialized, suggests that ecologically specialized species are especially prone to speciation. It is likely that such species are sensitive to slight local variations in the environment, and so diverge rapidly. However, it is also possible that the diversity of species is high in these groups because their specialization reduces interspecific competition. Certainly the most striking examples of adaptive radiation, such as the African cichlids, and the drosophilids, honeycreepers (Chapter 2), and Compositae (Carr and Kyhos 1981) of Hawaii, illustrate that speciation and adaptive divergence occur most luxuriantly when other competitors are few or absent. Under these circumstances, divergence may occur with such speed and magnitude that it has been termed QUANTUM EVOLUTION (Simpson 1944; Chapter 14).

## THE SIGNIFICANCE OF SPECIES AND SPECIATION

Speciation is usually not adaptive in itself, but is ordinarily a byproduct of adaptive differentiation under selection, with or without an impetus from genetic drift. In some cases, perhaps, speciation is the consequence of genetic drift alone. But speciation does have consequences for adaptation and long-term evolution.

Some paleontologists have argued that speciation may well be required for much evolution to occur at all: that established species are static, and that only the isolation of a small population from gene flow, and the destabilization of the genome during speciation, enables appreciable response to selection. This, the theory of punctuated equilibrium (Eldredge and Gould 1972, Stanley 1979), has been extremely controversial (Charlesworth et al. 1982). We will consider this issue in Chapter 14.

There is little doubt, however, that speciation contributes to adaptation in the long run, because the acquisition of reproductive isolation can enhance population fitness by restricting recombination. Asexual forms, as Hutchinson (1968) noted, are frequently organized into discrete units that can be defined by morphological or ecological criteria as "asexual species." In such forms, discrete phenotypes may coexist, each adapted to a different resource. But in sexual populations, although different genotypes may be adapted to different resources because of specific combinations of alleles at several loci, such genotypes are generally ephemeral because of recombination. In a sexual population, therefore, exceptional resources that require exceptional genotypes are out of bounds, and the population is more likely to consist of ecologically generalized genotypes, each only moderately adapted to a variety of resources, than of a spectrum of specialized genotypes (Roughgarden 1972). Indeed, multiple-niche polymorphism appears to be more frequent among asexual than sexual insects (Futuyma and Peterson 1985). But if genotypes adapted to different resources are reproductively isolated, the ecological and morphological diversity encompassed by a group of species can persist.

Local populations respond rapidly to selection. But local populations are ephemeral: as the environment changes, they become extinct, or migrate to new, ecologically suitable areas. Even if the rate of gene flow among populations is usually low, as it is in many species (Chapter 5), over longer periods populations are brought into contact, and much of the divergence they accomplished will be lost by interbreeding. If they have become different species, however, they can retain their diverse adaptations, and refine them even while sympatric (Futuyma 1986). So speciation, by enabling populations to retain evolutionary advances that otherwise would be reversed, can contribute to long-term evolutionary trends. As Mayr (1963, p. 621) puts it (a bit teleologically, to be sure), "Speciation, the production of new gene complexes capable of ecological shifts, is the method by which evolution advances. Without speciation there would be no diversification of the organic world, no adaptive radiation, and very little evolutionary progress. The species, then, is the keystone of evolution."

## SUMMARY

Speciation, the multiplication of species, consists of the evolution of genetic barriers to gene exchange among populations. Probably this occurs most frequently by polyploidy and by genetic divergence among spatially segregated (allopatric or parapatric) populations, but sympatric speciation is also at least theoretically possible. Species exist not by virtue of complete differentiation throughout the genome, but by virtue of reproductive barriers that have the same kinds of genetic bases as characters that vary within populations. Differentiation of populations into distinct species may occur rapidly or slowly, by natural selection, genetic drift, or a combination of the two. There are reasons to believe that small, localized populations may differentiate most rapidly into new species. Genetic barriers to gene flow (isolating mechanisms) usually arise as byproducts of genetic change rather than as mechanisms to prevent hybridization.

## FOR DISCUSSION AND THOUGHT

1. Speciation in plants often differs from speciation in animals; speciation by polyploidy is much more common, and the incompatibility of related plant species commonly has a chromosomal rather than a genic basis. Discuss why these differences exist.
2. Among African cichlids, there are far fewer species of *Tilapia*, which are ecologically generalized fishes, than of forms such as *Haplochromis*, which tend to be more specialized (Fryer and Iles 1972). How could one determine whether the disparity in species richness is caused by a difference in speciation rate or by competitive exclusion among ecologically similar species?
3. Alexander and Bigelow (1960) proposed that two species of crickets arose by allochronic speciation: as the climate cooled, a species that had bred year round became split into two moieties that passed the winter in different developmental stages and reached reproductive age at different times of the year. Mayr (*Systematic Zoology* 12:206, 1963) and Harrison (*Evolution* 33:1009, 1979) have disagreed. What kinds of data could test Alexander and Bigelow's hypothesis?
4. Divergence among geographically distinct populations is generally accepted as evidence for allopatric or parapatric speciation, yet the existence of sympatric species is not generally considered evidence for sympatric speciation. Why not? How could you tell whether sympatric species had arisen sympatrically or allopatrically? Is it possible to tell?
5. Would you expect speciation rates to differ in the recent past between phylogenetically "old" groups such as ferns or amphibians and "young" groups such as composites or mammals?
6. Theories of speciation by the founder effect depend on the principle that epistatic interactions among loci are prevalent and important, yet statistical analysis of genetic variance in quantitative characters (e.g., morphology) shows that the epistatic component of the variance in many traits is quite low. Does this apparent conflict mean that the founder effect is unlikely to be important?
7. Stanley (1979) has provided some evidence that long-term rates of morphological evolution have been higher in taxa that are rich in species than in taxa that have always had few species (such as aardvarks). What hypotheses might account for this pattern?
8. "Genetic distances" based on allozyme data (Chapter 4) have frequently been used to estimate time since speciation. For example, Nei (1971) estimated on this basis that about 500,000 years have elapsed since the formation of sibling species of *Drosophila* and about three times as many since the formation of non-sibling species. What assumptions are necessary if these estimates are correct?
9. Related species usually differ in chromosome arrangement, yet this does not necessarily mean that chromosome rearrangements caused speciation or were even associated with speciation. Why not?
10. Discuss the likelihood that hybrids between two diploid species could constitute or develop into a new, third species (e.g., Straw 1955, Ross 1958, Johnston 1969).

## MAJOR REFERENCES

Mayr, E. 1963. *Animal species and evolution*. Harvard University Press, Cambridge, MA. 797 pages. Contains a wealth of classical information bearing on speciation, and discussion of influential ideas.

Grant, V. 1981. *Plant speciation.* Second edition. Columbia University Press, New York. 563 pages. Extensive information on speciation and related issues in higher plants.

White, M.J.D. 1978. *Modes of speciation.* Freeman, San Francisco. 455 pages. Summarizes a great deal of more recent information on genetic and chromosomal aspects of speciation, emphasizing modes of speciation other than allopatric.

Templeton, A.R. 1981. Mechanisms of speciation: a population genetic approach. *Ann. Rev. Ecol. Syst.* 12:23-48. Detailed discussion of the genetics of speciation. See also papers by Carson and Templeton and by Barton and Charlesworth in *Ann. Rev. Ecol. Syst.*, Volume 15 (1984).

# Adaptation

# Chapter Nine

Adaptations—those features of organisms that, as Darwin said, "so justly excite our admiration"—are central to the study of biology. When a biochemist, physiologist, or behaviorist sets out to determine the function and operation of a feature, the very assumption that it has a function is usually tantamount to the belief that it has been shaped by natural selection to some particular end. Within evolutionary biology the study of adaptation is likewise a dominant theme. Ever since *The Origin of Species*, two great themes have pervaded our explanation of the features of organisms: genealogy, whereby the explanation is found in an organism's ancestry, and adaptation, whereby it is found in the organism's conditions of life.

In physiology, the word "adaptation" is often used to describe an individual organism's phenotypic adjustment to its environment, as in physiological acclimation. In evolutionary biology, however, an adaptation is a feature that, because it increases fitness, has been shaped by specific forces of natural selection acting on genetic variation. Sometimes the word refers to the process whereby a population is altered in such a way as to be better suited to its environment. As the etymology of the word implies, an adaptation suits its bearer *to* something; cryptic coloration, for example, is usually an adaptation to avoid predation. The analysis of adaptations, then, entails showing that the trait has been developed by natural selection, and specifying the nature of the selective agent or agents that have favored the trait.

## PROBLEMS IN RECOGNIZING ADAPTATION

There is some ambiguity in saying that a population, having acquired a trait by natural selection, is better adapted to its environment, because the relation of the population to its environment may not have been altered in any measureable way. An enzyme that enables an insect to digest its food more efficiently may enable it to produce more eggs, and so may be favored by selection, yet the population may not increase in density if it is limited by predation. Although the population's genetic characteristics are altered, the adaptation is a property of individual organisms. As we saw in Chapter 6, an increase in mean fitness ($\bar{w}$) of a population need not increase a population's abundance, growth rate, or long term persistence. Thus $\bar{w}$ is a summary description of the average relative fitness of individuals in the population, not of the absolute fitness of the population as such. Similarly, adaptedness is a measure of a genotype's capacity for survival and reproduction relative to that of other genotypes, and an adaptation is a property that confers such an increase in relative fitness.

Not all features of organisms are adaptations, so the question of how one determines whether or not a trait is adaptive is critical. We are led to suspect strongly that a feature is adaptive if it is complex, for complexity requires an organizing principle such as selection. Complexity is difficult to measure and is sometimes in the eye of the beholder, and not all adaptations are complex. Nevertheless, "biological intuition" can be a useful guide. It took many years of study to show that the ampullae of Lorenzini in sharks and rays (Figure 1) had a function—namely, enabling the animals to locate prey in the mud—yet their structure was an assurance that they must have some function (Maynard Smith 1978a). Although a complex structure or pattern is typically adaptive, variations in the individual components of the structure need not be. For example, a protein

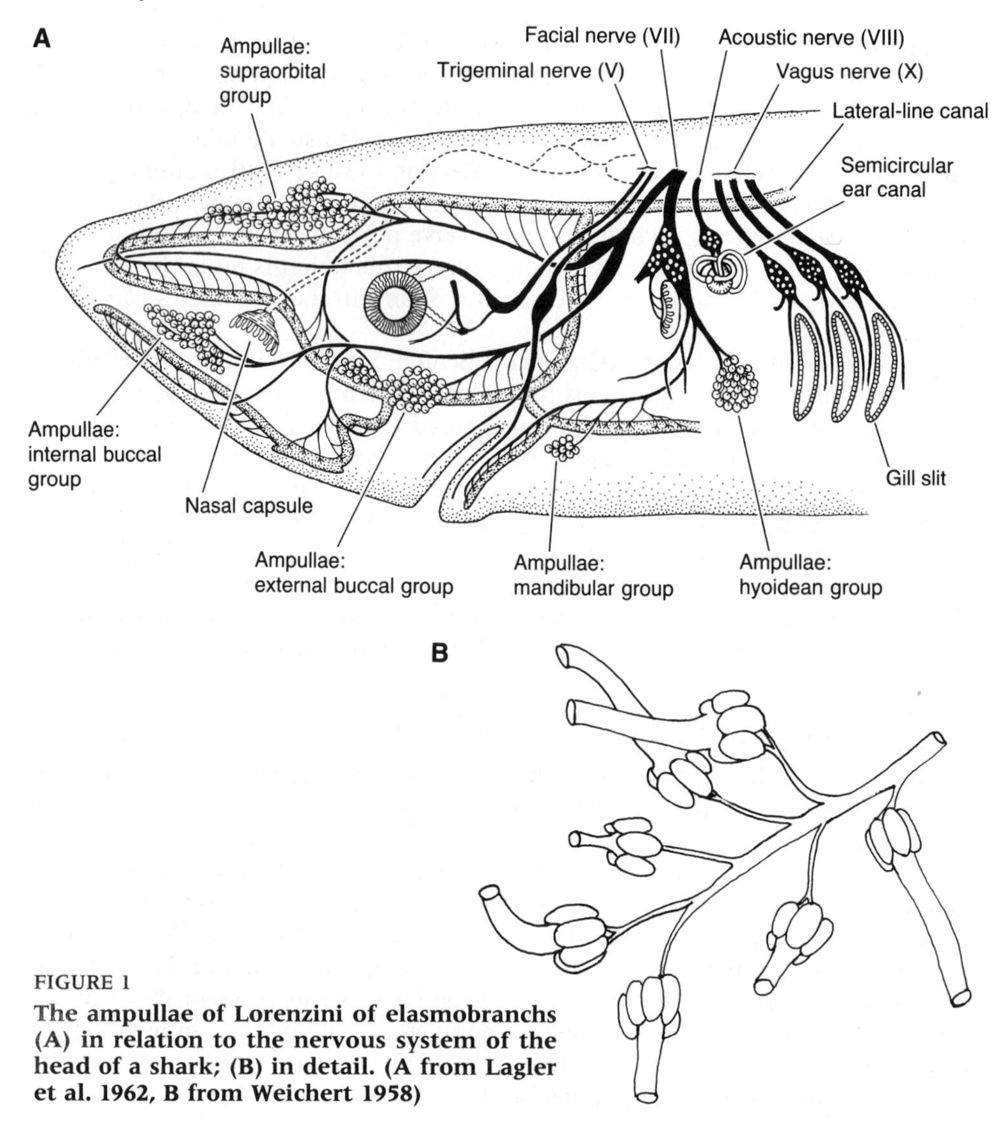


FIGURE 1

**The ampullae of Lorenzini of elasmobranchs (A) in relation to the nervous system of the head of a shark; (B) in detail. (A from Lagler et al. 1962, B from Weichert 1958)**

such as hemoglobin is an adaptation, but some of the amino acid differences among the hemoglobins of different species are likely to be neutral.

Demonstrating function can be difficult. In many cases, adaptation is strongly implicated by the correspondence between the form of a structure and the design that an engineer might specify for a particular function (Gans 1974); for example, the narrow, pointed wings of falcons conform to the aerodynamic specifications needed for rapid, sustained flight, whereas the short, rounded wings of forest hawks are appropriate for maneuverability and rapid acceleration.

The COMPARATIVE METHOD, widely used since Darwin introduced it, provides insights into adaptation by correlating differences among species with ecological factors. This method is most powerful when it draws on patterns of convergent evolution for its inferences; for example, one may infer that leaflessness in plants is often an adaptation to xeric conditions from its prevalence in many unrelated desert plants, or that a red, tubular corolla displayed by many unrelated plants is an adaptation to bird pollination. It is not sufficient to draw a correlation merely by counting species with one or another trait, because ecologically similar species may be closely related and therefore share a trait solely because of common ancestry. They will not provide independent evidence that the trait is an adaptation to that ecological circumstance. Therefore it is important to count the number of times a trait has independently arisen during phylogeny, and to determine if each of these evolutionary events transpired in a particular selective context (Clutton-Brock and Harvey 1984, Felsenstein 1985). Thus the comparative method is best used in conjunction with phylogenetic analysis (Chapter 10). The comparative method, in conjunction with considerations of optimal design, can yield predictions that constitute tests of hypotheses about adaptation. For example, if competition for mates is more intense among the males of polygynous than monogamous species, and if greater body size and weaponry such as canine teeth or antlers confer an advantage in fighting, one might predict that sexual dimorphism in these features would be more pronounced in polygynous species. Among genera of primates, deer, and many other groups, this prediction is upheld (Clutton-Brock and Harvey 1984). Even so, sexual dimorphism in body size is a phylogenetically conservative trait among primates, and has not evolved independently in different species solely in response to the mating system (Cheverud et al. 1985).

Such correlational studies provide inferences about the fitness value of traits, but direct evidence must come from experimental studies or from an explicit analysis of fitness. For example, the hypothesis that the color pattern of a butterfly is aposematic, providing protection against predation, can be tested by monitoring the predation rate on butterflies that differ in phenotype either because of genetic differences (as in the study of genetic polymorphism) or because they have been experimentally altered (e.g., by painting the wings). Experimental alterations of the phenotype have the advantage that differences in fitness may then be ascribed directly to the phenotypic trait in question, rather than to pleiotropic effects of a variable locus. It is sometimes even possible to evaluate experimentally the adaptive function of characteristics of extinct organisms. For example, Niklas (1983) measured the effectiveness of physical models of the reproductive structures of Paleozoic seed plants in capturing pollen in wind tunnels, and found that some phylogenetic trends in morphology were associated with increasing effectiveness of pollination.

In some cases, however, even experimental methods do not conclusively demonstrate that a trait evolved because of a particular function. The problem is that the *effects* of a character may differ from its function. For example, plant compounds such as terpenes and alkaloids have often been shown to repel or poison herbivorous insects, yet some authors (e.g., Jermy 1984) maintain that the compounds may have other physiological functions, or be merely waste products of plant metabolism, and that their effects on insects are only incidental. "Isolating mechanisms" have the effect of preventing gene flow between species, but there is seldom reason to believe that they evolved for this function (Chapter 8).

## THE ADAPTATIONIST PROGRAM

For much of the first half of this century, Darwin's theory of natural selection was not widely accepted. It became respectable only after Fisher, Wright, and Haldane showed that even a slight selective advantage could bring about fixation of an allele, and after ecological geneticists such as Dobzhansky and Ford showed that some natural polymorphisms were subject to appreciable selection. The view then became widespread that virtually every genetic difference, no matter how slight, has adaptive significance. In the 1960s, the debate over adaptation entered a new phase when V.C. Wynne-Edwards (1962) argued that dispersal of animals has evolved because it relieves population density and so benefits the population as a whole, rather than the individual members that disperse. Controversy then arose about whether such seemingly altruistic traits could evolve by natural selection at the individual level, or whether group selection—evolution by the differential extinction or proliferation of whole populations—was necessary. Among the most influential contributions to the debate were those of W.D. Hamilton (1964), who articulated the theory of kin selection as an explanation of altruistic behavior, and of G.C. Williams (1966), who argued that group selection is so weak a force that alternative explanations for altruism should be sought in selection at the level of the gene, the individual organism, or the kin group.

The enthusiastic response to these proposals ushered in an era of analysis of the adaptive value of puzzling traits, especially features of life histories and of social behavior. At their best, these analyses have led to deep insights, some of which we will consider later in this chapter. At its worst, the enthusiasm has produced wild, unsupported speculations about adaptation, with little consideration of either evidence or of alternative, nonadaptive explanations. To pick egregious examples of such "idle Darwinizing," as Lewontin (1977) has called it, some authors have proposed that the exposed mucosa of our lips serves as a sexual signal because it resembles the vulval labia, or that women tend to carry babies on the left arm because the children are soothed by the sound of the mother's heartbeat.

Gould and Lewontin (1979) applied the term "adaptationist program" to research that assumes that all characters are adaptive, that they are nearly optimal, and that the differences between species are invariably species-specific adaptations to different selective factors. Yet as Williams (1966) pointed out, "adaptation is a special and onerous concept that should be used only where it is really necessary." All of biology compels us to recognize that organisms are not optimally designed, that many features are not adaptive, and that species may differ for reasons other than natural selection:

—The trait may not be genetically programmed, but instead a direct consequence of the action of the environment or of learning. This is an especially important consideration in interpreting the behavior of humans or other animals in which learning is prominent. When "cultural inheritance" is important, there is no reason to suppose that organisms learn only what is in their best interests.

—The trait may be a simple consequence of the laws of physics or chemistry. Williams (1966) notes that when we see a flying fish return to the water, we might interpret this as adaptive behavior, since the fish could not survive long in air; but since the fish can actually only glide, its return to the water is surely a simple effect of gravity.

—The evidence that genetic drift plays an important role in evolution grows steadily (Chapters 5 and 15). If drift has affected evolution at the molecular level as much as it now appears, it can surely have affected morphological and behavioral features as well. In populations that consistently or frequently are small in size, selection cannot maintain allele frequencies at an adaptive peak unless it is quite strong. If, for example, we seek to explain the evolutionary loss or vestigialization of a feature, we must entertain both selectionist and non-selectionist hypotheses. Useless features such as the eyes of cave animals degenerate during evolution (Culver 1982); female mating behavior degenerated in parthenogenetic laboratory populations of *Drosophila mercatorum,* a normally bisexual species (Carson et al. 1982). In some cases, the loss of a structure may be adaptive because it saves energy or materials; wingless morphs of many insect species typically develop faster and are more fecund than winged morphs (Harrison 1980, Zera 1984). But in other cases a complex structure that is no longer used may degenerate because of the accumulation of selectively neutral mutations, as the existence of functionless pseudogenes (Chapter 15) indicates, or because of pleiotropic effects of genes selected for other functions (Prout 1964).

—Organisms cannot be decomposed into separate traits, for development is integrated and genes are pleiotropic (Gould and Lewontin 1979). Two species of *Drosophila* may differ in bristle number not because each has the bristle number appropriate for its ecology, but because the difference is a pleiotropic consequence of allelic differences that developed for other selective reasons or by genetic drift. Worse yet, the trait may not exist as such. It is pointless to ask why mammals have red blood if the property of redness is inseparable from the physical properties of hemoglobin. Gould and Lewontin (1979) suggest that the human chin is not an adaptive feature, and in a sense does not exist; its form is the consequence of the difference in the rate at which the mandible and the tooth-bearing part of the lower jaw have become shortened during human evolution.

—A trait may be genetically correlated, either because of pleiotropy or linkage, with another trait that is changing under selection. A common manifestation of pleiotropy, for example, is ALLOMETRIC GROWTH (Chapter 14), whereby a part of the body develops at a disproportionate rate relative to another part. Thus, *Tyrannosaurus rex* may have had absurdly small front legs not because they were adaptive, but because of a negative developmental correlation between the size of the body and the forelegs in theropod dinosaurs generally.

—Adaptive landscapes usually have multiple adaptive peaks, so that the course of evolution depends on the initial genetic composition of a population (Chapter 7). Species of grouse vary in the cryptic color pattern of the chicks (Figure 2), and a cryptic pattern is certainly adaptive; but the differences among species may stem from accidents of their genetic history, not because the optimal pattern differs from one species to another.

—Multiple adaptive peaks exemplify the historical contingency of evolution. The kinds of variations that can arise in any population are determined by evolutionary history. For example, every species has DEVELOPMENTAL CONSTRAINTS (Chapter 14) arising from the history of evolution of its developmental

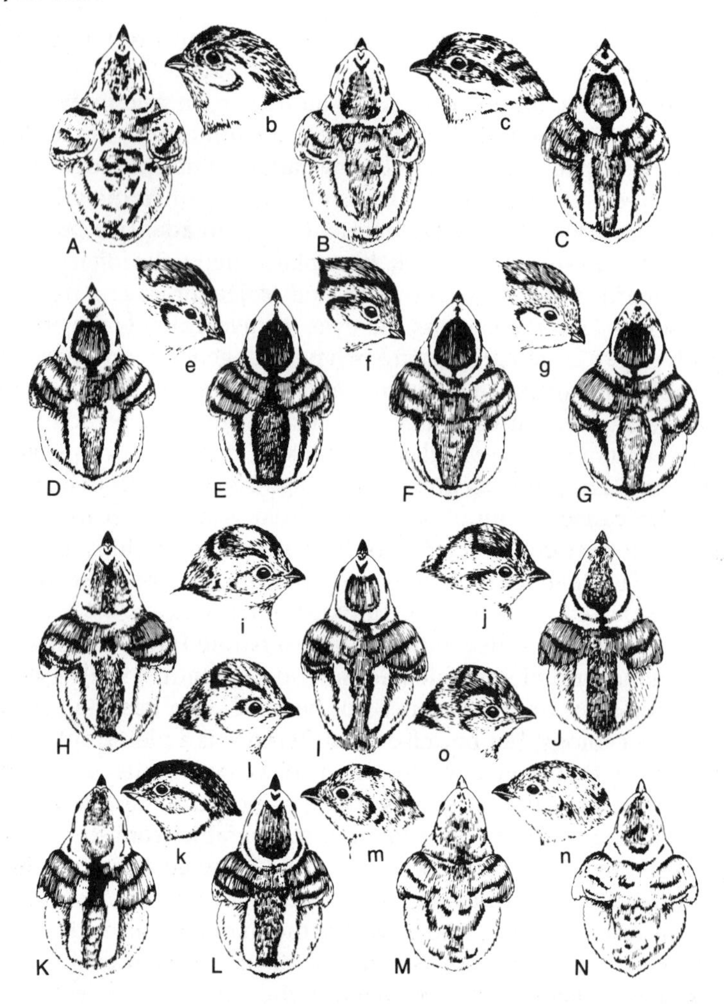


FIGURE 2

**Plumage patterns of newly-hatched grouse and ptarmigans (Tetraoninae). These down patterns are replaced within a few days by a juvenal plumage. The patterns doubtless provide cryptic protection, but it is quite possible that the differences among the species have little adaptive significance. (A) Sage grouse (*Centrocercus urophasianus*); (B,b) blue grouse (*Dendragapus obscurus*); (C,c) sharp-winged grouse (*Dendragapus falcipennis*); (D) spruce grouse (*Dendragapus canadensis*); (E,e) rock ptarmigan (*Lagopus mutus*); (F,f) willow ptarmigan (*Lagopus lagopus*); (G,g) white-tailed ptarmigan (*Lagopus leucurus*); (H) capercaillie (*Tetrao urogallus*); (I,i) black grouse (*Tetrao tetrix*); (J,j) hazel grouse (*Bonasa bonasia*); (K,k) ruffed grouse (*Bonasa umbellus*); (L,l) black-billed capercaillie (*Tetrao parvirostris*); (M,m) greater prairie chicken (*Tympanuchus cupido*); (N,n) sharp-tailed grouse (*Tympanuchus phasianellus*); (O) Caucasian black grouse (*Tetrao mlokosiewiczi*). (From *Grouse of the World* by Paul A. Johnsgard. Copyright 1983 by the University of Nebraska Press)**

system (Alberch 1982). It is evident that a Pegasus has never evolved because the wings of vertebrates invariably have evolved by modification of the front limbs, rather than sprouting *de novo* from the shoulders. In the bolitoglossine salamanders, there are 16 possible arrangements whereby the tarsal bones might be fused or divided, yet only three arrangements have been found as variations within populations, and each of these is the typical condition in one or another species: there appear to be developmental constraints on the kinds of variations that can arise and become fixed (Alberch 1983). Our understanding of developmental processes is so primitive that we have little grasp of what variations can and cannot arise.

—All organisms have anachronistic features, for the environment that an organism inhabits today is never identical to that in which the features evolved. The characteristics of the fruits of many tropical trees, for instance, seem to be adaptations for dispersal by large mammals that became extinct during the Pleistocene (Janzen and Martin 1982). A character that originally evolved for one function, or for no function at all, may be used by an organism for an entirely different function—but its current function cannot be viewed as its adaptive *raison d'être*. Our hands did not evolve their present form so that we could play the piano. Darwin pointed out in *The Origin of Species* that the sutures in the skull of young mammals "have been advanced as a beautiful adaptation for aiding parturition, and no doubt they facilitate, or may be indispensable for this act; but as sutures occur in the skulls of young birds and reptiles, which have only to escape from a broken egg, we may infer that this structure has arisen from the laws of growth, and has been taken advantage of in the parturition of the higher animals." Gould and Vrba (1982) have offered the term *exaptation* to describe features that now enhance fitness, but were not built by natural selection for their current role.

—It is a great oversimplification to say that features develop to "adapt the species to its environment," as if evolution by natural selection merely consisted of solutions to problems that the environment presents (Lewontin 1983). Organisms set up their own selection pressures, and engender their own evolution, by evolving into new ecological roles. Many organisms have only recently embarked on a new way of life, to which they are by no means optimally adapted. Water ouzels (*Cinclus*; Figure 3) and certain aquatic Hymenoptera (Proctotrupidae) are structurally much like their terrestrial relatives and show little morphological evidence of their aquatic habits (see Darwin's *Origin*, Chapter 6).

FIGURE 3

**The American dipper, or water ouzel (*Cinclus mexicanus*), which shows few special adaptations for its habit of foraging underwater in mountain streams. (Redrawn from Peterson 1961)**

In short, the main justification for assuming a priori that a trait is adaptive is not that this assumption is likely to be true, but that it is heuristic; it impels us to do research that will reveal more about the feature. But if such research fails to confirm the hypothesis of adaptation, it is unreasonable to insist that it must have some function yet to be discerned; it is most reasonable to consider alternative hypotheses that may shed light on other important issues in evolution, such as phylogenetic history and developmental processes. Still, many of the features that "excite our admiration" are indeed adaptive, so we turn now to recent developments in the analysis of adaptation.

## LEVELS OF SELECTION

Under ordinary natural selection, alleles increase in frequency if they enhance the fitness of their bearers, relative to that of genetically different individuals in the same population. By this mechanism, a trait cannot increase in frequency if it is harmful to the individual yet good for the population or species. Thus pure altruism, such as responding to excessive population density by becoming sterile or by committing suicide (as in the myth of lemmings marching to the sea to drown), cannot evolve. Yet many animals seem at first glance to have altruistic, cooperative behavior, and other features such as sexual reproduction and the process of mutation also seem to benefit the group rather than the individual.

A possible explanation for such traits is that even if natural selection within populations acts to decrease an allele's frequency, the allele might increase in the species as a whole if the rate of extinction of whole populations is lowered, or if the rate of proliferation of new populations is enhanced, by a high frequency of the allele (Figure 4). Thus the population or deme might be the unit of selection. Williams (1966), however, pointed out that such GROUP SELECTION (or INTERDEMIC SELECTION) will seldom be able to counteract the force of individual selection. The chief reason is that the rate of allele frequency change by individual selection must usually be vastly greater than the rate of change by group selection, because the number of individual organisms is much greater than that of populations and

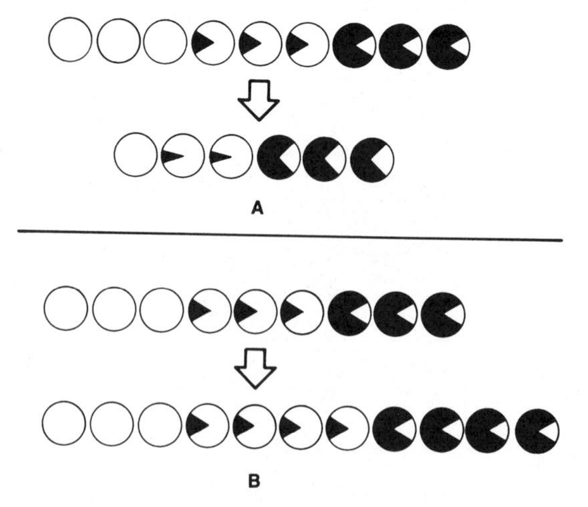


FIGURE 4
**Two possible modes of group selection. Each circle represents a population, with the proportion of dark and light representing the frequency of individuals of two genotypes, 1 (light) and 2 (dark). (A) The probability of survival of a population increases if genotype 2 is in high frequency, although individual selection reduces the frequency of this genotype within populations. (B) The probability of founding new populations increases if genotype 2 is in high frequency. At an individual level, the genotypes are selectively neutral.**

the rate of turnover of individual organisms is much higher. Moreover, even if an individually disadvantageous trait is temporarily fixed in a population by group selection, the genetic constitution is unstable, because if an individually advantageous allele again arises by mutation (or enters by gene flow), it will increase in frequency. Williams therefore argued that group selection should be an explanation of last resort. His arguments had a strong impact, but as we shall see, some recent models show that group selection may sometimes be a plausible mechanism of evolution. Let us first consider alternative explanations for features that seem not to provide benefit to individual organisms.

### The adaptation may be illusory

To illustrate this principle, consider that it is to a population's long-term advantage to be genetically variable, and hence capable of adapting to environmental change. Yet balanced polymorphisms do not exist for the sake of the future benefit they may provide in genetic flexibility; they exist because of certain modes of selection acting at the present time. The sickle-cell hemoglobin polymorphism is an incidental consequence of heterozygous advantage, not insurance for the future. A rather more complicated case is the question of whether mutation rates have evolved toward an optimum level. The mutation rate will be affected by any gene that influences DNA repair enzymes or sensitivity to mutagens such as ultraviolet light, and many genes are known to affect mutation rates (Chapter 3). An allele that increases the mutation rate at many loci, such as Treffer's mutator, *mutT*, in *Escherichia coli*, should ordinarily be eliminated from a population, because it will be inherited along with the mutations it causes, most of which are deleterious. By this reasoning, genes that suppress mutation should be advantageous, and the mutation rate should evolve toward zero, even if mutation is ultimately necessary for population survival.

Surprisingly, however, the frequency of the *mutT* mutator allele typically increases in frequency in chemostat cultures of *Escherichia coli* (Gibson et al. 1970), apparently because in this asexually reproducing organism, those clones that carry newly induced advantageous mutations, and which carry *mutT* as well, have a substantial advantage relative to clones that lack the mutator allele (Chao and Cox 1983). The mutator allele increases by hitchhiking with the favorable mutations it induces (Figure 5). The increase in frequency of the mutator allele depends on a low rate of recombination between the mutator and the advantageous mutations (Leigh 1973). Although mutator alleles may increase in laboratory populations, they appear to be rare in natural populations of *E. coli* (Chao and Cox 1983).

### Individual selection

Many features that at first glance appear advantageous at the population level may actually be advantageous to individual organisms (Williams 1966). For example, many organisms reduce their reproductive rate as population density increases. This can be naively interpreted as an adaptation to prevent overpopulation and exhaustion of resources. But if there is a high likelihood that the stress of high density will be alleviated later in an individual's lifetime, the net reproductive rate of an organism may be higher in the long run if it uses scarce

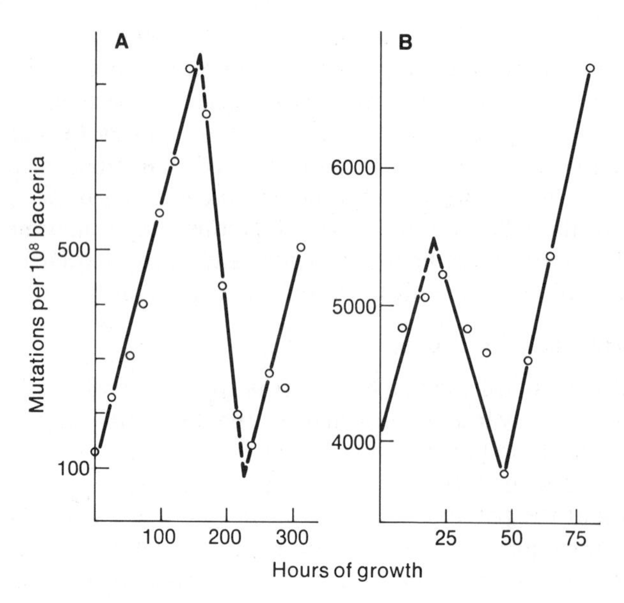


FIGURE 5

**Demonstration of "periodic selection" in *Escherichia coli*. This term refers to rapid, irregular changes in allele frequency caused by hitchhiking in clonal organisms. The population in (A) lacks a mutator allele; the population in (B) carries the mutator allele, and so has a higher mutation rate. The changes in genetic composition occur much more quickly in population (B); note the difference in scale of the axes. The populations are tested at intervals for the frequency of a neutral marker allele (the vertical axis). Its frequency increases if an advantageous mutation occurs in a cell that carries the marker allele, so this clone increases in frequency. When a more advantageous mutation occurs in a cell that lacks the marker allele, the unmarked clone increases in frequency, so the frequency of the marker decreases. These events happen more rapidly in the culture bearing the mutator allele. In clonally reproducing organisms, alleles such as neutral markers or mutator alleles therefore change in frequency by hitchhiking with advantageous mutations. (From Nestmann and Hill 1973)**

resources for its own growth and maintenance and delays reproducing, rather than squandering resources on offspring that are unlikely to survive. (The evolution of life histories will be treated later in this chapter.)

## Genic selection

Some traits are the consequence of selection at the level of the gene rather than the individual organism. MEIOTIC DRIVE (or SEGREGATION DISTORTION), whereby one allele is carried by more than half the gametes of a heterozygote, is an example (Lewontin 1970). For example, some recessive alleles (*t*) at the *T* locus in house mice (*Mus musculus*) cause homozygous males to be sterile—an obvious disadvantage. A male $+t$ heterozygote, however, produces 80–95 percent *t*-bearing sperm, so the *t* allele has a selective advantage and will increase when rare. If sterility and segregation distortion are the only effects of this allele, it should reach an equilibrium frequency of 0.70 (Lewontin 1962). A similar kind of advantage at the genic level is displayed by transposable genetic elements, which can proliferate within the genome and be transmitted in great numbers through the gametes (Chapters 8 and 15). Such elements can rapidly increase in frequency within a population even if they have some deleterious effects on the organism.

The sex ratio among offspring at birth is 1 in many species. This is not a necessary consequency of an X/Y sex-determining mechanism, for some organisms do not have a sex ratio of 1, and even in those that do, mutations are known that alter the segregation ratio of X and Y chromosomes. Nor is it obvious how an individual's fitness is enhanced by having equal numbers of sons and daughters rather than, say, mostly daughters.

A widely accepted explanation, proposed by Fisher (1930), entails frequency-dependent selection at the level of the gene. Since every offspring has one mother and one father, the average number of offspring per individual of a given sex is greater for the sex that is rarer in a population. Thus, if an allele is more frequent within the rarer sex than within the commoner sex, it will increase in frequency within the population as a whole. If in a male-biased population an allele *A* causes an individual to have more female than male offspring, this allele will be more frequent among the females in the next generation, and so will increase in frequency because of the greater average reproductive success of this sex. Conversely, an allele that biases an individual's brood toward males will increase in a female-dominated population. An allele that causes its bearers to "invest" equally in sons and daughters will consequently not be replaced by any allele that biases the sex ratio toward one sex or the other. The evolution of sex ratios and related phenomena is treated in detail by Charnov (1982).

## Kin selection

Suppose a mutation causes bright coloration in an otherwise cryptically colored, distasteful species of butterfly, and predators learn, by killing and tasting the mutant, to avoid all members of that species, both aposematic and cryptic. Such a mutation cannot increase in frequency, because the benefits are distributed equally both to other carriers of the allele and to carriers of alternative alleles, whereas the "cost" in mortality is borne only by carriers of the allele in question. If, however, predators learn to avoid only the brightly colored phenotype, the allele can increase in frequency because the cost in mortality has the consequence of benefiting individuals that carry other copies of the allele, that are identical by descent to those that are lost to predation. Thus the change in allele frequency depends on both the direct influence of the trait on the individual organism, and on its indirect effect: the increment (or decrement) in fitness that the trait bestows on related individuals that carry other copies of the allele.

This combination of direct and indirect effects, termed INCLUSIVE FITNESS (Hamilton 1964), may be expressed as

$$w_i = a_i + \sum_j r_{ij} b_{ij}$$

In this equation, $w_i$ is the inclusive fitness effect of genotype $i$, $a_i$ is the direct effect of the trait on $i$'s individual fitness, $b_{ij}$ is the effect of $i$'s trait on the fitness of individual $j$, and $r_{ij}$ is a coefficient of relatedness between $i$ and $j$: the fraction of $j$'s gametes that are identical by descent at this locus to the alleles carried by $i$

(Michod 1982). The summation sign indicates that inclusive fitness is calculated by taking all of *i*'s relatives into account. Thus an allele may increase in frequency if it is individually disadvantageous ($a_i < 0$), as long as it confers a sufficient advantage on the bearer's relatives. Hence this form of selection is known as KIN SELECTION (Maynard Smith 1964). The central theorem of kin selection, often known as "Hamilton's rule," is that a kin-selected trait can increase in frequency if $a_i < \sum_j r_{ij} b_{ij}$: if the trait's cost to its bearer in terms of individual fitness ($a_i$) is less than the benefit in fitness dispensed to relatives ($b_{ij}$), weighted by the coefficient of relatedness ($r_{ij}$) between the donor and the beneficiary of the action. Since half one's genes are carried, on average, by a son or daughter, half by a sibling, and one-fourth by a niece or nephew, sterility could theoretically evolve if this sacrifice in individual fitness increased, by an equivalent amount, the reproductive success of more than two offspring or siblings or more than four nieces or nephews.

The theory of kin selection is considerably more complex than the simple version of Hamilton's rule indicates (Michod 1982), for it must take into account various aspects of population structure, as well as differences among interacting individuals in their expected reproductive success. For example, parental care is a form of kin selection; a female mammal does not herself benefit from suckling her young, but she is enhancing the survival of her own genes, half of which, on average, are carried by each offspring. Although the reciprocal degree of relationship between parent and offspring is the same, we would ordinarily expect parents to sacrifice more for their offspring than vice versa, since the probable future reproductive output of an adult who has already reproduced is less than that of one of the offspring. But if the offspring have a low probability of successful reproduction, they may propagate their genes indirectly by helping their parents to rear their siblings. This may explain why some birds and mammals help their parents raise one or more broods, before setting out to reproduce on their own (Emlen 1984). This is also one of the chief theories of the evolution of eusocial insects (termites, ants, some wasps and bees), in which some individuals become sterile workers, rearing siblings or half-sibs rather than their own offspring (West-Eberhard 1975, Wilson 1975, Brockmann 1984). In many such insects, established queens have a prolonged, successful reproductive life, whereas the chance that a young solitary queen will successfully breed is very low. In some social wasps, moreover, workers can become reproductive if the queen dies, so by staying at the nest they have at least a slight expectation of reproducing.

Because the evolution of self-sacrifice requires that benefits be dispensed to kin, inbreeding, which tends to structure populations into kin groups (Chapter 5), generally (although not always) favors the evolution of such traits (Michod 1982). This is evident also in another class of models of kin selection, the family-structured models (Wade 1979, Michod 1982). Suppose a population is structured, as it is among young birds before they leave the nest, so that interactions among relatives are more frequent than among individuals at random. An allele *A* influences its bearer to be nice to its nestmates, perhaps by not competing too intensely for the food the parent brings. The survival of such individuals will presumably be somewhat less than that of their more competitive, well-fed siblings of genotype *aa*. Hence *within* each nest, the frequency of *A* decreases (Williams and Williams 1957). However, the average number of young fledged from nests that harbor carriers of *A* may be higher than the number fledged from nests in which all the birds (genotype *aa*) compete intensely, with most losing in the competition. If so, the productivity of nests with *A*-bearing birds is higher, so the allele *A* increases in the population as a whole (Figure 6). Thus kin selection may be viewed as consisting of two components (Wade 1980a): individual selection, which always opposes the evolution of altruism, and selection among families or kin groups, which may favor it.

Wade (1980b) has provided an example of kin selection from experiments on the flour beetle *Tribolium confusum*, in which larvae eat eggs, and so act as cannibals (Figure 7). In some experimental populations, larvae were offered eggs drawn from the population at random, while in others, the eggs were full- or half-sibs of the older larvae. Each generation was bred from both the cannibals and those that survived cannibalism. Over the course of eight generations, the rate of cannibalism declined in those populations in which larvae fed on their relatives, and the effect was more pronounced when the larvae were fed full sib eggs than half sib eggs, as the theory of kin selection predicts. However, cannibalism declined only in those populations that were divided into inbreeding subpopulations, not in those that had a random mating structure. Thus the results conformed to the theory that altruism (i.e., refraining from cannibalism) can evolve by the differential productivity of kin groups that differ in genetic composition.

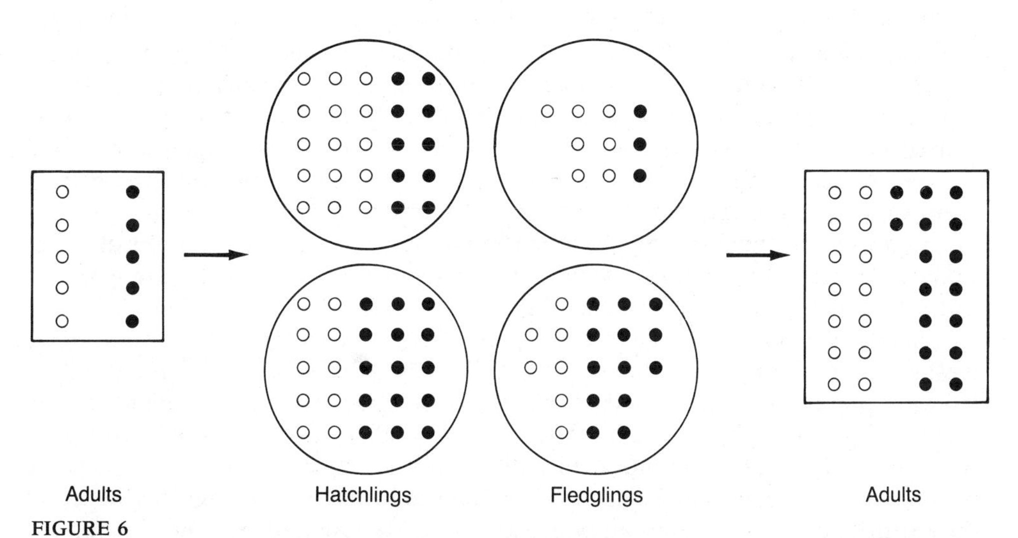


FIGURE 6

**A highly diagrammatic representation of a model of the evolution of cooperation by family-structured kin selection. The frequency of a cooperative genotype (●) varies among nests at hatching; although it may be selected against in interactions with the more competitive genotype (○), the productivity of families with a high frequency of the cooperative genotype is higher, so the genotype's frequency, measured at the adult stage, increases.**

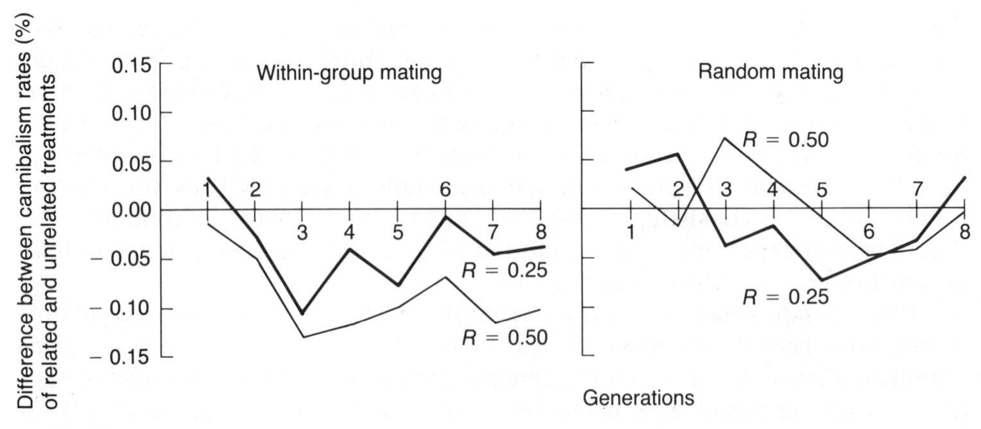


FIGURE 7

**The difference in average rates of cannibalism in experimental populations of *Tribolium* in which larvae were offered related eggs (full sibs, $R = 0.50$, or half-sibs, $R = 0.25$), compared to populations in which they were offered unrelated eggs (the zero line). Both the mating system and the degree of relationship affected the rapidity with which cannibalism declined due to genetic change. (From Wade 1980b)**

## GROUP SELECTION

We have now seen ways in which traits that appear to benefit the population rather than the individual can evolve, without invoking group selection, which has been considered a generally ineffective force of evolution (Williams 1966). However, mathematical models (reviewed by Maynard Smith 1976a, Wade 1978, Uyenoyama and Feldman 1980, D.S. Wilson 1983) have shown that group selection may take several forms, and that in some circumstances it may overwhelm the effect of individual selection against an individually disadvantageous allele. It is useful (Uyenoyama and Feldman 1980) to define a group as the smallest collection of individuals within a population, within which genotypic fitnesses do not depend on the genetic composition of any other group. Group selection, then, is variation in the rate of increase or of extinction among such groups, as a function of their genetic composition. The family-structured models of kin selection discussed in the previous section fit this definition: kin selection may sometimes be considered a special case of group selection.

Many models of group selection (e.g. Levin and Kilmer 1974, Aoki 1982) treat the hypothetical case in which the probability of extinction among semi-isolated populations is inversely related to the frequency of an allele that is selected against within the population. Such an allele can increase within populations, and so create variation in extinction rate among populations, only by genetic drift. In such models, group selection can overwhelm individual selection only if the rate of gene flow among populations is low and the populations are very small or the force of individual selection is very weak. Thus these models tend not to speak favorably for group selection.

In contrast, group selection can be strong when it results from differential productivity of groups. For example, D.S. Wilson (1980), among others, has advanced a model in which small groups of individuals (TRAIT GROUPS) settle at random into patches of suitable habitat after mating at random in a mating pool.

The frequency of an "altruistic" allele *A* that may reduce individual fitness will vary among groups as a consequence of random sampling. Because of the altruistic action of *A*-bearing individuals, trait groups with a high frequency of *A* have a higher proportion of survivors than groups with a low frequency of *A*, and so contribute more to the mating pool the next time random mating takes place. Consequently, *A* increases in frequency in the population as a whole, even though individual selection reduces its frequency within each trait group. This model is very similar to the family-structured models of kin selection, except that the members of a trait group are not related.

Group selection has been studied more in theoretical than in real populations, but a few examples may illustrate its effect. The recessive *t* alleles that cause sterility in male house mice are subject, as we have seen, to both genic and individual selection, which in concert should yield an equilibrium allele frequency of 0.70. In a large colony, however, the allele frequency appeared to be stable at 0.36. Using a computer program that simulated the processes of natural selection and genetic drift in such a population, Lewontin (1962) found that the discrepancy could be explained if the population was subdivided into small demes, each of about two males and four females. By genetic drift, some such demes lose their *t* alleles and so persist, whereas other demes experience an increase in the frequency of *t* so that all the males are homozygous and hence sterile. Such demes will become extinct. The extinction of demes with a high frequency of *t* reduces the allele's frequency in the population as a whole. This may not be the whole story, because many aspects of the complex biology of the *t* allele have not been taken into account, but it presents a plausible case in which selection at three levels—the gene, the organism, and the deme—acts to shape the genetic composition of a population.

Parasitologists have maintained that parasites typically have less harmful effects on their hosts if the association is old than if it is recent. There are several possible explanations for this, including the evolution of resistance by the host (Chapter 16). However, it is possible that under some circumstances a parasite may evolve lower virulence by group selection (D.S. Wilson 1983). If a host is weakened in proportion to the reproductive rate of the parasites within it, individual selection within the parasite group (i.e. the parasites inside a single host) favors high fecundity, but group selection favors low fecundity; the entire group of parasites may die if the host dies. For example, the myxoma virus, introduced into Australia to control European rabbits (*Oryctolagus cuniculus*) at first caused immense mortality. But within a few years, mortality levels were lower, both because the rabbits evolved resistance and because the virus had evolved lower virulence (Fenner 1965, May and Anderson 1983). Because the virus is spread by mosquitoes that feed only on living rabbits, the chance of transmission from rabbits that carry a high load of virulent virus genotypes is reduced, because such rabbits are likely to be the first to die.

In a panmictic population, Fisher's (1930) theory of sex ratio, considered earlier in this chapter, predicts that the sex ratio at birth will evolve toward 1. Some organisms, however, such as certain nectar-feeding mites (*Rhinoseius*) that are carried from flower to flower by hummingbirds (Trochilidae), have more female than male offspring. This can be explained by a trait-group model of group selection (Colwell 1981, Wilson and Colwell 1981). If the mites breed within the flower and one or a few males can inseminate all the females, groups of mites that contain a high frequency of an allele biasing the sex ratio toward females contribute more offspring to the dispersal pool than groups that lack such an allele. Wilson and Colwell (1981) present some indirect evidence that the group selection model applies to hummingbird-transmitted mites.

Partly because of the individual advantage of cannibalism, evolution within populations of the flour beetle *Tribolium castaneum* tends to cause a decrease in population size. Wade (1977) has shown that artificially imposed group selection for large population size in *Tribolium* can counteract the tendency of these populations to evolve toward smaller population size by individual selection (Figure 8). Thus both theoretical and empirical information indicates that group selection may be a more important force in evolution than has commonly been thought. However, not enough is known to say how often the conditions that favor group selection obtain in nature, and at the present time group selection has not yet won wide acceptance as a likely explanation of many evolutionary phenomena.

## THEORETICAL APPROACHES TO MODELING ADAPTATION

Nothing in the selection equations of classical population genetic theory (Chapter 6) tells us why the phenotype of a genotype should affect its fitness in any particular way. The problem in modeling the evolution of particular traits such as sex ratio or altruistic behavior is to construct reasonable models of the relation between phenotype and fitness. This is what is done, for example, in theories of the evolution of sex ratio, in which the fitness of a genotype that produces an excess of one sex is written as a decreasing function of the proportion of that sex in the population or group. All the models we have considered so far in this chapter have been developed mathematically from the allele frequency equations of population genetics, although I have presented only verbal descriptions of the models.

In recent years, numerous models of the evolution of certain adaptive traits have been developed in terms of phenotypes rather than genotypes. These models are commonly called "optimal models," because they attempt to determine which of a variety of hypothesized phenotypes would be fittest, or optimal, and assume that a genotype specifying such a phenotype would become fixed. In ignoring the complexities of Mendelian segregation, such models treat the population as if it were asexual, and assume that as long as there exists heritable variation for the trait, the details of inheritance can be ignored. For some questions, such as those entailing polymorphism or linkage disequilibrium, these models are clearly inappropriate.

Optimal models have been criticized on the grounds that organisms seldom if ever attain an optimal state, and that the models are difficult to test because if an organism's characteristics deviate from the predicted optimum, it is difficult to say which assumptions of the model are false. Proponents of optimal models (e.g., Maynard Smith 1978a, 1982, Pyke 1984) hold that these models, despite their name, do not assume that organisms are optimally adapted. Rather, organisms fall short of optimality because of various constraints that should be understood and incorporated into the models. Maynard Smith (1978a) points out that optimal models usually incorporate two kinds of constraints, by specifying both the variety of possible phenotypes on which selection can operate and the factors

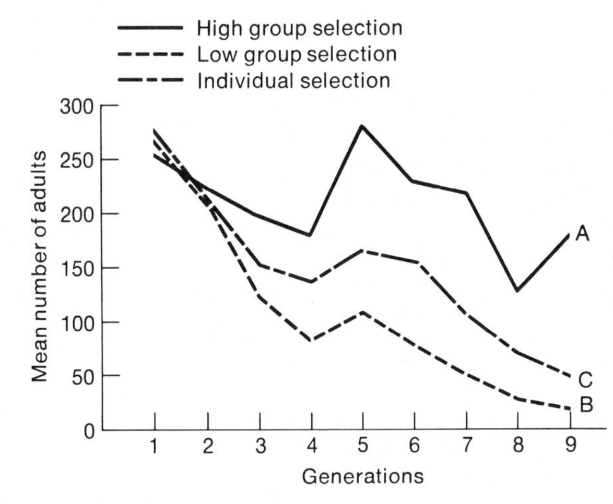


FIGURE 8
**Mean numbers of adult *Tribolium castaneum* beetles in populations evolving by group selection for high (A) and for low (B) population size, and by individual selection only (C). Individual selection (C) causes a decline in population size, an interesting effect in itself. This can be enhanced (B) or counteracted (A) by group selection, which was imposed by propagating new demes in each generation only from those demes that had the highest (A) or lowest (B) densities in the previous generation. In (C) all demes were propagated in each generation; there were no extinctions. (After Wade 1977)**

that are assumed not to change. For example, a model of the evolution of clutch size might assume that the size of the egg is free to vary within certain limits, but that the amount of material the organism can allocate to eggs is fixed. The assumptions of the model should take into account not only physical constraints, but those that arise from an organism's evolutionary history. In modeling the evolution of metabolism in mammals, it would be necessary to take into account not only the second law of thermodynamics, but also the inability of mammals to photosynthesize—which is a consequence of their evolutionary history.

### Optimal models: the evolution of optimal diets

Optimal models fall into two classes: frequency-independent and frequency-dependent models (Maynard Smith 1982). To exemplify frequency- independent models, let us consider the simplest form (MacArthur and Pianka 1966, Charnov 1976) of a model of an optimal diet for a predator that is faced with two or more kinds of potential food. The model is frequency independent because it assumes that the optimum for an individual forager is not affected by what other members of the population are eating (as it might be if they depleted the food supply). Among the numerous other assumptions are that the fitness of a forager increases linearly with its total intake of calories or some other measure of food value, and that a food item of a particular type takes a certain amount of time to handle, which detracts from the time the animal can spend searching for more food. The optimal diet, then, is one which maximizes the number of calories consumed ($E$) per unit time ($T$).

For simplicity, assume there are two kinds of food, which are encountered at rates $\lambda_1$ and $\lambda_2$ items per second, respectively, during $T_s$ seconds of searching time, that these types yield $E_1$ and $E_2$ calories per item, and that the predator, once it catches a prey item, must spend $h_1$ seconds handling an item of type 1 and $h_2$ on type 2. The theory predicts (Box A) that it would be optimal for the predator to eat only the type of prey that yields the most calories per unit handling time, if that kind of prey is sufficiently abundant. If we rank various types of prey in terms of their yield per handling time ($E/h$), then according to this model,

## A A Model of Optimal Diet

Prey items of types 1 and 2 yield a reward in calories of $E_1$ and $E_2$ per item, respectively, are encountered at rates $\lambda_1$ and $\lambda_2$, and require handling times $h_1$ and $h_2$ once captured (Charnov 1976; see text). A nonselective predator (a generalist) foraging for time $T_S$ will obtain $E = \lambda_1 T_S E_1 + \lambda_2 T_S E_2$ calories if it does not reject any prey, and the total time expended will be $T = (T_S) + (\lambda_1 T_S h_1 + \lambda_2 T_S h_2)$, where the terms in parentheses represent total search time and handling time respectively. Thus the average rate of intake will be

$$\frac{E}{T} = \frac{\lambda_1 E_1 + \lambda_2 E_2}{1 + \lambda_1 h_1 + \lambda_2 h_2}$$

Suppose prey type 1 is more profitable than type 2, in that it yields more calories per unit handling time ($E_1/h_1 > E_2/h_2$). Then the optimal diet should be type 1 only if

$$\frac{\lambda_1 E_1}{1 + \lambda_1 h_1} > \frac{\lambda_1 E_1 + \lambda_2 E_2}{1 + \lambda_1 h_1 + \lambda_2 h_2}$$

i.e., if the average rate of intake is greater for a specialist on type 1 than for a generalist. With a bit of algebra, this reduces to

$$\frac{\lambda_1 E_1}{1 + \lambda_1 h_1} > \frac{E_2}{h_2}$$

so specialization is favored if the specialist's average rate of intake is greater than the calories gained per time spent handling a prey item of the less profitable type. Note that the abundance of the more profitable item ($\lambda_1$) affects whether or not a predator should specialize on it, whereas the abundance of the less profitable item does not ($\lambda_2$ has dropped out of the inequality).

an animal should never specialize on a low-ranked food type regardless of its abundance, and a given food type should either be eaten whenever encountered or not at all (i.e., there should be no "partial preferences"). In experimental studies, mostly on birds and bees, animals have fulfilled these predictions to some degree, but seldom to the letter; in particular, they usually display partial preferences. However, this is but the simplest of a great array of models of optimal diet and optimal foraging, in which the assumptions vary from model to model (Pyke 1984). Optimal foraging models are important in accounting for the evolution of an important aspect of an animal species' ecological niche, namely its breadth of diet (Chapter 16). In contrast to speculations that interspecific competition may be the chief force of selection favoring specialized diets, optimal foraging models predict that specialization may arise simply in response to differences in quality and abundance of different kinds of food.

### Evolutionarily stable strategies

Frequency-dependent optimum models are those in which the ideal "strategy" (i.e., phenotype) of an individual depends on the "strategies" of other individuals in the population. The term "strategy" is natural in this context, because most of these models arise from the mathematical theory of games. As used in evolutionary theory, the central concept is an EVOLUTIONARILY STABLE STRATEGY, or ESS (Maynard Smith and Price 1973), defined by Maynard Smith (1982) as "a strategy such that, if all the members of a population adopt it, then no mutant strategy could invade the population under the influence of natural selection." A strategy is "a specification of what an individual will do in any situation in which it may find itself." A strategy may be pure, meaning that the individual always has the same phenotype, or it may be mixed, meaning that the phenotype of the individual varies, as is often the case in animal behavior. A mixed strategy is a description of an individual's variable phenotype, not a genetic polymorphism.

ESS models have been used mostly to analyze animal behavior (especially conflicts between individuals), sex ratio and related phenomena (Charnov 1982), and patterns of growth in plants. One of the earliest applications was the analysis of fighting behavior in animals. In many species, aggressive encounters consist largely of ritualized threat displays, the encounter seldom escalating to the point of physical damage to either animal (Figure 9). In the past, some authors assumed that ritualization evolved for the good of the species: a population of sharp-horned antelopes would quickly become extinct if every conflict between males over mates led to death. However, an ESS model shows that ritualization can evolve by individual advantage. Assume that there are two possible pure strategies: "Hawk," which escalates either until it is injured or the opponent retreats, and "Dove," which retreats as soon as its opponent escalates the conflict. Assume further that the winner of a conflict obtains a resource (such as food or a mate) that increases its fitness by an amount $V$, and that injury lowers an individual's fitness by an amount $C$. A game-theoretical model (Box B) shows that "Dove" is never an ESS (i.e., a "Hawk" genotype will be able to increase in frequency in a population of "Doves"), and that "Hawk" is an ESS if $V > C$, but not if $C > V$. If, however, there is a further strategy, the mixed strategy $I$ which specifies that an individual will play "Hawk" with probability $P$ and "Dove" with probability $1 - P$, then if $C > V$, $I$ is an evolutionarily stable strategy when $P = V/C$. This suggests that frequency or intensity of aggression ($P$) should be scaled to the value of the resource (in terms of fitness) relative to the cost of injury. This might explain the impression of naturalists that species with particularly dangerous weapons seem to have more ritualized aggressive behavior than those not so equipped.

ESS models have been used to analyze numerous problems: Should a parent care for its eggs or abandon them and have more offspring instead? If offspring benefit from a prolonged period of parental care, but parents benefit from turning their attention to another brood, what is the optimal resolution of the conflict? Is it advantageous for a male to display, and attract both females and attacks from other males, or to stay near displaying males and intercept females without displaying? A large field of behavioral ecology and sociobiology (E.O. Wilson 1975, Maynard Smith 1982, Krebs and Davies 1984) has developed around such questions. For many of the models, the empirical evidence is slight or indecisive, but in many cases there is qualitative agreement with predictions, and sometimes quantitative agreement as well. For example, pats of cow dung are the rendezvous for mating in the dung fly *Scatophaga stercoraria*, and the attractiveness of dung to

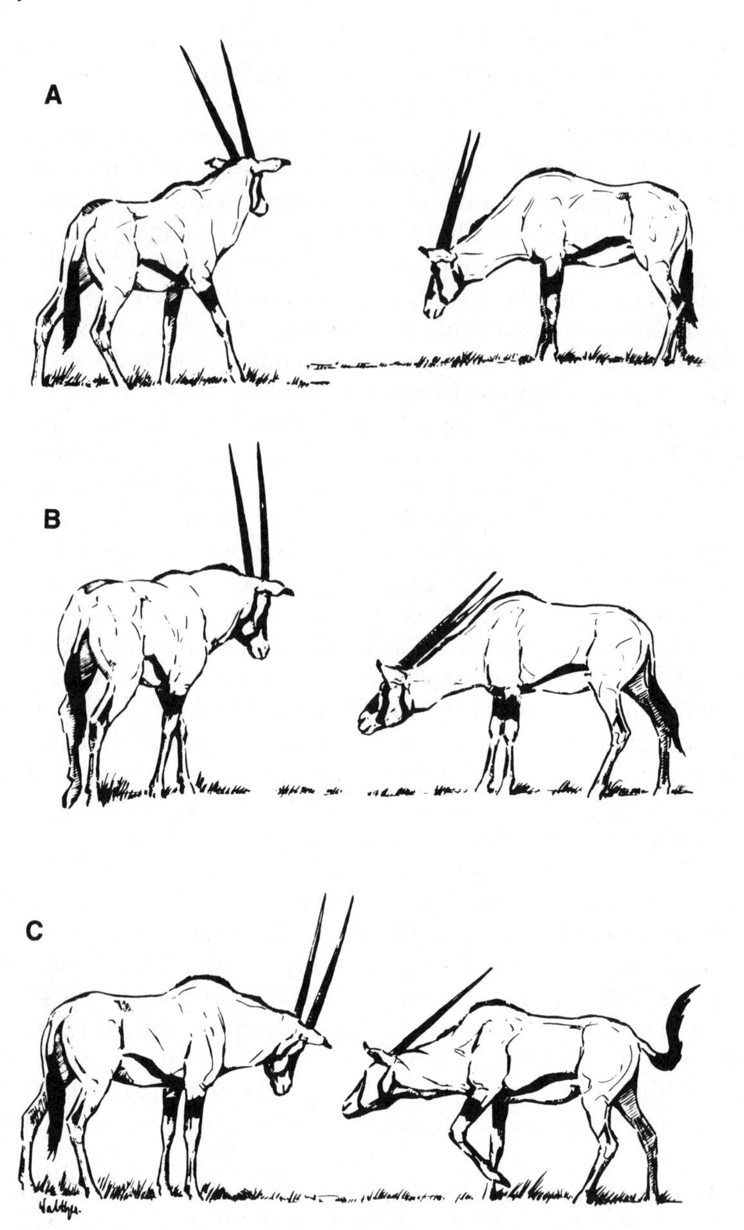


FIGURE 9

**Ritualized aggressive behavior in the East African oryx (*Oryx beisa*). (A) Subordinate male (right) responds to a dominant male by lowering its head. The horns are maintained in an upright position that enables him to counter an attack, typically by butting the aggressor. (B) As the dominant male intensifies the threat (by rotating his horns forward), the subordinate male assumes a more submissive attitude, which becomes so extreme (C) as to leave him defenseless against attack. The dominant male has intensified his display, but is unlikely to attack the submissive opponent. (From Walther 1984)**

## B *ESS Analysis of Animal Conflict*

Assume, following Maynard Smith (1982), that there are two strategies (phenotypes). Hawk (H) escalates until it is injured (with cost in fitness $C$) or until its opponent retreats (when Hawk gains the resource and has a gain in fitness $V$). Dove (D) retreats when threatened by H, and in that context neither gains nor loses in fitness. If two Doves meet, one gains the resource, so the average gain to an individual Dove in such encounters is $V/2$. A Hawk that encounters another Hawk wins with probability ½ and loses with probability ½, so its average "payoff" is $V/2 - C/2$. Thus we have a "payoff matrix:"

| | Opponent | |
|---|---|---|
| Payoff to | H | D |
| H | $\frac{1}{2}(V - C)$ | $V$ |
| D | 0 | $V/2$ |

The fitness of phenotype $i$ is $w(i)$; if $w(i) > w(j)$ when phenotype is rare, phenotype $i$ is an ESS. If $E(i,j)$ is the payoff to $i$ in a conflict with $j$, $E(i,i)$ the payoff to $i$ in a conflict with $i$, and so on, then for any two strategies $I$ and $J$

$$w(I) = w_0 + (1 - p)E(I,I) + pE(I,J)$$

$$w(J) = w_0 + (1 - p)E(J,I) + pE(J,J)$$

where $p$ and $1 - p$ are the frequencies of $J$ and $I$ in the population and $w_0$ is a "baseline" fitness. If $J$ is a rare mutant and $I$ is an ESS, $w(I) > w(J)$ by definition. From the expressions above, this will be true only if $E(I,I) > E(J,I)$ *or* if $E(I,I) = E(J,I)$ and $E(I,J) > E(J,J)$.

If Dove (D) were an ESS, $I = \text{D}$, $J = \text{H}$; but since $E(\text{D,D,}) < E(\text{H,D,})$ (that is, $V/2 < V$), D is not an ESS. Hawk (H) is an ESS if $V > C$ (because $E(\text{H,H}) = (V - C)/2$ is greater than $E(\text{D,H}) = 0$ if $V > C$). If $V < C$, neither H nor D is an ESS. However, suppose that $V < C$ and that there is a strategy $I$ that entails playing H with probability $P$ and D with probability $1 - P$. If genotypes vary in $P$, the fittest will be the one which plays each strategy often enough for its advantage to offset its disadvantage. That is, it must get an equal expected payoff from each randomly played strategy. From the payoff matrix, the expected payoffs from playing H and D with probabilities $P$ and $1 - P$ are respectively $P(V - C)/2 + (1 - P)V$ and $P(0) + (1 - P)(V/2)$. Equating these and solving for $P$, $P = V/C$. This is an evolutionarily stable strategy (Maynard Smith 1982).

females varies with its age. A male's chance of mating is increased by waiting at a pat of the ideal age, but decreases in proportion to the number of other males that congregate there. The solution is to adopt a mixed strategy, choosing inferior pats as the density of males on superior pats increases, so that the probability of successful mating is equalized over pats of different age. Parker (1974) provided evidence that the flies do just that.

## ADAPTATION: SPECIAL TOPICS

The remainder of this chapter treats several classes of traits that evolve under the influence of natural selection and have been the focus of a great deal of theoretical and empirical study in recent years. For reasons of space, I have had to treat other major topics, such as the evolution of social interactions, only by brief sketches of the theory in the preceding pages. The topics that follow are among those that have broad implications in evolution.

## THE EVOLUTION OF LIFE HISTORY CHARACTERISTICS

The "life history" of an organism refers to many features, including dispersal, seed dormancy (in plants), life span, the age at which reproduction begins, fecundity, frequency of reproduction, and parental care. We will consider here the evolution of the chief demographic characteristics that enter into the description of a species' population dynamics, namely the age-specific pattern of reproduction and mortality. We seek to explain why some organisms such as certain species of salmon and bamboo are SEMELPAROUS, reproducing only once, while others are ITEROPAROUS, reproducing repeatedly; why some, such as annual plants, reproduce early in life while others, such as trees, delay reproduction; why some have many eggs or seeds and others few; and why some are genetically programmed for senescence and death at an early age, and others at a later age (Williams 1966, Stearns 1976, 1977, 1980, Charlesworth 1980). In the past, some authors supposed that such features evolved for the good of the species; that fecundity evolved, for example, only to the extent that is necessary to balance mortality. Authors such as Lack (1954) and Williams (1966), however, recognized that natural selection must, by definition, always favor higher fecundity, as long as highly fecund genotypes do not suffer other disadvantages that lower their overall relative fitness. That is, they sought to explain life histories by the principles of individual selection.

### Age-specific reproduction and survival

Recall from Chapters 2 and 6 that if a population consists of several age classes, $l_x$ is the probability of surviving from birth to age $x$, $m_x$ is the average number of offspring produced by a female at age $x$, and the per capita rate of increase of the population is $r$, which is found from the equation

$$\sum_{x=1}^{L} l_x m_x e^{-rx} = 1$$

where the summation extends to $L$, the greatest age attained by any individual. The population growth rate $r$ is the mean of the various genotypes' values of $r$, weighted by the genotype frequencies. As we noted in Chapter 6, differences in $r$ are the selection coefficients that predict the rate of allele frequency change. Thus a genotype characterized by any value of $l_x$ or $m_x$ that increases $r$ should increase in frequency.

Charlesworth (1980) uses $P(x)$ to denote the probability that an individual will survive from the beginning to the end of the age interval $x$. He shows that an increase in $\log_e P(x)$ increases $r$ by an amount $s(x)/T$, where $T$ is the generation time of the population. The important element is $s(x)$, which is

$$s(x) = \sum_{y=x+1}^{L} l_y m_y e^{-ry}$$

where $y$ refers to each age class from $x + 1$ forward—i.e., all the ages that an individual of age $x$ might hope to attain. Thus $s(x)$, a measure of the selection coefficient associated with a gene that alters survival at age $x$, describes that fraction of the contribution to the population's growth that individuals of age $x$ can expect to make in the future, given their expectation of survival and reproduction ($l_y$ and $m_y$ for each age $y$ they might attain). Necessarily, then, $s(x)$ declines with age ($x$), since with advancing age individuals can look forward to contributing less to population growth. There are two reasons for this: individuals may not live to reproduce again, so $l_y$ decreases with age; and progeny born earlier in an individual's life contribute more to the rate of population growth (as long as $r \geq 0$) than those born later, because they reproduce sooner. Consequently, if a gene increases survival [i.e., increases $\log_e P(x)$] by acting at an early age class $x_i$, it will increase fitness ($r$) more than if it increases survival within a later age class $x_j$ by the same amount. The same is true for fecundity: an increment in fecundity at age $x$ increases $r$ by an amount $s'(x)$, which is similar to $s(x)$ in that it declines as a function of age.

Consequently, unless other factors enter into consideration, we should expect organisms to evolve to reproduce mostly early in life rather than later and to reproduce as soon after birth as possible. We can also explain the evolution of senescence, i.e. the degenerative changes that occur late in life. There is clearly no selection against alleles that cause degenerative effects after reproduction has ceased, for such genes have already been propagated to the next generation. But even alleles that have a degenerative effect late in the reproductive part of the life span will not be strongly selected against, because $s(x)$, the force of selection against them, is low. Thus if genes have advantageous effects early in life, but deleterious pleiotropic effects later in life, they will have a net selective advantage—but the consequence is degeneration, cessation of reproduction, and ultimately death (Williams 1957).

For example, Huntington's chorea, a severe disorder of the human nervous system caused by a dominant allele, generally causes death. But because it is not expressed until the age of 30 or 40, the net reproductive rate of choreics, 1.0308, is almost as high as that of the population as a whole, 1.0485 (Cavalli-Sforza and Bodmer 1971), and the allele is not eliminated as fast as it would be if it were expressed at an earlier age. Rose and Charlesworth (1981) and Luckinbill et al. (1984) found that when *Drosophila* populations were propagated only from eggs laid late in adult life, the populations evolved greater longevity and higher fecundity late in life, but that fecundity early in life decreased. Thus there appears to be a tradeoff—a negative genetic correlation—between fitness traits early and late in life, as the pleiotropy theory of senescence predicts. A consequence of this theory that has important medical implications is that since any gene with such effects is subject to the same principles of age-specific selection, senescence is likely to have numerous causes, rather than a single biochemical cause.

## Iteroparity and semelparity

To this point, the theory indicates that an "optimal" organism should concentrate all its reproduction into one big clutch at the earliest possible age: it should be semelparous. Iteroparity—repeated reproduction—can evolve, however, for several reasons. Contrast a semelparous genotype, with fecundity $m_s$, with an iteroparous genotype that has a constant probability of survival $P$ and a constant fecundity $m_i$ at each age. If the semelparous genotype's value of $r$ is to exceed that of the iteroparous genotype, its fecundity must be greater than the reproductive output achieved by the iteroparous genotype's repeated reproduction: $m_s > m_i/(1 - Pe^{-r})$ (Charnov and Schaffer 1973). Therefore an iteroparous life history may be favored when the probability of adult survival is high and the population's growth rate is low. In contrast, semelparity is favored when the likelihood of adult survival is low, as it is for annual plants, which typically are pioneering species that are soon displaced by other species in ecological succession.

In some organisms, such as trees and many fishes, body size and fecundity increase throughout life. If there is a tradeoff between "reproductive effort" at a given age and subsequent growth and fecundity, $r$ may be maximized by delaying at least some reproductive effort to later ages. If even a slight reproductive effort drastically reduces the prospect of subsequent survival and reproduction, as in salmon (*Oncorhynchus*) that make an exhausting journey upstream to spawn, then the optimal life history may be one of semelparous reproduction, deferred to an age at which large size and high fecundity have been reached (Gadgil and Bossert 1970). A tradeoff between reproductive effort and subsequent survival and growth has been documented by comparisons among closely related species (Figure 10), as well as within certain populations (e.g., Harper and White 1974, Snell and King 1977).

## Clutch size

A given reproductive effort may be distributed among many offspring, each of which begins life as a small egg or seed or is given little parental care, or among few offspring, each of which is provided with more yolk, endosperm, or parental

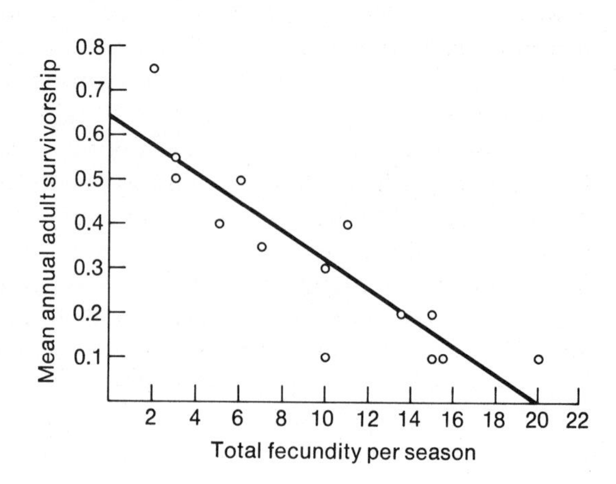


FIGURE 10

**An illustration of the cost of reproduction: fecundity versus survivorship in 13 species of lizards (one represented by two populations). (From Tinkle 1969)**

care. The optimal clutch size for a given reproductive effort maximizes the number of offspring that survive to reproductive age. When survival is not greatly affected by initial size, as is likely for many marine invertebrates and fishes with planktonic eggs that almost inevitably suffer high mortality, it may be optimal to lay numerous small eggs. But in many organisms, greater seed size or egg size confers greater juvenile survival; for example, the energy reserves in a seed affect the ability of a seedling to grow high enough to get enough light for photosynthesis, or to make a big enough root system to obtain sufficient water. In such cases, selection may favor genotypes that have fewer, larger offspring. Similarly, Lack (1954) found that starlings (*Sturnus vulgaris*) have the greatest number of surviving offspring if they lay no more than five or six eggs. The parents are unable to feed larger broods adequately, so the number of survivors from larger broods is less than the number from more modest clutches.

### Consequences of density-dependent population growth

Numerous factors besides those we have considered bear on the evolution of life histories, and the theories of life history evolution are correspondingly diverse (Stearns 1976). One important factor is the nature of population limitation. Usually $r$ declines from its maximum, $r_m$ (the intrinsic rate of natural increase), as population density increases (Chapter 2). This decline may, however, differ among genotypes (Figure 11) so that the genotype with the highest $r$ in a dense population may not have the highest $r$ when the population is sparse and is growing exponentially. For example, if competition is so intense that juveniles experience high mortality but adults, once established, have a long life expectancy, selection may favor iteroparity, delay of reproduction until the adult has

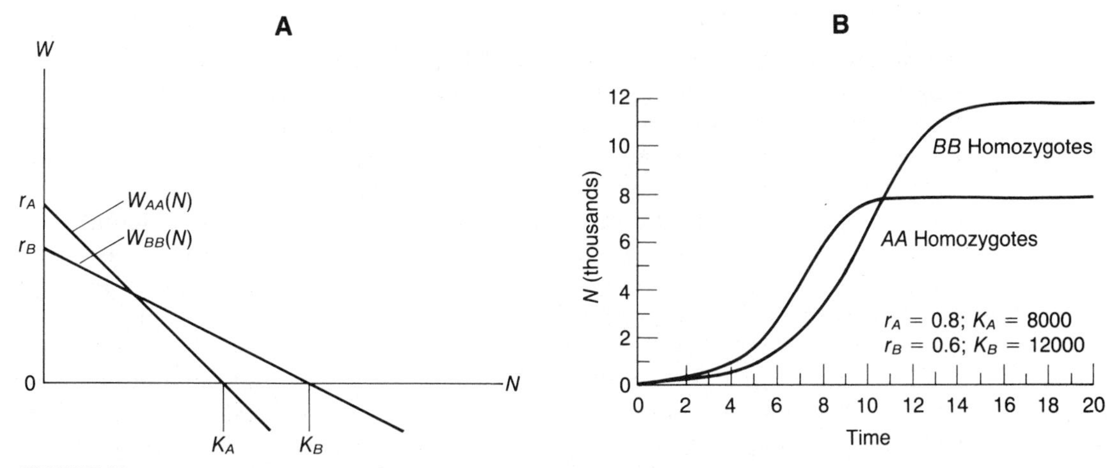


FIGURE 11

**A model of density-dependent natural selection. (A) for both genotypes *A* and *B*, the per capita rate of population growth, *r*, declines to 0 (at equilibrium density, *K*) as population density, *N*, increases. Genotype *A* has the higher fitness at low densities, whereas genotype *B* has higher fitness at high densities. (B) As a population consisting only of *AA* and *BB* homozygotes grows (e.g., a clonal population), genotype *BB*, with a higher equilibrium density, increases in frequency. (From Roughgarden 1971)**

reached large size, and small broods of large eggs or seeds that provide the hatchling or seedling with a competitive advantage. Such an organism will have a low intrinsic rate of natural increase ($r_m$).

MacArthur and Wilson (1967) suggested that populations that commonly experience exponential growth and are periodically reduced by catastrophic density-independent factors will tend to evolve life history traits that maximize the intrinsic rate of natural increase, and referred to these as *r*-SELECTED populations. Many weedy plants that colonize open patches of soil might fit this description. In contrast, a life history that optimizes competitive ability but is associated with a low intrinsic rate of natural increase may evolve in populations that typically are regulated near carrying capacity (*K*); they referred to these as K-SELECTED populations. For example, albatrosses (*Diomedea*) nest on oceanic islands in large colonies and probably compete intensely for food; some species lay only a single egg every two years and spend almost this long rearing the offspring to the point at which it can fend for itself. The life history traits that evolve when populations are regulated by density-dependent factors depend strongly, however, on the pattern of age-specific mortality; under some conditions, early prolific reproduction may be advantageous. Thus the simple categorization of life histories into *r*-selected and *K*-selected patterns does not adequately describe the diverse courses of life history evolution.

In simple models of density-regulated populations (e.g. Anderson 1971, Roughgarden 1971), evolution can increase population density, because genotypes that can best resist the limiting factor, say by finding food more efficiently, are most fit. When selection is frequency-dependent, however, as when individuals compete by territoriality or other mechanisms of interference, allele frequency change may decrease population density (Slatkin 1978).

## SEXUAL SELECTION

The difference between male and female is ultimately the difference in the size of their gametes. According to an ESS analysis (Parker et al. 1972, Maynard Smith 1978b), this difference may have evolved initially from the tradeoff in the reproductive advantage inherent in producing many small gametes and the advantage in survival of zygotes that develop from large eggs. Whatever its origin, a female typically invests more materials (and often more time and parental care) in each offspring than a male does, so the fitness of a female is likely to be reduced by unproductive mating more than that of a male (Trivers 1972). Perhaps for this reason, females are generally more selective in mating than males, which in some species may attempt to mate with extraordinarily inappropriate partners. In many species, then, the female chooses a mate from among contending males, although this pattern is reversed in some species such as seahorses (Syngnathidae) and phalaropes (Phalaropodidae), in which the male cares for the young.

Sexual selection, the term Darwin used to describe "the advantage which certain individuals have over others of the same sex and species solely in respect of reproduction," is therefore typically more intense among males, which commonly vary more in reproductive success than females (Wade and Arnold 1980). Sexual selection is especially intense in polygamous species, in which it commonly results in the evolution of pronounced sexual dimorphism in the color pattern, plumes, horns, or other features used in display (West-Eberhard 1983; Figure 7 in Chapter 8). Sexual selection may be INTRASEXUAL or INTERSEXUAL.

Intrasexual selection arises from contests among males for access to females or to favorable mating sites, as in elephant seals (*Mirounga angustirostris;* Cox and Le Boeuf 1977) and mountain sheep (*Ovis canadensis;* Geist 1971), in which large size or large horns confer an advantage in battle. The preference of females for certain males over others constitutes intersexual selection; for example, females of an African widowbird (*Euplectes progne*), a kind of finch in which the males have extremely long tail feathers, mate selectively with longer-tailed males (Andersson 1982a). Sexual selection in the weevil *Brentus anchorago* (Figure 12) appears to be both intrasexual and intersexual: males joust with their extremely elongated snouts, and long-snouted males are more successful both because they win access to females and because females mate preferentially with them (Johnson 1982). Although most discussions of sexual selection focus on animals (Blum and Blum 1979, O'Donald 1980, Bateson 1983, Thornhill and Alcock 1983), sexual selection surely occurs in plants as well (Willson and Burley 1983), in which intrasexual selection for reproductive success as a male may result in the production of far more flowers, acting as pollen parents, than can ever bear fruit.

The theory of intrasexual selection is straightforward: a character is favored insofar as it leads to success in competition among males. Sexual selection by female choice is rather more complex. In some cases, female preference for certain males is adaptive because the male provides something, such as parental care, that enhances the female's reproductive success. Males of some species of scorpionflies (e.g., *Hylobittacus apicalis*) capture insects and offer them to females, whose fecundity is thereby increased; females prefer males that offer large meals (Thornhill 1980). But in species in which the male contributes only gametes, and in which virtually all females mate, the female's fitness is not influenced by her choice of mate. How, then, does female preference for an exaggerated male trait evolve, and cause the male's trait to evolve in concert?

The answer to this question was suggested by Fisher (1930) and has been worked out by Lande (1981b) in a quantitative genetic model and by Kirkpatrick (1982) in a two-locus model. Their conclusions are much the same, so we will consider here the outlines of Kirkpatrick's model. For simplicity, consider a haploid organism in which males of types $T_1$ and $T_2$ have frequencies $t_1$ and $t_2$

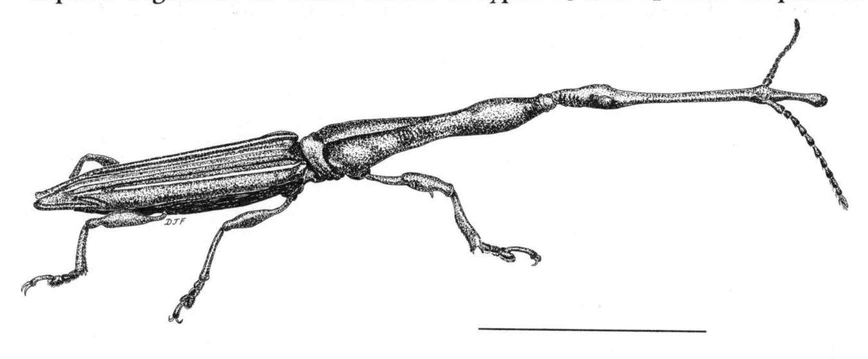


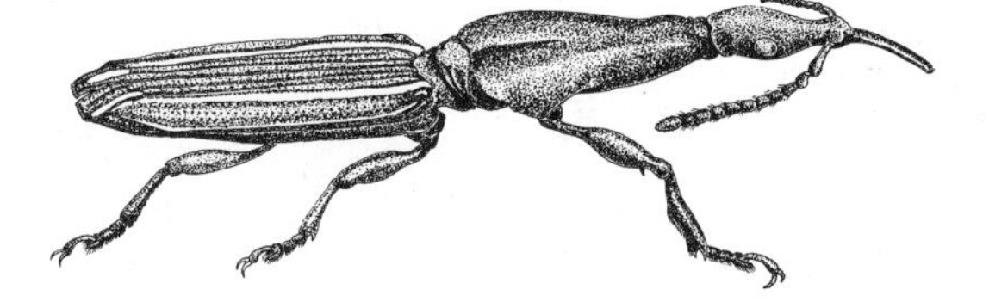

FIGURE 12

**Male (above) and female (left) of a sexually dimorphic tropical American beetle (*Brentus anchorago*) in which the elongated head of the male is subject to sexual selection. Bar = 1 cm.**

respectively; $T_2$ has some exaggerated trait such as a long tail that increases the risk of predation, so that its viability is $1 - s$ relative to $T_1$. Females of type $P_1$ (with frequency $p_1$) do not discriminate in mating, but those of type $P_2$, with frequency $p_2$, prefer $T_2$ males relative to $T_1$ males. If alleles $P_1$ and $P_2$ do not affect females' survival or fecundity, they are selectively neutral.

Because $T_2$ males are both acceptable to $P_1$ females and actively preferred by $P_2$ females, $T_2$ increases in frequency. Because the offspring of a $P_2 \times T_2$ mating inherit both the $P_2$ and $T_2$ alleles, a genetic correlation between female preference and male trait arises through linkage disequilibrium: $P_2$ becomes associated with $T_2$, and $P_1$ with $T_1$. As $T_2$ increases in frequency, so does $P_2$, according to the relation $\Delta p_2 = \Delta t_2[D/(t_2)(1 - t_2)]$, where $D$ is the coefficient of linkage disequilibrium. Thus the frequency of $T_2$-preferring females increases by "hitchhiking," and as it does, the reproductive advantage of $T_2$ males becomes even greater, and $T_2$ can increase even if $T_2$ males suffer a severe disadvantage in viability. In fact, $T_2$ may increase even if its increase lowers population size. Ultimately $p_2$ and $t_2$ arrive at a line of equilibria (Figure 13): various combinations of the gene frequencies $p_2$ and $t_2$ that are not changed by selection. However, the allele frequencies may change along this line of equilibria by genetic drift. The model is much the same if $P_1$ females actively prefer $T_1$ males and $P_2$ females prefer $T_2$ males. Under this model, then, genetic drift leads to divergence among populations in both female preference and male trait, and can result in speciation (Chapter 8).

Using a model in which both male trait and female preference are polygenically inherited, Lande (1981b) has described a similar line of equilibria for various combinations of the two characters. If, however, the genetic variance in female preference is great enough, assortative mating may generate such a strong genetic correlation between female preference and male trait that the equilibria become unstable. In this case, any exaggeration of the male's trait is favored by the females' preference, and female preference for ever more exaggerated males evolves virtually without limit. Fisher (1930) described this continual escalation as "runaway" sexual selection.

Runaway sexual selection is a fascinating example of how selection may proceed without adaptation. It may reduce population size, and so can be detrimental to the population as a whole (Lande 1980b, Kirkpatrick 1982). It can be shown that females prefer exaggerated males not because they provide genes that

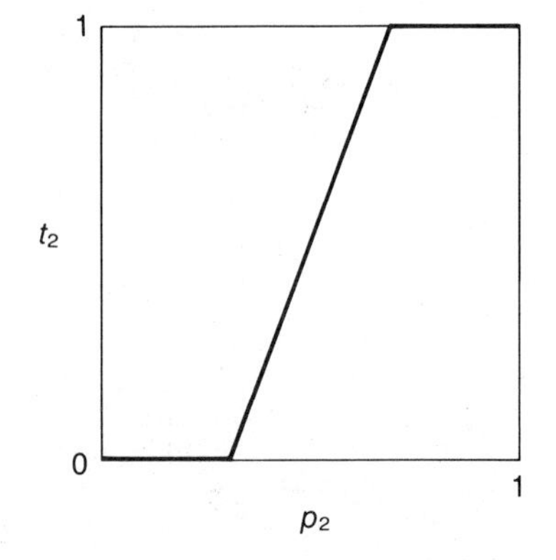


FIGURE 13

**A line of stable equilibria for the joint frequencies $p_2$, the frequency of an allele affecting female choice of mate, and $t_2$, an allele for a male character preferred by females of type $P_2$. Any point on the line is an equilibrium determined by sexual selection; movement along the line can occur by genetic drift. (After Kirkpatrick 1982)**

will enhance the survival or fecundity of the offspring, but merely because the daughters of exaggerated males tend to prefer males with those same characteristics. In these models, then, the evolution of female preference is not an adaptive process.

## THE EVOLUTION OF RECOMBINATION AND SEX

Among the most intriguing and difficult problems in evolutionary theory are those of the origin and maintenance of recombination and sexual reproduction, and the question of the conditions that favor the evolution of phenomena such as self-fertilization and parthenogenesis (Williams 1975, Maynard Smith 1978b, Lloyd 1980). In many species, genetic variation for recombination rate and mode of reproduction exists, so these features are evolutionarily labile. For example, Chinnici (1971) and others have shown that the crossover frequency between pairs of loci can be altered by artificial selection in *Drosophila melanogaster* (Figure 14) without affecting the frequency of crossing over elsewhere on the chromosome. The frequency of self-fertilization varies genetically within and among populations of many plants (Jain 1976). Parthenogenetic and sexually reproducing genotypes are found within many species of plants and animals.

Explanations for the maintenance of sex and recombination can be cast in group-selectionist or nongroup-selectionist terms. A common group-selectionist explanation is that recombination enhances the ability of populations to adapt to changing environments, and that asexual populations have a higher extinction rate. Suppose that in a population in which *A* and *B* have high frequency, the rare mutants *a* and *b* interact to give high fitness in an altered environment. If

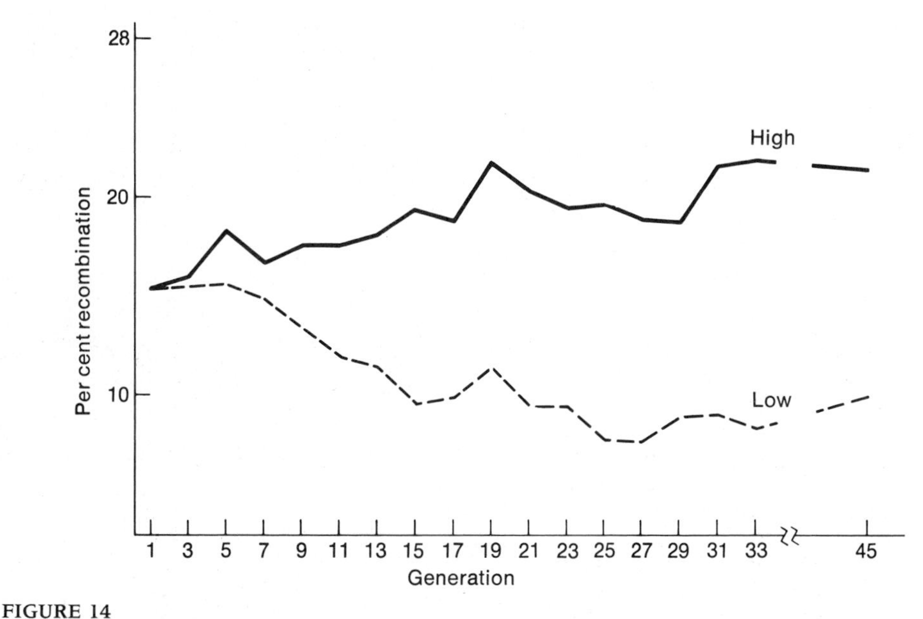


FIGURE 14

**Changes in the frequency of recombination between the loci *sc* and *cv* in laboratory stocks of *Drosophila melanogaster* selected for high and low recombination between these loci. (After Chinnici 1971)**

the population is finite, *a* and *b* will exist primarily in the repulsion combinations *aB* and *Ab*; *ab* will be rare, and the population is in linkage disequilibrium with $D < 0$. Recombination, by permitting the immediate formation of the *ab* combination, enables $D$ to exceed zero, hastening the rate of adaptation (Felsenstein 1974; see Chapter 7). In an asexual population, in contrast, the *ab* combination will be formed only when one mutation occurs in a direct descendant of an individual that bears the other (Figure 15). However, Crow and Kimura (1965) argue that recombination may not enhance the rate of evolution in small populations, because mutations will then occur so rarely that one mutation will be nearly fixed before another occurs.

Some artificial selection experiments have demonstrated that recombination enhances the response to selection; for example, Carson (1959) found a more rapid response to selection for motility in chromosomally monomorphic *Drosophila* populations than in populations polymorphic for inversions that reduce the frequency of crossing over. The taxonomic distribution of obligate parthenogenesis is consonant with the idea that parthenogenetic forms have high extinction rates: most such forms, such as the dandelion *Taraxacum officinale*, are closely related to sexual species, often retain accoutrements of sexual reproduction such as petals and pollen, and clearly have arisen only recently from sexual ancestors. In only a few cases, such as the rotifers of the order Bdelloidea, has a purely parthenogenetic group achieved such great morphological and taxonomic diversity as to imply that it has a long history.

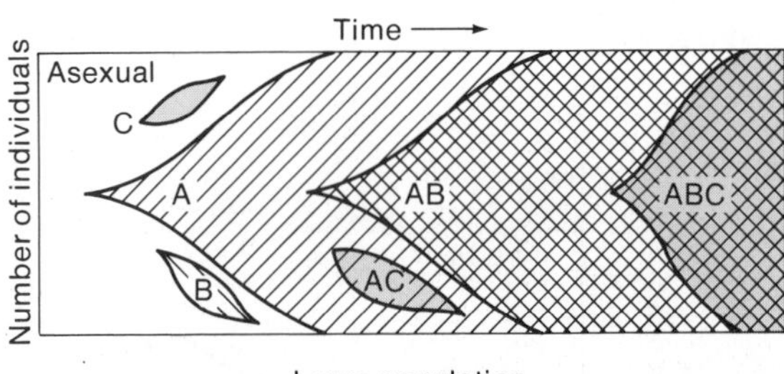


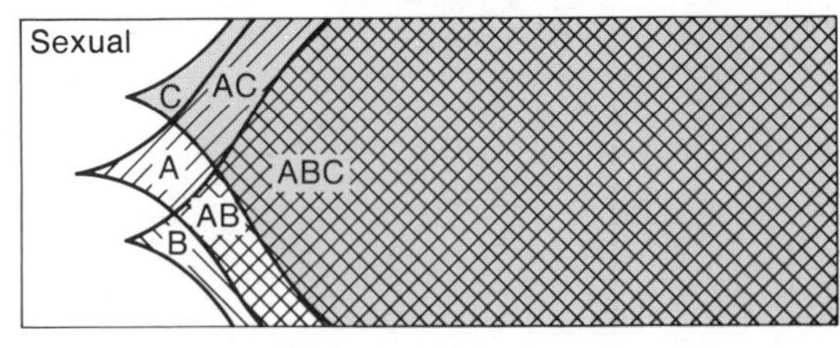


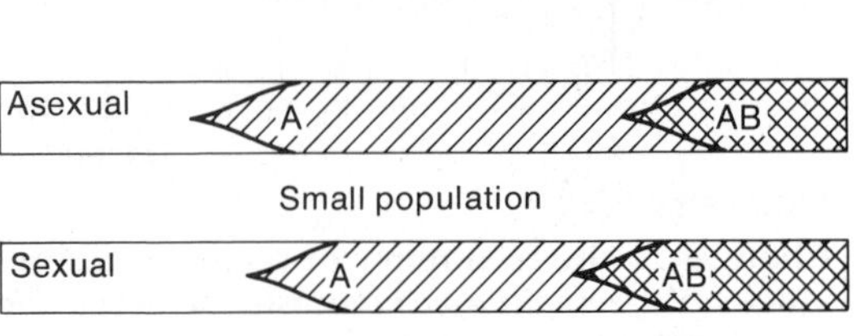


FIGURE 15

**One view of the effect of recombination on the rate of evolution. If evolutionary rates and the precision of adaptation depend on new mutations A, B, and C that are advantageous in concert, then adaptedness is achieved more rapidly in a sexual than in an asexual population because the mutations are assembled by recombination. This effect is lost in small populations, however, because mutations occur so rarely. This view of evolution assumes that favorable mutations limit the rate of evolution. (From Crow and Kimura 1965)**

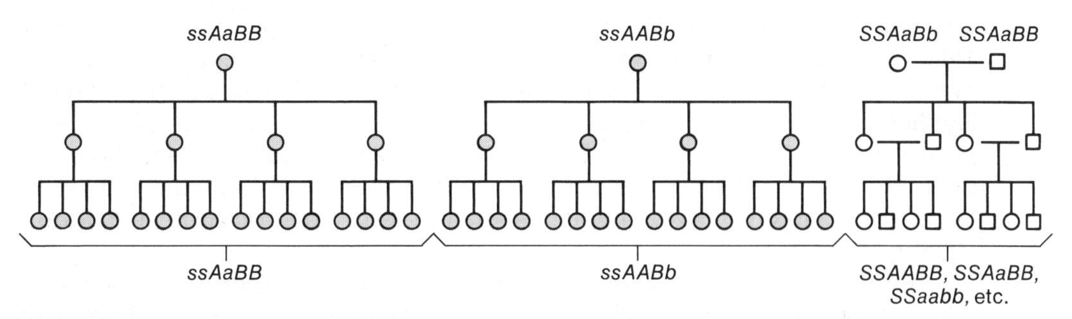


FIGURE 16

**The advantages and disadvantages of an allele *S*, which codes for sexual reproduction, compared to those of an allele *s* coding for asexual reproduction. Each female produces four equally fit offspring, but the frequency of the sexual allele drops rapidly from ½ in the first generation to ⅕ by the third generation. If, however, the environment then changes, so that only *aabb* genotypes survive, allele *S* will persist but *s* will be eliminated. Circles represent females, squares males.**

### The advantages and disadvantages of sex and recombination

Group selection, however, is a shaky explanation for recombination. The genetic variation that exists for recombination rates could readily result in the evolution of reduced recombination if it were favored by individual selection, and this is likely often to be the case. The question must be phrased as follows: If one allele increases the rate of crossing over or promotes sexual reproduction, and another reduces crossing over or causes parthenogenesis, which allele will increase in frequency?

Considering first just recombination in a species that is committed to sexual reproduction, assume that alleles affecting recombination rate do not affect survival or fecundity. They change in frequency, then, only by association (hitchhiking) with favorable or unfavorable allele combinations at other loci. In a stable environment, recombination is almost always selected against, because it breaks down favorable gene combinations, so that an allele fostering recombination is associated with less fit combinations. Accounting for the maintenance of sexual reproduction is even more difficult. In a species that has equal numbers of female and male offspring, the growth in numbers of a parthenogenetic genotype that produces only daughters is about twice that of a sexual genotype, because male offspring do not make more offspring (Figure 16). Thus an allele for parthenogenetic reproduction has an intrinsic twofold advantage.

Theoreticians have therefore looked for ongoing advantages of recombination that can maintain sexual reproduction in a population. One factor that may play a role is short-term environmental fluctuations that favor recombinant genotypes. This model explains recombination only if the fluctuations in various environmental factors follow a very particular pattern, or if the environment continually presents new challenges (Maynard Smith 1978b). Another model (Williams 1975) proposes that in a mosaic of different microenvironments, competition between the genetically uniform progeny of an asexual genotype and the diverse progeny of a sexual genotype will favor one or another of the latter. Frequency-dependent selection in a uniform environment may have the same effect (Price and Waser 1982). For example, Antonovics and Ellstrand (1984) found that the fitness of an average plant of a minority genotype of a grass, planted amidst large numbers of another genotype, was twice as great as when it was placed in competition primarily with other members of the same genotype. If, however, adaptation to each microenvironment is conferred by specific multi-locus combinations of alleles, it seems likely that recombination will be disadvantageous (Maynard Smith 1978b).

Although factors that favor recombination can be envisioned, it is uncertain whether the advantage they may impart to sexual reproduction is sufficient to counteract its twofold disadvantage. Intense selection for recombinant genotypes is necessary, and this may be possible only in highly fecund organisms (Williams 1975), in which the reproductive excess is so great as to permit a high intensity of selection (Chapter 6). Parthenogenesis and other mechanisms that reduce recombination are relatively rare in organisms that inhabit biologically complex environments, such as tropical forests (Levin 1975, Glesener and Tilman 1978), in which selective pressures may be so diverse and variable as to favor recombination. But so little information exists on the magnitude of the advantage of sex that the maintenance of sexual reproduction has not yet been fully explained.

**Inbreeding and outbreeding**

Pure self-fertilization, the most extreme form of inbreeding, ultimately results in a population of homozygous genotypes and no recombination (Chapter 5). Although populations of habitually inbreeding plants typically are genetically diverse, they may be expected, like asexual forms, to suffer high rates of extinction because of their reduced ability to generate new genotypes. But this long-term effect at the population level does not explain why either inbreeding or outbreeding systems evolve and are maintained.

Imagine that in a population of outbreeding but self-compatible plants an allele arises that in homozygous condition causes all the plant's ovules to be self-fertilized, but that the plant's pollen is nonetheless dispersed to other plants. Such an allele automatically has a strong selective advantage (Fisher 1941), because it contributes two gene copies (in the ovule and pollen) to the following generation by selfing, as well as single copies through fertilization of other plants. Selfing clearly can have other, ecological, advantages as well: fertilization is assured if the population is sparse or pollinators are scarce. Self-fertilization is, indeed, common among colonizing species of plants, and in environments such as the Arctic that are unfavorable for pollinating insects (Jain 1976).

The major disadvantage of self-fertilization is the inbreeding depression arising from homozygosity for deleterious recessive alleles, and it is presumably for this reason that self-incompatibility and other outcrossing mechanisms have evolved in numerous plants. If a population has already become highly inbred, however, this disadvantage is reduced, because much of the mutational load will have been purged by natural selection. There is, indeed, some evidence that inbreeding depression and its converse, heterosis, are less pronounced in habitually inbreeding species of plants than in outcrossing species (Jinks and Mather 1955). Thus alleles for self-compatibility or self-fertilization may be disadvantageous in large outbred populations, but advantageous in small or otherwise inbred populations.

## SUMMARY

Adaptation can be perceived at various levels from the gene to the population, corresponding to selection at each of these levels. There is considerable controversy about the prevalence of group-level adaptations because these require group selection, which often is weaker than selection at the level of the gene or the individual organism. Adaptation is an onerous concept, and the adaptive value of a trait should be demonstrated rather than assumed, for numerous factors other than adaptation can influence the evolution of a trait. Several methods of analyzing adaptation, some explicitly genetic and others not, have contributed to our understanding of features such as social behavior, life histories, and genetic systems.

## FOR DISCUSSION AND THOUGHT

1. Discuss the importance of distinguishing between the adaptive nature of a feature and the adaptive nature of variations in a feature.
2. Is the genetic variability in a population an adaptation?
3. Adaptations are features that enhance fitness. It has sometimes been claimed that fitness is a tautological and hence meaningless concept. According to this argument, adaptation arises from the "survival of the fittest" and the fittest are recognized as those which survive; consequently there is no independent measure of fitness or adaptedness. Evaluate this claim (see Sober 1984).
4. Adaptive explanations have been offered for human behaviors such as incest taboos, xenophobia (dislike of strangers), homosexual behavior, adolescent rebellion against authority, and capitalistic economic systems. How might such hypotheses be tested? What alternative explanations might be offered?
5. Dawkins (1976) has strongly advocated the view that an organism is merely a gene's way of making more genes, so that selection at the level of the gene provides the most insightful way of understanding evolution. Sober and Lewontin (1982) claim that this is an excessively reductionistic view that obscures the importance of interactions among genes and of selection at the level of phenotypes. Discuss the merits of each argument.
6. Kin selection can explain the evolution of cooperative behavior among closely related individuals. How can we explain "altruistic" behavior that is directed to unrelated individuals? For example, in many species of ants a colony has numerous, presumably unrelated queens, yet workers are sterile and rear presumably unrelated offspring.
7. There is evidence that in some social animals, individuals can recognize and direct their altruistic behavior preferentially toward kin (Lacy and Sherman 1983). How might such recognition be accomplished, and what are its evolutionary consequences?
8. Consider the evolution of reduced virulence in a parasite that reduces its hosts' viability by reproducing within the host. Under what circumstances would such evolution be attributed to individual selection, kin selection, or group selection?
9. Some authors argue that female choice of the most vigorously displaying or most ornately adorned males may be adaptive, because such males are likely to have "good genes" that contribute to their vigor, so females that choose such males would have vigorous progeny. One problem with this hypothesis is that according to Fisher's fundamental theorem (Chapter 6), the heritability of fitness in a population at allele frequency equilibrium is zero. Evaluate these arguments; see Zahavi (1975), Maynard

Smith (1976b), Andersson (1982b), Kirkpatrick (1982), Arnold (1983), Kodric-Brown and Brown (1984).

10. Under what circumstances would you expect the evolution of high versus low tendency for dispersal out of a population? What are the roles of individual and group selection in the evolution of dispersal? See Van Valen (1971) and Balkau and Feldman (1973).

## MAJOR REFERENCES

Williams, G.C. 1966. *Adaptation and natural selection.* Princeton University Press, Princeton, NJ. 307 pages. A clear and insightful essay on the nature of individual and group selection.

Wilson, E.O. 1975. *Sociobiology: The new synthesis.* Harvard University Press, Cambridge, MA. 697 pages. An important, in part controversial, synthesis of ideas and information on the evolution of social behavior.

Krebs, J.R., and N.B. Davies (eds). 1984. *Behavioral ecology: An evolutionary approach.* Second edition. Sinauer Associates, Sunderland, MA. 493 pages. A comprehensive set of essays on the evolution of ecological and behavioral traits, mostly in animals.

Gould, S.J., and R.C. Lewontin. 1979. The spandrels of San Marco and the Panglossian paradigm: A critique of the adaptationist programme. *Proc. Roy. Soc. Lond. B* 205: 581-598. A cogent account of the pitfalls in assuming adaptation without sufficient reason.

# Determining the History of Evolution

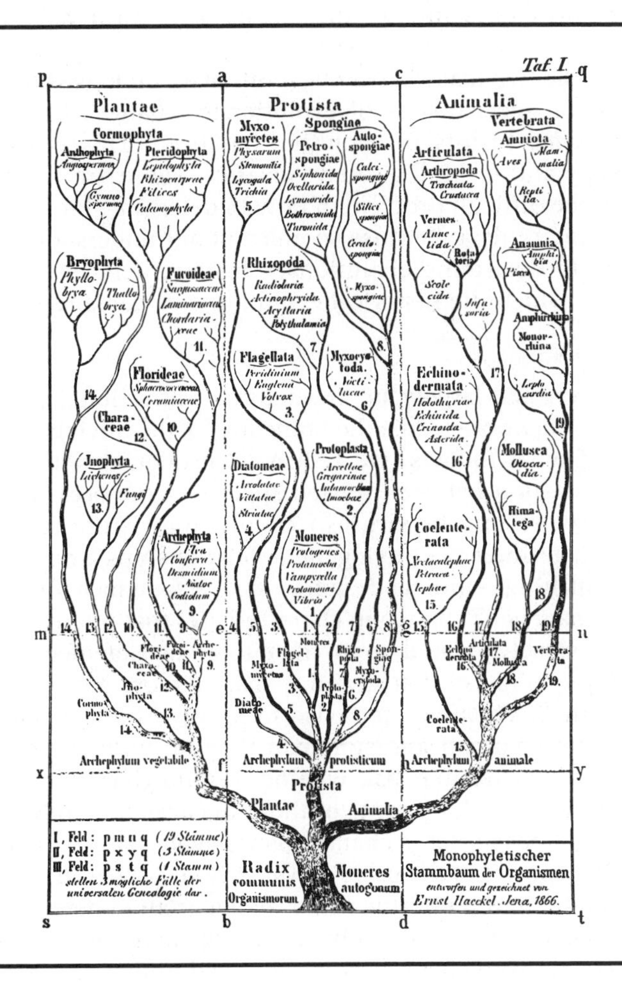


# Chapter Ten

The study of biology may be divided into two modes. "Functional biology" asks how an organism works—how does it develop and maintain itself? The other approach to biology, the historical approach, asks "How and why has life come to be this way?" The phenomena we study in biology are the products of historical development, in which past events may have determined the present state of affairs. Just as a political scientist cannot explain the present patterns of national boundaries or of tensions among peoples without reference to the history of, say, colonialism, a biologist may require a knowledge of the past to understand why the kidney of marine fishes seems designed more for life in fresh than in salt water, or why tapirs inhabit the forests of tropical Asia and America but not Africa.

One of the two major tasks of evolutionary biology is to determine the genetic and ecological mechanisms of evolutionary change. The other is to determine what the actual history of evolution has been. This is largely the province of paleontology and of biological systematics. These are the sciences that come to grips with the difficult problem of how to determine biological history from often very fragmentary data. Because so much of our understanding of biology rests on our ability to determine evolutionary history, it is essential to understand the methods by which the historical evolution of organisms can be inferred.

## DEFINITIONS

Evolution consists largely of ANAGENESIS—directional change within a single lineage—and CLADOGENESIS, the branching of the phylogenetic tree by speciation. (In most of what follows, we will not consider RETICULATE EVOLUTION arising from hybridization between related species or the occasional transfer of genetic material between taxa by viruses.) To infer the history of evolution of a group of organisms is to infer the phylogenetic relationships among the species—the pattern of branching—and to describe, for each branch of the tree, the rate and pattern of anagenetic change in characteristics that interest us. To say that species A is more closely related to B than to C (a statement of their phylogenetic relationships) is to say that A and B share a more recent common ancestor than they do with C. Thus we may wish to know if humans and chimpanzees share a more recent common ancestor than they do with the orangutan; this having been established, we may ask whether certain characteristics such as brain size have evolved faster along some of these lineages than others.

In every lineage, some characteristics evolve substantially and others less so. Every living species is a mosaic of ANCESTRAL (sometimes called primitive) characteristics inherited with little or no change from remote ancestors, and relatively DERIVED characteristics, which have undergone recent change. The large brain of humans is a derived characteristic compared to that of other primates, while our pentadactyl (five-fingered) limb is a primitive or ancestral state, in the sense that the number of digits has remained unchanged since our amphibian ancestors of the Devonian. A character (e.g., number of digits) can have one or more CHARACTER STATES (e.g., 5 digits as in humans, 4 as in sheep, 1 as in horses). An ancestral character state is sometimes termed PLESIOMORPHIC (Hennig 1979); a derived state may be termed APOMORPHIC or "advanced," a term that should not be taken to imply a value judgment about evolutionary "progress" or adaptiveness. In Figure 1, unprimed and primed letters (a, a′, etc.) denote hypothetical

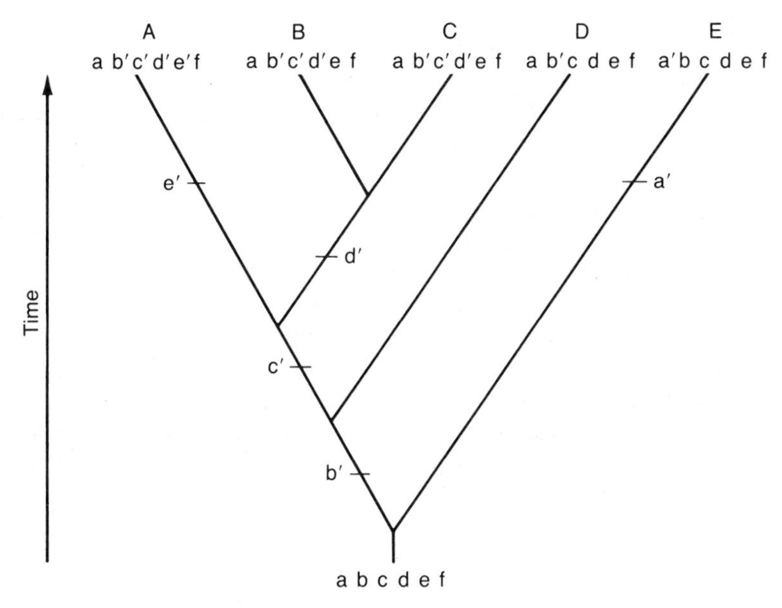


FIGURE 1

**A hypothetical phylogenetic tree, or cladogram, of five taxa A–E derived from a common ancestor, with the evolution of six characters a–f indicated by transitions from the ancestral (plesiomorphic) to derived (apomorphic) states (indicated by primes). In this example, each character changes once at most, and taxa with more recent common ancestors share more derived character states than taxa with more remote common ancestors. Each taxon retains some ancestral character states.**

states of each of five characters. An ancestral character shared by several species is termed SYMPLESIOMORPHIC; a derived character shared by two or more species is SYNAPOMORPHIC. Thus character state f in the diagram is symplesiomorphic for the entire group; a is symplesiomorphic for species A, B, C, and D—which share the derived character state b′ as a synapomorphy.

Note that a characteristic can undergo change long after a lineage has branched off from its "sister group." For example, although the divergence of gibbons from other apes may be ancient, some of the distinctive features of gibbons may well have evolved recently (e.g., character a′ in Figure 1). Therefore no *species* is primitive, although many of its characteristics may be primitive. Frogs are no more "primitive" than humans, for both are descended from Devonian amphibians (and from Precambrian protists!); the split between the lineage that ultimately evolved into frogs and the one that ultimately evolved into humans occurred when reptiles emerged in the Carboniferous, and both lineages have been evolving new characteristics ever since, although at different rates.

## CLASSIFICATION

The study of historical evolution requires some familiarity with taxonomy (the practice of classification) and systematics (the study of relationships among organisms) (Simpson 1961, Wiley 1981).

A major task of systematics is to classify individual organisms into species (see Chapter 8). Once recognized, species are combined into the more inclusive groups of the Linnaean hierarchy (genera, families, orders, classes, phyla or divisions, with intercalated categories such as subfamily and superfamily). The levels of classification, such as family, are called CATEGORIES; each set of real organisms that fill a category is called a TAXON (plural: taxa). For example, the Felidae (cats) and Apiaceae (carrots and related plants) are taxa in the category "family." "Higher taxa" are those above the species level. Nomenclature—the application of names to species and higher taxa—is governed by purely legalistic sets of rules. For example, certain leaf-feeding beetles (Chrysomelidae) that until recently were referred to the genus *Galerucella* are now placed in the genus *Pyrrhalta,* because the name *Pyrrhalta,* first used in 1866, has historical priority over *Galerucella,* which was not introduced until 1873.

A final task of systematics is to determine the evolutionary relationships among taxa, and to describe patterns of evolutionary change. But this raises the question of what criteria are used to define taxa; why are species placed in one group rather than another? This is the subject of sometimes rancorous controversy among several schools of thought.

## CONTENDING SCHOOLS OF SYSTEMATICS

Most, although not all, taxonomists agree that classification should somehow reflect evolutionary history, but this means different things to different people. For most systematists of the post-Darwinian era, a taxon, such as the class Reptilia, is understood to convey the fact that its members are derived from a common ancestor. Traditional systematists, sometimes called "evolutionary systematists," refer to this as a MONOPHYLETIC group, meaning that it has arisen from a single stem, namely a group of equal rank (the Reptilia arose from the class Amphibia) or lower rank (perhaps from Carboniferous amphibians of the order Diadectiamorpha).

A CLADE is the entire portion of a phylogeny that is descended from a single ancestral species, and therefore is a monophyletic group. It might be supposed that the higher taxa in current classifications are equivalent to clades, but this is sometimes not the case. In some cases, a taxon is composed of the descendants of unrelated ancestors that evolved similar features by convergent evolution and for this reason have been mistaken as relatives. Such a taxon is POLYPHYLETIC. For example, many botanists believe that the old order Amentiferae, including the oaks, walnuts, and willows, is an "unnatural" polyphyletic taxon, defined by a characteristic (the catkin) that evolved independently in these three groups. Some recent classifications place these plants in three presumably unrelated orders. Almost everyone agrees that taxa should not be polyphyletic.

Some taxa are not equivalent to clades because they are defined by characters that have evolved to different levels of organization. A group of species at the same level of organization is referred to as a GRADE (Figure 2). A grade may not equal a clade either because the characters have been attained independently by PARALLEL EVOLUTION in several closely related lineages, or because members of the clade that have evolved further are removed to another taxon. For example, some mammalian characteristics, such as the reduction of the lower jaw to a single bone, may have arisen repeatedly among the therapsid reptiles. Because this characteristic largely defines the class Mammalia, it was suspected that this

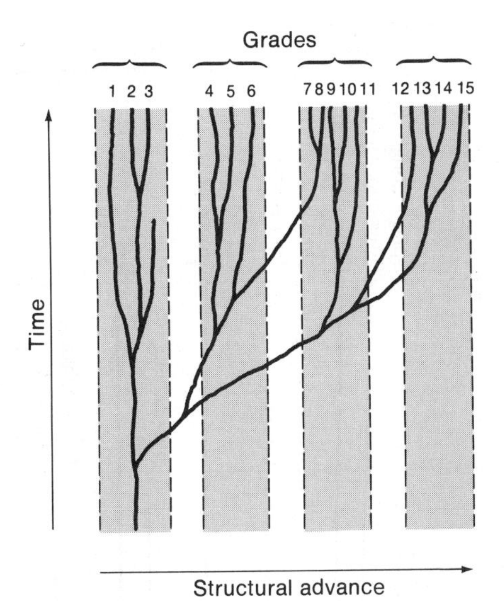


FIGURE 2

**Grades and clades. A group of species (e.g., 1, 2, 3) with a recent common ancestor forms a clade; a group with the same level of structural organization (e.g., 7–11) forms a grade. Members of a clade may belong to different grades because of differential evolutionary rates. (Modified from Simpson 1961)**

class had therefore arisen from several different therapsid stocks by parallel evolution, which if true would mean that mammals are a grade rather than a clade (see Simpson 1959, Lillegraven et al. 1979). Recent fossil evidence, however, supports the view that the Mammalia are monophyletic (Hopson and Crompton 1969, Kemp 1982).

Some taxa are grades because some of their descendants have become more "advanced," and are removed into their own taxa. For example, among the archosaurian reptiles that included the crocodilians and the dinosaurs, one lineage diverged dramatically and gave rise to birds (Figure 3). These are accorded their own class, Aves, because they are so different from their reptilian relatives. Thus this traditional classification attempts to reflect both common ancestry (putting turtles, lizards, snakes, and crocodilians together) and amount of evolutionary divergence in morphology; it attempts to reflect both clades (the class Aves) and grades (the class Reptilia). Mayr (1981) has articulated the reasons for this dual function of traditional classification. Note that the erection of a separate class for birds emphasizes the dissimilarity between birds and crocodilians, rather than their phylogenetic relationship. A taxon such as the class Reptilia that does not include all the descendants of an ancestor has been termed a PARAPHYLETIC group (Wiley 1981).

In the 1950s and 1960s, two new schools of systematics arose. PHENETICISTS (see Sneath and Sokal 1973) argue that a classification will be most informative if it is based on the overall similarity among species, measured by as many characteristics as possible, even if such a classification does not exactly reflect common ancestry. Pheneticists have developed elaborate numerical methods (and so call themselves "numerical taxonomists") for grouping species on the basis of overall similarity, and portraying the similarity as a PHENOGRAM (Figure 4A). The more characteristics two species have in common, the higher the level at which they are joined in a phenogram. Such a diagram, however, may not portray the true phylogeny, for two reasons. First, if a character state evolved independently

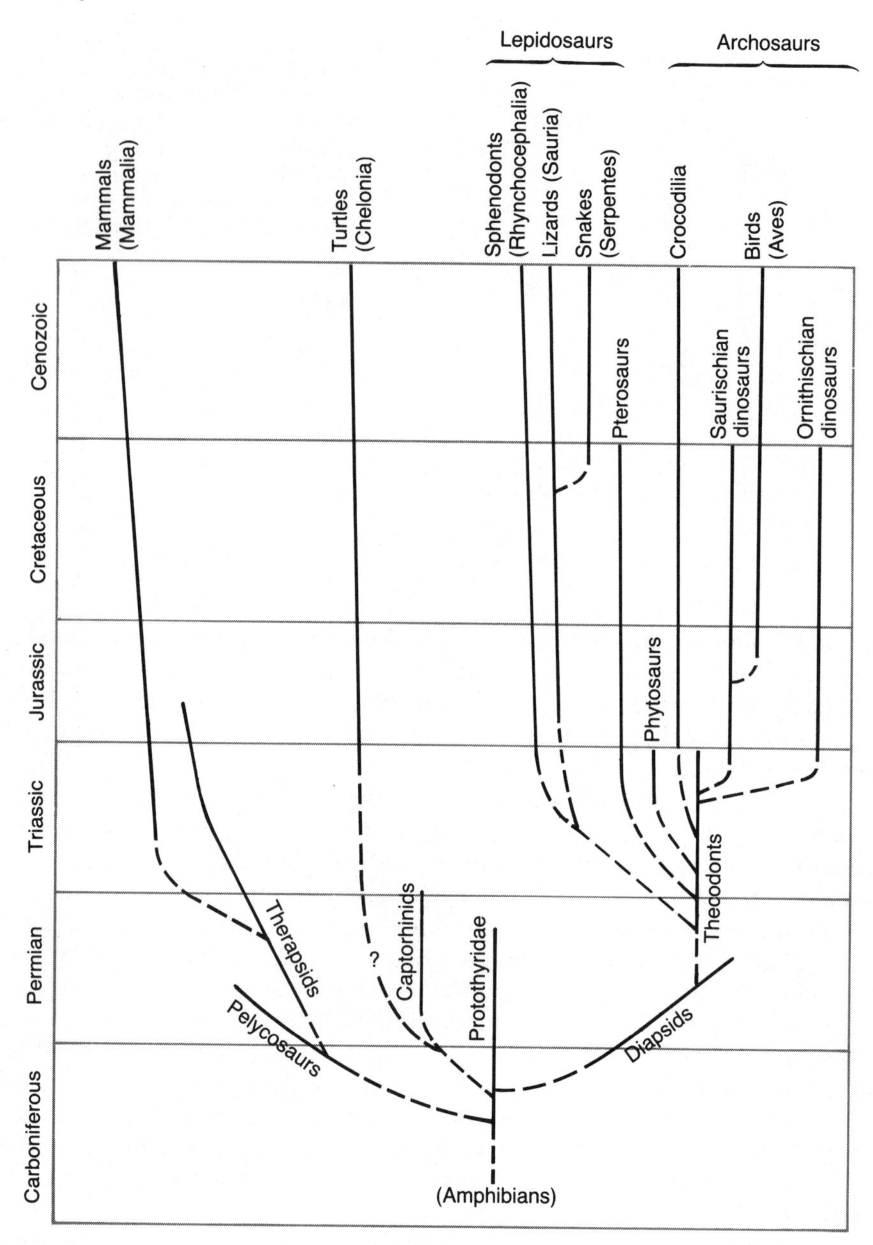


FIGURE 3

**A highly oversimplified phylogeny of the Reptilia and their derivatives. All the taxa included are placed in the class Reptilia except for the mammals and birds. Segregation of these into separate classes makes the Reptilia a paraphyletic taxon. Solid lines indicate approximate distribution in the fossil record; broken lines indicate postulated phylogenetic relationships. Many extinct groups are omitted. (Based on Romer 1966, Carroll 1982, Benton 1983, 1985)**

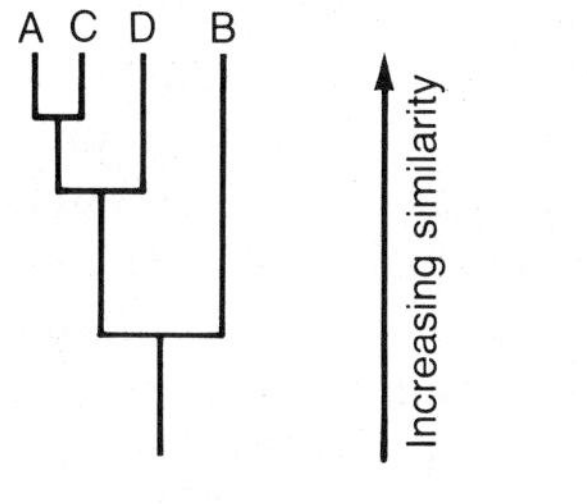

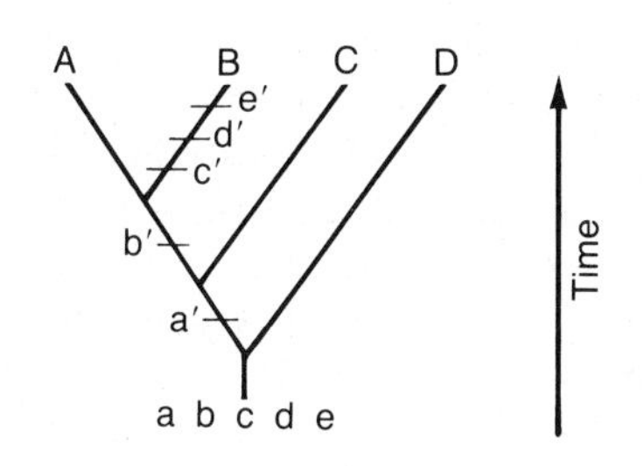


**B** Cladogram

**C** Character states of the taxa

| Taxa | 1 | 2 | 3 | 4 | 5 |
|---|---|---|---|---|---|
| | Characters | | | | |
| A | a′ | b′ | c | d | e |
| B | a′ | b′ | c′ | d′ | e′ |
| C | a′ | b | c | d | e |
| D | a | b | c | d | e |

**D** Matrix of shared character states

| | A | B | C | D |
|---|---|---|---|---|
| A | – | 2 | 4 | 3 |
| B | | – | 1 | 0 |
| C | | | – | 4 |
| D | | | | – |

**E** Matrix of shared derived character states

| | A | B | C | D |
|---|---|---|---|---|
| A | – | 2 | 1 | 0 |
| B | | – | 1 | 0 |
| C | | | – | 0 |
| D | | | | – |

FIGURE 4

**Comparison of a phenogram (A) with a cladogram (B), based on the hypothetical states of four taxa (C). The phenogram is constructed from the total number of character states shared by any pair of taxa (D), whereas the cladogram is constructed only from the derived character states (marked by primes) shared by pairs of taxa (E). The character state transitions are marked on the cladogram. Note that the evolutionary tree implied by the use of shared derived characters differs from that implied by the use of all shared characters.**

in two different lines, it will appear to indicate common ancestry when in fact the species are not derived from an immediate common ancestor. We will return to this problem below.

The other reason for a discrepancy between a phenogram and a true phylogenetic tree is differential rates of evolution. Suppose, for example, that the real phylogeny of species A–D is as portrayed in Figure 4B, in which the origin of derived character states is marked along the segments of the phylogenetic tree by primed letters. Because B has evolved more rapidly (in characters c, d, and e) than either A or C have, A and C appear more closely related than they actually are because they have retained more primitive character states in an unchanged condition. Based on the resulting phenogram (Figure 4A), a taxonomist might well place species A and C in one taxon, and B in another, which would not represent the true phylogeny. (The hypothetical species B, for example, might represent birds, while A and C might represent crocodilians and lizards respectively.)

The CLADISTIC SCHOOL of systematics, following many of the principles laid down by the entomologist Willi Hennig (1979), holds that a classification should express the branching (cladistic) relationships among species, regardless of their degree of similarity or difference. The relationships among species are portrayed in a CLADOGRAM, which can be considered an estimate of the true phylogenetic tree (Figure 1). As we shall see, a cladogram is not based on overall similarity of species, and so may differ substantially from a phenogram. Cladists argue that each taxon in a classification should be STRICTLY MONOPHYLETIC, meaning that it will include *all* the descendants of a particular ancestor. Paraphyletic groups are not admitted in such a scheme. Thus to express the common ancestry of birds and crocodilians, these groups might be ranked as, say, orders within a subclass of the class Reptilia. Such a classification would reflect the branching order of evolution by stressing the common origin of birds and crocodilians, but would not express differences in rates of evolution (the more pronounced divergence of birds than crocodilians from ancestral reptiles). Wiley (1981) has summarized the principles of cladistics, and Cracraft (1983) has described the use of cladistic classifications in studying evolution.

Under any philosophy of classification there are ambiguities. The absolute rank of taxa is rather arbitrary: a taxonomist may divide a group of species into two genera if he or she is impressed by differences, or combine them into one genus if the similarities are emphasized. Some authorities have included the tiger and other large cats in the genus *Felis* with the small cats, whereas other authorities have segregated them as the separate genus *Panthera*.

Moreover, no system of classification adequately expresses the relationships among ancestral and derived taxa at different geological times. Placing the Eocene *Hyracotherium* and its relatives in their own family would not show that they were the ancestors of the Rhinocerotidae (rhinoceroses) and Equidae (horses), but neither would placing them in either of these two families. Nor is it satisfying to place *Hyracotherium* in the Equidae because it leads to the horses, and to place its close Eocene relatives in a different family.

These classificatory problems are not in themselves part of the study of evolution, but classifications are an important part of the vocabulary of evolutionary studies (Cracraft 1983). For example, rates of evolution are often expressed in terms of the origination and extinction of families or other higher taxa (Chapter 12). Also, lack of familiarity with a classification can lead one astray. For example, the African apes (chimpanzees, gorillas) and the orangutan are traditionally placed in the Pongidae, and humans are assigned to the Hominidae. One might therefore surmise that humans diverged from the apes before the apes diverged from each other, but there is a great deal of evidence that the divergence of humans from African apes occurred after the orangutan lineage diverged (see below, and Chapter 17). In a cladistic classification, humans and African apes might constitute one subfamily, and the orangutan another.

## DIFFICULTIES OF PHYLOGENETIC INFERENCE

Whether or not a phylogenetic tree is expressed in the classification, it constitutes a hypothesis about evolutionary history. Like other scientific hypotheses, phylogenetic trees cannot be proven with absolute certainty. But it is possible to adduce evidence for or against a phylogenetic hypothesis.

If there were no convergence and if each lineage diverged from other lineages at a constant, clock-like rate, the degree of difference between any two species would be directly proportional to the time elapsed since they have become separate gene pools, i.e., since their common ancestor speciated. Consequently the phylogeny would be accurately inferred from the overall degree of difference among species. But if convergent evolution has occurred, the correspondence between phenotypic difference and age of divergence will be lowered. This will also be true if evolutionary rates have varied. To infer phylogenies, therefore, convergence and inconstant rates of evolution must be taken into account.

### Variation in evolutionary rates

Figure 5 illustrates the evolution of two tooth characteristics in horses. Each character evolved faster at some times than others and more rapidly in one lineage than another; and in each lineage the two characters often differed in their rates of change. The detailed fossil record of the radiolarian protozoan *Pseudocubus vema* (Figure 6) shows that size evolved at a highly inconstant rate, with three periods of rapid increase interspersed with periods of up to a million years in which little change occurred (Kellogg 1973).

The term MOSAIC EVOLUTION refers to differences among characters in their rate of evolution within a lineage. Compared to Carboniferous amphibians, for example, frogs have an ancestral number of aortic arches (2), but a derived number of fingers (4); humans have a derived number of aortic arches (1), but an ancestral number of digits (5). Humans (*Homo sapiens*) and chimpanzees (*Pan troglodytes*) differ strikingly in morphology, but their DNA and proteins are extraordinarily similar, suggesting that morphological and biochemical evolution have proceeded at different rates (King and Wilson 1975). Mosaicism of evolution is the rule rather than the exception; higher taxa emerge not by coherent transformation of all or even most of their features, but by sequential changes in various traits (Schaeffer 1956). In all but its feathers and a few other bird-like features, for example, *Archaeopteryx* had the reptilian characters of the small dinosaurs that were its close relatives (Figure 14 in Chapter 11).

Slowly evolving, or CONSERVATIVE, characters are frequently shared in a similar condition by all or most of the members of a major group. The form of the incisors in modern rodents as different as beavers (*Castor*) and kangaroo rats (*Dipodomys*) is much like that in the earliest fossil rodents. Having once arisen, such characters probably change little because they serve for adaptation to a great variety of environmental conditions. Rodent incisors are a GENERAL ADAPTATION

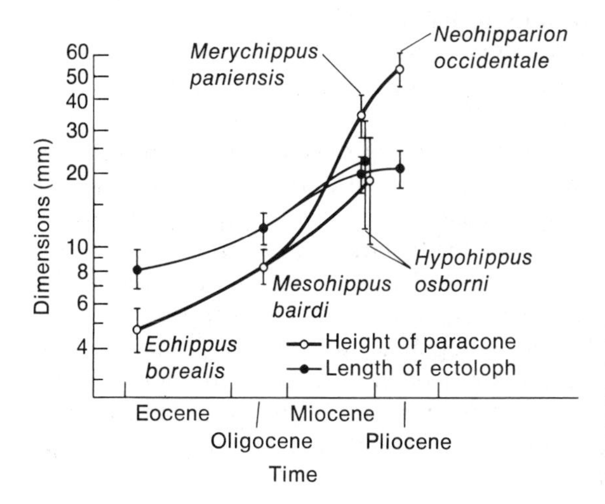


FIGURE 5

**Changes in two dental characters in a phylogeny of five genera of horses. The rates of change vary between characters, among phylogenetic sequences, and among time periods. (After Simpson 1953)**

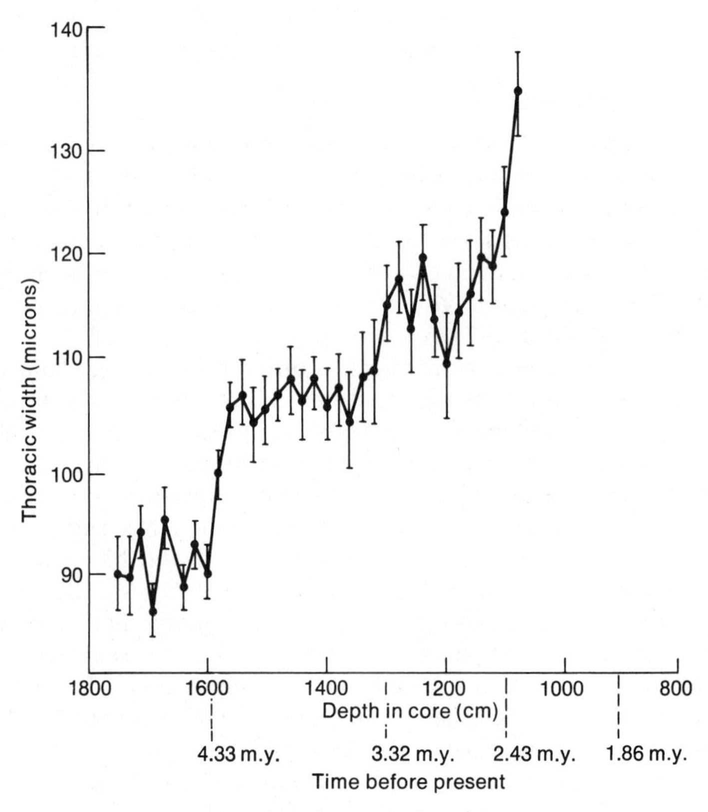


FIGURE 6
**Change in size of the radiolarian protozoan *Pseudocubus vema*, for which the fossil record is highly detailed. Points represent means, vertical bars the amount of variation. The rate of evolution was higher at some times than others, and fluctuations are evident even over short periods. The rate of evolution at any time is not correlated with the amount of phenotypic variation. (Adapted from Kellogg 1973)**

(Brown 1959) to most of the conditions rodents face. But the form of the legs varies more among rodents, and has evolved more rapidly: for example, the long hind legs and reduced number of toes in kangaroo rats are SPECIAL ADAPTATIONS to desert conditions.

Some characters may be conservative because of developmental canalization (Chapters 7 and 14). Their alteration may require drastic and unlikely remodeling of the pattern of development. The arrangement of ovules in angiosperms—on a central column or on the ovarial partitions, for instance—is a rather conservative character that appears to be independent of the plant's ecology. Some conservative characters seem so trivial that it stretches the imagination to suppose that they are adaptive. The species in the fly family Sepsidae, for instance, share a single anteriorly directed bristle on the rear margin of the posterior thoracic spiracle. Among bees, one of the distinguishing characteristics of the large family Halictidae is the curved basal vein of the forewing, a vein that is straight in the related but distinct family Andrenidae (Figure 7).

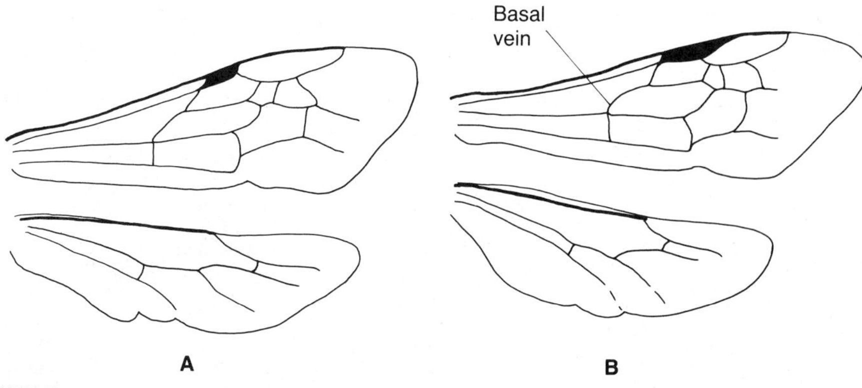


FIGURE 7

**The wings of bees of the families Andrenidae (A) and Halictidae (B). The basal vein, which is straight in Andrenidae and curved in Halictidae, is a chief distinguishing feature between these large families, and therefore represents a character that has remained constant during the long evolution of each of these groups. (From Borror et al. 1981)**

## Homology and homoplasy

A character state is HOMOLOGOUS in two species when it has been inherited by both from their common ancestor. On the other hand, if a character state has evolved more than once—if it is possessed by two species but was not possessed by all the ancestors intervening between them and their most recent common ancestor—it exhibits HOMOPLASY. One form of homoplasy is CONVERGENT EVOLUTION, which is quite common because different species are often subject to similar selection pressures. The mouthparts of true bugs (Hemiptera) and of biting flies such as mosquitoes (Diptera) have become modified into piercing–sucking beaks, but the independent origin of this modification is evident in their different anatomical construction (Figure 8). They are more similar in name than in structure.

Convergent evolution often results in such similar features that we would be hard put to identify them as convergent if other features did not indicate that the groups are unrelated. Herbaceous plants have evolved repeatedly from woody ancestors; the leaflessness and growth form of New World cacti and some Old World Euphorbiaceae are extremely similar adaptations to arid conditions; leglessness has evolved in snakes and in several unrelated groups of lizards. Hole-nesting birds, whatever their phylogenetic affinities, typically lay white rather than colored or speckled eggs. Rensch (1959) lists dozens of other patterns of convergent adaptation.

Convergent evolution grades into two other forms of homoplasy: PARALLEL EVOLUTION and EVOLUTIONARY REVERSAL. Ideally, "convergent evolution" describes cases in which similar phenotypes have evolved by different developmental pathways, whereas "parallel evolution" refers to independent developmental modifications of the same kind. Because related species have similar developmental programs, parallelism is frequent among closely related species. For example, the elongated body of burrowing salamanders has evolved numer-

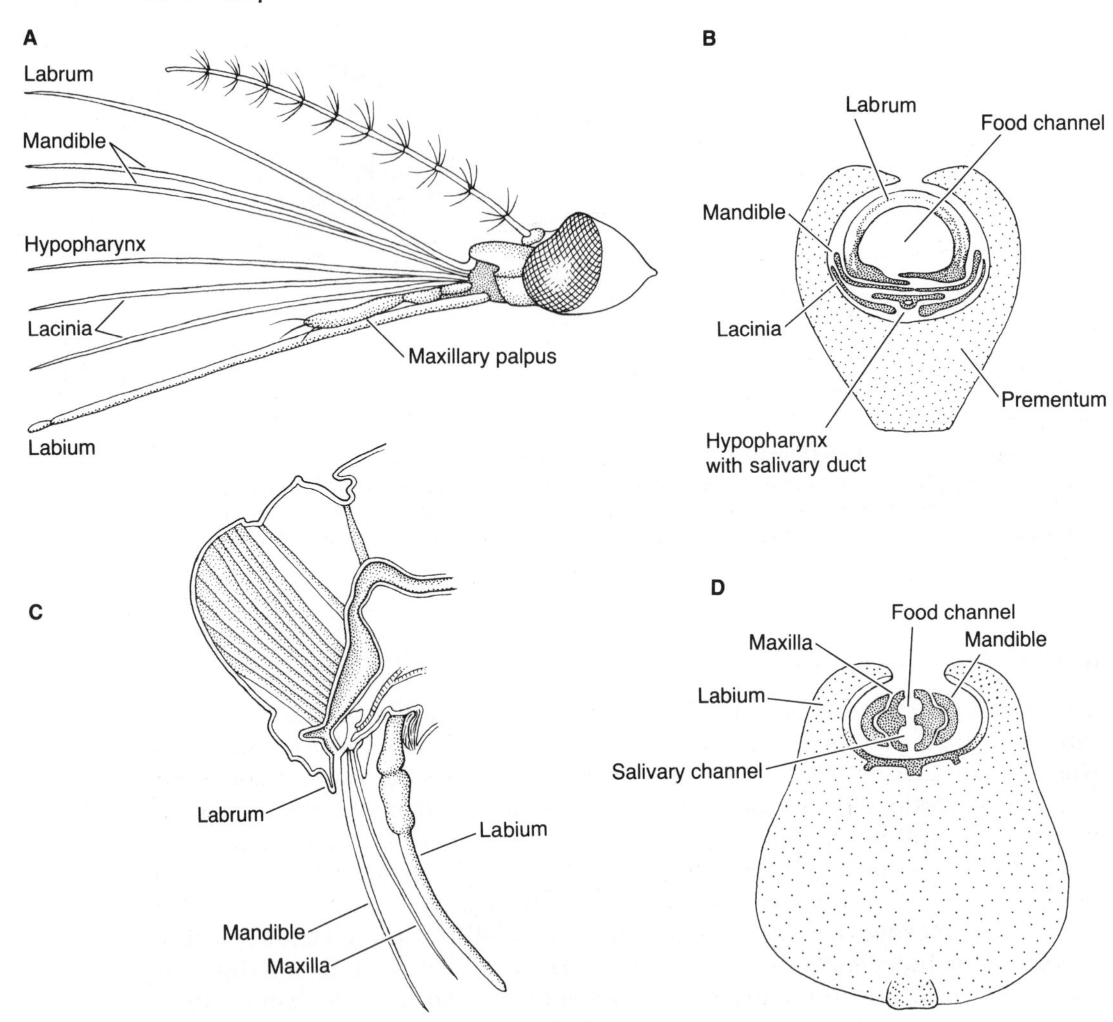


FIGURE 8

**Convergent evolution: piercing-sucking mouthparts in the insect orders Diptera (mosquito) and Hemiptera (cicada). (A) Lateral view of the head of a mosquito (mouthparts separated for visibility). (B) Cross section of mosquito mouthparts. (C) Lateral view of the head of a cicada, in sagittal section. (D) Cross section of cicada mouthparts. The labrum encloses the food channel in the mosquito, but is abbreviated in the cicada; the maxillae form separate piercing organs (laciniae) in the mosquito, but enclose the food and salivary channels in the cicada. (From Atkins 1978)**

ous times by an increase in the number of vertebrae (parallelism), but in a few cases by increase in the length of individual vertebrae (convergence). A similar pattern of banding on the wings of many moths has arisen repeatedly, apparently by similar modifications of a basic "ground plan" that almost all moths share (Figure 9). Often, of course, we do not know enough about the developmental

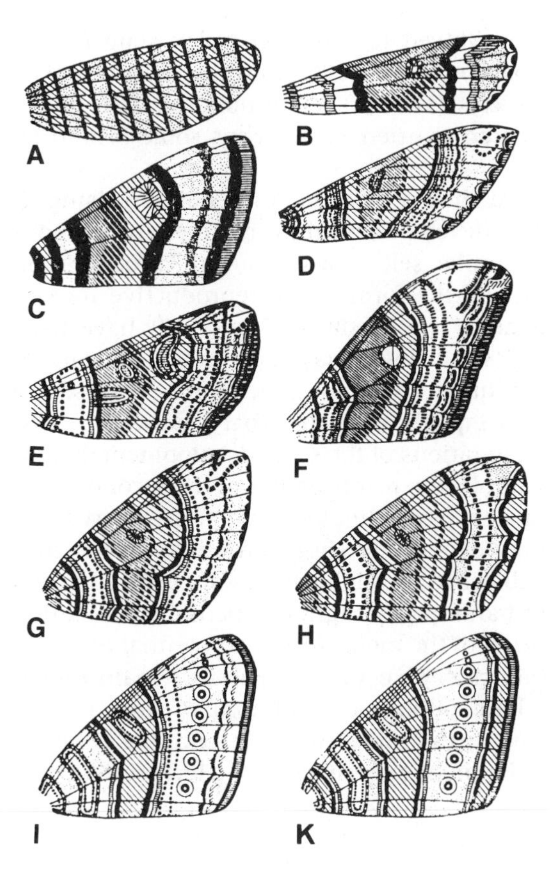


FIGURE 9

**Parallel evolution in the wing patterns of Lepidoptera (moths and butterflies). Similar patterns of pigmentation have evolved repeatedly. (A) Tineidae and Tortricidae; (B) Pyralidae; (C) Arctiidae; (D) Sphingidae; (E) Noctuidae; (F) Saturniidae; (G,H) Geometridae; (I,K) Nymphalidae. (From Rensch 1959)**

basis of a feature to judge whether it has arisen by convergence or parallelism in different species.

Parallel evolutionary tendencies are evident in almost every major group. Female winglessness has evolved repeatedly in the moth family Geometridae, as in other families of moths; colonial, social behavior has arisen quite a few times among bees and in other aculeate Hymenoptera as well; cuckoos seem inclined toward nest parasitism. Each of the traditional suborders of rodents arose several times from the generalized early rodents, the protrogomorphs (Wood 1959).

Reversals are likewise common in evolution. The fossil record of elephants, for example, shows that the general trend toward greater size was reversed in several lines that evolved dwarf species, and reversals in the structure of the teeth accompanied the change in body size in each instance (Maglio 1972). Quite often a complex character may degenerate and return to its original state; thus winglessness in insects can be either a primitive condition as in silverfish (Thysanura) or a derived condition as in lice, fleas, and the many wingless species in almost every insect order that has evolved from winged ancestors. To be sure, there is a degree of irreversibility in evolution, as expressed by DOLLO'S LAW, which states that complex structures, once lost, are unlikely to be regained in their original form. However, the structure of the eye in snakes is different enough from that of other vertebrates to suggest that it evolved to a new complex state from a reduced condition in a burrowing ancestor (Porter 1972). Of the several molars possessed by primitive Carnivora, the cats (Felidae) retain only the first, but in the lynx the second molar has reappeared (Kurtén 1963). Thus some lost structures may indeed be regained.

That parallel evolution and reversal should be common is not surprising. If related species have similar patterns of development, they are likely to be modified in similar ways if subjected to similar selection pressures. Throckmorton (1965) has suggested that peculiarities of the form of the reproductive tract in species of *Drosophila,* whose phylogenetic relationships (Figure 10) have been determined by chromosome analysis (see below) may have arisen repeatedly when identical genes carried in low frequency in some species are expressed in homozygous condition in others. It is equally plausible that parallelisms and reversals may be genetically different alterations of the same developmental pathway in each of several species. For example, the reappearance of the second molar in the lynx probably did not require a slew of new mutations to form a new tooth *ex nihil.* Whether a tooth develops may depend on slight variations in the concentration of some tooth-inducing substance. A slight mutation could suppress tooth development in the Felidae, and an equally slight compensatory mutation in the lynx could allow the manifestation of the molar-forming potential that may be present but suppressed in the other cats. That such potentials remain immanent in developmental systems for long periods of evolutionary time is well known (Chapter 14). Frogs have lacked teeth in the lower jaw since the Jurassic, but it is possible to induce their development experimentally (Hecht 1965), and

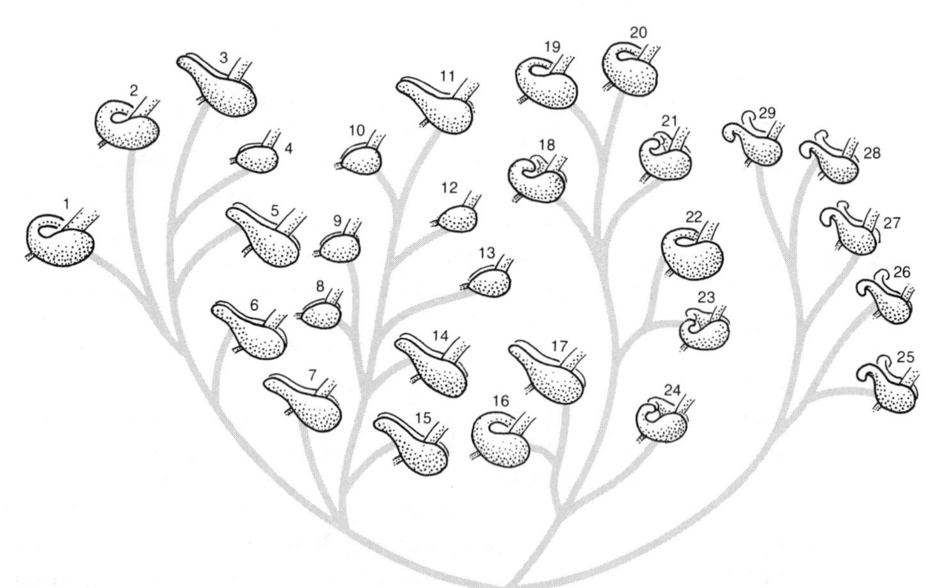


FIGURE 10

**Independent origins of forms of the ejaculatory bulb in species of the *Drosophila repleta* group. The phylogeny of these species has been inferred from chromosomal evidence. Note the identical bulb shape in, for example, species 3, 5, 14, 17; 4, 10; and 1, 16, 19. (Redrawn from Throckmorton 1965)**

one South American tree frog, *Amphignathodon*, has re-evolved true teeth in the lower jaw (Noble 1931).

### The usefulness of the fossil record

Many people suppose that evolutionary relationships among species can be read directly from the fossil record, but this is seldom the case. For most groups, especially those that do not fossilize readily, the paleontological record is too fragmentary to be useful. Even in groups with a good fossil record, there are seldom evenly graded series of fossils between old and young forms (Chapter 12). There are often reasons to believe that a fossil species is not the direct ancestor of a more recent species, but rather is a related lineage that became extinct without issue. Therefore a derived character state may appear early in the fossil record, in an extinct lineage, while the related lineage leading to a living group may retain a primitive character state until much later. To use an absurd example, birds appear earlier in the fossil record (Jurassic) than do snakes (Cretaceous), but this does not mean that all the features of birds are more primitive than those of snakes, nor that snakes are descended from birds. Granted, primitive characters such as the single occipital condyle of reptiles often do appear earlier in the fossil record than derived features such as the double occipital condyle of mammals. But only if a character state is ubiquitous among the fossil members of a group thought to have included the ancestors of a modern taxon can the fossil's features safely be assessed as the ancestral state (Schaeffer et al. 1972). Fossils can provide corroboration of relationships: for example, reptiles appear before mammals, and there are numerous intermediates between the two groups. But relationships cannot be inferred solely from temporal sequences of fossils.

## INFERRING PHYLOGENY FROM MORPHOLOGICAL DATA

The problem of phylogenetic inference is to join species into monophyletic groups that reflect common ancestry, without being misled by homoplasy and differential evolutionary rates. Hennig (1979) was among the first to point out that the key to recognizing a monophyletic group is that its members share UNIQUELY DERIVED CHARACTER STATES: synapomorphies that have arisen only once. For example, the amnion is a synapomorphy that unites reptiles, birds, and mammals into the Amniota. Ostriches and penguins share synapomorphies such as feathers that combine them into the monophyletic group Aves within the Amniota.

If we can distinguish derived from primitive character states, and if each transition has been unique (i.e., there is no homoplasy), inferring the phylogeny is rather simple. In Figure 4B, for example, two derived characters a′ and b′ are shared by the monophyletic group A + B, and only one derived character, a′, is shared by the more inclusive monophyletic group A + B + C. Species C shares a more remote common ancestor with A and B than A and B do with each other, and consequently C has fewer derived characteristics in common. Whereas a phenogram (Figure 4A) incorporates information on all shared characters, both primitive and derived, the cladistic approach of using only shared derived characters provides a direct inference of the phylogeny. However, homoplasy may lead us astray; leaflessness, for example, is a derived character that might lead us to suppose that cacti and Euphorbiaceae share an immediate common ancestor, whereas the structure of their flowers argues against a close relationship.

Numerous ways of finding the "best," or most likely, phylogenetic tree in the face of such conflicting data have been proposed (reviewed by Felsenstein 1982). At present, the most widely used method (Edwards and Cavalli-Sforza 1964, Farris 1970, 1983, Wiley 1981) uses the criterion of PARSIMONY: that tree is chosen which requires us to postulate the smallest number of evolutionary changes throughout the entire tree. This is equivalent to minimizing the number of convergent events or reversals. For example, the females of both fall cankerworm moths (*Alsophila pometaria*) and the wasps known as velvet ants (Mutillidae) are wingless; but to suppose that this is a uniquely derived synapomorphy, i.e., that they inherited this condition from a common ancestor, we would have to postulate that the fall cankerworm is convergent with other moths, and the velvet ant with other wasps, in innumerable other characteristics. This would violate the criterion of parsimony. We conclude, then, that female winglessness is a convergent condition. The criterion of parsimony is the basis of the widely used Wagner method (Kluge and Farris 1969), which arranges taxa into a network that minimizes the number of character state changes (Figure 11). Such a network can be transformed into a phylogenetic tree if there is reason to believe that the network consists of two groups of related species, connected to each other via a stem from their common ancestor (Figure 11E). Various computer algorithms have been devised to find the most parsimonious of the many possible trees that may be constructed for even a few taxa.

In contrast to the parsimony methods, which attempt to minimize the number of evolutionary changes that must be assumed, some authors (e.g., Meacham and Estabrook 1985) favor COMPATIBILITY methods, which assume that the most likely phylogeny is the one that is fully compatible with the largest number (or "clique") of individual characters. Thus the phylogenetic tree in Figure 11B, which is compatible with four characters, would be preferable to that in Figure 11C, which is compatible with only two.

The phylogenetic trees found by parsimony methods and compatibility methods will seldom be the same, and there is intense dispute over which methods are preferable. Felsenstein (1982) has criticized both approaches on the grounds that they do not include statistical methods for evaluating whether one proposed phylogeny is really more likely than another to be correct. He has also suggested that if one assumes that convergences and reversals are scattered at random over characters, parsimony will be the better approach; if only certain characters are prone to homoplasy, compatibility analysis, by discarding these characters, may be preferable. The experience of taxonomists has been that most characteristics are subject to some degree of convergent evolution, but it is often possible to identify some characters that are especially likely to converge within each group. If characters that are known from past experience to be prone to convergence are excluded a priori from phylogenetic analysis, the parsimony method will be stronger.

### Distinguishing ancestral from derived character states

If a monophyletic group is one that shares uniquely derived character states, it is necessary to be able to distinguish derived from primitive character states, i.e., to infer the direction (polarity) of evolution. Several lines of evidence can be used (Hennig 1979, Stevens 1980).

**A** Character states of the taxa

| | Character | | | | | |
|---|---|---|---|---|---|---|
| Species | 1 | 2 | 3 | 4 | 5 | 6 |
| A | a′ | b′ | c | d | e | f |
| B | a | b | c′ | d | e′ | f′ |
| C | a′ | b′ | c | d | e′ | f′ |
| D | a | b | c′ | d′ | e | f |
| E | a | b′ | c | d | e′ | f′ |

**B** Tree implied by characters 1–4, assuming primed states are derived

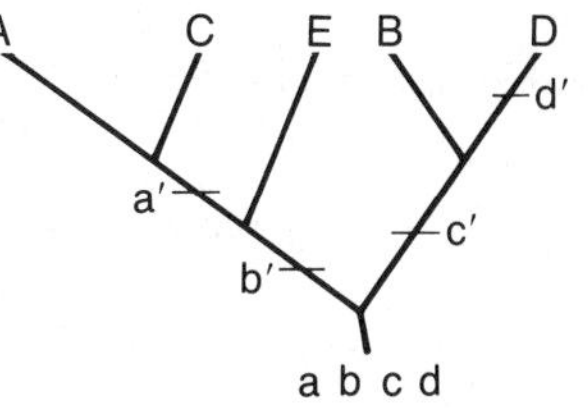


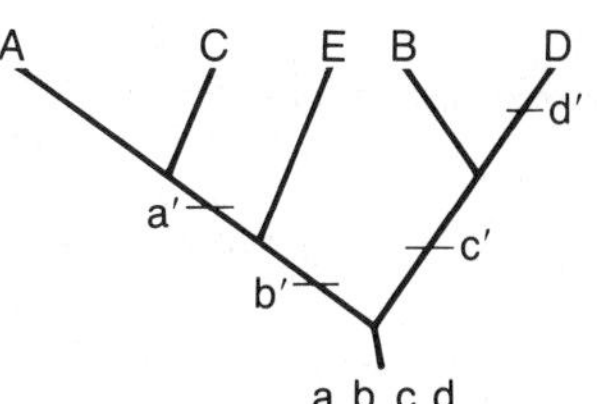

**C** Tree implied by characters 5 and 6, assuming primed states are derived

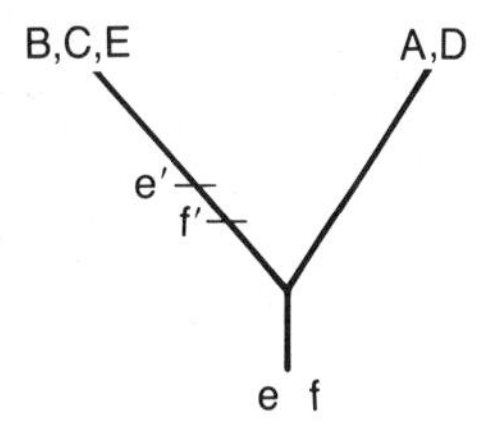


**D** Most parsimonious undirected tree (Wagner network)

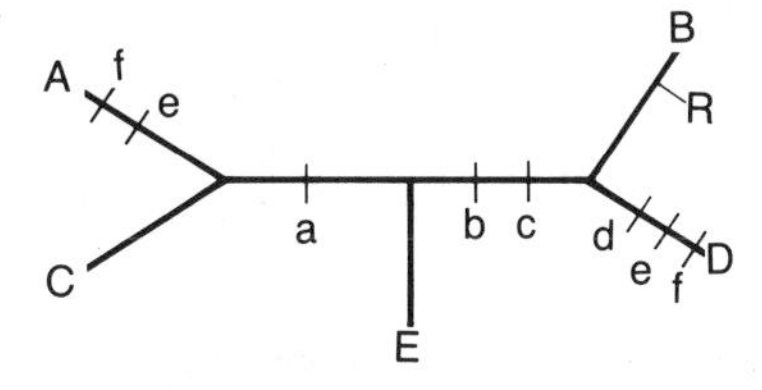


**E** Most parsimonious directed tree (Wagner tree), by assuming B an outgroup, rooting tree at point R

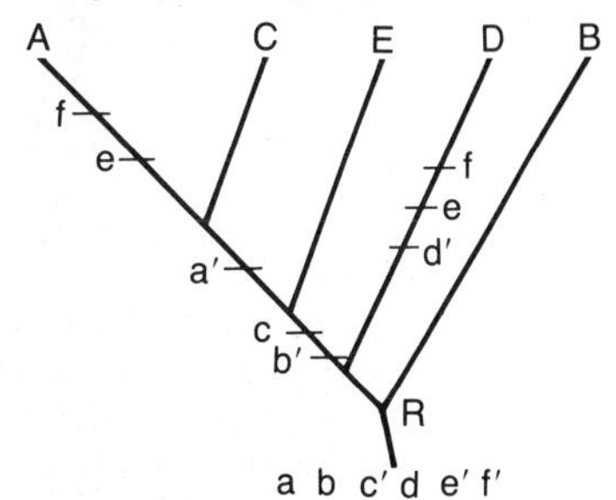


FIGURE 11

**An example of cladistic inference of phylogenetic relationships, using the criterion of parsimony. (A) Hypothetical data on six characters of five species. If it is assumed that primed character states are derived, characters 1–4 imply one tree (B), but characters 5 and 6 imply another (C) that is incompatible with the relationships implied by characters 1–4. Some characters are therefore homoplasious. By assuming a minimal number of character state changes, an undirected network (D) of minimal length can be constructed. The diagram indicates the branches along which each character changed, but does not specify the direction of change (e.g., *a* indicates a change in state but does not specify a → a′ or a′ → a). If there is some independent reason to believe that species B is less closely related to any of the other species than they are to each other, the common ancestor of B and the other species is placed at point R, yielding a directed (rooted) phylogenetic tree (E). This implies that the ancestral character states were a, b, c′, d, e′, f′, and that characters 5 and 6 (e, f) each changed twice. (After Felsenstein 1982)**

When a character exists in three or more states, it is often possible to order these in a nondirected MORPHOCLINE or TRANSFORMATION SERIES such as A ↔ B ↔ C, implying that B is an intermediate stage between A and C. For example, inversions in the chromosomes of *Drosophila* (Chapter 3) are identified by changes in the sequences of individually identifiable bands. If these form a sequence ABCDEFG in a "standard sequence," the sequences AB*FEDC*G ↔ ABCDEFG ↔ A*EDCB*FG ↔ AED*FBC*G, in which the italicized letters represent inversions, constitute a transformation series. By this logic, the evolution of chromosome inversions in many groups of *Drosophila* has been inferred (Figure 12). But the problem in such a transformation series (or in a simple two-state series, such as presence or absence of wings) is to identify which character state is ancestral.

The single most useful criterion in judging which character state is primitive

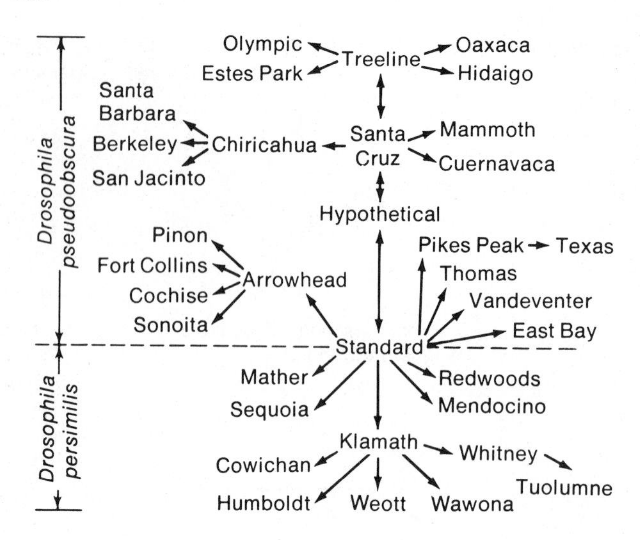


FIGURE 12

**An undirected phylogeny of the arrangements of the third chromosome in *Drosophila pseudoobscura* and *D. persimilis*. Each of the named arrangements differs from its neighbor in the phylogeny by a single inversion; only one step (*Hypothetical*) has not been found in natural populations. The species share the *Standard* arrangement, which has been postulated to be ancestral to the others. (From Anderson et al. 1975)**

within a group of species is its condition in related forms outside the group (Stevens 1980). If species A, B, and C are known to share a more recent common ancestor with each other than they do with species D, E, or F, then by the criterion of parsimony the character state that is ubiquitous among D, E, and F is most likely to be primitive within A, B, and C. For example, among the butterflies (all of which clearly have a common ancestor), the brush-footed butterflies (Nymphalidae) and the monarchs (Danaidae) have four functional and two highly reduced legs, whereas the swallowtails (Papilionidae) and the sulphurs (Pieridae) have six functional legs. The six-legged condition is assumed to be ancestral because this is the condition in other Lepidoptera (moths) and indeed in other orders of insects. The moths and the other insect orders are OUTGROUPS which enable us to tell which condition is primitive within the "ingroup" (the butterflies) that we are analyzing. We therefore infer that reduced anterior legs are a derived character state shared by Nymphalidae and Danaidae, leading us to postulate that these families share a more recent common ancestor than either does with the Papilionidae or Pieridae. This logic has been incorporated into computer programs that produce the shortest networks by the parsimony criterion: the members of the ingroup are arranged to minimize the number of evolutionary steps required to derive their advanced character states from the primitive character state represented by the outgroups that are included in the analysis (Figure 11).

An example of the use of the "outgroup" criterion in establishing phylogeny is the analysis of relationships among the *grimshawi* group of 101 species of *Drosophila* endemic to the Hawaiian Islands (Carson and Kaneshiro 1976, Carson

1983). Inversions in the gene order of the chromosomes have been used to establish transformation series in the manner described above. The gene order of one species, *D. primaeva,* has been considered as the ancestral condition because it is similar to that of the *robusta* group (an outgroup) of continental species, from which the Hawaiian radiation is presumably derived. Thus a phylogeny of these Hawaiian species has been established with considerable confidence (Figure 13).

An exceptionally complete fossil record may provide useful information for phylogenetic analysis, but the outgroup method remains necessary for its interpretation. For example, the fossil record of horses, showing a progressive reduction in the number of digits, confirms that among the living members of the order Perissodactyla the one-toed condition of horses is derived relative to the multitoed condition in rhinoceroses and tapirs. In this case the fossil record confirms what we can deduce equally well from the outgroup criterion, since the multiple-toed condition is general among the orders of mammals. In contrast, the Litopterna, an extinct order of South American horse-like mammals, included one-toed forms in the Miocene, whereas only three-toed genera are found later, in the Pliocene and Pleistocene (Romer 1966). Despite the temporal sequence, the outgroup criterion holds that the three-toed condition is primitive, and indeed three-toed litopterns occurred both before and after the one-toed form.

Embryology has long been used for phylogenetic analysis, and many authors have supposed that it provides direct evidence on the polarity of evolutionary changes (see analyses by Kluge and Strauss 1985, de Queiroz 1985). The idea that early embryological features represent evolutionarily primitive states and later embryological features more derived states arose first in Haeckel's (1866) famous BIOGENETIC LAW: "ontogeny recapitulates phylogeny." Haeckel's belief that each embryonic stage represents the adult stage of one of its ancestors is flatly wrong (Gould 1977). Butterflies are not descended from larviform ancestors. Quite often VON BAER'S LAW is closer to the truth: the early developmental stages of a characteristic tend to be more similar among related species than the later stages, so that the characteristics that differentiate the taxa are embryologically later accretions on a fundamentally similar developmental plan.

Although von Baer, who wrote in 1828, was not an evolutionist, his "law" was later invested with an evolutionary intepretation, namely that characters which depart further from the embryonic condition in some species than others represent a derived condition. For example, in cattle as in other mammals, the metacarpal bones develop first as separate elements; later they fuse into the cannon bone in cattle, but not in most other mammals. The rudiments of teeth develop in anteaters, but are later resorbed. The "ontogenetic method" of determining character evolution would therefore hold that the cannon bone of cattle and the toothlessness of adult anteaters are derived characters. This is so, but in fact we judge this to be true only because unfused metacarpals and adult teeth are widespread among other mammals. The outgroup criterion shows that von Baer's law often does not hold. Early developmental stages may have their own special adaptations that represent derived rather than primitive characters; for example, the cotyledons of a plant do not represent ancestral leaves, and the horny beaks of tadpoles and the comb-like milk teeth of young bats are special adaptations to the juvenile environment. Early stages of an ancestor's ontogeny are sometimes lost from a descendant's ontogeny; for example, earthworms have

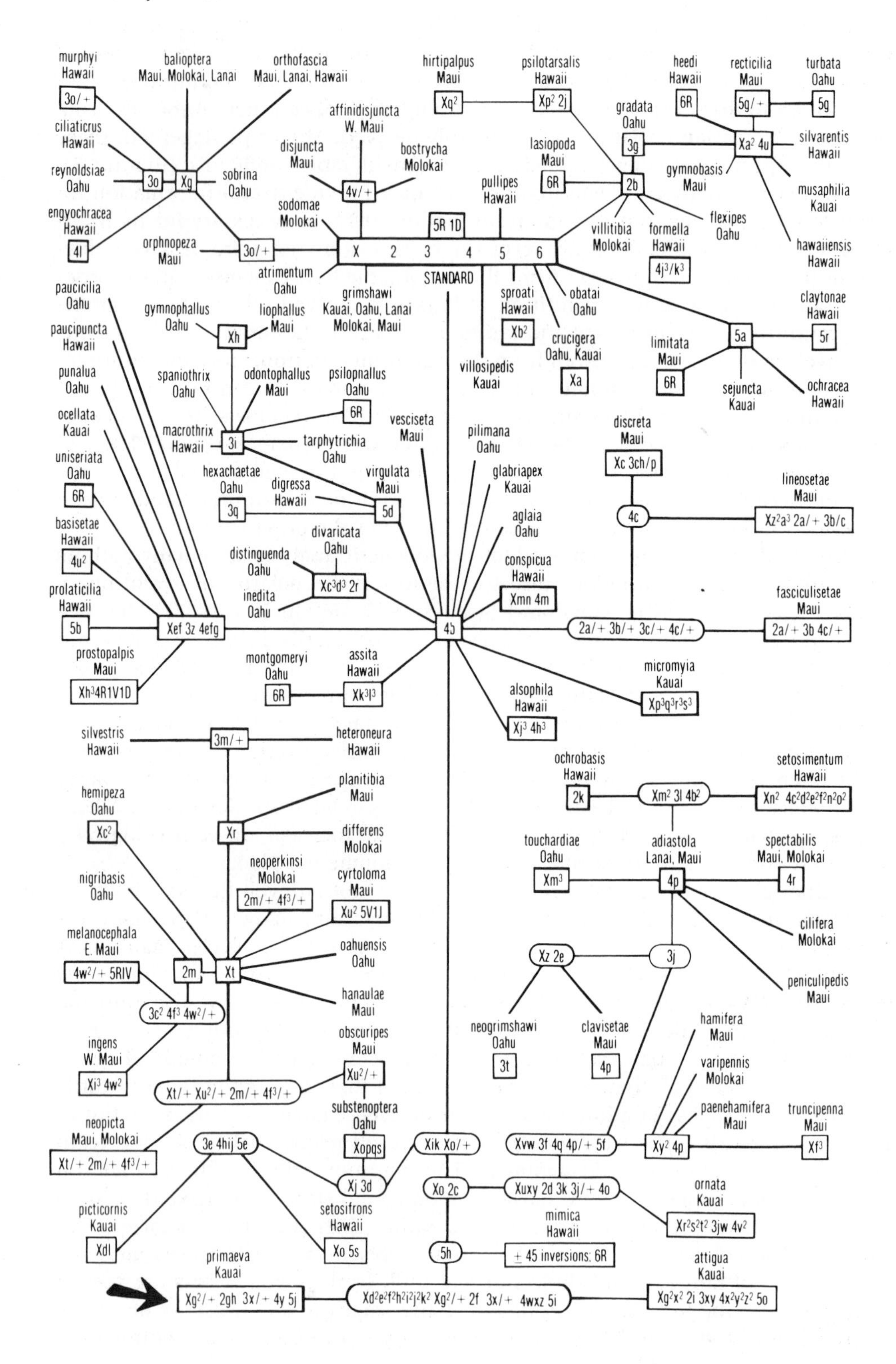
murphyi Hawaii
3o/+
balioptera Maui, Molokai, Lanai
orthofascia Maui, Lanai, Hawaii
ciliaticrus Hawaii
reynoldsiae Oahu
3o
Xg
sobrina Oahu
engyochracea Hawaii
4l
orphnopeza Maui
3o/+
affinidisjuncta W. Maui
disjuncta Maui
bostrycha Molokai
4v/+
sodomae Molokai
hirtipalpus Maui
$Xq^2$
psilotarsalis Hawaii
$Xp^2$ 2j
lasiopoda Maui
6R
pullipes Hawaii
5R 1D
X 2 3 4 5 6
STANDARD
atrimentum Oahu
grimshawi Kauai, Oahu, Lanai Molokai, Maui
heedi Hawaii
6R
recticilia Maui
5g/+
turbata Oahu
5g
gradata Oahu
3g
$Xa^2$ 4u
silvarentis Hawaii
gymnobasis Maui
2b
musaphilia Kauai
villitibia Molokai
formella Hawaii
$4j^3/k^3$
flexipes Oahu
hawaiiensis Hawaii
sproati Hawaii
$Xb^2$
obatai Oahu
crucigera Oahu, Kauai
Xa
villosipedis Kauai
claytonae Hawaii
5r
limitata Maui
6R
5a
sejuncta Kauai
ochracea Hawaii
paucicilia Oahu
paucipuncta Hawaii
gymnophallus Oahu
Xh
liophallus Maui
punalua Oahu
spaniothrix Oahu
odontophallus Maui
psilopnallus Oahu
6R
ocellata Kauai
macrothrix Hawaii
3i
tarphytrichia Oahu
vesciseta Maui
pilimana Oahu
discreta Maui
Xc 3ch/p
uniseriata Oahu
6R
hexachaetae Oahu
3q
digressa Hawaii
virgulata Maui
5d
glabriapex Kauai
aglaia Oahu
lineosetae Maui
4c
$Xz^2a^3$ 2a/+ 3b/c
basisetae Hawaii
$4u^2$
divaricata Oahu
distinguenda Oahu
$Xc^3d^3$ 2r
inedita Oahu
conspicua Hawaii
Xmn 4m
fasciculisetae Maui
prolaticilia Hawaii
5b
Xef 3z 4efg
4b
2a/+ 3b/+ 3c/+ 4c/+
2a/+ 3b 4c/+
prostopalpis Maui
$Xh^3$4R1V1D
montgomeryi Oahu
6R
assita Hawaii
$Xk^3l^3$
micromyia Kauai
$Xp^3q^3r^3s^3$
alsophila Hawaii
$Xj^3$ $4h^3$
silvestris Hawaii
3m/+
heteroneura Hawaii
ochrobasis Hawaii
2k
$Xm^2$ 3l $4b^2$
setosimentum Hawaii
$Xn^2$ $4c^2d^2e^2f^2n^2o^2$
planitibia Maui
hemipeza Oahu
$Xc^2$
Xr
differens Molokai
touchardiae Oahu
$Xm^3$
adiastola Lanai, Maui
4p
spectabilis Maui, Molokai
4r
nigribasis Oahu
neoperkinsi Molokai
2m/+ $4f^3$/+
cyrtoloma Maui
$Xu^2$ 5V1J
cilifera Molokai
melanocephala E. Maui
$4w^2$/+ 5RIV
2m
Xt
oahuensis Oahu
Xz 2e
3j
$3c^2$ $4f^3$ $4w^2$/+
hanaulae Maui
peniculipedis Maui
ingens W. Maui
$Xi^3$ $4w^2$
obscuripes Maui
$Xu^2$/+
neogrimshawi Oahu
3t
clavisetae Maui
4p
hamifera Maui
varipennis Molokai
Xt/+ $Xu^2$/+ 2m/+ $4f^3$/+
substenoptera Oahu
paenehamifera Maui
truncipenna Maui
neopicta Maui, Molokai
Xt/+ 2m/+ $4f^3$/+
3e 4hij 5e
Xopqs
Xik Xo/+
Xvw 3f 4q 4p/+ 5f
$Xy^2$ 4p
$Xf^3$
Xj 3d
Xo 2c
Xuxy 2d 3k 3j/+ 4o
ornata Kauai
$Xr^2s^2t^2$ 3jw $4v^2$
picticornis Kauai
Xdl
setosifrons Hawaii
Xo 5s
mimica Hawaii
5h
± 45 inversions: 6R
primaeva Kauai
$Xg^2$/+ 2gh 3x/+ 4y 5j
$Xd^2e^2f^2h^2i^2j^2k^2$ $Xg^2$/+ 2f 3x/+ 4wxz 5i
attigua Kauai
$Xg^2x^2$ 2i 3xy $4x^2y^2z^2$ 5o

FIGURE 13

**A proposed phylogeny of 103 Hawaiian species of *Drosophila* in the "picture-winged group," based on chromosome inversions. These are indicated by the notations in the boxes (rectangular for observed cases, rounded for hypothetical intermediates), with the identity of the affected chromosome (X, 1, 2 . . .) indicated by numbers. The sequence in *D. grimshawi* (near top) has been taken as an arbitrary standard. The chromosomes of *D. primaeva* (lower left: large arrow) resemble those of some North American species, and may represent an ancestral condition. Almost every species is endemic to only one of the Hawaiian Islands, as indicated. (From Carson 1983)**

lost the trochophore larval stage that is primitive in annelids. The terminal stages of an ancestor's ontogeny are sometimes lost from the ontogeny of a descendant, which may therefore depart less from the juvenile condition than its ancestor did. For example, metamorphosis from the larval to the adult form is the primitive condition in salamanders, judging from its ubiquity among the many families of salamanders; paedomorphosis (see below) in those salamanders that retain some larval characteristics throughout life is clearly a derived condition.

To say that the primitive condition within an ingroup is that condition which is found among various outgroups is not the same as saying that common characters are the most primitive. To be sure, there is sometimes such a correspondence. For example, among the monkeys and the apes only the owl monkey (*Aotus*) in the family Cebidae is nocturnal, and its nocturnal habit is doubtless a derived state; but we deduce this from the fact that the diurnal habit of all the Cebidae except *Aotus* is shared by virtually all monkeys in other families. Most vertebrates have jaws, but this does not mean that the lack of jaws in cyclostomes (lampreys and hagfish) is derived; in fact these are the only vertebrates that retain the primitive condition. Thus it is the distribution of a character among outgroups, not merely the number of species possessing it, that identifies it as a primitive trait.

The reader may have noticed that in most modern approaches to phylogenetic analysis, inferences about the polarity of evolution and the relationships among taxa are drawn solely from the pattern of distribution of features among taxa, without reference to the adaptive significance of the features, or indeed to whether they have any adaptive function at all. In the past, some authors assumed that because a feature is adaptively "important" (e.g., feathers), it must be "important" for determining relationships as well. Other authors maintained just as vehemently that the more adaptively important a feature is, the less trustworthy it is as an indication of relationship, because it would be especially prone to evolve convergently in different groups. It is indeed true that many adaptive characters display convergent evolution (e.g., radial symmetry in coelenterates and echinoderms), but some (e.g., feathers, amnion) have evolved only once. But we know this only because of the pattern of distribution of these features relative to other features. (For example, it is obvious from numerous features besides feathers that birds are a monophyletic group.) A character that serves an unusual ecological function, such as the raptorial forelegs of a praying mantis (Mantidae), is often a derived rather than an ancestral character, but again we judge this from the more general distribution of the less specialized character (e.g., the nonraptorial forelegs of most insects). It is possible that understanding the adaptive function of a feature may help in phylogenetic inference, but no one has yet specified how.

### Establishing homology and convergence

Before any phylogenetic methods can be applied, it is necessary to confront what can sometimes be a vexing problem: How do we know that the characters of two species are homologous, that is, that they have been derived (with modification) from a characteristic in a common ancestor? Sometimes this is easy; it is quite evident, for example, that the leg bones of humans and monkeys are homologous, despite minor differences in size, shape, and muscle insertions. But it is less evident that the bones of a bird's forelimb are homologous to those of a reptile's. The position of a structure relative to other features is often used (Wiley 1981). Even though the bones of a bird's wing are highly modified relative to those of a reptile's foreleg, the spatial relationships among the bones identify them as humerus, carpals, and so forth. Very often the embryological development of a feature provides better evidence of homology than its final form; this is the basis on which the ear ossicles of mammals are considered homologous to the jaw elements of reptiles—a conclusion that is supported by intermediate fossils.

Often a similar character state in two species is revealed to be convergent rather than homologous by detailed anatomical analysis; the construction of the beak of a mosquito, for example, is quite different from that of a bug (Figure 8). Knowing this, a systematist who set out to determine the relationships among orders of insects would not use the character "beak," but would define the state of each of the component mouthparts as a separate character. Sometimes, however, the details of anatomy or embryology do not help, as when an ancestral character has been lost and the question is whether it was lost more than once. In this case one must rely on an argument from parsimony, as in the case of winglessness in moths and wasps described earlier.

A serious problem in phylogenetic analysis is determining whether different measurements are really independent characters, because parsimony analysis depends on the number of shared derived characters. Because of genetic correlations (Chapter 7), characters may change in concert, and so constitute only a single characteristic. For example, many groups of salamanders have undergone PAEDOMORPHOSIS (Chapter 14) whereby descendant species retain throughout their reproductive lives a suite of correlated features that are larval characteristics in their ancestors. If all these many characters are treated separately, parsimony analysis will classify several paedomorphic families of salamanders as a single monophyletic group. But if the entire suite of paedomorphic traits is treated as a single character, other traits (which then assume greater relative importance) show that the several paedomorphic families do not form a monophyletic group (Hecht and Edwards 1977).

To summarize, it is possible to eliminate some of the "noise" from a phylogenetic analysis by excluding a priori characters that are known to be prone to convergence (e.g., leaflessness in desert plants), or by recoding characteristics that are shown to be convergent by anatomical or embryological analysis. One then performs an analysis (e.g., parsimony or compatibility analysis) that will give a "most likely" phylogenetic tree that will reveal some of the remaining homoplasies.

## PHYLOGENETIC INFERENCE FROM MACROMOLECULES

An increasingly valuable source of information for phylogenetic inference is provided by proteins and nucleic acids. The amount of information is potentially enormous; for example, every base pair in a DNA sequence is a separate character. Several kinds of data, and several kinds of analysis, have been used (reviewed by Fitch 1977a, Wilson et al. 1977, Felsenstein 1982, 1985b).

The most direct approach is to determine directly the nucleotide sequence of one or more homologous genes in the species whose phylogeny is sought. Partial sequence data can be obtained from the use of restriction enzymes, which cleave DNA at particular oligonucleotide sequences (Chapter 3). A segment of DNA, such as the mitochondrial DNA, can be cleaved into fragments in each of several species. For each such restriction enzyme, comparison of the fragments reveals the extent to which the species differ in the distribution of the enzyme-specific nucleotide sequences.

A less direct, and technically much more tedious, estimate of differences in nucleotide sequence can be obtained from the amino acid sequence of homologous proteins such as cytochrome *c*. By reference to the RNA code, the minimal number of nucleotide substitutions required to account for each amino acid difference can be inferred. As with direct estimates of nucleotide sequences, this is a minimal estimate because reversals (e.g., A → G followed by G → A) cannot be detected. Thus even if the rate of substitution were constant, sequence differences between species would not be linearly related to time since they diverged from a common ancestor, but rather would tend to level off with time. Techniques for correcting for reversals have been suggested.

In addition to direct sequence data, several indirect measures of molecular divergence have been used. Separated strands of DNA of each of two species will form duplexes (can be "hybridized") in vitro. The stability of such duplexes is greater the more similar the nucleotide sequences of the hybridizing strands. This stability can be measured by the temperature required to destabilize the duplexes.

Many authors have used dissimilarity of allozyme frequencies over a number of loci as a measure of the "genetic distance" among species (reviewed by Avise and Aquadro 1982, Avise 1983, Buth 1984). Nei's (1971) measure *D* (Chapter 4) is the most frequently used index of difference, although there are others (Thorpe 1982, Felsenstein 1985b). Electrophoretic data are most useful for very closely related species.

An indirect measure of difference in amino acid sequence of homologous proteins is the "immunological distance" found by comparing the protein's relative affinity to antibodies against that of the protein of a standard species. Immunological distances among vertebrates, found by comparing their serum albumins, are rather well correlated with "genetic distances" found with electrophoretic data (Sarich 1977, Highton and Larson 1979, Wyles and Gorman 1980). The "albumin immunological distance" is moderately well correlated with divergence times estimated from paleontological data (Wilson et al. 1977). Nevertheless, discrepancies occur and are to be expected if only a single protein is used.

Molecular sequence data can be used to estimate phylogenies by any of several cladistic techniques, i.e., those that search for *derived* characters, which for sequence data are nucleotide substitutions (reviewed in Fitch 1977a, Felsenstein

1982). Indirect measures of molecular difference—"distance" measures such as Nei's *D*—are based on an overall value of similarity or dissimilarity. As with phenetic measures of overall morphological similarity, these distance measures will correlate with time since divergence only if the rate of divergence is approximately constant. Hence the question is raised of whether or not nucleotide substitutions occur at a constant rate—of whether or not there is a "molecular clock."

### Molecular clocks

Over long periods of time, molecular evolution might occur at an apparently constant rate even if it is caused by natural selection, simply because of an averaging of selection coefficients over millions of generations (Lewontin 1974a). However, the theory of genetic drift (Chapter 5) explicitly predicts that purely neutral mutations will be substituted in an equilibrium population at a rate equal to the rate of mutation per generation (Kimura 1983a). Empirical tests of the constancy of divergence at the molecular level have been of two kinds. One test is to plot a measure of genetic distance between species, or the number of nucleotide differences that are either determined directly or are inferred from amino acid differences in one or more proteins, against the time of divergence as estimated from the fossil record. A strong linear correlation would imply a constant rate of divergence per year. For example, an analysis of this kind using combined data from the amino acid sequences of seven proteins from a variety of mammal species indicated that nucleotide substitutions have occurred at a rate only about twice as variable as that expected of a purely random process (Fitch 1977b; Figure 14 in Chapter 5). The correlation between molecular distance and paleontological times of divergence can then be used as a calibration to estimate the time of divergence between forms that have a poor fossil record.

On this basis, advocates of the molecular clock hypothesis have suggested that one unit of immunological distance based on serum albumin corresponds to 0.54–0.58 Myr (millions of years) since divergence. For the genetic distance *D* based on electrophoretic data, a distance of $D = 1$ is often quoted as corresponding to about 18 Myr, although this depends on which enzymes are studied, since some evolve faster than others. However, widely varying estimates of the relation between *D* and divergence time have been used by different authors (Avise and Aquadro 1982), and there are few independent estimates of time since speciation that allow an accurate calibration of electrophoretic divergence. The genetic distance *D* between pairs of fish species on either side of the Isthmus of Panama, which arose 2–5 Myr ago, conformed approximately to the clock hypothesis (Vawter et al. 1980), but the genetic distance between pairs of sea urchins conformed less well (Lessios 1981). Some of the argument about the constancy or inconstancy of the molecular clock arises because paleontological data are often inadequate to provide precise estimates of divergence times by which to calibrate molecular differences (Radinsky 1978a, Novacek 1982).

The second way of testing for a constant rate of divergence is the RELATIVE RATE TEST (Wilson et al. 1977), which is independent of paleontological data. If molecules diverge at a constant rate, the "distance" (e.g., in molecular substitutions) should be equal from an outgroup to each of the members of an ingroup (Figure 14). For example, Wilson et al. used nonprimate mammals as an outgroup

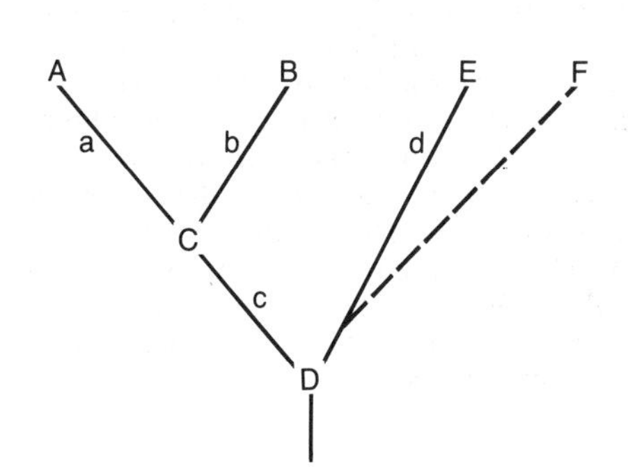


FIGURE 14

**The relative rate test for constancy of the rate of molecular divergence. Sequences (or other data) are obtained for the living species A and B and for the outgroup species E (perhaps also F). C and D represent common ancestors. The genetic distance (e.g., in terms of nucleotide differences) between A and E is $D_{AE} = a + c + d$. That between B and E is $D_{BE} = b + c + d$. If the rate of nucleotide substitution is constant, $a = b$, so $D_{AE} = D_{BE}$. If constancy of rate holds throughout the tree, the distance between any pair of species that have D as a common ancestor will equal that between any other pair of species (e.g., $D_{AE} = D_{BF}$).**

to compare the rate of divergence between the lineage leading to Old World monkeys (Cercopithecoidea) and that leading to humans and apes (Hominoidea). The rate of evolution of amino acid sequences has been slower for some proteins in the Hominoidea than the Cercopithecoidea, but faster for other proteins, so that for the combination of five proteins, the two groups of primates have diverged at similar rates. Immunological data gave a similar result.

The apparent constancy of divergence is a constancy with respect to time. If taxa vary in generation time, however, we might expect the number of substitutions per million years to be higher in organisms with short generation time than long generation time, because mutations are generally thought to occur at a roughly constant rate per generation (Chapter 3). Wu and Li (1985) applied the relative rate test to nucleotide sequence data for 11 genes in rodents (mouse or rat) and humans, using cattle and other mammals as outgroups. They discovered that in most of the genes, rodents have diverged more from the outgroup than humans have, implying a higher evolutionary rate that Wu and Li ascribed to the rodents' shorter generation time. The disparity in rate was greater for synonymous than for nonsynonymous nucleotide substitutions. To account for this, Wu and Li propose that rodents have larger effective population sizes ($N_e$). Because slightly deleterious mutations are more likely to be eliminated by selection in large than small populations (Chapter 8), and because nonsynonymous substitutions are more likely to have deleterious effects than synonymous substitutions, a smaller proportion of the nonsynonymous substitutions are likely to be fixed by genetic drift in rodents than in primates. Whatever the reason for the disparity, these results conform to other evidence that the rate of evolution of several proteins has declined in the hominoid primates compared to other vertebrates (Goodman et al. 1982).

## Examples of phylogenetic inference from molecular data

Many of the phylogenetic trees derived from molecular data concur with those previously established from morphological information. For example, one of the first such phylogenetic trees, based on amino acid sequences of the slowly evolving protein cytochrome *c*, resembles the traditional phylogeny in all but a few

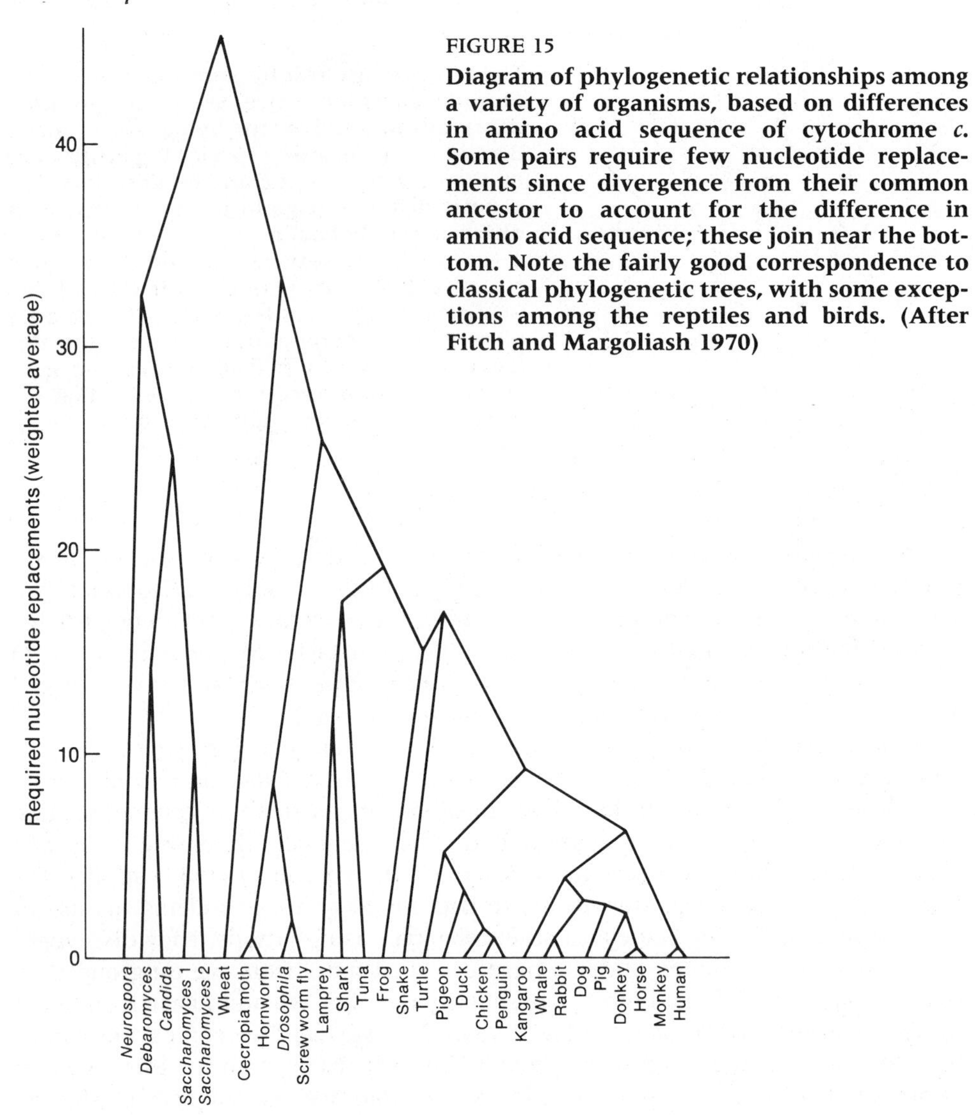


FIGURE 15

**Diagram of phylogenetic relationships among a variety of organisms, based on differences in amino acid sequence of cytochrome *c*. Some pairs require few nucleotide replacements since divergence from their common ancestor to account for the difference in amino acid sequence; these join near the bottom. Note the fairly good correspondence to classical phylogenetic trees, with some exceptions among the reptiles and birds. (After Fitch and Margoliash 1970)**

details (Figure 15). (The position of the turtle, for example, is almost certainly in error, but after all this tree is based on only a single protein.)

In some cases, morphological and molecular data are in conflict. For example, the chachalaca (*Ortalis*), a neotropical bird that is similar morphologically to chickens and pheasants, traditionally has been placed with them in the order Galliformes. Immunologically, however, its lysozyme is more similar to that of a duck (order Anseriformes) than to that of a chicken. One might dismiss the limited data provided by a single protein, but three other proteins give the same result, suggesting that the relationships of the chachalaca may need to be reassessed (Wilson et al. 1977).

A

B

C

FIGURE 16

**Representative plethodontine salamanders. (A) *Plethodon neomexicanus.* (B) *Ensatina eschscholti.* (C) *Aneides lugubris. Ensatina* is readily recognized by the constriction at the base of the tail, and this species of *Aneides*—which departs furthest from the ancestral condition represented by *Plethodon*—has an expanded skull and truncated digits. (Photographs by E.D. Brodie/BPS)**

The tribe Plethodontini is a group of terrestrial North American salamanders comprising the genera *Plethodon, Ensatina,* and *Aneides* (Figure 16). Wake (1966) had postulated that *Plethodon* resembles the ancestor of the group, that *Ensatina* is a morphologically divergent derivative of the ancestral stock, and that *Aneides,* which is morphologically adapted for arboreal locomotion, is a specialized derivative of *Plethodon* (meaning, incidentally, that *Plethodon* is a paraphyletic taxon). Using other salamanders as an outgroup, Larson et al. (1981) confirmed Wake's hypothesis by a cladistic analysis based on morphological characters (Figure 17A), and found that phylogenetic trees based on immunological and electrophoretic data confirmed the morphological analysis in most respects (Figure 17B,C). Assuming that one unit of immunological distance corresponds to 0.58 Myr of separation, and one unit of genetic distance in the electrophoretic data corresponds to 14 Myr, the authors concluded that *Plethodon* is between 48 and 64 Myr old (Eocene to Paleocene) and that *Aneides* arose from *Plethodon* between 24 and 38 Myr ago (in the Oligocene). Wake (1966) had already suggested, based on biogeographical and paleontological data, that *Aneides* separated from *Plethodon* in the Oligocene, and the recent discovery of both *Plethodon* and *Aneides* fossils in slightly younger (lower Miocene) rocks confirms this belief. If the divergence times calculated from the molecular data are correct, all the species of *Aneides* arose within a period of 20–24 Myr during the late Oligocene and Miocene. There was a burst of adaptive change and morphological divergence within *Aneides* while the genus *Plethodon* remained virtually unchanged.

Many relationships among humans, the orangutan (*Pongo pygmaeus*), and the

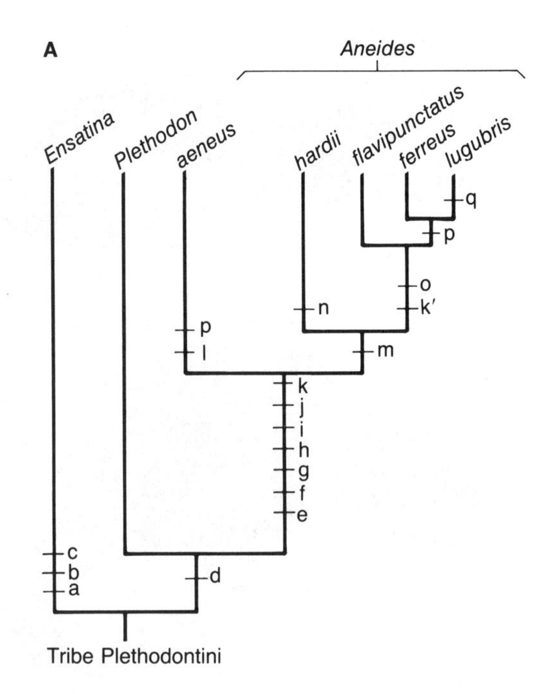


FIGURE 17

**(A) Cladogram of salamanders of the tribe Plethodontini (Figure 16), based on 17 morphological characters. Letters mark transitions from the ancestral character states implied at the base of the tree. *Plethodon* retains ancestral states except for character d. Only one character (p, in lineages leading to *Aneides aeneus* and the ancestor of *A. ferreus* and *A. lugubris*) exhibits parallelism. (B) Phylogenetic tree of some of these species, based on immunological distance data from serum albumin. (C) Phylogenetic tree based on genetic distances calculated from enzyme electrophoresis. The cladogram (A) is based on shared derived characters, whereas the other phylogenetic diagrams are based on overall similarity versus difference. The phylogeny based on morphology differs from the other two estimates only in the position of *A. lugubris*. (After Larson et al. 1981)**

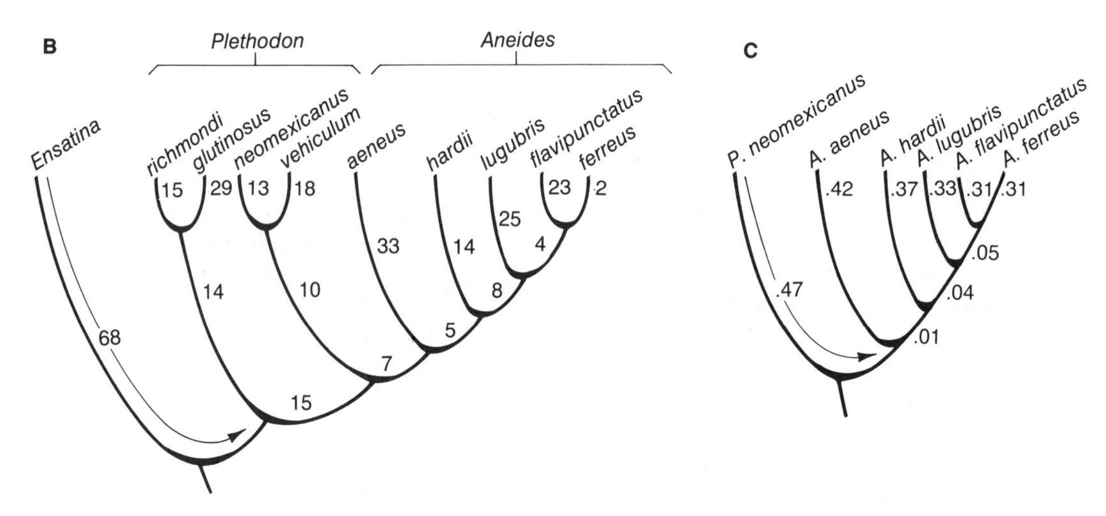


African apes (gorilla, *Gorilla*, and chimpanzee, *Pan*) have been suggested by different authors (Figure 18; see Sibley and Ahlquist 1984 for review). Kluge (1983), using a parsimony analysis (Wagner tree), primarily of morphological characters, proposed that humans diverged before the orangutan and the African apes diverged from each other (tree 5 in Figure 18). Most other authors have concluded that the orangutan diverged before the split between humans and African apes, but numerous molecular analyses have been unable to resolve the relationships among human, gorilla, and chimpanzee (tree 4). Clearly these three lineages branched closely together in time.

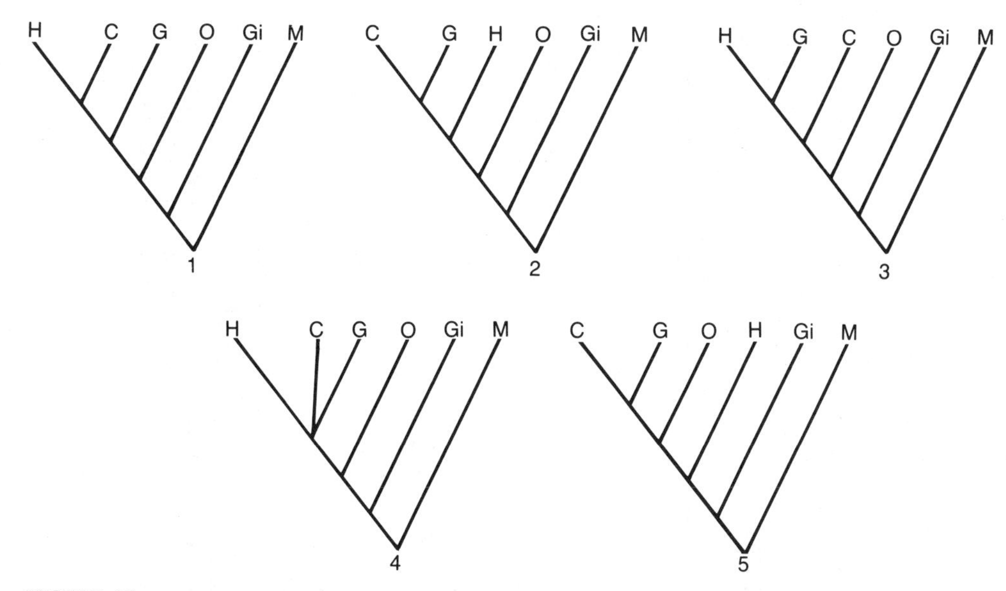


FIGURE 18

**Some phylogenetic relationships that have been proposed for the Hominoidea. H, human (*Homo*); C, chimpanzee (*Pan*); G, gorilla (*Gorilla*); O, orangutan (*Pongo*); Gi, gibbons (Hylobatidae); M, Old World monkeys (Cercopithecoidea), an outgroup. Tree 5 has been suggested by some analyses of morphology, tree 2 by an analysis of restriction enzyme data, tree 1 by hybridization of single-copy DNA.**

Templeton (1983) claimed that tree 2, in which *Homo* is the sister group of the African apes, is most likely, using data from restriction enzyme analysis of mitochondrial DNA and hemoglobin genes. For example, if a certain DNA fragment is present in the *Eco-R1* digest of species 1 and is absent in species 2, we know that species 1 has the sequence GAATCC, but that species 2 has different bases at the first two nucleotide sites, where this enzyme cleaves. Because more mutations will alter a specific sequence than will create it from another sequence, convergence due to losses of specific nucleotide sequences should be more common than convergence due to gains of such sequences. Using data from each of 19 restriction enzymes and taking the gibbon as an outgroup, Templeton constructed a separate maximum parsimony cladogram from each of the 19 enzymes by assuming that convergent gains are less likely than convergent losses. By compatibility analysis, he concluded that phylogeny 2 was consistent with more of these cladograms than any other phylogeny, and provided a statistical analysis to show that it was significantly more consistent.

Templeton's analysis has been criticized by Nei and Tajima (1985), who argue that the maximim parsimony procedure does not give the most reliable estimate of the phylogeny if nucleotide substitutions are fixed at so high a rate that convergences are abundant. Since, moreover, several of the restriction enzymes recognize similar DNA sequences, they do not provide independent data. When the redundant data are excluded, trees 1 and 2 become equally likely. Tree 1, in fact, is supported by sequence data from several proteins, by chromosome band-

TABLE I

**Distance values among primates determined from DNA-DNA hybridization** [a]

| | Human (*Homo sapiens*) | Pygmy chimpanzee (*Pan paniscus*) | Common chimpanzee (*Pan troglodytes*) | Gorilla (*Gorilla gorilla*) | Orangutan (*Pongo pygmaeus*) | White-handed gibbon (*Hylobates lar*) |
|---|---|---|---|---|---|---|
| Pygmy chimpanzee | 1.9 $n = 7$ (1.65 / 11) | | | | | |
| Common chimpanzee | 1.8 $n = 12$ (1.63 / 64) | 0.7 $n = 6$ (0.73 / 6) | | | | |
| Gorilla | 2.4 $n = 10$ (2.27 / 69) | 2.3 $n = 6$ (2.37 / 11) | 2.1 $n = 10$ (2.21 / 69) | | | |
| Orangutan | 3.6 $n = 11$ (3.62 / 24) | 3.7 $n = 9$ (3.56 / 13) | 3.7 $n = 10$ (3.56 / 20) | 3.8 $n = 10$ (3.54 / 22) | | |
| White-handed gibbon | 5.2 $n = 10$ (4.76 / 17) | 5.6 $n = 5$ (4.87 / 6) | 5.1 $n = 10$ (4.82 / 18) | 5.4 $n = 12$ (4.72 / 24) | 5.1 $n = 15$ (4.83 / 20) | |
| Cercopithecid monkeys | 7.7 $n = 9$ (7.36 / 22) | 8.0 $n = 6$ (7.01 / 11) | 7.7 $n = 9$ (7.30 / 22) | 7.5 $n = 5$ (7.18 / 28) | 7.6 $n = 4$ (7.43 / 15) | 7.4 $n = 3$ (7.25 / 8) |

[a]The values not included in parentheses are those published by Sibley and Ahlquist (1984). The values in parentheses, based on larger samples, are by Sibley and Ahlquist (in preparation) and were kindly provided by Charles G. Sibley.

Each entry is the average delta $T_{50}H$ value for a number ($n$) of comparisons among individual organisms. Delta $T_{50}H$ is the depression in the temperature (in degrees C) required to achieve 50 percent dissociation of hybrid DNA duplexes, relative to the temperature required to achieve this for native (single-species) DNA. Most of the DNA in these tests is single-copy. Sibley and Ahlquist present standard errors (ranging from 0.05 to 0.2) for these mean values. The variations are ascribed to experimental error.

ing patterns, and by Sibley and Ahlquist's (1984) data on DNA-DNA hybridization (Table I).

Sibley and Ahlquist's data illustrate particularly well the relative rate test for constancy of the molecular clock. In the phylogeny proposed by Sibley and Ahlquist, each taxon listed on the left side of Table I is an outgroup relative to all the taxa above it, and so should have the same distance value to each of these taxa if divergence has occurred at a constant rate. This is very nearly true in all cases; for example, taking the orangutan as an outgroup relative to human, pygmy chimpanzee, common chimpanzee, and gorilla, the distance values (3.6, 3.7, 3.7, and 3.8 respectively) are the same, within the range of experimental error. Sibley and Ahlquist (1983) have used this technique to estimate relationships among many families of birds, and find that these too pass the relative rate test for constancy of divergence rate. It is interesting that DNA hybridization data do not show an effect of generation time.

Sibley and Ahlquist (1984) propose that DNA hybridization values must almost inevitably show a uniform rate of divergence because this will be "the inevitable statistical result of averaging over billions of nucleotides and millions of years." If they are correct, we humans are the chimpanzee's closest relatives (see also Chapter 17).

## SUMMARY

The analysis of virtually all questions about the history of evolution requires that the phylogenetic history of groups of species—the relationships among them—be inferred. Fossils are sometimes helpful, but are usually inadequate for this task. Various techniques, however, make it possible to infer phylogenetic relationships among living species. The chief obstacles to phylogenetic analysis are unequal rates of evolution and convergent evolution, but these problems can be minimized if the direction of evolution of many characters can be specified. Macromolecular data are a rich source of information for phylogenetic analysis.

## FOR DISCUSSION AND THOUGHT

1. Reproductive characters such as those that serve as isolating mechanisms are important for distinguishing species, but are not necessarily any more useful than other characters in assessing the phylogenetic relationships among species and higher taxa. Why not?
2. Much (but by no means all) of the higher classification of organisms, developed in pre-Darwinian times, has remained rather little changed by the introduction of evolutionary thought into systematics. Why?
3. Higher taxa that are defined by shared uniquely derived character states often include species that do not have the definitive character states. For example, the insect suborder Hemiptera (true bugs) is distinguished by partially sclerotized forewings, yet many hemipterans such as bedbugs possess no wings at all. Explain how their affinities can be determined.
4. Is it possible to draw a phylogenetic tree to describe genealogical relationships among intraspecific groups such as subspecies? What are the difficulties of such an attempt?
5. This discussion of phylogeny has assumed that diversity arises by irreversible splitting of lineages between which interbreeding ceases. How can a phylogenetic diagram accomodate cases (as in many plants) in which a new species arises by hybridization?

6. Review the history of a taxonomically difficult group to determine why authorities differ in their judgments of phylogenetic affinities and how they justify their conclusions. An illuminating case is that of the frogs. Compare treatments by Griffiths (1963), Inger (1967), Kluge and Farris (1969), and Savage (1973).
7. How can one tell whether a reduced character is vestigial (a derived state) or incipiently evolving (an ancestral state with reference to a more elaborated condition in other species)?
8. Some systematists have argued vehemently that classifications should not include paraphyletic taxa (e.g., taxa such as the Reptilia from which some descendants, such as birds, have been excised). They claim that paraphyletic groups "do not exist" because they are defined by plesiomorphic characters (e.g., scales rather than feathers) instead of by synapomorphic characters. Is this debate about scientific issues or about conventions of classification?
9. A few cladists (who have come to be known as "transformed cladists" or "pattern cladists") argue that the goal of systematics should be to construct classifications based on synapomorphies, but that it is neither necessary nor desirable that evolution be assumed in order to construct such classifications (Patterson 1982, Nelson and Platnick 1984). Discuss.
10. It is commonly asserted that sexual differences in behavior must have existed in the ancestor of the human species, since they commonly exist in other primates. Some authors would go further and assert that such a genetically based pattern of behavior has been inherited by modern humans with little change. Apply the principles of this chapter to evaluate these arguments.

## MAJOR REFERENCES

Hennig, W. 1979. *Phylogenetic systematics.* University of Illinois Press, Urbana. 263 pages. A translation of Hennig's 1966 book from the German, in which many of the principles of cladistic analysis and classification are established.

Stevens, P.F. 1980. Evolutionary polarity of character states. *Ann. Rev. Ecol. Syst. 11:* 333–358. A useful discussion of the criteria for establishing the direction of evolutionary change.

Eldredge, N., and J. Cracraft. 1980. *Phylogenetic patterns and the evolutionary process.* Columbia University Press, New York. 349 pages. Describes the applications of phylogenetic analysis to the study of evolution.

Felsenstein, J. 1982. Numerical methods for inferring evolutionary trees. *Quart. Rev. Biol.* 57: 379–404. The most general review of the theoretical bases of the methods of phylogenetic inference.

Wiley, E.O. 1981. *Phylogenetics: The theory and practice of phylogenetic systematics.* Wiley, New York. 439 pages. Describes the practices of cladistic systematists, especially the use of parsimony.

# The Fossil Record

# Chapter Eleven

The phylogenetic analysis of living organisms provides indirect evidence of many aspects of their evolutionary history. The only direct evidence of evolutionary history, however, is provided by the fossil record. It is in the sedimentary rocks deposited in ages past that we find not only glimpses of the ancestry of living organisms, but the traces of innumerable others that strutted their hour and then were heard no more. The record spans almost inconceivably vast time: the earliest archaeological record of human agriculture is about 12,000 years old, yet this is less than 1/8000 of the time since *Tyrannosaurus rex* walked the earth, and *Tyrannosaurus* was a latecomer to the evolutionary play. Whole communities of organisms quite different from those of today have arisen and perished since the days of the dinosaurs; continents have moved, sea level has risen and dropped, climates have changed, glaciers have scoured the continents.

In this light, statements about evolution that at first seem farfetched become entirely plausible. Evolutionary biologists are forever invoking unknown environmental changes to explain extinctions or the origin of adaptations; yet we know that enormous changes in both the physical and biotic environment have occurred continually for billions of years. The time available for evolution has been great enough for the most improbable of events, such as rare mutations, to have occurred repeatedly.

The record of the rocks is not a book that can be opened to the page of choice. It is distressingly incomplete, and in view of the factors that hinder the paleontologist's efforts, it is actually remarkable how good the fossil record is. Organisms without hard skeletons can be fossilized only under exceptionally favorable conditions. Even organisms with fossilizable parts are unlikely to be fossilized unless they occupy habitats like swamps or estuaries where their remains can become buried in sediments. These sediments must become compacted into rock and persist without metamorphosis or erosion for millions of years if we are to discover their contents, and they must become exposed in localities accessible to investigation. The very rarity of many fossil species, many of which are known from only single specimens, hints at the vast number of which we have no record. It is unusual to find a continuous sequence of strata spanning more than a few hundreds of thousands of years; even a moderately continuous sequence usually has gaps of thousands of years. Except under unusual circumstances, even a single bedding plane contains a "time-averaged" sample of organisms that lived years, decades, or centuries apart. Moreover, it is difficult to establish that strata in different localities were deposited at the same time; the correlation between sites may have an error of 100,000 years or more. Jablonski et al. (1986) discuss these and other limitations of the fossil record.

## DATING THE PAST

Long before Darwin convinced the scientific world of the reality of evolution, geologists had established the *relative* chronology of the geological periods, i.e., their chronological order, on the basis of the principle of superposition—that younger rocks are deposited on top of older rocks. Strata can often be assigned to geological periods, or to stages within periods, on the basis of their fossil contents—index fossils of species that persisted for so short a time that they are the signatures of the age in which they lived. Not until the early twentieth century, when radioactive decay was discovered, did it become possible to obtain *absolute* ages of rocks. Radioactive dating is based on the constant rate of decay of unstable parent nuclides into daughter nuclides. For example, half the potassium-40 nuclides incorporated into a rock during its formation will decay to argon-40 in 1300 million years; half the remaining potassium-40 nuclides will decay in the following 1300 million years, and so on. Thus the ratio of parent to daughter nuclides measures the age of the rock. The different half-lives of various unstable isotopes provide independent evidence of age. Because the estimation procedure has an error of a few percent, the precision of the estimate decreases with increasing age. Radioactive dating cannot be performed on the sedimentary rocks that contain fossils, but on igneous rocks only. The age of a sedimentary rock is found by dating igneous formations between which it is bracketed. Eicher (1976) provides an introduction to dating methods. The geological time scale presented in Table I provides a framework for the material in this and the next chapter.

The oldest rocks that have been discovered on earth have been radiometrically dated at 3.8 billion (3800 million) years. However, dating of meteorites and of moon rocks, together with other astronomical evidence, indicates that the earth and the solar system came into existence about 4.6 billion years ago.

## THE HISTORY OF THE EARTH

The environment in which organisms have evolved has undergone vast changes, in which astronomical influences, the dynamics of the earth itself, and the activities of organisms have all played a role. Before life evolved, and probably for a considerable time thereafter, the earth had a reducing atmosphere, without free oxygen; the evolution of photosynthetic organisms about 3.2 billion years ago created the oxidizing atmosphere. On the continents, organic soils are the product of communities of terrestrial vegetation, dating from the Silurian, about 438 million years (Myr) ago. Fischer (1984) discusses these and other biological effects on the earth's environment.

The arrangement of the seas and land masses has changed enormously over time because of PLATE TECTONICS. The earth's crust consists of plates about 100 km thick that move about over the mantle by the process of sea-floor spreading, whereby material from the mantle arises along oceanic ridges and pushes the plates to either side. Plates may therefore diverge along their margins; one plate may plunge under another where they converge; or two plates may move laterally with respect to each other along a fault, such as the San Andreas fault in California. Mountain belts can be produced where plates meet, and continents may be fragmented when a zone of divergence arises beneath them. The migration, or "drift," of plates and their associated continents has had enormous influences on sea levels, ocean currents, climates, and the geographic distribution of organisms (Gray and Boucot 1979, Bambach et al. 1980, Hallam 1983).

Recent evidence (Bambach et al. 1980) suggests that in the late Cambrian there were six continents, each straddling the equator. The largest of these, Gondwana, moved south and lay across the south pole, while two of the smaller continents coalesced into the land mass Laurussia, which became united with Gondwana in the Carboniferous. The continent known as China, as well as two others that fused into a single continent, moved northward and remained as island continents very near the northern edge of Laurussia. By Permian time, then, there was a single large land mass (Pangaea) at the south pole, extending northward along one face of the earth to the two small island continents. An enormous world ocean stretched from pole to pole, and at the equator was twice

TABLE I

**The geological time scale**

| Era | Period | Epoch | Millions of years from start to present | Major events |
|---|---|---|---|---|
| CENOZOIC | Quaternary | Recent (Holocene) | 0.01 | Repeated glaciations; extinctions of large mammals; evolution of *Homo*; rise of civilizations |
| | | Pleistocene | 2.0 | |
| | Tertiary | Pliocene | 5.1 | Radiation of mammals, birds, angiosperms, pollinating insects. Continents nearing modern positions. Drying trend in mid-Tertiary. |
| | | Miocene | 24.6 | |
| | | Oligocene | 38.0 | |
| | | Eocene | 54.9 | |
| | | Paleocene | 65.0 | |
| MESOZOIC | Cretaceous | | 144 | Most continents widely separated. Continued radiation of dinosaurs. Angiosperms and mammals begin diversification. Mass extinction at end of period. |
| | Jurassic | | 213 | Diverse dinosaurs; first bird; archaic mammals; gymnosperms dominant; ammonite radiation. Continents drifting. |
| | Triassic | | 248 | Early dinosaurs; first mammals; gymnosperms become dominant; diversification of marine invertebrates. Continents begin to drift. Mass extinction near end of period. |
| PALEOZOIC | Permian | | 286 | Reptiles, including mammal-like forms, radiate; amphibians decline; diverse orders of insects. Continents aggregated into Pangaea; glaciations. Major mass extinction, especially of marine forms, at end of period. |
| | Carboniferous (Pennsylvanian and Mississipian) | | 360 | Extensive forests of early vascular plants, especially lycopsids, sphenopsids, ferns. Amphibians diverse; first reptiles. Radiation of early insect orders. |
| | Devonian | | 408 | Origin and diversification of bony and cartilaginous fishes; trilobites diverse; origin of ammonoids, amphibians, insects. Mass extinction late in period. |
| | Silurian | | 438 | Diversification of agnathans, origin of placoderms; invasion of land by tracheophytes, arthropods. |
| | Ordovician | | 505 | Diversification of echinoderms, other invertebrate phyla, agnathan vertebrates. Mass extinction at end of period. |
| | Cambrian | | 570 | Appearance of most animal phyla; diverse algae |
| PRE-CAMBRIAN (SINIAN) | Vendian | | 670 | Origin of life in remote past; origin of prokaryotes and later of eukaryotes; several animal phyla near end of era. |
| | Sturtian | | 800 | |

(Numerical ages based on Harland et al. 1982.)

as wide as the Pacific Ocean is at present. Consequently, climates then were very different; much of eastern Pangaea was hot and humid, although there were glaciers in southern Pangaea in the Carboniferous and early Permian. During the Paleozoic, there were several major TRANSGRESSIONS, when the sea level rose and EPICONTINENTAL SEAS spread over parts of the continents. These transgressions, in the late Cambrian, late Ordovician, and early Carboniferous, were followed by REGRESSIONS, when sea level dropped. Sea level seems to have reach an all-time low during the regression at the end of the Permian (Hallam 1984).

Following the Permian, Pangaea separated into a different set of continents than those from which it had been formed. During the Permian, the eastern part of what is now North America was in contact with Europe and Africa, against which South America also abutted (Figure 1). During the late Triassic, Asia and Africa started to separate, and this was followed in the early Jurassic by the beginning of a westward movement of North America away from Africa and South America, although it was still connected to Europe in the north. At this time, then, North America and Eurasia formed a single northern continent, Laurasia, and the southern land mass formed a single continent known as Gondwanaland. By the late Jurassic, a narrow seaway had formed between North America and Eurasia, connecting the Arctic Ocean to the Tethys Sea, a broad seaway that separated Eurasia from Gondwanaland (Figure 1). During the Jurassic and Cre-

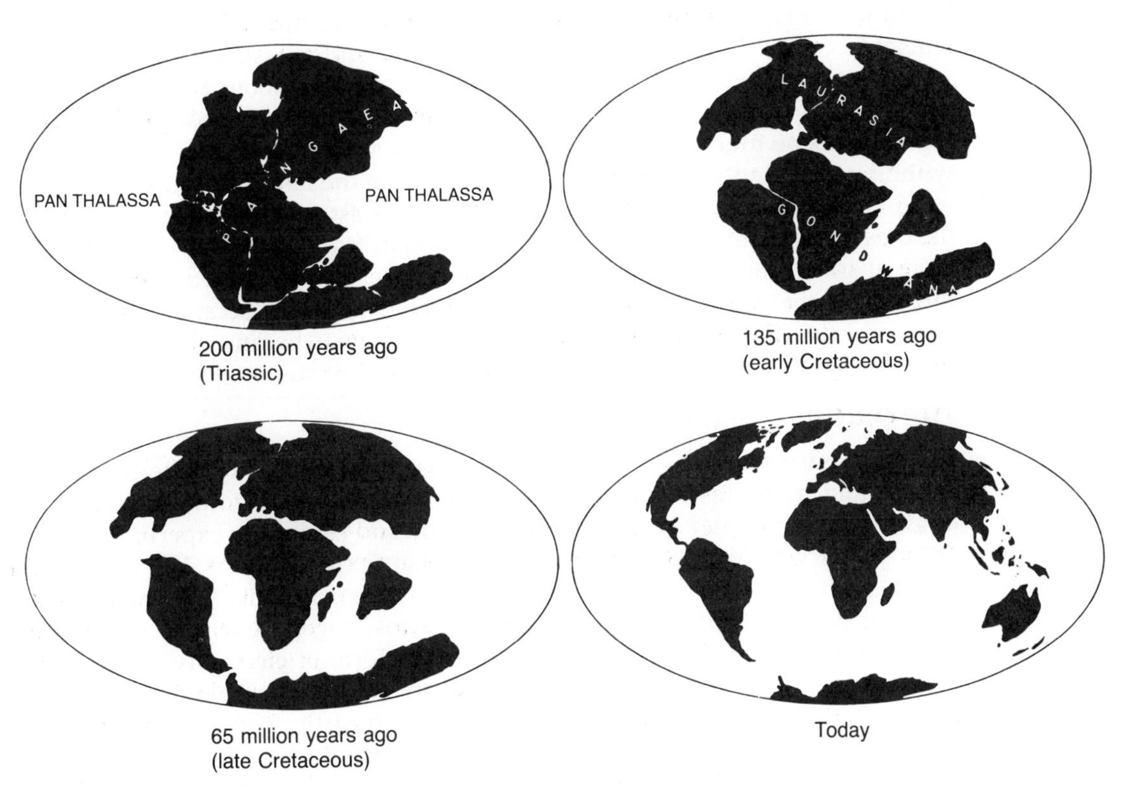


FIGURE 1
**Disposition of major land masses in the mid-Triassic, early and late Cretaceous, and the present. (After Cloud 1978)**

taceous, extensive epicontinental seas covered much of western North America, Canada, and central Eurasia.

By the mid-Cretaceous, Gondwanaland had begun to break up into Africa, South America (which however remained in contact with Africa in the north), and a land mass consisting of Australia, Antarctica, and India. By the late Cretaceous, India had broken free and was moving northward, ultimately to collide with Asia; South America had separated completely from Africa but was still narrowly joined in the south to the Antarctica-Australia land mass; and North America had moved so far to the west that it was separated fully from western Europe, but had made contact with northeastern Asia to form the Bering land bridge in the region of Alaska and Siberia. There was a massive regression of epicontinental seas in the late Cretaceous, exposing a great deal of land. During the early Tertiary, the Atlantic Ocean was considerably narrower than at present; indeed the westward movement of the Americas is still in progress. Finally, during the late Pliocene, the Panamanian isthmus arose, connecting North and South America and dividing the marine biota of the Caribbean from that of the eastern Pacific.

For much of this history, the climate of the earth was considerably warmer and less latitudinally stratified than it is today (Frakes 1979). Colder periods that included glaciations seem to have occurred only in the late Precambrian, the Carboniferous, the Permian, and the Pleistocene. The most recent cooling trend began in the early Tertiary. By Oligocene time the northern latitudes were considerably cooler and drier than they had been, and grasslands and deciduous forests became widespread. In the Pleistocene, starting about two million years ago, the climate changed drastically; the mild climate of the polar and subpolar regions became much colder, the polar ice caps formed, and glaciers spread and withdrew repeatedly over the northern portions of the northern continents. There were numerous minor glacial episodes, and at least four major ones, of which the most recent, called the Wisconsin glaciation in North America, withdrew little more than 10,000 years ago. During the glacial episodes, sea level dropped throughout the world by as much as 100 meters as water became locked in ice caps, and the climate in tropical and subtropical areas became drier; during the interglacial episodes, climates became warmer and wetter, and sea level rose.

## THE ORIGIN OF LIFE

The mechanisms by which living matter arose from nonliving materials are not to be found in the fossil record; organic chemical reactions do not fossilize. The origin of life is a matter of informed speculation and laboratory experiment (Orgel 1973 and Dickerson 1978 provide reviews). Many experimenters, beginning with Miller and Urey (1953) and Oparin (1953), have experimentally shown that the building blocks of macromolecules—amino acids, sugars, purines, and pyrimidines—will form in abundance if energy in the form of electricity or ultraviolet light is discharged into a reducing atmosphere (of, for example, $H_2$, $CH_4$, $NH_3$, and $H_2O$) like that postulated for the prebiotic earth. The amino acids will spontaneously form short polypeptides, but the conditions for spontaneous formation of nucleotides and nucleic acids appear to be considerably less likely. Monomers can become concentrated and form stable polymers by adsorbtion into mineral particles such as clay, or by evaporation or freezing. Polypeptides, moreover, will aggregate into colloidal droplets or coacervates that, some have suggested, could serve as "proto-cells" within which growth and replication might occur. These coacervates can indeed catalyze a number of organic reactions.

In the presence of RNA polymerases, short strands of RNA will become assembled from nucleotides even in the absence of a preexisting RNA template (Eigen et al. 1981); and RNA molecules will become replicated in a cell-free system of RNA templates, free nucleotides, and polymerases. Moreover, mutation and natural selection occur in such mixtures; errors in copying occur, and some RNA sequences become replicated more rapidly than others. However, it has not yet been possible to demonstrate in the laboratory the de novo origin of nucleic acid-coded polypeptides that can act as nucleic acid polymerases. Thus the origin of life has still not yielded to the efforts of chemists. It has been suggested that terrestrial life may have had an extraterrestrial origin, but this would merely push the problem back in time. But the earliest fossil evidence of life is about 3.4 billion years old, so the time available for chemical processes to yield primitive life forms, in an earthly environment of which we have very imperfect knowledge, is about a thousand million years. There is no reason to argue that the inability of chemists to synthesize life de novo in a mere thirty years of experimentation is evidence against the origin of life on earth.

## PRECAMBRIAN LIFE

The earliest fossil indication of life is in South African rocks dated at 3.4–3.1 billion years old, which contain forms that resemble bacteria, including Cyanobacteria (the blue-green bacteria or "algae"), and stromatolites—mound-like structures that are still formed in parts of Australia by Cyanobacteria. The earliest known organisms, then, were prokaryotes, apparently capable of photosynthesis. (This does not imply that the first organisms were photosynthetic; they were probably heterotrophic.) By about two billion years ago, photosynthetic activity had created an oxygen-rich atmosphere, which must have led to the demise of many anaerobic early organisms. Some such organisms, such as methanogenic bacteria and their relatives, still persist in anaerobic environments, and are so different from other bacteria in the sequence of their rRNA that Woese (1981) has suggested that they have constituted a separate, anaerobic, lineage for 3.5 billion years (Figure 2).

The Cyanobacteria and other prokaryotes appear to have held sway for almost two billion years (see Schopf 1983 on Precambrian life). The earliest known eukaryotes, probably green algae, are in 0.9 billion-year-old rocks, although there is some evidence that they go back to about 1.5 billion years.

The origin of eukaryotes is a major event in the history of life, for it marks the evolution of chromosomes, meiosis, and organized sexual reproduction. Not without challenge (e.g., Raff and Mahler 1972), Margulis (1970) has argued from biochemical and structural evidence that the eukaryotic cell evolved in part by symbiosis: that mitochondria and plastids, and perhaps other organelles as well, are descended from prokaryotes that became incorporated as intracellular symbionts. Both the circular form and the nucleotide sequence of mitochondrial and plastid DNA, for example, have similarities to those of bacteria. How eukaryotic chromosomes and meiosis evolved, however, is entirely mysterious.

Almost certainly the several kingdoms into which many authors classify

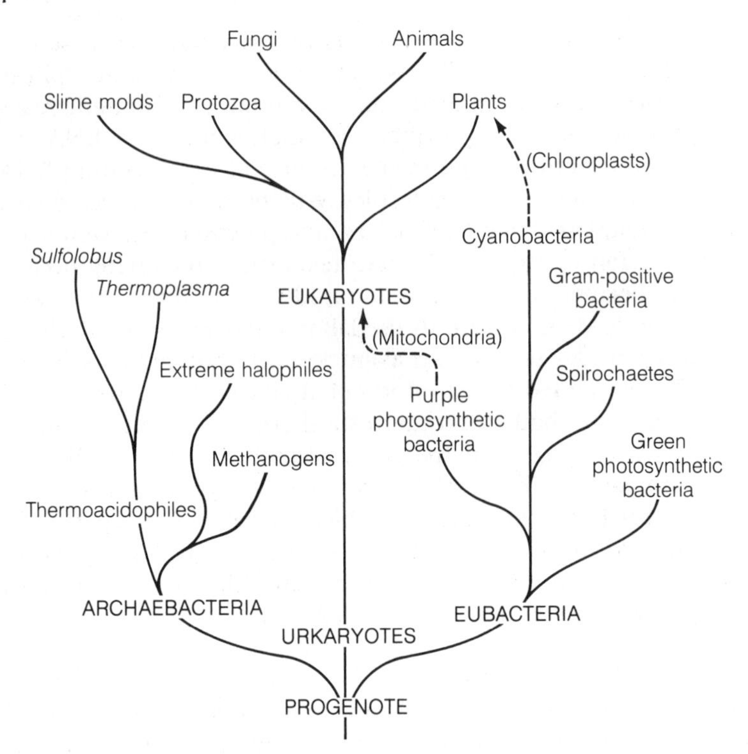


FIGURE 2

**A possible phylogeny of major groups of organisms, based on differences in the nucleotide sequences of 16S ribosomal RNA. Archaebacteria are as distinct from other prokaryotes (termed Eubacteria in this classification) as the latter are from eukaryotes. The broken lines represent derivation of mitochondria and chloroplasts from bacteria that are thought to have been intracellular symbionts of eukaryotes. (After Woese 1981)**

eukaryotes—fungi, several kingdoms of protozoans and of algae, plants, and animals—became differentiated during the Precambrian, but their fossil record is far too fragmentary to document their origins. Only the kingdom Animalia has a promising early fossil record. The first complex, multicellular animals, known as the Ediacaran fauna (Cloud and Glaessner 1982), are dated at about 640 Myr B.P. (million years before present). These fossils include traces of burrows and tracks and a number of soft-bodied animals, some of which have been interpreted as polychaete annelid worms, coelenterates, and possibly soft-bodied arthropods. Because they differ considerably from later members of these groups, however, their identification with these phyla has been questioned. Certainly the Ediacaran fauna includes other organisms that seem not to fall into the later phyla, and it seems likely that this fauna represents an early animal radiation that was largely extinguished (Seilacher 1984).

## THE PALEOZOIC ERA

Phanerozoic time embraces all of earth's history since the end of the Vendian period of the Precambrian, and is divided into the Paleozoic, Mesozoic, and Cenozoic eras. Because the eras and periods of the Phanerozoic were originally

FIGURE 3

**(Left) A trilobite, *Paradoxides davidis,* of the Cambrian. (Photograph courtesy of R. Levi-Setti) (Right) A brachiopod, *Gotatrypa orbicularis,* of the Silurian. (Photograph courtesy of P. Copper)**

defined by their distinct fossil faunas, the borders between them are marked by major episodes of extinction or by the diversification of major new groups of animals.

The appearance of abundant, diverse forms of invertebrates marks the beginning of the CAMBRIAN period (about 590 Myr B.P.). During the Cambrian, all the animal phyla that have fossilizable skeletons appear, many in profusion (see Valentine 1977 and House et al. 1979 on metazoan phylogeny and paleontology). These include the Arthropoda, represented by the trilobites (Figure 3), the Brachiopoda (lamp shells; Figure 3) the Mollusca (including gastropods, bivalves, and cephalopods), the Porifera, a great variety of classes of Echinodermata, and several of the "minor phyla" (e.g., Nemertea and Pogonophora) that apparently have existed at low diversity from the Cambrian to the present. Cnidarians and annelids, perhaps present in the Ediacaran fauna, were well established in the Cambrian. They are among the remarkably well preserved soft-bodied animals found in the Burgess Shale of British Columbia (ca. 530 Myr), which includes animals in about ten extinct phyla that are known only from this formation (Conway Morris and Whittington 1979; Figure 4). The Burgess Shale also includes the earliest known chordate (*Pikaia*), an animal that bears some resemblance to *Amphioxus.* The origin of vertebrates, however, is not well recorded by fossils. The earliest vertebrate remains, in 510 million-year-old marine deposits of the upper Cambrian (Repetski 1978), are fragments of the external armor of ostracoderms, a group of jawless, finless "fishes" that are more abundantly represented in the Ordovician.

It is considered likely that all the animal phyla became distinct before or during the Cambrian, for they all appear fully formed, without intermediates connecting one phylum to another. Thus our understanding of the phylogenetic relationships among the phyla, which are a matter of some dispute (Figure 5), is based on inferences from their anatomy and embryology. The rapid origin of the

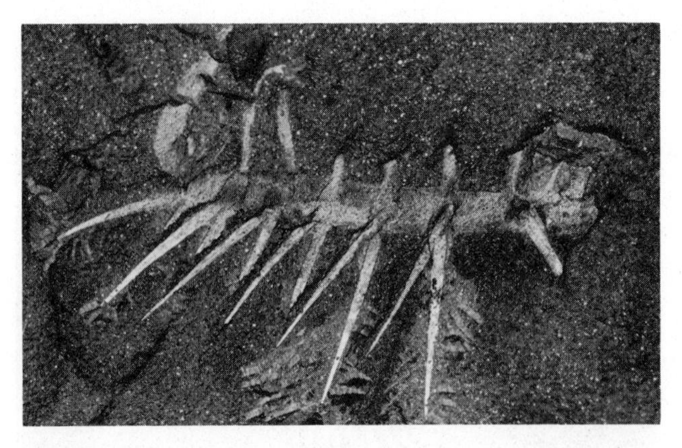

FIGURE 4

**Two peculiar animals from the Burgess Shale of British Columbia, a Cambrian deposit dated at ca. 530 Myr. These animals are not assignable to extant phyla, and are thought to represent phyla now extinct. (Left) *Hallucinogenia sparsa* had seven pairs of spines, with which it may have walked, and dorsal tentacles that perhaps were used for feeding. (Right) *Wiwaxia corrugata*, a flattened, crawling animal that probably fed by scraping. The peculiar scales and spines are unlike those of any other organism. (Photographs courtesy of S. Conway Morris)**

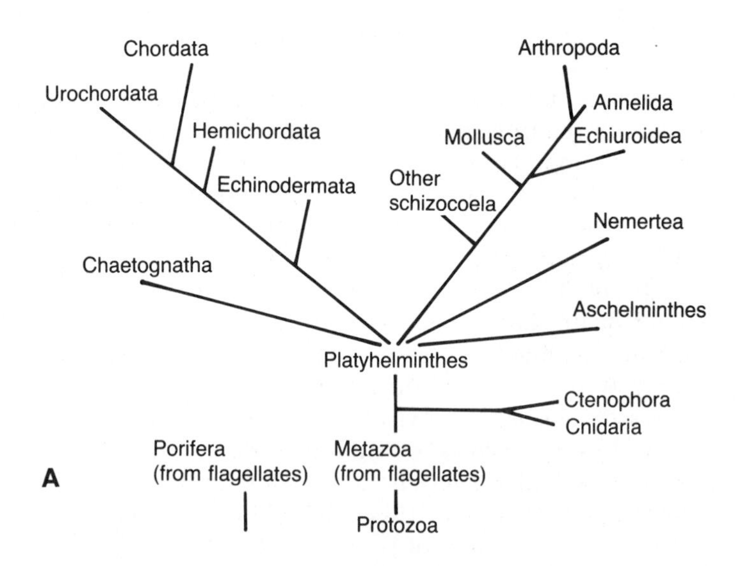


FIGURE 5

**Several authors' suggestions of the phylogenetic relationships among animal phyla. None of these is an explicit cladistic analysis; each relies primarily on the author's proposition that certain characters provide more phylogenetic information than others. Modern techniques have not yet been used to infer the phylogeny of the phyla. [After Valentine 1977. Based on interpretations by (A) L.H. Hyman 1940, (B) J. Hadzi 1963, (C) L. von Salvini-Plawen 1969, and (D) G. Jägersten 1972. See also Nielsen 1985.]**

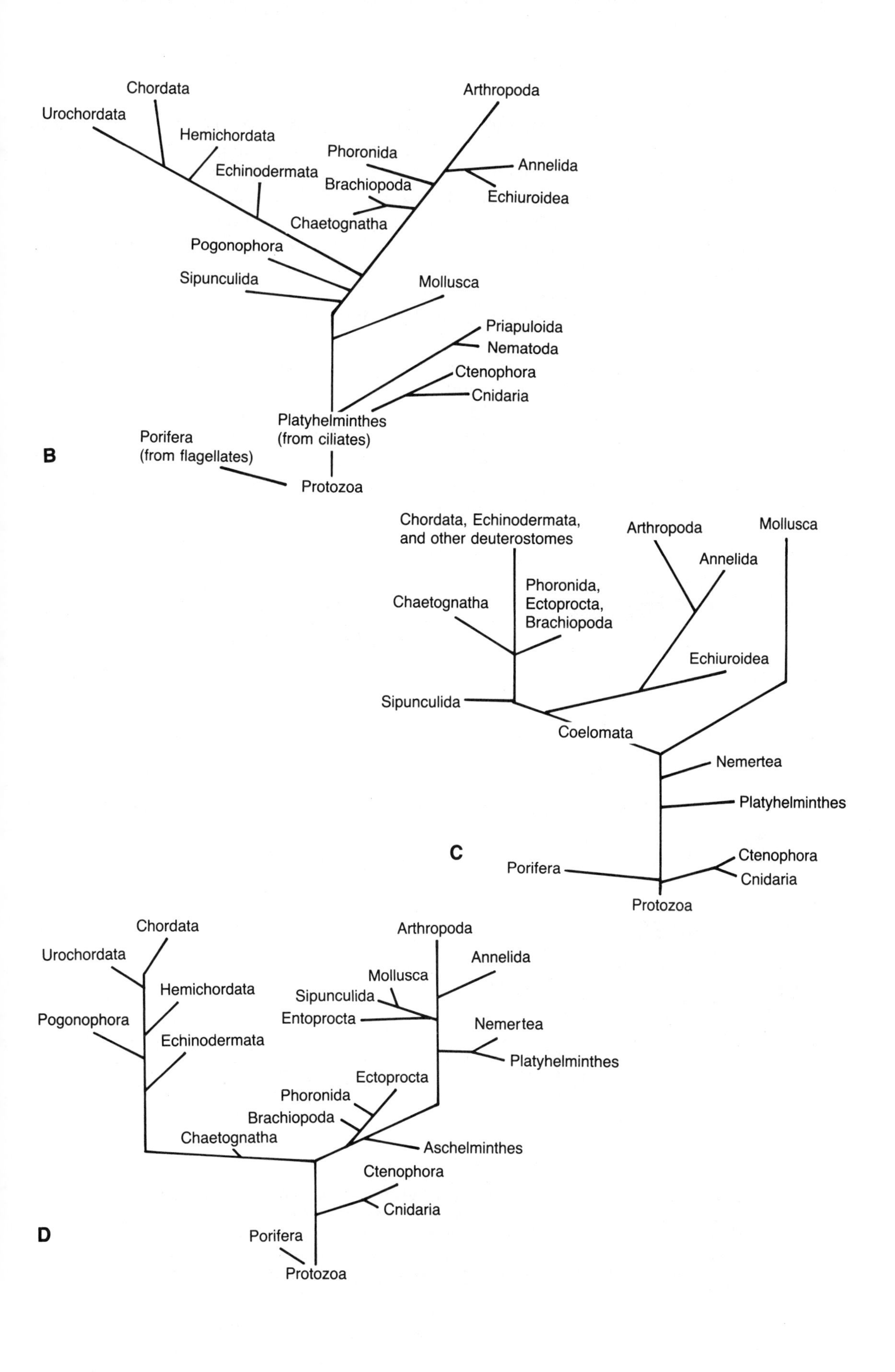
Chordata
Urochordata
Hemichordata
Echinodermata
Arthropoda
Phoronida
Annelida
Brachiopoda
Echiuroidea
Chaetognatha
Pogonophora
Sipunculida
Mollusca
Priapuloida
Nematoda
Ctenophora
Cnidaria
Platyhelminthes
(from ciliates)
Porifera
(from flagellates)
B
Protozoa
Chordata, Echinodermata,
and other deuterostomes
Arthropoda
Mollusca
Annelida
Phoronida,
Ectoprocta,
Brachiopoda
Chaetognatha
Echiuroidea
Sipunculida
Coelomata
Nemertea
Platyhelminthes
C
Ctenophora
Porifera
Cnidaria
Protozoa
Chordata
Arthropoda
Urochordata
Annelida
Mollusca
Hemichordata
Sipunculida
Pogonophora
Entoprocta
Nemertea
Echinodermata
Platyhelminthes
Ectoprocta
Phoronida
Brachiopoda
Chaetognatha
Aschelminthes
Ctenophora
Cnidaria
D
Porifera
Protozoa

animal phyla, apparently transpiring in the 100 Myr between the Ediacara and Burgess Shale faunas, has been considered one of the great problems of evolution. Some authors (e.g., Durham 1971) have argued that their appearance was preceded by a long evolution that left few traces in the fossil record; but most (e.g., Cloud 1976, Gould 1976) hold that they diversified rapidly after a late Precambrian origin. Sepkoski (1978) has argued that the rate of increase in animal diversity is no greater than might be expected of an exponential increase in taxa radiating to fill an ecological vacuum.

During the ORDOVICIAN (505–438 Myr B.P.), many of the animal phyla radiated into a great profusion of classes and orders. Ectoproct bryozoans and reef-building corals appeared. There was an especially marked increase in diversity of suspension-feeding coelomate groups such as bivalves and echinoderms; by the end of the Ordovician, 21 classes of echinoderms had come into existence, most of which became extinct at the end of the period. The earliest known reasonably complete vertebrate fossils (Figure 6A), the heavily armored jawless ostracoderms (class Agnatha), are found in the Ordovician (see Romer 1966 and Olson 1971 for the vertebrate fossil record).

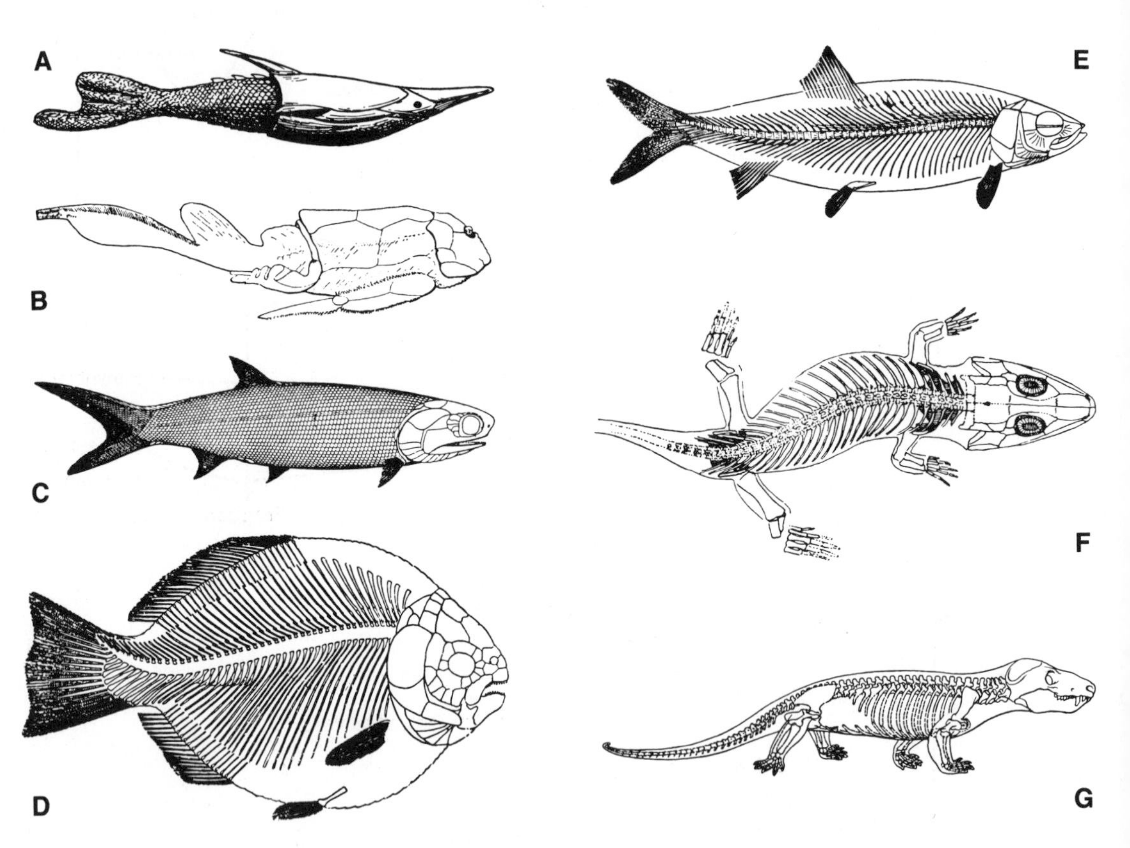


FIGURE 6

**Extinct representatives of some vertebrate groups. (A) *Pterapsis*, a Devonian ostracoderm; (B) *Bothriolepis*, a Devonian placoderm; (C) *Palaeoniscus*, a Permian chondrostean; (D) *Dapedius*, a Jurassic holostean; (E) *Leptolepis*, a Jurassic teleost; (F) *Diplovertebron*, a Carboniferous seymouriamorph amphibian with reptilian features; (G) *Cynognathus*, a Triassic therapsid (mammal-like reptile). (From Romer 1960)**

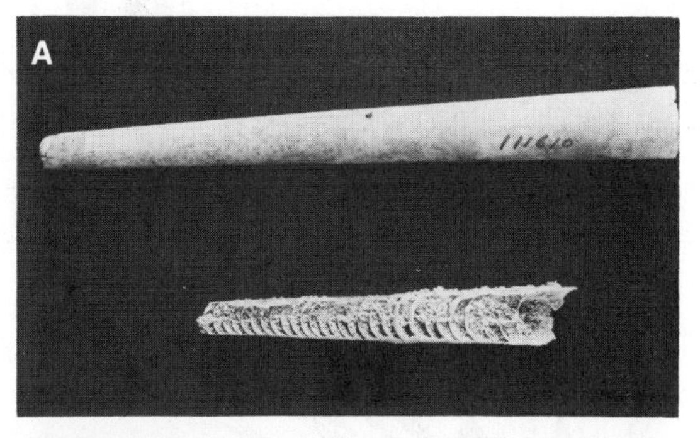


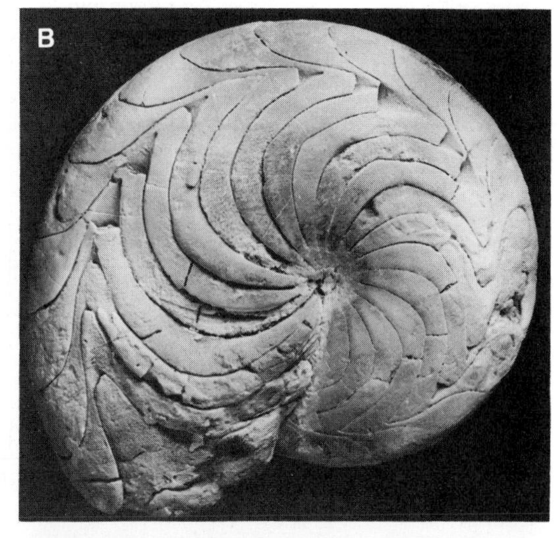

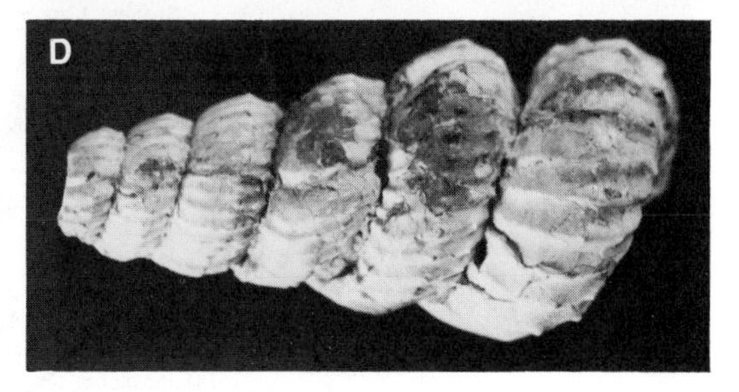


FIGURE 7

**A few of the diverse forms of ammonites (cf. also chapter frontispiece). (A) *Mooreoceras* sp. (Permian). (B) *Imitoceras rotatorius* (Mississippian). (C) *Harpoceras serpentinum* (Jurassic). (D) *Turrilites splendidus* (Cretaceous). (Photographs courtesy of National Museum of Natural History/The Smithsonian Institution)**

At the end of the Ordovician, numerous families and orders became extinct, in concert with a pronounced drop in sea level; during the SILURIAN (438–408 Myr B.P.), diversity increased once again. The agnathans diversified during this period, and there arose a new class of vertebrates, the placoderms, which had jaws and, in some cases, fin-like structures (Figure 6B). The first evidence of terrestrial life appears in the Silurian: scorpions, millipedes, and small, simple, dichotomously branched vascular plants such as *Cooksonia* that resembled in some ways the modern psilopsids and club mosses.

The DEVONIAN (408–360 Myr B.P.) was a period of great adaptive radiation of corals and trilobites, and saw the emergence from the bactritids of the Ammonoidea, shelled squid-like cephalopods (Figure 7) that began in the Devonian the first of several periods of enormous diversification. Agnathans and placoderms reached the peak of their diversity; only a few placoderms survived beyond the Devonian into the Permian, and the agnathan fossil record stops at the end of the Devonian, although agnathans are represented among living organisms by the lampreys and hagfishes, which do not have fossilizable skeletons. During the

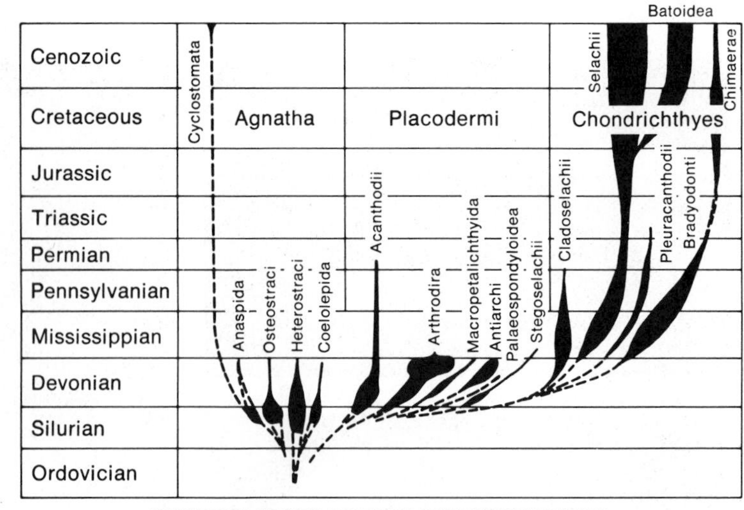


THREE CLASSES OF FISHLIKE VERTEBRATES

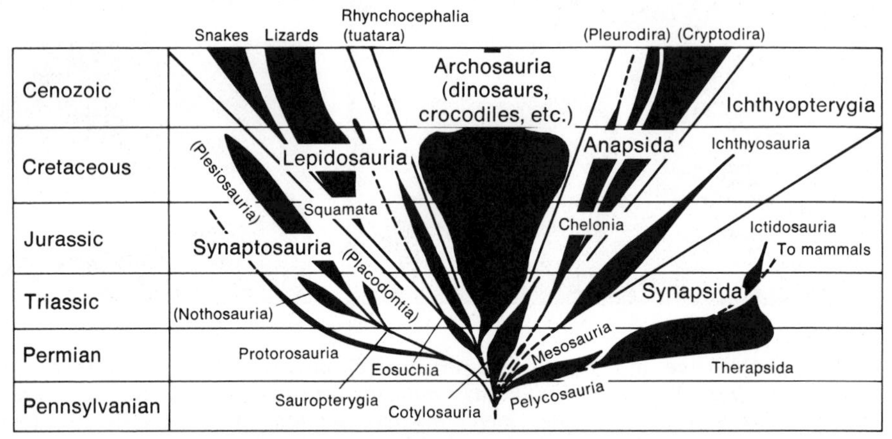


REPTILES

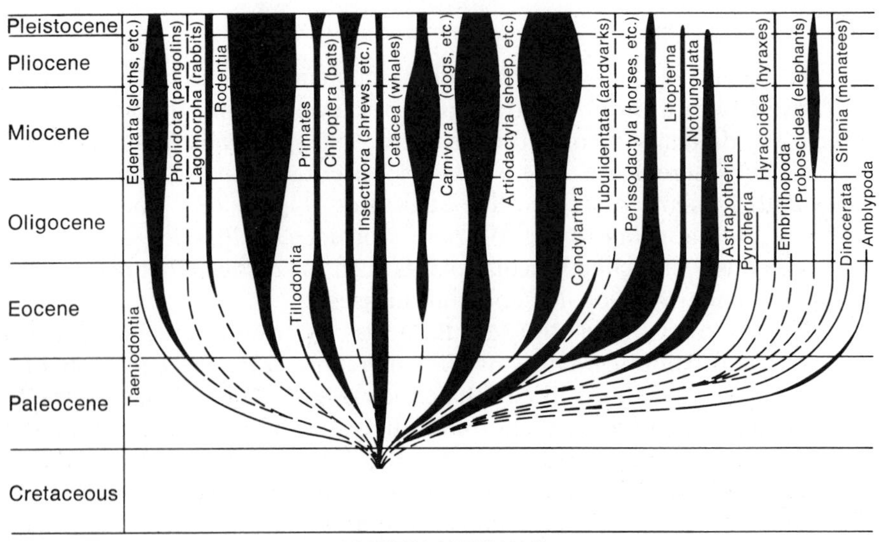


PLACENTAL MAMMALS

FIGURE 8

**Chronological distribution and very approximate changes in diversity of some major groups of vertebrates. Width of the bar is proportional to diversity observed in the fossil record. Broken lines indicate inferred but not observed presence at that time. Lineages are joined to indicate hypothesized phylogenetic relationships. (A) Three classes of fish-like vertebrates. (B) Reptiles. (C) Placental mammals. See Figure 3 in Chapter 10 for a more recent reptile phylogeny. (From Romer 1960)**

Devonian, the "Age of Fishes", the sharks (class Chondrichthyes) arose (Figure 8A), although they were rather unlike modern sharks which, along with the rays, emerged during a later radiation in the Jurassic and Cretaceous. The cartilaginous skeleton of the Chondrichthyes is thought to have been derived from the bony skeleton of their ancestors, probably placoderms. Certainly the true bony fishes, class Osteichthyes, arose earlier in the Devonian than the sharks.

The ancestors of the Osteichthyes are unknown; moreover, both major subclasses of bony fishes, the Choanichthyes and the Actinopterygii, have their distinctive character when they first appear, the former somewhat earlier that the latter. The Actinopterygii are the fishes that are most familiar to us. The Devonian representatives were in the Chondrostei (Figure 6C), the group that today is represented by sturgeons. The Choanichthyes, with internal nostrils and lobe-like paired fins with an internal skeleton, include the lungfishes and the crossopterygians, both of which became diverse in fresh water in the Devonian. Today the lungfishes are represented by only three species and the crossopterygians by one, the coelacanth *Latimeria chalumnae* (Figure 9). The fossil record of crossopterygians extends only to the Cretaceous, and the crossopterygians had been thought entirely extinct until *Latimeria* was dredged from the Indian Ocean in the 1930s. Both *Latimeria* and the lungfishes are so similar to Mesozoic forms that they are considered "living fossils."

It is in the Devonian that an abundant terrestrial biota is first recorded. Amphibians arose at this time, probably from crossopterygians; *Ichthyostega*, of the latest Devonian, had fully formed limbs, but its skull was similar to that of crossopterygians (Figure 10). Bryophytes (liverworts) appeared, as did a great variety of vascular plants, including ferns, lycopsids (club-mosses of today), sphenopsids (represented today by the horsetails), and, toward the end of the period, the first gymnosperms (for terrestrial paleobotany, see Knoll and Rothwell 1980, Stewart 1983). A fossil springtail (order Collembola) is the first representative of the insect-like arthropods (although many authors do not consider springtails true insects). The later part of the Devonian is marked by a substantial

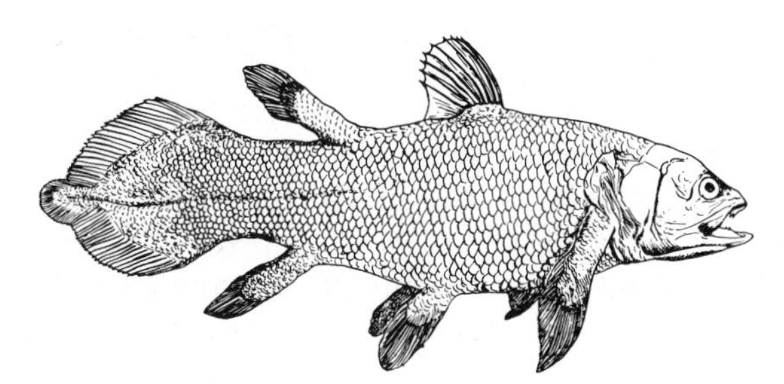

FIGURE 9

**The only living crossopterygian, the coelacanth *Latimeria chalumnae*. (From Gregory 1951)**

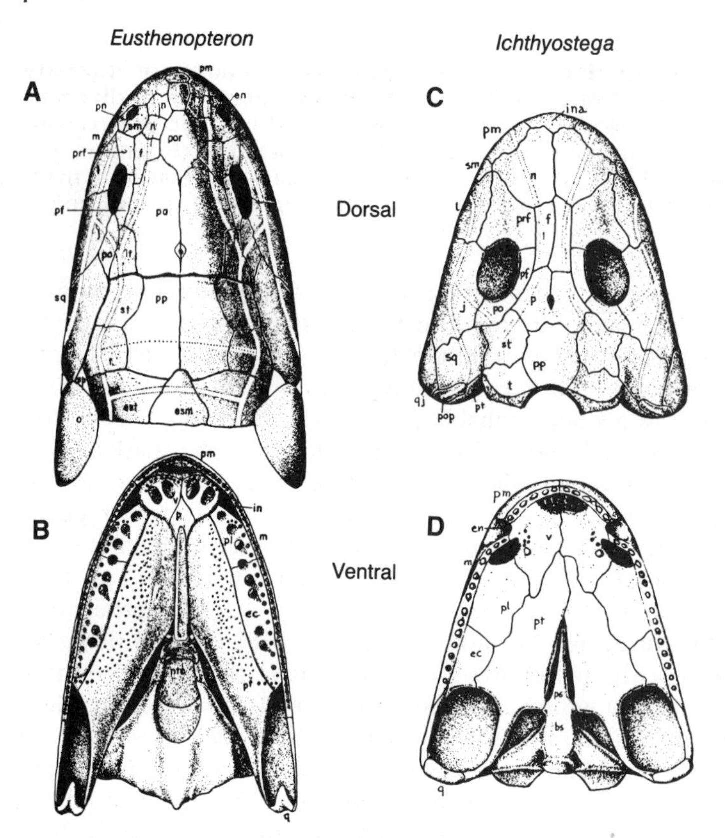


FIGURE 10

**Dorsal and ventral views of the skulls of *Eusthenopteron* (A, B), a crossopterygian fish of the Upper Devonian, and *Ichthyostega* (C, D), the oldest known amphibian (uppermost Devonian or earliest Carboniferous). Although the skull of *Ichthyostega* has many definitive characteristics of amphibians, it is similar in certain critical details to the crossopterygian skull. (From Romer 1966)**

mass extinction of marine invertebrates, apparently coinciding again with a drop in sea level (Hallam 1984).

The coal beds of the CARBONIFEROUS period (360–286 Myr B.P.), sometimes divided into the Mississippian (360–320) and the Pennsylvanian (320–286), are the legacy of the extensive swamp forests of tropical and subtropical Pangaea. These were dominated by ferns, seed ferns, and tree-sized lycopsids and sphenopsids (Figure 11). A profusion of primitive orders of insects appeared, including Orthoptera (the grasshoppers and their relatives of the modern era), the Blattaria (roaches), Ephemerida (mayflies), Homoptera (the order that includes cicadas), and a group of primitive "paleopterous" orders that did not survive beyond the Paleozoic (see Carpenter 1976, Carpenter and Burnham 1985 on fossil insects). The amphibians, many of them large (more than 15 feet) lumbering animals and all of them quite unlike the modern amphibians, underwent an adaptive radiation that continued through the Permian, at the end of which most became extinct. In the early Pennsylvanian, the first reptiles, the protorothyrids, emerged from an amphibian stock, the diadectiamorphs (Figure 6F), which were so reptilian in character that they have sometimes been classified as reptiles.

PERMIAN deposits (286–248 Myr B.P.) yield the first fossils of most of the insect orders that have not yet been found in the Carboniferous: the extinct Protodonata (including giant dragonflies), the Odonata (true dragonflies), Plecoptera (stoneflies), heteropteran Hemiptera (true bugs), Neuroptera (antlions and allies), Mecoptera (scorpionflies), Trichoptera (caddisflies), Coleoptera (beetles), and Diptera (true flies). Few intermediates among these orders, that might help to establish phylogenetic relationships, are known as fossils. A variety of reptile groups (Figure 8B) evolved from the protorothyrids, including the pelycosaurs, of which *Dimetrodon*, with its large dorsal sail, is the best known. Closely related to the pelycosaurs was a dominant group of Permian and Triassic reptiles, the therapsids (Figure 6G), within which mammalian skeletal features arose.

In fresh waters, the Permian was a time of substantial radiation of the chondrostean bony fishes, and the emergence of a new grade of bony fishes, the Holostei (Figure 6D), represented today only by the bowfin (*Amia*) and the garpikes (*Lepisosteus*). In the sea, the ammonites continued to proliferate wildly, but their proliferation was abruptly curtailed by one of the most spectacular events in the history of life: the greatest mass extinction that living things have yet suffered. At the end of the Permian, at least 52 percent of the families of skeleton-bearing marine invertebrates became extinct, and it has been estimated that as many as 96 percent of all species met their end. The trilobites, which had been declining in diversity for some time, were extinguished completely, and the diversity of groups such as anthozoan corals, ammonites, ostracods, crinoids, and articulate brachiopods declined precipitously. Curiously, there was relatively little extinction of fishes or of terrestrial life, but the change in the marine invertebrate

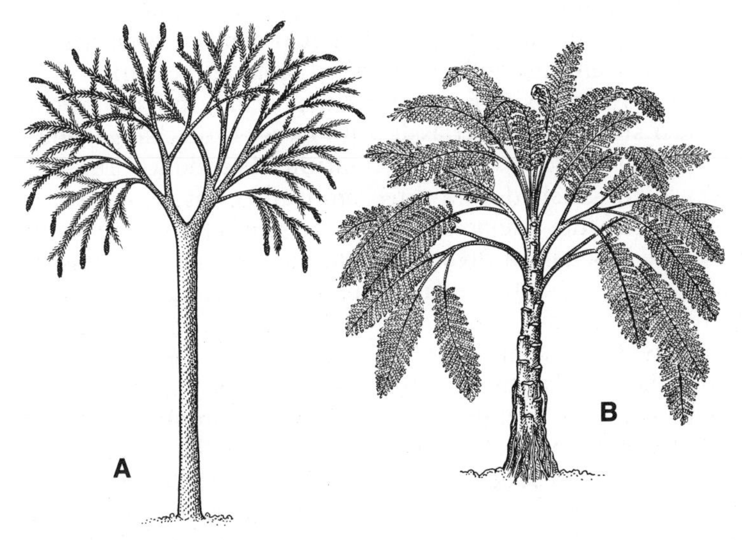


FIGURE 11

**Reconstruction of two extinct vascular plants. (A) *Lepidodendron*, a Carboniferous lycopsid. (B) *Medullosa*, a Carboniferous pteridosperm. All pteridosperms (seed ferns) are now extinct. (From Delevoryas 1962)**

fauna was so thorough as to mark the end of the Paleozoic era and the beginning of the Mesozoic.

## THE MESOZOIC ERA

During the Mesozoic, the world continent Pangaea began to split by plate tectonic processes into separate continents—although the separation was not marked until the Jurassic and later (Figure 1). In consequence, the world biota, which had been rather homogeneous in the late Paleozoic and early Mesozoic, became increasingly PROVINCIALIZED: increasing isolation among land masses and their associated continental shelves fostered the independent development of regional biotas (Valentine 1973, Hallam 1974, Valentine et al. 1978). In the marine realm, diversity increased rapidly as the survivors of the great Permo-Triassic extinction radiated, in part no doubt filling ecological niches that had been vacated.

In the TRIASSIC period (248–213 Myr B.P.), the ammonites underwent a second great proliferation, modern (scleractinian) corals arose and proliferated, and numerous invertebrate groups, such as bivalves, increased in diversity while others, such as brachiopods, showed little recovery from the low stature to which they had been reduced. Some invertebrate groups entered into new ways of life; for example, Paleozoic bivalves and echinoids (sea urchins) had mostly been epifaunal (surface-dwelling), whereas in the Triassic some began to evolve an infaunal (burrowing) habit, a trend that became much more pronounced in the Jurassic and Cretaceous. On land, gymnosperms (including the Ginkgoales, represented today by the "living fossil" *Ginkgo biloba*) and seed ferns achieved dominance over the lycopsids and sphenopsids, and the first fossils of Lepidoptera (moths) and Hymenoptera (wasps) are found—although these orders may have arisen earlier. The reptiles entered upon the spectacular radiation that gives to the Mesozoic its popular name, the "Age of Reptiles." These included the turtles, the first of which have all the characters of the order; the marine ichthyosaurs and plesiosaurs; and the archosaurs (see Figure 3 in Chapter 10).

The archosaurs are a group of orders that evolved from the Thecodontia, in which the teeth were inserted into sockets. Among the derived groups were the crocodile-like phytosaurs (Figure 10 in Chapter 12), the true crocodilians, which have persisted since the late Triassic with rather little change, and two orders of dinosaurs that arose independently from the thecodonts. The order Saurischia (Figure 12), distinguished by a three-pronged pelvis, arose in the late Triassic and persisted into the Cretaceous. The Triassic representatives of this diverse group were the bipedal, carnivorous, theropods. Some were less than a meter in length, but by the late Cretaceous, some theropods such as *Tyrannosaurus* were more than 15 meters long.

Late in the Triassic, at about the same time as the first dinosaurs evolved, some of the therapsids—the mammal-like reptiles that had arisen in the Permian—approached the mammalian condition so closely that some of them are considered the first mammals. Among these are the haramyids, the likely ancestors of the multituberculate mammals of the later Mesozoic (Figure 15), and *Kuehneotherium*, a genus related to the probable ancestors of the marsupials and placental mammals. Because most of these early mammals are known only from a few teeth and other fragments, there is considerable uncertainty about the details of the relationships among them, and between them and later mammals

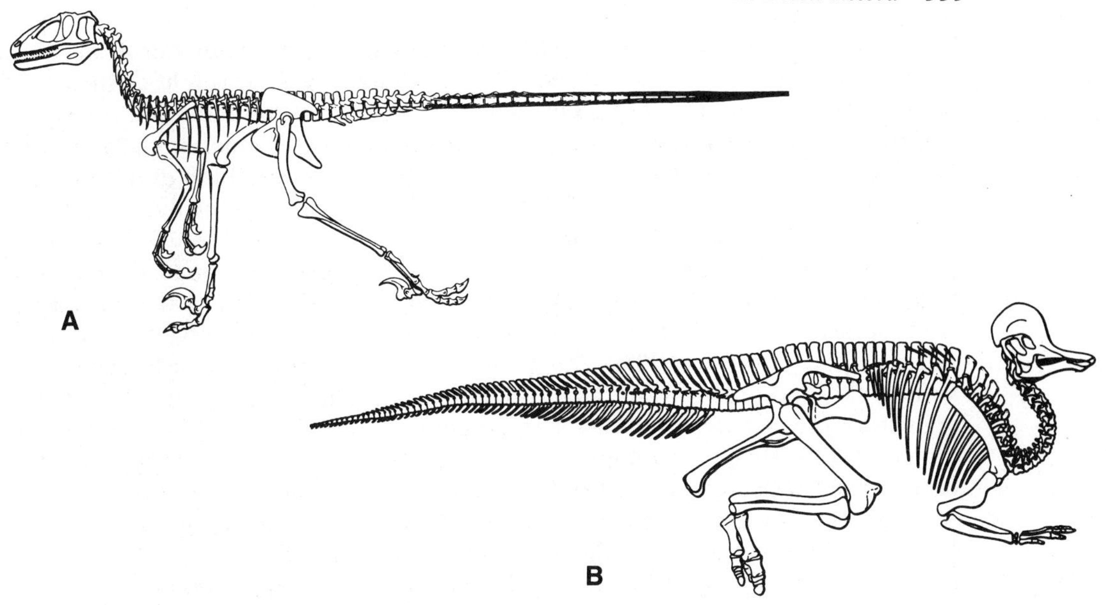


FIGURE 12
**(A) A saurischian dinosaur, *Deinonychus*, an early Cretaceous carnivorous theropod allied to the stock from which birds probably arose (see Figure 14). The powerful claw on the second toe is thought to have been used to disembowel prey. (B) An ornithischian dinosaur, *Corythosaurus*, a late Cretaceous herbivorous hadrosaur. Although not evident in this illustration, the hadrosaurs were primarily bipedal. The form of the pelvis distinguishes the ornithischian and saurischian dinosaurs. (From Colbert 1980)**

(for Mesozoic mammals and therapsids, see Lillegraven et al. 1980, Kemp 1982).

The JURASSIC period (213–144 Myr B.P.) began after a substantial mass extinction that eliminated most of the ammonites and many other marine invertebrates. The ammonites underwent yet another adaptive radiation, and many invertebrate groups increased in morphological diversity, acquiring sturdy shells that probably served as protection against the more effective predators—echinoids, crabs, predatory gastropods, and predatory fishes—that appeared at this time. Planktonic Foraminifera evolved from benthic ancestors, and a new group of large, peculiar bivalves, the rudists, arose, forming reefs from their skeletal secretions. A new grade of ray-finned bony fishes, the teleosts (Figure 6E), arose from holostean ancestors, and began an enormous adaptive radiation that continued almost unabated until the present, when they constitute the dominant fishes.

The amphibian groups that had prevailed became extinct by the end of the Triassic, but the first primitive frogs appear in the mid-Jurassic, and salamanders are represented by the end of the period. The first fossil lizards are Jurassic, and among the archosaurs, the radiation that had begun in the Triassic led to some extraordinary forms. The thecodonts gave rise to an adaptive radiation of pterosaurs, flying reptiles in which the wing membrane was supported by an immensely elongated fourth digit. The bipedal theropod dinosaurs continued to diversify, and gave rise to quadrupedal, mostly herbivorous sauropods. Although not all the sauropods were large, they included among their number *Apatosaurus* (also known as *Brontosaurus*), up to 67 feet in length with a weight estimated at 30 tons; the more slimly built *Diplodocus*, the longest of dinosaurs at more than 87 feet; and doubtless the heaviest terrestrial animal of all time, *Brachiosaurus*, 80 feet long and probably 50 tons in weight. These were saurischian dinosaurs; but in the Jurassic, another order of dinosaurs, the Ornithischia, evolved from thecodont ancestors (Figure 12). The ornithischians, characterized by a four-pronged pelvis, were herbivores; some were bipedal, but most were quadrupedal. The bipeds included the duck-billed dinosaurs (hadrosaurs), many of which had the skull bones extended into a variety of curious crests which probably served for social display (Hopson 1975). The quadrupeds included the Jurassic *Stegosaurus*, famous for its dorsal rows of large triangular plates that were probably thermoregulatory organs; the Cretaceous ankylosaurs, which had the body encased in armor; and the diverse ceratopsians, of which *Triceratops* is the most familiar. The ceratopsians appeared late in the Cretaceous, and became extinct at its end.

*Archaeopteryx lithographica*, from the mid-Jurassic, is the earliest known bird (Figure 13). Except for its feathers and a few osteological features, it has none of the anatomical peculiarities that typify later birds. It shares numerous synapomorphies with the small, bipedal theropod dinosaurs of the order Saurischia (Ostrom 1976, Thulborn and Hamley 1982; Figure 14), so most authorities now agree that birds are descended directly from dinosaurs. Among living organisms, crocodilians—the only other surviving descendants of the archosaurs—are the closest relatives of birds.

Although their Jurassic remains are few, by the end of the period diverse groups of archaic mammals had appeared, most of which did not survive long into the Cretaceous. The most notable group was the Multituberculata (Figure

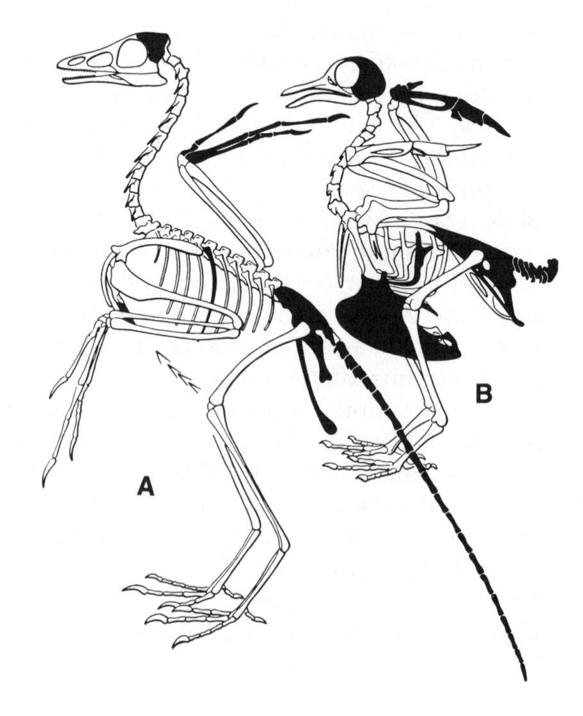


FIGURE 13

**Comparison of the skeletons of (A) *Archaeopteryx*, the earliest known bird, and (B) a modern pigeon (*Columba*). Some homologous regions are in black. Relative to *Archaeopteryx*, the modern bird has an expanded brain case, reduced and fused digits in the wing, an enlarged breastbone, a reduced tail, and a coalescence of the pelvic bones with each other and with the neighboring vertebrae. (From Colbert 1980)**

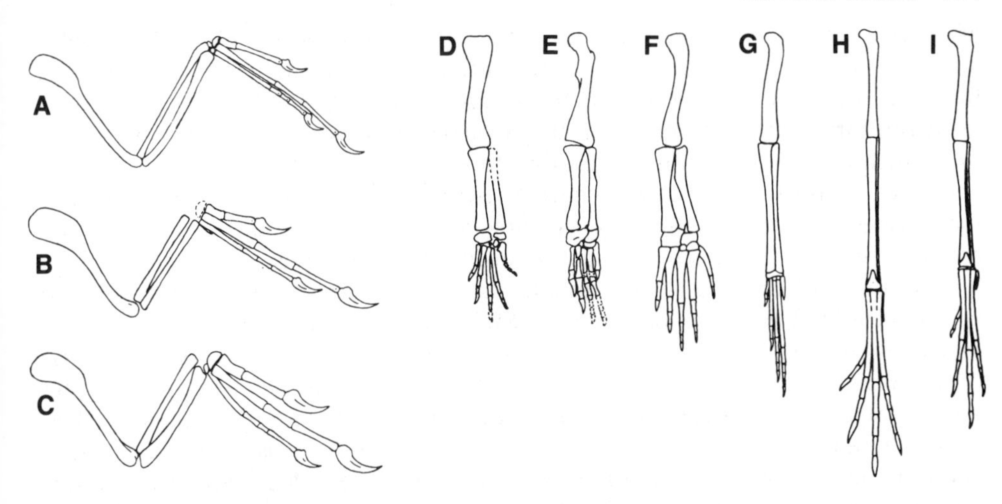


FIGURE 14
**Skeletal features of *Archaeopteryx* compared to other archosaurs. Left, the forelimb of *Archaeopteryx* (A) and two theropod dinosaurs, *Ornitholestes* (B) and *Deinonychus* (C). Right, the hind limb of *Archaeopteryx* (H) compared to those of the theropod dinosaur *Compsognathus* (I) and those of several pseudosuchians (D–G), a group of more primitive archosaurs. (From Ostrom 1976)**

15), which appeared during the Upper Jurassic, diversified into a variety of herbivorous rodent-like forms in the Cretaceous and early Tertiary, and became extinct during the Eocene epoch.

During the CRETACEOUS period (144–65 Myr B.P.), marine invertebrate diversity continued to increase, ammonites continued to diversify and then dwindled toward the end of the period, and both the marine and terrestrial biota became increasingly provincialized. Modern sharks and rays evolved, and teleosts continued their rise to dominance. The dinosaurs continued to diversify, but all

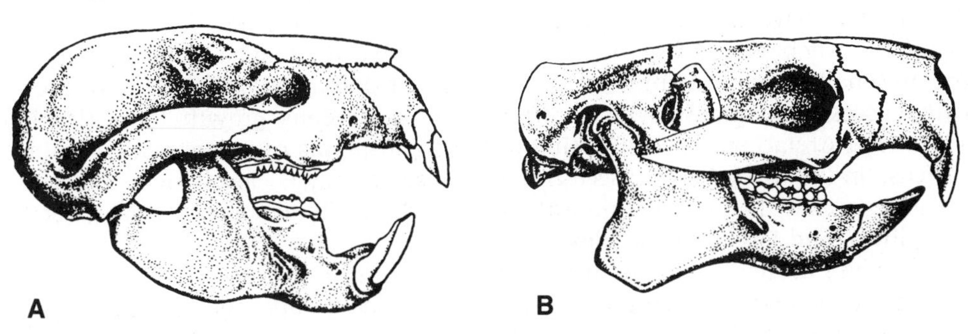


FIGURE 15
**An example of convergent evolution. (A) A Paleocene multituberculate, *Taeniolabis*. (B) An Eocene rodent, *Paramys*. Rodents replaced the structurally and ecologically similar multituberculates, but whether or not they excluded them by competition is not known. (From Romer 1960)**

of them became extinct at or before the end of the period. The first snakes appeared in the Cretaceous, and began a radiation that would become pronounced in the Miocene. A few Cretaceous birds, such as the aquatic *Hesperornis*, retained teeth but otherwise differed from *Archaeopteryx* by possessing modern avian features such as a reduced tail, a well-developed, keeled sternum, and the reduced and fused hand bones that are characteristic of the wings of modern birds. At the very end of the Cretaceous, the first few bird fossils appear that may be ascribed to modern orders, such as the order of the pelicans.

By the mid-Cretaceous, the mammals included the therians, which cannot be characterized as either marsupials (Metatheria) or placentals (Eutheria), and may well include the common ancestors of these two groups. By the Upper Cretaceous, the Metatheria and Eutheria had become distinct. The upper Cretaceous marsupials included members of the family of the modern opossum *Didelphis*, although not *Didelphis* itself. Upper Cretaceous eutherians included a considerable variety of indistinctly different groups, many of which have traditionally been placed in the order Insectivora because of their generalized early-placental traits that the modern insectivores (e.g., shrews and hedgehogs) retain in a plesiomorphic condition. Other Upper Cretaceous fossils are assigned to the Primates and to the extinct order Condylarthra (Figure 16), but at this time these orders show only minor differences from the "insectivores."

There was, thus, the beginning of a modern cast to terrestrial animal life in the Cretaceous, and the same was true of the flora, for it is in the early Cretaceous that the flowering plants, the angiosperms, evolved from gymnosperm ancestors (Hickey and Doyle 1977, Doyle 1978). The magnolia family and its relatives such as the Winteraceae have long been postulated, on anatomical grounds, to present ancestral angiosperm features, and the Winteraceae, found in 105–110 million-year-old strata (Walker et al. 1983), are indeed among the oldest angiosperms known (the oldest certain angiosperm remains—pollen—are dated at 109–114 Myr). The earliest known fossil flowers have most of the features that have long been considered primitive among the angiosperms. Most of the features that in concert define the angiosperms, such as vessels, enclosed ovules, and the foreshortened strobilus that is the anatomical ground plan of a flower, had arisen separately among various gymnosperm groups, among which parallel evolution was extensive (Niklas et al. 1980). The combination of these features in one such lineage may have facilitated speciation and adaptive radiation, for the angiosperms quickly began an exponential increase in diversity that has continued until the present. The earliest angiosperms were insect-pollinated, and may have been early-successional plants with a shrubby or herbaceous growth form (Doyle 1978). The Cretaceous diversification of the angiosperms coincides with a major increase in the diversity of insect groups whose ecology is closely tied to that of plants, especially Lepidoptera (moths and butterflies), Hymenoptera (wasps, bees, and allies), and Diptera (true flies).

The end of the Cretaceous was marked by a major drop in sea level and the second largest extinction in the history of life. Some groups, such as ammonites and dinosaurs, had been declining in diversity and disappeared altogether at the end of the period; others experienced an abrupt decline which in some cases, such as the rudist bivalves, led to complete extinction. On land, all vertebrates larger than about 25 kilograms in body weight seem to have become extinct. The

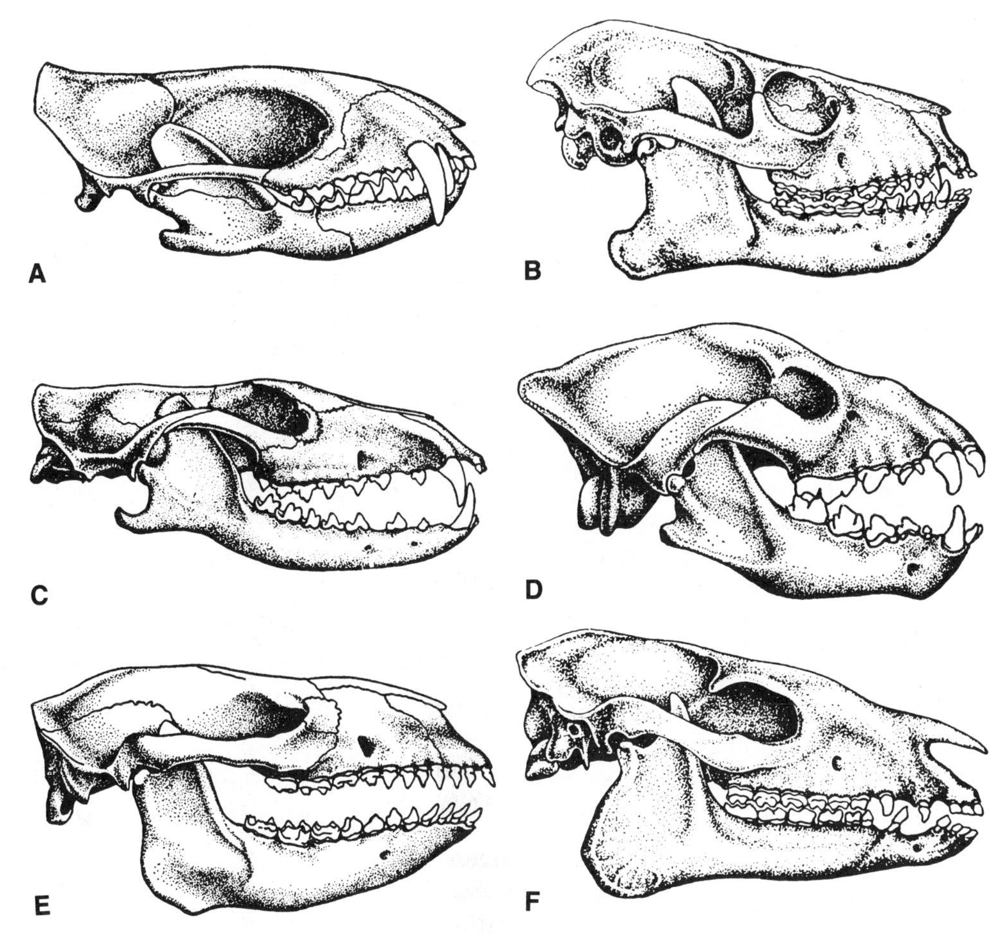


FIGURE 16

**Skulls of placental mammals of several orders, illustrating similarities among their early representatives. (A) *Deltatheridium* (Cretaceous), a representative of a basal placental stock often classified as the order Insectivora. (B) *Notharctos* (Eocene), order Primates. Note the postorbital bar and flattened cheek teeth; otherwise similar in many respects to Insectivora. (C) *Sinopa* (Eocene), a hyaenodontid (order Carnivora). The cheek teeth have shearing edges. (D) *Hyaena* (late Tertiary), a modern member of Carnivora. The teeth are enlarged and reduced in number, but otherwise the skull differs little from earlier carnivores. (E) *Hyopsodus* (Eocene), order Condylarthra. A member of a group that was closely related to the ancestors of the ungulates. The skull is similar to that of generalized insectivores and primates, although the cheek teeth are flattened. (F) *Hyracotherium* (Eocene), order Perissodactyla. The ancestor of the horses. Except for the incipient development of a space behind the canine tooth, the skull resembles that of condylarths (compare with the skulls of later horses shown in Figure 16 of Chapter 14). (From Romer 1966)**

mass extinction was most marked among marine plankton and benthic invertebrates, was considerably less so among small terrestrial vertebrates, and was effectively undiscernible among fishes and terrestrial plants. The end-Cretaceous extinction marks the end of the Mesozoic era.

## THE CENOZOIC ERA

Some paleontologists divide the Cenozoic era into the Paleogene (65–24.6 Myr B.P.) and Neogene (24.6 Myr B.P.–present) periods; more traditionally it is divided into the Tertiary (65–2.0 Myr B.P.) and Quaternary. In the TERTIARY period, the angiosperms diversified enormously, and came to dominate most of the world's forests. Herbaceous angiosperms, derived by convergent evolution from many different woody families, became prominent in the middle of the Tertiary, for during the Oligocene epoch, the climate became considerably cooler and drier, creating conditions favorable for herb- and grass-dominated savannahs. Many of the modern families of plants had acquired their distinguishing features by the first Tertiary epoch, the Paleocene, and most of the modern insect families of which there is any substantial fossil record appear to have become differentiated at this time. A mid-Cretaceous (100 Myr B.P.) fossil, *Sphecomyrma* (Figure 17), signals the evolution of the ant family from wasps; by the time ants are found in abundance, in Baltic amber from the Oligocene (38–24.6 Myr B.P.), a profusion of genera that still exist today had evolved (E.O. Wilson 1971).

Among the vertebrate classes, some modern families differentiated in the Paleocene (65–54.9 Myr B.P.), but terrestrial deposits from this epoch are sparse; it is during the Eocene (54.9-38 Myr B.P.) that modern families appear in moderate numbers amidst a great many other families that have since become extinct. There are Paleocene records of groups such as Cyprinidae (carps) and Dasypodidae (armadillos), but the Eocene records of modern families are far more diverse, including, for example, Centrarchidae (basses), Bufonidae (toads), Boidae (boas), Ardeidae (herons), and Leporidae (hares). Not until the Oligocene and especially the Miocene (24.6-5.1 Myr B.P.), however, are there substantial numbers of the vertebrate genera that persist today.

The differentiation of most of the orders of placental mammals occurred in the Paleocene (Figures 8C and 16), although the sparse deposits of that epoch do not record it in detail. Some of the orders, such as the Chiroptera (bats, first recorded in the Eocene), first appear as fossils that are similar in most respects to their modern representatives. Many of the early Tertiary placentals, however, are generalized mammals that, although displaying the beginnings of adaptive radiation into diverse modes of life, are not easily divided into orders. Some of them are assigned to the Insectivora, a grade (rather than a clade) that is essentially devoid of the apomorphic characters that distinguish most of the other orders. The Edentata are probably an early derivative of the insectivores; armadillos appear in the Paleocene. Of the numerous extinct South American edentates, the last were the giant ground sloths of the Pleistocene, which extended into North America. The Condylarthra (Figure 16) were generalized herbivores of the Paleocene and Eocene; some of them, such as the Eocene *Hyopsodus*, were so primitive in character that they are sometimes classified as insectivores or even as primates.

The condylarths were the stock from which many of the orders of hoofed mammals (ungulates) evolved. Four of these orders were restricted to South America, and became extinct by the Pleistocene or before. These included the Litopterna, some of which resembled camels and horses, and the Notoungulata, which ranged from rat-sized to rhinoceros-sized animals. The condylarths also

**FIGURE 17**
***Sphecomyrma freyi*, a mid-Cretaceous North American ant that appears to bridge the gap between modern myrmecioid ants and the tiphioid wasps from which the ants are thought to have arisen. The node on the waist (pedicel) is a distinguishing features of ants, but some features, such as the form of the antennae (thread-like rather than elbowed as in typical ants), are wasp-like. Most of the features of *Sphecomyrma* match those that had been hypothesized for the ancestors of ants before this specimen was discovered. Several other Cretaceous species of sphecomyrmine ants have since been described. (Photograph courtesy of E.O. Wilson)**

gave rise to the Perissodactyla, Artiodactyla, and Proboscidea. Among the Perissodactyla, *Hyracotherium* of the Lower Eocene, sometimes classified as a condylarth, is the ancestor of the horses, which radiated throughout the Tertiary into a great diversity of browsing forest animals and rapidly running grazers. The same condylarth stock gave rise to the titanotheres (Figure 15 in Chapter 12) of the Eocene and Oligocene, and the rhinoceroses, a diverse group in the Miocene and Pliocene; one of the extinct rhinoceroses, *Baluchitherium*, was 18 feet high at the shoulders and was the largest land mammal known. The earliest even-toed ungulates (Artiodactyla), in the Eocene, are not easily distinguished from the generalized insectivore or condylarth stock except by their distinctive astragalus (tarsal) bone; nor is it possible to assign the Eocene artiodactyls to the suborders—pigs, camels, and ruminants—into which they are easily classified today. Camels and their relatives appeared in considerable diversity in North America in the Eocene, but most of the other modern groups of artiodactyls date from the Oligocene or later. The order radiated greatly in the Miocene, when a variety of peculiar, horned relatives of the giraffes appeared, and in the Pliocene, when the Bovidae (antelopes, goats, cattle) underwent a striking diversification. Among the ungulates, finally, the elephants (Proboscidea) appeared in the Upper Eocene and very rapidly evolved into a great diversity of forms. At least 24 genera of proboscideans are recognized, mostly from the Miocene and Pliocene. Among the last of them were *Mastodon* and *Mammuthus* (the mammoths) of the Pleistocene; only two species of proboscideans, in two genera, survive at present.

The basal insectivore stock also gave rise to the Carnivora and the Primates; members of both groups are known from the Paleocene, but are difficult to distinguish from the insectivores. The carnivores have undergone a succession of radiations. Early carnivores are often placed in the order Creodonta; the first of these were the arctocyonids of the Paleocene, which were replaced in the Eocene by mesonychids, hyaenodonts, and oxyaenids (Figure 16). Most of these became extinct at the end of the Eocene, being replaced by modern carnivores such as weasels (Mustelidae), cats (Felidae), and dogs (Canidae), all of which began to diversify in the late Eocene or Oligocene. The first seals (Pinnipedia) appeared in the Miocene and bears (Ursidae) in the Pliocene; both groups are considered descendants of the dogs (Canidae).

Rodents are probably derived from the Paleocene insectivoran stock. They appeared first in the Eocene, and by the Oligocene had diversified greatly; some were the size of large pigs. The diversification of the rats and mice of the family Muridae began in the Oligocene; during the Pliocene they rose to their present status as the most species-rich family of mammals. During the Paleocene, also, one of the morphologically most primitive of mammalian orders, the Primates, arose from the insectivoran stock. The earliest primates, similar in some respects to lemurs and tarsiers, were only slightly distinct from the insectivores. The lemuroids and tarsioids radiated throughout much of the world in the Paleocene and Eocene; the Old World monkeys began to differentiate in the Oligocene, and the earliest fossils of the modern South American monkeys are Miocene in age. The superfamily Hominoidea, consisting today of apes and humans, may have as its earliest representative the cat-sized *Aegyptopithecus* of the Oligocene; during the Miocene, the hominoids are represented by the dryopithecines, from which the great apes and hominids are believed to have arisen. The first true hominids, the australopithecines, are known from almost four-million-year-old Pliocene deposits of Africa. The later details of hominid evolution will be treated in Chapter 17.

The final geological period, the QUATERNARY, began with the Pleistocene epoch about two million years ago; the last 10,000 years or so are referred to as the Recent or Holocene. Aside from genera that became extinct without issue, the Pleistocene biota consists of forms so similar to Recent species that they are placed in the same genera and often in the same species. The salient feature of the Pleistocene was the drastic cooling and fluctuation of climate. During several minor and at least four major glacial episodes, the climate throughout much of the world became cooler and drier, and the distributions of animal and plant species shifted toward the tropics; for example, Arctic forms such as spruce (*Picea*) and musk-ox (*Ovibos*) occurred in the southern United States. During the warm interglacials, many species were distributed far to the north of their present range; England had elephants, hippopotami, and lions. During the glacials, many species that had moved to the south were restricted to "refuges"—local pockets of favorable habitat—so that their ranges were fragmented, and genetic divergence to the race or species level occurred. In parts of the tropics, the dry climate of the glacials probably fostered genetic divergence among populations of rain forest organisms, isolated in local areas of high precipitation (Simpson and Haffer 1978, Prance 1982). As the last glaciation (the Wisconsin) retreated about 10,000 years ago, species spread out from their refuges at different rates over the glaciated northern landscape, and formed new ecological associations; thus some of the plant communities of today, such as the spruce-fir association of the northern United States and southern Canada, may be no more than a few thousand years old (Davis 1976).

Curiously, the violent fluctuations of the Pleistocene climate seem to have had rather little effect on the rate of evolution or the rate of extinction. For example, virtually all Pleistocene beetle fossils are identical to living species (Coope 1979), and most Pleistocene mammals show only minor differences from their living descendants (Kurtén 1968). An increase in extinction rate has been discerned only for Atlantic invertebrates, especially in tropical regions, and for the "megafauna" of large birds and mammals, especially in North and South America. Many of these creatures—sabertooth cats, mammoths, giant bison, giant sloths and many others—became extinct about 11,000 years ago. Possibly by 18,000 years ago, and certainly by 13,000 to 12,000 years ago, humans crossed the Bering land bridge into the New World—shortly before the extinction of much of the megafauna. The extinctions may well have been caused directly by human hunting (Martin and Klein 1984). Whatever its cause, the extinction of these abundant large animals was so recent that the species that interacted with them have not had time to adapt to their absence. The fruits of many trees, for example, appear to be adapted for dispersal by large mammals that no longer exist (Janzen and Martin 1982).

Whether one is studying the genetic structure of populations, the adaptations or geographic distributions of species, or the interspecific interactions that influence the structure of communities, it is well to bear in mind that history casts its shadow on the present. Whatever genetic or ecological equilibria may have existed before the Pleistocene were probably destabilized by its vicissitudes; there has not been enough time since then for genetic and ecological equilibria to be fully reattained; and hardly had the glaciers retreated when a major new disruption of ecological communities began. The advent of human agriculture about 12,000 years ago began yet another reshaping of the terrestrial environment. For the last several thousand years, deserts have expanded under the impact of overgrazing, forests have succumbed to firing and cutting, and climates have changed as vegetation is modified or destroyed. At present, under the impact of a population of about 5 billion people and of modern technology, tropical forests with their richness of species face almost complete annihilation, temperate zone forests and prairies have been eliminated in much of the world, and even marine communities suffer pollution and overexploitation. In the next several hundred years one of the greatest mass extinctions of all time will come to pass unless we act now to prevent it (Ehrlich and Ehrlich 1981).

## SUMMARY

This chapter has been an overview of salient features of the origin and extinction of major taxa over the 3.5 billion years of the fossil record of life. These facts raise numerous questions of interpretation, some of which are treated in the following chapter.

## FOR DISCUSSION AND THOUGHT

1. "Living fossils" such as the coelacanth *Latimeria chalumnae* present two questions. Why have they undergone so little morphological evolution? How have they persisted so long after their close relatives have become extinct? Discuss.
2. In what sense, if any, can groups with many species be considered more "successful" than groups with fewer species? Is high diversity evidence of better adaptation?
3. Cite instances in which the evolution of a group of organisms provides a stimulus, or opportunity, for the evolution of other groups. Does diversity beget diversity?
4. Higher taxa (e.g., phyla, classes) persist longer in the fossil record than lower taxa (e.g., families, genera). Why?

5. Creationists often claim that the fossil record provides no evidence for the evolution of higher taxa from preexisting forms. Examine cases cited in this chapter that refute this claim. How can we account for the many cases in which, indeed, no intermediate fossils have been found?
6. After land masses became separated by continental drift, progressive differentiation of their biotas occurred (provincialization). Was this because gene flow among conspecific populations ceased, or because species could not disperse readily from one region to another?
7. How might you account for the observation that evolutionary rates seem not to have been accelerated by the Pleistocene changes in climate?
8. How might you account for the fact that most of the existing families of insects and reptiles are older than most of the existing families of mammals? Does this mean that insects and reptiles evolve more slowly than mammals?

## MAJOR REFERENCES

Hallam, A. (editor). 1977. *Patterns of evolution, as illustrated by the fossil record.* Elsevier, Amsterdam. 591 pages. Highly informative, interpretive essays by many authors on the fossil record of particular groups.

Knoll, A.H. and G.W. Rothwell. 1980. Paleobotany: perspectives in 1980. *Paleobiology* 7: 7–35. An excellent review of current knowledge of patterns of evolution of terrestrial plants.

Stewart, W.N. 1983. *Paleobotany and the evolution of plants.* Cambridge University Press, Cambridge. 405 pages. One of several textbooks on paleobotany.

House, M.R. (editor). 1979. *The origin of major invertebrate groups.* Academic Press, New York. 515 pages.

Carpenter, F.M., and L. Burnham. 1985. The geological record of insects. *Ann. Rev. Earth Planet. Sci.* 13: 297–314. (The fossil record is also briefly treated in H.B. Boudreaux, 1979. *Arthropod phylogeny, with special reference to insects,* Wiley, New York, 320 pages.)

Romer, A.S. 1966. *Vertebrate paleontology.* Third edition. University of Chicago Press, Chicago. 468 pages. Somewhat out of date, but still the major single reference on the subject.

# The History of Biological Diversity

# Chapter Twelve

The account of evolutionary history presented in the preceding chapter with little interpretation raises a host of questions. How has diversity changed overall during evolution? Does diversity reach an equilibrium, increase without limit, or fluctuate randomly? What causes changes in diversity? Are there trends in long-term evolution? How do we account for the geographical distribution of species? What genetic processes are involved in the origin of higher taxa?

## CHANGES IN DIVERSITY

In attempting a quantitative analysis of changes in diversity over time, the paleontologist faces both methodological and conceptual difficulties. Biases in the fossil record can give an illusion of an increase in diversity even if it has remained constant (Raup 1972). For example, at least for the last 100 Myr or so, young rocks are less likely than old rocks to have been destroyed by the ravages of time, so the last occurrences of a group or species are more likely to be found than the first occurrences, and the geographic coverage of young rocks will be greater. Moreover, if an extant group is known from just a single fossil, its geological range is from that time to the present. If it were not extant, its geological range would be very brief instead of covering all subsequent time. This effect has been termed THE PULL OF THE RECENT (Raup 1979). Another problem is raised by the imprecise time estimates given in the literature compilations for many groups (e.g., "Lower Silurian"); thus estimates of, say, the extinction rate per million years will often be similarly imprecise.

Moreover, artifacts of taxonomy raise serious problems. "Species" in paleontology are fossils that differ enough (often quite subtly) to be given different names. If two such forms occur in the same deposits, they probably represent different biological species, populations that do not interbreed. But a single evolving lineage is often given different names as it changes in morphology. Thus when a single gene pool changes into a new CHRONOSPECIES—when it changes enough to acquire a new name—one nominal "species" has become "extinct" and another has "originated." These events—PSEUDOEXTINCTION and PSEUDOSPECIATION—are not the same as true extinction (annihilation of a lineage) or speciation (splitting), but it is often difficult to tell from paleontological compilations what proportion of the species names represent chronospecies. This is one reason why many paleontological treatments of diversity tabulate genera or other higher taxa rather than species. A single gene pool may evolve into a new chronospecies, but all the members of a genus will not evolve in concert into another genus. However, the usage of higher taxa has problems too, for the criteria for naming organisms as distinct genera and families are arbitrary, and differ from one group to another (although the criteria will often be reasonably consistent among researchers on a particular group). For example, it is likely that species are more finely divided into genera in morphologically complex than in morphologically simpler groups (Schopf et al. 1975). Moreover, many of the higher taxa now recognized are not strictly monophyletic groups in the cladistic sense (Chapter 10), but paraphyletic groups or even grades rather than clades.

Whether we count species or higher taxa, the number $N$ at any time $t$ may change at a rate $dN/dt$. Like the number of individuals in a population, this value is a function of the rate of input or origination ($O$) and the rate of loss or extinction ($E$). The origination rate $O$ is the number of new taxa per preexisting taxon per

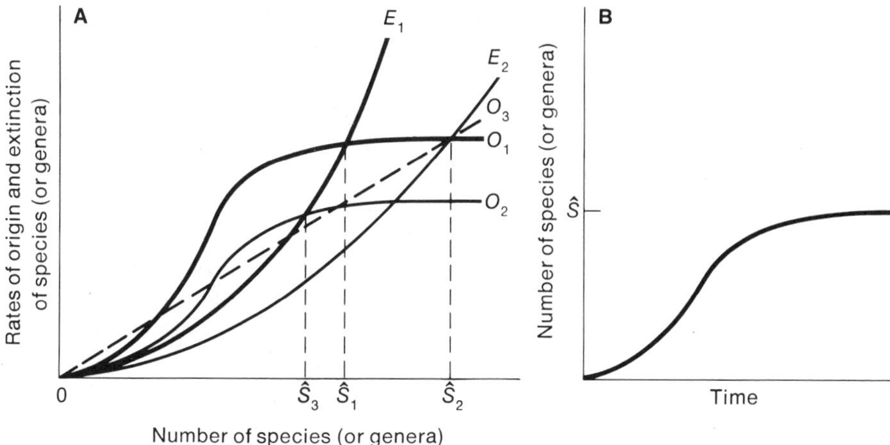


FIGURE 1

**(A) Both the number of new species (or genera, or higher taxa) originating per unit time and the number of extinctions per unit time must increase as the number of species already present rises. If rates of extinction ($E$) increase more rapidly with species number than do origination rates ($O$), an equilibrium number of species ($\hat{S}$) will exist when $O = E$. Whether origination rates are constant ($O_3$) or decline as diversity increases ($O_1$, $O_2$), $\hat{S}$ will be lower if extinction rates are greater ($\hat{S}_1 < \hat{S}_2$) or if origination rates are lower ($\hat{S}_3 < \hat{S}_1$). (B) The change in species diversity over time, according to the equilibrium model.**

unit time; the rate of extinction $E$ is determined by the probability of extinction during the period $dt$. Thus the average rate of change in diversity is

$$dN/dt = ON - EN = N(O - E)$$

If $O = E$, diversity is constant, but it will remain constant over a long term only if a negative feedback process holds it at equilibrium. DIVERSITY-DEPENDENT controlling factors exist if, as diversity increases, the rate of origination declines and/or the extinction rate increases (Figure 1).

The rates of origination and extinction calculated from fossil occurrences depend on taxonomic level. Because a higher taxon such as a family does not become extinct until all its constituent species are extinct, extinction rates fluctuate less over time for higher than for lower taxa. Similarly, because the acquisition of character differences marked enough to warrant recognition of a family usually takes longer in evolution than the evolution of generic characters, the origination rate at the family level is lower and less variable than the origination rate of genera.

After each major extinction, diversity has increased rapidly. On land, the diversity of vascular plants increased from the Devonian to the Permian, dropped at the end of the Permian, and then rose to a plateau that was maintained throughout the Mesozoic until the Upper Triassic, when angiosperms began a diversification that has continued until the present (Niklas et al. 1980). The number of plant species in individual fossil assemblages increases in parallel with the global increase in diversity, as does the variety of growth forms (Niklas 1986). The fossil record of insects is poor, but some of the largest orders, such as the Lepidoptera and Hymenoptera, did not diversify until the late Mesozoic and Cenozoic; it is almost certain that the enormous diversity of insects, especially those that are specialized feeders on various angiosperms, has increased since the Mesozoic. The diversity of genera of mammals has risen constantly and exponentially throughout the Cenozoic, although the increase in number of families is much less marked (Gingerich 1977).

Marine invertebrates have left the most complete fossil record, and have been most intensively analyzed (Figure 2). Diversity increased steadily from the Cambrian to the Ordovician, and then after a mass extinction increased to a roughly steady plateau throughout the rest of the Paleozoic, interrupted only by a Devonian mass extinction. Following the great late Permian extinction, diversity increased throughout the Mesozoic and Cenozoic to an all-time maximum (Sepkoski et al. 1981, Sepkoski 1984). It has been estimated that the average number of marine invertebrate species was about 40,000 in the Paleozoic and Mesozoic,

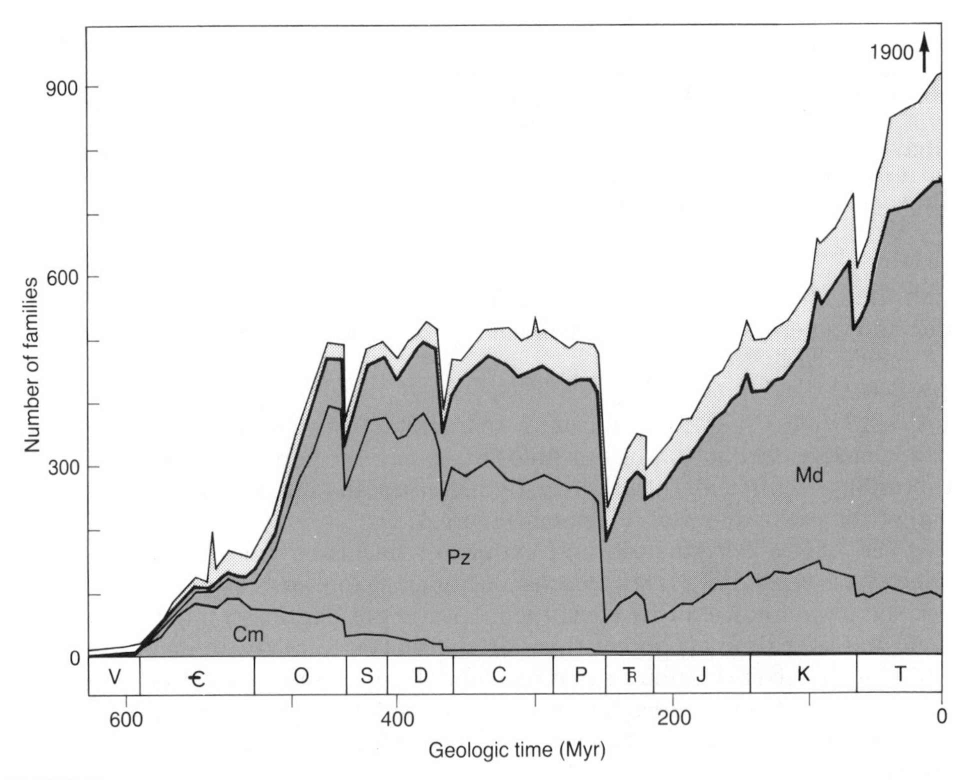


FIGURE 2

**The history of the taxonomic diversity of marine animal families during the Phanerozoic. The uppermost line is the total number of families. The lightly stippled area below this line represents those that are rarely preserved. The remaining area, consisting of well-preserved families with a more reliable fossil record, is divided into the three "evolutionary faunas" (Cm, Cambrian fauna; Pz, Paleozoic fauna; Md, modern fauna) distinguished by Sepkoski. The figure "1900" at upper right is the approximate number of animal families in the oceans today, including many soft-bodied groups that are only rarely preserved as fossils. The symbols along the bottom represent geological periods. (After Sepkoski 1984)**

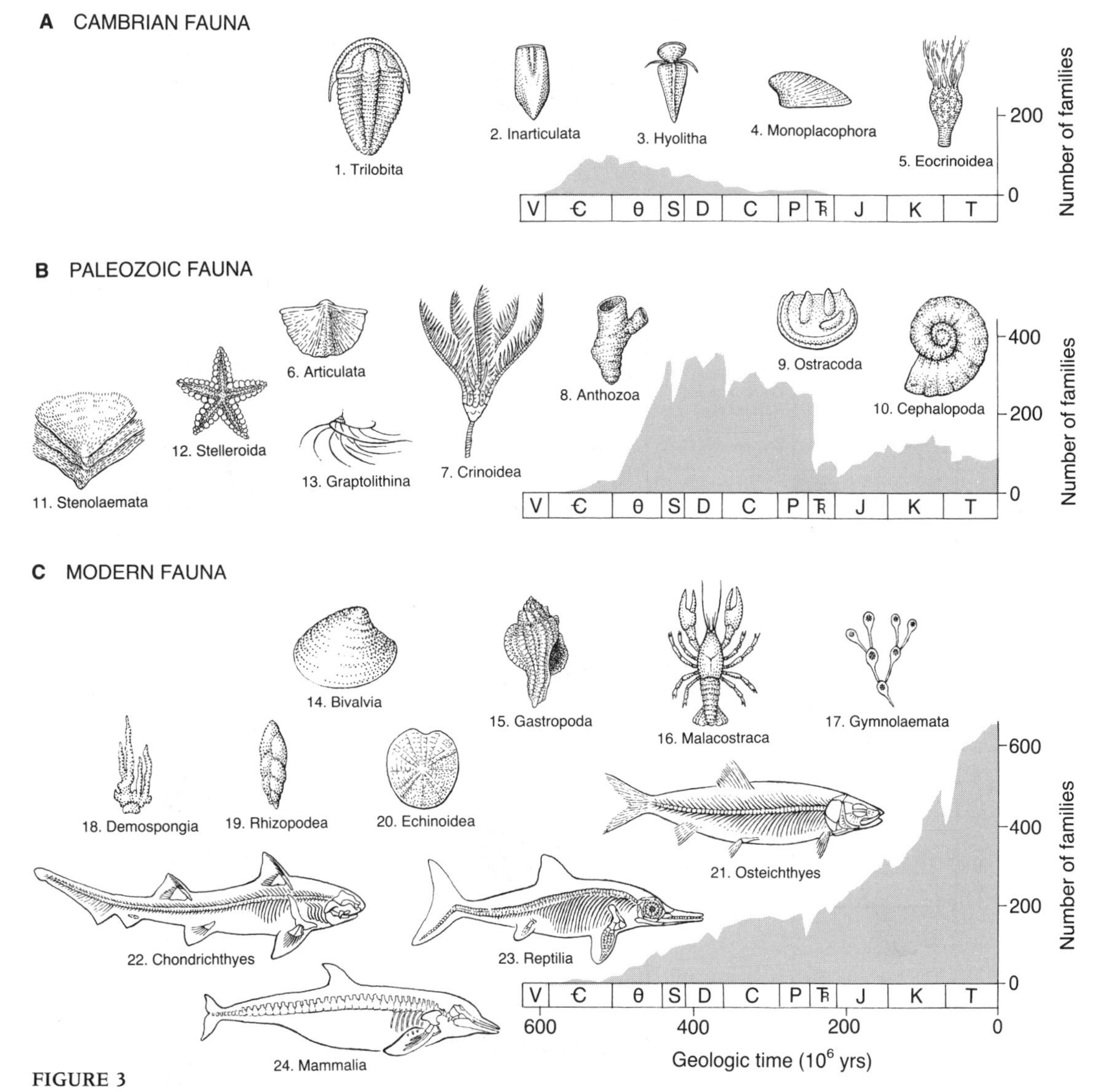


FIGURE 3

**The history of diversity of each of the three "evolutionary faunas" (Figure 2) in the marine fossil record, with illustrations of major forms. (From Sepkoski 1984)**

and about 240,000 in the Cenozoic (Valentine et al. 1978). Sepkoski (1981) has statistically distinguished three major, taxonomically different, "evolutionary faunas" that have dominated the seas sequentially through Phanerozoic time (Figure 3).

Much of the late Mesozoic and Cenozoic increase has been attributed to increasing provincialization (Valentine 1973, Valentine et al. 1978); regional biotas differentiated as they became isolated by plate tectonic processes and as the Cenozoic cooling of climate led to greater latitudinal differentiation of communities. However, even within communities, invertebrate species diversity in-

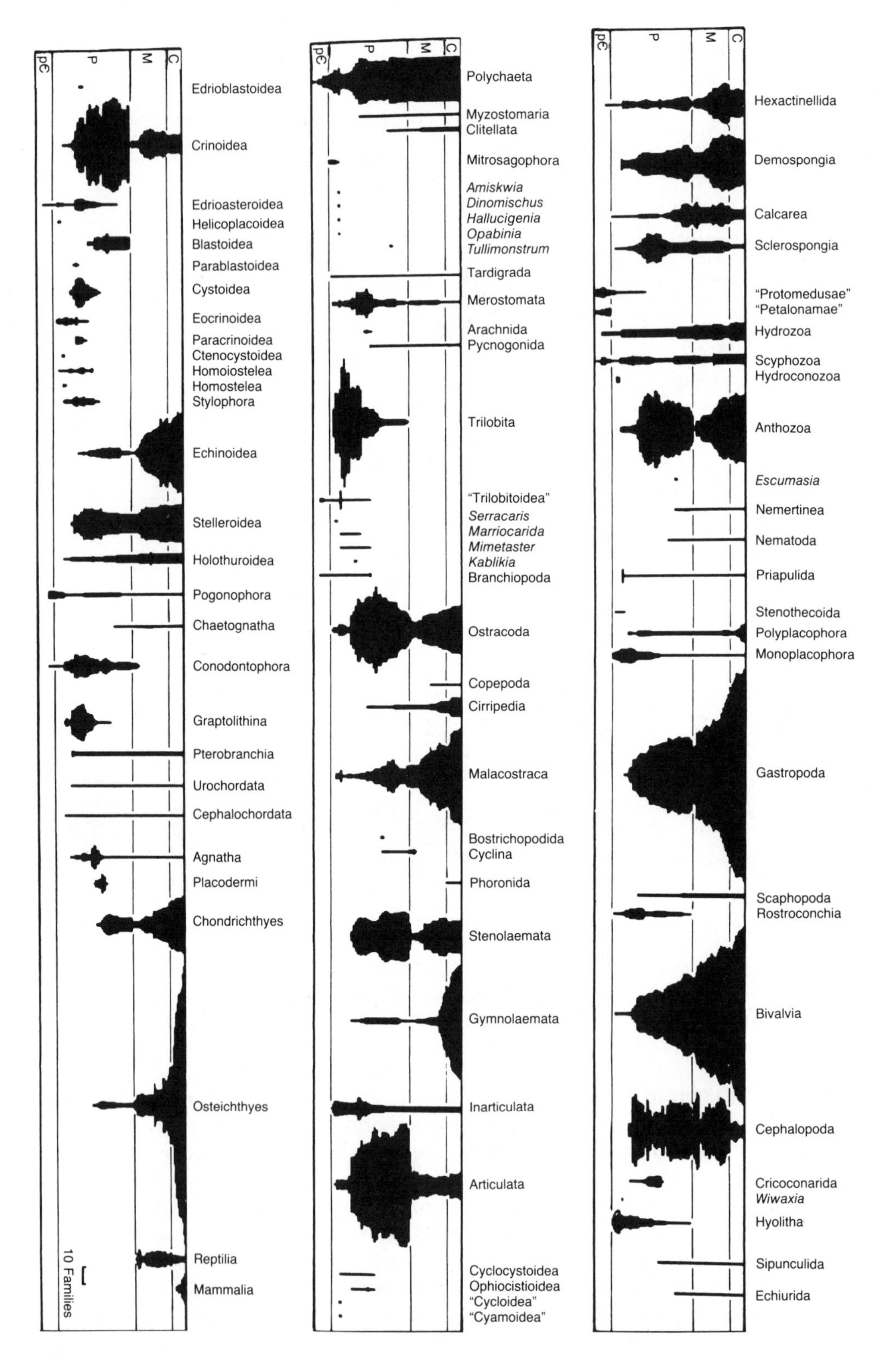
pЄ
P
M
C
Edrioblastoidea
Crinoidea
Edrioasteroidea
Helicoplacoidea
Blastoidea
Parablastoidea
Cystoidea
Eocrinoidea
Paracrinoidea
Ctenocystoidea
Homoiostelea
Homostelea
Stylophora
Echinoidea
Stelleroidea
Holothuroidea
Pogonophora
Chaetognatha
Conodontophora
Graptolithina
Pterobranchia
Urochordata
Cephalochordata
Agnatha
Placodermi
Chondrichthyes
Osteichthyes
Reptilia
Mammalia
10 Families
Polychaeta
Myzostomaria
Clitellata
Mitrosagophora
Amiskwia
Dinomischus
Hallucigenia
Opabinia
Tullimonstrum
Tardigrada
Merostomata
Arachnida
Pycnogonida
Trilobita
"Trilobitoidea"
Serracaris
Marriocarida
Mimetaster
Kablikia
Branchiopoda
Ostracoda
Copepoda
Cirripedia
Malacostraca
Bostrichopodida
Cyclina
Phoronida
Stenolaemata
Gymnolaemata
Inarticulata
Articulata
Cyclocystoidea
Ophiocistioidea
"Cycloidea"
"Cyamoidea"
Hexactinellida
Demospongia
Calcarea
Sclerospongia
"Protomedusae"
"Petalonamae"
Hydrozoa
Scyphozoa
Hydroconozoa
Anthozoa
Escumasia
Nemertinea
Nematoda
Priapulida
Stenothecoida
Polyplacophora
Monoplacophora
Gastropoda
Scaphopoda
Rostroconchia
Bivalvia
Cephalopoda
Cricoconarida
Wiwaxia
Hyolitha
Sipunculida
Echiurida

creased during the early Paleozoic, and again between the Mesozoic and Cenozoic, so that late Cenozoic communities are about twice as rich in species as Paleozoic communities (Bambach 1977). Bambach (1983) has classified each species in certain fossilized communities into a "guild" describing its way of life (e.g., "epifaunal predator," "infaunal suspension feeder"), and has concluded that most of the increase in community diversity is a consequence of the addition of organisms with new ways of life rather than an increase of diversity within guilds.

Individual higher taxa vary greatly in their patterns of change (Sepkoski 1981; Figure 4). Among classes of marine invertebrates, some, such as the starfish (Stelleroidea), increased rapidly to a high level of diversity and maintained that level thereafter; some, such as trilobites (Trilobita), rose to a peak and then dwindled to extinction; others radiated, were reduced moderately (e.g., sea urchins, Echinoidea) or drastically (e.g., anthozoan corals), and then reradiated to a high diversity level. This occurred several times in the history of the ammonites (Figure 5). Interesting patterns are displayed by groups that apparently have been low in diversity since their origin (e.g., tooth shells, Scaphopoda), and by others that have persisted in low diversity long after most of the group became extinct. These include some of the "living fossils" such as the coelacanth (*Latimeria chalumnae*), the horseshoe crab (*Limulus*), and the ginkgo (*Ginkgo biloba*).

## IS DIVERSITY REGULATED?

Just as one may ask if the density of a population fluctuates randomly or is regulated around an equilibrium by density-dependent factors, one may ask if the diversity of species is regulated around an equilibrium. Even if it is, of course, the equilibrium itself may change in time as environmental conditions change, so inconstancy of species diversity is not in itself evidence against regulation of diversity. Evidence for regulation would be provided by changes in diversity that differ from random expectations, or from direct evidence that origination or extinction rates depend on diversity.

Raup et al. (1973) performed a computer simulation of changes in diversity, by assuming that in each time period a lineage could remain unchanged, branch into two, or become extinct. They assumed random fluctuations in speciation and extinction events, the only constraint being that the mean probability of speciation equaled that of extinction. Some of their randomly generated "clades" showed the same kinds of patterns that appear in the fossil record (Figure 6), so they concluded that many of the historical changes in diversity of particular taxa may be attributable to long successions of random events. However, their simulations did not result in certain patterns observed in the fossil record, which may therefore call for special explanations. None of the random clades increased as fast as some groups have, such as the therapsids; nor did they observe what they dubbed the "coelacanth effect," the long-term survival of one or a few species in

FIGURE 4

**Diversity profiles of higher taxa of marine animals in the fossil record. The scale marks off the Precambrian (PC) and the Paleozoic (P), Mesozoic (M), and Cenozoic (C) eras. The number of families is indicated by the width of the profile. The taxa include phyla (e.g., Priapulida), classes (e.g., Anthozoa), orders (e.g., Ostracoda), and several genera (e.g., *Amiskwia*) of uncertain affiliation. (From Sepkoski 1981)**

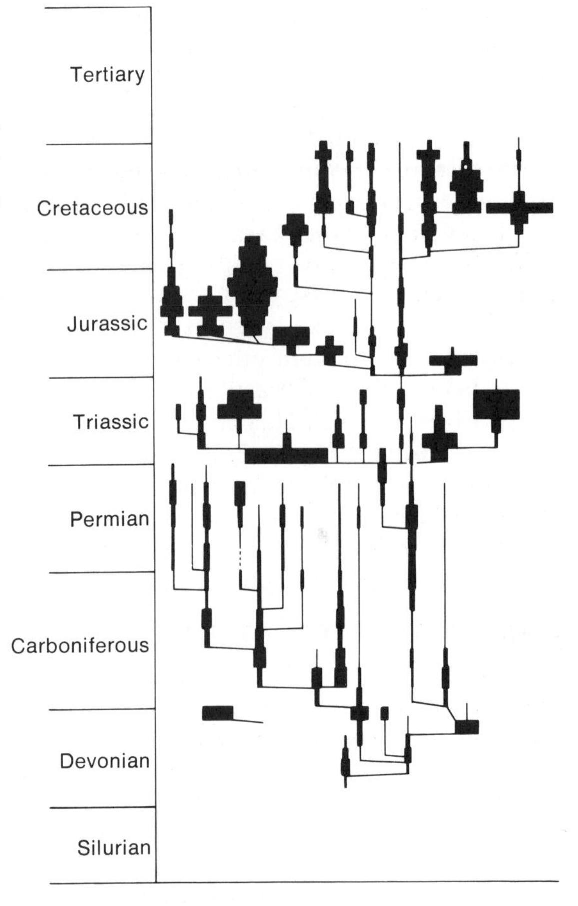


FIGURE 5

**The history of 41 superfamilies of ammonite cephalopods, showing mass extinctions at the end of Devonian, Permian, Triassic, and Cretaceous periods. In several instances a single group persisted, diversifying again after a mass extinction. The width of each bar is proportional to the number of genera known for that time period. (After Newell 1967)**

a group that otherwise became extinct long before. Strathmann and Slatkin (1983) note that the probability is very low that a phylum that originated in the Cambrian should still exist today if it has few species (e.g., the Brachiopoda), if speciation and extinction rates have varied at random around a constant mean. They suggest that small phyla may be the remnants of a much larger number that once existed (which in view of the Ediacaran and Burgess Shale faunas might be true), or that speciation and extinction rates are regulated by diversity.

The rate of origination of new taxa is correlated with the extinction rate both across groups (Stanley 1979) and within some groups over time (Webb, 1969, Mark and Flessa 1977) Over the last 12 Myr, for example, extinction rates have tended to fluctuate in concert with origination rates among North American mammals, so that diversity has remained fairly constant. Such correlations may show that density is regulated—that communities are so saturated with species that extinction rates rise to match increases in originations; they could mean that biological characteristics of species have common effects on the likelihood both of speciation and extinction (Stanley 1985); or they could merely mean that extrinsic changes in the environment influence both processes.

Ecologists have found that the number of species in a group is a fairly

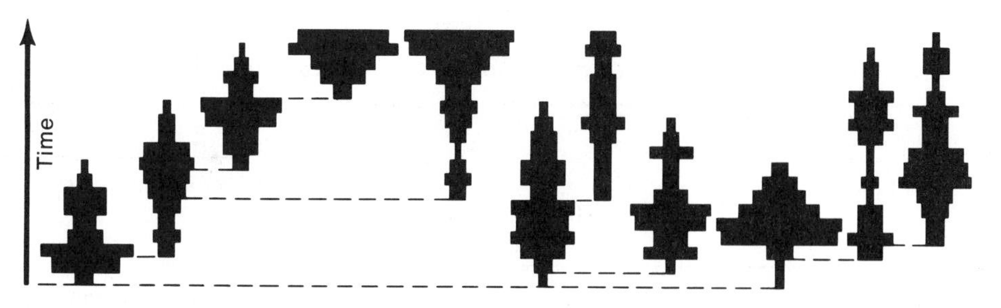


FIGURE 6

**Phylogeny and temporal variation in diversity of 11 lineages (clades) generated on a computer by random speciations and extinctions. The width of a bar reflects the number of species. Many of the changes in diversity actually observed in the fossil record are mimicked by these random clades. (After Raup et al. 1973)**

consistent function of geographic area (Chapter 2). Large areas include more habitats that support different species, and can harbor large populations, which will have a lower risk of extinction than small populations. The species diversity of marine invertebrates throughout most of the Phanerozoic has often changed in concert with fluctuations in sea level, which have affected the total area of continental seas (Simberloff 1974, Sepkoski 1976, Hallam 1984)—but the correlation is far from perfect. These various lines of evidence nevertheless suggest that for at least some taxa in some communities, biological processes regulate species diversity near an equilibrium.

Stanley (1975, 1979) has assumed that the number of species in a recently arisen group increases exponentially, as does the size of the population, following the equation $N_t = N_0e^{Rt}$ (see Chapter 2). The per capita rate of increase, $R$, is $O - E$, the rate of speciation less the rate of extinction. For each of several groups of bivalves and mammals, the time $t$ since origin is known from the fossil record; the initial number of species, $N_0$, is assumed to be 1; $E$ is calculated from the average life span of fossil species; and $N_t$ is the number of existing species. Solving for $O$ (Table I), Stanley has concluded that speciation rates have been higher in families of mammals than of bivalves.

Sepkoski (1978, 1979, 1984; see also Kitchell and Carr 1985) has extended the analogy between the growth of a population and that of the diversity of taxa in an analysis of the diversity of families of marine invertebrates. During the late Precambrian and the early Cambrian, the growth in diversity appears to have been exponential, with no evidence of acceleration at the Precambrian–Cambrian boundary as previous authors had often supposed. For the entire Paleozoic, diversity fits a logistic model (see Chapter 2) fairly well, for although there were continual originations and extinctions, diversity fluctuated around a fairly constant level from the late Cambrian through the Permian. This implies that as diversity increases, origination rates decline and/or extinction rates rise because of interactions among species, possibly including competition. Sepkoski has extended the logistic model by treating each of the three "evolutionary faunas" that he distinguished (Figure 2 and 3) as a separate "population" of families, each of which has characteristic values of $O$ and $E$ that are functions of total diversity. The model, then, is analogous to a model of competition among three

TABLE I

**Estimated rates of speciation *O*, extinction *E*, and increase in diversity *R***

| | *t* (million years) | *N* (species) | *R* | $\bar{R}$ | *E* | *O* |
|---|---|---|---|---|---|---|
| | | | | (per million years) | | |
| Bivalvia | | | | 0.07 | 0.17 | 0.24 |
| Veneridae | 120 | 2400 | 0.06 | | | |
| Tellinidae | 120 | 2700 | 0.07 | | | |
| Mammalia | | | | 0.20 | 0.50 | 0.70 |
| Bovidae (cattle, antelopes) | 23 | 115 | 0.21 | | | |
| Cervidae (deer) | 23 | 53 | 0.17 | | | |
| Muridae (rats, mice) | 23 | 844 | 0.29 | | | |
| Cercopithecidae (OW monkeys) | 23 | 60 | 0.18 | | | |
| Cebidae (NW monkeys) | 28 | 37 | 0.13 | | | |
| Cricetidae (mice) | 35 | 714 | 0.19 | | | |

(After Stanley 1975)

competing species. The rise and decline of family diversity in each of the three faunas can be accounted for by this model. Both the origination and extinction rates of families appear to depend on total diversity (Figure 7).

Among the three successive faunas that Sepkoski recognizes, the per capita rate of increase during both their initial exponential growth and during their recovery from mass extinctions was greatest for the "Cambrian fauna" and least for the "Modern fauna." Nevertheless, the three faunas attained progressively greater levels of diversity. Sepkoski (1979) notes that many of the Paleozoic forms appear, both from their morphology and the range of habitats they occurred in, to have been more ecologically specialized than their Cambrian predecessors, and that the later faunas display a greater diversity of ways of life and, presumably, of resource use. Bambach (1983), as noted above, has also provided evidence that much of the increase in diversity during the Phanerozoic has come about through increased occupancy of "ecospace," i.e., the variety of resources used. It is likely that the diversity of resources itself has increased, in that species are resources for other species that act as predators, parasites, or symbionts. For example, most angiosperms are hosts of specialized herbivorous insects, which in turn are hosts to thousands of species of parasitic insects that are commonly very host-specific. Almost every abundant organism provides a resource for other species, such as the many beetles and other arthropods which are specialized inhabitants of the nests of various species of termites and ants. But even though it is likely that diversity begets more diversity to some degree, the fossil record has not yet been analyzed well enough to determine if this is so.

## PATTERNS OF ORIGINATION

In the early days of paleontology it was presumed that new forms proliferated most rapidly during times of intense geological activity that brought with them

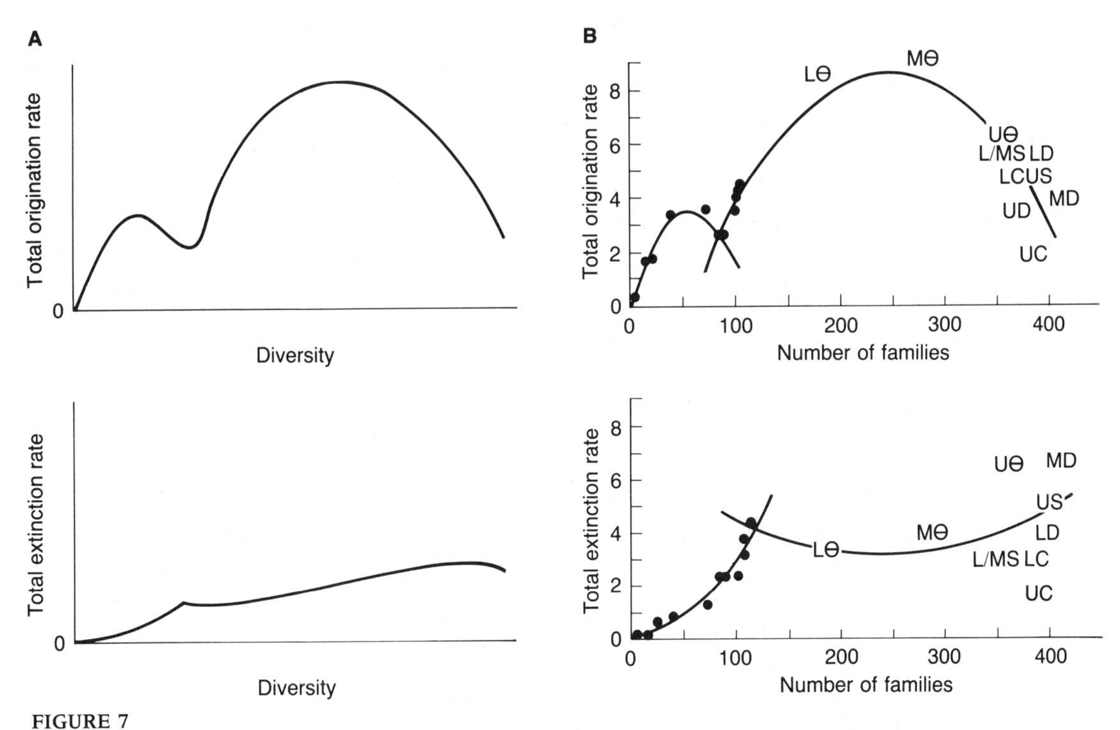


FIGURE 7

**(A) Model of the origination rate (above) and extinction rate (below) of taxa as a function of diversity, assuming diversity-dependent rates of these processes in each of two interacting groups of taxa (evolutionary faunas). (B) Total rates of origination and extinction of pre-Permian Paleozoic marine animal families, plotted against familial diversity at various pre-Permian times. The solid points represent several times during the Cambrian; letters (e.g., LC, Lower Carboniferous) represent data for post-Cambrian times. The data resemble the theoretical curves in (A), and suggest that origination and extinction rates are diversity-dependent. (From Sepkoski 1979)**

major changes in the physical environment. But although periods of diversification are related to times of major geological change, it is not clear that changes in the physical environment themselves are the major stimulus to evolution. Rather, as Simpson (1944, 1953) argued, diversification and the evolution of major new forms of life appear to result from the invasion of a new ADAPTIVE ZONE, by which Simpson meant a group of similar ecological niches different from those occupied by other groups. For example, different species of owls differ in what they eat and where they nest, but as a group they differ from hawks, and may be said to occupy a different adaptive zone. Once a lineage acquires the characteristics that enable it to occupy a new adaptive zone, the other species to which it gives rise by speciation can diverge and fill a variety of niches within that zone. Thus owls, having acquired the features required of a nocturnal bird of prey, range from thrush-sized to eagle-sized birds, and differ accordingly in the kinds of prey they favor. Such adaptive radiations may happen very rapidly on a geological time scale, as the radiation of the mammalian orders within about 12 Myr in the early Tertiary illustrates (see Stanley 1979).

This argument implies that empty ecological niches exist. It is a truism among ecologists that ecological niches cannot be identified independently of the species that occupy them, for the number of niches depends on how finely the species are able to divide resources among them. Nevertheless, it is hard to resist the conclusion that some ecological opportunities await the evolution of new species to take advantage of them. For example, sea snakes are diverse in the tropical parts of the Indian and Pacific Oceans, but are absent from the Atlantic; fish-eating and vampire bats inhabit the New World tropics, but not those of the Old World. It is hard to prove that opportunity awaits sea snakes in the Atlantic or blood-drinking bats in Africa, but the recent expansion of some species' ranges, without obvious competitive exclusion of other species, suggests that ecological vacancies may persist for millions of years without any species' evolving to fill them. The cattle egret (*Bubulcus ibis*; Figure 8), which throughout the Old World hunts insects stirred up by grazing ungulates, is thriving in both North and South America after having dispersed over the Atlantic Ocean in the late nineteenth century. The egret associates mostly with domestic cattle, but there were large native ungulates on the plains of both continents until recently. Flying insects, a potential food source for large volant predators, existed long before pterodactyls, birds, or bats evolved. Thus vacant adaptive zones seem to exist—although new species do not evolve just because opportunity awaits them. Nature does not so abhor a vacuum as to create species just because they would fit in. If the extinction rate is high relative to the speciation rate, communities will not be saturated with species. Walker and Valentine (1984) have suggested that extinction rates are usually high enough to create numerous ecological vacancies.

A lineage may enter an adaptive zone and proliferate either because it was preadapted for niches that become available, or because it evolves "key innovations"—critical new adaptations that enable it to use resources from which it was previously barred. In many cases, as Simpson (1953) suggested, the evolution of a key adaptation is the "ticket to entry" into a new adaptive zone which, once invaded, is the stage of adaptive radiation. The evolution of flight by the archosaurs that gave rise to birds, of ever-growing incisors by the ancestors of rodents, and of insect pollination by the ancestral angiosperms exemplify Simpson's point.

FIGURE 8

**The cattle egret, *Bulbucus ibis*, which a few decades ago invaded the New World from the Old World, and has become established without obvious competition from other species. (After Henry 1971)**

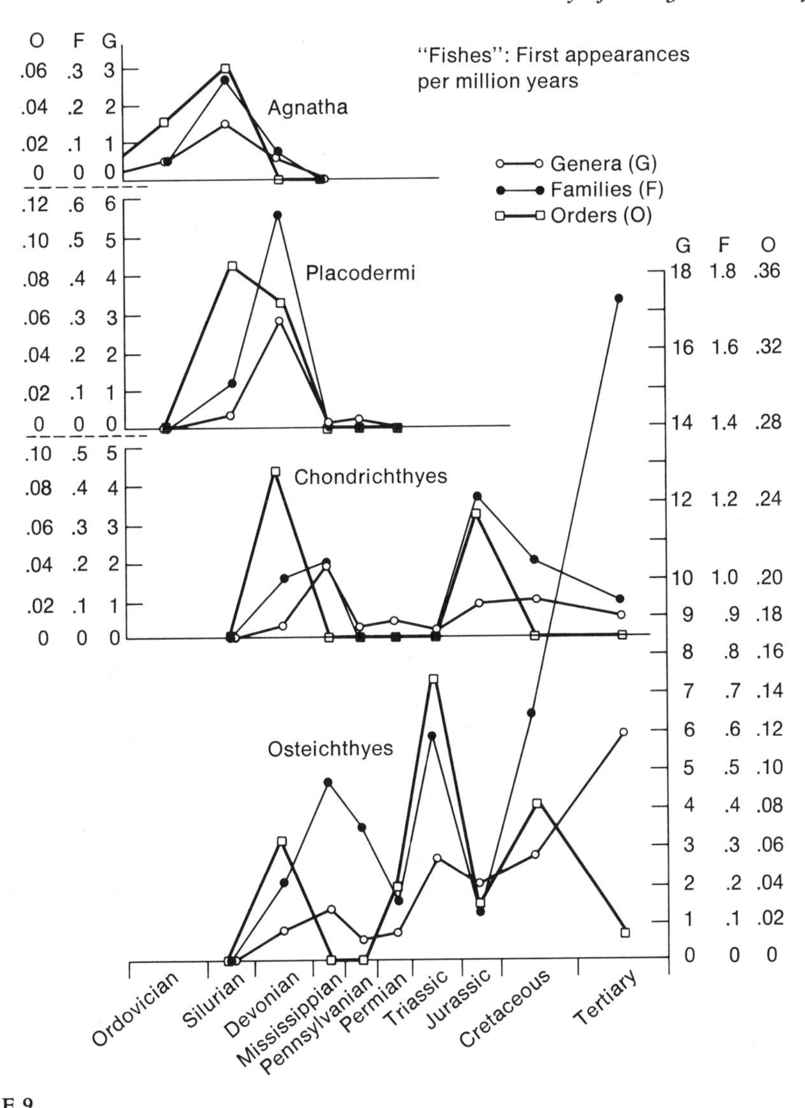


FIGURE 9

**Rates of origination of new orders, families, and genera in each of the four classes of fish-like vertebrates, from the Ordovician to the Tertiary. The three major peaks in the Osteichthyes (bony fishes) correspond roughly to the rise of the subclasses Chondrostei, Holostei, and Teleostei. (After Simpson 1953)**

The various classes of fish-like vertebrates underwent a sequence of adaptive radiations (Figure 9), and each group has a distinctive complex of morphological features—the acquisition of a bony skeleton by the agnathans or of jaws by the placoderms, for example—that may account for its rise to dominance. The several peaks of diversification of the Osteichthyes correspond to subclasses in which successive improvements in the structure of the tail and jaw mechanism were associated with new locomotory and feeding activities (see Romer 1966 for details). Van Valen (1971) has suggested that the emergence of about half the orders of mammals was accompanied by special adaptations for feeding or escaping predators; but in other orders, such as the Carnivora, diversification, although perhaps including an improvement in the efficiency of ancestral features (such as shearing teeth), was not tied to major new morphological traits. Similarly, the modern carnivore families radiated in the Oligocene not because they acquired different key innovations, but, perhaps, because they filled niches that had been vacated by the extinction in the Eocene of archaic carnivores such as oxyaenids (Radinsky 1982).

Although it is common to read that the "evolutionary success"—usually meaning the great number of species—of a group is a consequence of some special adaptive feature, one should be careful about such claims, for they are seldom if ever tested. It is certainly plausible to suppose that the key to "success" of the mammals is homeothermy or a large brain, or that the good fortunes of the angiosperms are tied to the flower or to the protection of the ovules in carpels. But there really is no indisputable evidence that this is so, and each group has other features that might equally well explain its diversity. The plasticity of branching patterns and growth form in angiosperms, for example, might be the key to their competitive superiority, not their reproductive characters (Niklas et al. 1980). It is very hard to test hypotheses about the causes of unique historical events, such as the rise of the angiosperms.

There are many examples of ECOLOGICAL REPLACEMENT in the fossil record, which have often been assumed to exemplify competitive exclusion of a less "efficient" by a more "efficient" group. Ecological replacements may be mediated by predation. For example, the reproductive structures of coralline encrusting algae appear to be more resistant to grazing by marine herbivores than those of solenoporacean algae. With the radiation of sea urchins and other grazers in the Mesozoic, the competitive balance between the two groups of algae appears to have shifted in favor of the corallines, which diversified more than 100 Myr after they first originated (Steneck 1983). Competition has often been invoked to explain the rise of one group and the extinction of another. For example, the rodents are often thought to have been superior to the multituberculates, which had rodent-like teeth and ways of life (Figure 15 in Chapter 11). In the Eocene of the Rocky Mountains, where early rodent and late multituberculate fossils occur, the diversity of the one group drops as that of the other rises.

However, to assume that the diversification of one group caused the decline of another merely because of their temporal sequence would be to fall into the logical fallacy of *post hoc, ergo propter hoc*—the supposition that because event B followed event A, it was caused by event A. In some cases it is evident that a group proliferated only after an ecologically similar group had already become extinct, leaving vacant an adaptive zone. This may well be the more common course of events. For example, the Crocodilia invaded their present adaptive zone shortly *after* the phytosaurs (Figure 10) had already become extinct, and only then evolved their special adaptations for aquatic life (Colbert 1949). Clams and brachiopods are superficially similar, and the great decline of brachiopods at the end of the Permian was followed by a great radiation of clams, but this replacement seems not to involve competitive exclusion, as has long been thought. The clams were not greatly affected by the late Permian extinction, and proliferated from a relatively high post-Permian level of diversity, while the brachiopods

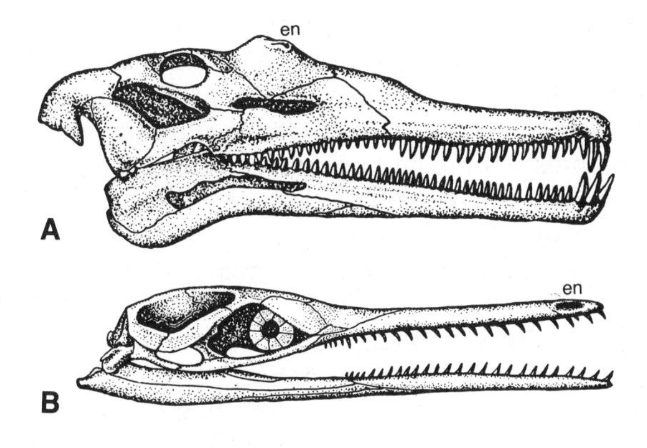


FIGURE 10

**Ecological replacements in time. (A) *Machaeroprosopus*, a phytosaur. (B) *Geosaurus*, a crocodilian. Note the differences in skull structure, especially the position of the external nares (en). (From Romer 1960)**

never recovered from the low diversity to which they had been reduced (Gould and Calloway 1980).

There is actually very little evidence that newly evolved groups achieve greater diversity by competitively excluding preestablished groups, although possible exceptions are exemplified by terrestrial plants among which, for example, the gymnosperms gradually waned as the angiosperms waxed (Niklas 1986). The more common pattern implies that established groups preempt resources, and that their extinction releases rapid adaptive radiation and morphological evolution in groups that had previously been less diverse (Raup 1984, Jablonski 1986a). The spectacular radiation of the mammals after the end-Cretaceous extinction of dinosaurs and many other reptiles is only one of numerous cases of proliferation after mass extinction events. Major changes in the physical environment such as regressions may therefore trigger higher diversification—not directly, but indirectly, by increasing extinction rates.

Maynard Smith (1976; also Stenseth and Maynard Smith 1984) has suggested on theoretical grounds that under some circumstances, ecological interactions among species may cause evolution to cease, because the interaction with one or another species would make most changes disadvantageous. In this model, external perturbations are necessary to move the system of interacting species away from its equilibrium, and hence to stimulate a renewal of evolutionary change. Raup (1984) has likewise suggested that were it not for major extinction events, the evolution of new forms of life would generally be suppressed by older forms of life that were well adapted to preempt resources. In all likelihood, then, extinction has played a major role in the diversification of life.

## PATTERNS OF EXTINCTION

Although it is generally estimated that far more than 99 percent of the species that have ever lived are extinct (Romer 1949, Simpson 1952), very little is known about the immediate causes of extinction, even of species that have become extinct in historic time (Simberloff 1986). Certainly changes in both the physical and biotic environment of a species can cause extinction, but the relative importance of various factors is unknown. Ecologists have documented instances of population extinction by competitive exclusion, as in the introduced parasitoid wasps

*Aphytis* (Chapter 2), by the incursion of predators such as *Cichla ocellaris* in Gatún Lake (Chapter 2), by weather (Ehrlich et al. 1972), and by the combination of weather and interspecific interactions (Jaeger 1980; see Chapter 5). Pathogenic organisms can also be important; within the last few decades the American chestnut (*Castanea dentata*) has been almost eliminated by the introduced fungus *Endothia parasitica*, which is a rather benign parasite on Asiatic species of chestnut; and the Dutch elm disease *Cerastosomella ulmi* is rapidly devastating the American elm (*Ulmus americana*). The destruction of habitats constitutes the major threat to species at present.

Simpson (1953) and Williams (1975) have pointed out that extinctions may vary in kind, depending on the form of the environmental change. Some changes, such as volcanic eruptions, simply cannot be countered by evolutionary change. Other changes can in principle be survived by resistant genotypes, but may occur so rapidly and in such magnitude that there is no time for such genotypes to arise; Simpson cites the case of the extinction of bird species on Caribbean islands by the predatory mongoose (*Herpestes auropunctatus*). In such cases the species survives only if a resistant genotype is already present when the catastrophe occurs. A few blight-resistant American chestnut trees may already have been present when the fungus swept through; if so, the survival of the species depends on these trees. Other changes, such as the increasing concentration of pollutants that we are now experiencing, may occur slowly enough for mutation and recombination to generate resistant genotypes before the environment becomes completely intolerable.

However, many of the environmental changes that cause extinction evoke no adaptive response, simply because they do not impose any new selective pressures. For example, in resource-limited species many individuals fail to survive because of insufficient resources. Genotypes that can use other resources would have an advantage if it were not for countervailing pressures such as competition with other species. Thus the species is always under pressure to expand its niche but cannot. If the limiting resource diminishes, the population decreases, but individuals are dying at a greater rate for the same reason as before. The cause of mortality, to which the species has already adapted as far as possible, remains the same—so there is no selection for new genotypes, even if the resource becomes so rare that the population becomes extinct.

Even without genetic changes, some species survive a change in the environment while others do not. Survival may be conferred by properties of individual organisms (e.g., broad temperature tolerance, seed dormancy) or by species-level properties such as breadth of geographic distribution. A broadly distributed species is likely to survive environmental changes that are not equally widespread.

## THE DISTRIBUTION OF EXTINCTION RATES

Paleontologists have commonly distinguished "normal" or "background" extinctions from "mass extinctions" such as that near the Permo-Triassic Boundary. It is primarily for the mass extinctions that geological data suggest possible causes. Whether or not mass extinctions represent qualitatively different phenomena from background extinctions, with entirely different causes, is disputed. Raup and Sepkoski (1982) plotted the number of extinctions of marine animal families,

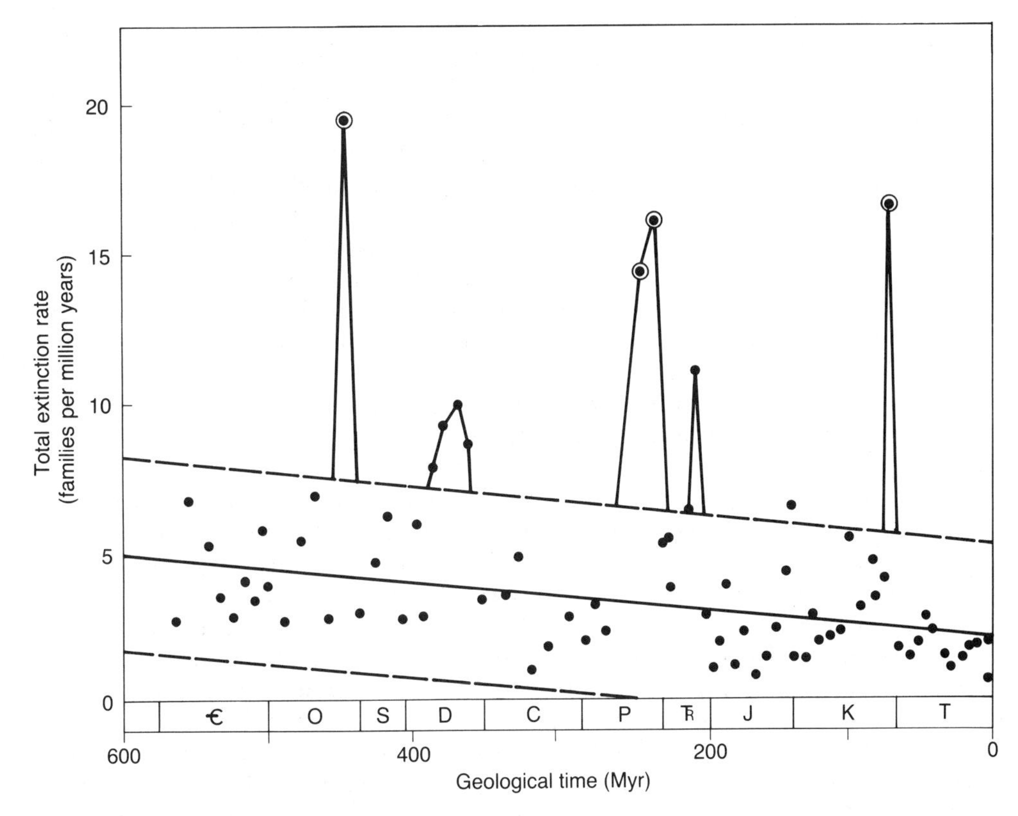


FIGURE 11

**Total extinction rate of marine animal families throughout the Phanerozoic. The points representing fewer than eight extinctions per million years are described by the solid regression line; the regression line indicates an apparent decline in "background" extinction rate per time. Circled points deviate significantly from this "background" cluster. (After Raup and Sepkoski 1982)**

resolved to within 7.4 Myr on average, throughout the Phanerozoic (Figure 11). Five episodes of exceptionally high extinction rate stood out at or near the close of the Ordovician, Devonian, Permian, Triassic, and Cretaceous periods. Under different statistical assumptions, however, the distinction between background and mass extinctions becomes blurred, and there may be a continuum of extinction rates from low to very high (Quinn 1983; see also the associated response by Raup et al.). Perhaps the most interesting feature of Figure 11 is that the background extinction rate has apparently declined steadily since the Cambrian: about 710 families did not become extinct that would have if the Cambrian extinction rate had been sustained. This lessening of the extinction rate can account for the net increase in the number of marine animal families during the Phanerozoic.

The drop in the rate of "normal" extinctions might be expected if evolution has led to a steady improvement in the adaptation of organisms to the vicissitudes of the environment. But since species cannot adapt to future changes in the environment, and since novel environmental changes have occurred throughout history, there is no necessary reason to expect species to become more resistant to extinction. If, for example, a common cause of extinction in the past has been the destruction of species' habitats, the survival or extinction of a species may not depend on intricate adaptations, efficiency, competitive ability, or even genetic flexibility, but on the kind of habitat it occupied. Extinction could then well be random with respect to a species' adaptations (Gould and Eldredge 1977).

The decline in background extinctions may not reflect progressive adaptation, however. Paleontologists have described an increasing number of species per family from early to late Phanerozoic time; since extinction of a family requires extinction of all its species, the extinction rate of families will decline even if the probability of extinction of a species is constant (Flessa and Jablonski 1985). The number of species per family could increase with time merely because extinctions that are randomly distributed over families will extinguish small groups sooner than large ones.

Van Valen (1973) has provided evidence that *within* most taxonomic groups, the probability of extinction of a genus or family is independent of its prior duration. He found that the logarithm of the number of taxa that survive for $t$ years, plotted against $t$, is a straight line with negative slope (Figure 12), as expected if older taxa are no more likely to survive at a given time than younger taxa. This pattern implies as the evolution of a group proceeds, it becomes neither more nor less resistant to new changes in the environment. To account for the data, Van Valen proposed the RED QUEEN HYPOTHESIS. This holds that the environment is continually deteriorating because of the continual evolution of other species (competitors, predators, parasites), so that like the Red Queen in *Through the Looking Glass,* each species has to run (i.e., evolve) as fast as possible just to hold its own. Species, then, fail to do so at a stochastically constant rate. The evidence of such continual coevolution is not very great (see Chapter 16), but a constant rate of extinction does not require coevolution; it could equally occur if many different kinds of environmental changes, each bearing a small risk of extinction, occur continually.

Although each major taxon may have a characteristic extinction rate, this rate differs among taxa. The average duration of a marine invertebrate genus has been 11.1 Myr during the Phanerozoic (Raup 1978), but there is considerable variation among invertebrate groups. The average longevity of a genus has been 16–20 Myr in brachiopods (Williams and Hurst 1977), 7.3 Myr in ammonites (Ward and Signor 1983), 78 Myr in bivalves (Simpson 1953). By comparison, the average duration of a genus of Carnivora has been 8 Myr (Simpson 1953). Species of Jurassic bivalves lasted 20 Myr on average (Hallam 1976), whereas an average species of Cretaceous ammonite survived for only 500,000–700,000 years (Ward and Signor 1983). Stanley (1979, 1985) has summarized data of this kind.

It is frequently thought that species meet extinction because they are "overspecialized," but this proposition is difficult to test, for every species is specialized in some way to some extent. Paleontologists have often contrasted the longevity of morphologically unusual organisms with that of their less flamboyant relatives, on the presumption that the latter were less ecologically specialized. During periods of background extinction, the longevity of ammonite genera was not related to their morphological complexity, although the longest lasting forms

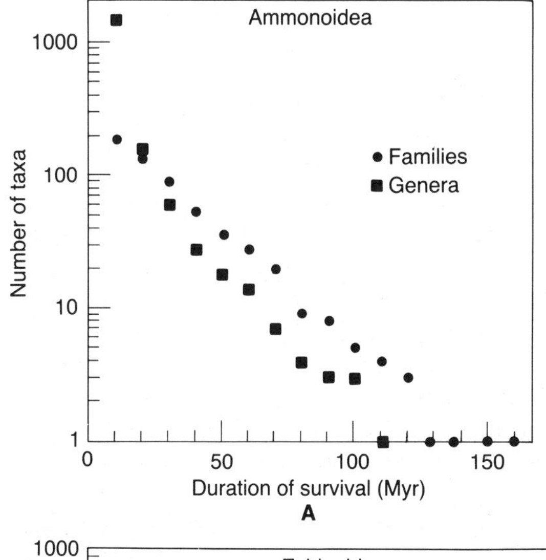


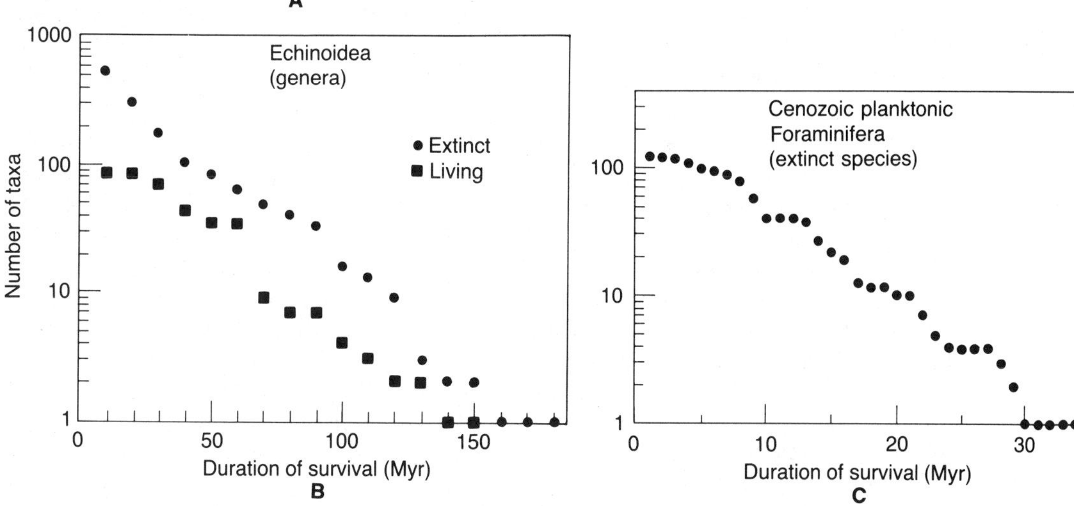


FIGURE 12

**Taxonomic survivorship curves for (A) families and genera of Ammonoidea, (B) genera of Echinoidea (sea urchins), and (C) extinct species of planktonic Foraminifera. Each point represents the number of taxa that persisted in the fossil record for a given duration, irrespective of when the taxa originated. The approximate linearity of the plots suggests that the probability of extinction is independent of the age of a taxon. (After Van Valen 1973)**

shared a morphological structure that is thought to have enabled them to live at greater depths (Ward and Signor 1983). Among Paleozoic hydrozoans, the longevity of morphologically complex, presumably more specialized genera was greater than that of morphologically simple genera during intervals between mass extinctions (Anstey 1978). Among benthic molluscs and some other invertebrates the mode of larval development appears to have an important influence on extinction rates. Species with planktotrophic larvae (i.e., a planktonic feeding stage) are prevalent in shallow waters, and appear to be ecologically generalized forms with wide geographic distributions. Those with nonplanktotrophic, less mobile larvae have narrower geographic and ecological ranges and are more prevalent in deeper waters (Jackson 1974, Jablonski and Lutz 1983). The background extinction of planktotrophic forms is less than that of nonplanktotrophs, presumably because ecologically generalized, widespread species are less susceptible to extinction by local environmental changes (Hansen 1980, Jablonski and Lutz 1983).

The features that impart resistance to extinction, however, appear to differ during episodes of mass extinction.

## MASS EXTINCTIONS

Perhaps no issue in paleontology has been as controversial as the search for causes of mass extinctions. For one thing, there is some debate about whether they are qualitatively different from "normal" extinction rates, or are simply quantitatively extreme cases in a continuum of extinction rates (Quinn 1983). If the latter is true, mass extinctions may fit comfortably into the uniformitarianism that has guided geology ever since Charles Lyell, and represent merely high rates of the same processes that have caused extinction throughout the Phanerozoic. If they are qualitatively different events, however, they may have had truly unusual causes—catastrophic, nonuniformitarian, interruptions of the normal course of earthly events. There is the further problem of whether each of the five major mass extinctions had a different cause or whether a single explanation holds for all.

At one time or another, almost every conceivable catastrophe, terrestrial or extraterrestrial, has been advanced to explain mass extinctions—often without evidence. The most recent catastrophic theory—for which there is considerable evidence—has been offered specifically for the extinction at the end of the Cretaceous. Alvarez et al. (1980) suggested that an asteroid or large meteorite collided with the earth at that time, with a force so great as to throw a pall of dust into the atmosphere that darkened the skies and lowered temperatures. Marine phytoplankton, incapable of normal photosynthesis, would suffer, and so then would the rest of the marine food chain. Their evidence was the existence of various chemical anomalies at the Cretaceous-Tertiary Boundary in many sites throughout the world, especially a concentration of iridium and other heavy metals that have higher concentrations in meteorites than on earth. Abundant evidence has been cited both for and against this hypothesis (see summaries by Jablonski 1984, 1986b, Van Valen 1984). There is some dispute as to whether or not the iridium anomaly at various localities has the same age. Moreover, the time span over which the mass extinction took place—whether or not there was a single extinction event—is greatly disputed. The impact hypothesis requires that the extinctions be instantaneous on a geological scale—on the order of tens of years to at most a few thousand years. Yet several authors (e.g., Clemens et al. 1981, Officer and Drake 1983) have argued that most of the groups that declined at the end of the Cretaceous dwindled slowly toward their nadir over a period of many thousands of years. For some, such as the ammonites and dinosaurs, the end of the Cretaceous merely marked the end of a long decline. In contrast, however, there are cases such as the brachiopods in a Danish locality, at which deposits that permit unusually fine temporal resolution have revealed synchronous extinctions of many species (Surlyk and Johansen 1984). But the species in a single locality may well become extinct synchronously because of a local event, while extinctions occur at different times elsewhere.

In contrast to the impact hypothesis, some authors have explained the end-Cretaceous extinction by the normal geological events that are known to have occurred at that time: mountain-building, massive volcanic activity, and especially a major regression of sea level that reduced the area of continental shelf and effectively eliminated the epicontinental seas. T. Schopf (1974) and Simberloff (1974) argued that a similar regression at the end of the Permian accounted for the great extinction at that time; species diversity dropped to a new equilibrium in accord with the ecological species-area "rule" $S = cA^z$ (Chapter 2). However, Jablonski (1986b) has pointed out that because the shallow waters around islands would provide refugia for shallow-water species during regressions, lowering of sea level is unlikely to account for mass extinctions. Moreover, if the end-Cretaceous mass extinction was as rapid as many paleontologists believe it to have been, it is unlikely to have been caused by a regression.

As of this writing, it appears that a majority of geologists may agree that an extraterrestrial body struck the earth at the end of the Cretaceous, and that at least some major groups of organisms became extinct rather abruptly; but there is still no clear consensus on whether or not an extraterrestrial impact caused the extinction. Whether or not the various mass extinctions during the Phanerozoic had a common cause is even less certain. Raup and Sepkoski (1984) presented tentative evidence that mass extinctions may have occurred with a regular period of about 26 Myr, which would imply a common, extraterrestrial cause. If this controversial suggestion is true, mass extinctions may have occurred under extraordinary environmental conditions, to which the past adaptations of species provided little preadaptation.

There is evidence that in some groups the survivors of mass extinction tended to be the more ecologically and morphologically generalized species; among planktonic Foraminifera, the only survivors of major extinctions were a few morphologically simple species, whose form and wide geographic range suggest that they were ecologically generalized (Cifelli 1969). Morphologically complex bryozoans suffered greater extinction during Paleozoic mass extinctions than simpler forms, even though their background extinction rate was lower in between mass extinction events (Anstey 1978). Jablonski (1986a,b) has found that during intervals between mass extinctions, the survival of species of molluscs is enhanced by a broad geographic range and by the planktotrophic larval habit, and the survival of higher taxa ("clades") is greater for clades that are rich in species. In contrast, these properties did not enhance survival through the end-Cretaceous extinction. Species-rich clades did not have a higher likelihood of survival, perhaps because their included species tend to be stenotopic. The breadth of geographic distribution of a clade appears to have been the major factor enhancing likelihood of survival (Figure 13). The end-Cretaceous extinction of marine forms was more severe in tropical than in high latitudes, so clades with a primarily tropical distribution were especially vulnerable.

In the adaptive radiations that occurred after mass extinctions, ITERATIVE EVOLUTION sometimes took place: some members of a later adaptive radiation took on morphological features of some members of an earlier one. For example (Figure 14), Cretaceous planktonic Foraminifera included morphological types termed globigerine, turborotalid, hastigerine, and globorotalid, of which the globigerine is morphologically simplest. Only globigerines (and possibly only one species of them) survived the end-Cretaceous extinction, after which hastigerine, turborotalid, and globorotalid forms again evolved. By late Oligocene time, all the forms except the globigerines had again become extinct, but by mid-Miocene times, new representatives of all these morphological types had again evolved.

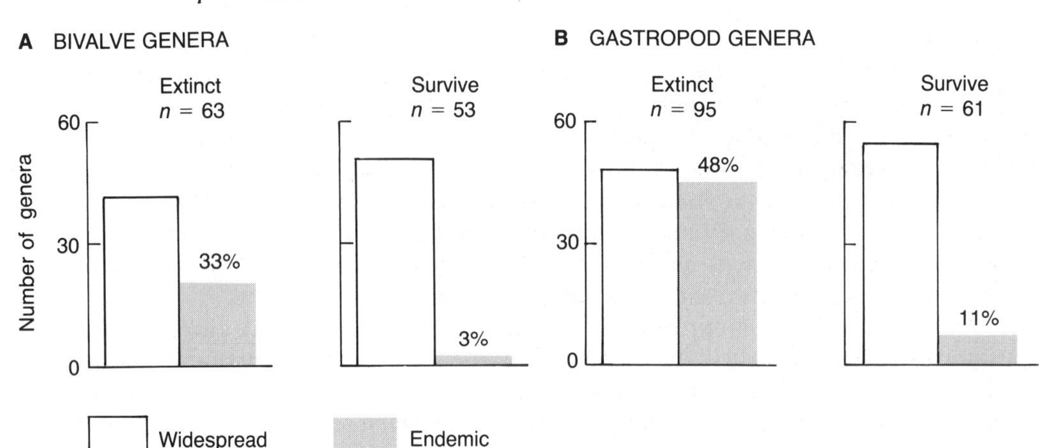


FIGURE 13

**Among both bivalves (A) and gastropods (B), the proportion of genera that survived the extinction event at the end of the Cretaceous was greater for geographically widespread forms than for genera with narrow geographical distributions (endemics). The histograms plot the percentage survival and extinction of bivalves and gastropods of the Gulf and Atlantic coastal plain of North America, distinguishing those restricted to the region from those that were more widespread. (From Jablonski 1986b)**

Each reemergence of the globorotalid morphotype displayed a trend toward evolution of a sharp keel (Cifelli 1969).

But iterative evolution is more the exception than the rule. The chief impact of the mass extinctions was the obliteration of many forms of life whose like has never reappeared (Strathmann 1978). The extinction of numerous sessile echinoderm groups in the Ordovician diminished echinoderm diversity forever; were it not for the end-Cretaceous extinction we might still have rudist reefs instead of coral reefs. The great periods of extinction were intervals in which the ecological stage was reset for whole new evolutionary dramas only partly touched by what had gone before.

## TRENDS IN EVOLUTION

If there were such a thing as evolutionary progress, it certainly was set back repeatedly by mass extinctions. But whether or not any progress can be discerned in evolution is rather dubious; at most it is possible to describe trends within individual taxonomic groups.

Discussions of the direction of evolutionary change often use the word *trend* in two distinct but related senses. It may mean a consistent, monotonic directional change within a single phyletic line in one or more characteristics, as in the increase of cranial capacity in the hominid line. Or it may mean a prevailing inclination of independent change in many related lineages—parallel evolution—as in the tendency of many mammals to evolve toward larger size, or the tendency of the sutures in ammonite shells to become increasingly complex.

Temporal sequences in the fossil record sometimes show progressive changes in one or more features within a phyletic line; in the titanotheres, for example,

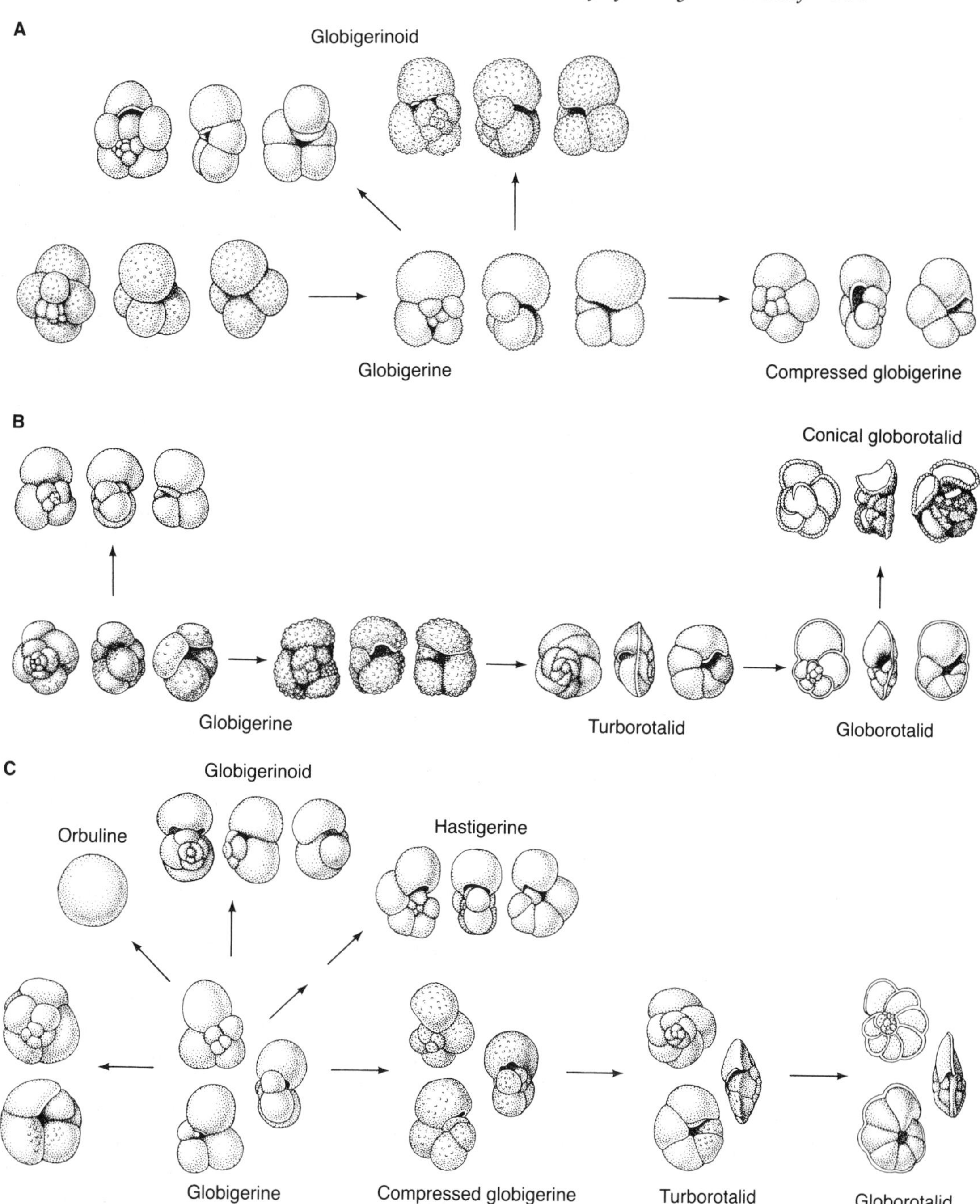


FIGURE 14

**Iterative evolution in planktonic Foraminifera. (A) Early Paleocene forms, showing presumed evolutionary transitions between morphological types. (B) Keeled forms appeared in the late Paleocene. (C) Middle Miocene forms that evolved after extinction of almost all planktonic Foraminifera in the Oligocene. Note the morphological similarity of the keeled types to the Paleocene forms. Each triplet of figures presents different views of the same organism. (From Cifelli 1969)**

FIGURE 15

**Reconstruction of four fossil species of titanotheres, showing progressive increase in size and positive allometry of the horn. (Redrawn from Stanley 1974; after Osborn 1929)**

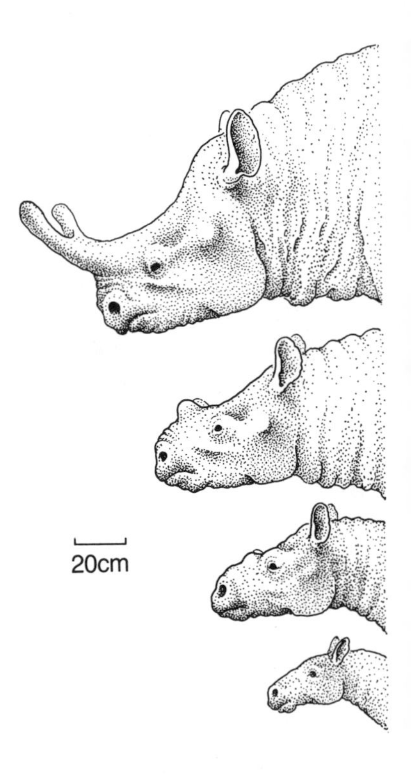


body size and the size of the nasal horn increased (Figure 15), and this trend proceeded in parallel in three subfamilies. This is but one instance of a trend toward increasing body size that is common in mammals and other groups; it has been named COPE'S RULE. Brain size has increased, relative to body size, in many groups of mammals and birds (Radinsky 1978), but, like body size, this is not an invariable trend. In no case does the fossil record show a constant rate of progression in any direction; as discussed in Chapter 14, periods of change are typically interspersed with long periods of apparent stasis.

For any clade, the most common pattern by far is not unidirectional parallel change in all branches, but divergent evolution—adaptive radiation. Cope's rule may have applied to elephants in general, but dwarf forms evolved in several elephant lineages, and reversals in the structure of the teeth accompanied the change in body size in each instance (Maglio 1972). Even a small group such as the Falconiformes exhibits not directional change, but adaptive radiation: falcons that rapidly chase their prey, eagles and hawks that hunt by soaring, ospreys and fishing eagles that dive for fish, the long-legged secretary bird that runs after its prey, vultures and caracaras that feed on carrion, and even an African vulture that feeds mostly on the fruit of the oil palm. This being the case, some of the trends we may perceive are illusions, created only by looking retrospectively from the present. It is possible, for example, to look back from the modern horse (*Equus*) through each intermediate form to the Eocene "dawn horse" *Hyracotherium*, and to discern, then, trends in the reduction of toe number and increase in height of the cheek teeth, length of the face, and body size. But to do so is to single out only one of the numerous equid lineages that can be traced back to *Hyracotherium*.

When the entire equid clade is considered, it presents not a straight line to the modern horse, but a complex branching pattern in which some lineages became adapted for browsing rather than grazing, some retained ancestral features while others acquired new ones, some decreased rather than increased in size (Figure 8 in Chapter 14).

If adaptive radiation is the norm for each group, it is certainly the character of life taken as a whole. Every attempt to find a universal trend in evolution has failed. It is possible to imagine, for example, that the evolution of the nervous system from the simple system of flatworms to the complex brain of mammals and especially of humans represents a trend toward higher consciousness or greater behavioral flexibility; but to see such a trend is to ignore the innumerable animal groups (and plants!) in which no such trend has occurred. The degree to which behavior is learned rather than instinctual is as much a special adaptation to the unpredictability of the environment and the social organization in which the individual functions as the ability to acclimate to changes in salinity is a special adaptation to variations in salinity—and is no more likely to evolve universally. Morphological complexity, including the complexity of the nervous system, has certainly increased since the Precambrian, as has the diversity of complex forms, but some Paleozoic animals rivaled those of today in complexity, and reduction in complexity has occurred in many individual groups.

Whether trends be illusory or real, many authors in the past have attempted to see a purpose or goal in evolution. ORTHOGENESIS, a word that may simply mean a documented pattern of unidirectional change in a lineage, was invested by some early authors with the idea of goal-directedness or predestination—as if, for example, Silurian ostracoderms were destined by necessity to evolve toward the final goal, the emergence of *Homo sapiens*. This mystical idea places the motive force of evolution in an Aristotelian final cause, rather than an efficient cause such as natural selection, which is not sentient and has no vision of the future. There is not the slightest evidence that evolution has a purpose or goal (Simpson 1944), and indeed the notion that evolution has a purpose is incompatible with scientific thought. Except in human affairs, the future cannot determine the present.

### Causes of evolutionary trends

If we abandon vitalistic or teleological explanations of evolutionary change, how can we explain the trends that have in fact occurred? One possibility (Chapter 14) is that developmental constraints make some changes more likely than others; for example, the number of tarsal segments in insects has invariably decreased rather than increased in evolution, and the number of digits in tetrapod vertebrates has almost always decreased. Another explanation of trends is directional selection. We can imagine either that the environment changes directionally in time, so that directional evolution occurs as a lineage tracks its environment; or that the environment is static, but that there exists an optimal phenotype for that environment toward which the lineage evolves.

There is little evidence that long-term evolutionary trends are caused by tracking of steady, long-term changes in the physical environment. It is possible, for example, that the evolution of horses' teeth matched, step by step, the expansion of the grasslands in the mid-Tertiary. At a very coarse level of resolution this is true (Woodburn and MacFadden 1982), but horse evolution was far too complex, and the resolution of the fossil record is too poor, for us to say whether the evolution kept precisely in step with the change in environment. It is somewhat more likely that progressive changes in the biotic environment impose pressure for steady change. Thus coevolution of prey with increasingly effective predators or of predators with increasingly well defended prey could result in trends, as could coevolution between competing species or between mutualistically interacting species. This is the subject of Chapter 16; suffice it to say at this point that evidence both for and against this view has been advanced.

New selective pressures can come to bear even if the ecological structure of the environment remains unchanged, for organisms by their own evolution affect what selective pressures they will experience. When a lineage first enters a new adaptive zone, it does not have all the features that would be optimal for that way of life, but by adopting it the lineage comes under selection toward this optimum. Thus a bird of prey that becomes specialized for hunting birds in open country will experience selection for long, narrow wings that, like those of a falcon, are adapted for speed; one that becomes specialized for hunting birds in forests will be selected for short rounded wings that provide greater maneuverability. If any trend is apparent in the history of life, it is the emergence of various specialized lines from generalized ancestors, as in the adaptive radiation of the mammals, and it is certainly possible to document in some cases the replacement of earlier forms by those that, by the criteria of optimal design, were more efficient in their particular way of life. Later ungulates were clearly capable of running faster than earlier ones (Bakker 1983); Cambrian echinoderms, if we may judge from anatomical evidence, had lower feeding efficiency that those that succeeded them (Paul 1977).

To these traditional explanations of trends, based on individual selection, a new explanation has been added by the proponents of punctuated equilibrium (Gould and Eldredge 1977, Stanley 1975, 1979; Chapter 14). Suppose that some individual lineages in a clade repeatedly evolve one trait—smaller body size, for example—under individual selection. An equal number of lineages evolve another trait, such as larger size. If small-sized lineages have a higher speciation rate or a lower extinction rate than large-sized lineages, then the clade as a whole will shift directionally in size over time (Figure 3 in Chapter 14). Trends could thus be caused by *species selection*. It is possible, of course, that species selection could reinforce natural selection within species; but is is also conceivable, if the rate of evolution under natural selection is low enough, that differential speciation or extinction rates could reverse the trend that natural selection alone would yield. If species selection is a reality, then any trait which enhances the rate of speciation or lowers the rate of extinction would increase in frequency among the species in a clade. So would associated traits that in themselves have no influence on susceptibility to speciation or extinction.

From the evidence so far available, the most likely candidates for characters that have been subject to species selection are those associated with ecological specialization (Vrba 1980) and gene flow (Hansen 1982, Jablonski and Lutz 1983). Among many benthic invertebrates such as gastropods, species with nonplanktotrophic larvae tend to have more limited dispersal and more restricted geographic and ecological distributions than species with planktotrophic larvae, and consequently should have higher rates of both speciation and extinction. In fact the extinction rates of nonplanktotrophic forms have been higher, there is some evidence that they have speciated faster, and there has been a long-term trend toward a greater prevalence of nonplanktotrophic species in several groups.

## SUMMARY

After each of the several mass extinctions in the history of life, diversity has rapidly increased, and has reached successively higher equilibrium levels with the passage of geologic time. Much of the long-term increase in diversity may be the consequence of exploiting a greater variety of resources, in part, perhaps, because species constitute resources for other species. Rapid evolution and diversification have typically occurred when a lineage evolves adaptations for utilizing new resources, and so enters a new "adaptive zone." This often appears to have been facilitated by the prior extinction of other taxa. Extinction, a prominent feature of evolutionary history, sometimes resets the stage for novel evolutionary events. The causes of both "mass" and "background" extinction are not entirely understood.

Long-term trends in the evolution of phenotypic characteristics are less prominent in the history of life than adaptive radiation. Those trends that can be documented can be attributed to natural selection within lineages and perhaps to species selection.

## FOR DISCUSSION AND THOUGHT

1. Is there any way in which we might recognize major innovations in existing organisms that could lead to future diversification? Can we identify existing species that are the progenitors of future higher taxa in new adaptive zones?
2. Argue for and against the proposition that unfilled niches exist.
3. Is there any reason to suppose that the probability of a species' undergoing speciation depends on the existing levels of species diversity?
4. If there is any tendency for species to evolve to be specialized, and if specialized species have a high risk of extinction, it would appear that the overall course of evolution is toward greater susceptibility to extinction. Does this contradict the supposition that natural selection results in greater adaptation of organisms to their environments?
5. In the past, some authors argued that extinction of groups such as trilobites or dinosaurs has been caused by "racial senescence," by which they meant that a taxon, like an individual organism, naturally goes through stages of birth, growth, maturity, and senescence. Does this analogy have any value? What are the uses and dangers of such analogies? See Gould (1972) for an analysis of a case history.
6. Bakker (1977) and Stanley (1979) have suggested that the extinction of higher taxa during periods of mass extinction might be caused by a reduction of the rate of speciation below the background extinction rate, rather than an elevation of the extinction rate. Why might this occur, and what evidence might bear on the hypothesis?
7. If the impact hypothesis for the end-Cretaceous extinction is true, and a pall of dust reduced insolation, one would expect the extinction rate of some kinds of organisms to be higher than others. What predictions would you make, and how does the evidence conform to your predictions?
8. In *The Origin of Species*, Darwin said that "new species become superior to their predecessors; for they have to beat in the struggle for life all the older forms, with which

they come into close competition. We may therefore conclude that if under a nearly similar climate the eocene inhabitants of the world could be put into competition with the existing inhabitants, the former would be beaten and exterminated by the latter . . ." Discuss Darwin's scenario in light of the information in this chapter.

9. Evaluate the role of competition in governing levels of species diversity and the evolution of new taxa.
10. Discuss the proposition that by chance, taxa that had been minor constituents of the world biota may survive during mass extinctions, so that the course of evolution after such events is not predictable from the evolution that preceded these events.

## MAJOR REFERENCES

Simpson, G.G. 1944. *Tempo and mode in evolution.* Columbia University Press, New York. This book and its successor, *The major features of evolution* (Columbia University Press, 1953), are seminally important interpretations of the fossil record. Some of the information is out of date, but these books are still among the most incisive analyses of evolution from a paleontological viewpoint.

Stanley, S.M. 1979. *Macroevolution: pattern and process.* Freeman, San Francisco. 332 pages. Contains much useful information on rates of diversification and extinction, used to develop controversial arguments for punctuated equilibrium and species selection.

Valentine, J.W. (editor). 1985. *Phanerozoic diversity patterns: profiles in macroevolution.* Princeton University Press, Princeton, New Jersey. 441 pages. Treats much of the material of this chapter in depth.

Elliott, D.K. (editor). 1986. *Dynamics of extinction.* Wiley, New York. Essays on extinction.

Raup, D.M., and D. Jablonski (editors). 1986. *Patterns and processes in the history of life.* Springer-Verlag, Berlin and New York. Proceedings of a workshop on rates and patterns of evolution through geologic time.

# Biogeography

# Chapter Thirteen

Biogeography is the study of the geographic distributions of organisms. It attempts to explain how species and higher taxa came to be distributed as they are and why the taxonomic composition of the biota varies from one region to another.

Biogeography is intimately related to both ecology and geology, for the answers to some biogeographic questions are more ecological and for others more historical. An understanding of present patterns of distribution depends strongly on a knowledge of historical changes in climates, geography, and species' distributions, but the relationship between geology and biogeography is a mutualistic one. The geologist's evidence of the past distribution of continents and land bridges is essential to the biogeographer's task, and patterns of biotic similarity often provide evidence for geological events. For example, the Great Lakes of North America presently drain into the Atlantic via the St. Lawrence River. But the composition of their fish fauna is very similar to that of the Mississippi River, indicating that until recently the lakes drained into the Gulf of Mexico.

Biogeography played a critical role in the origins of evolutionary theory; for example, part of the inspiration for Darwin's conviction of the reality of evolution came from his observations of the distribution of similar species of birds and tortoises in the Galápagos Islands and from the similarities and differences between fossil and Recent South American mammals. Darwin used biogeography extensively as evidence for evolution in *The Origin of Species*, and Alfred Russel Wallace spent much of his life developing the concepts of biogeography.

The study of geographic distributions may be roughly divided into historical biogeography and ecological biogeography. To some extent these approaches conflict, for historical biogeographers tend to see distributions largely as the consequence of past events such as continental drift, while ecological biogeographers tend to invoke contemporary factors such as interspecific interactions or the distribution of habitats. Sometimes the differences in emphasis are a matter of spatial scale; historical events may explain why some groups but not others occur in a certain large geographic region, while ecological factors may explain their localized distribution. In most cases both historical and ecological processes must be invoked to understand the patterns.

## THE IMPORTANCE OF PHYLOGENETIC ANALYSIS

Historical biogeography is completely reliant on phylogenetically accurate taxonomy. There is no point in trying to explain the distribution of a taxon unless its members really constitute a monophyletic group. For example, the peculiar Bornean lizard *Lanthanotus* was once considered a close relative of the Gila monster, *Heloderma*, of the southwestern United States and Mexico. This distribution appeared to require an extraordinary explanation until reanalysis of *Lanthanotus* showed that it is really related to the monitor lizards (Varanidae), which are broadly distributed throughout the Old World tropics. The hypothesis of continental drift has long been advanced to account for the distribution of the ratite birds, which include the African ostrich (Struthionidae), the South American rheas (Rheidae) and tinamous (Tinamidae), and the emu (Dromiceiidae) and cassowaries (Casuariidae) of the Australian region (Figure 1). Some ornithologists argued against this, claiming that these birds had converged from different ancestors, and so had evolved independently in situ. Recently, evidence from

FIGURE 1

**The living families of ratite birds. (A) Struthionidae (ostrich), Africa. (B) Rheidae (rhea), South America. (C) Dromiceiidae (emu), Australia. (D) Casuariidae (cassowary), Australia and New Guinea. (E) Apterygidae (kiwi), New Zealand. (F) Tinamidae (tinamou), tropical America. Despite their disjunct distribution, these birds, which except for the tinamou are flightless, are a monophyletic group. Not drawn to scale. (From Van Tyne and Berger 1959)**

morphology and from DNA-DNA hybridization has indicated that the ratites are indeed monophyletic (Cracraft 1974, Sibley and Ahlquist 1981), so the continental drift hypothesis is now favored. Phylogenetic analysis, finally, is integral to the methodology of historical biogeography, as I will explain below.

## GEOGRAPHIC PATTERNS

Although a few species are virtually cosmopolitan in distribution, the geographic range of every species is limited to varying degrees, and most higher taxa are likewise ENDEMIC—restricted—to a particular geographic region. The level of endemism depends on taxonomic rank; thus the wood warblers (Parulidae) are endemic to the New World, the warbler genus *Vermivora* is restricted to North and Central America, and the species *V. bachmani* is found in only a few localities in eastern United States.

Wallace and other early biogeographers recognized that many endemic taxa had more or less congruent distributions. For example, South and Central America have a unique fauna that includes edentates (anteaters and relatives), platyrrhine primates, caviomorph rodents (guinea pigs and allies), suboscine birds (antbirds, for example), various families of catfishes, and many other groups. Some of these extend north to a greater or lesser extent, but the faunal elements from central Mexico southward are distinctive enough for Wallace to have christened this area the "neotropical realm." The BIOGEOGRAPHIC REALMS that Wallace designated (Figure 2) are still widely recognized today, although some workers divide them more finely. The fauna of Central America, for example, has enough endemic forms to be designated a realm by some biogeographers. The realms do not correspond exactly to continental boundaries, but sometimes meet along a major barrier to dispersal, such as the Sahara Desert.

Within major realms, barriers to dispersal such as mountain ranges or regions of abrupt environmental change such as the edges of deserts often mark the boundaries between associations of species. Students of plant geography and ecology have been particularly active in dividing the flora into biomes, such as the northern hardwood forest and the boreal coniferous forest that meet in northern North America and bear with them distinct associations of animal species.

An important revelation of the early descriptive work was that although a particular type of habitat might occur in several widely scattered places in the world, the species in that habitat are typically more closely related to nearby

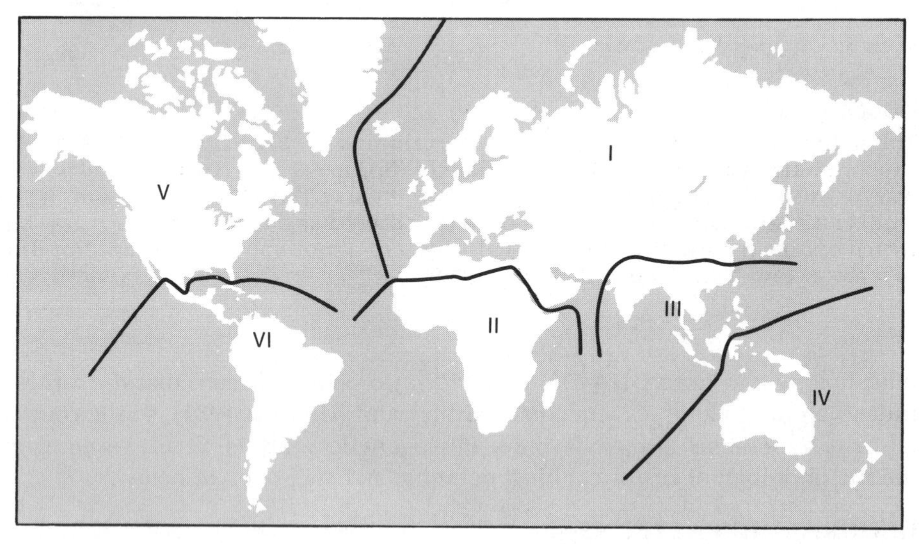


FIGURE 2

**The zoogeographic regions recognized by A.R. Wallace: Palearctic (I), Ethiopian (II), Oriental (III), Australian (IV), Nearctic (V), and Neotropical (VI). Note that the borders (which are rather arbitrary) do not necessarily demarcate the the continents.**

species in other habitats than to species that inhabit the same habitat in other realms. Very often, however, corresponding habitats will have species that are convergently similar in their adaptations. Thus in the temperate zone, a regime of cool wet winters and hot dry summers—a "Mediterranean climate"—fosters the development of a chaparral-like, shrubby, fire-resistant vegetation which is similar in aspect in California, coastal Chile, the Mediterranean region, southern Australia, and southern Africa, but which is made up of a taxonomically distinct flora in each of these areas. Superficially similar jumping rodents have evolved in the deserts of North America (kangaroo rats, Heteromyidae), central Asia (jerboas, Dipodidae), and Australia (hopping mice, Muridae); each is related to nonjumping rodents that occupy different habitats in the same region. Observations of this kind were among the strongest early arguments for the reality of evolution.

Several major patterns of geographic distribution may be described for higher taxa. Many taxa are primarily or entirely limited to a single continent or biogeographic realm: for example, the Bromeliadaceae (pineapple family) is neotropical, although a few species extend into the United States and one occurs in Africa. Australia and nearby islands constitute a region of rich endemism of forms such as marsupial mammals and the plant family Myrtaceae (*Eucalyptus* and its relatives). Many groups, such as voles (Microtinae) and maples (Aceraceae), have a holarctic distribution, encompassing Eurasia and North America. Some groups are virtually cosmopolitan, such as the skinks (Scincidae) and the pigeons (Columbidae), which occur in all but the coldest parts of all the continents. A few higher taxa, in contrast, are very narrowly endemic: the reptile order Rhynchocephalia is represented today by only a single species, the tuatara (*Sphenodon punctatus*; Figure 3) on islands along the coast of New Zealand, and the family Todidae (small birds related to kingfishers) is restricted to the Caribbean islands.

Taxa that occur in widely separated localities are said to have DISJUNCT DISTRIBUTIONS. Typically, the group is represented in each such area by different species or genera. Many disjunct distributions follow one of several patterns. Some groups are circumtropical, such as cuckoos, trogons, and parrots. Many occur only on two or more of the southern land masses. Africa, Australia, and South America all have lungfishes, ratite birds, the plant family Proteaceae, and quite a few other groups. Australia and South America share marsupials, chelydid

FIGURE 3

**The tuatara, *Sphenodon punctatus*, the sole surviving member of the ancient family Sphenodontidae. It has a relictual distribution, limited to small islands along the coast of New Zealand. (From Bellairs 1970)**

FIGURE 4

**The South American sideneck turtle *Podocnemis expansa.* A similar species, in the same genus, occurs on the island of Madagascar. (New York Zoological Society photo)**

turtles, the southern beeches (*Nothofagus*), and others. Africa and South America share cichlid and characid fishes, the gymnosperm family Podocarpaceae, and pelomedusid turtles. The genus *Podocnemis* (Figure 4) in this family occurs in both South America and Madagascar! AMPHITROPICAL DISTRIBUTIONS (Raven 1963) are rare among animals except in cold-water marine forms such as seals, but they are fairly common among plants, in which congeners or even the same species occur on either side of the tropics. For example, *Sanicula crassicaulis* in the carrot family (Apiaceae) occurs in California and southern Chile, and creosote bushes (*Larrea*) occur in the deserts of western North America and southern South America. Another common disjunct pattern is that of an eastern Asia-eastern North America distribution, as exemplified by alligators (*Alligator sinensis* and *A. mississippiensis*), skunk cabbage (*Symplocarpus foetidus*), and the tulip tree (*Liriodendron tulipifera*). Finally, some peculiar distributions are unique or shared by few taxa. Cacti are American except for one species in Madagascar. Tapirs (Figure 5) are limited to tropical America and Malaya. The Camelidae are Asian (camels) and South American (llama and relatives). The lizard family Iguanidae is diverse in North and South America but has a few species in Madagascar, Fiji, and Tonga, and the boas have a similar distribution (South and Central America, Madagascar, and Polynesia). How have all these patterns of distribution come about?

## CAUSES OF GEOGRAPHIC DISTRIBUTIONS

A possible explanation of a taxon's distribution is that the group achieved its present range by DISPERSAL from the region in which the clade originally evolved. This is not a controversial interpretation when the members of a taxon are distributed over a large contiguous area. For example, the Formicariidae (antbirds; Figure 6) are distributed throughout South and Central America, and we should encounter no opposition if we argue that the first formicariid evolved in some localized part of South America, and that there has been range expansion and speciation in the neotropics since then. But when a group occurs on either side of a barrier or is otherwise disjunctly distributed, two explanations are possible. The ratite birds in the southern hemisphere, for example, could either have dispersed to each of the continents from a CENTER OF ORIGIN (Africa, for example), or the existing distribution may arise from a formerly continuous distribution that has become fragmented by some external factor. This is the hypothesis of VICARIANCE. Vicariance can be caused either by the extinction of intervening populations or by the separation of one land mass or body of water into two, so that members of a continuously distributed biota become separated

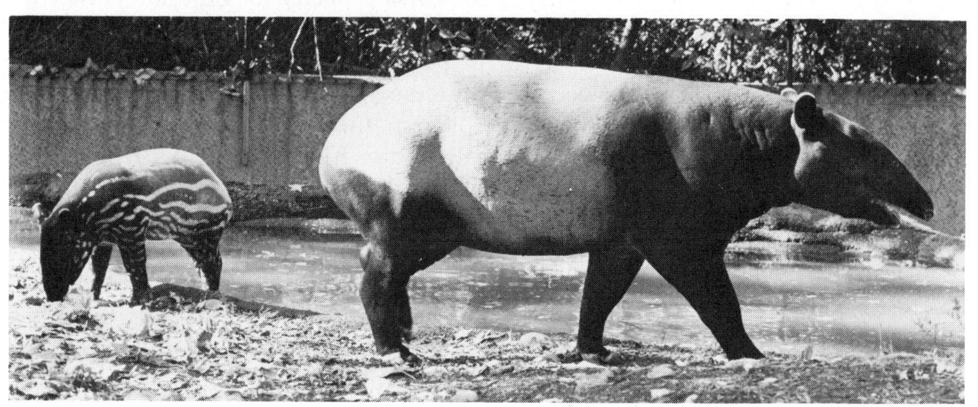


FIGURE 5
**(A) South American tapir, *Tapirus terrestris*. (B) Malayan tapir, *Tapirus indicus*. (New York Zoological Society photo)**

and evolve independently. Both dispersal and vicariance are known to have occurred.

## Dispersal

The first strong argument in favor of continental drift was presented in 1912 by Alfred Wegener, a German meteorologist, but it was not until the 1960s that the evidence and the mechanism of plate tectonics became acceptable to geologists. Consequently, although some biogeographers relied on continental drift in the early part of this century to explain distributions, the majority took the present positions of the continents as fixed, and were forced to explain disjunct distri-

FIGURE 6
**The barred antshrike (*Thamnophilus doliatus*), a member of a family (Formicariidae) of suboscine birds found throughout the Neotropics. (From Haverschmidt 1968)**

butions by dispersal. This often resulted in fantastic ad hoc hypotheses that geology could not support, such as narrow land bridges spanning the oceans. The reality of plate tectonics and moving continents, though, is just as amazing as the subsidence of imaginary land bridges, so we shouldn't feel too smugly superior to our predecessors. Although plate tectonics accounts for the distributions of some taxa, dispersal accounts for the distributions of others.

The ability to disperse over long distances varies greatly from group to group. Encysted protozoans, tardigrades, and rotifers, the spores of fungi, and many other small organisms can withstand adverse conditions and can be blown long distances by wind, so it is not surprising that many such forms are widely distributed. The seeds of plants vary in size and capacity for dormancy; mechanisms of dispersal over long distances include water, transport in bird guts, or attachment to the feet, feathers, or fur of birds or mammals. Some freshwater organisms, such as snails, are occasionally dispersed long distances as eggs in mud that adheres to the feet of aquatic birds, but others, such as freshwater fishes that cannot tolerate salt water, must usually swim from stream to stream. Geological processes that connect and disconnect different drainage systems affect their distribution. Seeds of land plants and small terrestrial organisms such as land snails, earthworms, and lizards can be carried across broad expanses of salt water by rafts of floating vegetation and debris. Large rafts of debris that emerge from rivers after heavy rains have been found far from shore. Small flying insects can be carried long distances by wind, and birds and bats, although they generally adhere closely to their home ranges and to fixed migration routes, are capable of occasional dispersal over broad expanses of ocean. Many birds, however, are very sedentary, and some birds of the forest seldom cross even short unforested gaps.

Very often the geographic distribution of taxa corresponds to their dispersal abilities. Bats, for example, are the only mammals native to Hawaii and New Zealand; in contrast, frogs and salamanders, which are incapable of surviving in salt water, are absent from most oceanic islands. However, the capacity for dispersal is only one of many factors that affect the likelihood that new populations will be founded. Even if the dispersing individuals can survive where they arrive, they may disperse and be unable to find mates. Thus in plants, hermaphroditism, self-compatibility, apomixis, and vegetative reproduction are common features of colonizing species, and in animals, parthenogenesis, internal fertilization (permitting a single fertilized female to found a colony), and a tendency to travel in flocks enhance the chances of establishment. Once arrived, an invading species is often unable to increase in numbers because it is not adapted to the local physical environment or the native species. Entomologists have made numerous attempts to introduce parasitic or predatory insects to control agricultural pests; if we may judge by the very low proportion of these that have become successfully established, most colonizations probably fail in nature as well.

### Vicariance

A disjunct distribution can arise by the extinction of intervening populations, presumably because of changes in the environment. For example, during Pleistocene glacial periods, cold-adapted northern species were distributed at lower latitudes and altitudes than they are today, and warm-adapted North American species were restricted to REFUGIA (refuges) in the southernmost United States.

With the retreat of the glaciers, the distribution of cold-adapted species became limited to northern North America, and, farther south, to the tops of mountains. During the glacials, lakes covered much of what is now desert in the American southwest. Desert pupfishes (*Cyprinodon*) and other aquatic organisms were once widely dispersed throughout the Death Valley region, but now occur only in isolated springs.

Changes in the distribution of land have produced many vicariant patterns. Before the Pliocene, Central America consisted only of a chain of islands; when the Panamanian isthmus arose, Caribbean populations of marine species became isolated from Pacific populations, and many forms have since differentiated to the subspecies or species level. The plate tectonic movement of land masses has been immensely important, but obviously it can only explain the distribution of taxa that already existed when the disjunction of continental land masses took place. For example, the Mesozoic breakup of Gondwanaland (Figure 1 in Chapter 11) can explain the distribution of ancient groups such as ratites in Australia, Africa and South America, and marsupials in South America and Australia. But the Elephantidae and Camelidae, to cite just two examples, evolved after the continents were well separated. Thus the Elephantidae are believed to have originated in Africa in the Pliocene and dispersed on foot throughout Eurasia and over the Bering Land Bridge that connected Siberia to North America, where they occur as fossils (mammoths) in the late Cenozoic. The Camelidae originated in North America in the Eocene and by the Pleistocene had dispersed into Eurasia via the Bering bridge and into South America via the Central American isthmus; they have since become extinct everywhere except Asia, northern Africa, and South America.

## EVIDENCE USED IN HISTORICAL BIOGEOGRAPHY: PALEONTOLOGY

By far the most useful evidence for explaining the distribution of a group is a good fossil record. It is often critical for determining when a group arose, and consequently whether or not it could have been fragmented by plate tectonics. The past distribution of a group often sheds light on whether its present distribution came about by dispersal or vicariance. For example, the Camelidae arose in the Eocene, much too late to explain their present distribution by the breakup of Pangaea. They were extremely diverse in, and limited to, North America throughout the Tertiary, so even though North America was connected to Eurasia by the Bering bridge during this period, their absence from Eurasia in the Tertiary indicates that they arose in North America and dispersed into Eurasia only recently. Similarly, we need not invoke continental drift to explain the distribution of tapirs in tropical America and Malaysia, or of the tulip tree *Liriodendron* in eastern Asia and eastern North America. Tapirid fossils are found in North America, Europe, and Asia from the Oligocene to the Pleistocene, and *Liriodendron* is part of a mesic flora that extended from eastern Asia across northern North America until quite recently. These vicariant distributions represent the RELICTS (remnants) of once widespread groups.

In contrast, the fossil record of many older groups indicates that plate tectonics can account for their present distribution. The breakup of Gondwanaland in particular resulted in many vicariant distributions on the southern continents.

The pipid frogs of South America and Africa are known from the Cretaceous of both continents. Marsupials surely originated on Gondwanaland and became isolated in South America and Australia; early Tertiary marsupial fossils have been found in Antarctica, which was still joined to South America and Australia in the late Cretaceous. The turtle genus *Podocnemis* was distributed over South America, Africa, Eurasia, and North America during the late Cretaceous and early Tertiary; its present distribution in South America and Madagascar is surely a consequence of continental drift followed by widespread extinction. The fossil record of many other groups with a disjunct southern continent distribution, such as the iguanid lizards, the boas, and the plant families Podocarpaceae and Proteaceae, shows that they are old enough to have been carried on the pieces of Gondwanaland.

## EVIDENCE USED IN HISTORICAL BIOGEOGRAPHY: SYSTEMATICS

In the absence of an adequate fossil record, the history of a group's present distribution can often, although not always, be inferred from phylogenetic analysis. In the past, biogeographers have relied on several lines of evidence that may prove untrustworthy. For example, it was often supposed that the region in which a group is most diverse was its center of origin from which it dispersed to other locales, on the supposition that the longer the time a group has been in a place, the more species will have originated (the age-and-area hypothesis of Willis 1922). However, this argument fails because a group may undergo adaptive radiation after colonizing a new region, or the present distribution may be relictual. Primitive primates, the lemurs, are most diverse in Madagascar, but this is a relictual distribution; lemuroid primates were widespread in the early Tertiary.

### Inferring dispersal from phylogenetic relationships

It is often possible to infer that a group arose in one area and dispersed to another if an accurate phylogeny of the group has been determined by cladistics or similar methods. If members of various branches of the phylogeny occur in area A, and if the species in area B are a sister group of one of the taxa in area A, A is likely to be the source area (Figure 7A). This principle is illustrated by the mammal faunas of North and South America, which evolved in isolation from one another until the Pliocene, when the Central American land bridge became complete. Fossils help to show which groups moved in which direction, but similar conclusions may be drawn from the relationships among living taxa.

The northern groups that moved into South America include rabbits, squirrels, cricetid rodents, bears, mustelids (weasels and skunks), cats, and deer. In North America and Eurasia, each of these groups is represented by many genera, clearly indicating a long history of adaptive radiation. In contrast, the South American species are in almost all cases very closely related to North American forms. South American rabbits, for example, are in the same genus (*Sylvilagus*) as the North American cottontails, and one of the South American cats (the mountain lion *Felis concolor*) is the same species as the North American form. It clearly entered South America very recently. The procyonids (raccoons and relatives) and the cricetid rodents, on the other hand, have been in South America longer, for they include some endemic neotropical genera. Most authors believe they

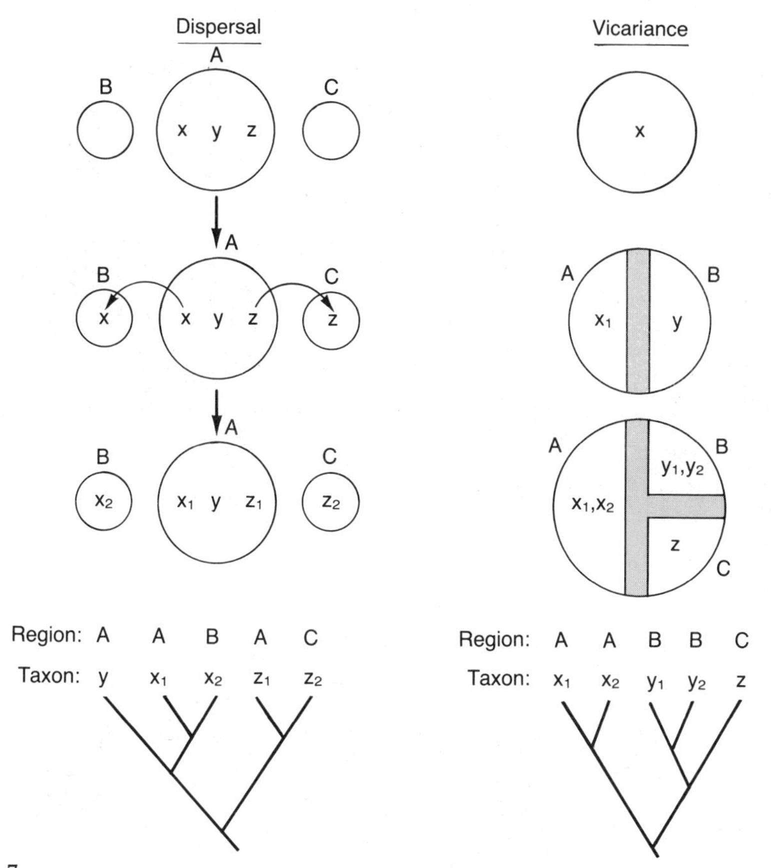


FIGURE 7

**Cladistic relationships as indicators of biogeographic history. (Left) If some members of a taxon have recently dispersed from a source area, A, to different areas, B and C, they are likely to be cladistically related to some of the species in area A (cladogram below, left). (Right) A vicariant history of successive separation of faunas is likely to yield cladistic relationships among taxa that parallel the order in which the areas became separated.**

moved into South America in the Miocene by dispersing across the Central American chain of islands (Simpson 1980, Reig 1981). Platyrrhine monkeys such as capuchins and caviomorph rodents such as guinea pigs, on the other hand, are endemic South American groups that have obviously been diversifying in South America for a long time. It is not clear whether they entered from North America early in the Tertiary, or were separated by continental drift from Africa, where distantly related forms exist at present (Figure 8).

Fewer South American groups entered North America than vice versa (Simpson 1980, Brown and Gibson 1983). Of the many mammalian groups that experienced spectacular adaptive radiation in South America during the Tertiary, only three families now have representatives in North America, and all are recent immigrants. Of the numerous Tertiary families of South American marsupials, only several families of opossums survive, and one species of opossum (*Didelphis*

FIGURE 8

**A problem in zoogeography and systematics. American porcupines (e.g., the neotropical prehensile-tailed porcupine *Coendou prehensilis*, shown at top) and Old World porcupines (e.g., the Malayan porcupine *Hystrix brachyurum*, lower photo) may be unrelated and convergently similar. If they do form a monophyletic group, it is difficult to account for their distribution. (From Walker 1975)**

*virginiana*), which is closely related to South American forms, now occurs throughout much of North America and is still moving northward. The edentates were diverse in South America; only three well-differentiated groups, the anteaters, tree sloths, and armadillos, survive. Ground sloths related to South American forms existed in North America in the late Pliocene and Pleistocene, and one of the South American species of armadillos (*Dasypus novemcinctus*) extends into the southern United States. Only one of the caviomorph rodents, the porcupine *Erethizon dorsatum*, occurs in North America; it is in a different genus from the South American porcupines, but it nevertheless clearly has a South American ancestry. On each American continent, then, the mammal fauna includes AUTOCHTHONOUS groups—those that evolved there—and ALLOCHTHONOUS groups that entered by dispersal.

### Inferring vicariance from phylogenetic relationships

"Vicariance biogeography" has recently developed as a major occupation of the cladistic school of systematics (Nelson and Platnick 1981). Its central principle is that if the region occupied by a widespread taxon has been successively fragmented into separate habitable areas, the cladistic relationships among members of the taxon should mirror the history of fragmentation (Figure 7B). Thus if a landmass A was divided, say by plate tectonics, into regions A′ and B, and if B was then divided into regions B′ and C, the taxa that occupy B′ and C have had relatively little time to diverge, and so will be cladistically more closely related to each other than to those in A′. Thus, the taxa in regions B′ and C′ are sister groups that bear an equal cladistic relationship to the taxa in A′.

In one of the first applications of these principles, Brundin (1965) inferred the phylogeny of the genera in several subfamilies of midges (Chironomidae), flies whose larvae live primarily in mountain streams of both the north and south temperate zones (Figure 9). Within two subfamilies, the most ancient bifurcations in the phylogeny are between groups that live in the north versus the south temperate zone. The sister groups of south African midges occur either in the north temperate zone or in Australia and South America. The sister groups of New Zealand midges are not localized in Australia, but occur in both South America and Australia or in South America only. Numerous genera are shared between South America and Australia; the closest relative of every one of the nine Australian genera is South American. Brundin concluded that the separation of boreal and southern forms occurred early in the Mesozoic; this was followed by the early isolation of Africa from the rest of Gondwanaland in the early Cretaceous. Next, New Zealand became separated from the complex made up of Australia, Antarctica, and South America. Finally Australia, and later South America, became separated from Antarctica. Geological evidence supports this sequence of events as inferred from the midge phylogeny.

In the absence of detailed phylogenetic information, some authors have proposed vicariant events based on the congruent pattern of disjunct distributions of many unrelated taxa, such as the African-Australian-South American pattern that many groups share. Vicariance biogeographers argue that a "generalized track," i.e., a more or less congruent distribution of many taxa, can only be produced by vicariance. They claim that dispersal alone would not yield such congruence because rare long-distance dispersal events in taxa with very different dispersal abilities are not likely to yield a correlated pattern. However, other biogeographers argue that when dispersal occurs incrementally over contiguous land masses, such as the invasion of Central America by North and South American organisms after the Isthmus of Panama arose, it too can yield congruent distributions (Savage 1982).

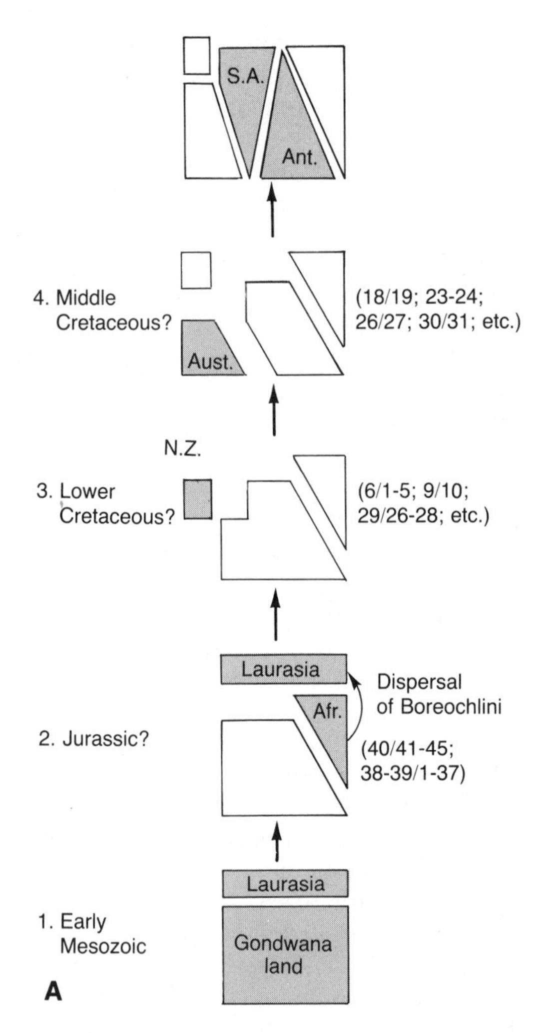


FIGURE 9

**Vicariance biogeography of chironomid midges of the subfamilies Podonominae and Aphroteniinae, as interpreted by Brundin from cladistic analysis. (A) The proposed history of separation (by continental drift) of regions inhabited by the midges, based on the distribution of sister groups in the phylogeny illustrated in (B). In (A), the times of separation of the land masses are suggested by geological information. To the right of each figure are listed sets of sister groups that imply each vicariant event; for example, the sister group of taxon 40 (in South Africa) is the set of species groups 41 through 45, found in South America and Australia. This implies separation of Africa from Gondwanaland before South America and Australia became separated. In (B), each taxon at the top of the cladogram is a species group, numbered above in a rectangle showing geographic distribution. The dots above the numbers indicate the number of species in each species group. (B from Brundin 1965)**

## HISTORY AND THE COMPOSITION OF REGIONAL BIOTAS

As the preceding discussions should make clear, the taxonomic composition of the biota of any region has been affected by both ancient and more recent history. South America, for example, has (1) some elements that are the remains of the early Mesozoic Gondwanaland biota and are shared with other southern continents (e.g., lungfish, pipid frogs, many families of plants and insects); (2) autochthonous groups that arose and diversified in the region after it became isolated in the late Mesozoic by continental drift (e.g., antbirds, edentates); (3) some forms that diversified after their progenitors entered by dispersal from North America during the Mid-Tertiary (e.g., cricetid rodents); (4) some that entered from North America during the Quaternary (e.g., mountain lion); and even (5) some species that have arrived by dispersal during historic time (e.g., the cattle egret; Figure 8 in Chapter 12). Moreover, the succession of glacial and interglacial periods in the Pleistocene has left its mark on the diversity and distribution of

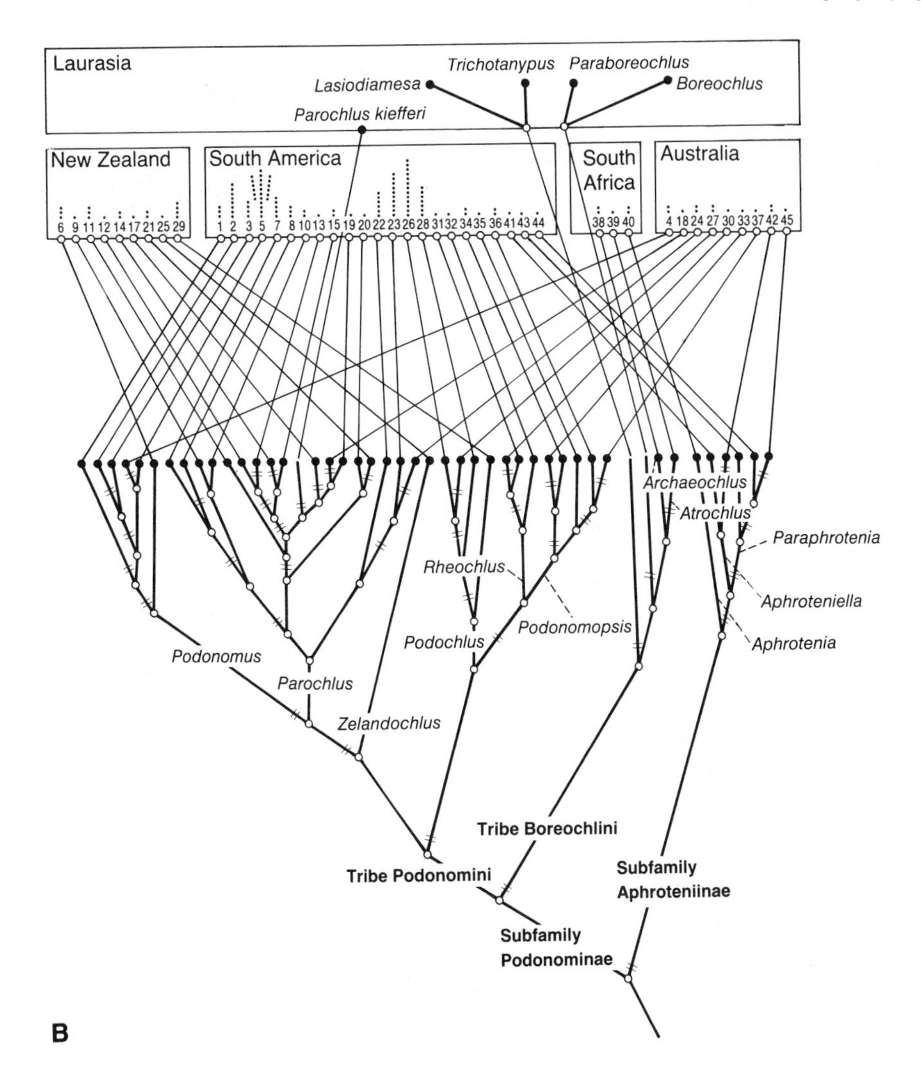


species. In North America, for example, numerous taxa are thought to have survived the glacials in southeastern and southwestern refugia. This may account for the distribution of species such as the scrub jay (*Aphelocoma coerulescens*) and the burrowing owl (*Speotyto cunicularia*), found now in western North America and peninsular Florida. Some such pairs of populations have diverged enough to be termed different subspecies, while others apparently have become different species. The eastern and western diamondback rattlesnakes (*Crotalus adamanteus* and *C. atrox*) may have developed in this way (Figure 10). Moreover, some northern forms have persisted in the south, perhaps as relicts of the ice age; thus there are beeches (*Fagus*) in the mountains of Guatemala and yews (*Taxus*) in Florida.

If we direct our attention to a local assemblage of species, such as the plants of an Adirondack forest or the birds of Cuba, questions arise that are the province both of historical biogeography and of ecology. We are interested in knowing why certain species occur in the locality and others do not. Some potential

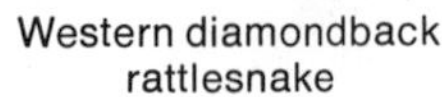


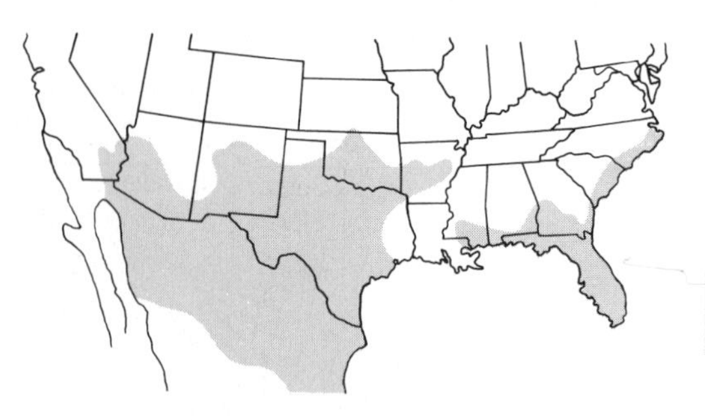

Eastern diamondback
rattlesnake

FIGURE 10

**The western and the eastern diamondback rattlesnakes, *Crotalus atrox* and *C. adamanteus*, closely related species that differentiated in southwestern and southeastern refugia during the Pleistocene. (Photographs from Klauber 1972; map after Conant 1958)**

candidates for inclusion in the community are absent for historical reasons: they have not been able to disperse to the region, or were formerly there but have become extinct. Others may have access to the locality, but are absent because they are not adapted to physical aspects of the environment, or are excluded by predators, parasites, competing species, or the absence of species on which they depend. These ecological reasons for the species composition of communities are more the province of ecological than of evolutionary studies. Nevertheless, the task of the biogeographer is to distinguish historical from ecological causes of distributions.

## ARE COMMUNITIES IN EQUILIBRIUM?

Many ecologists subscribe to the idea that under a given set of physical conditions such as temperature and rainfall, there is an upper limit to the number of species that can coexist and form a stable community. As climates change over geological time, the theoretical equilibrium diversity of species would change as well, with a time lag, because the processes that reestablish equilibrium take time. These processes would include immigration from similar habitats of species that are adapted to the new conditions; extinction of species because of interactions with the physical environment or with other species; and evolutionary adjustment of species to the new physical and biotic conditions. We must ask whether there is a theoretical maximum to the number of coexisting species, and whether present communities are at their maxima.

These questions are the subject of a large ecological literature (e.g., Brown 1981, Brown and Gibson 1983, Pianka 1983 for review). One line of evidence that communities approach a predictable equilibrium structure is that the number of species (e.g., of birds or plants) is often closely correlated with area, whether of islands within an archipelago or of regions within a continent (Chapter 2). The number of species $S$ is related to area $A$ by the relation $S = cA^z$, where $c$ and $z$ depend on the specific taxon and region under consideration. The fact that $S$ is predictable implies that immigration and/or extinction rates do not vary at random and that immigration of species does not increase without limit. The equilibrium predicted by this theory of island biogeography follows simply from the suppositions that the number of new immigrant species will decline, and the number of extinctions will increase, with the number of species in the locality; extinction rates are assumed to be higher in small areas simply because the likelihood of extinction depends on population size. The relationship of species diversity with area does not depend, then, on the supposition that interspecific interactions such as competition and predation set an upper limit to the number of coexisting species. According to the theory of island biogeography, the equilibrium number of species would increase if the immigration rate rose or the extinction rate declined because of changes in the physical environment. This theory therefore predicts a noninteractive species equilibrium (Wilson 1969).

Interactions among species may cause extinctions, establishing an "assortative species equilibrium" (Wilson 1969), a stable association of species. One line of evidence that such assortment exists is that related species (of birds, for example) sometimes have complementary distributions, suggesting that competitive exclusion limits the number of coexisting species (e.g., Diamond 1975). There is considerable controversy, however, about how to determine whether the frequency of such complementary distributions is any greater than if species were distributed at random with respect to each other (Connor and Simberloff 1984, Gilpin and Diamond 1984). Certainly some species' distributions are affected by interactions with other species, including competitors, but the incidence of these effects is hard to quantify. Paleontologists' observations that both local and global species diversity seem to have long-term plateaus in geological time (Chapter 12) and that the rate of diversification of a group is frequently correlated with the prior extinction of other species also imply that diversity attains an equilibrium.

Wilson (1969) has suggested that the assortative equilibrium established by interspecific interactions may move slowly over evolutionary time to an "evolutionary species equilibrium." Diversity could be augmented by the process of speciation, or by improved adaptation to the physical and biological environment, so lowering the rate of extinction. That speciation can increase diversity is implied by cases like the Galápagos finches: each of the major islands in the archipelago has several species that are thought to have attained sympatry after diverging on different islands. In contrast, Cocos Island is isolated from the archipelago and has only one species of finch (Chapter 8). There is less evidence that local adaptation enhances diversity by reducing extinction rates, and coevolution between species may be expected frequently to cause extinction rather than coexistence (Chapter 16).

Most communities are probably open to the addition of more species than they presently have, for several reasons. First, some organisms, such as herbivorous insects (Lawton and Strong 1981) may suffer little competition, so that this factor at least may not limit the number of coexisting species. Second, the "missing species" may simply have not yet reached the area; for example, species may still be expanding out of Pleistocene refuges. The evidence against this hypothesis is that most species, even those with apparently limited dispersal capacity, seem to be able to expand their range over broad expanses of favorable habitat with surprising rapidity. Thus the muskrat (*Ondatra zibethica*), introduced into Czechoslovakia in 1905, has already spread to easternmost Siberia, and species such as the red-backed salamander (*Plethodon cinereus*) have recolonized vast expanses of previously glaciated terrain. On the other hand, the limits to the distribution of some species appear to reflect history. Most species of North and South American freshwater fishes are absent from Central American streams, which have comparatively few species of fish. Some species of bog-inhabiting beetles—species that fly very little and disperse slowly—reach their northern limit along the southernmost frontier of the Wisconsin glaciation, even though apparently suitable habitats exist immediately to the north of this border (Reichle 1966). Many mountaintops in western North America have fewer species of mammals than they "should" have for their area, apparently because they have not been recolonized after local extinction (Brown 1971).

Third, present communities may have most of the existing species that could coexist, but new forms might arise by evolution that could utilize resources in such a way as not to exclude any preexisting species. We can envision both minor and major evolutionary changes that might add species to a biota. For example, the tree species in a deciduous forest vary greatly in the number of species of caterpillars that feed on them (Futuyma and Gould 1979), and we can imagine that, say, oak-feeding moths could give rise to new species feeding on the less heavily utilized plants. Clearly the evolution of major new taxa in the past, such as bees and birds, must have filled many vacancies in the economy of nature. As I emphasized in Chapter 12, however, it may be a very long time before such major taxa arise and fill the ecological vacuum.

Patterns of convergence suggest that in some instances, particular "guilds" of organisms—those that use a particular resource—have evolved to an approximate equilibrium. The nectar-feeding guild of birds, for example, is represented by hummingbirds (Trochilidae) and coerebid honeycreepers in the New World tropics, by sunbirds (Nectariniidae) in the African and Asian tropics, by honeyeaters (Meliphagidae) in Australia, and by honeycreepers (Drepanididae) in the Hawaiian Islands; all are convergently similar in having a long, thin beak that enables them to probe deep into flowers. Different North American deserts have similar associations of seed-eating rodents, with corresponding niches filled by species that are not necessarily closely related (Figure 11 in Chapter 2). Some communities have repeatedly achieved similar structure over evolutionary time; for example, bivalve species in the Silurian apparently partitioned the benthic habitat in ways surprisingly similar to the unrelated, modern bivalve fauna of Massachusetts (Levinton and Bambach 1975). Although individual groups of organisms may converge, it has been difficult to show that unrelated biotas in their entirety have achieved similar overall diversity and structure (Orians and Paine 1983).

Despite these indications that communities may approach an equilibrium, few ecologists would argue that any community is saturated with species. It is puzzling, then, that of the great number of species that humans have transported from one part of the earth to another, very few have become established except in highly disturbed environments. For example, numerous Eurasian plants now flourish along North American roadsides, but few indeed can be found in mature forests or prairies. To explain the failure of introduced species to invade mature communities, we might suppose they are not well adapted to the new climate, but the success of "weedy" species shows that this is not invariably true. An introduced species may be susceptible to local competitors or predators, but a priori it seems just as reasonable to expect the native species to be poorly adapted to a newcomer. It has been suggested, therefore, that communities of species have coevolved to fill available niches, and so by diffuse interactions exclude invading species (see Brown and Gibson 1983). A problem with this interpretation, though, is that most of the species in a local community have rather independent geographic distributions and their ranges have shifted independently as climate has changed (Davis 1976). Thus communities are not tightly integrated entities, and their members cannot be expected to have evolved such fine adaptations to each other as to create a seamless community fabric (Futuyma 1986). The question of how species invade new areas and become assembled into communities is far from a satisfactory solution.

## REGIONAL VARIATIONS IN SPECIES DIVERSITY

The complex interplay of historical and ecological factors comes into full focus in attempting to account for differences in the diversity of species in different parts of the world. We may take latitudinal gradients in species diversity as a preeminent example of this problem. To be sure, some major taxa, such as sandpipers, ducks, chironomid midges, and ichneumonid wasps, reach their highest diversity in boreal or temperate regions. But the majority of higher taxa are far more diverse in the tropics than at higher latitudes. A research station of 3600 acres in the lowland rain forest of Costa Rica has at least 1500 species of vascular plants, including 440 species of trees; only 150 species of trees occur in the much larger Great Smoky Mountains National Park in the southern Appalachians. Almost every major group of animals, in both the terrestrial and marine environment, is likewise more diverse in the tropics.

Ecologists have advanced numerous explanations for this gradient (reviewed by Pianka 1983), most of them based on the assumption that diversity is near equilibrium in most regions. It has been proposed, for example, that the ecological niches of competing species can be narrower, so that resources are more finely divided, in the supposedly more constant tropical environment than in the more fluctuating climate of the higher latitudes; but tropical climates seem to be just as variable (in rainfall) as elsewhere. There is some support for the idea that more species can maintain minimal viable population sizes in regions of high primary productivity than low (Brown 1975). The proximate cause of high diversity of some animal groups, such as insects and birds, may well be the greater variety of plants on which they feed and forage; but we then must account for the species diversity of plants.

The alternative to ecological explanations may be historical; one possibility is that the higher latitudes are impoverished in species because they have not recovered from the Pleistocene. Against this view, one may argue (Brown and Gibson 1983) that temperate zone species could have survived the Pleistocene in refugia just as tropical species did, that the climate of the tropics was by no means constant during the Pleistocene, and that not many species became extinct during the Pleistocene anyway.

However, the Pleistocene glaciations are not the only important historical event. Not until the late Oligocene did the present pronounced latitudinal stratification of temperature develop; until then, most of the world had a tropical or subtropical biota. It is quite possible that even 30 million years has not been enough time for many warm-adapted groups to evolve the adaptations necessary for surviving freezing temperatures and associated seasonal changes. Plants must become deciduous or cold-hardy; insectivorous birds must migrate or change their foraging habits to feed on concealed hibernating insects or on seeds; fruit- and nectar-feeding birds have little option but to migrate. It is likely, then, that some environments—very cold or very dry ones, for example—are simply so harsh that they cannot be invaded except by lineages that acquire a complex suite of adaptations, and that such evolutionary changes occur rather rarely. Those groups that have achieved such adaptations are quite diverse; for example, there are many species of winter-hardy seed-eating finches and insectivorous hibernating bats in the temperate zone. For a comparable example, life in sandy beaches requires special morphological adaptations that few groups of macro-invertebrates have acquired, and those only recently (Stanley 1979). It is conceivable, then, that given enough time, most tropical groups might spawn lineages that could invade the higher latitudes and give rise to new radiations of cold-adapted species, so that species diversity at high latitudes could some day equal that of the tropics. But it is hard to think of a test of this hypothesis.

## THE ORIGINS OF DOMINANT GROUPS

Biogeographers have long wondered whether "dominant" taxa—those that are geographically widespread and rich in species—are more likely to arise in some habitats or geographic regions than others. They have imagined that some conditions might favor the development of competitive superiority, enabling such groups to spread into new communities and displace the residents. Matthew (1915), for example, maintained that dominant groups of mammals had originated in and spread from the temperate zone. Darlington (1957), in contrast, held that almost every major group of vertebrates had spread out from the Old World tropics, where they had acquired competitive superiority in the crucible of intense interactions with a highly diverse biota. Stebbins (1974), reflecting on the great diversity of the tropical flora, wondered if the tropics were the "cradle" in which major new groups of plants evolved because of interspecific interactions, or a "museum" for the survivors of groups that had originated elsewhere. He opted for the museum.

In light of recent information, much of this controversy is moot. Most of the world has been tropical or subtropical in climate for most of the Phanerozoic, and many major groups were so widely distributed over vast land masses that have since been fragmented by plate tectonics that they can hardly be said to have had a center of origin. It is unlikely that any particular geographic region is a special site of origin of dominant groups, and indeed such a notion receives little support from the fossil record. But the possibility that some ecological circumstances foster the evolution of competitive superiority remains alive.

For example, innumerable continental species have become established on islands, and have often caused the extinction of endemic island species; the reverse seems seldom if ever to have happened. This could imply that species that have survived competition in diverse communities may be competitively superior to those in less diverse communities. The only quantitative test of this hypothesis so far seems to be Wilson's (1965) analysis of exchanges of ant species among Melanesian islands. More species have invaded small islands from large

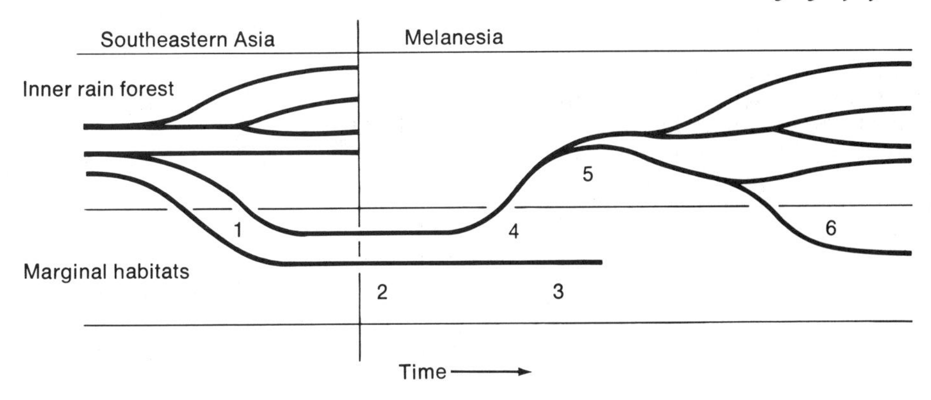


FIGURE 11

**The taxon cycle in Melanesian ants, according to Wilson. In southeastern Asia, species adapt to marginal habitats (1), then invade similar habitats on Melanesian islands (2), where they either become extinct (3) or give rise to species adapted to the rain forest (4), where they may diversify (5). Some of these species may again adapt to marginal habitats (6) and colonize other islands. (From Wilson 1959)**

islands than vice versa, but Wilson could not reject the hypothesis that this merely reflected the greater number of potential colonists on the larger islands. More analysis will be needed, however, to evaluate this hypothesis fully.

The possibility that superior competitors arise in unstable environments—which typically have low rather than high species diversity—is raised by the TAXON CYCLE that Wilson (1961) described for Melanesian ants and that seems also to hold for island birds (Greenslade 1968, Ricklefs and Cox 1972). According to Wilson, islands are colonized by ecologically generalized species adapted to ecologically marginal habitats (second growth, for example). Meeting little resistance from a species-poor native fauna, they invade the primary forest and evolve into ecologically specialized species. These species tend not to colonize other islands, but eventually become extinct and are ultimately replaced by new colonizing species that repeat the cycle (Figure 11). However, there is no direct evidence that the extinctions are caused by direct competition with new invaders; whether or not the latter are truly superior in competition remains to be determined.

The fossil record of marine benthic invertebrates indicates that origination and extinction rates—turnover rates—have been higher in tropical than temperate latitudes, and in offshore, supposedly stable habitats than in nearshore, physically less stable habitats. For example, genera of brachiopods and Foraminifera endemic to tropical waters are more numerous and younger than cosmopolitan genera, which have been less susceptible to extinction but also have a lower rate of origination (Figure 12; Stehli et al. 1969). This pattern does not support the hypothesis that dominant groups arise in areas of high species diversity where they have been tested against diverse competitors and predators. Successful invaders might therefore come from less species-rich, unstable inshore or high latitude communities. Recent paleontological evidence (Jablonski et al. 1983) reveals a pattern reminiscent of Wilson's taxon cycle: both in the Cambrian-Ordovician and the late Cretaceous, a succession of new community types, con-

FIGURE 12

**Age of genera of Cretaceous Foraminifera in low latitudes (0°–50° N, top) and high latitudes (north of 50° N, bottom). The diverse fauna of the low latitudes contained far more "young" genera, which suggests greater rates of origination in or near the tropics. (From Stehli et al. 1969)**

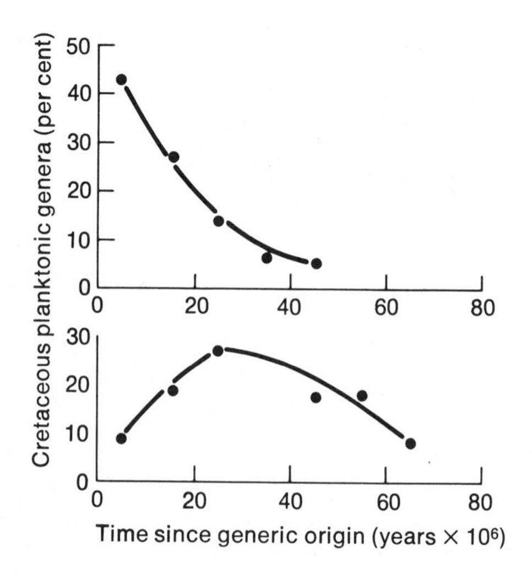


sisting of newly evolved higher taxa, appeared first in nearshore habitats, and progressively spread to deeper waters, replacing older taxa as they expanded. It has not yet been determined whether this pattern is attributable to higher evolutionary rates within nearshore lineages or to lower extinction rates that would give them more time to evolve novel adaptations. But at least in this case it appears that less diverse communities in more unstable environments may be a cradle of major new forms of life that arise from ecologically generalized species.

## SUMMARY

Phylogenetic inference and the fossil record can be used to determine the historical processes responsible for the geographical distributions of taxa. Both dispersal and vicariance, the disjunction of populations by plate tectonics or the extinction of intermediate populations, have influenced the distributions of species. The biota of any region often has a complex history of endemic diversification, vicariance, and invasion of different groups from one or more other regions at various times in the past. Geographic variation in the diversity of species has been affected by all these processes, and so has a historical component, but is probably influenced as well by ecological factors operating at present. There is some evidence that communities in unstable environments may be the locus of origin of major new dominant taxa.

## FOR DISCUSSION AND THOUGHT

1. By what mechanisms could amphitropical plant distributions have arisen?
2. Except for birds and bats, there are almost no native land vertebrates in New Zealand. The fauna includes one frog, a few lizards, *Sphenodon*, and several flightless birds (kiwis and moas, the latter recently extinct). But there are no snakes, freshwater fishes, or terrestrial mammals. Account for this peculiar situation.
3. Southwood (1961) presented evidence that plant species with a longer fossil record and wider geographic distribution are hosts for more insect species than plant species with a more limited temporal and spatial distribution. Discuss the implications of this observation for the question of whether or not communities are in evolutionary equilibrium.

4. In some cases it can be shown that species are physiologically incapable of surviving temperatures beyond the border of their range. Do such observations prove that cold regions have low diversity because of their harsh physical conditions?
5. There is considerable evidence that geographically widespread species tend to have higher local population densities than geographically more restricted species (see Brown 1984). What are the possible causes and implications of this correlation?
6. From Sanders's data on the species diversity of benthic communities, Slobodkin and Sanders (1969) proposed the "time-stability hypothesis," which states that diversity is greatest in the most stable and long-persistent environments. What is the relation between this hypothesis and the historical versus ecological explanations for regional variation in species diversity?
7. What is the significance of Darwin's observation that different habitats in the same geographic region harbor related species, whereas similar habitats in different parts of the world tend to harbor unrelated species?
8. Suppose we wish to decide whether the biota of a land mass such as New Zealand was derived primarily by dispersal from another region such as Australia, or by divergence after separation of formerly contiguous land masses (vicariance). What is the significance of missing elements (e.g., freshwater fishes)?
9. Discuss the proposition that a congruent geographic distribution of many taxa can arise by vicariance but not dispersal.

## MAJOR REFERENCES

Brown, J.H., and A.C. Gibson. 1983. *Biogeography*. Mosby, St. Louis, MO. 643 pages. A comprehensive textbook of biogeography, with excellent coverage of both historical and ecological aspects of animal and plant distributions.

# The Origin of Evolutionary Novelties

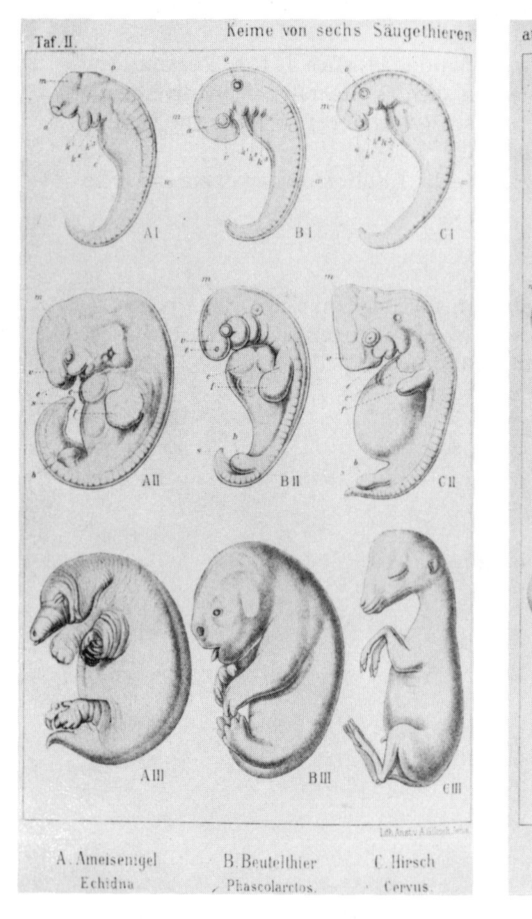


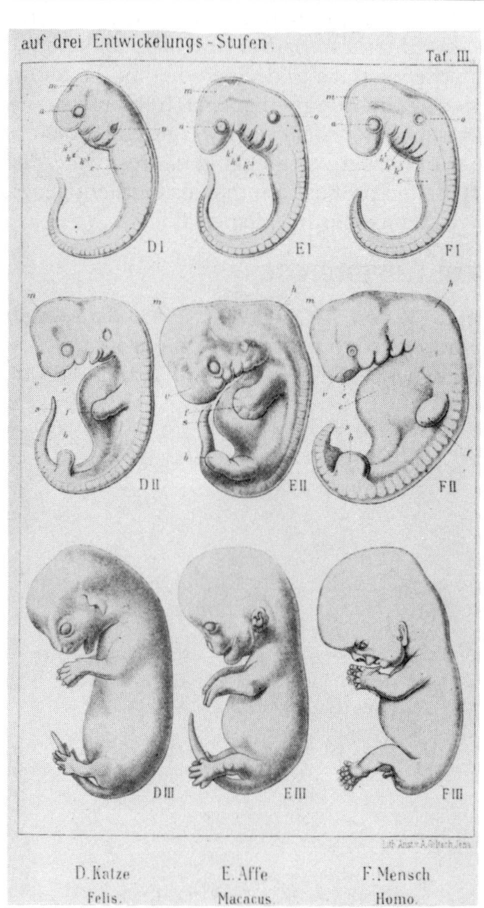


# Chapter Fourteen

Among the most challenging topics in evolutionary biology is the evolution of those differences among organisms that are so great as to distinguish higher taxa. By what mechanisms have the differences among families, orders, classes, and phyla evolved? Few other topics in evolution have been as endlessly debated, or as difficult to resolve.

We are concerned with what is generally called MACROEVOLUTION or "evolution above the species level" (Rensch 1959). The question of whether or not macroevolution entails different processes from MICROEVOLUTION, or genetic change within populations and species, was a contentious issue earlier in this century, and has been revived recently, especially by some paleontologists and developmental biologists. Several decades ago the controversy included the ideas that new features emerge by major reorganization, or SALTATION, rather than by progression through slight intermediate stages; that new features emerge by Lamarckian mechanisms; and that evolutionary trends are caused by internal, "autogenetic" drives rather than by natural selection (Mayr 1982a). Much of the effort of the leading figures in the Modern Synthesis, especially of Simpson, Rensch, Mayr, and Dobzhansky, was devoted to showing that these ideas are erroneous: that the mechanisms of microevolution adequately account for macroevolutionary phenomena, and that macroevolutionary changes are compounded of microevolutionary events. However, the question of saltational versus gradual change is once more a topic of debate.

The chief issues with which we shall be concerned are the rates at which novel features evolve, continuous versus discontinuous change, the genetic and developmental bases of morphological change, the adequacy of natural selection as an explanation for morphological evolution, and the factors that constrain the variety of features that may evolve.

## RATES OF EVOLUTION

Simpson (1944, 1953) distinguished PHYLOGENETIC rates of evolution, by which he meant the rates at which single characters or complexes of characters evolve within individual lineages, from TAXONOMIC rates of evolution, rates at which taxa with different characteristics replace each other. The taxonomic rate of evolution of a single lineage will be high if its characters evolve so rapidly that a paleontologist divides it into many chronospecies (Chapter 8); each "species" soon becomes "extinct" (pseudoextinction) and is replaced by another "species" simply because successive stages are given different names. In this case, the taxonomic rate of evolution will be correlated with the phylogenetic rate of evolution for those characters used to define different chronospecies. But a high taxonomic rate of evolution will also be calculated for groups in which real extinction, and replacement by newly evolved forms, occurs at a high rate. The taxonomic rate of evolution will thus be inversely proportional to the extinction rate, and will not necessarily be correlated with the phylogenetic rate of evolution.

Using taxonomic rates, Simpson calculated that bivalves have evolved at about one tenth the rate of mammals in the order Carnivora, for the average duration of a bivalve genus is about 78 Myr and that of a carnivore genus about 8 Myr. This contrast is clearly a comparison of extinction rates. "Species" of Cenozoic mammals appear to persist for about 1 Myr (Stanley 1979), while it is not unusual for a fossil "species" of marine invertebrate to have persisted for

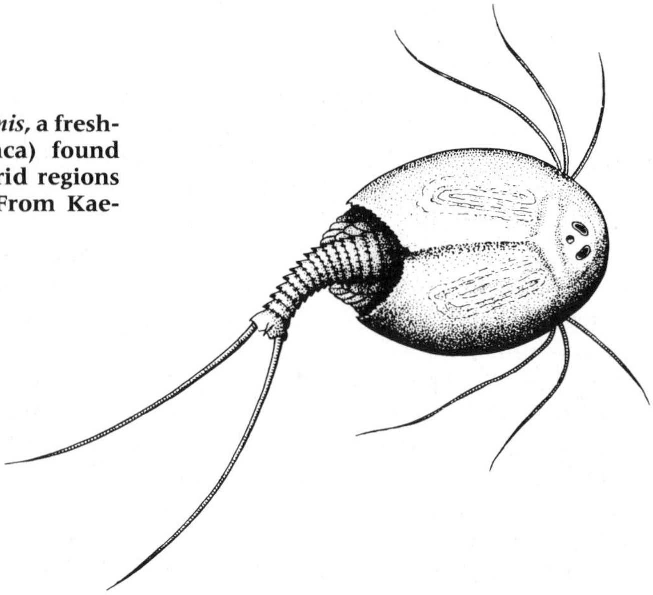

FIGURE 1

**The tadpole shrimp *Triops cancriformis*, a freshwater crustacean (order Notostraca) found primarily in temporary pools in arid regions of Eurasia and northern Africa. (From Kaestner 1970)**

more than 12 Myr (Hoffman 1982). Since fossil "species" are defined by morphological criteria, this means that most of the characters that can be measured in such fossils persist without substantial change for long periods. Extremely sluggish evolution is displayed by "living fossils. " The living coelacanth (*Latimeria chalumnae*; see Figure 9 in Chapter 11) is very similar to the coelacanths of the late Carboniferous (250 Myr ago) and the tadpole shrimp *Triops cancriformis* (Figure 1) is indistinguishable from 180 million-year-old Triassic fossils that are given the same species name. This is the oldest known living species of animal (Stanley 1979).

In contrast, new higher taxa have often evolved so rapidly that their ancestry cannot be traced well in the fossil record. Very rapid evolution is especially typical of adaptive radiations; for example, numerous families of ammonoids appear to have evolved within perhaps only 7 or 8 Myr from only a few genera that survived the mass extinction at the end of the Permian (Stanley 1979). Some of the most rapid examples of evolution are in the elephants, once a very diverse group; for example, the genera *Elephas* (Indian elephant), *Loxodonta* (African elephant), and *Mammuthus* (mammoths) seem to have arisen from *Primelephas* within about a million years during the Pliocene (Maglio 1973). Simpson coined the term QUANTUM EVOLUTION for cases of rapid, substantial evolution entailing a shift into a new adaptive zone (Chapter 8). He suggested that once a threshold has been passed in the acquisition of an adaptation to a new way of life, strong directional selection rapidly shapes the feature into more efficient form. The pulley-like tarsal bone that defines the order Artiodactyla, for example, is an adaptation for running that seems to have evolved very rapidly, and may be responsible for the great adaptive radiation of this order into deer, pigs, camels, antelopes, and other forms.

## Rates of evolution of single characters

Various measures of rates of character evolution (phylogenetic rates) have been proposed, including the *darwin*, defined by Haldane (1949) as a change of a factor of *e* (the base of natural logarithms, 2.718) per Myr. For example, the average rate of change of the height of molars in horses during the Tertiary (Figure 5 in Chapter 10) was 40 millidarwins, or 4 percent per Myr; the average rate of increase in size from the smallest to the largest of the ceratopsian dinosaurs (such as *Triceratops*) was 60 millidarwins (a factor of $6 \times 10^{-8}$ per year). In contrast, artificial selection in laboratory experiments can often change characters at rates exceeding 60,000 darwins for short periods, and introduced species often have evolved within human history at rates exceeding 400 darwins (Gingerich 1983). For example, house sparrows (*Passer domesticus*), introduced into North America from Europe about 100 years ago, have become geographically differentiated into races that are adapted in size and coloration to different North American environments (Johnston and Selander 1964). Some of their skeletal dimensions have diverged from those of European populations at rates of 50–300 darwins.

The low rates of change typically described from fossils are average rates, based on deposits separated by many thousands or millions of years. More continuous fossil records show that a low average rate of change masks frequent, rapid advances and reversals (Gingerich 1983; Figure 6 in Chapter 10). Kurtén (1959), for example, showed that characteristics of some late Cenozoic mammals fluctuated rapidly, attaining rates as high as 12 darwins for short periods. Charlesworth (1984a) has shown that the rate of evolution in single characters is highly variable in almost every case for which a detailed fossil record is available.

Can morphological rates of evolution be accounted for by population genetics? Lande (1976b) and Charlesworth (1984b) have addressed this question by quantitative genetic models. Given the variance in a character, an estimate of its heritability ($h_N^2$), and the change in the mean over the course of $t$ generations, it is possible to estimate the intensity of directional selection (i.e., the proportion of the population that fails to reproduce) that would be required each generation to change the mean by the observed amount. In contrast, if the character is subject to weak stabilizing selection but the population is finite, the mean may change by genetic drift at the loci that contribute to the character. Lande has calculated the effective population size $N^*$ at which the mean has more than a 5 percent chance of changing at a given rate purely by genetic drift.

Assuming, from data on living species, that $h_N^2 = 0.5$ and is constant, Lande applied this theory to some paleontological data, with striking results (Table I). For two measurements of the molars in each of several periods in horse evolution, the average rate of evolution, if caused by constant directional selection, would have required only about two selective deaths per million individuals per generation, even though the teeth are one of the more rapidly evolving and ecologically important features of horses. If the effective population size $N^*$ was any less than $10^4$ or even $5 \times 10^6$, genetic drift alone would suffice to explain the average rate of change. Similar conclusions have been reached by applying this theory to other fossil lineages as well (Lande 1976b, Reyment 1982, Bakker 1983, Charlesworth 1984b). Even rapid, irregular changes such as those exhibited by the radiolarian *Pseudocubus vema* (Figure 6 in Chapter 10) would require only weak selection (Charlesworth 1984b). Similarly, threshold characters such as digit number evolve at rates that can readily be explained by the theory of quantitative genetics (see Chapter 7). Many such characters have been documented in the fossil record; for example, the number of toes varied within populations of the horse *Pliohippus* during the transition from three toes to one. A small crest on the teeth of some individuals of the Oligocene horse *Mesohippus* became a fixed feature of its descendant *Parahippus*, and gradually became enlarged into a major feature of the teeth of later horses (Simpson 1953).

TABLE I

**Quantitative genetic analysis of evolutionary changes in two dimensions of the third upper molar of the horse lineage *Hyracotherium* to *Neohipparion* (see Figure 8)**

**a. Basic data**

| Species | Mean paracone height ($\log_e$) | Mean ectoloph length ($\log_e$) | Age $\times 10^6$ years |
|---|---|---|---|
| A. *Hyracotherium borealis* | 1.54 | 2.11 | 50 |
| B. *Mesohippus bairdi* | 2.12 | 2.48 | 30 |
| C. *Merychippus paniensis* | 3.53 | 2.99 | 15 |
| D. *Neohipparion occidentale* | 3.96 | 3.03 | 8 |
| Average standard deviation: | .055 | .052 | |

**b. Amounts of change in units of phenotypic standard deviations**

| Transition | ln paracone height | ln ectoloph length | Elapsed generations |
|---|---|---|---|
| A–B | 10.6 | 7.1 | $10 \times 10^6$ |
| B–C | 25.6 | 9.8 | $5 \times 10^6$ |
| C–D | 7.8 | 0.8 | $1.75 \times 10^6$ |
| A–D | 44.0 | 17.7 | $16.75 \times 10^6$ |

**c. Minimum selective mortality per generation required to explain transitions, assuming no genetic drift**

| Transition | ln paracone height | ln ectoloph length |
|---|---|---|
| A–B | $4 \times 10^{-7}$ | $3 \times 10^{-7}$ |
| B–C | $2 \times 10^{-6}$ | $8 \times 10^{-7}$ |
| C–D | $2 \times 10^{-6}$ | $2 \times 10^{-7}$ |
| A–D | $1 \times 10^{-6}$ | $4 \times 10^{-7}$ |

**d. Effective population sizes below which transitions may be explained by genetic drift alone**

| Transition | ln paracone height | ln ectoloph length |
|---|---|---|
| A–B | $2 \times 10^5$ | $4 \times 10^5$ |
| B–C | $1 \times 10^4$ | $1 \times 10^5$ |
| C–D | $6 \times 10^5$ | $5 \times 10^6$ |
| A–D | $2 \times 10^4$ | $1 \times 10^5$ |

From Lande (1976), based on data in Simpson (1953). See text for explanation.

## PUNCTUATED EQUILIBRIUM

The deposits in which successive fossils of an evolving lineage are found are often separated by gaps of at least 50,000–100,000 years, so the short-term fluctuations in a species' characteristics are seldom evident. Lineages with a coarse fossil record commonly display little substantial change for millions of years. It is not uncommon, moreover, for their morphologically distinguishable descendants (with different species names) to appear suddenly, with virtually no evidence of intermediate transitional forms, after long periods of little change (Figure 2). Except in unusually complete fossil sequences, the impression is one of apparent STASIS for long periods, "punctuated" by periods of very rapid shift to a new stable morphology.

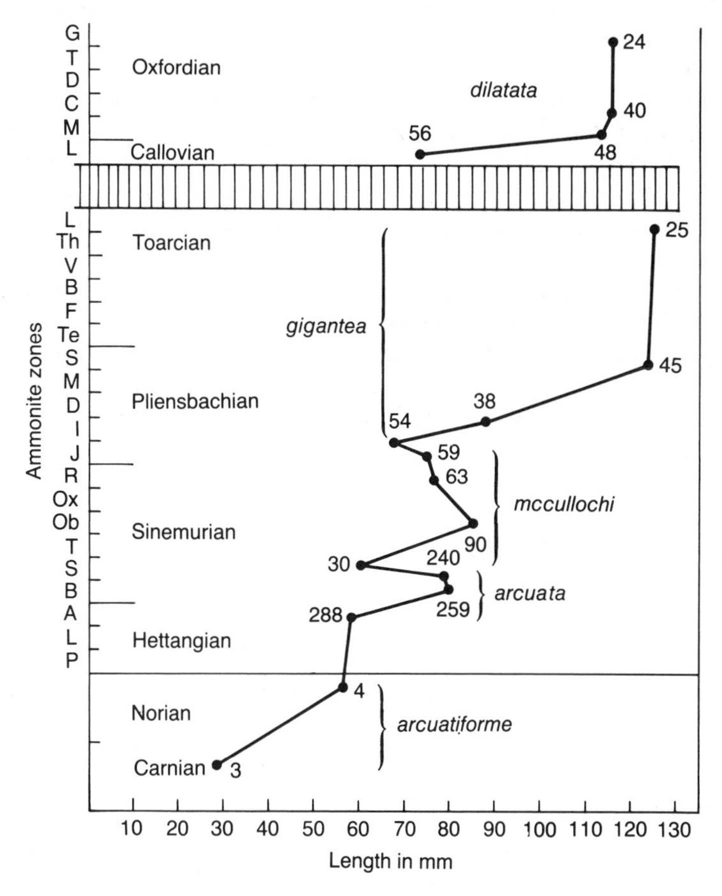


FIGURE 2

**Punctuated change without biological speciation (bifurcation) in Jurassic oysters of the genus *Gryphaea*. Length of the left valve is plotted for samples (number of specimens indicated) in a stratigraphic series of zones of about one million years' duration, distinguished by different ammonite faunas. Sequential members of the same lineage are given different species names (chronospecies). Rapid changes are evident in *dilatata* and in the transition from *mccullochi* to *gigantea*. (From Hallam 1982)**

Eldredge and Gould (1972) termed this pattern PUNCTUATED EQUILIBRIUM, and contrasted it with PHYLETIC GRADUALISM: steady anagenetic change. To explain the pattern, they invoked Mayr's theory (1954) of peripatric speciation (Chapter 8), and proposed that most evolutionary change transpires rapidly in small, localized populations in concert with the acquisition of reproductive isolation (i.e., true speciation)—an idea that Mayr himself (1954, 1963) had adumbrated. Having achieved reproductive isolation, the new form expands from its site of origin into the range of the unchanged parent species, and becomes sufficiently abundant and widespread to be recovered in the fossil record. In this theory, then, most evolutionary change is associated with and contingent on speciation (i.e., bifurcation of lineages; Figure 3). Eldredge and Gould (1972) and Stanley (1975) went on to argue that because individual species are static, long-term trends in morphology are the consequence not of anagenetic change within single lineages, but of selection among species. Character states associated with low rates of extinction or high rates of speciation will become prevalent within a clade, and establish a long-term trend, even if the direction of morphological change during the speciation process varies at random with respect to the trend (Figure 4). Stanley (1975) boldly concluded that "Macroevolution is decoupled from microevolution," and Gould and Eldredge (1977) argued that "anagenesis is only accumulated cladogenesis filtered through the directing force of species selection."

The theory of punctuated equilibrium includes the traditional neo-Darwinian processes of gradual evolution: within speciating populations, characters shift gradually, although rapidly, under the influence of genetic drift and individual selection. The proposition that characters change by discontinuous macromutational jumps (see below) is entirely different, and is not necessarily part of the

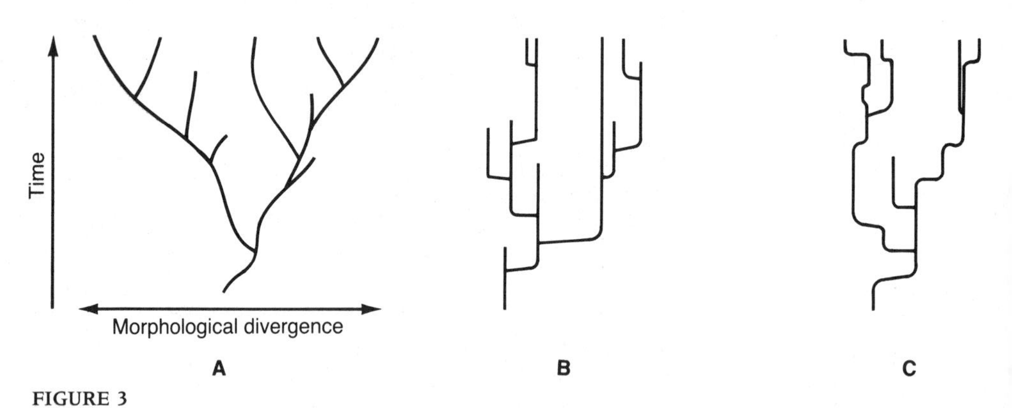


FIGURE 3

**Three scenarios for the long term evolution of a morphological character in a lineage. (A) An idealized version of phyletic gradualism: change is rather slow and steady, and is not accentuated during speciation. (B) An idealized case of punctuated equilibrium: change occurs rapidly, and only during speciation (bifurcation). (C) A neo-Darwinian view: substantial changes are rapid when they occur, so the fossil record appears punctuated. Morphological change is not necessarily concentrated in speciation events.**

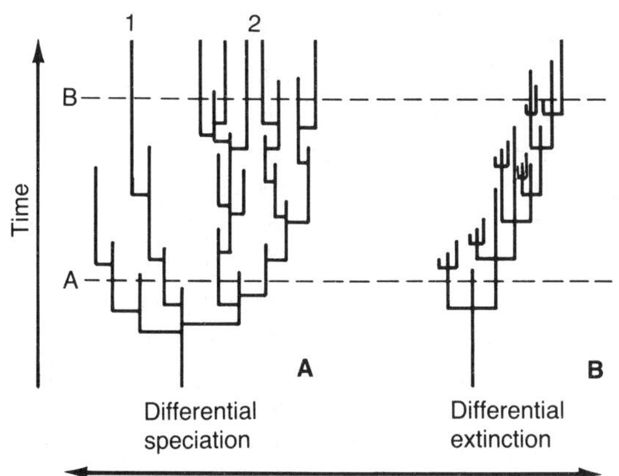


FIGURE 4

**Evolutionary trends arising from two forms of species selection, according to the hypothesis of punctuated equilibrium. (A) Differential speciation: species on the right of the phylogeny have higher speciation rates than those on the left, so this kind of species (e.g., of larger size) comes to dominate the clade. (B) Differential extinction. Most species give rise to both smaller and larger species, but larger species survive longer. In both clades, the average value of the character is greater for species at time B (upper dashed line) than at time A (lower dashed line). (After Gould 1982)**

theory of punctuated equilibrium. Moreover, the morphological changes claimed to occur during punctuations are rather modest, and are not meant to explain the origin of higher taxa. One of the original illustrations of punctuated equilibrium theory was a decrease in the number of rows of eye facets from 18 to 15 in the trilobite genus *Phacops* (Eldredge 1971). The descendant was not different enough to be placed in a new genus.

The theory expounded by Eldredge and Gould (1972) and by Stanley (1975, 1979), and the observations on which it is based, have been highly controversial. They raise several important questions: If stasis is real, how do we account for it? Is there evidence that speciation is necessary for substantial evolutionary change, and if so, why? Is macroevolution really decoupled from microevolution?

### Punctuated equilibrium: Pro and con

As a theoretical explanation for why speciation should be necessary for evolutionary change, Eldredge and Gould (1972) followed Mayr (1963) in arguing that coadapted gene pools resist genetic change and that a shift from one adaptive peak to another is facilitated by the destabilizing effect of small population size (the founder effect). Stanley (1979), in contrast, argued that local populations of a widespread species are subject to conflicting selection pressures so that gene flow among such populations prevents selection from changing allele frequencies substantially. These arguments have been strongly opposed by population geneticists, who argue that the founder effect is usually ineffective in shifting populations to new adaptive peaks (Lande 1980c, Barton and Charlesworth 1984) and that gene flow is seldom sufficient to counteract strong selection (Charlesworth et al. 1982). Substantial geographic variation within species, often over short distances, demonstrates that evolution can proceed without speciation. Whereas advocates of punctuated equilibrium propose that evolution is most rapid in single, small, localized populations, Wright's theory (1977) of shifting balance (Chapter 6) holds that progressive adaptive evolution is most likely in widespread species made up of many local populations among which there is a low level of gene flow. Wright (1982) believes that evolution is concentrated in occasional rapid events that correspond not to speciation but to changes in ecological conditions (especially the availability of new ecological niches).

One must conclude that the theory and data of population genetics do not support the notion that evolution of characters requires speciation. Moreover, to show in the fossil record that rapid change is accompanied by speciation, it would be necessary to show that an unchanged ancestral form persists sympatrically along with a modified descendant. Few if any such examples have been well documented. Nevertheless, speciation may play a role in anagenetic advance. Local populations of a widespread species develop diverse adaptations to their respective environments, but the species as a whole will become fixed only for alleles that are uniformly advantageous throughout its range and are spread by gene flow. But such "generally adaptive" (Brown 1959) traits probably seldom arise; they are likely to be uniformly advantageous only if changes in ecological conditions are widespread. Moreover, adaptive advances in local populations may be ephemeral in geological time, because local populations become extinct and because shifts in the geographical distribution of habitats may bring about interbreeding among formerly isolated, differentiated populations. Attainment of reproductive isolation permits a population with divergent characters to become sympatric with the parent species, without losing its identity by interbreeding; thus speciation, although not necessary for the evolution of new characters, may permit their retention (Futuyma 1986).

The data of the fossil record permit several interpretations. There is some debate, first of all, about whether some sequences display gradualism or stasis with punctuation; Figure 5, for example, is interpreted by its author (Gingerich 1976; see also Bookstein et al. 1978) as an instance primarily of phyletic gradualism, but by Gould and Eldredge (1977) as a pattern of punctuated equilibrium. That is, the distinction between gradual and punctuated change is not well defined. The more general problem is that if fossil sequences appear to display rapid shifts from one static morphology to another, speciation may have occurred, or there may simply have been stepwise anagenetic advance in a single lineage, without speciation (Figure 3). Geologically "instantaneous" events, in the eyes of a paleontologist, may take thousands of years—during which considerable change can occur that a population geneticist sees as slow and gradual (Stebbins and Ayala 1981). Turner (1986) points out that a modest change in a character under natural selection will almost inevitably take less than 50,000 generations if a change in optimal phenotype is even slightly favored by natural selection. For example, a rare dominant allele with a selective advantage of 0.01 will become almost fixed in a population of 1000 individuals within 3160 generations. Evolution by natural selection will usually transpire more rapidly than a coarse fossil record can document. Thus, says Turner, "population genetics predicts that evolution by natural selection will be punctuational!"

## Stasis

Perhaps the most interesting problem posed by the fossil record is stasis. Granted that many characters fluctuate slightly but rapidly and are not truly static, the relative constancy of morphology for millions of years nevertheless seems surprising, in view of the extreme inconstancy of the environment. Not only fossils, but extant forms provide evidence of stasis. For example, protein differences

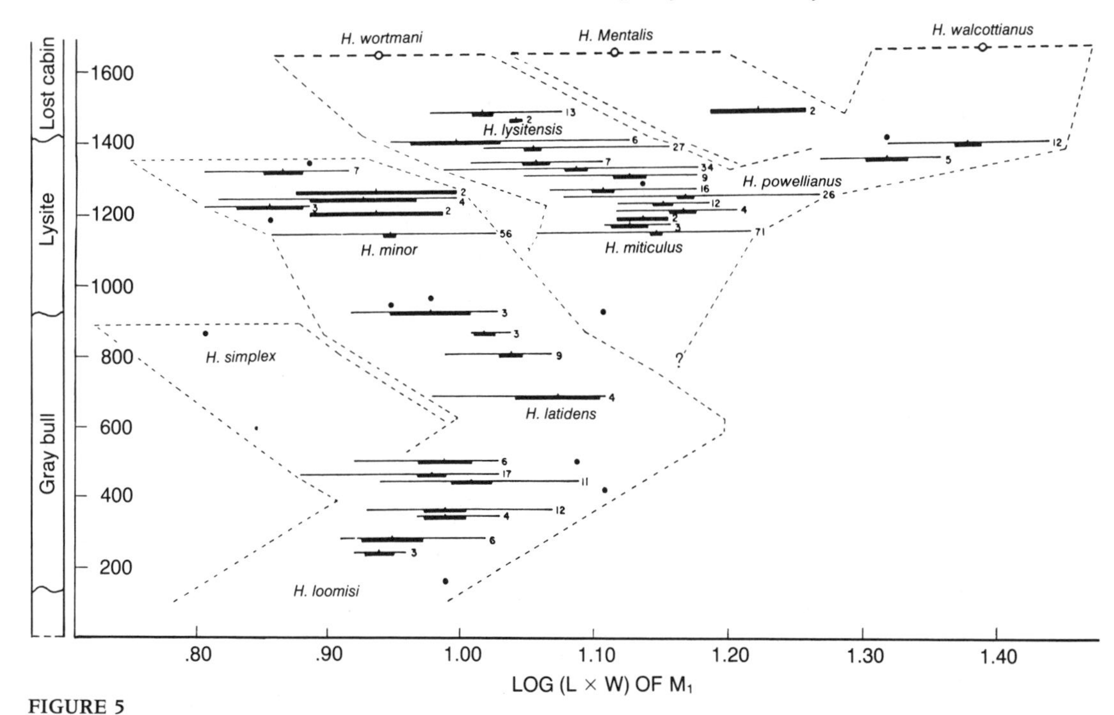


FIGURE 5

**Evolution of the first upper molar of the condylarth *Hyopsodus* in early Eocene deposits in the Big Horn Basin of Wyoming. The mean of each sample is shown with the standard error (horizontal bar) and the range (horizontal line). Sample sizes are indicated at the right of each distribution; points are single specimens. The dotted envelopes show Gingerich's interpretation of the data as reflecting both gradual anagenetic change and speciation. These data have also been interpreted by other authors as an example of punctuated equilibrium. (From Gingerich 1976)**

among some species of the salamander genus *Plethodon* suggest (assuming relative constancy of molecular evolution; Chapter 10) that the most distantly related species diverged at least 60 Myr ago. Yet the numerous species in the genus are morphologically almost indistinguishable except in size, coloration, and minor differences in skeletal dimensions (Wake et al. 1983). Abundant speciation has occurred, yet substantial morphological divergence has occurred only in one lineage that became adapted to an arboreal adaptive zone and is recognized as a distinct genus (*Aneides*). Stasis in *Plethodon* is surely not attributable to gene flow, which is extremely low in these salamanders.

From the standpoint of population genetics, the most likely explanation of stasis is stabilizing selection—but how could selection favor the same morphology in the face of environmental change? It is precisely in this context that we must recognize (cf. Chapter 2) that organisms are not merely the passive objects of environmental influences, but that they define and create their environments (Lewontin 1983). Having evolved to use a certain microhabitat or certain kinds of food, a lineage may be buffered from more macroscopic changes, of which it is, so to speak, blissfully unaware. *Drosophila* in a bottle cannot escape artificial selection for tolerance of high temperatures, but *Drosophila* in a forest can escape by finding cool microhabitats. *Plethodon* salamanders, inhabiting moist forest litter and employing a very generalized feeding mechanism to capture small invertebrates, may have experienced few novel selection pressures for as long as moist forests and litter arthropods have existed (Wake et al. 1983). The environment of any particular locality changes continually over geological time, but habitats persist; they shift about in space, and their associated species shift with them, while the populations from which the shifting habitats are colonized simply become extinct. For example, the distribution of North American species such as spruces (*Picea*) shifted in concert with that of their habitat throughout the Pleistocene (Chapter 13). Northern populations were extinguished during glacial episodes and southern populations became extinct during the warm interglacial periods; there are no warm-adapted spruces along the Gulf of Mexico where they once grew.

### Species selection and the hierarchical nature of evolution

Speciation does not appear necessary for anagenetic change, but it nevertheless may play an important role in long-term evolution. The variation among a group of species, in one or more characters, can be much greater than within a single panmictic unit, because recombination within a population constrains the variance within rather narrow limits (Chapter 7). Reproductive isolation enables the variance to increase, so that a group of species can occupy several adaptive peaks, whereas a single population can occupy only one (Hutchinson 1967; Figure 6). Thus speciation is a prerequisite for adaptive radiation into different sympatric niches. If a new optimum phenotype is favored by changes in ecological conditions, one or another member of a group of species is more likely to evolve toward the optimum than a single species might, because the genetic variance of the

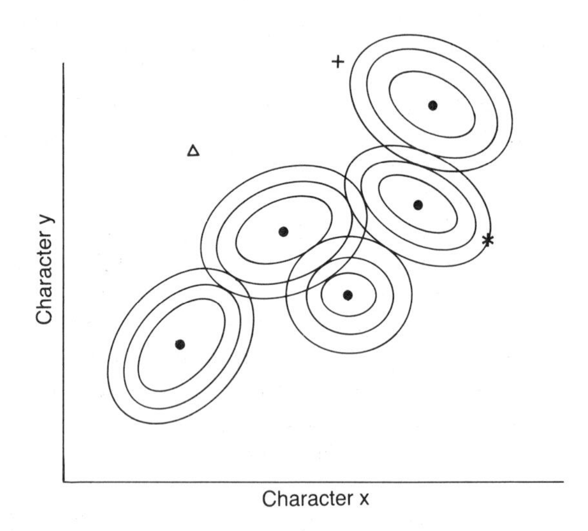


FIGURE 6

**An illustration of how adaptation to new environmental conditions may occur in one or another of a group of species, but would be less likely if only a single species exists. For each of five species, the variation in two characters is represented as a bivariate normal distribution, with the three concentric rings representing the frequency of individuals within 1, 2, and 3 standard deviations from the mean. In some of the species the characters are genetically correlated. The range of the species in aggregate includes more phenotypic space than any one species can include, because of recombination. Each of the three symbols represents a new optimum phenotype that would be favored because a new resource appears. The triangle is an optimum unlikely to be achieved both because it falls well outside any species' range of variation and because it is orthogonal to the correlation between the characters in the closest species. The optimum represented by the cross is more likely to be achieved because a slight expansion of the range of variation in one of the species will enable selection for this optimum to occur. The optimum represented by the star will be achieved rapidly because the variation and covariation in one of the species happens to include it.**

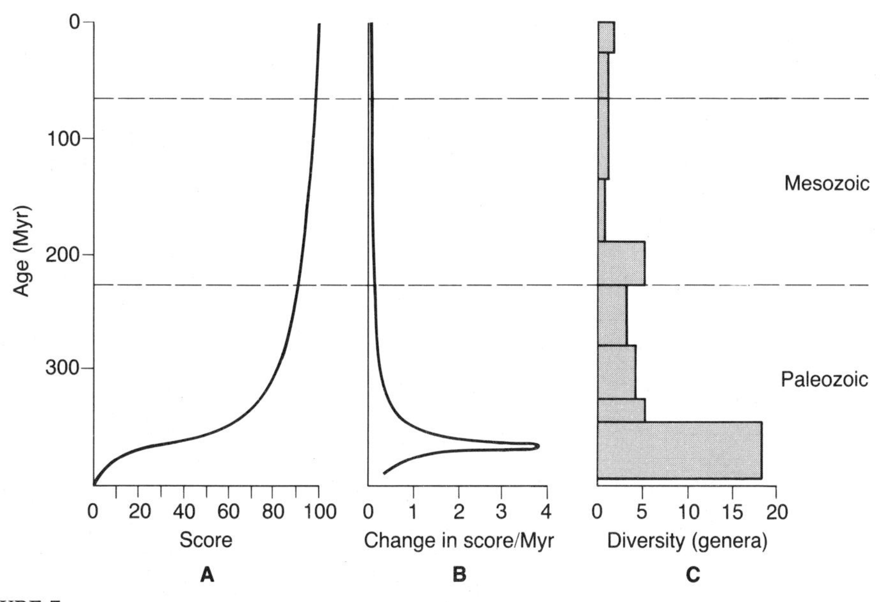


FIGURE 7

**Rates of evolution in lungfishes. The score given in (A) and (B) is an average over genera of the degree of "advancement" from the ancestral condition of several skeletal characters. The rate of change was high early in the history of the group, and very low in the late Paleozoic and thereafter. The diversity of lungfishes was likewise highest early in their history. (From Stanley 1979; A and B after Simpson 1953)**

group as a whole is greater than that of any one species (Arnold and Fristrup 1982). Hence speciation could facilitate anagenesis by multiplying the variety of gene pools within which natural selection can act. In addition to facilitating anagenesis, speciation may contribute to long-term trends if species selection in the strict sense proposed by Eldredge, Gould, and Stanley is a reality. That is, the duration or speciation rate of a lineage may be correlated with one or more characters that, because of the correlation, increase in frequency among the species in a clade (cf. Chapter 12 for possible examples).

The most strongly suggestive evidence that speciation enhances long-term evolutionary rates (Stanley 1979) is that many of the "living fossils" represent clades that have had low species diversity throughout most of their long histories; in the lungfishes, for example, morphology changed rapidly early in the clade's history when it was diverse, but very slowly after its diversity declined (Figure 7). However, Douglas and Avise (1982) found no difference in the total amount of morphological divergence among the numerous species of *Notropis* (minnows) than among the few species of *Lepomis* (sunfishes)—two genera of about the same age. This suggests that speciation does not necessarily facilitate morphological evolution.

Figure 8 is a greatly simplified representation of the phylogeny of the horse family, Equidae. No character showed a steady trend; most features displayed reversals in one or more lineages, some became fixed without further change

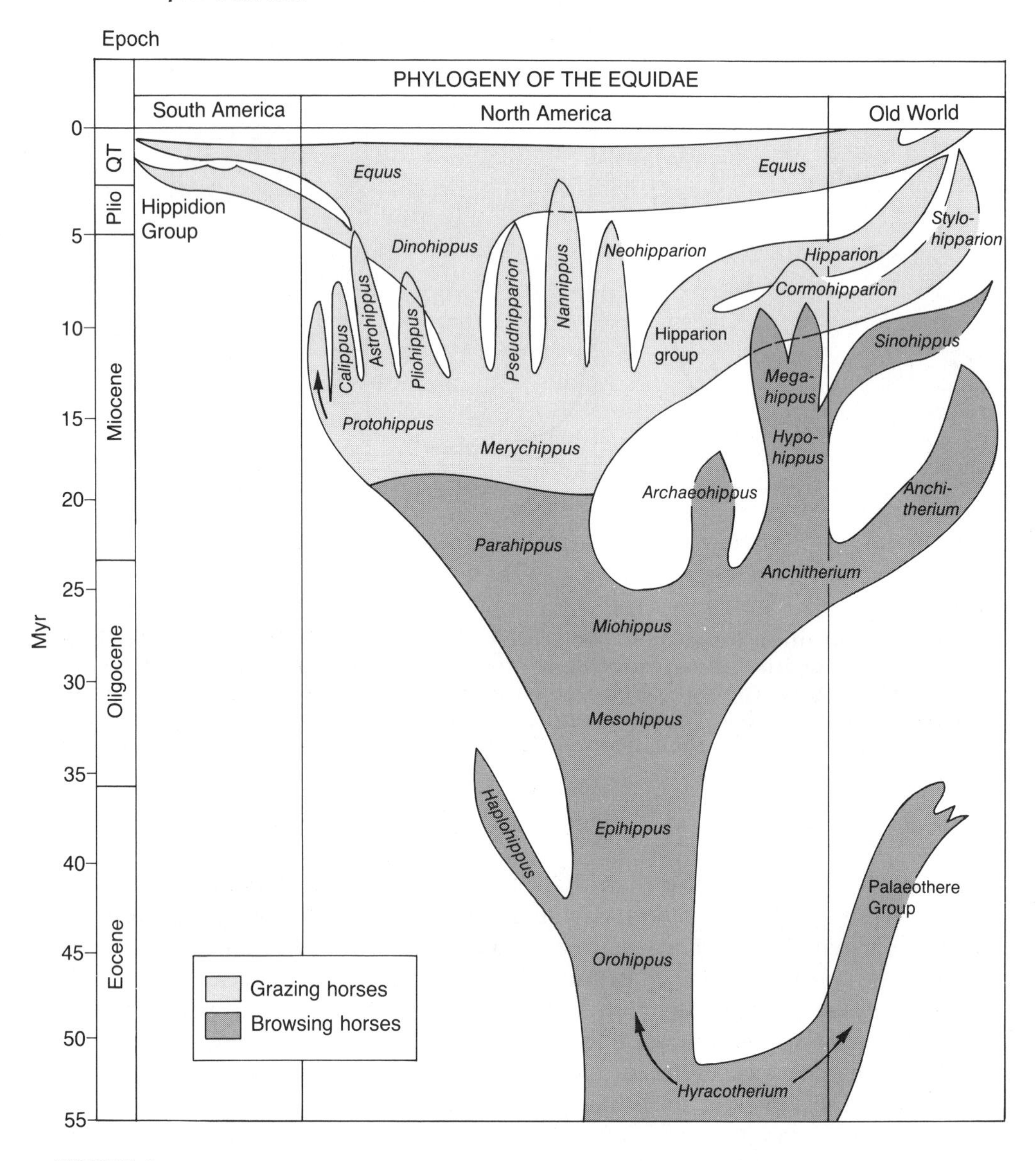


FIGURE 8

**The phylogeny of the horse family, Equidae, as currently understood. Note that the grazing habit evolved in the Miocene, in only one of the lineages extant at that time. (From MacFadden 1985)**

until extinction, and the rate of evolution of every character varied greatly (Simpson 1953). Trends, in toe number for example, can be traced from *Hyracotherium* to *Equus,* but *Equus* is the only genus that happens not to have become extinct. Our notion of trends in the evolution of horses would be quite different if, say, *Anchitherium* or *Stylohipparion* were extant. We do not know why *Equus* alone survived—whether it was because of the structural features that distinguish it from other genera, or because it was "lucky" that its habitat has persisted.

All the morphological changes in the history of the Equidae can be accounted for by the neo-Darwinian theory of microevolution: genetic variation, natural selection, genetic drift, and speciation. Yet this theory, because it does not in itself include extinction, does not explain why the only living equids have one toe on each foot. The current complexion of the Equidae is as fully a consequence of extinction versus survival of lineages as it is of adaptive change within populations. In this sense, macroevolution is indeed decoupled from microevolution, for microevolutionary events have no long-term consequences if they are obliterated by extinction. To understand evolution in the long run, it is as necessary to study the history of speciation events and extinction events as it is to study the processes of microevolution. A comprehensive view of evolution, then, must be hierarchical in its structure (Gould 1982), encompassing both the dynamics of species and higher taxa, and of the genotypes and genes they include.

## REGULARITIES OF PHENOTYPIC EVOLUTION

We turn now to the origin of novel characters, primarily morphological characters of the kind that define higher taxa. We seek to understand the genetic and developmental mechanisms by which novel phenotypes arise. It will be helpful, first, to draw on comparative anatomy and embryology for a description of some of the common kinds of transformations that have occurred in evolution. It must be emphasized that the entire discussion of this subject requires that we be able to recognize evolution from primitive (plesiomorphic) to derived (apomorphic) characters, and so requires phylogenetic analysis (Chapter 10) as its foundation.

### Biochemical changes

At the biochemical level, many of the differences among species and higher taxa are to be found in differences in the amino acid sequences of enzymes and proteins that affect their binding capacities, reaction rates, or temperature optima (Hochachka and Somero 1973). An alteration of fructose-diphosphatase appears to be a mechanism by which bumblebees have evolved homeothermy, which enables them to live in colder regions than most other bees. The reaction catalyzed by this enzyme is coupled to the reaction ATP $\rightarrow$ ADP + $P_i$ + heat. Cleavage of ADP generates AMP, which in most species inhibits fructose-diphosphatase and consequently the catalysis of ATP. The fructose-diphosphatase of bumblebees is not inhibited by AMP, so the heat-yielding reaction proceeds at a high rate.

Physiological adaptations are often accomplished by evolutionary changes in the regulation of levels of enzymes; for example, detoxification of DDT in some resistant strains of houseflies is accomplished by an increased concentration of DDT-dehydrochlorinase which is present even in susceptible flies and presumably catalyzes other reactions under normal conditions (O'Brien 1967). Larvae of the black swallowtail butterfly (*Papilio polyxenes*) feed on parsley and related plants, and are able to degrade the toxic compounds (psoralens) in these plants. Other Lepidoptera that do not specialize on these plants have the same ability to detoxify psoralens, but it is less well developed (Ivie et al. 1983). Changes in enzyme regulation undoubtedly affect the distribution among tissues of pigments, lipids, lignins, and other constituents. For example, the thick layer of waxes on the epidermis of desert plants is largely a quantitative change in materials widespread among plants.

Major new biochemical pathways seldom come into existence. As every biochemist knows, the biochemical characteristics of organisms are far less diverse than morphological features. Many physiological adaptations entail not biochemical but behavioral or structural changes, as of the thickness of fur in cold-adapted mammals. The basic biochemical pathways and even the kinds of cells that make up an animal are almost invariant throughout the Metazoa; and the evolution of changes in morphology involves changes in the developmental patterning of cellular mechanisms, not of the cellular mechanisms themselves (Gerhart et al. 1982).

### Alterations of morphological features

Truly new morphological features—those for which we can find no hint of a predecessor—are unusual. Most of those that do provide us with the greatest puzzles had their origin very early in evolution: the paired appendages of vertebrates and the wings of insects, for example. More often, distinctive features can be traced by anatomical or embryological studies to quite different ancestral features. For example, the middle ear bones of mammals are modified from certain jaw elements of reptiles; these in turn are homologous to parts of the gill arches of fishes, and perhaps to elements of the branchial basket of agnathans. Such transformations often include changes in the number of elements, in their size, shape and position, in their association with other parts of the body, and in the degree of differentiation of serially homologous elements (repeated elements based on the same developmental plan).

An organ may become elaborated during evolution by increase in size, as in the enlargement of the cerebral hemispheres of mammals compared to those of reptiles, or by an enhancement of the complexity of its form, as in the greater dissection of the lungs into bronchioles and alveoli as one passes from amphibians to some reptiles and thence to mammals. Conversely, reduction in size or complexity, and even complete loss, is a common trend, probably the single most common trend in morphological evolution. Throughout the angiosperms one encounters instances of the loss of leaves and petals, and reduction of the number of ovules; among insects the loss of wings, of certain tarsal segments, and of numerous structural features associated with the genitalia is commonplace; the history of the vertebrate skeleton is in large part one of simplification, from the numerous skull bones of fishes, for example, to the few in mammals. Reduction and loss is so common that phylogenetic systematists can often assume as a rule of thumb that the absence of a trait is a derived state relative to its presence. In some cases reduction is more apparent than real, because ancestrally separate elements may become fused during development. This is characteristic, for example, of many parts of a bird's skeleton (Figure 9). In other cases embryonic elements that become separate in an ancestor fail to separate in a descendant.

SERIALLY HOMOLOGOUS ORGANS, those that are repeated within an individual organism, evolve in number, position, and degree of differentiation from each other. Their number appears to depend on the *pattern* by which cells do or do not differentiate into a particular kind of tissue (Goldschmidt 1938, Waddington 1956a, Sondhi 1963, Wolpert 1982). The number and geometric arrangement of petals in a flower, digits on a foot, or scales on a snake depend on the mechanisms that dictate that cells in certain sites rather than others differentiate into petals, digits, or scales. The number of serially homologous elements may increase in

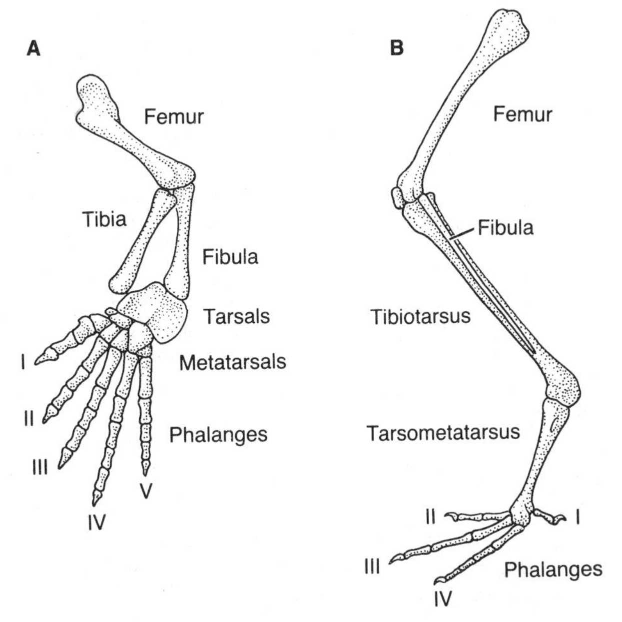


FIGURE 9

**Fusion and reduction of parts. (A) The skeleton of the hind limb of a reptile and (B) that of a bird. In the bird, some of the tarsals have become fused with the tibia and others with the metatarsals; the number of digits and phalanges is reduced, and the fibula is reduced in size. (A after Romer 1956)**

evolution, as have the number of vertebrae of snakes, body segments of millipedes, and ovules in plants such as lilies. More often the number of elements is reduced. Reduction in the number of vertebrae, aortic arches, digits, and teeth is one of the common themes in the comparative anatomy of the vertebrates, and most derived families of angiosperms have fewer stamens and carpels than groups such as magnolias that retain the ancestral condition. Evolutionary changes in number commonly occur when the number of parts is "indeterminate," i.e., large and variable, as in the many stamens of magnolias and of mimosaceous legumes such as *Acacia*. When the number is smaller, it is often more "determinate;" the elements vary less in number and position, both within and among species. Thus stamen number is virtually fixed at ten in papilionaceous legumes such as peas and at six in the mustard family Brassicaceae. Stebbins (1974) refers to the greater evolutionary lability of variable multiple features as "modification along the lines of least resistance."

Among the most important regularities in evolution is the transformation of the homogeneous serially homologous structures of an ancestor into diverse structures serving different functions in a descendant. In numerous plants, some or all leaves are transformed into tendrils or spines (Figure 10). Whereas the appendages of trilobites were rather homogeneous in form, the adaptive radiation of the Crustacea and other arthropods is largely a consequence of the differentiation of appendages into various mouthparts and locomotory and reproductive

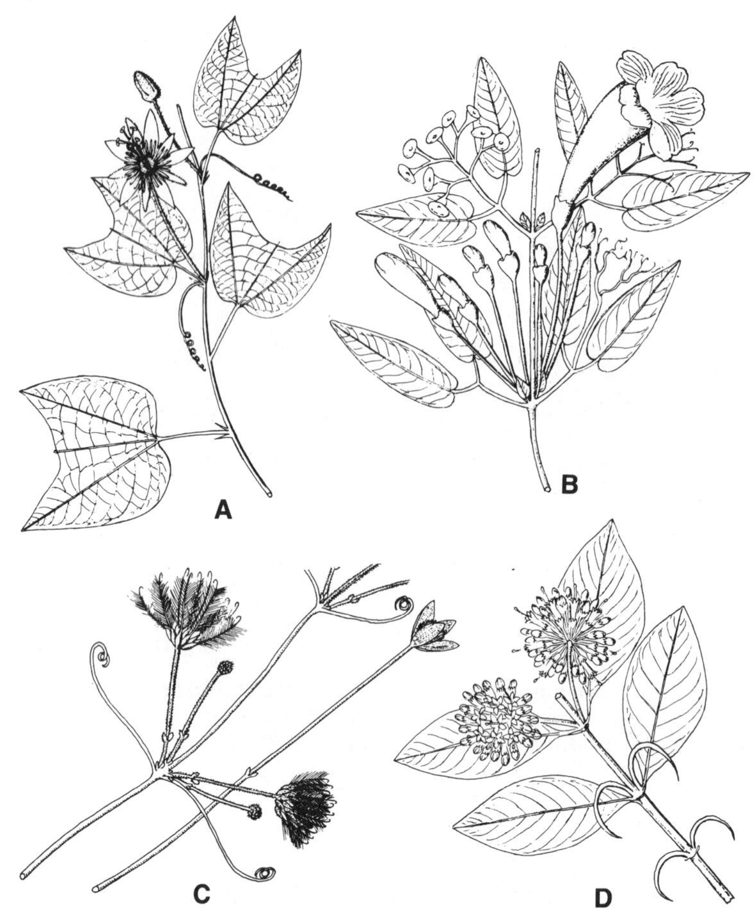


FIGURE 10

**Adaptive differentiation of serially homologous structures: adaptations for climbing in vines. (A) Stipules, normal in form at the base of the lowermost leaf, modifed into tendrils in Passifloraceae. (B) Terminal leaflets of three-lobed leaves modified into tendrils and into holdfasts in Bignoniaceae. (C) Leaves, subtending inflorescences, modified into tendrils in Ranunculaceae. (D) Peduncles of inflorescences, modified into hooks in Rubiaceae. (From Hutchinson 1969)**

organs. Differentiated structures sometimes, although rarely, become more homogeneous during evolution; for example, the teeth of toothed whales, unlike those of most other mammals, are homogeneous in form. Differentiation of serially homologous structures often follows or accompanies an evolutionary transition from indeterminate to determinate number: the highly differentiated teeth of mammals are virtually fixed in number within species, whereas the teeth of most reptiles are more numerous, variable in number, and little differentiated.

## ALLOMETRY AND HETEROCHRONY

Much of morphological evolution can be described in terms of changes in the shape of one or more individual elements; elongation of the phalanges "accounts for" the shape of a bat's wing, for example, while an increase in the length and thickness of the central toe relative to the side toes describes one of the major trends in horse evolution. Almost all such changes in shape can be expressed mathematically as a change in duration or rate of growth of one dimension or one body part relative to that of others. Differential rates of growth of different measures of an organism are referred to as ALLOMETRIC GROWTH. The most common expression for allometric growth (from Huxley 1932) is $y = bx^a$, where $y$ and $x$ are two measurements taken on a number of specimens. For example, $y$ and $x$ might refer to the height and width of a tooth, or to the weight of the brain and the body. (In many studies $x$ is indeed a measure of body size.) This allometric equation has the form of a straight line if plotted on logarithmic axes or if it is transformed to the relation $\log y = \log b + a \log x$. Thus the constant $\log b$ is the intercept, and the constant $a$ (the allometric coefficient) describes the slope of the relationship. If $a = 1$, then $y$ is a constant proportion of $x$, and an increase in $x$ (perhaps body size) entails no change in shape. If $a$ does not equal 1, shape changes during development (Figure 11); for example, during human ontogeny the head grows more slowly ($a < 1$) and the legs more rapidly ($a > 1$) than the body (Figure 12). *Intraspecific* allometry may be described by ontogenetic data (measures on individuals of different ages) or "static" data (individuals of similar age but different size). *Interspecific* allometry (e.g., Figure 13) describes differences, usually measured in adults, among species. For reasons that are not fully understood, the allometric coefficients often differ among these three relationships (Cock 1966, Gould 1966, Riska and Atchley 1985).

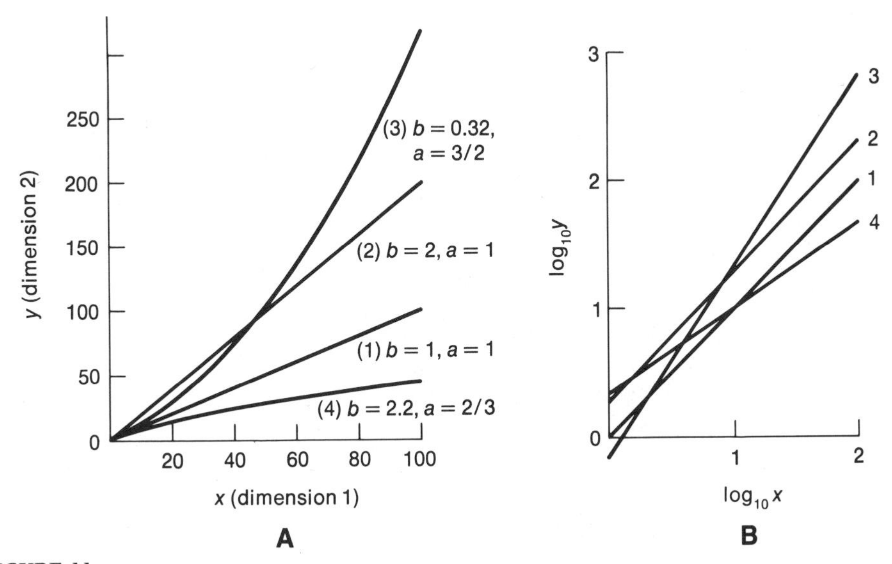


FIGURE 11

**Allometric growth. (A) Arithmetic plot of the lengths, $x$ and $y$, of two structures or dimensions. Curves 1 and 2 show isometric growth ($a = 1$); dimension 2 equals dimension 1 in curve 1 ($b = 1$) and is twice as great in curve 2 ($b = 2$). Curves 3 and 4 show positive ($a > 1$) and negative ($a < 1$) allometry respectively. (B) Logarithmic plots of the same curves. Curves 1 and 2 both have slope 1, but differ in intercept (compare with Figure 13). The slope is greater than 1 in curve 3, and less than 1 in curve 4.**

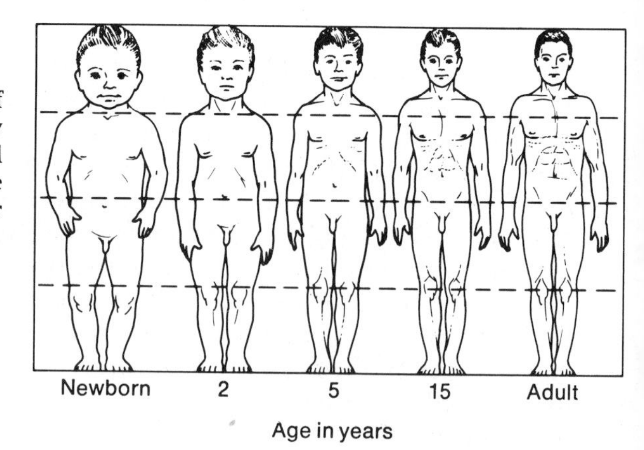


FIGURE 12

**Allometric growth in humans. Individuals of different ages, drawn at equal heights, show proportionately less rapid growth of the head and more rapid growth of the legs than of the body as a whole. (Redrawn from Sinclair 1969)**

Many allometric relationships are adaptive. For example, organs like the intestine, whose function is surface-dependent, often grow faster than does body mass. The ratio of the organ's surface area to the volume of body it serves is constant only if $a \approx 3/2$, because the area of the organ increases as the square of its length $y$, while the volume (and weight) of the body increases roughly as the cube of body length $x$.

Two allometrically related features will display a genetic correlation (Chapter 7) if genotypes vary in shape ($a$) or in the age at which they stop growing along a uniform allometric curve. Allometric coefficients vary within populations, and

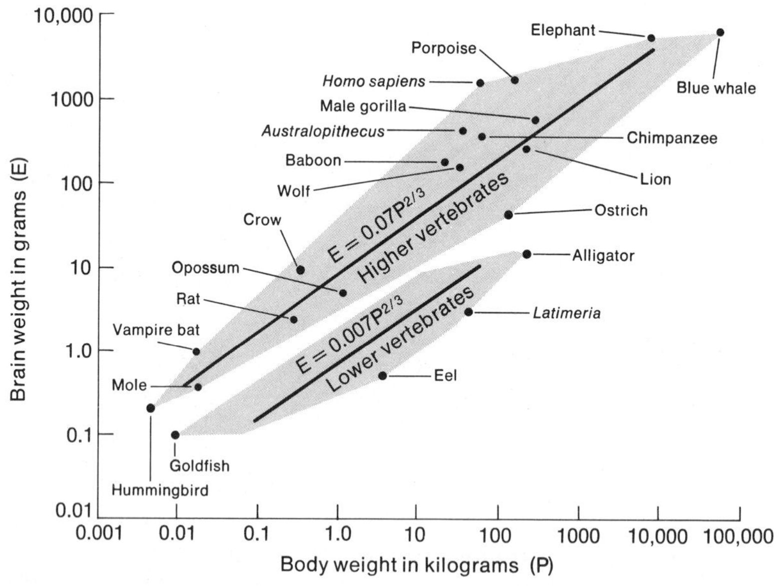


FIGURE 13

**Interspecific allometry, illustrated by the relationship between brain weight and body weight in vertebrates. The axes are logarithmic, so larger species have relatively larger brains. Those of endotherms ("higher vertebrates") are 10 times as large as those of ectotherms of comparable size. (Redrawn from Jerison 1973)**

the variation has both genetic and environmental components (Atchley and Rutledge 1980). Evolutionary changes in the time of development of an organism's features are referred to as HETEROCHRONY (Gould 1977). Heterochronic changes can be of several kinds (Alberch et al. 1979). Suppose that in an ancestral form, $x$ represents body size, $y$ the size of some feature and that this feature begins to develop at age $\alpha$ and ceases to grow at age $\beta$ (Figure 14A). During this interval $y$ and $x$ grow at rates $k_y$ and $k_x$. If the period of development of feature $y$ (e.g., until the onset of sexual maturation) is extended by some amount $\Delta\beta$ during

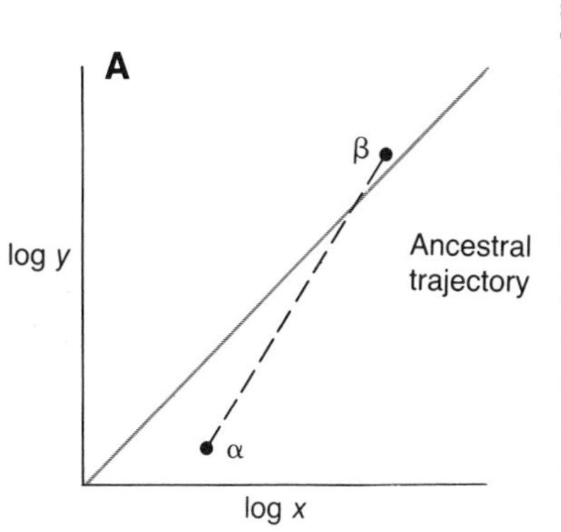


FIGURE 14

**Diagrammatic representation of some forms of heterochrony. (A) The ontogenetic change in an ancestor is represented by the broken line expressing the growth of two structures, with values $y$ and $x$ plotted on logarithmic axes. The slope exceeds 1 (solid line), so there is positive allometry. Growth proceeds from age $\alpha$ to age $\beta$. (B) Hypermorphosis in a descendant: extending development from age $\beta$ to $\beta + \delta\beta$ results in peramorphosis—in this case a greater ratio $y/x$ at maturity. (C) Progenesis in a descendant: development ceases at age $\beta - \delta\beta$, leading to a juvenile morphology at maturity (paedomorphosis). (D) Acceleration of the rate of development of $y$ (i.e., increasing $a$) leads to peramorphosis. (E) By neoteny, a decrease in the rate of development of $y$ relative to $x$, a juvenile condition (paedomorphosis) of character $y$ but not of $x$ is expressed in the adult.**

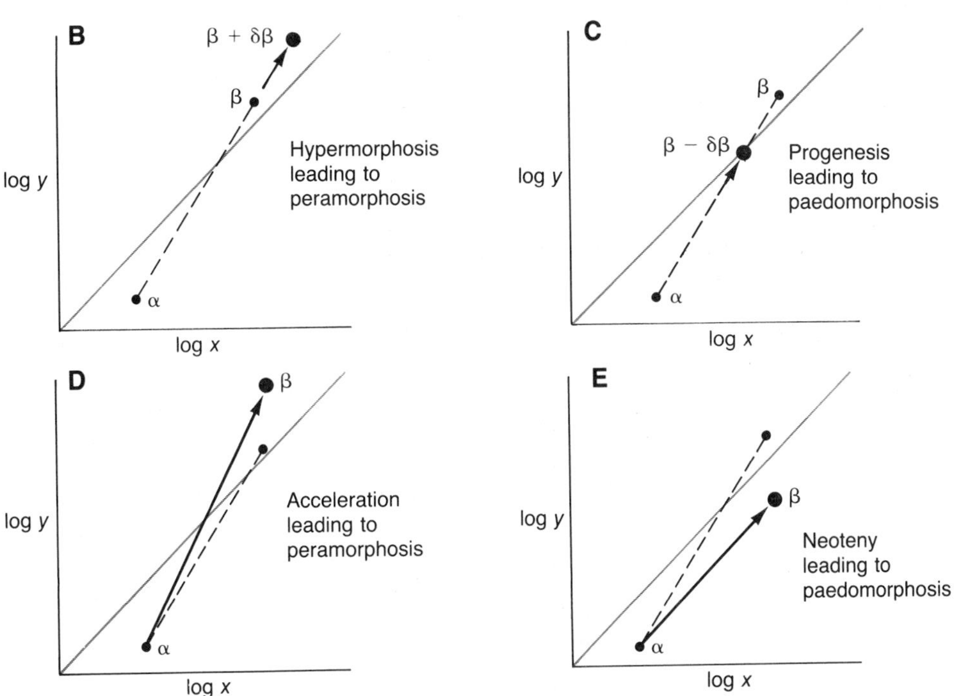

evolution, its size at maturity will increase. The descendant will have the same shape (expressed by the ratio $y/x$) as its ancestor if $a = 1$, but a different shape ($y$ greater relative to $x$) if $a > 1$ (Figure 14B). During its ontogeny, however, the descendant will pass through a juvenile stage in which its shape resembles the adult form of its ancestor—the phenomenon of RECAPITULATION made famous by Haeckel's pronouncement that "ontogeny recapitulates phylogeny." Alberch et al. use the term HYPERMORPHOSIS for the process by which development is extended during evolution, and PERAMORPHOSIS for its morphological consequence, the more exaggerated shape ($y/x$) of the descendant compared to the ancestor. If, however, development is truncated during evolution by early maturation (at age $\beta - \Delta\beta$), the adult descendant will be smaller and character $y$ will be less fully developed (Figure 14C). The morphological expression of this process of PROGENESIS is PAEDOMORPHOSIS: an adult organism with features typical of the juvenile of its ancestor. This is "reverse recapitulation," the very opposite of Haeckel's dictum.

Suppose that the period of growth ($\alpha$ to $\beta$) remains unchanged, but that the allometric coefficient $a$ increases by an increase in $k_y$ relative to $k_x$. Like hypermorphosis, this process of "acceleration" again results in peramorphosis: the shape of the descendant is exaggerated relative to that of the ancestor (Figure 14D). Conversely, if $k_y$ decreases, the descendant attains the same adult size as its ancestor, but the feature $y$ has a relatively juvenile condition (Figure 14E). This evolutionary process is termed NEOTENY; like progenesis, it results in paedomorphosis. Table II summarizes these phenomena.

### Examples of heterochrony

The allometric consequences of evolutionary changes in body size are sometimes spectacular; for example, in the largest species of deer, the extinct Irish elk (*Megaloceros giganteus*; Figure 15), allometric growth resulted in monstrously large antlers (Gould 1974). This appears to be an example of peramorphosis. During the evolution of the line leading to the modern horse (Figure 16), the length of

Table II
**Processes and results of heterochrony**

| Control parameter | Evolutionary change | Process | Morphological expression | Phylogenetic phenomenon |
|---|---|---|---|---|
| $k_x$ | $+\Delta k_x$ | Proportional giantism | | Recapitulation |
| | $-\Delta k_x$ | Proportional dwarfism | | Reverse recapitulation |
| $k_y$ | $+\Delta k_y$ | Acceleration | Peramorphosis | Recapitulation |
| | $-\Delta k_y$ | Neoteny | Paedomorphosis | Reverse recapitulation |
| $\beta$ | $+\Delta\beta$ | Hypermorphosis | Peramorphosis | Recapitulation |
| | $-\Delta\beta$ | Progenesis | Paedomorphosis | Reverse recapitulation |
| $\alpha$ | $+\Delta\alpha$ | Postdisplacement | Paedomorphosis | Reverse recapitulation |
| | $-\Delta\alpha$ | Predisplacement | Peramorphosis | Recapitulation |

Modified from Alberch et al. (1979)

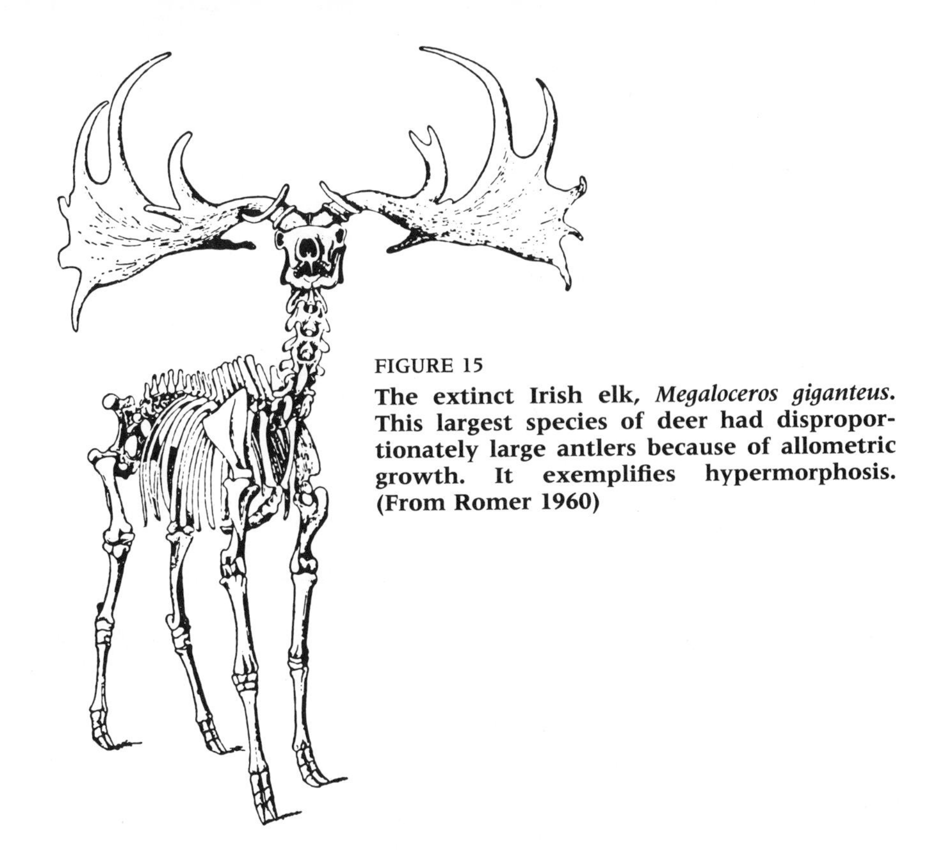

FIGURE 15

**The extinct Irish elk, *Megaloceros giganteus*. This largest species of deer had disproportionately large antlers because of allometric growth. It exemplifies hypermorphosis. (From Romer 1960)**

the facial portion of the skull relative to the braincase increased along an allometric curve as body size increased. Some of the changes in the skull, however, were not merely allometric but entailed reorganization; for example, the tooth row moved forward when high-crowned teeth evolved (Radinsky 1984).

Paedomorphosis resulting from neoteny is prevalent in many groups of salamanders, in which juvenile features such as gills are retained in the sexually

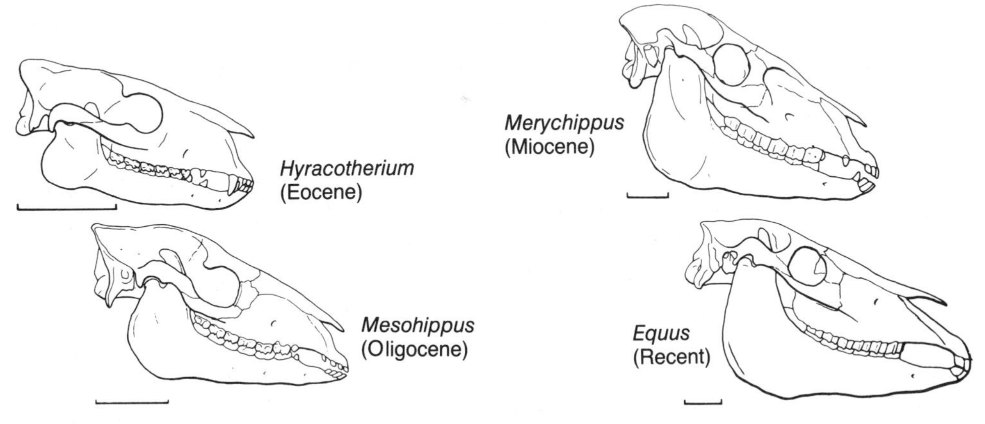


FIGURE 16

**Skulls of four stages (*Hyracotherium, Mesohippus, Merychippus, Equus*) of the lineage leading to the modern horse. *Merychippus* and *Equus* have high-crowned teeth, an anteriorly displaced tooth row, and a longer face compared to the older forms. The figures are drawn to the same size (bars = 5 cm). (From Radinsky 1984)**

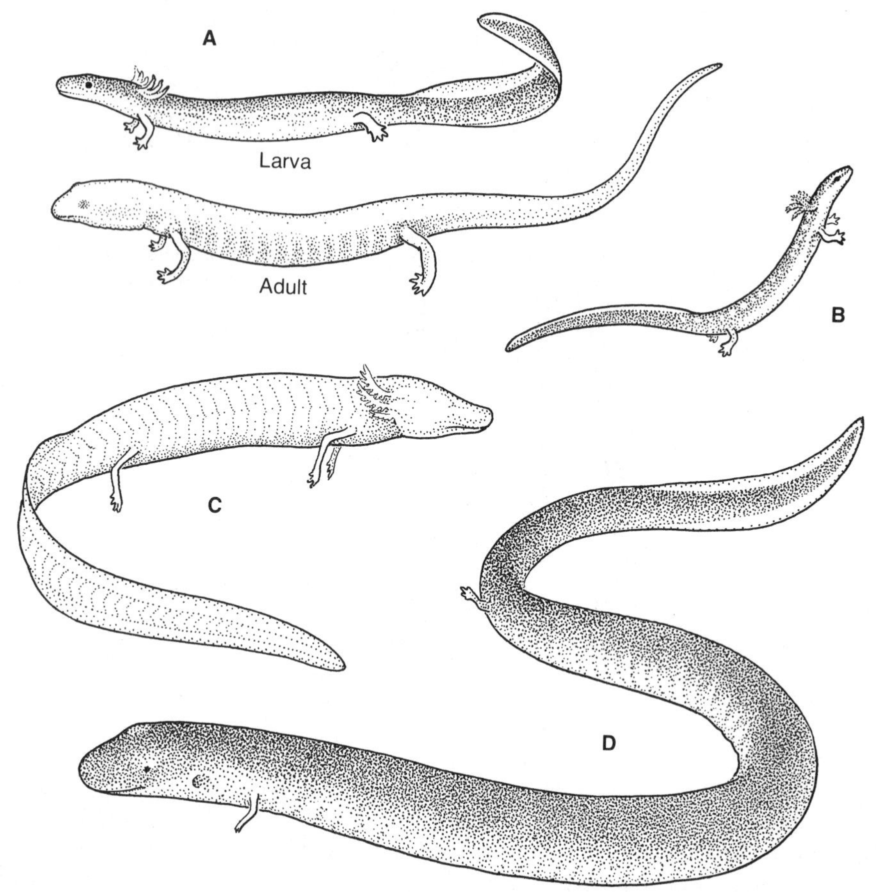


FIGURE 17

**Neoteny in salamanders. (A) Larva and non-neotenic adult of *Typhlotriton spelaeus*. (B) Adult *Eurycea neotenes*, with gills as in larval *Typhlotriton*. (C) A highly modified neotenic derivative of *Eurycea*, the cave salamander *Typhlomolge rathbuni*. (D) An unrelated neotenic form, *Amphiuma means*, in which the gills are retained but are internal. (A and B redrawn from Conant 1958; C and D redrawn from Noble 1931)**

mature adult (Figure 17). But not all the features of neotenic salamanders retain the juvenile state (Tompkins 1978); evolution is mosaic even when it occurs by neoteny. Many neotenic lineages (such as *Amphiuma*; Figure 17D) have evolved additional species-specific features that are not the consequences of neoteny.

The external features of the flower of the larkspur *Delphinium nudicaule*, which has become adapted for hummingbird pollination, illustrate mosaic evolution clearly. Overall, the flower resembles the bud of ancestral bee-pollinated species and so is paedomorphic; but the nectar-bearing petal develops more rapidly, and achieves a peramorphic condition (Guerrant 1982). Both paedomorphosis and peramorphosis are evident in the same flower (Figure 18).

Progenesis—paedomorphosis achieved by rapid maturation at a small size—appears to be typical of some parasitic crustaceans (Gould 1977), and is likely to

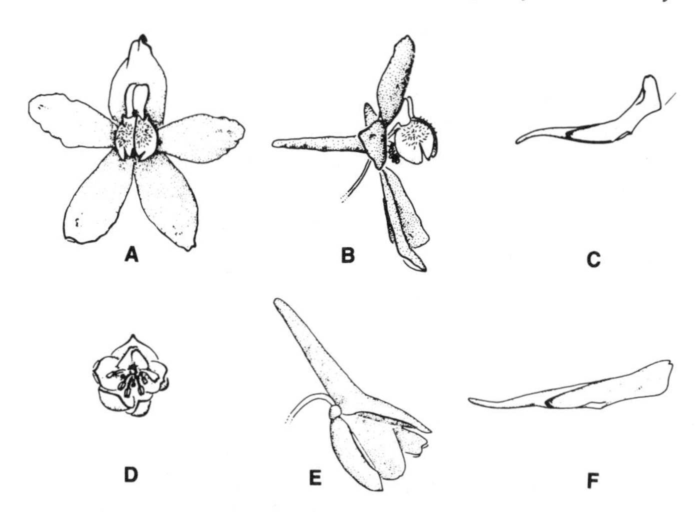


FIGURE 18

**Heterochrony in the evolution of flower form in *Delphinium*. (A, B, C) Frontal view, lateral view, and left upper petal of *D. decorum*, representing the ancestral condition. (D, E, F) Frontal view, lateral view, and left upper petal of *D. nudicaule*, representing the derived condition. The mature flower of *D. nudicaule* resembles the bud of *D. decorum*, and so is neotenous; the upper petal of *D. nudicaule* is relatively elongated, a peramorphic condition. (From Guerrant 1982)**

account for the evolution of an extinct group of echinoderms, the isorophids, small animals that resembled the juveniles of cyathocystids from which they arose (Sprinkle and Bell 1978). Some very small species of salamanders resemble the juveniles of larger related species; for example, many skeletal elements are absent or are not fully ossified (Alberch and Alberch 1981, Hanken 1984). In these cases, progenesis has an organism-wide effect on the form of numerous characters.

Many authors (e.g., Garstang 1922, Gould 1977) have suggested that some major groups of animals have arisen from neotenic or progenetic ancestors. For example, features associated with flightlessness in birds such as rails and ostriches have a juvenile aspect; insects may have arisen from the six-legged larval condition in millipede-like ancestors; chordates may have evolved from the tadpole larvae of tunicates. Humans retain in adult life some of the features (e.g., the short face and relatively large cranium) of juvenile apes, and have frequently been cited as examples of neoteny (Gould 1977).

## THE ORIGIN OF HIGHER TAXA

If we compare chordates, echinoderms, and annelids, it is evident that their structural organization is fundamentally different. The "blueprint" on which each phylum is constructed has been termed a *Bauplan* (plural, *Baupläne*). Within each phylum, classes display important variations in *Bauplan*, representing what might be termed *Unterbaupläne.* Orders within a class—bats and whales, for example—differ considerably as well. Many *Baupläne*—such as those of the animal phyla—are not bridged by any known intermediate forms, either living or extinct. How, then, have different *Baupläne*, features that distinguish higher taxa, evolved?

Some authors, such as Schindewolf (1936) and Goldschmidt (1940), argued that many of the distinctive features of higher taxa evolved by single saltational mutations: that the first bird, for example, hatched with essentially birdlike features from the egg of a typical reptile. Goldschmidt believed that neo-Darwinian processes accounted for microevolution within species, but that higher taxa arose by "systemic mutation," reorganization of chromosomal material in its entirety. Goldschmidt's view of the genome was not accepted when he proposed it, and derives no support from modern genetics (Charlesworth 1982, Templeton 1982).

The neo-Darwinian theory holds that the characters of higher taxa evolve mosaically, as individual characters or as groups of correlated traits (cf. Chapter 10), and that most such characters evolve gradually through intermediate stages. Some discontinuous changes are also recognized, as in color patterns that are controlled by single genes or threshold characters with a polygenic basis. In the neo-Darwinian view, one seeks to explain the origin not of a higher taxon as an entity, but of the particular features that characterize the taxon.

The saltational theory derives its support from the apparent absence of intermediate stages among many of the higher taxa (cf. Olson 1981), from the difficulty its adherents find in accounting for the selective advantage that a new structure might have in its incipient stages, and from the existence of mutations that have dramatic, discontinuous effects on the phenotype. In true flies (Diptera), for example, the third thoracic segment is smaller than the second, and the wings are modified into small balancing organs (halteres). Certain homeotic mutations of the *bithorax* gene complex in *Drosophila* cause the third segment to resemble the second, with winglike halteres (Figure 19; see below).

The neo-Darwinian argument for gradual transformation is based in part on the numerous cases of intermediate forms (transformation series) between higher taxa, both in the fossil record (Chapter 11) and among living organisms (Chapter 10). For example, a brilliant tropical butterfly (order Lepidoptera) differs greatly

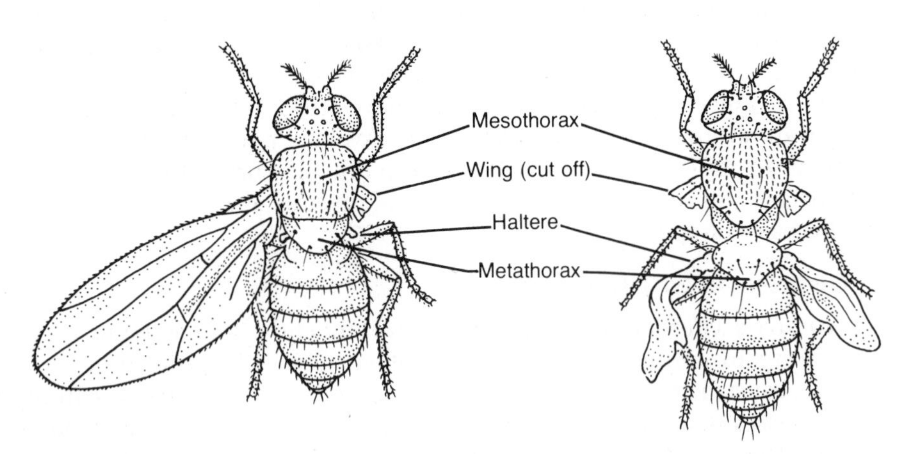


FIGURE 19

**A wild-type (left) and a *bithorax* mutant (right) of *Drosophila melanogaster*. In the mutant the metathorax has a mesothoracic form, including the development of halteres into winglike structures. (Adapted from Waddington 1956b)**

from other insects. It has large wings covered with small, colorful scales, mouthparts reduced to a coiled proboscis, and, in some instances, four rather than six functional legs. But there exist butterflies with functional forelegs, and the size and color of the wings vary gradually among species, from the dullest moths to the glamorous morphos. The most primitive moths, the Micropterygidae, have a full set of mouthparts, including chewing mandibles, and their maxillae are only slightly elongated into a rudimentary proboscis. Their wing venation is much like that of primitive caddisflies (order Trichoptera), some of which have mothlike wing scales. The Lepidoptera clearly arose from Trichoptera-like ancestors by mosaic evolution of several characters, at least some of which evolved through intermediate stages. In the fossil record, the neo-Darwinian position is supported by instances of intermediates between higher taxa, such as *Archaeopteryx,* essentially a small dinosaur with feathers and a few other avian features; of graded series through time, as of the horses; and of polymorphic characters that became fixed in later populations. We have already noted the small cusp on the teeth of some specimens of *Mesohippus* that became a major part of the tooth later in the evolution of the horses.

From a genetic point of view, the neo-Darwinian position is supported by the observation that most mutations with pronounced phenotypic effects, such as the homeotic mutants of *Drosophila,* have such disruptive pleiotropic effects on development that they greatly reduce viability. Fisher (1930) argued that an organism, like a delicate machine, is so intricately constructed that major alterations will disrupt its function. Such an argument from analogy is not proof, of course, and the possibility remains that single mutations of large effect can be advantageous (Maynard Smith 1983). Nevertheless, many (but not all) of the morphological differences between closely related species are polygenic (Chapter 8), and so have been built up from gene substitutions of small effect.

The evidence of comparative morphology and of genetics makes the saltationist view of Goldschmidt and Schindewolf untenable. But it is equally fallacious to view each of the character differences between taxa as an independent event stemming from allele frequency changes at numerous loci. It appears likely that the transformation of suites of developmentally integrated characters can sometimes have a rather simple genetic basis. For example, polyploidy in plants brings about an immediate change in numerous physiological and biochemical characteristics that might permit invasion of new habitats (Levin 1983). Failure to metamorphose in neotenic axolotls (*Ambystoma mexicanum*) appears to be caused by a low concentration of thyroxin, which in non-neotenic salamanders induces coordinated changes in many characters during metamorphosis. A single gene substitution in this species can determine whether or not metamorphosis takes place (Tompkins 1978). This should not be construed as the evolution of a developmental pattern *de novo* by a single mutation; it is instead a rather simple shift along a developmental trajectory that presumably was built up during millions of years of vertebrate evolution.

Some abstract models of development, such as the models of heterochrony and allometry discussed previously, suggest that complex changes in the geometric form of organisms may arise from simple changes in developmental rates. D'Arcy Thompson (1917) pioneered this approach by illustrating the simple mathematical transformation by which the shape of one species could be derived

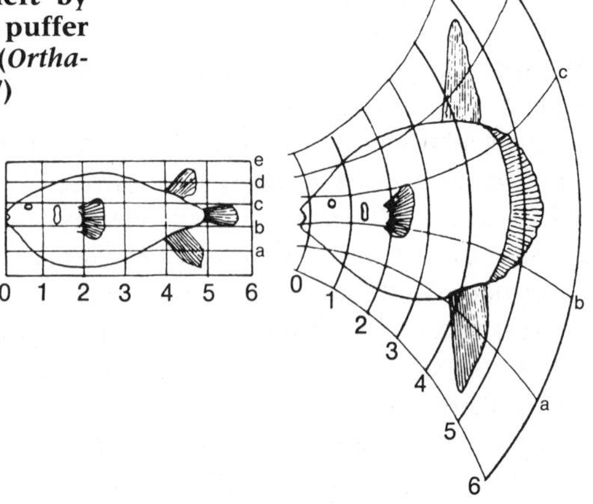


FIGURE 20

**D'Arcy Thompson's method of transformations. Replacing the coordinates at left by those at right transforms the shape of a puffer (*Diodon*) into that of an ocean sunfish (*Orthagoriscus* = *Mola*). (From Thompson 1917)**

from another. If a figure of, say, a porcupine fish is inscribed in the Cartesian coordinates *x*, *y*, it can be transformed into the shape of an ocean sunfish by replacing the *x* coordinates with hyperbolas and the *y* coordinates by a system of concentric circles centered near the head (Figure 20). Raup (1962, 1966) has shown that most of the diverse shapes of snail shells that exist could theoretically by generated by varying only four basic parameters (Figure 21): the shape of the generating curve (the cross-sectional outline of the shell aperture), the distance

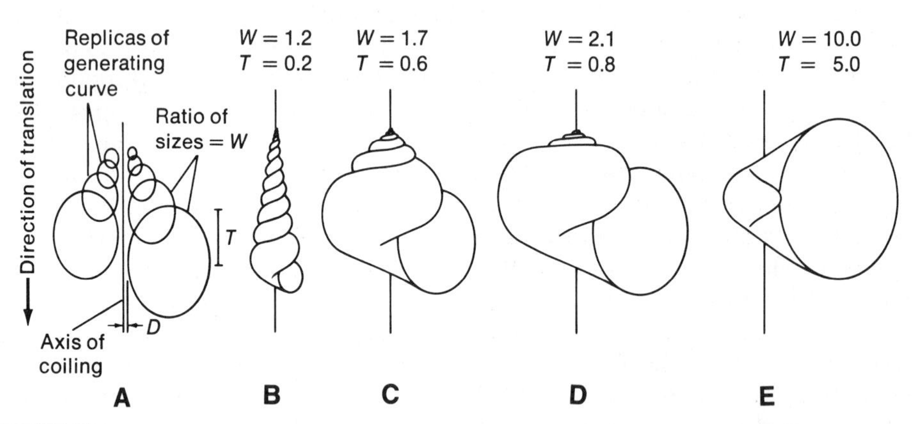


FIGURE 21

**Hypothetical snail forms (B–E) drawn from computer-generated plots. They were derived from the information in (A), which shows the form of the generating curve, the axis of coiling, the size ratio *W* of successive generating curves, the distance *D* of the generating curve from the axis, and the proportion *T* of the height of one generating curve that is covered by the succeeding generating curve in one full revolution. These hypothetical forms closely resemble known species. (After Raup 1962)**

between the generating curve and the axis about which it spirally revolves during growth, the rate at which the generating curve (aperture) is enlarged, and the rate at which it is translated along the axis. It is conceivable that developmental transformations under such mathematically simple "rules" have a rather simple genetic basis.

## THE ADAPTIVE CONTEXT OF EVOLUTIONARY INNOVATIONS

If new traits develop by natural selection, and if natural selection depends only on present rather than prospective advantage, how can the first slight manifestations of a trait be adaptive? Can there be adaptive advantage to featherlike structures that are insufficient for flight? And if the function of a complex organ depends on the congruence of interdependent parts, how can the several parts evolve if they do not arise in concert?

These questions have not been entirely resolved (Frazzetta 1975), but they are not insuperable problems. We have noted (Chapters 6 and 7) that even slight phenotypic differences are sometimes subject to strong selection. Moreover, the parts of a complex organ are often not strongly interdependent in the early stages of its evolution. The function of a vertebrate eye requires precision of form of lens, iris, retina, humors, muscles, and nerves; but an incipient eye can be quite advantageous with only a few of these features, as the less elaborate photoreceptors of many invertebrates bear witness. Eyes could well have evolved by successive slight alterations (Eakin 1968).

Mayr (1960) has stressed that the major changes in organisms' features usually follow from intensification, diminution, or change of function. Intensification of function is the adaptive basis for the progressive elaboration of horns as weapons in ungulates, and for the high-crowned molars of horses and many rodents that feed on tough, silica-laden grasses. Conversely, many evolutionary changes stem from the reduction or loss of a structure's function, as in wingless insects, legless snakes, or parasitic plants without chlorophyll. In some cases, the loss of a structure is doubtless directly adaptive; snakes can presumably move and burrow better without legs. In other cases, as with the vestigial eyes of cave fishes, accumulation of neutral degenerative mutations appears to be a sufficient explanation (Wilkens 1971, Culver 1982).

Possibly the most common basis of evolutionary transformations is a change in a characteristic's function. The wings of auks and several other birds are used in the same way in both air and water, and in penguins they have become entirely modified for underwater flight. In animals a change in behavior often precedes changes in morphological and physiological features; for example, the Galápagos marine iguana (*Amblyrhynchus cristatus*) feeds on subtidal algae, but is hardly more physiologically adapted for diving than its more terrestrial relatives (Dawson et al. 1977).

Quite often a structure serves both new and old functions, as in many plants (maples, *Acer*, and stick-tights, *Desmodium*, among others) in which the ovary wall both protects and disperses the seeds. Thus an evolutionary innovation may have developed because of an original function, to the point where a new function becomes possible. The ability of an electric eel (*Electrophorus electricus*) to kill prey and defend itself by electric shocks is an elaboration of the much weaker electric fields generated by other gymnotid knife-fishes, which use their electricity for orientation and communication in murky waters.

A structure that takes on a new function is often said to have been a PRE-ADAPTATION. This does not mean that it evolved in anticipation of future need (Bock 1959), but merely that its function has changed more rapidly than its structure. The kea (*Nestor notabilis*), a New Zealand parrot that rips through the skin of sheep with its sharp beak to feed on fat, has a beak much like that of other parrots. Thus many parrots could be said to be preadapted for carnivory. Progenitors of a group usually display preadaptations for that group's distinctive adaptations; for example, epiphytic plants such as bromeliads, which do not have access to wet forest soil, developed from arid-adapted ancestors.

The origin of flight in birds illustrates the role of preadaptation in the evolution of a major adaptive shift (Ostrom 1976, Caple et al. 1983, Padian 1985). *Archaeopteryx* shares with coelurosaur dinosaurs numerous synapomorphies, including features that are often thought to be typical only of birds, such as hollow bones and fused clavicles (the "wishbone"). These and other features that provide the lightness and structural strength of skeleton required for flight were already characteristic of the agile, bipedal, carnivorous coelurosaurs. A critical prerequisite of flight is the ability to generate lift by moving the forelimbs down and forward. *Deinonychus* and related coelurosaurs, uniquely among the reptiles, had long forelimbs capable of exactly these movements (Figure 14 in Chapter 11); equipped with strong claws, the forelimbs were used to grasp prey. If the surface area of the forelimbs were even slightly expanded, their down-and-forward motion would have enabled the animal to stay momentarily aloft following a running leap into the air. Feathers, the major feature of *Archaeopteryx* that distinguishes it from coelurosaurs, may be the outcome of selection for expanded surface area.

An evolutionary change in one structure often sets in motion selection for changes in other structures. In most fishes the premaxillary and maxillary bones, which bear teeth, must perform the dual functions of collecting and manipulating food; deeper within the gullet the pharyngeal bones bear teeth that help hold the prey. In the cichlid fishes, the articulations and musculature of the pharyngeal bones make them more versatile, so they can manipulate prey. Thus the premaxillary and maxillary bones are freed for food gathering. Liem (1973) has proposed that the liberation of these bones from their primitive functional constraints has resulted in the enormous adaptive radiation of this family, in which species vary greatly in structure and feeding habits (Figure 13 in Chapter 8).

Incipient changes of some characteristics may well be nonadaptive, with new adaptive functions emerging only later. Features that have little selective advantage in an animal of small size may become enlarged by allometric growth as body size evolves, and become modified for some function. In some cases, variations in a developmental pattern may become selectively neutral if body size is sufficiently small (Wright 1982). In very small, progenetic species of the salamander genus *Thorius*, for example, there is extraordinarily great variation in the development of many skeletal elements (Hanken 1984), perhaps because the physical force required for muscular function is not as great as in larger species. It is easy to imagine that if a species of *Thorius* evolved larger size, certain skeletal elements would become fixed in a different pattern from that of other plethodontid salamanders.

## GENETICS, DEVELOPMENT, AND EVOLUTION

Genetic descriptions of phenotypic differences and mathematical models of changes in shape, such as models of allometric growth, provide rather abstract pictures of the changes in developmental processes that might have transpired in evolution. But useful as they are, these abstract descriptions do not tell us the mechanisms by which changes in DNA are translated into changes in morphology. We have no idea of what we would have to do at a molecular or cellular level to transform *Drosophila melanogaster* into *D. simulans,* much less a fly into a flea. In all of biology, the mechanisms of development are the area of greatest ignorance, but they are central to major questions in evolution. Why are some characters more variable than others? Why do some mutations but not others have deleterious pleiotropic effects? These questions of developmental biology are central to the problem of why some characters can evolve more readily than others.

Much of the progress in developmental biology bears only a tenuous, hypothetical relation to evolutionary studies. The mechanisms by which some genes exert their morphological effects are known, but chiefly through the study of rather drastic mutations; seldom does a geneticist determine the mechanism by which a gene difference between related species causes their difference in morphology. Similarly, few studies in experimental embryology describe the mechanisms that cause differences between related species. Geneticists and developmental biologists are fully occupied with the enormously difficult problems that are their proper province; the evolutionary aspects of development must be studied by evolutionary biologists who understand and apply the techniques of molecular and developmental biology.

## THE GENETIC AND DEVELOPMENTAL BASIS OF MORPHOLOGICAL EVOLUTION

In asking what genetic changes cause macroevolutionary change in morphology, let us first dispose of simplistic answers. Simple changes in the amount of DNA, for example, have little predictive value. Closely related species can differ severalfold in genome size, and the amount of DNA bears little relation to morphological complexity or the time of origin of a group (Chapter 15). Nor are morphological changes generally caused by changes in chromosome structure. To be sure, some examples of position effects are known; for example, some genes in *Drosophila* cease to function when they are moved by a chromosome inversion into a heterochromatic region. But the vast majority of structural alterations of chromosomes have no evident phenotypic effect (Chapter 4), and similar species often differ greatly in chromosome configuration. A particularly striking example (Figure 22) is that of two phenotypically similar species of small deer: *Muntiacus reevesii* with $2n = 46$, and *M. muntjac* with $2n = 6$ or 8 in different populations.

As far as known, morphological change seldom involves the evolution of entirely new enzymes or proteins. Moreover, the histologically recognizable types of cells are highly conservative in evolution, differing hardly at all throughout the Metazoa (Gerhart et al. 1982). Rather, most of morphological evolution appears to consist of changes in the spatial organization of cell types within the developing organism, the time at which cell types and tissues differentiate, and the geometric form of tissues and organs.

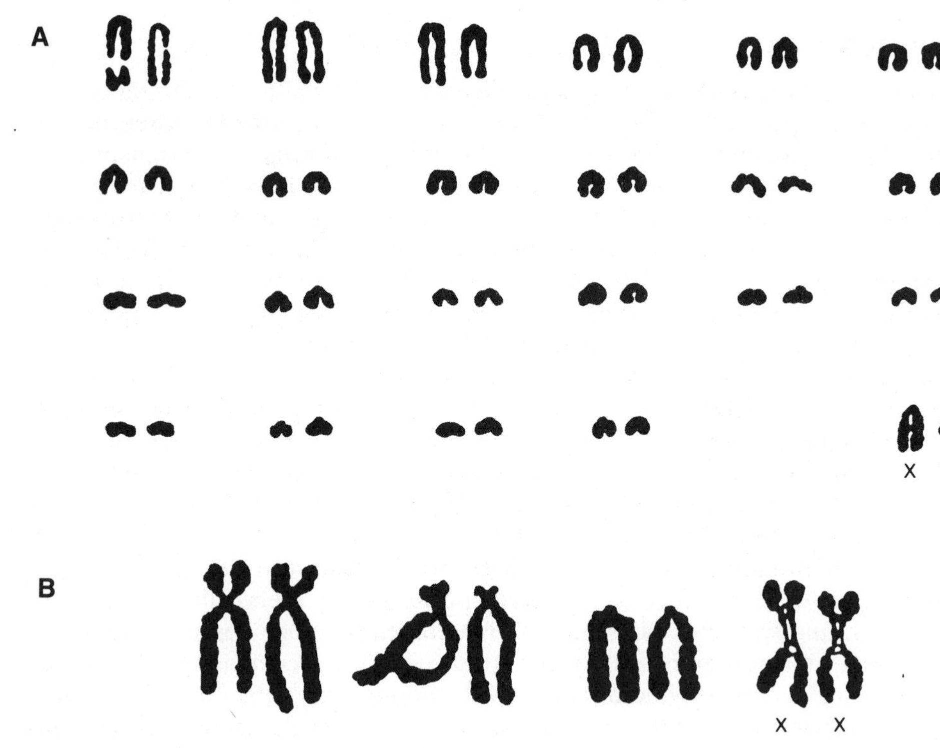


FIGURE 22

**Chromosome complements of two closely related and phenotypically similar species of muntjac deer. (A) *Muntiacus reevesi* (2*n* = 46). (B) *M. muntiacus* (2*n* = 8). (From White 1978)**

The phenotype at any stage in ontogeny is the consequence of interactions among tissues and cells, and between the organism and physical factors; the physical stress of muscular contraction, for example, is often necessary for the development and maintenance of the normal form of a bone. Moreover, each step in development depends on prior developmental conditions. Even the earliest steps in the differentiation of most embryos depend on the spatial organization of factors in the cytoplasm of the egg, established during oogenesis; these factors apparently include thousands of different sequences of maternal messenger RNAs and a cytoskeletal matrix that seems to organize their location (Raff and Kaufman 1983). Consequently the egg cytoplasm is as crucial a foundation for an embryo's development as is the embryo's genome. But this cytoplasmic foundation is itself under genetic control. For example, a polymorphism in a nuclear gene determines the direction of coiling in the snail *Lymnaea peregra*, but the gene acts by altering the cytoplasm of the egg so as to determine the direction of cleavage; thus the coiling of progeny is determined entirely by the maternal genotype (a maternal effect). Ultimately, then, all developmental events can be altered by changes in the genes and in that sense are caused by genes. But development is EPIGENETIC: it so depends on prior developmental events that it cannot be understood merely in terms of primary gene action.

### Differentiation of cells and tissues

Some genes, such as those coding for immunoglobulins in mammals, undergo rearrangements in different somatic cells, and so influence cell differentiation. Most of the nuclear genome, however, is homogeneous throughout the soma, and cytodifferentiation occurs through the differential expression of genes in response to signals, presumably of a chemical nature. Some of these signals are hormones; for example, many tissues of amphibians undergo metamorphic change in response to an increased concentration of thyroxine. Thyroxine is produced by the thyroid gland in response to release of thyrotropin by the pituitary, which in turn is stimulated by the production of thyrotropin-releasing hormone by the hypothalamus of the brain. In some amphibians that have recently evolved to become neotenic, such as the axolotl *Ambystoma mexicanum*, the thyroid and pituitary glands can function normally, and metamorphosis can be induced by injection of thyroxine. Thus neoteny has evolved through a cessation of the hypothalamic releasing mechanism. But treatment with thyroxine does not induce metamorphosis in salamanders with a longer evolutionary history of neoteny (Figure 17), in which the tissues have lost the ability to respond to thyroxine (Dent 1968). The genetic and molecular basis of this change in competence is not known.

Tissue differentiation often depends on induction by neighboring tissues, presumably through the diffusion of chemical signals—although in no instance are the identity of the chemicals, the mechanism triggering their production, and the mechanisms of response fully understood. For example, the lens of the vertebrate eye develops under successive inductive influences from endoderm, heart mesoderm, and the developing optic cup (Raff and Kaufman 1983). The optic cup itself develops in response to induction by prechordal mesoderm during gastrulation. In an eyeless mutant of the axolotl, the optic cup fails to develop because it is incapable of responding to the mesodermal inducer. In some cave populations of the Mexican characid fish *Astyanax mexicanus*, the eyes are degenerate because of a genetically reduced capacity of the optic cup to induce lens development (Cahn 1959). In this as in many other instances, it is possible to explain an evolutionary event that results from the loss of a developmental mechanism (i.e., failure to produce or respond to an inducing influence); it is more difficult to understand how the ancestral developmental system evolved in the first place.

### Morphogenesis

The acquisition of form of an organ, or of the whole organism, is termed MORPHOGENESIS. Morphogenesis and tissue differentiation together determine structure: cytodifferentiation in a leg primordium results in bone, muscle, and so forth, while factors governing morphogenesis determine the shape and size of the appendage. Several mechanisms are entailed in morphogenesis (Wessels 1982). Some structures are formed from the migration of individual cells; the melanocytes (pigment cells) of vertebrates, for example, are derived from the neural crest. The size and shape of structures formed from masses of mesenchymal cells can be influenced by the rate and duration of mitosis, the spatial orientation of mitosis, and the density of cell packing. Any or all of these could influence allometric growth, but there is little information on the subject. A mutation in the house mouse that prevents the face from elongating into a snout apparently acts by increasing the density of cell packing, possibly by increasing adhesion among cells. Another mechanism of morphogenesis is cell death; for example, the digits of the limbs of amniotes are formed by the death of cells in the embryo's paddle-shaped hand or foot between the growth points of the digits. There is less interdigital cell death in the duck's foot than in the chicken's, resulting in the duck's more fully webbed foot (Raff and Kaufman 1983).

### Pattern formation

Among the most important developmental factors in evolution are those that determine the number and spatial arrangement of structures such as petals in a flower, digits on a foot, or scales on a snake. These patterns imply mechanisms whereby cytodifferentiation and morphogenesis occur in certain sites rather than others. Differences in the spatial pattern of mitosis, for example, appear to determine whether the petals of a flower are fused or unfused, a character that differentiates many families of plants. If the tissues (intercalary meristems) between the petal primordia grow by mitosis when the petal primordia grow, the petals are fused; if not, the petals are separate (Figure 23; Stebbins 1974).

Early in the history of experimental embryology, the concept of GRADIENTS in inducing substances was developed. For example, cells at the posterior edge of the developing wing of a chick induce the development of three digits (II, III, and IV) which do not differentiate if the posterior edge is removed. If posterior wing tissue is grafted onto the anterior edge, a mirror-image limb is formed, with the digit sequence IV-III-II-III-IV. This is interpreted to mean that successively anterior digits develop in response to successively lower concentration of some substance produced by the posterior region. Wolpert (1982) suggests that cells gain POSITIONAL INFORMATION from the concentration gradient, and differentiate on the basis of their position. Evolutionary changes in pattern, then, would come about through genetic changes in cells' responses to their position.

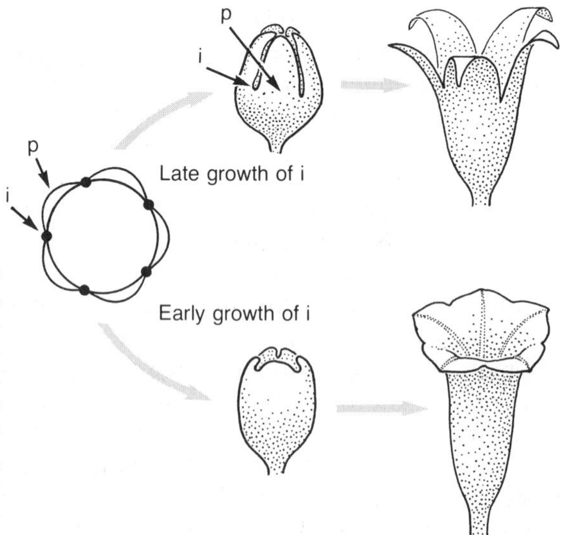


FIGURE 23

**Effects of developmental timing on form. The developing flower, shown from above at left, has petal primordia (p) separated by intercalary cells (i). If the intercalary cells begin dividing only after the petal primordia grow, the petal lobes of the mature flower (side view, at right) are well separated. If the intercalary cells develop along with the petal primordia, a largely fused (sympetalous) corolla is formed. (Adapted from Stebbins 1974, after Payer 1857)**

Alternatively, evolution might occur by alteration of the concentration gradient. In this view, the inducing substance forms a PREPATTERN (Stern 1968), to which the ultimate pattern of structures is a response. The tarsus of a female *Drosophila*, for example, has an underlying prepattern for the development of a male sex comb (a dense patch of modified bristles), since male epidermal cells on an otherwise female tarsus develop a sex comb even though a female's tarsus ordinarily lacks one.

Rather complex prepatterns can, theoretically, be generated quite simply. The mathematician Alan Turing (1952) showed that if two substances that react to give an inducing substance diffuse across an area (or "field"), the reaction can generate a standing wave pattern in the concentration of inducer. The pattern depends on the size and shape of the field, on the rates of diffusion of the precursors, and on the kinetics of their reaction. Changes in the arrangement of the wave pattern or in the total concentration of inducer could then change the pattern of the structures that develop in response to the inducer (Figure 24). Sondhi (1963) invoked this model to explain alterations in the pattern of head bristles in *Drosophila* selected for higher bristle number. The new bristles appeared consistently at specific positions (Figure 25), as if selection had raised the concentration of an inducer that is present in a specific pattern in wild-type flies but which is insufficient to induce bristles in certain positions. In some stocks whole groups of bristles were shifted to new sites as if the prepattern had been shifted, as in Turing's model. An interesting aspect of Sondhi's work was that certain

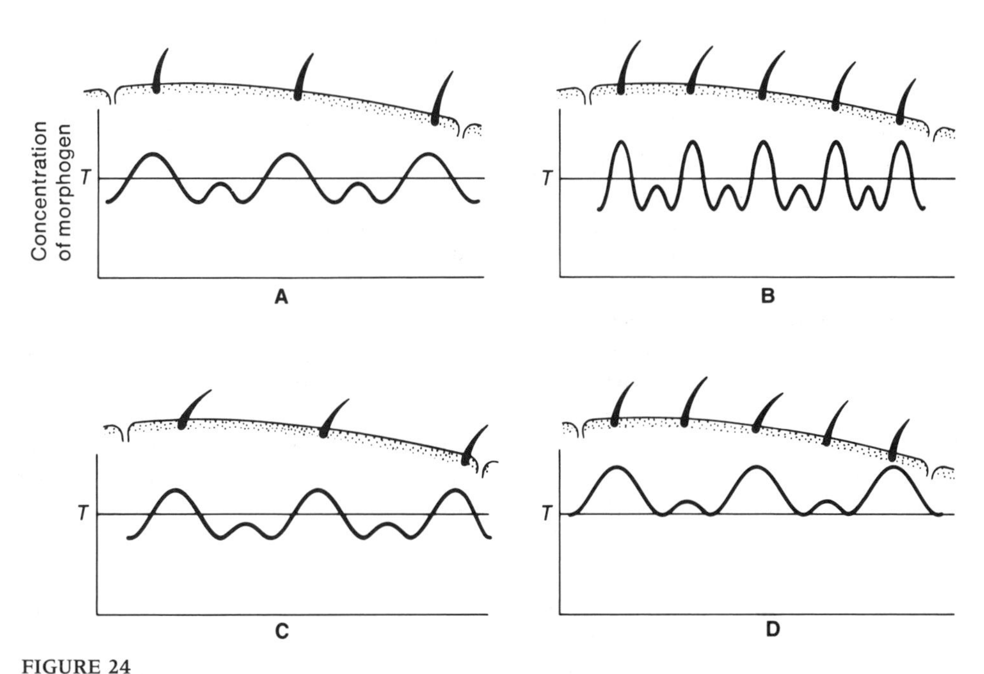


FIGURE 24

**Hypothetical changes in an underlying prepattern (A) of a morphogen, or material that induces bristle formation. If the concentration of morphogen exceeds a threshold *T*, a bristle is formed. Change in the kinetics of morphogen synthesis can change the spacing pattern (B) or the position of the morphogen peaks (C); increasing morphogen concentration can also change the number of bristles (D).**

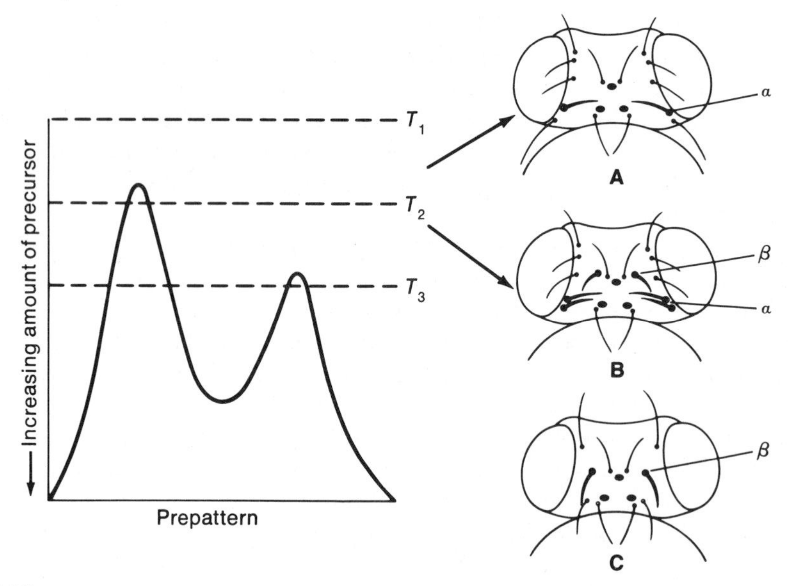


FIGURE 25

**Sondhi's model of origin of a neomorphic ("new") pattern. If the first peak of the prepattern is between threshold levels $T_1$ and $T_2$, it produces the wild-type pattern of bristles and ocelli (A) in *Drosophila subobscura*. If the second peak exceeds threshold $T_3$, additional bristles are formed (B): bristle α is doubled, and new bristles (β) arise. These new bristles are unknown in normal drosophilids, but have a counterpart in *Aulacigaster leucopeza* (C), a member of a related family. (After Sondhi 1962)**

new patterns, although never observed in normal *Drosophila,* resembled those of different, related families of flies (Figure 25).

Turing's theory has been extended mathematically by Murray (1981) in a model of coat colors in mammals. Supposing that melanin is formed by a chemical reaction of the kind envisaged by Turing, Murray predicts that uniform pigmentation would develop if the field is either very small or very large, as it might be if the prepattern were formed either very early or very late in development. Above a minimum field size, a one-dimensional pattern of stripes will develop; as the field size increases, a two-dimensional pattern of spots emerges (Figure 26). Thus on a surface with a tapering cylindrical form, such as a leg or a tail, spots may develop in the thick basal area, but only bands will develop near the tip—a pattern observed in many mammals. Murray's theory suggests that there should exist no striped mammals with spotted tails.

## HOMEOTIC MUTATIONS IN DROSOPHILA

Some of the greatest progress in developmental genetics is being made in the study of homeotic mutants of *Drosophila melanogaster.* The evolutionary significance of this work lies in its possible (but not fully demonstrated) relationship to the evolutionary history of the Arthropoda.

Arthropods are derived from annelid-like ancestors. Each segment of an annelid's body contains intrasegmental longitudinal muscles that insert on invaginations at each end of the segment. Each segment of an arthropod's trunk

FIGURE 26

**An example of possible constraints on the phenotype arising from physical chemistry. (A) Theoretical patterns of pigmentation arising from a mathematical model of the kinetics of a biochemical reaction in successively larger fields a, b, and c. The pattern of one-dimensional bands in a small field (or in the narrower region of a larger field) gives way to a two-dimensional pattern of spots in the larger field. (B) Markings on the tails of mammals resemble the theoretical patterns. a, zebra (*Equus burchelli*); b, genet (*Genetta genetta,* a viverrid carnivore); c, cheetah (*Acinonyx jubatus*); d, jaguar (*Panthera onca*). (From Murray 1981)**

corresponds to the posterior portion of one annelid segment, joined to the anterior portion of the succeeding segment (Figure 27A); the muscles thus run from one segment to the next. Each arthropod segment is a double unit from a phylogenetic point of view.

In the Onychophora (e.g., *Peripatus*), peculiar animals long thought to represent intermediates between annelids and arthropods, the mouth is preceded by an antenna-bearing segment (the protocephalon) and followed by a trunk made up of uniform segments with primitively jointed legs. In arthropods, the legs have a characteristically jointed form that is also evident in the several pairs of mouthparts (mandibles, maxillae, labium) that lie behind the mouth (Figure 27B). This part of the arthropod head, the gnathocephalon, is thus derived from trunk segments. The remaining body segments almost all bear legs in myriapods such as millipedes. In "primitive" (apterygote) insects such as silverfish, the trunk is

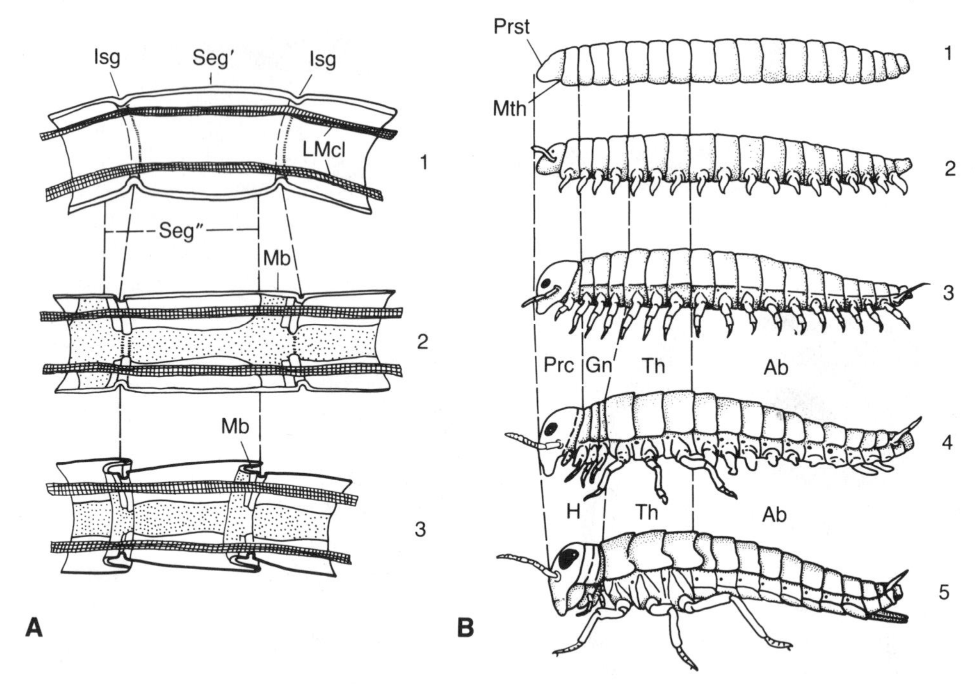


FIGURE 27

**Evolution of arthropod segmentation, based on comparative anatomy and embryology. (A) Hypothesized evolution of secondary segmentation of arthropods (3) from the primary segmentation seen in annelids (1). Cut-away view of interior of body wall. The primary segment (Seg′) is delimited by intersegmental folds (Isg) on which the longitudinal muscles (LMcl) insert. In arthropods, the secondary segments (Seg″) are delimited by intersegmental membranes (Mb) formed toward the posterior of the primary segments. Each secondary segment corresponds to parts of two primary segments. (B) Hypothesized evolution of the segmentation of insects (5) from a wormlike ancestor (1), through intermediate steps resembling Onychophora (2) and myriapods (3 and 4). H, head; Th, thorax; Ab, abdomen; Mth, mouth; Prst, primitive head; Prc, protocephalon; Gn, gnathocephalon. (After Snodgrass 1935)**

differentiated into a three-segmented thorax with legs and an abdomen whose appendages are vestigial or absent except on the terminal segments, where they are modified into genital structures. In the winged (pterygote) insects, finally, the abdomen retains only genitalic appendages, and the second and third thoracic segments (the mesothorax and metathorax) bear wings. Winglike flaps on the prothorax (the first thoracic segment) are seen on certain Carboniferous insect fossils. In the order Diptera (the true flies), the metathoracic wings have become modified as small "balancers" called halteres. With this phylogenetic background in mind, we turn to the development and genetics of segmentation in *Drosophila* (Lewis 1978, Raff and Kaufman 1983, North 1984, Slack 1984).

In the embryonic development of *Drosophila*, a layer of cells—the blastoderm—forms on the surface of the egg. After several invaginations (gastrulation), transverse furrows develop, marking off 17 segments. At this time, the fate of cells that will give rise to both larval and adult structures has already been fixed. The adult structures are formed from groups of cells, the imaginal discs, that do not become morphologically differentiated until the pupal stage.

The first five segments of the embryo give rise to the head (Figure 28); the gnathocephalon is formed from the mandibular (Ma), maxillary (Mx), and labial (Lb) segments. The eyes and antennae of the adult appear to develop from the procephalic segment (Pc). Following the head are three thoracic ($T_1$ through $T_3$) and eight abdominal ($A_1$ through $A_8$) segments, and a caudal segment (C). Most of these segments can be distinguished quite early in larval development by slight differences in morphology, and all are readily recognizable in the adult.

The fundamental pattern of segmentation is governed by genes whose action is deduced from their effects when mutated or deleted. The mutant *knirps* has only one large abdominal segment in place of $A_1$ through $A_7$, and the mutant *Rg(pbx)* has a single abdominal-type segment in place of Mx through $A_1$. The normal alleles at these so-called "gap loci," then, establish major domains of differentiation. Several "pair rule loci" then divide these domains further. For example, the mutant *ftz* has the anterior portion of a given segment fused with the posterior portion of the succeeding segment so that the intervening two regions are deleted. Thus if $T_1$ consists of anterior and posterior compartments $T_{1a}$ and $T_{1p}$, and $T_2$ of compartments $T_{2a}$ and $T_{2p}$, segments of the form ($T_{1a}$ + $T_{2p}$) develop in *ftz*. The mutant *engrailed* replaces posterior compartments with anterior

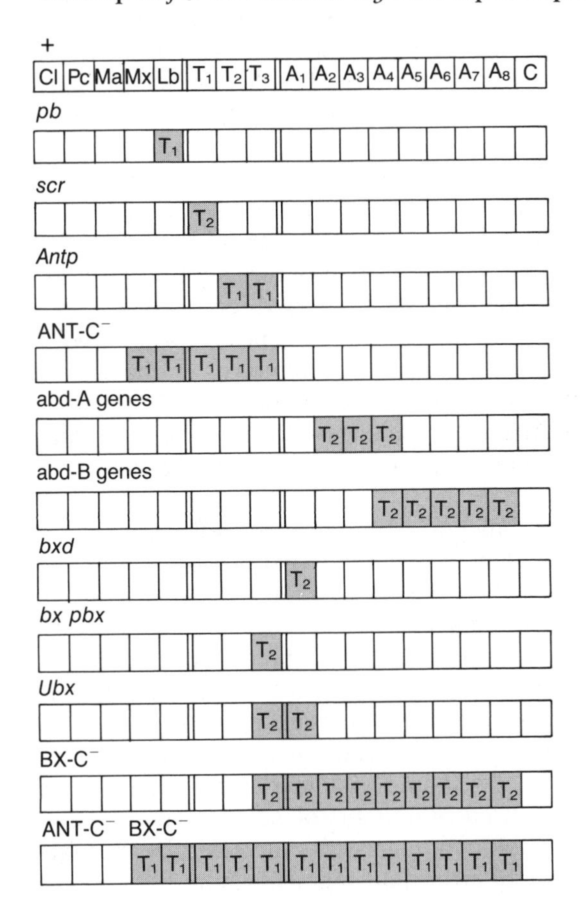


FIGURE 28

**Diagram of homeotic transformations of segments caused by loss mutations or deletions of genes in the ANT-C and BX-C gene complexes of *Drosophila melanogaster*. Normal segment identity (+) at top; mutant genes indicated above each of the other diagrams. Affected segments are shaded and their new identity is indicated. Segment abbreviations as in the text. (Adapted in part from Raff and Kaufman 1983)**

compartments, so that each segment has two anterior compartments arranged as mirror images.

The genes with homeotic effects are arranged in two clusters, the Antennapedia complex (ANT-C) and the Bithorax complex (BX-C). Within each cluster, the order of the genes corresponds fairly well with the order of the segments they control. Within ANT-C, the mutant *proboscipedia (pb)* converts labial palps (segment Lb) into prothoracic legs ($T_1$); *Scr* transforms $T_1$ into $T_2$, and *Antp* converts both $T_2$ and part of $T_3$ into $T_1$. When the entire ANT-C region is deleted, all the segments from Mx through $T_3$ are transformed into $T_1$-like segments (Figure 28). Thus most of the loci in ANT-C are normally responsible for the differentiation of one or more gnathocephalic or thoracic segments from a fundamentally prothoracic "default" plan.

The BX-C complex consists of the three regions *Ubx, abd-A,* and *abd-B,* each comprising several loci. Regions *abd-A* and *abd-B* govern segments $A_2$ through $A_4$ and $A_4$ through $A_8$, respectively. Mutations of loci in each of these regions convert one or more abdominal segments into mesothorax ($T_2$). Each of the several genes in the *Ultrabithorax (Ubx)* region governs a compartment from $T_2$ through $A_1$. The mutant *bxd,* for example, converts $A_1$ to $T_2$; *bx* and *pbx* together convert $T_3$ to $T_2$, creating a four-winged fly. Deletion of the entire BX-C converts each of the segments $T_3$ through $A_8$ into $T_2$, the "default" state from which the normal alleles of various loci in the BX-C promote one or more segments to other states. An interesting mutant in the BX-C, *Contrabithorax (Cbx),* converts $T_2$ to $T_3$, thereby replacing wings with halteres. The function of the normal allele at this locus may be to repress the action of the normal alleles of genes *bx* and *pbx* in segment $T_2$ so that they do not convert it into $T_3$.

The mRNA transcripts of some of these loci have been localized primarily in the specific segments affected by mutations at the loci that code for them (North 1984, Slack 1984, Harding et al. 1985). The transcript of the *ftz* locus (one of the "pair rule loci") is distributed in the blastoderm in bands corresponding to the positions at which segments later become evident; there is direct evidence of a prepattern for the later morphological pattern. Most of the genes associated with segmentation contain a sequence of 180 nucleotides, the "homeo box," that varies only slightly from one gene to another. This repeated element is consistent with the hypothesis that the evolutionary differentiation of serially homologous structures has been accomplished in part by the serial repetition and divergence of genes.

Both the molecular and the developmental patterns, then, reflect serial homology, a common developmental plan among segments that is modified by the addition of segment-specific switch mechanisms. It is interesting that some of the mutants have, in part, an ATAVISTIC quality, causing some apparently ancestral patterns to be revealed. When mutated or deleted, *Ubx* eliminates the difference between mesothorax and metathorax that distinguishes Diptera from other pterygotes, *Antp* yields homogeneous thoracic segments resembling those of apterygotes, ANT-C and BX-C together promote a reversion to the myriapod pattern in which there is no distinction between thorax and abdomen. The origin of the segmented mouthparts from legs is reflected in mutants of ANT-C. The division of the segments into anterior and posterior compartments, revealed by mutants such as *ftz,* reminds us of the evolution of arthropod segments from the annelid pattern, although a real correspondence of the mutants to that pattern is far from certain.

The development of differentiated segments appears to be controlled hierarchically; for example, the ANT-C and BX-C controls are necessary for the differentiation among gnathocephalon, thorax, and abdomen. Given such differentiation, genes such as *bx* and *pbx* act to differentiate meta- from mesothorax. This suggests that the evolution of features peculiar to a developmental unit, such as the halteres of the metathorax, requires first that the unit have acquired a distinct developmental identity. The metathorax must exist as such, under separate developmental control from, say, abdominal segments, before it becomes capable of evolution in its own right. Thus the metathorax of a fly is homologous to that of a grasshopper, but it is not homologous to any specific segment in a millipede, in which no differentiation between thorax and abdomen has evolved. Homology therefore evolves by the acquisition and fixation of developmental control (Wagner 1986).

## CONSERVATISM AND CHANGE IN DEVELOPMENTAL PROGRAMS

Atavistic alterations, or "throwbacks" to ancestral form, occur as mutations or as nongenetic developmental anomalies that can often be experimentally induced (Hall 1984). Whales occasionally sport rudimentary hind limbs, and horses sometimes develop side toes. In birds, the fibula ordinarily develops only into a small splint, whereas the tibia, fused with the tarsal bones, forms the major bone of the chicken drumstick. Hampé (1960) found that when he inserted a slip of mica between the developing tibia and fibula of the chick, the fibula grew to the same length as the tibia. The enlarged fibula induced the tarsals to develop separately from the tibia, producing a pattern very similar to that of *Archaeopteryx* (Figure 29).

Atavistic anomalies are evidence that some developmental mechanisms are highly conservative during evolution. Experimental embryology provides many examples of such conservatism. For example, the dermis of vertebrates induces

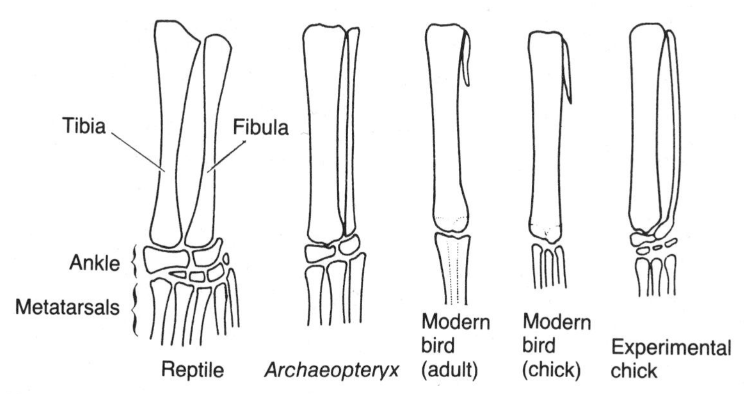


FIGURE 29

**The development of the leg and ankle bones of Hampé's experimental chicks (right), compared to their state in reptiles, *Archaeopteryx*, and the normal condition in modern birds. (From Frazzetta 1975)**

the differentiation of epidermal structures. The embryonic epidermis of a lizard, when grafted onto mouse dermis, develops scales in a pattern typical of mouse hairs (Dhouailly 1973); the "signal" from the dermis is conservative enough to evoke a response across class lines. In toothed vertebrates, teeth form from epidermis in response to induction by jaw mesoderm. Kollar and Fisher (1980) reported that jaw epithelium from chick embryos developed teeth when placed on the jaw mesoderm of mice—a striking example of the conservation of a response to induction that has not been expressed for more than 80 million years.

Some developmental mechanisms, then, retain their integrity even if they have not been phenotypically expressed for millions of years. The genome therefore retains a latent capacity for phenotypes that may be expressed as evolutionary reversals (Chapter 10). The genes governing these mechanisms must have other developmental functions, for if they were nonfunctional they would degenerate by mutation and genetic drift (Chapter 15). We can understand in some instances why structures appear in the embryo, later to be lost, for they serve critical developmental roles; for example, the notochord of vertebrates, although no longer serving its original function of structural support, makes a temporary appearance because it induces the development of the nervous system. Developmentally critical mechanisms, then, are not free to evolve rapidly. Indeed, they are partly responsible for our ability to recognize homology (Wagner 1986).

Nevertheless, the structures of different organisms can be homologous even though their developmental pathways have come to differ substantially; form can be more conservative than the developmental route by which it is achieved. For example, digits are formed by cell death in interdigital regions in amniotes, but by cell division at digital growth points in amphibians. The pattern of cartilage formation by which the bones of a salamander's limb are formed is utterly different from that of other tetrapods, yet few would argue that the limbs of salamanders are not homologous to those of frogs (Hinchliffe and Griffiths 1983). Much as alleles affecting a polygenic character can change in frequency by genetic drift even though the character is maintained constant by stabilizing selection (Chapter 7), a developmental pathway may evolve while the form of its product is retained. Similarly, the cytoskeleton of the ciliate protozoan *Tetrahymena* is constant in form among species, but the proteins of which it is composed differ greatly (Williams 1984). Many of the parts of an organism, then, are so integrated that their form retains an identity through evolutionary time despite changes in their molecular constitution or developmental pathway.

## EVOLUTIONARY CONSTRAINTS AND PHENOTYPIC GAPS

We may approach an overall perspective on macroevolution by asking why all imaginable intermediates among organisms do not exist. Neo-Darwinian explanations, in contrast to the theory of saltations, are of several kinds. Sexual reproduction, by constraining the variance in polygenic traits, can cause gaps between closely related species (Maynard Smith 1983; Chapter 7). Gaps among higher taxa are often merely the historical consequence of extinction. The phylogenetic divergence of different lineages from a common ancestor almost inevitably results in gaps; for example, there has never been an intermediate between a modern horse and a modern rhinoceros. Similarly, characters are correlated among species because of common ancestry (Raup and Gould 1974). All living homeotherms that possess only a right aortic arch have feathers, and those with only a left aortic arch have mammary glands—not because these characters are functionally related, but because birds and mammals have inherited these features from their remote ancestors.

Many gaps exist because of discontinuities in adaptation to the environment: discrete peaks in the adaptive landscape. The convergence of sympatric mimetic butterflies into several discrete color patterns is an example of such ADAPTIVE CONSTRAINTS.

The causes of gaps about which we know least are those constraints on variation that arise not from environmental selection but from internal factors: DEVELOPMENTAL CONSTRAINTS (Alberch 1982, Maynard Smith et al. 1985). Population genetics describes the fate of variations once they arise, but does not predict which variations will arise and which will not. The study of developmental constraints focuses on this critical issue.

Some developmental constraints may arise from physical or chemical principles that simply forbid the origin of certain variations. For example, if Murray's (1981) extension of Turing's prepattern theory is correct, spots cannot develop in too small a developmental field (page 430), and small mammals with spots should have evolved only rarely if at all.

Another kind of developmental constraint describes variations that can arise but which so disrupt the function of the organism that they are invariably selected against—headless *Drosophila* larvae carrying the *bicaudal* mutation, for example. In such cases, there may be little distinction between developmental constraints and adaptive constraints. Developmental constraints certainly come into play when an advantageous feature is caused by alleles whose pleiotropic effects on other characters are highly deleterious.

Undoubtedly the majority of developmental constraints are not inherent in physical principles or in gene action, but are products of evolution and so are historically contingent and taxon-specific. We have noted that a highly variable character in one taxon is frequently almost invariant in another. Nevertheless, unexpressed genetic variation sometimes underlies phenotypically invariant characters, suggesting that canalization of the development of the character has evolved (Waddington 1956a).

Moreover, the developmental system has come to be differentially canalized in different taxa, so that different variations arise. For example, reduction of digits, a common event in the evolution of amphibians, usually entails partial loss of preaxial digits in frogs but partial or complete loss of postaxial digits in salamanders. The preaxial digits of salamanders are the first to differentiate during ontogeny, whereas the central and postaxial digits differentiate first in frogs. When an inhibitor of mitosis was applied to developing limbs (Figure 30), the phalanges and digits that failed to differentiate were the latest to develop: the postaxial ones in a salamander, but the preaxial ones in a frog (Alberch and Gale 1985). It appears, then, that the developmental differences that have evolved in these groups determine the variations that can arise and become fixed during evolution.

Genetic correlations among characters that prevent them from evolving independently are another form of evolved developmental constraint. Olson and Miller (1958) and Riedl (1977, 1978) proposed that genetic and developmental

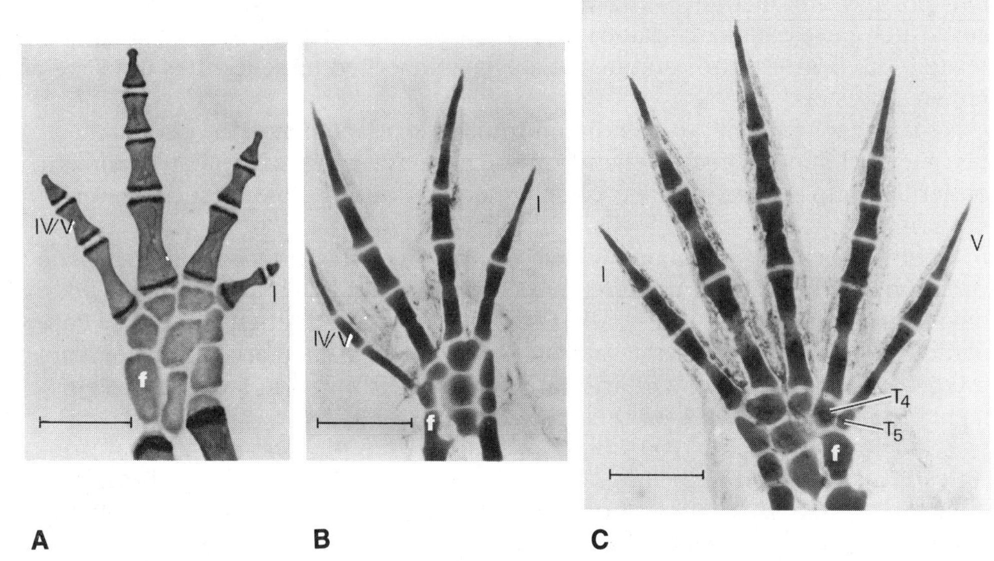


FIGURE 30

**(A) Dorsal view of the left hind foot of the four-toed salamander, *Hemidactylium scutatum* (Plethodontidae); the four-toed condition is normal for this species. (B) Left hind foot of the axolotl *Ambystoma mexicanum* (Ambystomatidae), treated by an inhibitor of mitosis in the limb bud stage. (C) Right hind foot of the same axolotl; untreated control, showing the normal five-toed condition. The experimentally treated foot is smaller than the control, lacks a postaxial toe and some phalanges, and resembles the condition in *Hemidactylium*. (From Alberch and Gale 1985; photograph courtesy of P. Alberch)**

correlations evolve by natural selection. Since as much as 85 percent of the genome may act to shape development, one trait or another would be very likely to suffer a maladaptive developmental accident if each gene were expressed independently. Riedl argues that if traits are brought under a system of hierarchical control so that they develop in integrated sets, there will be effectively fewer independent traits, and hence fewer opportunities for developmental errors. Thus major genetic "switches" ought to trigger the coordinated action of groups of other genes. The loci in the ANT-C and BX-C gene complexes of *Drosophila* appear to be switches of this kind; for example, genes that shape the structure of a normal leg have the same effects on legs that develop in place of antennae. Such integration should be expected, especially for functionally related traits such as the length of the upper and lower jaws of a mammal.

There is some evidence that functionally related traits are integrated in just this way. Cheverud (1982) measured 56 dimensions of the skull of female rhesus macaques (*Macaca mulatta*) and their offspring, and calculated phenotypic correlations ($r_P$), genetic correlations ($r_G$), and environmental correlations ($r_E$) among them (see Chapter 7). On the basis of embryology and functional morphology, he assigned these traits to functional groups; for example, various dimensions of the jaws constitute the "masticatory set." As predicted by the theory of morphological integration, both genetic and environmental correlations were generally higher among traits within functional groups than between them (Table III).

Table III

**Average correlations among rhesus macaque skull measurements assigned to the same ("within F-sets") and different ("between F-sets") functional complexes**

| Kind of correlation | Average correlation within F-sets | Average correlation between F-sets | Percent correctly classified[a] |
|---|---|---|---|
| Phenotypic ($r_P$) | 0.269 | 0.105 | 92 |
| Environmental ($r_E$) | 0.281 | 0.093 | 75 |
| Genetic ($r_G$) | 0.270 | 0.138 | 58 |
| Genetic (neuro.)[b] | 0.370 | 0.110 | 96 |
| Genetic (facial)[b] | 0.123 | −0.042 | 74 |

From Cheverud (1982)

[a]"Percent correctly classified" describes the proportion of traits that cluster together in the "correct" F-set according to the kind of correlation (phenotypic, environmental, or genetic) used.

[b]Genetic correlations for the neurocranial and orofacial groups of traits separately considered. Each of these was broken down into several F-sets.

## DEVELOPMENTAL INTEGRATION AND MACROEVOLUTION

Developmental integration imposes constraints on evolution, but has other consequences as well (Riedl 1978). If developmentally integrated traits are controlled by genetic "switches," mutation of the switches may cause development to proceed in a different, but harmonious, channel. Such integrated systems are likely to display a limited, recurring repertoire of variations, giving rise therefore to parallel evolution and to atavistic variants that reveal in a recapitulatory way the ancestral foundations of the developmental program. If each of the virtual infinity of an organism's characters were independently variable and subject to environmental selection, most similarities among species would be attributable to similarity of function rather than common ancestry; but the individual elements of highly integrated developmental systems are not readily changed without disrupting function. Thus homology, the trace of common ancestry that makes phylogenetic analysis and evolutionary classification possible, "is the consequence of epigenetic fixation beyond recent functional requirements" (Riedl 1977).

Developmental integration, although constraining the paths of evolution, makes the evolution of complex systems possible. The difference in the length of a leg between two vertebrates can be described in terms of differences in the dimensions of bones, muscles, blood vessels, and nerves—but the individuals do not differ at separate loci controlling each of these structures. Rather, a single control system alters them in concert. If a change in leg length required a separate mutation for each of the elements of the leg, the probability that an individual would inherit the constellation of mutations necessary for a properly formed but longer leg would be vanishingly small. Given developmental integration, however, coordinated change in the elements of the leg becomes not only theoretically probable, but physically observable. Heterochronic evolutionary changes such as progenesis represent the extreme in coordinated change owing to developmental integration. Thus the major problem of macroevolution, the evolution of complex traits that cannot function unless their several parts evolve in concert, may be partly solved by the hierarchical, integrated nature of development.

## NEO-DARWINISM AND ITS CRITICS

In both the popular and professional literature (e.g., Ho and Saunders 1984), the neo-Darwinian theory that emerged from the Modern Synthesis has been variously criticized as incomplete, inadequate as an explanation of evolution, or just plain wrong. The latter charge can be dismissed; there is simply too much evidence in the theory's favor. Whatever elaborations of evolutionary theory develop in the future, they must be compatible with and include random (i.e., not adaptively directed) mutations, individual selection, and the rest of population genetics, just as biochemical theories must be compatible with the mechanisms of physics and chemistry.

Whether or not the theory adequately explains all of evolution depends on one's personal judgment of what constitutes explanation. The theories of physics explain the weather in principle, but the empirical information that the theory requires to explain fully why there was a thunderstorm in Ithaca, New York on August 2, 1985 is lacking. Similarly, neo-Darwinian theory explains in principle the evolution of complex organs, but for lack of information we do not have a verifiable population genetic theory of the origin of feathers.

Neo-Darwinian theory might well be incomplete. It would be rash to maintain that new factors may not be discovered through, say, the study of molecular mechanisms (Chapter 15). But most discoveries of the last 30 years or so have only elaborated and given substance to the form of theories that existed by the 1940s. Our knowledge of the mechanisms of mutation has grown greatly, but mutations are as random now as formerly supposed. Genetic drift has taken on greater significance as a factor of evolution that was imagined 30 years ago, but the fundamental theory of genetic drift has not changed.

It is true that classical neo-Darwinism emphasized some modes of explanation, levels of analysis, and questions to which inquiry was directed, at the expense of others. Evolutionary theory is currently being expanded (Gould 1982) in all these respects. The favored mode of explanation has been the operation of natural selection within populations; factors such as genetic drift and the extinction of populations and species were recognized, but only recently have received more emphasis. The traditional level of analysis has been the study of gene action and gene frequency change; there is now a greater emphasis on higher levels of biological organization such as development and the constraints imposed by phylogenetic history. The importance of development and historical contingency has always been recognized, but until recently it has been the kind of formal, polite recognition accorded to a stranger at an otherwise intimate party.

The power of neo-Darwinism lies in its generality of explanation. But like most general theories, it is highly abstract. It gains full explanatory power when concepts such as gene frequencies and selection are given empirical content by applying them to real features of real organisms: behavior, life histories, breeding systems, physiology and morphology. When this is done, however, new questions appropriate to those particular features emerge and context-specific factors must be added to the theory. The evolution of behavior raises questions about issues such as kinship, and requires us to analyze factors such as learning. Likewise morphology, that neglected subject that was once the chief substance of biology, requires us to understand not only the abstract theory of gene frequencies, but those developmental and historical factors that determine the forms of morphological evolution.

The theory of neo-Darwinism is a theory of mechanisms. But it was erected to explain the existence of a history of evolutionary change. While the study of mechanism has held center stage, the study of history—through paleontology, systematics, and morphology—has been slighted. But just as the actual history of human affairs is as viable and rich a subject as the sociology and political science that address its mechanisms, so the study of evolutionary history raises questions and hypotheses with their own rich intellectual content, and describes patterns of diversification, extinction, and historically contingent change that present to us intrinsic interest and grandeur.

## SUMMARY

Morphological evolution proceeds at highly variable rates. The theory of population genetics adequately explains morphological change and stasis, and does not require a correspondence between morphological evolution and speciation. Higher taxa emerge through mosaic evolution, whereby individual characters and developmentally integrated sets of characters evolve quasi-independently, each typically proceeding through intermediate stages for which selective advantages, especially changes in function, can generally be envisioned. Because many characters, especially functionally related characters, are integrated during ontogeny, coherent and substantial morphological change can sometimes arise from fairly simple genetic changes in critical control mechanisms of development. The developmental integration that evolves in the course of phylogeny both constrains the expression of genetic variation so that some paths of evolution are more likely than others, and enables evolutionary change in complex characters.

## FOR DISCUSSION AND THOUGHT

1. In Chapter 6 of *The Origin of Species*, Darwin remarked, "If it could be demonstrated that any complex organ existed, which could not possibly have been formed by numerous successive, slight modifications, my theory [of evolution by natural selection] would absolutely break down." Discuss this statement in light of this chapter. Darwin's Chapters 6 and 7 make fascinating reading in this context.
2. Paedomorphosis has been advanced as a mechanism whereby lineages can "escape from specialization" and embark on new evolutionary paths. Can you find evidence in support of this idea? Do new higher taxa evolve from specialized or unspecialized ancestors? How could you tell?
3. Why should the number of serially homologous parts be more variable if they are many than if they are few?
4. Discuss ways in which paedomorphosis by acceleration (progenesis) can be distinguished from paedomorphosis by retardation (neoteny).
5. Explain how natural selection can shift the development of structures forward during ontogeny so that they appear before they have any function. For example, human embryos have calluses on their feet.
6. Why is the average rate of evolution of a character so much slower over long periods than over short periods?
7. From comparative anatomy and/or paleontology, trace the evolutionary history of the

features of a modern group such as an order of mammals or a family of angiosperms. Do derived features arise from prexisting structures in all cases? Is there evidence that their transformation has been gradual?

8. What is the current understanding of the mechanisms by which gene action is regulated, and how can such understanding be applied to morphological evolution? See Davidson and Britten (1973), Dawid et al. (1982), MacIntyre (1982).
9. The generalization that early ontogenetic stages are evolutionarily conservative sometimes does not hold. For example, modes of gastrulation differ among and within phyla (compare frogs and birds); the gastrulae of oligochaete annelids develop directly into the adult form rather than into a trochophore larva as in polychaetes. How can such drastic remodeling of early development be explained?
10. Certain substantial evolutionary changes in morphology can be plausibly attributed to simple genetic changes, sometimes at a single locus; neoteny in *Ambystoma* is one example (p. 421), and Raff and Kaufman (1983) cite others. Interpret these cases in terms of Goldschmidt's views on the one hand and those of neo-Darwinians such as Fisher on the other.

## MAJOR REFERENCES

Raff, R.A. and T.C. Kaufman. 1983. *Embryos, genes, and evolution. The developmental-genetic basis of evolutionary change.* Macmillan, New York. 395 pages. A developmental interpretation of evolution that provides a good introduction to developmental genetics.

Rensch, B. 1959. *Evolution above the species level.* Columbia University Press, New York. 419 pages. An old but excellent treatment of the evolution of new features, discussed in part from a developmental viewpoint.

Bonner, J.T. (editor) 1982. *Evolution and development.* Springer-Verlag, Berlin. 356 pages. Proceedings of a conference on the subject, with contributions from molecular, developmental, and evolutionary biologists.

Gould, S.J., and N. Eldredge. 1977. Punctuated equilibria: The tempo and mode of evolution reconsidered. *Paleobiology 3:* 115-151. One of the more detailed explications of the theory of punctuated equilibrium. A more extended defense of the hypothesis is provided by Stanley, S.M. 1979. *Macroevolution: Pattern and process.* Freeman, San Francisco.

Charlesworth, B., R. Lande, and M. Slatkin. 1982. A neo-Darwinian commentary on macroevolution. *Evolution 36:* 474-498. One of the more comprehensive critiques of punctuated equilibrium.

# Evolution at the Molecular Level

# Chapter Fifteen

The general principles of evolution explain variation in morphology, physiology, biochemical pathways, behavior, life histories—the classes of characteristics to which evolutionary biologists have devoted much of their attention. Each such class of characteristics calls for special study in its own right, and for special extensions and modifications of the general theory that embraces them all. So it is with yet another dimension of the diversity of life: the structure, organization, and function of the hereditary material. Fascinating as an object of study in itself, the genetic material is, however, more. It is the repository of information for phenotypic characters, and so we look to its evolution as a source of insight into the evolution of morphology, behavior, biochemistry—in short, of organisms. Abstracted from the organism, sequences of DNA are neither more nor less interesting than bones or flowers; viewed in the context of the organism, they provide fundamental insights into life and its history.

## THE USES OF MOLECULAR INFORMATION IN EVOLUTIONARY STUDIES

The molecules that make up an organism include products of biochemical reactions—lipids, steroid hormones, alkaloids, carbohydrates, and many more—that have been the subjects of numerous comparative evolutionary studies. Our focus, however, will be on macromolecules: DNA sequences, RNA sequences, and proteins, the latter being the proximal agents of most biochemical diversity. We will be most concerned with sequences of nucleotide bases or of amino acids, or with less detailed contrasts that reflect sequence differences with greater or lesser precision. The evolutionary literature that makes use of such information is of several kinds.

First, molecular variants are tools for addressing traditional problems in population genetics (see Chapters 4 through 8). Protein differences revealed by techniques such as electrophoresis are used to study gene flow, genetic variation, and natural selection. By direct sequencing of DNA or by the use of restriction enzymes, it is now possible to identify base pair polymorphisms that provide more refined information for these purposes. For example, where two forms of mice (*Mus musculus* and *M. domesticus*) meet in Denmark, electrophoresis revealed little introgression of the nuclear genes of *domesticus* into the *musculus* population; yet the restriction enzyme patterns of mitochondrial DNA indicated that there had been flow of mitochondria between them (Ferris et al. 1983). In theory, selectively neutral mitochondrial variants will reach similar frequencies in two hybridizing populations, even if the frequency of hybridization is low and there is strong selection against heterozygotes at loci in the nuclear genome (Takahata and Slatkin 1984).

A second use of macromolecular data is in the inference of phylogenetic relationships (Chapter 10). Amino acid and nucleotide sequences in particular provide very large numbers of characters, and so are being used with increasing frequency to determine branching patterns in phylogeny. As for population genetic studies, molecular data in systematics are used as tools for addressing traditional evolutionary questions.

The third kind of molecular evolutionary literature takes the genome as an object of interest in itself. Without reference to organismal-level features, we may ask how and at what rate DNA sequences evolve. This is a chief subject of this chapter.

Finally, we may ask how changes at the level of the DNA effect changes in the properties of organisms. In large part, this amounts to asking how changes in macromolecules affect their function, and how such functional changes are translated into evolutionary changes in physiology, development, and morphology. This too is part of the subject matter of this chapter.

## TECHNIQUES

The recent spectacular advances in the techniques of molecular biology have made possible extraordinarily detailed descriptions of the genetic material. Some of the techniques were described briefly in Chapter 3; recent textbooks of genetics should be consulted for further details. It is now possible to "clone" DNA sequences of interest by placing them in bacteria, and to build up a "library" of an organism's genes. Within a DNA sequence, the relative location of certain short base-pair sequences can be determined by restriction enzymes that cut the DNA at specific sequences five or six base pairs (bp) long (Figure 1). It is considerably more tedious, but nevertheless now routine, to determine the complete base-pair sequence of fragments several thousand base pairs (kilobases, kb) long. Cloned genes can be radioactively labeled and "hybridized" with preparations of intact chromosomes to determine the location(s) of a sequence. For much of this chapter, limitations of space preclude more detailed descriptions of the molecular techniques used in individual studies; the focus instead will be on some general conclusions that are emerging from molecular studies, and their evolutionary interpretation. It must be emphasized that important new discoveries are being

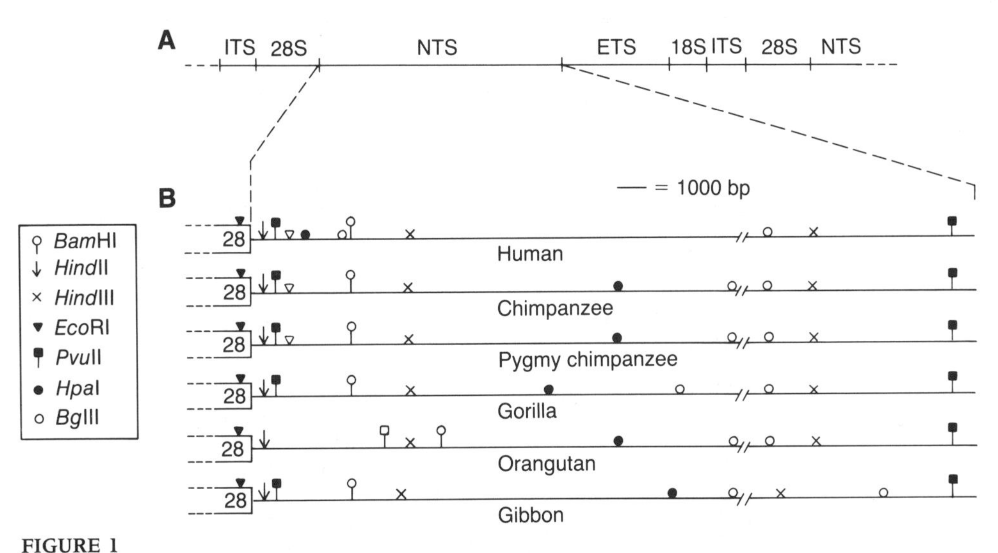


FIGURE 1

**Restriction enzyme maps of the nontranscribed spacer regions in the rRNA genes of human and apes. (A) The organization of one of the repeated units, in which the transcribed region carries sequences for 18S and 28S rRNA. (B) The sites recognized by seven restriction enzymes are indicated for the nontranscribed region of each of several species. The break in the sequence indicates a portion that has not been studied; the inverted triangles (▽) are regions that are heterogeneous in length among different copies. Note the several restriction sites that, although they are unique to one or another species, are present in most or all copies in the genome. (After Arnheim 1983)**

TABLE I

**Variation in nucleotide sequence in 11 *Adh* genes of *Drosophila melanogaster***

| Region | 5′ flanking sequence | Adult leader (exon 1) | Intron 1 | Larval leader | Exon 2 | Intron 2 | Exon 3 | Intron 3 | Exon 4 | 3′ untrans- lated region | 3′ flanking sequence |
|---|---|---|---|---|---|---|---|---|---|---|---|
| Average number of nucleotides | 63 | 87 | 620 | 70 | 90 | 65 | 405 | 70 | 264 | 178 | 767 |
| Number of polymorphic sites[a] | 3 | 0 | 11 | 1 | 1 | 2 | 4 | 5 | 9 | 2 | 5 |
| Percent polymorphic | 4.7 | 0 | 1.8 | 1.4 | 1.0 | 3.1 | 1.0 | 7.1 | 3.5 | 1.1 | 0.6 |

From Kreitman (1983)

[a]In addition to the nucleotide polymorphisms, there were short insertion/deletion polymorphisms at two sites in intron 1, one site in the 3′ untranslated region, and 2 sites in the 3′ flanking sequence.

made at such a great rate that some of the material in this chapter will be out of date before it is published.

## VARIATION IN SINGLE DNA SEQUENCES

Even the coarse description of DNA provided by restriction enzyme mapping has revealed considerable intraspecific variation at the nucleotide level (Avise and Lansman 1983, Nei 1983). For example, Langley et al. (1982) used eight restriction enzymes to study variation in a 12 kb region containing the alcohol dehydrogenase (*Adh*) locus of *Drosophila melanogaster,* and found four polymorphic sites in a sample of 18 chromosomes. Because recognition sequences for a particular enzyme are only sparsely distributed throughout the DNA, this figure represents an average heterozygosity of about 0.004 per nucleotide site; that is, a fly may be expected on average to be heterozygous at about four out of every thousand positions.

A more detailed study was performed by Kreitman (1983), who determined the complete sequence of the *Adh* locus and its flanking sequences for 11 individual genes taken from *Drosophila melanogaster* collected on four continents. This locus is polymorphic for two electrophoretically distinguishable alleles, $Adh^F$ and $Adh^S$, that show parallel latitudinal clines in allele frequency on several continents and produce allozymes that differ in activity (Chapter 6). Kreitman's sample of five $Adh^F$ and six $Adh^S$ alleles revealed a single nucleotide substitution accounting for the observed difference in the amino acid sequences of the enzymes coded for by the two alleles (threonine versus lysine at one position). Forty-two other sites varied in the entire sequence, which on average was 2721 bp long; at several sites there were variations in length owing to polymorphism for deletions or insertions of short sequences. Silent polymorphisms (i.e., synonymous codons) were abundant, as was variation in the introns (Table I).

Kreitman noted that the $Adh^F$ genes were less variable than the $Adh^S$ genes, and have such similarity of sequence as to suggest that the $Adh^F$ allele (the threonine-lysine substitution) arose recently from a single mutation of an $Adh^S$ ancestral gene (Figure 2). Two close relatives of *Drosophila melanogaster* possess only the $Adh^S$ allele, strengthening this inference. Thus the $Adh^F$ allele has had less time to accumulate variation by mutation. Kreitman also noted that the "phylogeny" of the 11 genes implied by certain sites was incompatible with other sites, and inferred that at least two instances of intragenic recombination have occurred in the history of these genes (Figure 2). Hudson and Kaplan (1985) analyzed the data further, and concluded that at least five and perhaps as many as 150 intragenic recombination events have occurred.

## RATES OF SEQUENCE EVOLUTION

Both amino acid sequencing of proteins and direct nucleotide sequencing of homologous genes in different species have shown that some DNA sequences evolve at much higher rates than others (Chapters 5 and 10). The sequences of nonfunctional or nearly nonfunctional polypeptides, such as the C peptide that is discarded from preproinsulin when this molecule is processed to form insulin, evolve at higher rates than functionally constrained proteins (Table II; see also Chapter 5). Synonymous changes in codons accumulate faster than changes that cause amino acid substitutions. The genetic code is most degenerate at the third

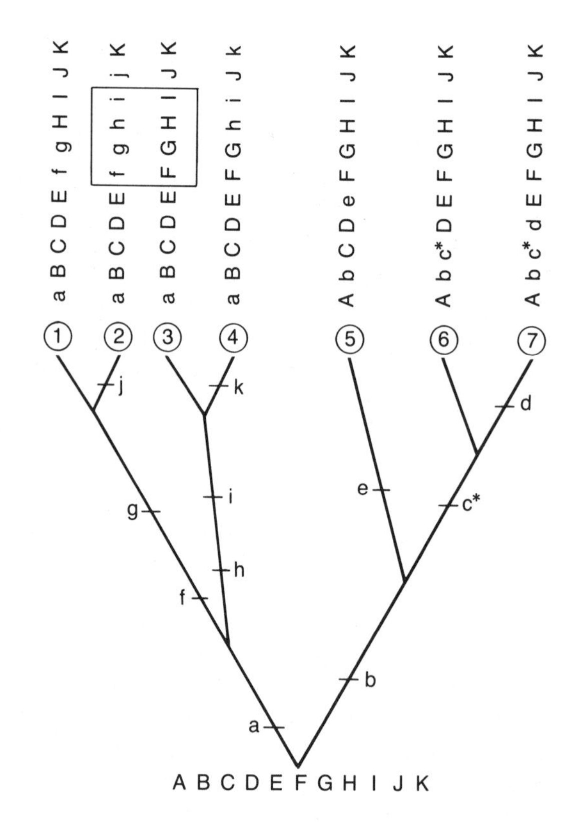


FIGURE 2

**A schematic example of inference of mutational events in the history of a sample of genes. A through K represent variable base pairs in a longer DNA sequence. The change *c** alters an amino acid, so the protein product of genes 6 and 7 is a different electromorph from that of genes 1–5. The greater homogeneity of sequence of 6 and 7 than of 1–5 reflects the recency of the electrophoretically detectable mutation. The variation in sequence among genes 1 through 4 may be explained by postulating that either the pair of mutations f and g or the pair h and i occurred twice. More parsimoniously, we may infer that each mutation occurred only once, as shown on the diagram, but that intragenic recombination between sites G and H gave rise to the four combinations. Note that the genealogy of genes may be inferred by the same principles used to infer phylogenetic relationships among species.**

codon position and least at the second, and the rate of divergence at these positions in the codons of translated sequences corresponds to this variation in degeneracy (Table II). The rate of base-pair divergence is greater in introns than in exons (Jeffreys 1982, Bodmer and Ashburner 1984). These observations, as we noted in Chapter 5, support the neutralists' argument that the fixation of most nucleotide substitutions is the consequence of random genetic drift rather than natural selection. We have also noted that averaged over a number of homologous genes, nucleotide sequences appear to diverge at an approximately constant rate in different lineages, although some evidence of greater rates in organisms with short rather than long generation times suggests that the rate of fixation may be approximately constant with respect to numbers of generations rather than absolute time (Chapters 5 and 10).

Over sufficiently long time, multiple substitutions can occur at the same position, so that the observed number of differences between two species that share a remote common ancestor will be less than the number of substitutions that have transpired. For mitochondrial DNA of mammals, the relation between sequence difference and time since divergence is linear for about 5–10 million years (Myr), because most substitutions within this time occur at different sites (Figure 3). The mitochondrial DNA of mammals evolves at five to ten times the rate of nuclear DNA, perhaps because the DNA polymerase that replicates mitochondrial DNA lacks the proofreading activity of the polymerases that replicate the nuclear genome (Brown et al. 1979, Brown 1983). In both mitochondrial and

TABLE II

**Numbers and rates of base pair substitutions in several genes, inferred from amino acid sequences[a]**

| Comparison | Approximate time of divergence (years) | Number of nucleotide differences, per site, at codon position | | | | Nucleotide substitutions per year in each lineage ($\times 10^{-9}$) at codon position | | | |
|---|---|---|---|---|---|---|---|---|---|
| | | 1 | 2 | 3(total) | 3(synonymous) | 1 | 2 | 3(total) | 3(synonymous) |
| Human/rat | $8 \times 10^7$ | | | | | | | | |
| Pregrowth hormones | | 0.26 | 0.18 | 0.53 | 0.44 | 1.6 | 1.1 | 3.3 | 2.8 |
| Preproinsulins | | | | | | | | | |
| A+B chains (insulin) | | 0.04 | 0.00 | 0.46 | 0.38 | 0.25 | 0.00 | 2.7 | 2.4 |
| C peptide | | 0.18 | 0.27 | 0.95 | 0.77 | 1.1 | 1.7 | 6.0 | 4.8 |
| Chicken/rabbit | $3 \times 10^8$ | | | | | | | | |
| β globins | | 0.30 | 0.19 | 0.64 | 0.54 | 0.5 | 0.32 | 0.40 | 0.34 |
| Chicken/rat | $3 \times 10^8$ | | | | | | | | |
| α tubulins | | 0.025 | 0.005 | 0.58 | 0.47 | 0.015 | 0.003 | 0.36 | 0.29 |
| *S. purpuratus/P. miliaris* (sea urchins) | $10^8$ | | | | | | | | |
| Histone H2β | | 0.09 | 0.02 | 0.48 | 0.43 | 0.056 | 0.013 | 0.30 | 0.27 |
| Histone H3 | | 0.008 | 0.008 | 0.47 | 0.41 | 0.005 | 0.005 | 0.29 | 0.26 |

Based on Kimura (1983)

[a]Approximate divergence times are based on the fossil record; the divergence time given for the sea urchins is less well established than the others. Kimura provides standard errors for the estimated numbers of substitutions, which are inferred from amino acid sequences under a model that estimates the proportion of synonymous substitutions.

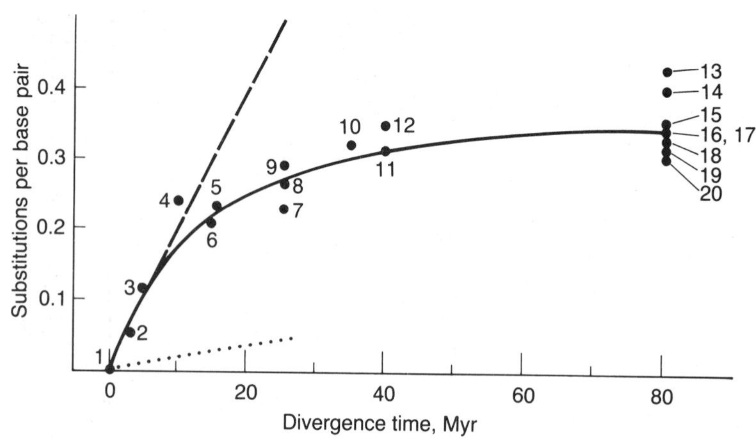


FIGURE 3

**The number of base pair differences per site between mitochondrial DNA of pairs of species is plotted against the estimated time since their most recent common ancestor. Sequence difference is calculated from restriction enzyme data, divergence time from paleontological data and protein differences. The dashed line estimates the rate of substitution by linear extrapolation from the initial slope of the curve. The dotted line estimates the rate of substitution in single-copy nuclear DNA. Because of the high substitution rate in mitochondrial DNA, multiple substitutions at the same sites after 10–15 Myr cause the curve to level off. (Point 1 = variation among individual humans; 2 = goat and sheep; 3–9 = pairs of primates; 10–12 = pairs of rodents; 13–20 = various rodent–primate species pairs.) (From Brown et al. 1979)**

nuclear DNA, transitions account for far more base-pair substitutions than transversions (Brown 1983, Li and Gojobori 1983).

Some sequences of DNA appear to be conserved for extraordinary spans of time. Part of the sequence of a gene that affects circadian rhythms in *Drosophila* is found as a multiply repeated sequence in the DNA of birds and mammals, although its function in vertebrates is unknown (Shin et al. 1985). Several of the genes (*ftz, Ultrabithorax, Antennapedia*) that affect segmentation in *Drosophila* (see Chapter 14) share a sequence, coding for 60 amino acids, that appears similar to DNA sequences in frogs and humans (Gehring 1985). Some similarity of sequence, which however may be convergent rather than homologous, is also found in a protein of yeast (Shepherd et al. 1984). An interesting contrast in evolutionary rates is provided by those regions of genes at which transcription by RNA polymerases is initiated (Arnheim 1983). RNA polymerase I transcribes only DNA sequences that code for ribosomal RNA (rRNA genes). The initiation site of the rRNA genes of *Drosophila,* human, and the frog *Xenopus* is very different in nucleotide sequence. RNA polymerase II transcribes many of the genes that code for messenger RNA, so it might be expected that the initiation site of such genes would not be free to vary. In fact a CONSENSUS SEQUENCE for the initiation site—an "average" DNA sequence that shows only minor variations—can be identified among various protein-coding genes both within and among phylogenetically diverse organisms. That the function of these sites is conserved is shown by experiment; the transcription system of humans can transcribe the silk protein gene of silkworms (*Bombyx mori*) in vitro, and RNA transcripts of sea urchin (*Strongylocentrotus*) genes are produced when sea urchin DNA is injected into *Xenopus* oocytes.

## EVOLUTIONARY CHANGES IN THE LOCATION AND NUMBERS OF GENES

The linkage relationships among genes have long been known to evolve by chromosome rearrangements such as inversions, translocations, and the fusion and fission of chromosomes (Chapter 3). The total amount of DNA is increased greatly when polyploidy occurs. But other mechanisms of change in the location and number of DNA sequences have been revealed by the study of repeated DNA (Davidson and Britten 1973, Dover et al. 1982, Arnheim 1983). In all eukaryotes studied to date, FAMILIES of DNA sequences with identical or very similar sequences have been revealed. The number of elements (copies) in a gene family ranges from two to more than $10^6$; the human genome, for example, contains the *Alu* family of more than 500,000 copies, among which there is only minor variation in nucleotide sequence. The number of families per genome often measures in the hundreds, and the members of a family may either be clustered on a single chromosome or, as in the *Alu* family, interspersed among other genes throughout the chromosomes. A unit consisting of the 18S and 28S rRNA genes together with associated spacers is repeated in tandem several hundred times along one chromosome of *Xenopus*, whereas in humans this gene family is clustered on five different chromosomes. Differences in copy number of repeated sequences account in large part for interspecific differences in the size of chromosomes. In the salamander genus *Plethodon*, species differ more than threefold in DNA content per cell, and the differences are paralleled by differences in chromosome size. Remarkably, the relative sizes and arm lengths of the various chromosomes do not differ among species of *Plethodon*, apparently because changes in amount of repetitive DNA have been evenly distributed among and within chromosomes (MacGregor 1982). Gene families raise the question of how copy number changes in evolution, how the copies are moved to new locations, what effects copy number has, and what governs the similarity of sequence among family members.

## UNEQUAL CROSSING OVER AND THE EVOLUTION OF DUPLICATE GENES

When unequal crossing over occurs between single genes on homologous chromosomes, one strand carries a deletion of the gene and the other a tandem duplication (Chapter 3). Unequal crossing over between two duplicated sequences gives rise to a single copy on one strand and three copies on the other. It is likely that the greater the number of tandem copies, the more likely unequal crossing over becomes, because a copy on one strand can pair with any of the copies on the other. Thus variation in copy number is generated, and a chromosome with a particular number of copies may be fixed either by genetic drift or natural selection (Ohno 1970, Ohta 1980, Arnheim 1983). Duplicate genes may diverge in sequence and function under the influences of mutation, genetic drift, and natural selection.

Duplicate loci within a species are referred to as PARALOGOUS GENES, whereas corresponding genes in different species are ORTHOLOGOUS. Thus the phylogeny of genes can differ from the phylogeny of the species that bear them. Methods of phylogenetic inference (Chapter 10) can be used to trace the phylogeny of genes both within and among species. For example, a maximum parsimony analysis of amino acid sequences indicates that myoglobin and the several hemoglobin chains of vertebrates have originated by a succession of gene duplications from an ancestral globin (Figure 4; Goodman et al. 1982). The β chains of human and

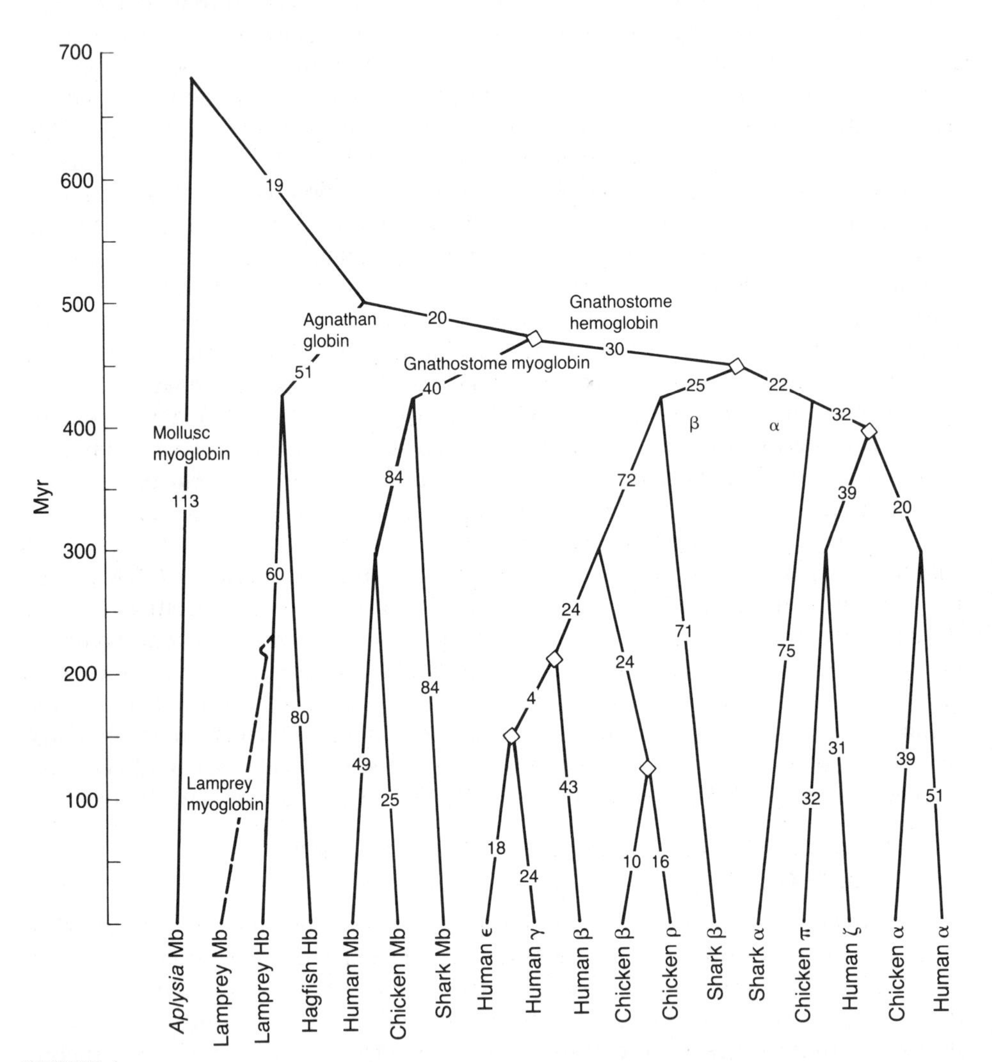


FIGURE 4

**A phylogeny of the myoglobins (Mb) and hemoglobin chains (Hb and Greek letters) of a mollusc (*Aplysia*) and several vertebrates, based on a cladistic analysis of amino acid sequences. Branch points marked with diamonds represent gene duplications, the descendant genes being paralogous. Branch points without diamonds mark the most recent common ancestors of different taxa. Divergence times of taxa (scale at left) are based on the fossil record. Numbers along the branches are numbers of nucleotide replacements, estimated from amino acid differences. The lamprey myoglobin was placed on the tree on the basis of information from a different study. (After Goodman et al. 1982)**

chicken, for example, are considered orthologous, whereas the α and β chains of humans are paralogous. The globins have proliferated by unequal crossing over considerably more than Figure 4 indicates; recall (from Figure 4 in Chapter 3) that the human genome contains at least 11 hemoglobin-like nucleotide sequences, each consisting of three exons and two introns. All the α-like sequences are tightly clustered on chromosome 16, and all the β-like sequences on chromosome 11. Some members of the gene family have diverged in function: myoglobin is expressed in muscle and hemoglobin in erythrocytes, and the expression of several of the hemoglobin genes differs among embryonic, fetal, and adult stages in mammals. Direct evidence that unequal crossing over occurs in the globin clusters is provided by cases of individual humans who carry three rather than two α sequences on one of their chromosomes, and by the anomaly known as Lepore hemoglobin, in which the N-terminal part of the δ chain is fused to the C-terminal part of the β chain, owing to deletion of the DNA between these regions.

In addition to the functional members of the globin family, there exist nonfunctional members, or pseudogenes (denoted ψ). Pseudogenes have numerous substitutions compared to their functional relatives, including frameshift mutations and termination codons that prevent translation into functional polypeptides. The nucleotide sequences of the three pseudogenes in the human globin family show homology to both the exons and the introns of the functional genes. Sequences of this kind are sometimes called "traditional" pseudogenes. Numerous structural genes have related pseudogenes.

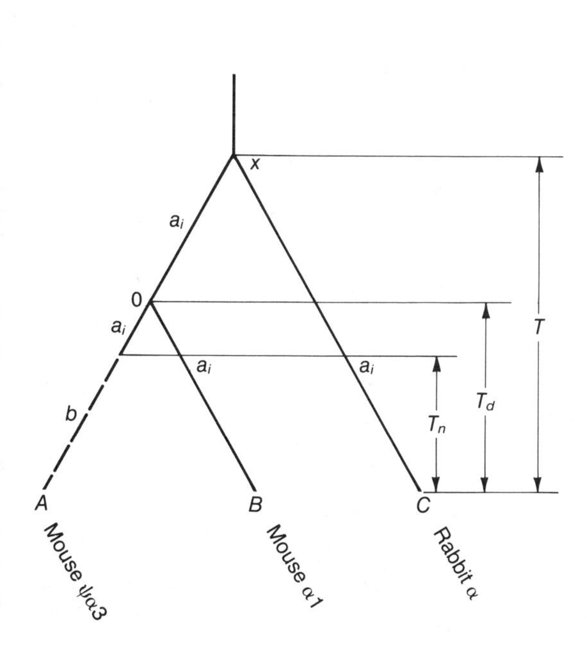


FIGURE 5

**Analysis of the evolution of pseudogenes. At codon position *i*, the average rate of substitution in a functional sequence is $a_i$. The rate of evolution after a gene becomes nonfunctional (at time $T_n$) is *b*. *T* is estimated from the fossil record. The number of nucleotide differences between two sequences (e.g., *B* and *C*) at codon position *i* is $d_{BCi}$. For each of the three codon positions, $d_{AB} = 2aT_d + (b - a)T_n$, $d_{AC} = 2aT + (b - a)T_n$, and $d_{BC} = 2aT$. Therefore $a_i$ can be estimated as $d_{BCi}/2T$, and the time since gene duplication as $T_d = [d_{ABi} - (d_{ACi} - d_{BCi})]/2a_i$. With more algebra, it can be shown that $T_n \cong [(d_{AC} - d_{BC}) - (d_{AC3} - d_{BC3})]/(a_3 - a)$, and $b = (a_3[d_{AC} - d_{BC}] - a[d_{AC3} - d_{BC3}])/([d_{AC} - d_{BC}] - [d_{AC3} - d_{BC3}])$, where absence of the subscript 3 refers to an average of *d* over codon positions 1 and 2. That is, it is possible to estimate the time at which a pseudogene became nonfunctional, and its rate of evolution thereafter, by a comparison of its sequence divergence to that of functional genes, and contrasting its divergence to the difference between the rate of evolution at third base positions and first and second positions. (From Li et al. 1981)**

By assuming that there exists a typical rate of evolution at each of the three codon positions in a coding sequence, Li et al. (1981) estimated the time at which certain globin genes became nonfunctional, and the rate at which their sequences evolved after loss of function (Figure 5; Table III). The ψα3 pseudogene of mouse and the ψα1 pseudogene of human both appear to have become nonfunctional about four Myr after they arose by gene duplication, whereas the ψβ2 pseudogene of rabbit apparently was silenced almost immediately after it arose. (These estimates are imprecise, however.) A more complex history is that of the β globins of the goat (*Capra hircus*; Li and Gojobori 1983; Figure 6). The β gene was duplicated about 42 Myr ago, and one copy became a pseudogene (ψβ) about 6 Myr later. The β-ψβ pair was then duplicated about 15 Myr ago, so the goat now has two functional genes ($\beta^A$, $\beta^C$) and two pseudogenes ($\psi\beta^X$, $\psi\beta^Z$). In between these events, moreover, the β gene gave rise by duplication to yet another functional gene, γ. These analyses also showed that the rate of sequence evolution in a pseudogene after it has become nonfunctional is much higher than in functional genes.

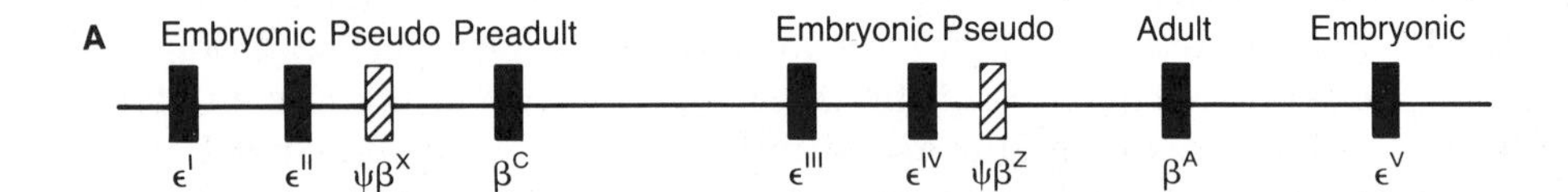


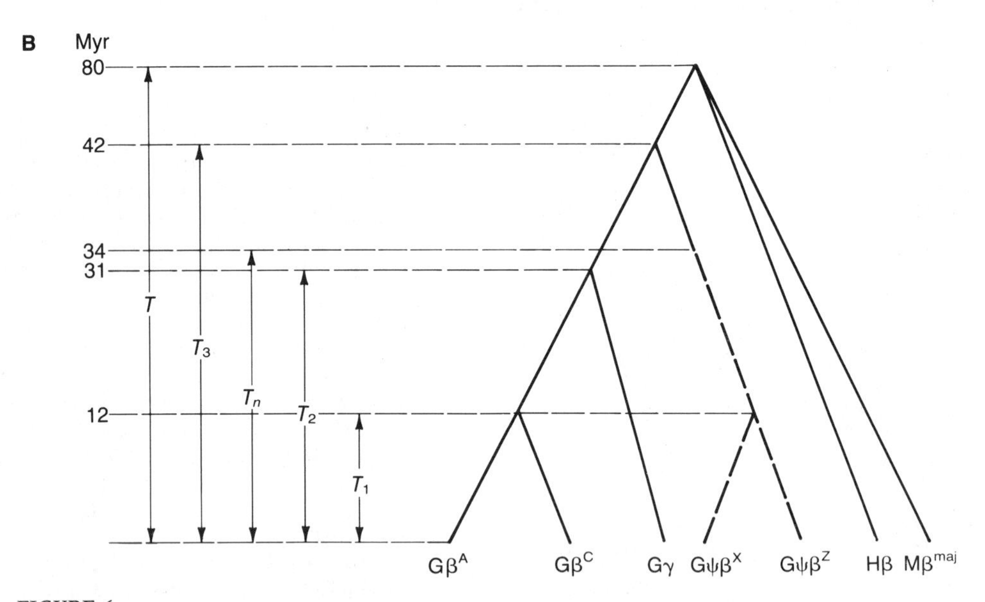


FIGURE 6

**(A) Diagram of the part of the β globin gene family of the domestic goat. (B) A genealogy of some of the members of the goat (G) β globin family, inferred by reference to the β globin of human (H) and mouse (M). The divergence time of the mammalian orders to which these species belong is dated at about 80 Myr from the fossil record; the dates of gene duplication ($T_3$, $T_2$, $T_1$) and nonfunctionalization ($T_n$) are estimated by the method exemplified in Figure 5. (Modified from Li and Gojobori 1983)**

TABLE III

**Estimated times of origin and rates of nucleotide substitution for three globin pseudogenes (see Figure 5)**

| Pseudogene (Sequence A) | Functional gene (Sequence B) | Functional outgroup gene (Sequence C) | Rate of nucleotide substitutions (*a*) per site per year, in functional genes ($\times 10^{-9}$) at codon position | | | Time since gene duplication[a] ($T_d$) | Time since loss of function of pseudogene[a] ($T_n$) | Rate of nucleotide substitution (*b*) since loss of function ($\times 10^{-9}$) |
|---|---|---|---|---|---|---|---|---|
| | | | 1 | 2 | 3 | | | |
| Mouse $\psi\alpha 3$ | Mouse $\alpha 1$ | Human $\alpha$ or rabbit $\alpha$ | $0.69 \pm 0.26$ | $0.69 \pm 0.26$ | $3.32 \pm 0.73$ | $27 \pm 6$ | $23 \pm 19$ | $5.0 \pm 3.2$ |
| Human $\psi\alpha 1$ | Human $\alpha$ | Mouse $\alpha 1$ or rabbit $\alpha$ | $0.74 \pm 0.27$ | $0.67 \pm 0.26$ | $2.51 \pm 0.61$ | $49 \pm 8$ | $45 \pm 37$ | $5.1 \pm 3.3$ |
| Rabbit $\psi\beta 2$ | Rabbit $\beta 1$ | Human $\beta$ or mouse $\beta^{maj}$ | $0.71 \pm 0.27$ | $0.51 \pm 0.22$ | $2.09 \pm 0.51$ | $44 \pm 8$ | 44 | 3.6 |
| | Average | | 0.71 | 0.62 | 2.64 | | | 4.6 |

From Li et al. (1981)

[a]Times $T_d$ and $T_n$ are in Myr.

Like gene duplication, polyploidy increases the number of copies of a gene, and as in the case of gene duplication, duplicate copies diverge in function. The fish families Salmonidae (trout, salmon, and relatives) and Catostomidae (suckers) are both descended from ancestors that became tetraploid about 50 Myr ago, and in both groups some of the duplicate genes have diverged in function (Ferris and Whitt 1979, Allendorf and Thorgaard 1984). Ferris and Whitt (1979) have studied the divergence of expression of duplicate genes in numerous species of catostomids by measuring the difference in activity, in each of several tissues, of the isozymes produced by duplicate copies. In about 40 percent of the comparisons, the duplicate loci have not diverged in expression; in many other cases, one isozyme is more active than the other throughout all the tissues tested; in about 19 percent of the cases, different isozymes are expressed differentially in different tissues, suggesting that they are independently regulated. By mapping divergence in isozyme expression on a phylogenetic tree of the Catostomidae, Ferris and Whitt showed that various duplicated loci have acquired differential expression at different times. Almost all the enzymes studied show a loss of expression of one duplicate copy in one or another species; in all, about half of the enzyme–species combinations reflect such a loss of function. Assuming that a nonfunctional (null) allele at one locus does not reduce fitness because enzyme function is provided by the duplicate locus, Li (1980) has shown that fixation of null alleles by random genetic drift can explain the loss of gene function in the Catostomidae. The rate of loss of gene expression will decrease rapidly, though, if the duplicate copies assume different functions and so are maintained in the population by natural selection.

## MOBILE GENETIC ELEMENTS

Possibly the most startling discovery in molecular genetics is the existence of numerous sequences of DNA that can be duplicated and inserted into many sites in the genome. These mobile elements (reviewed in Finnegan et al. 1982, Campbell 1983, Shapiro 1983) include episomes, which can replicate even when they are not inserted into the chromosome, and other sequences, termed TRANSPOSABLE ELEMENTS or TRANSPOSONS, that replicate only when inserted. Transposable elements are recognizable by their characteristic sequence structure, which typically includes terminal repeats at either end; their location is signaled, moreover, by short direct repeats that flank the inserted sequence and are created in the "target" DNA by the process of insertion (Figure 18 in Chapter 3). Inserted sequences are excised at a very low rate (typically about $10^{-10}$–$10^{-9}$/gamete), and the rate of transposition, although higher, is still quite low (about $10^{-5}$–$10^{-4}$/gamete). Under some conditions the transposition rate increases substantially. For example, the $F_1$ offspring of a cross between females of *M* strains of *Drosophila melanogaster*, which lack *P* elements in the chromosomes, and males from *P* strains, which possess them, have *M*-type cytoplasm and some *P*-bearing chromosomes (see Chapter 8). The rate of transposition of *P* elements is more than 20-fold higher in these offspring than in flies with *P*-type cytoplasm (Engels 1983), which indicates that *P* strains have some property that regulates the transposition rate. In prokaryotes and possibly in eukaryotes some inserted elements can suppress further insertion of other elements in their immediate vicinity (TRANSPOSITION IMMUNITY). In both prokaryotes and eukaryotes, *trans*–acting REPRESSION of insertion has been described: inserted sequences can decrease the rate at which other sequences are inserted elsewhere in the genome (reviewed by Charlesworth and Langley 1986). The several families of transposable elements in a species may account for 10 percent or more of the total DNA. *Drosophila melanogaster* has at least 30 families of transposable elements that in aggregate make up about half the middle repetitive DNA.

Many kinds of transposable elements are replicated via reverse transcription, and are called RETROTRANSPOSONS (Rubin 1983, Baltimore 1985). Like retroviruses, with which they share common structural features, these elements replicate by transcription of the inserted DNA sequence into an RNA that is then reverse transcribed into DNA (denoted cDNA) by the enzyme reverse transcriptase. The gene (*pol*) for this enzyme is carried by retroviruses and by the *Ty* transposable element of yeast, and a similar DNA sequence makes up part of the *copia* element of *Drosophila*. The DNA sequence of many families of repeated DNA, such as the *Alu* family in mammals, indicates that these sequences originated by reverse transcription (Arnheim 1983, Baltimore 1985). A family of transposable elements may be found inserted at many sites on the chromosomes; in *Drosophila*, any particular site is typically occupied at a low frequency. For example, Montgomery and Langley (1983) determined the distribution of three families of transposable elements in 20 X chromosomes derived from a wild population of *D. melanogaster*. In all, 158 insertions were found with an average of 7.9 elements per chromosome. The elements were found at 80 cytologically distinguishable sites, most of which were occupied only once. These data suggest that there exist numerous sites at which elements may be inserted, perhaps at random. Some laboratory studies, however, suggest that some regions of the chromosomes are more susceptible to insertion than others (see Charlesworth 1986).

## EFFECTS OF TRANSPOSABLE ELEMENTS

Some transposable elements carry genes with adaptive phenotypic effects (Calos and Miller 1980, Kleckner 1981); for example, genes for drug resistance and for the metabolism of novel substrates frequently reside in the episomes and other mobile elements of bacteria. In most such instances, however, these genes appear to be host genes that have been captured by transposable elements. Most transposable elements (such as *copia* in *Drosophila*) do not in themselves carry genetic information for the organism's phenotype; the only information they carry appears to be that required for their own replication. Because the number of copies of an element can increase by transposition, the proportion of the genome consisting of a transposable family can increase without, at first surmise, any limit. Any variant that can transpose at a higher frequency than others will comprise an increasing proportion of the family of elements, and so may be said to have a selective advantage. This is a prime example of selection not at the level of the individual organism, but at the level of the "gene," or replicator (Dawkins 1976). These sequences do not exist because they serve the organism, but because they autonomously propagate themselves. They have been termed SELFISH DNA (Doolittle and Sapienza 1980, Orgel and Crick 1980), and may be viewed as parasites of the genome in which they reside (Hickey 1982). Selfish DNA is different from functionless DNA that is replicated along with functional genes but does not increase in proportion; Dover (1980) has suggested that this be called "ignorant DNA."

Transposable elements persist in spite of their effects on organisms, not because of them. Their chief effect (not function) on organisms is to cause mutations. Because they carry promoters for RNA transcription, they are thought occasionally to activate repressed host genes downstream from them (Rubin 1983), but they can also abolish gene function by interrupting the coding sequence or a control region. Many of the classical mutations in *Drosophila*, such as *bithorax* and various mutations of the *white* locus that affects eye color, are caused by insertion of transposable elements (Rubin 1983).

Chromosome rearrangements such as inversions and deletions are frequently caused by recombination between two members of a family of transposable elements (Figure 7). The deleted sequence that is then associated with one of the copies may be inserted with it elsewhere in the genome. (If insertion does not occur, one copy of the transposable element has been deleted from the genome.) Sequences transported this way are often many thousands of base pairs long, and it is often such a huge sequence that, upon insertion into another gene, abolishes its function. Many of the mutations of the *white* locus of *Drosophila* are caused by insertions as much as 14 kb long; others are caused by deletions of several thousand kb (Zachar and Bingham 1982). Like most mutations, those caused by transposition are likely to be deleterious, but some advantageous ones have been observed in chemostat cultures of bacteria (Chao et al. 1983, Hartl and Dykhuizen 1984).

The reverse transcription of retroviruses and retrotransposons has given rise to a special class of pseudogenes called PROCESSED PSEUDOGENES (Walsh 1985a). For example, the nucleotide sequence of the $\psi\alpha 3$ pseudogene of the mouse *Mus musculus* is homologous to that of the $\alpha$ globin gene, except that the sequences corresponding to the introns are precisely lacking, exactly as expected if the pseudogene arose by reverse transcription from a mature mRNA transcript. Many such pseudogenes have been described in the genomes of human, mouse, and other mammals. They are usually flanked by short direct repeats and are displaced in the genome away from their functional relatives, unlike traditional pseudogenes, which are usually adjacent to functional copies. Because the promoters of genes transcribed by RNA polymerase II (i.e., most genes) are upstream from the transcribed sequence, processed pseudogenes formed from such mRNAs cannot themselves be transcribed into RNA and so are unlikely to generate additional copies. For genes transcribed by RNA polymerase III (such as genes for the tRNAs), however, the promoter for the RNA polymerase lies within the transcribed region, so this sequence will be retained in the reverse-transcribed pseudogene, which may therefore give rise to more copies. The *Alu* sequence contains such a promoter.

Transposable elements can sometimes multiply within the genome and spread through populations of their host with extraordinary rapidity. For example, *P* elements are absent from all stocks of *Drosophila melanogaster* collected before the 1950s (Engels 1983), whereas natural populations throughout the world now carry them. Both mathematical theory (Charlesworth 1986) and biological evidence indicate that the elements were not simply lost from the old laboratory stocks. If *P* were an ancient feature of *D. melanogaster*, we might expect to find it in closely related species such as *D. simulans*, but this and the other close relatives of *D. melanogaster* lack *P* elements (Brookfield et al. 1984). More distantly related species groups of *Drosophila*, however, do carry *P* elements and may have been

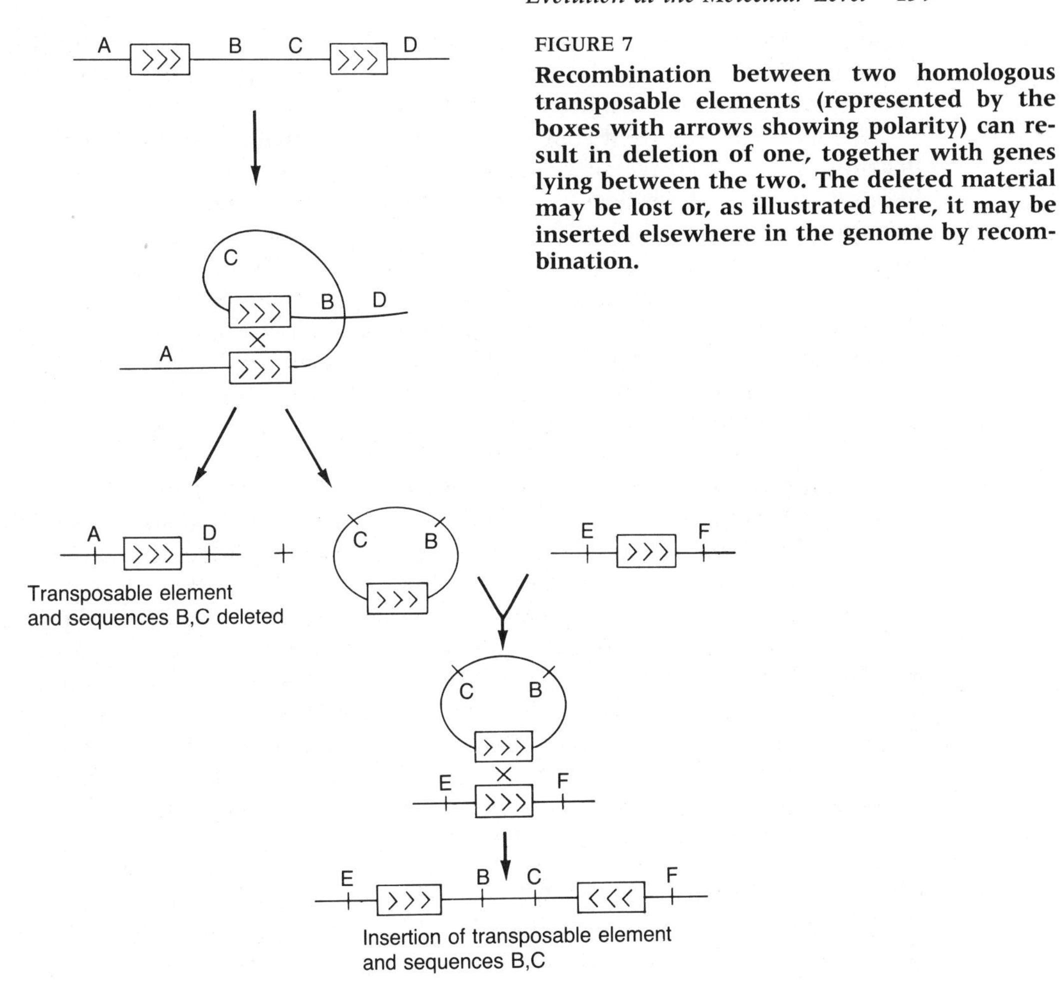


FIGURE 7

**Recombination between two homologous transposable elements (represented by the boxes with arrows showing polarity) can result in deletion of one, together with genes lying between the two. The deleted material may be lost or, as illustrated here, it may be inserted elsewhere in the genome by recombination.**

the source of infection (Daniels et al. 1984). If transposable elements can multiply so quickly, we are led to ask what, if anything, limits the number of copies—and what, for that matter, controls the amount of repetitive DNA in general.

## EVOLUTION OF THE SIZE OF THE GENOME

The DNA content per haploid genome varies enormously among organisms, even among closely related species (Figure 3 in Chapter 3). In itself, the amount of DNA has few discernible effects on an organism's phenotype, except for its influence on cell size and the rate of cell division. Both the time between mitotic divisions and the duration of meiosis increase with the amount of DNA (Bennett 1982, Rees et al. 1982). Species with high C values (a term used to describe DNA content) frequently develop more slowly than those with low C values; among plants, ephemeral species with very rapid development have low C values, whereas species with high C values are perennials (Bennett 1982). The salamander *Plethodon vehiculum*, with a C value almost twice that of *P. cinereus*, reaches about the same adult size, but with half the number of cells (MacGregor 1982).

Little research has been done on the implications of these relationships for the evolution of life histories, but it seems likely that selection on the rate of development will affect the evolution of DNA content.

It might be supposed that divergence of populations in the number and distribution of repetitive sequences could reduce chromosome pairing in their hybrids and so reduce fertility, leading to speciation. Although there is some evidence that pronounced differences in DNA content can interfere with chromosome pairing (Flavell 1982, Rees et al. 1982), the effect is surprisingly slight; hybrids between related species of grasses that differ by as much as 50 percent in DNA content have virtually normal chromosome pairing, chiasma formation, and segregation (Rees et al. 1982). It has been postulated that homologous DNA sequences along the chromosomes pair normally, while interstitial repetitive sequences that differ in length are projected in unpaired loops (Figure 8).

There is little evidence that small differences in the number of copies in a gene family have a substantial effect on fitness simply because of total DNA content. Rather, amplification of the number of copies is likely to affect fitness because of variation in the quantity of the specific gene product (e.g., an enzyme) of that family. If an excess of a gene product reduces fitness, selection may favor mechanisms of DOSAGE COMPENSATION, whereby the activity of one or more gene copies is reduced (Allendorf 1979). There is evidence, for example, that in some tetraploid fish such as carp (*Cyprinus carpio*), levels of RNA and enzymes have been regulated down to the levels expressed in related diploid species (Leipold and Schmidtke 1982). Ribosomal RNA, for instance, is degraded at a higher rate in the tetraploid species. Conversely, selection may favor a high copy number of genes whose product is required in large quantities, such as rRNA, for which hundreds of genes are carried in the genome.

### Models of the evolution of copy number

If excess copies reduce fitness, the frequency in a population of a chromosome with $n$ gene copies is affected by the rate at which it arises by unequal crossing over, and the rate at which it is eliminated by natural selection. At equilibrium, the mean copy number is given by a complex expression that is analogous to the equilibrium frequency ($\hat{q} = \sqrt{u/s}$; see Chapter 6) of a deleterious allele maintained in a population by recurrent mutation (Charlesworth et al. 1986; see also Crow and Kimura 1970). If the effect of the number of copies is so small that it may be ignored (as is likely for noncoding sequences such as satellite DNA), one chromosome or another, bearing some number of copies, will approach fixation by genetic drift. Charlesworth et al. (1986) model the case in which the effective population size is so small that a chromosome is fixed before another variant arises by unequal crossing over. If copy number is low, and if recombination is high enough to generate one-copy chromosomes with appreciable frequency, there is a high probability that a one-copy chromosome will ultimately be fixed by genetic drift. Unequal crossing over is less likely to generate variation in copy number from a single gene copy than from a multigene family, so a population is likely to remain at the single-copy state for much longer than at any other copy number. Thus in chromosome regions with high rates of recombination, single gene copies are more likely than repeated sequences, unless these are favored by natural selection.

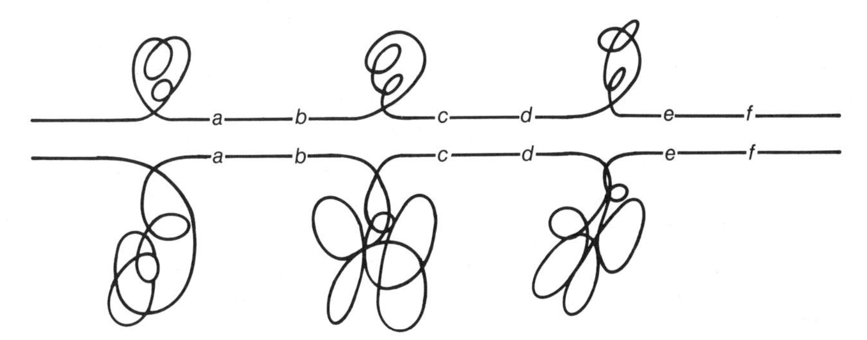


FIGURE 8

**A model showing how pairing of corresponding sequences (*a, b, . . . f*) on homologous chromosomes may be achieved even if they differ in the length of interstitial repeated sequences. (From Rees et al. 1982)**

By this same reasoning, highly repeated sequences are expected to build up in chromosome regions with low recombination, because variants with lower copy number arise less frequently by unequal crossing over, and a chromosome with multiple copies remains intact long enough to be fixed by genetic drift. Thus the absorbing boundary state of a one-member gene family is less likely to arise and to be fixed. This is one possible explanation for the observation that middle and highly repetitive DNA (e.g., satellite DNA) is especially abundant in the vicinity of centromeres, at the ends of chromosomes, and on Y chromosomes—all of which have low rates of crossing over (Charlesworth et al. 1986).

The size of gene families that proliferate via transposable elements is determined by rates of transposition and deletion. For example, Walsh (1985a) has modeled changes in numbers of processed pseudogenes of the kind that are inserted by reverse transcription but cannot generate further copies of themselves. He assumes that they do not affect fitness because they will seldom be inserted into functional genes (which comprise a small fraction of the genome). If a gene family has $k$ functional members, the rate of origin of pseudogenes is $\nu$ per copy per generation, and the rate of deletion of pseudogenes is $\delta$, the change in the average number ($n$) of pseudogenes per generation is $dn/dt = \nu k - \delta n$, so the equilibrium number is $\hat{n} = k(\nu/\delta)$. The equilibrium will be reached after about $5/\delta$ generations. With random (Poisson) variation around this mean, the probability that a particular gene family has no associated pseudogenes is $e^{-k\nu/\delta}$. Thus if $\nu/\delta$ is greater than 5, the probability that a single active gene ($k = 1$) has no associated pseudogenes is less than 0.006. The proportion of functional genes in a family is $\pi = k/(k + n) = 1/[1 + \nu/\delta]$. From the few preliminary estimates of $\pi$ that have been determined for several gene families, $\pi$ averages about 0.1, so $\nu/\delta$ may be calculated to range from 1.7 to 15. Therefore it is likely that few genes lack associated processed pseudogenes. Walsh postulates, on this basis, that processed pseudogenes may constitute about 9 percent of the total DNA.

## Controlling selfish DNA

Self-replicating transposable elements, as we have seen, can be viewed as parasites in the host's genome; like traditional parasites, they raise the question of what prevents them from increasing without bound. Their increase has two components: increase in copy number within a single host's genome, and increase within the host population by transmission from parents to offspring. As Hickey (1982) has pointed out, evolution of copy number in asexually reproducing species requires that clones with variant copy numbers change in frequency. If transposable elements decrease the fitness of their hosts, fixation of a clone with low copy number is likely. It is perhaps for this reason that prokaryotes have less middle repetitive DNA than eukaryotes. In sexual populations, on the other hand, recombination among sites occupied by a transposable element combines together insertions that have arisen in different individuals, so the distribution of copy number in the population behaves much like a polygenic character (Chapter 7).

If transposable elements had no effects on fitness, the mean number ($\bar{n}$) per diploid genome would increase approximately as $\Delta\bar{n} = \bar{n}(u - v)$ per generation, where $u$ and $v$ are the rates of transposition and excision, respectively; thus if $u >> v$, as seems to be the case, the equilibrium number would approach $2m$, where $m$ is the number of sites per haploid chromosome complement that a transposable element could occupy (Charlesworth 1986). That is, the genome would be saturated with transposable elements. The random distribution of *copia*-like elements among chromosome sites in *Drosophila melanogaster* (Montgomery and Langley 1983) suggests that the number of available sites might be very large.

However, transposable elements appear generally to have deleterious effects on organisms because of the mutations they induce. Assuming that fitness declines with the number of such elements in the genome, the mean copy number will tend toward an equilibrium between the rate of origin by transposition and the rate of elimination by selection. This equilibrium will be stable, though, only if the deleterious effect per element increases with the number of elements already present in the genome (Charlesworth 1986). Deleterious mutations in *Drosophila* can have such synergistic effects (Mukai 1969), so it is possible that selection can regulate the number of transposable elements. Perhaps this accounts for the low frequency with which sites are occupied in chromosomes from natural populations of *Drosophila* (Montgomery and Langley 1983, Ajioka and Eanes, in preparation).

The number of transposable elements per genome could also attain a low equilibrium value if the transposition rate per copy were regulated; i.e., if it were to decline as copy number increases (Charlesworth and Charlesworth 1983, Langley et al. 1983). The question, then, is how such regulation might evolve. At the genic level of selection (i.e., among variant sequences of a transposable element), any sequence with a higher transposition rate has a selective advantage, by definition of selection. We may, however, conceive of several mechanisms by which mutant elements with altered transpositional properties can increase in frequency, relative to nonmutant elements with transposition rate $u$ (Charlesworth and Langley 1986).

Transposition immunity, whereby an element excludes other elements from inserting within its immediate neighborhood on the same chromosome (i.e., in *cis* configuration), may be viewed as competition for chromosome regions, rather like competition for space among territorial animals. A mutant that can exclude nonmutant elements will have a selective advantage as long as the mutant does not exclude other mutant copies. However, the selection coefficient in favor of the mutant is approximately

$$s \approx \frac{\bar{n}u(u - v)}{L} \ln\left(1 + \frac{\rho}{2u}\right)$$

where $\bar{n}$ is the average number of copies per genome, $v$ is the excision rate, $\rho$ is the length of the protected territory and $L$ is the total map length of the host's genome. The selective advantage is likely to be exceedingly small—about $\bar{n}u^2$, where $u$ is about $10^{-4}$—unless $\rho$ is very large or $L$ is very small. If the mutant excludes other mutant copies as well as nonmutant copies, a selective advantage to transposition immunity is even less likely. In general, then, transposition immunity appears unlikely to evolve.

Charlesworth and Langley (1986) have also modeled the evolution of transposition repression, supposing that a mutant sequence reduces the transposition rate throughout the genome (in both *cis* and *trans* configurations). If the mutant reduces the transposition rate not only of competing sequences but of like sequences, it is exceedingly unlikely to increase in frequency. This is analogous to the evolution of altruism or spite among organisms (Chapter 9): a genotype that alters the reproductive rate (either positively or negatively) of other individuals in the population will not increase in frequency if both like and unlike genotypes receive equal benefit (or harm).

The most likely circumstance, then, under which repression of transposition will evolve is if selection at the level of the organism eliminates newly inserted elements together with the deleterious mutations with which they are associated. The force of selection will be especially strong if the mutants are highly deleterious and dominant (for they will then be eliminated immediately). A mutant element that represses transposition may then have a selective advantage because genomes that carry it suffer fewer deleterious mutations. Charlesworth and Langley calculate that if a fraction $p$ of insertions of newly transposed elements cause dominant lethal or sterile mutations, a mutant element with the capacity for transposition repression will be selectively advantageous if $p > 2/(\bar{n} + 4)$. This requires that if the mean copy number is 50, at least 3.7 percent of the insertions must cause dominant lethal or sterile mutations. The chromosome breaks and deletions caused by transposable elements often have dominant lethal or sterile effects, and in $P \times M$ dysgenic crosses in *Drosophila melanogaster*, they appear to occur frequently enough to provide an advantage to transposition repression. Even under these most favorable of circumstances, however, the selective advantage of transposition repression has the same order of magnitude as the transposition rate (probably about $10^{-4}$), and so it is very slight. Moreover, the higher the recombination rate and the greater the size of the host genome, the more members of a family of transposable elements will be dissociated from other members that, inserted into functional genes, are eliminated by selection. Thus selection for repression of transposition is weaker in species with high recombination (Charlesworth and Langley 1986).

Selection at the organismal level favors not only transposable sequences that repress transposition, but also whatever properties of the organism itself can achieve repression. But in this respect, too, the coefficient of selection for a host allele that represses transposition will be of the same order of magnitude as the transposition rate, and so may be counteracted by genetic drift unless the population is large. At the same time, selection at the genic level among variant transposable sequences clearly favors the most selfish of selfish DNAs: those whose transposition is repressed neither by other elements nor by the host. There may well be waged within an organism an evolutionary battle between the host and elements in its own genome, like the coevolutionary "arms race" in which species of predators and prey, and of hosts and parasites, are thought to be engaged (Chapter 16). The theory we have discussed suggests that it is a battle in which the selfish DNA has the advantage.

## THE EVOLUTION OF GENE FAMILIES

Possibly the most remarkable feature of gene families is that even those consisting of many thousands of members are highly homogeneous in nucleotide sequence. For example, about 400 copies of a sequence carrying the 18S rRNA and 28S rRNA genes and a nontranscribed spacer (NTS) are distributed over five chromosomes in the human genome (Figure 1). Each of these units has a recognition site for the restriction enzyme *Hpa*I near the beginning of the NTS that is not found in chimpanzees or other apes (Arnheim 1983). Because this restriction site is found only in humans, it must have originated since the divergence of humans from apes. Either the same mutation has arisen independently and been fixed in all 400 genes—which is most unlikely—or the mutation has been spread from one copy to other members of the family. In one of the first reports of this phenomenon (Brown et al. 1972), the several hundred tandemly arranged rRNA genes of the clawed frog *Xenopus laevis* were shown to be homogeneous in nucleotide sequence (as inferred from restriction sites), and to differ from the nucleotide sequence, equally homogeneous, of these genes in *Xenopus borealis*. Note that homogeneity is among gene copies within an individual organism, not merely among individuals of a species. Divergence of species, then, is accompanied by CONCERTED EVOLUTION (Zimmer et al. 1980) of the members of a gene family. Concerted evolution of many gene families has been described in numerous taxa (Dover et al. 1982).

One might expect that the multiple genes of any given family would evolve independently of one another, so that comparisons between different copies within a species (paralogous comparisons) would show the same amount of divergence as comparisons between homologous loci in different species (orthologous comparisons). That paralogous comparisons are more homogeneous than orthologous comparisons implies that the members of a gene family within a species are not evolving independently. We can explain this by postulating that all members of a gene family within each of the species descended from a common ancestral nucleotide sequence after the several species diverged from each other. Thus there must exist mechanisms that homogenize the members of some gene families. The homogenizing mechanisms are believed to be transposition, unequal crossing over, and gene conversion.

### Concerted evolution by transposition

The homogenizing action of transposition may be illustrated by those processed pseudogenes that are not capable of forming additional copies by reverse transcription (Walsh 1985a). New copies are formed by reverse transcription from the active gene or genes, while previously generated copies are subject to deletion. If the deletion rate is high enough compared to the rate of mutation of the nucleotide sequence, a family of processed pseudogenes will be homogeneous in sequence because most will have arisen recently by transposition from a common ancestor, the functional gene. Families of processed pseudogenes seem to show only about 5–15 percent divergence in sequence, suggesting that the deletion rate may be fairly high (more than $2 \times 10^{-7}$ if the mutation rate per nucleotide site is $5 \times 10^{-9}$).

Slatkin (1985) has taken a similar approach in calculating the probability that two copies of a transposable element have the identical nucleotide sequence, assuming that the excision rate is high, so that the average number of copies per genome, $n$, is low. As transposable elements turn over by transposition and deletion, variation in sequence arises by mutation (at rate $u$), but is lost by genetic drift, which increases the probability of identity by descent. The theory of genetic drift (Chapter 5) tells us that at equilibrium, the probability of identity by descent for two gene copies at a locus, the inbreeding coefficient, is $F = 1/(1 + 4N_e u)$, where $N_e$ is the effective population size. Slatkin finds that the probability of identity for two transposable element copies is a similar expression, but with $2N_e n$ in place of $4N_e$. Because multiple copies of a transposable sequence offer more opportunity for mutation, they will vary more in sequence, the greater their number is. This calculation assumes that selection does not act on the sequences.

### Concerted evolution by unequal exchange

Families of nonmobile genes may be homogenized by unequal crossing over and gene conversion (Figure 9). The two mechanisms are often treated together as "unequal exchange," because both appear to arise from nonreciprocal recombination events (Chapter 3). However, they differ in some important ways (Nagylaki and Petes 1982). Unequal crossing over changes the number of tandem copies on a chromosome, whereas gene conversion changes identity but not number. Unequal crossing over affects only tandemly repeated genes, but gene conversion is known to occur between distantly located genes on the same chromatid, on homologous chromosomes, and on nonhomologous chromosomes (Figure 10). Unequal crossing over produces complementary gametes (e.g., chromosomes bearing one and three gene copies by unequal exchange between two tandem loci), with no bias as far as is known. Some loci, however, show BIASED GENE CONVERSION: given a conversion event between sequences $A$ and $A'$, $A$ has a conversion bias of $\delta$ if the probability that $A'$ is converted is $\frac{1}{2} + \delta$. About half the loci studied in yeast display biased gene conversion, with $\delta$ ranging up to 0.25, and in some fungi, up to 0.50 (Nagylaki and Petes 1982). The mechanism of conversion bias is poorly understood. The frequency of unequal exchange (whether biased or unbiased) appears generally to be about $10^{-5}$ to $10^{-2}$ per site (Ohta and Dover 1984).

The rate of intrachromosomal gene conversion (Figure 10A) seems to be higher than the rate of gene conversion between either homologous or nonhomologous chromosomes (Dover et al. 1982), but whether or not the conversion rate among genes on nonhomologous chromosomes is lower than on homologous

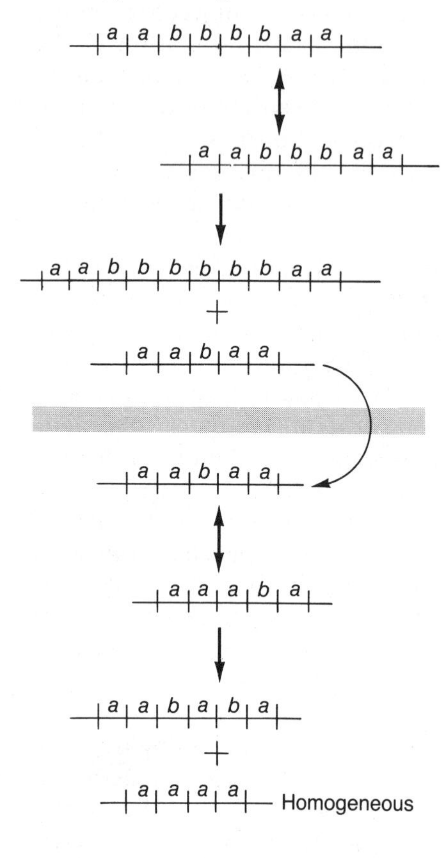


FIGURE 9

**Meiotic events in two generations (separated by the shaded bar), illustrating the effects of unequal crossing over (A) and gene conversion (B) on a gene family that includes two variant sequences *a* and *b*. Chromosomes in the population differ in the distribution of *a* and *b*. A chromosome altered in the first generation is followed (curved arrows) to the second generation when, by a second recombination or conversion event, a homogeneous gene family is produced. In the diagram of gene conversion, two repeats are converted from *a* to *b* by copying from the elevated portion of the other chromosome. Note that unequal crossing over alters the number of repeats, but gene conversion does not. (After Arnheim 1983)**

chromosomes is unclear. Several families of genes are known to consist of "subfamilies" of homogeneous sequences (e.g., *aaaa* and *a'a'a'a'*) that have not been homogenized into a single sequence. In some instances, the different subfamilies are located on nonhomologous chromosomes, as expected if the rate of gene conversion among them is lower than among homologous chromosomes;

but in other cases each of the subfamilies is found on each of several chromosomes (Dover 1982).

Both unequal crossing over and gene conversion contribute to concerted evolution. For example, the nucleotide sequence is homogeneous among rRNA genes in *Drosophila melanogaster*, but the number of copies varies considerably among different X chromosomes taken from a single population (Dover et al. 1982). Unequal crossing over is doubtless responsible for the variation in copy number, but it is unlikely to explain the homogeneity of sequence among rRNA genes on the X and Y chromosomes, because these chromosomes do not undergo crossing over. Concerted evolution entails events at the gene level and at the population level. Consider gene conversion (Figure 11) within a chromosome that does not undergo crossing over. A mutant sequence *b* converts one or more copies of the nonmutant sequence *a*. In subsequent generations, subsequent conversion events on descendant copies of the chromosome give rise to chromosomes that vary from entirely *b* to entirely *a*. On some *chromosomes*, then, *b* has been fixed in the gene family. Fixation in the *population* requires that chromosomes carrying only *b* copies replace all others, either by natural selection or genetic drift. If there is recombination along the chromosome, most chromosomes

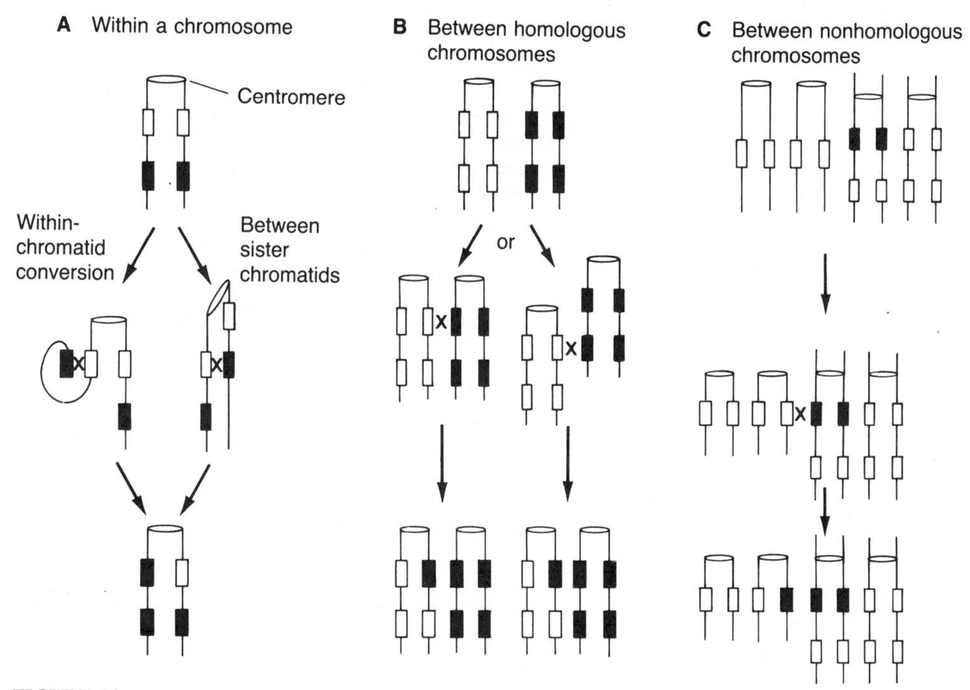


FIGURE 10

**Possible modes of gene conversion. Boxes on each chromatid represent DNA sequences that are sufficiently similar for pairing; black and white boxes represent nucleotide sequence variants; crosses represent gene conversion events. In unbiased gene conversion, white would be converted to black as often as vice versa; in biased conversion, conversion to one sequence (e.g., white to black, as shown here) is more frequent. These events can occur both in somatic cells and in meiosis. (A after Nagylaki and Petes 1982)**

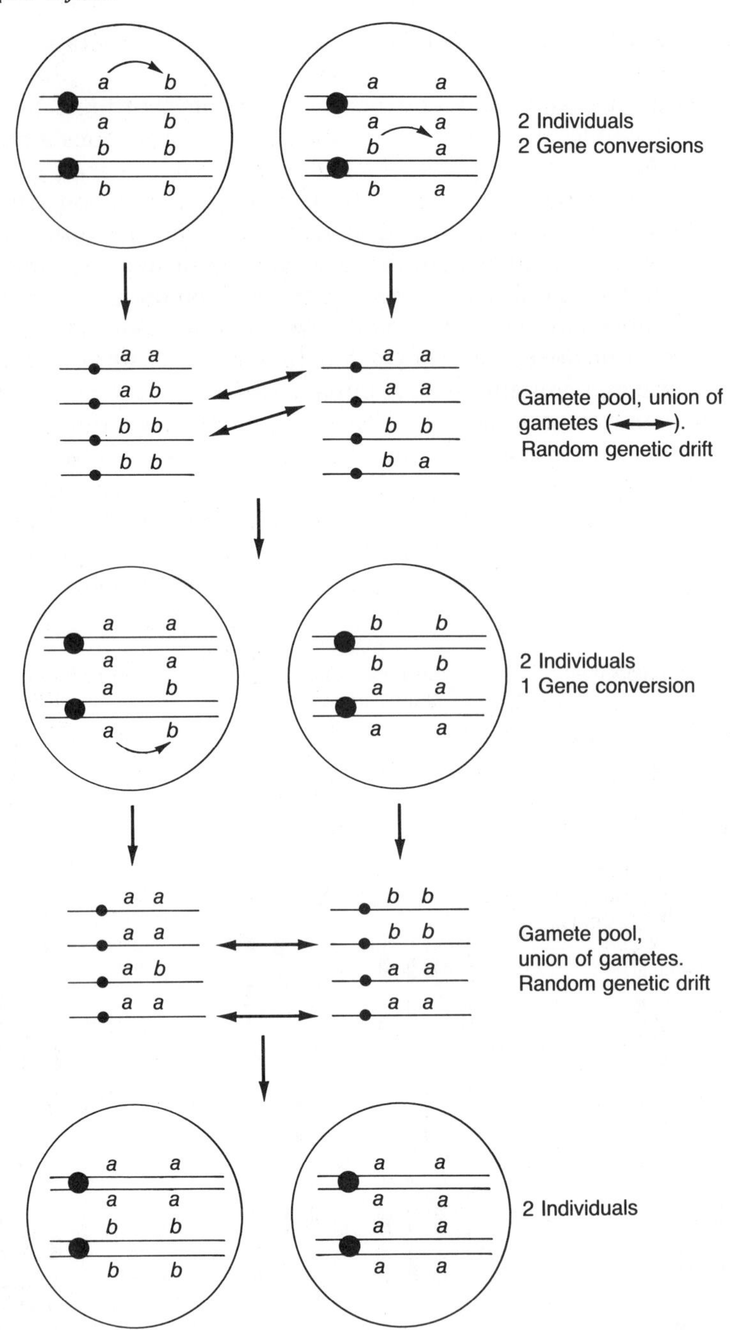


FIGURE 11

**The effects of intrachromosomal gene conversion and random genetic drift in a population of two individuals. Within chromosomes, the proportion of variants *a* and *b* is altered by gene conversion. The diagram illustrates diploid individuals with a pair of homologous chromosomes, each with two chromatids; there is no crossing over. Changes in the frequency of different *a–b* arrays are a population-level process, here depicted as random genetic drift by sampling of gametes. This illustration shows the average identity of copies on the same chromosome increasing (to complete identity), but the population still has different arrays.**

will carry a mixture of *a* and *b* variants. Fixation of *b* at all loci might occur if at each locus, variant *b* drifts to fixation or is favored by selection, but *interchromosomal* gene conversion can affect the rate of fixation, especially if *b* has a conversion bias.

## Models of concerted evolution by gene conversion

A considerable body of mathematical theory has been developed to describe concerted evolution, chiefly by gene conversion (e.g., Ohta 1980, 1985, Nagylaki and Petes 1982, Walsh 1983, 1985b, Nagylaki 1984a,b, Ohta and Dover 1984, Slatkin 1985). The formulations are complex, but the general conclusions may be summarized along the following lines. We consider a gene family with $n$ copies, each of which may mutate at rate $u$ to a different sequence. Crossing over between adjacent loci occurs at rate $\beta$, and the overall recombination rate among loci then depends on $\beta$ and the number of loci. Conversion of one gene into another occurs at rate $\lambda$ per gamete, and may be either interchromosomal or intrachromosomal, depending on the model. Conversion may be unbiased, or one sequence may have a conversion bias $\delta$, again depending on the model. The effective population size is $N_e$. In some models, the fitness of an organism depends on the number of mutant versus nonmutant sequences, but for the moment we assume there is no selection. We are interested in knowing how many generations ($T$) are required for the proportions of various sequences to reach equilibrium in the population. This equilibrium may be described by three coefficients of identity (Figure 12): the probability of identity of genes at the same site on two chromosomes taken at random from the population ($\hat{f}$), at any two sites on a randomly chosen chromosome ($\hat{g}$), and at two different sites on two randomly chosen chromosomes ($\hat{h}$). Complete fixation of a mutant sequence throughout the members of a gene family (i.e., at all sites) and throughout a population would mean that all three coefficients equal 1.

Mutation acts to reduce the identity coefficients, while genetic drift acts to fix one or another allele at any site; $\hat{f}$ is determined by the balance between these factors, and decreases as either $u$ or $N_e$ increase. Now consider the effect of intrachromosomal, unbiased gene conversion (Nagylaki 1984a). Gene conversion is necessary to spread a mutational change from one site to others on the same chromosome, so without it, $\hat{g}$ and $\hat{h}$ would be zero. But the higher the rate of

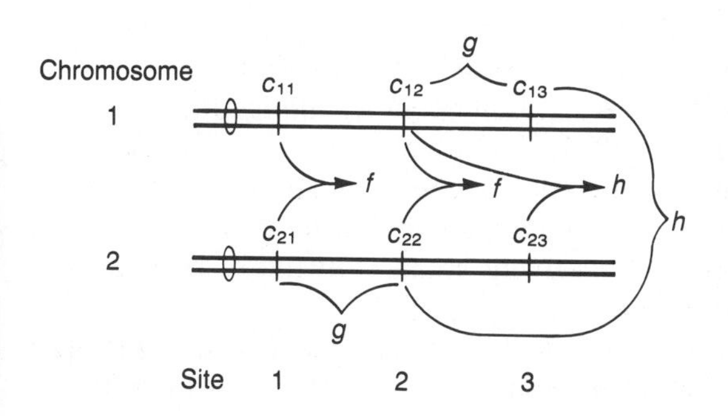


FIGURE 12

**Identity coefficients for describing concerted evolution. Two randomly chosen chromosomes from the population have members of a gene family at sites 1, 2, and 3, so for example the copy on chromosome 2 at site 1 may be named $c_{21}$. $f$ is the identity coefficient between genes at the same site on different chromosomes (e.g., between $c_{11}$ and $c_{21}$ or $c_{12}$ and $c_{22}$), $g$ is that between sites on the same chromosome (e.g., $c_{12}$ and $c_{13}$), and $h$ is that between different sites on different chromosomes (e.g., $c_{13}$ and $c_{22}$). The theoretical description of concerted evolution uses the average values ($\bar{f}, \bar{g}, \bar{h}$) over all such pairs.**

unbiased gene conversion ($\lambda$), the more likely it is that a variant sequence is converted to any of a number of other sequences, thus slowing down the rate at which the mutants drift toward fixation in the population. As a general rule $\hat{f}$, $\hat{g}$, and $\hat{h}$ all decrease if $\lambda$ is very large. Similarly, the identity coefficients are lower, the greater the size ($n$) of the gene family. Finally, the effect of recombination among sites is generally to decrease the identity coefficients. If there were no crossing over, a chromosome on which all sites had been converted to the mutant sequence could fluctuate in frequency and eventually reach fixation, just like a single allele. If $n$ sites freely recombine, fixation requires that $n$ independent loci be fixed for the same sequence, which is clearly less likely. By these same lines of reasoning, it can be shown that the rate at which equilibrium is attained generally increases with $n$ or $N_e$, but it decreases with higher conversion rate $\lambda$ (even though any given sequence is farther from fixation at equilibrium). Recombination may either increase or decrease the time to reach equilibrium, depending on the values of other parameters.

The numerical values obtained for these rates and identity coefficients depend on the particular molecular mechanism by which gene conversion is supposed to occur, but if realistic values are assigned to the various parameters, the equations show that a gene family can become fairly homogeneous for a mutant sequence in a reasonably short span of evolutionary time (several million generations or less).

Greater homogeneity of sequence is attained, at a higher rate, if a sequence has a conversion advantage over other sequences (Nagylaki and Petes 1982, Walsh 1985b). However, most mutant sequences are unlikely to have a conversion advantage over a prevalent sequence, because most variants will have arisen at one time or another in the past, and the ones with the greatest conversion advantage will already have come to prevail (Nagylaki and Petes 1982). This is the same reasoning that leads us to expect that most mutations that affect fitness should be deleterious, because advantageous mutations have already been fixed by natural selection. Here we apply the same principle, but selection, at the level of the gene, is expressed as differences in conversion advantage.

The dynamics of concerted evolution by interchromosomal gene conversion between homologous chromosomes differ in several respects from those under intrachromosomal gene conversion (Nagylaki 1984b). Increasing the rate of gene conversion again reduces $\hat{f}$ and the rate at which the equilibrium is attained, but in this case it increases $\hat{g}$: homogeneity among different sites is enhanced. The recombination rate does not influence the time to equilibrium ($T$). Compared to intrachromosomal gene conversion, interchromosomal gene conversion results in less homogeneity of sequence at equilibrium ($\hat{f}$ and $\hat{g}$ are lower) and the equilibrium takes longer to attain, because interchromosomal gene conversion, like recombination, sets back any progress toward homogeneity along the chromosome. If $\lambda = 10^{-6}$ and $N_e = 10^6$, a family of five genes will reach equilibrium in about 24 million generations, and a family of 30 genes in about 149 million generations (Nagylaki 1984b). These calculations assume that neither conversion bias nor natural selection are acting. A relatively small conversion bias could greatly hasten the approach to equilibrium.

Comparisons of concerted evolution in *Drosophila melanogaster* and its relatives (Dover et al. 1982) have revealed that some sequences have diverged in concert faster than others. For example, although the nontranscribed spacers in the rRNA genes differ between, and are homogeneous within, the most closely related of the species, the coding regions of these same genes do not differ among even the most distantly related species in the group, and so have not undergone concerted divergence. These disparities imply that the homogenizing processes are different, or that other factors such as natural selection are operating. Circumstantial evidence that biased gene conversion may be important has been found in a 20,000-member gene family that is homogeneous within, but different between, species of mice (*Mus*) that may have diverged less than two million years ago (Dover 1982). A gene family of this size is unlikely to undergo concerted evolution so rapidly without conversion bias.

### Concerted evolution and natural selection

Virtually nothing is known about the effect on fitness of variation in the sequences for which gene families become homogeneous, but it is to be expected that some mutations in such genes affect fitness and phenotypic characteristics. Dover (1982) has introduced the term MOLECULAR DRIVE for the fixation in a population of a variant sequence throughout a gene family. He postulated that homogenizing mechanisms, especially biased gene conversion, could cause species to diverge in phenotypic characteristics, even if the variant phenotype were not selectively advantageous. Molecular drive, he suggested, could even cause speciation if a population were fixed for a variant that causes hybrid sterility (i.e., one that lowers fitness when heterozygous; see Chapter 8).

For a single such locus, a conversion advantage of one sequence (say *a*) over others is conceptually equivalent to meiotic drive (Chapter 9): it represents a selective advantage at the level of the gene. Genic selection will tend to fix such an allele in a population if it is neutral with respect to individual selection, and will hasten the fixation of an allele favored by individual selection. The question then is whether or not a conversion bias can fix an allele that is disadvantageous at the individual level (such as an allele that lowers fertility when heterozygous). From studies of meiotic drive (e.g., the *t* locus in mice; Chapter 9), we know that an allele can increase in frequency even if it is selected against at the organismal level, if genic selection (e.g., the conversion advantage) is stronger than individual selection (the selection coefficient *s*).

The efficacy of both genic and individual selection depend on the effective population size $N_e$, because genetic drift affects allele frequencies more than selection if $N_e$ and $s$ are small (if $4N_es << 1$; see Chapter 6). Suppose the genotypes *AA, Aa,* and *aa* have fitnesses 1, $1 - s$, and $1 - 2s$ respectively, and *a* has a conversion bias $\delta$. For fixation of the mutant *a* at *n* sites in the genome, we must consider the magnitude of both $\delta$ and $s$ in relation to the effective population size (Walsh 1985b). The spread of the mutant among sites by gene conversion occurs only within individual organisms. Among these sites, the proportion of mutant copies changes in part at random (because *A* copies are converted to *a* and vice versa) and in part by genic selection. Because the "population" of sites undergoing change is only *n*, a large conversion bias $\delta$ is required to prevail over the random changes. At the population level, however, selection at the individual level can discriminate among $2N_e$ gene copies at each locus, and so may be efficacious even if the selective disadvantage ($s$) of a mutant copy is small. Thus even a small selective disadvantage will prevent a mutant with a conversion bias from being fixed throughout a gene family if the population is large relative to the size of the gene family; selection predominates if $N_e s/n > 2\delta$. Similarly, selection against alleles that are disadvantageous in heterozygous condition (and so might contribute to speciation) strongly opposes gene conversion, which is therefore unlikely to cause speciation (Walsh 1985b).

## ADAPTIVE EVOLUTION FROM A MOLECULAR PERSPECTIVE

The mechanisms and rates of change in DNA are fascinating, but are far removed from the phenotypic characteristics of organisms. Few connections have been made between the evolution of the genome and the evolution of the phenotype. Indeed, much of the evolution of the genome—silent substitutions, evolution of pseudogenes, evolution of nontranscribed sequences—has little if any phenotypic effect. To trace a path from the evolution of DNA to the evolution of the phenotype, it is necessary to examine changes in structural genes and in genes that regulate their expression. Ultimately, a molecular description of evolution requires the equivalent, at the molecular level, of comparative anatomy and embryology: an analysis of how proteins evolve in structure and function, of how their organization in biochemical pathways evolves, and how their expression in different tissues, or at different times in development, is regulated.

## EVOLUTION OF GENES AND PROTEINS

Most proteins are divisible into several or many DOMAINS: compact, continuous regions of the molecule that are spatially distinct from each other. Very often the domains have different functions, or perform the same function more or less independently; for example, each of the three domains of ovomucoid, a protein in chicken egg white that inhibits trypsin, has a trypsin binding site. In many, but not all, proteins there is some correspondence between the protein domains and the exons that make up the gene. For example, introns separate the coding regions for the three domains of ovomucoid. However, each protein domain corresponds to two exons; thus there is partial but not complete correspondence between exons and domains (Li 1983; Figure 13). It has been suggested, then, that in such cases primordially separate genes have become fused, in that they are transcribed together into an RNA that is subsequently spliced into a single message. In the process, new enzymatic functions may arise. For example, the first step of the tryptophan biosynthesis pathway in most bacteria is catalyzed by an enzyme consisting of two subunits, coded for by two genes that we may call 1A and 1B. The second step is catalyzed by an enzyme coded for by gene 2, which lies immediately downstream from 1B. In *Escherichia coli* and some other enteric bacteria, genes 1B and 2 have fused, so that a single large polypeptide is produced. This binds with the product of 1A to give a single complex protein with different enzymatic properties (Crawford 1982).

Duplication of genes, in whole or in part, may provide proteins with new structures and functions (Li 1983). The considerable similarity of amino acid sequence of the three domains of chicken ovomucoid, for example, suggests that they have diverged from a single bipartite gene (containing one intron) that became triplicated and subsequently organized into a single transcription unit. Many proteins, moreover, contain internally repeated sequences of amino acids,

Domain I

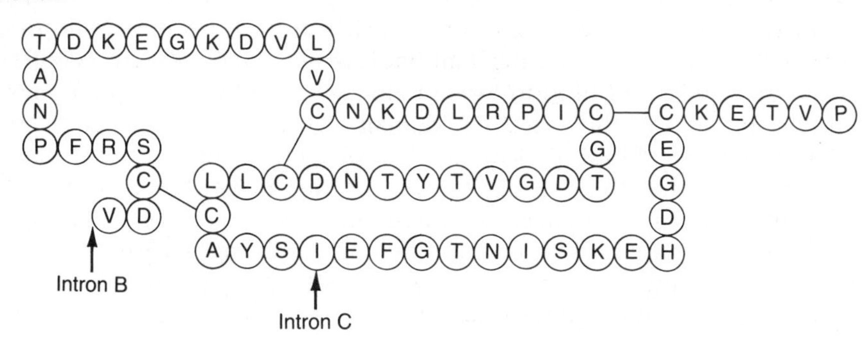


Domain II

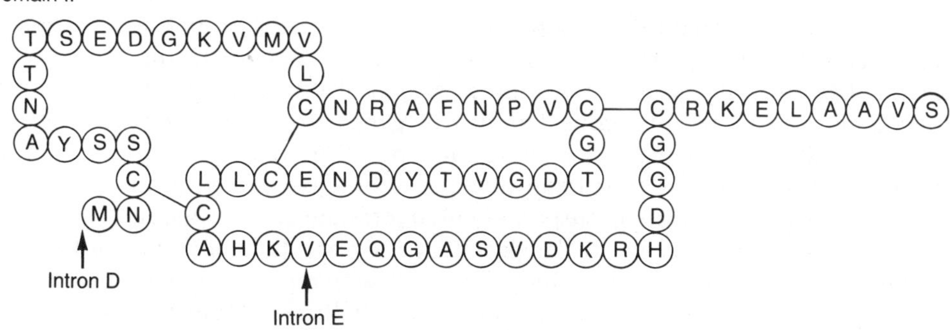


Domain III

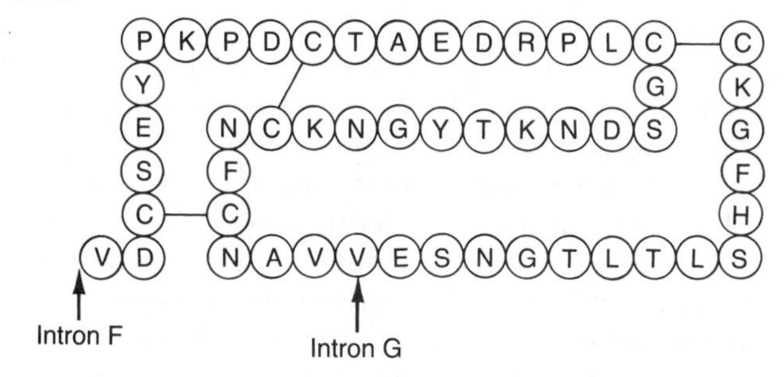


FIGURE 13

**The three functional domains of the chicken ovomucoid protein. Note the overall similarity in structure (e.g., the cysteine-cysteine bonds C—C) and the fairly high similarity between amino acids (denoted by letters) at corresponding positions. The coding region of the gene has introns (D, F) that separate the regions for these domains, but each domain corresponds to two exons separated by introns (C, E, G) at corresponding points. (From Li 1983)**

implying that the gene has become elongated by successive internal duplications. For example, 21 of the more than 50 exons in the collagen gene of chicken have been sequenced. Each consists of 45, 54, 99, 108, or 162 base pairs. Each of these is a multiple of a 9-bp sequence coding for an amino acid trio glycine-X-Y, where X and Y represent different amino acids. These variations in exon length probably arose by unequal recombination. In this instance, protein domains have much the same function, but in other cases domains that arose by exon duplication have diverged. The immunoglobulins (antibodies) of vertebrates are complex molecules that recognize and bind to foreign molecules (antigens). Antigen recognition is the function of the "variable" domains of the protein, whereas the "constant" domains effect binding. Sequence homology between these two kinds of domains indicates that they probably arose by duplication of a primordial gene. Each kind of domain, moreover, is encoded in the several exons of multiple genes, and variation in antibodies among different cell lineages arises within a single organism by fusing different exons by unequal crossing over into a multitude of combinations (Hunkapiller et al. 1982).

### Divergence in protein function

We have noted throughout this book instances in which the properties of a protein are altered by changes in amino acid sequence. Divergence in the sequence of genes that arose by gene duplication is also well known, as in the vertebrate hemoglobins and in numerous families of genes that code for tissue-specific isozymes. Much of the vast biochemical repertoire of organisms appears to have arisen by structural and functional divergence of successively duplicated genes. For example, many of the proteases of eukaryotes are similar enough in parts of their amino acid sequence to suggest that they are homologous (Barker and Dayhoff 1980). These enzymes include the digestive proteases trypsin, chymotrypsin, elastase, carboxypeptidase, and phospholipase. Moreover, several enzymes that play proteolytic roles in blood clotting and in the breakdown of blood clots are related in sequence to the digestive proteases (Figure 14A). Various of these proteins include functional "modules" of several types (Figure 14B), coded by exons that appear in various combinations among the genes that code for these proteins. Thus new genes have evolved by EXON SHUFFLING, the joining of different exons into new combinations (Doolittle 1985, Patthy 1985).

Jensen (1976) has proposed that substrate-specific enzymes may have arisen by gene duplication and divergence from less substrate-specific ancestral enzymes. He notes that many enzymes are quite versatile: L-fucose isomerase, for example, utilizes as a substrate not only L-fucose, but also L-xylose and D-arabinose. Quite often an organism has both broadly and narrowly reactive enzymes that perform related functions; the fungus *Aspergillus*, for instance, has not only a generalized amidase, but the substrate-specific enzymes acetamidase and formamidase. Jensen suggests that entire new biochemical pathways may arise by "recruiting" for a new function variant enzymes in an analogous pathway. For example, the pathways for synthesis of lysine, isoleucine, and leucine are similar to each other and to the tricarboxylic acid cycle (Figure 15). Jensen's hypothesis predicts that the enzymes at corresponding steps in these pathways should have homologous amino acid sequences; whether or not this is so is not known. Jensen cites examples in which the same enzyme catalyzes corresponding reactions in two or more parallel pathways.

### Evolution of enzyme regulation

Biochemical adaptation may be achieved by changes in the structure of an enzyme, by changes in its regulation, or both. Little is known about the evolution of

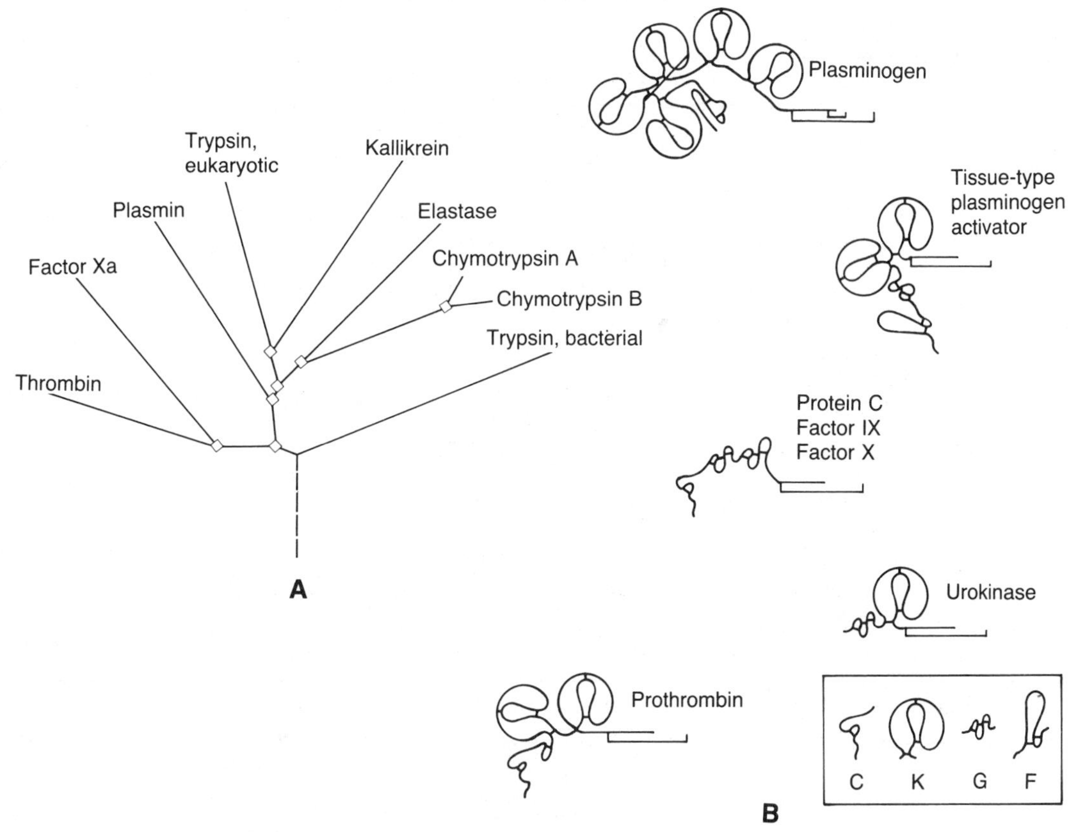


FIGURE 14

**(A) An evolutionary tree of diverse proteases thought to have evolved by gene duplication (diamonds) and subsequent divergence. The tree is based on similarities in amino acid sequences of vertebrate proteins (other than bacterial trypsin), but many of the proteins are not restricted to vertebrates. Thrombin, factor Xa, plasmin, and kallikrein are proteins associated with blood clotting. (B) Diagrams of the structures of some proteins in which the protease region (represented by the shaded bar) is similar in amino acid sequence to the proteases in (A). These proteins have nonprotease modules of types C, K, G, and F, coded for by exons associated with the protease-coding regions. The same kinds of modules occur in various combinations, indicating that the genes for these proteins have been formed by assembly (shuffling) of several types of exons. (A after Barker and Dayhoff 1980, B after Patthy 1985)**

enzyme regulation save that it occurs, as the studies of tissue-specific enzyme expression in tetraploid fish illustrate (see above). The importance of regulatory evolution may be exemplified by the adaptation of herbivorous insects to toxic compounds in their host plants. For example, certain species of the carrot family (Apiaceae) contain toxic linear furanocoumarins such as xanthotoxin. Not surprisingly, these plants are fed on primarily by specialized species of insects, such as larvae of the black swallowtail butterfly, *Papilio polyxenes* (Berenbaum 1983). This species rapidly degrades ingested xanthotoxin to nontoxic molecules. The

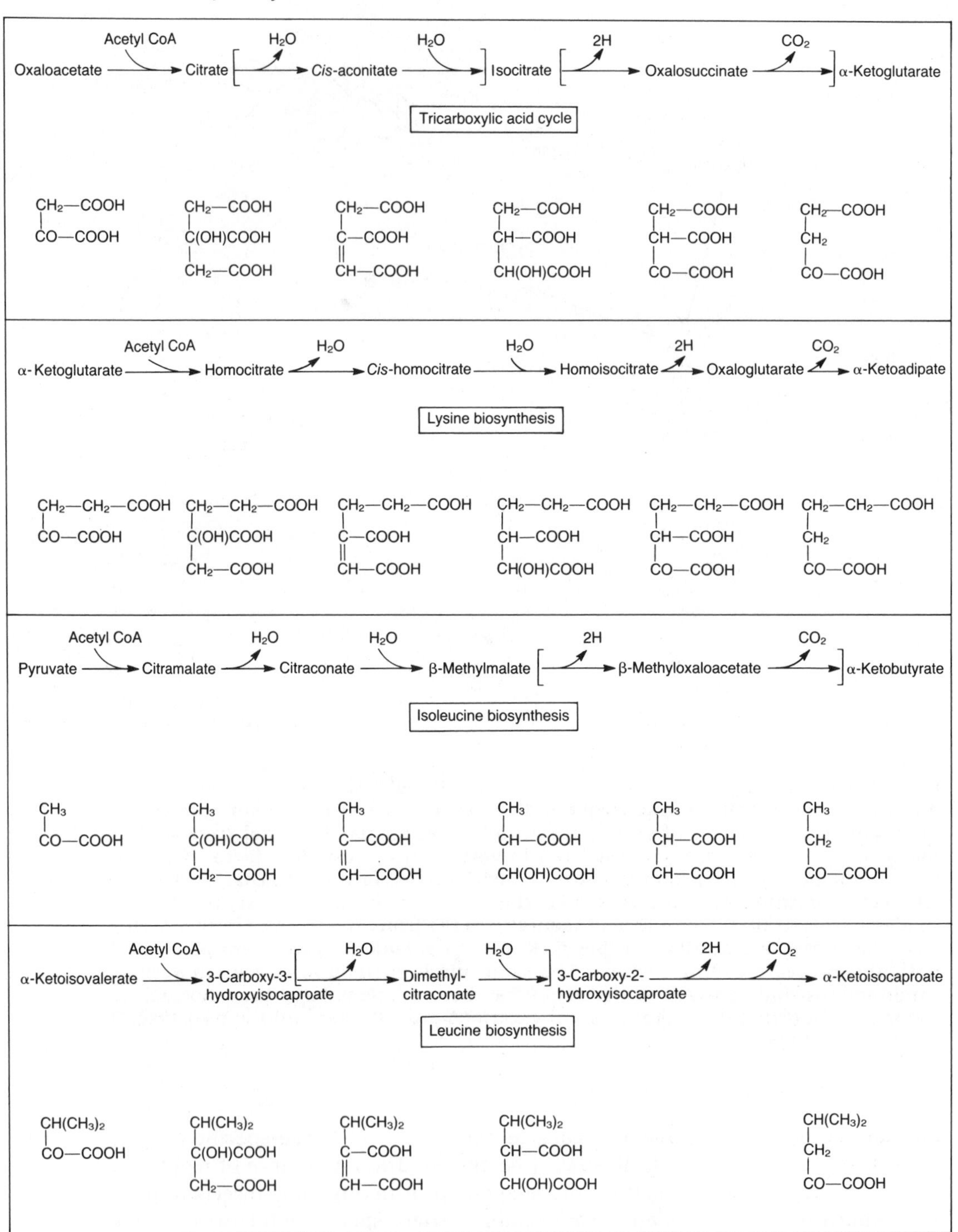
Acetyl CoA
Oxaloacetate
Citrate
H2O
Cis-aconitate
H2O
Isocitrate
2H
Oxalosuccinate
CO2
α-Ketoglutarate
Tricarboxylic acid cycle
α-Ketoglutarate
Homocitrate
Cis-homocitrate
Homoisocitrate
Oxaloglutarate
α-Ketoadipate
Lysine biosynthesis
Pyruvate
Citramalate
Citraconate
β-Methylmalate
β-Methyloxaloacetate
α-Ketobutyrate
Isoleucine biosynthesis
α-Ketoisovalerate
3-Carboxy-3-hydroxyisocaproate
Dimethyl-citraconate
3-Carboxy-2-hydroxyisocaproate
α-Ketoisocaproate
Leucine biosynthesis

FIGURE 15

**Analogous reaction sequences in four major biochemical pathways. Instances in which free intermediates may not be formed are in brackets. The similarity of reaction sequences suggests the hypothesis that the enzymes mediating corresponding steps may be homologous (i.e., paralogous), having acquired specialized functions after gene duplication. (From Jensen 1976)**

southern armyworm (*Spodoptera frugiperda*), a generalized feeder that is not well adapted to xanthotoxin, can also degrade the molecule, but at a much lower rate (Ivie et al. 1983). The detoxifying activity is distributed throughout the tissues in both species, but it is much higher in the midgut of *P. polyxenes*, enabling rapid detoxification. The specialized feeding habit of this species, then, has been accompanied by a change in the activity and tissue specificity of a biochemical mechanism that appears to be widespread in the Lepidoptera.

The evolution of a complex biochemical function is likely to require changes both in enzyme structure and regulation. The origin of such changes by mutation has been extensively studied in bacteria by selecting mutants with the capacity to grow on a novel substrate as their source of energy (e.g., Clarke 1974, Hall 1982, 1983, Mortlock 1982). For example, in *Escherichia coli* metabolism of lactose is normally accomplished by the *lac* operon, in which an operator regulates transcription of the genes *lacZ* (which codes for β-galactosidase) and *lacY* (which codes for lactose permease; Chapter 3). When lactose leaks into the cells it is hydrolyzed by β-galactosidase into glucose and galactose, which are used for energy, and into allolactose, which derepresses the operator, enabling transcription of *lacZ* and *lacY*. The permease that is then synthesized enables lactose to enter freely, permitting growth of the cell.

Using an *E. coli* strain in which most of the *lacZ* gene had been deleted, Hall (1982, 1983) selected for ability to grow on lactose as an energy source. A chemical additive that induces *lacY* was provided so that lactose could enter the cells. Mutants that could grow on lactose were obtained; they proved to have mutations in an entirely different operon, the EBG ("evolved β-galactosidase") operon, distant on the chromosome from the *lac* operon. These were double mutants. They carried a structural mutation of an enzyme (the *ebg* enzyme) that enabled it to hydrolyze lactose; this enzyme is different from the normal β-galactosidase. Transcription of the gene (*ebgA*) for this enzyme is ordinarily repressed by an operator gene (*ebgR*) that is not induced by lactose. Thus growth on lactose required a mutation in *ebgR* as well. Some *ebgR* mutants were constitutive (i.e., permanently derepressed); others were induced by lactose. So far, however, growth depended on the permease-inducing additive in the medium.

When Hall selected his mutant lactose-utilizing strains for the ability to grow on a second sugar, lactulose, he obtained further mutants that, by virtue of a second alteration of the *ebg* enzyme, could grow on both lactose and lactulose. These mutants had several fortuitous properties, such as the ability to metabolize a third sugar, lactobionate. One of these mutants also incidentally metabolized lactose to allolactose. Because this sugar is an inducer of the *lacY* gene, this genotype was capable of synthesizing lactose permease without the aid of the additive. Thus an entire system of lactose utilization had evolved, consisting of changes in enzyme structure enabling hydrolysis of the substrate; alteration of a regulatory gene so that the enzyme can be synthesized in response to the presence of its substrate; and evolution of an enzyme reaction that induces the permease needed for entry of the substrate. One could not wish for a better demonstration of the neoDarwinian principle that mutation and selection in concert are the source of complex adaptations.

## HORIZONTAL GENE TRANSFER

Except for gene flow (hybridization) among very closely related species, the genetic changes that transpire in one species are seldom transmitted to other species. However, some molecular data suggest that transfer of genetic information between widely different taxa has occurred occasionally in the course of evolution. Very early in evolutionary history, the eukaryotic cell incorporated prokaryotes as intracellular symbionts: mitochondria and chloroplasts have such structural similarities to bacteria, including the structure of their genomes, that we must view a eukaryote as a compound organism (Margulis 1981). The intimacy of association becomes complete when part or all of an intracellular symbiont's genome becomes included in that of its host, as it does for viruses and plasmids. Viruses and plasmids can carry genes that affect the phenotype of the host, and sometimes act as agents of gene transfer between unrelated species. For example, *Agrobacterium tumefaciens*, a bacterial associate of dicotyledonous plants, induces crown gall tumors in the host plant when part of the bacterial genome is carried by a plasmid into the plant's cells, where it probably becomes incorporated into the genome (Drummond 1979). The amino acid sequence of the enzyme superoxide dismutase in the bacterium *Photobacterium leiognathi*, a symbiont of the ponyfish (*Leiognathus splendens*), is more similar to the enzyme of fish than of other bacteria, suggesting transfer of genetic material from the host to the symbiont (Bannister and Parker 1985). The most startling case of possible transfer of genes between species is presented by the leghemoglobin molecule of soybean (*Glycine max*) and some other legumes. The exon–intron structure, and the amino acid sequence in part, are so similar to the hemoglobins of vertebrates and certain invertebrates that it seems possible the gene was transferred from an animal (Jeffreys 1982). Whether or not gene transfer between unrelated taxa will prove, as seems likely, to be merely an occasional curiosity of evolution is not yet known.

## MOLECULAR BIOLOGY AND EVOLUTIONARY BIOLOGY

In biology, to a greater extent than in many other sciences, there exists an ebb and flow of tension between two outlooks that are often termed reductionist and antireductionist. Both philosophers and biologists have discussed this issue at some length (see, for example, Simpson 1964, Chapters 5 and 6; Monod 1971; Ayala and Dobzhansky 1974; Dobzhansky et al. 1977, Chapter 16; Levins and Lewontin 1985). At one extreme is the ultimately reductionist view that all biological phenomena can be explained and predicted by the study of physics and chemistry; at the other, the view that biological phenomena cannot be understood except by the study of organisms and the ecological communities in which they function. Although few if any biologists subscribe to either extreme view, some favor one or the other mode of explanation. Reductionism seeks an explanation of phenomena at "higher" levels in terms of phenomena at "lower" levels, ultimately molecular and atomic; in principle, it is said, we might describe the behavior of a bird in terms of the activities of neurons, these in terms of enzymes and hormones, these in terms of DNA, and the behavior of DNA in terms of chemistry and biophysics. What Simpson (1964) calls *compositionist* explanation makes reference to the context in which each part of the organism, and the organism as a whole, functions: the behavior of DNA depends on its structural and chemical environment, and that of a bird on its ecological environment.

The question of reductionism in biology arises in several domains (Dobzhansky et al. 1977). In the *ontological* domain, we ask if physical and chemical processes, and only physical and chemical properties, underlie all biological phenomena. Virtually all biologists affirm that this is so. In the *methodological* domain, we ask whether answers are to be found only by investigations at lower levels, or whether reference to higher levels (the compositional approach) is also necessary. Most biologists agree that both approaches are required, but this is most vigorously affirmed by those who study whole organisms. In the *epistemological* domain, we ask whether the theories and laws of one science (e.g., biology) are special cases of those described by another science (e.g., chemistry): whether the one science can be "reduced" to the other. For such reduction to be successful, it is necessary to show that all the theories and laws of the science to be reduced are logical consequences of the theories and laws of the other science, and that all the terms of the one discipline can be redefined in the terms of the other.

Neither of these criteria is satisfied if we attempt to reduce biology to chemistry, which has no vocabulary for and does not predict phenomena such as self-replication, morphogenesis, and natural selection. As Simpson (1964) says, all known material processes and explanatory principles apply to organisms, but only a limited number of them apply to nonliving systems. The failure of epistemological reduction of biology to chemistry implies, further, a failure of methodological reductionism. The study of chemistry is necessary, but not sufficient, to explain biological phenomena. If we wish to explain the migration of warblers, we need to understand their hormones—a chemical study, but one that nevertheless provides little insight unless we can understand the organization of the nervous system on which the hormones act. But even this understanding, although it may someday tell us *how* a warbler is stimulated to migrate, will not tell us *why* warblers migrate (and why crows do not). For this we require insight into evolutionary history—into the history of natural selection, and hence into the history of the species' ecology.

As biology is not reducible to chemistry, so the biology of organisms is not reducible to molecular genetics and biochemistry. The spectacular advances in these fields clearly provide information that bears on the questions asked by any biologist, but this fascinating and profoundly important information is not, and will not be, sufficient to answer all questions. All but the simplest questions about organisms require more than DNA sequences for their answer; our knowledge of an organism's DNA does not predict the dynamics of meiosis, much less the ontogeny of a moth or the number of species in a community. These questions are as fascinating, as mysterious, and as important—even in their merely utilitarian applications—as ever; they hold their proper places in the biological sciences; they require methods and concepts appropriate to their respective biological disciplines. We hear on occasion that the future of biology lies in the molecular realm, that all else is stamp collecting. But the richness and importance of biological phenomena at all levels of organization, the immense gap of understanding between molecular and organismal phenomena, even the necessity of turning to evolutionary theory to explain molecular phenomena such as the homogeneity of gene families, belie this claim.

A familiar refrain in evolutionary circles is Theodosius Dobzhansky's pungent remark that "nothing in biology makes sense except in the light of evolution." Without evolutionary theory, we can describe how the parts of organisms function, but we are left with endless descriptions of detail without this single theory that provides them with unity and coherence. It is remarkable that, with little change in the traditional equations of evolutionary genetics, molecular phenomena that were unknown a decade ago have been integrated into the same theory that has served for understanding the genesis of organisms' morphology, physiology, and behavior. As at these higher levels of biological organization, phenomena at the molecular level require and are given explanation, unity, and coherence not only by reduction to submolecular forces, but by the compositionist theory of evolution.

## SUMMARY

Within populations, sequences of DNA vary in sites that alter gene products and in sites that do not. The rate of sequence evolution varies among genes, and is greater in sequences that lack function. Families of related genes arise by transposition and unequal recombination; by these mechanisms as well as gene conversion, they frequently evolve in concert. Transposable elements affect organisms chiefly by causing mutations; they may persist primarily by their capacity to transpose ("selfish DNA") rather than by providing advantages to the organism. Fusion and internal duplication of DNA sequences may give rise to enzymes with new functions. Adaptive mutational change in the structure and regulation of enzymes is an important feature of evolution. These and other revelations at the molecular level are readily incorporated into well established population genetic models of evolutionary change.

## FOR DISCUSSION AND THOUGHT

1. How would the evolution of repression of transposition be affected if the mechanism of repression extended across families of transposable elements, compared to the case in which it is specific to one such family?
2. What are the similarities and differences between the evolution of the regulation of transposition and the evolution of mutation rates in general?
3. How might one determine if any phenotypic adaptations of organisms originated as mutations caused by transposable elements?
4. Provide an explanation of Takahata and Slatkin's (1984) conclusion that mitochondria may be transmitted between hybridizing species even if there is strong selection against hybrids.
5. There is some evidence that codons which, under random mutation, are likely to mutate to synonymous codons occur in coding regions more frequently than expected under a random model, compared to codons that are likely to cause an amino acid change if mutated. Provide an evolutionary explanation.

6. Introns are generally lacking in the genes of mitochondria and prokaryotes, but are typical features of eukaryotic genes. Suggest a hypothesis to account for the evolution of introns.
7. From information in this chapter, discuss the role that mathematical theory plays in directing empirical research toward important phenomena that have not yet been fully described.
8. How might the size of a gene family affect the allele frequencies of deleterious mutations at each of the individual loci?
9. Read one or more of the theoretical papers describing the evolutionary consequences of molecular phenomena (e.g., transposable elements or concerted evolution), and discuss the modifications of traditional population genetic theory that have been necessary to describe them. Has a new theory been required for these phenomena? (An exercise for the mathematically able.)
10. In what ways has our understanding of evolution at the organismal level been affected by molecular discoveries in the last 10 years?

## MAJOR REFERENCES

Each of the following books is a collection of essays by leading workers in the area of molecular evolution.

Dover, G.A. and R.B. Flavell (eds.). 1982. *Genome evolution.* Academic Press, London. 382 pages.

Nei, M. and R.K. Koehn (eds.). 1983. *Evolution of genes and proteins.* Sinauer Associates, Sunderland, MA. 331 pages.

MacIntyre, R.J. (ed.). 1986. *Molecular evolutionary genetics.* Plenum, New York. 610 pages.

# The Evolution of Interactions Among Species

# Chapter Sixteen

One of the greatest challenges to evolutionary biology is understanding how interspecific interactions have influenced rates of evolution and patterns of adaptive radiation, and how evolution affects interspecific interactions and thereby the structure of ecological communities. This is a tall order, for it requires a synthesis of two complex and incomplete theories: the genetical theory of evolution and the ecological theory of community structure. The synthesis of these theories has only barely begun. The purpose of this chapter is to point out questions that are only now being asked, and to describe problems that are currently among the most exciting in evolutionary biology. Research in this area has implications for our understanding of historical changes in diversity (Chapter 12) and of biogeography (Chapter 13), as we have seen.

## COEVOLUTION

The word COEVOLUTION was coined by Ehrlich and Raven (1964) in their description of the probable influences that plants and herbivorous insects have had on each other's evolution. The word has been used in various ways, and there is no general agreement on its definition (Futuyma and Slatkin 1983b). It has been broadly defined (Roughgarden 1976) as evolution in which the fitness of each genotype depends on the population densities and genetic composition of the species itself and the species with which it interacts. Other authors (e.g., Janzen 1980) are more specific, requiring that each of two or more species change in genetic composition in response to a genetic change in the other. Thus a trait in one species has evolved in response to a trait of another species, which trait itself has evolved in response to the trait in the first. This definition implies that two or more lineages, e.g., a predator and a prey, evolve specifically and reciprocally in response to each other. Such evolution raises a host of questions: Will evolution promote coexistence or destabilize the interaction and lead to extinction? Will antagonistic species evolve indefinitely in an evolutionary "arms race," or will they arrive at an evolutionary equilibrium? Can evolution be rapid enough to save a species from extinction by competition or predation?

Such problems are most easily thought of in terms of a single pair of species, such as a prey and a predator. However, most species interact with a variety of prey or predators, and it is doubtful that all of them will evolve in the same way in response to an evolutionary change in any one species. Often, then, a trait in, say, a prey species will evolve in response to a class of predators rather than to any one specific predator. The reciprocal evolutionary interactions among classes of species, often termed DIFFUSE COEVOLUTION, must often be affected by conflicts among the adaptations appropriate to each pairwise interaction.

Just as adaptation is an onerous concept that should not be invoked except when warranted by evidence (Chapter 9), so perhaps coevolution should not be invoked save when necessary. Before assuming that the traits that influence an interspecific interaction have been shaped by coevolution, we should ask if they can be adequately explained without reference to the particular interaction in question. For example, related coexisting species often do not overlap completely in their diet. It is possible in such instances that under the pressure of past competition, they evolved differences in diet in response to each other's presence. But ecological theory tells us that under some conditions, excessive overlap in resource utilization by two competing species will lead to the extinction of one of them by competitive exclusion. Therefore the only coexisting species may be those that had already evolved different diets before they encountered each other. Thus the properties of coexisting species might be explained purely by ecological processes of immigration and extinction (i.e., during the assembly of communities) rather than by coevolution (Case 1982, Schluter and Grant 1984, Rummel and Roughgarden 1985).

## THE EVOLUTION OF RESOURCE UTILIZATION

Let us take this approach in asking what determines how many species a given species uses as resources. We are asking, in effect, what determines the number of host species of a parasite, the number of prey species taken by a predator or herbivore, or the number of hosts (e.g., plants) used by a "mutualist" (e.g., a mycorrhizal fungus). The evolution of generalization or specialization is partly affected by the difference among potential resource species, relative to the tolerance of the focal species' genotypes; two resources that are sufficiently different cannot both be used effectively by even the most generalized genotype. But even over the spectrum of usable resources, generalized genotypes will have an advantage under some circumstances and specialized genotypes under others (Levins 1968, Pyke et al. 1977). Although under some special circumstances a population may have a broad niche by virtue of a polymorphism for individually specialized genotypes (multiple niche polymorphism; see Chapter 6), we will consider here the circumstances under which selection results in fixation of either a generalized or a specialized genotype.

Selection is more likely to favor genotypes that specialize on common than on rare resource species (cf. optimal foraging theory, Chapter 9); if no one resource species is abundant, generalized genotypes will be favored. These considerations may explain why widespread, abundant species of trees harbor more species of insects than do rare or localized tree species (Figure 1; Southwood 1961), and why mycorrhizal fungi form more specialized associations with particular tree species in low-diversity temperate zone forests than in high-diversity tropical forests. Optimal foraging theory predicts that a consumer can "afford" to specialize on a superior resource if it is common, but not if it is rare (Emlen 1966, MacArthur and Pianka 1966), and indeed some species display more specialized preferences as superior resources increase in abundance (Schoener 1971). Similarly, generalization is favored if each individual resource fluctuates considerably in abundance relative to the time scale of utilization, or if a particular resource varies in quality (e.g., energy content; Caraco 1980, Real et al. 1982). A hummingbird, for example, feeds year round and must be able to use a seasonal succession of flowering plants for nectar; in contrast, some solitary bees are dormant except for a brief period each year during which they obtain nectar and pollen from one or a few plant species, and many such bees are quite specialized (Feinsinger 1983). Numerous other factors can affect the evolution of specialization; for example, a particular host plant may provide to an herbivorous insect not only food but also protection against predators and a rendezvous for mating (Colwell 1985).

We therefore expect allopatric populations to diverge in resource use if the relative abundance of various resources differs among localities. For example, an Arizona population of the Colorado potato beetle (*Leptinotarsa decemlineata*) prefers

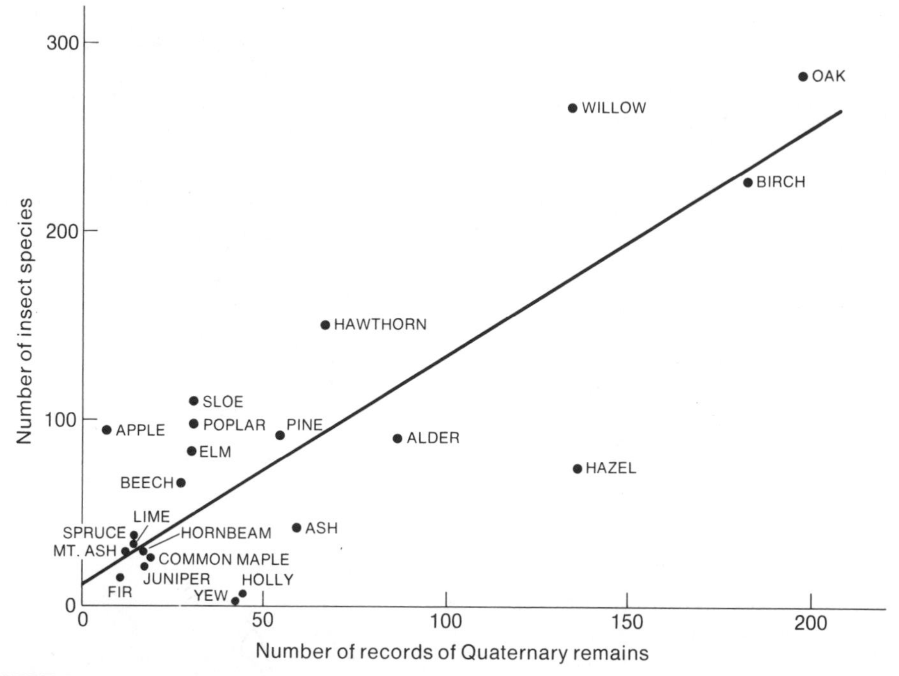


FIGURE 1

**The relationship between the number of insect species associated with a tree species in Britain and the historical abundance of the tree, as measured by the number of records of Quaternary fossil remains. It is unclear whether the relationship depends simply on how widespread the tree is, or whether the length of time it has been available has affected the number of insect species it harbors. (After Southwood 1961)**

and is physiologically adapted to a locally abundant host plant which other populations of the species do not prefer (Hsiao 1978). Thus species that have evolved in different regions will commonly already differ in resource use if they become sympatric; not all niche differences between sympatric species are the result of coevolutionary responses to their interaction.

## COEVOLUTION OF COMPETING SPECIES

According to the classical Lotka-Volterra model of interspecific competition, stable coexistence of two or more competing species is not possible unless intraspecific competition within each species is stronger than interspecific competition (Chapter 2). This theory assumes, of course, that the species do indeed compete for limiting resources. Although field experiments have shown that interspecific competition affects the populations of many species, it is by no means universal (Connell 1983, Schoener 1983).

If competition exists only by virtue of the common use of resources, we can represent the "niche" of each species by a "utilization function" (Figure 2A) that describes the frequency with which each kind of resource (e.g., size of insects eaten by a lizard) is eaten. Competition between the species increases as $d$, the difference between the means of their utilization functions, decreases, or as $w$,

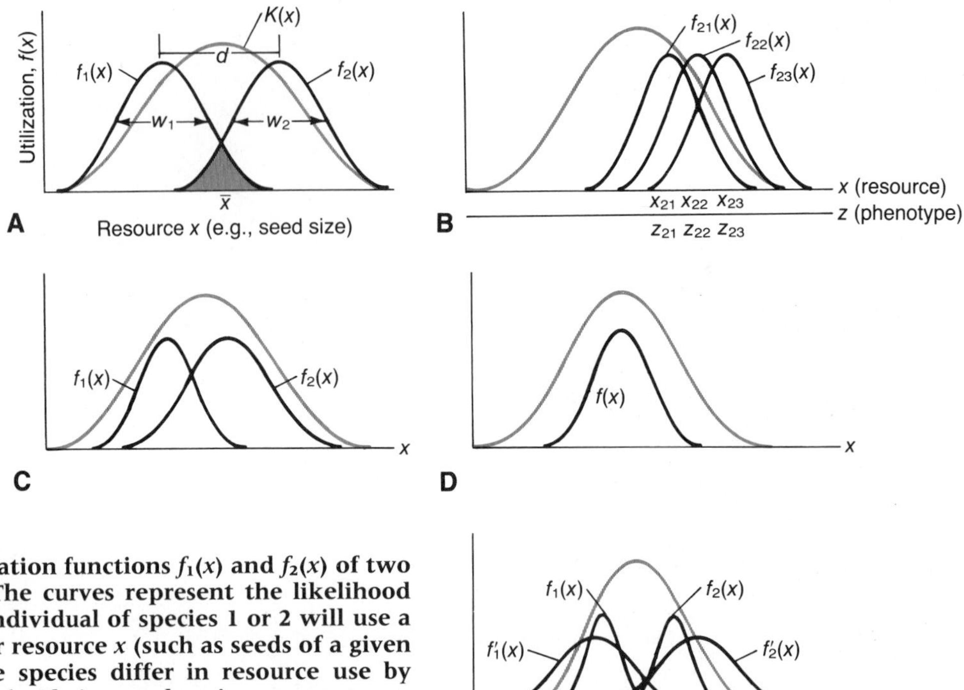


FIGURE 2

**(A) Utilization functions $f_1(x)$ and $f_2(x)$ of two species. The curves represent the likelihood that an individual of species 1 or 2 will use a particular resource $x$ (such as seeds of a given size). The species differ in resource use by amount $d$. Their overlap in resource use (shaded) depends on $d$ and on the width ($w_1$, $w_2$) of the species' utilization functions. The curve $K(x)$ represents the abundance of the resources in all five panels. (B) The utilization functions for three phenotypes ($z_{21}$, $z_{22}$, $z_{23}$) of species 2; these might represent individual birds with different beak sizes that differ in the average size ($x_1$, $x_2$, $x_3$) of seeds they consume. Note that phenotype $z_{21}$ has access to more resources than the others. (C) Competition is asymmetrical if one species has a broader utilization function ($f_2(x)$) than the others. (D) The optimal utilization function toward which a solitary species will evolve. (E) If two species have narrow utilization functions ($f_1(x)$ and ($f_2(x)$)), they may diverge in resource use. If they have broad utilization functions ($f_1'(x)$ and $f_2'(x)$), they may converge from the positions shown and evolve greater overlap.**

the "niche breadth," increases. MacArthur and Levins (1967), who first provided this formulation, suggested that species could coexist only if their degree of nonoverlap ($d/w$) exceeded some critical value, the "limiting similarity." It might be further assumed that some anatomical feature of an individual is closely correlated with its optimal food, so that the size of a lizard's mouth or of a bird's beak may be an index of the size of insect or seed that it eats. If this assumption is valid, the degree of difference in resource use among coexisting species may be indicated by differences in their morphology. Surprisingly little information bears on this assumption. In some cases (D.S. Wilson 1975), larger species merely feed on a greater range of resources than smaller ones, rather than specializing on larger food items. Pulliam (1985) used the rapidity with which a finch husks a seed as a measure of the efficiency with which finches handle seeds of different sizes (Figure 3A). Large seeds are handled more efficiently by finches with large

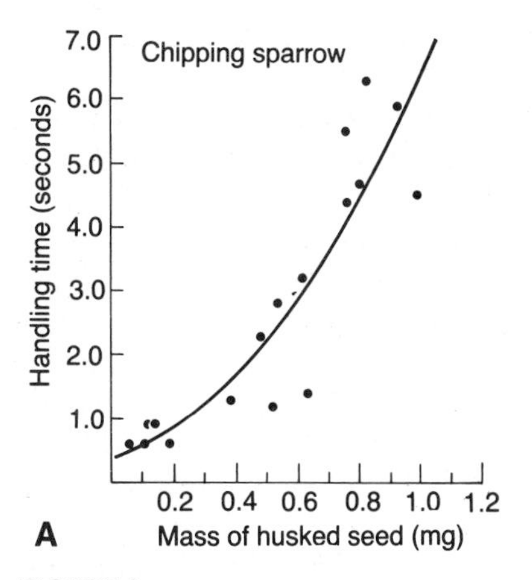


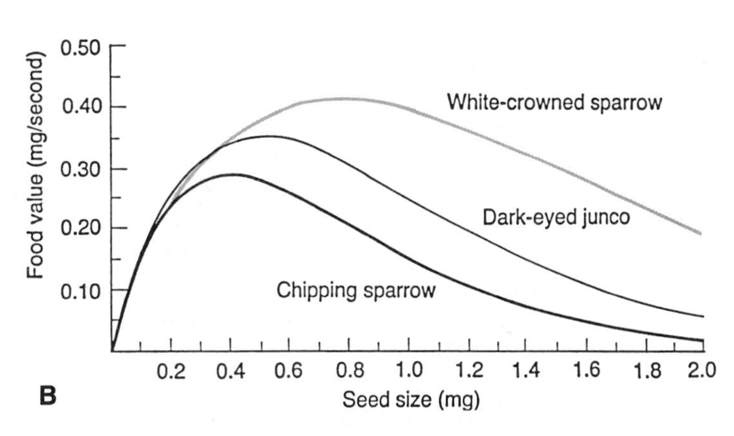


FIGURE 3

**Analysis of the efficiency with which species use different resources. (A) The time required for chipping sparrows (*Spizella passerina*) to husk and consume seeds of different sizes. (B) As seed size increases, its food value to a bird increases because of its greater energy content, but decreases because of greater handling time; thus seeds of intermediate size have maximal food value, measured as mg consumed per second. Beak size, and hence rapidity of husking, is least for chipping sparrows, greater for juncos (*Junco hyemalis*), and greatest for white-crowned sparrows (*Zonotrichia leucophrys*), so optimal seed size increases accordingly. The food value of large seeds is greater for the larger-beaked species because they handle them more rapidly. Small seeds, however, are handled equally rapidly by, and so have the same food value for, all three species. Thus greater efficiency of handling large seeds does not lower efficiency in handling small seeds. (From Pulliam 1985)**

beaks, but beak size does not affect the efficiency with which small seeds are handled. Taking into account handling efficiency and the energy content of seeds, Pulliam calculated that under conditions of food scarcity, finches of different sizes would overlap broadly in the range of seeds they could profitably consume (Figure 3B).

Sympatric species (e.g., of finches) often differ in features such as body size or beak size, and such differences have frequently been interpreted to mean that the species would not otherwise coexist. Many authors have assumed, further, that the differences evolved to reduce interspecific competition. However, Simberloff and his colleagues (e.g. Simberloff and Boecklen 1981, Simblerloff 1983), noting that a community of species is generally a sample from a larger pool of species that vary in size, argue that the differences among sympatric species are often no greater than would be expected if the species were sampled at random from the pool. If they are correct, neither competitive exclusion of excessively similar species nor evolutionary responses to competition need be invoked to explain the differences among species. Numerous investigators have contested this claim that communities of related species conform to a random model (see papers in Strong et al. 1984 for an introduction to the controversy). In one test, Schoener (1984) compared the size ratios of sympatric pairs of bird-eating hawks throughout the world to the distribution of size ratios that would be expected if each pair were drawn at random from the worldwide pool of 47 species. The observed size ratios were significantly greater than random (Figure 4), as expected if competition were important.

If coexisting species do indeed differ more than by chance in resource use, the difference may be attributable to ecological factors, evolution, or both. Species may evolve different utilization curves when allopatric; the only ones that can then successfully invade a community are those that are different enough to avoid immediate exclusion by competition. Alternatively, niche differences could evolve in response to interspecific competition, an instance of character displacement (Brown and Wilson 1956). Both the ecological and the evolutionary theories are highly complex, however. Under some circumstances, such as temporal fluctuation in the abundance of resources, there may be no limit to the similarity of competing species (Armstrong and McGehee 1980, Chesson and Warner 1981, Abrams 1983).

The coevolution of competing species has been modeled by several authors, usually by examining the evolution of a trait such as beak size and assuming a correspondence between this trait and the resources (e.g., seed size) that the animal uses. The "niche" of each phenotype $i$ is specified by its location along the resource axis and by its width (the within-phenotype niche breadth $w_i$). In the absence of competing species, the mean phenotype of a single species will evolve to match the peak of the resource distribution, i.e., the most abundant kind of food (Roughgarden 1976). Slatkin (1980), using a quantitative genetic model (Chapter 7), assumed that the intensity of competition between any two phenotypes does not depend on which species they represent, but only on their

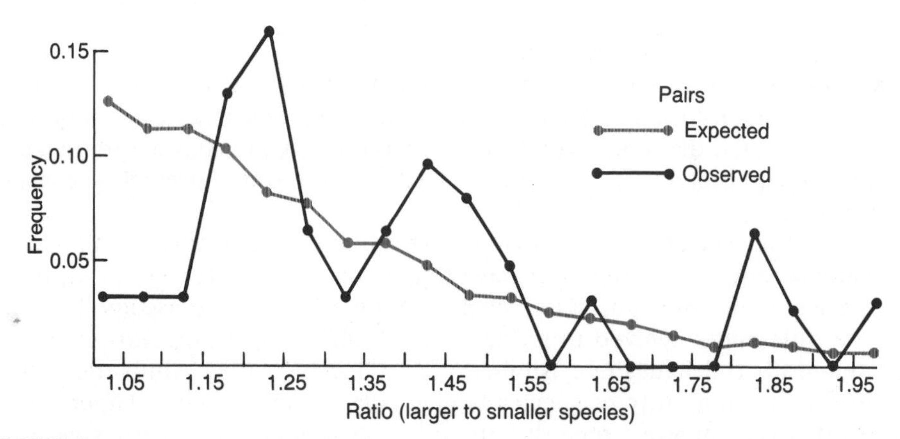


FIGURE 4

**Sympatric pairs of species of bird-eating hawks (genus *Accipiter*) are more different in body size than expected at random. Because hawks of different sizes feed on different kinds of prey, this implies either that excessively similar species cannot coexist (the competitive exclusion principle) or that they diverge in response to competition (character displacement). The black line represents the proportion of pairs of sympatric species, worldwide, with a given ratio of body sizes. The shaded line is calculated from all possible pairs of the world's 47 species of *Accipiter*. Note that few observed pairs of sympatric species have a low ratio of body sizes, although many such pairs occur in the random sample. (From Schoener 1984)**

degree of niche overlap. The fitness of a phenotype is proportional to the abundance of its resources, and so is affected by both the form of the resource distribution (curve $K(x)$ in Figure 2) and the amount of competition with other phenotypes of both the same and the other species. The mean phenotype of each species therefore evolves to maximize mean fitness. Slatkin found that the amount of character displacement is ordinarily very slight, and that the species will sometimes converge rather than diverge. The reason is that selection for divergence imposed by interspecific competition is counteracted by the scarcity of resources at either end of the spectrum relative to the middle (Figure 2E). Convergence is especially likely if the resource spectrum is narrow compared to the within-phenotype niche breadth.

Taper and Case (1985) confirmed these results, but extended the model to include genetic variation in degree of specialization so that the within-phenotype niche breadth could evolve. In this model, moreover, the resource distribution is not fixed; each resource along the axis is treated as a growing population, so its abundance is determined by its rate of renewal relative to the rate at which it is consumed. Taper and Case found that if the initial overlap between species is high, the species converge; if not, they diverge appreciably. Depending on initial conditions, therefore, competing species may either diverge or converge. If they converge, the intensification of competition may well lead to the extinction of one of the species. Coevolution, therefore, does not necessarily foster coexistence and higher species diversity in a community.

Coevolution among competitors is likely to cause extinction if competition is asymmetrical (Figure 2C). Roughgarden (1983b) has modeled this case, based on his studies of *Anolis* lizards in the West Indies. Large lizards can eat small insects and so impose strong competition on small lizards, but small lizards cannot eat large insects. Large insects are relatively rare, so according to Roughgarden's model a large species of lizard is selected for smaller body size, and so converges toward the smaller species. The small species evolves still smaller size in response, but nevertheless may be driven to extinction (Figure 5). Roughgarden believes that this model accounts for the observation that almost all the anoles that are the sole species on their islands have the same body size, except for several cases that may represent intermediate stages in this process.

Rummell and Roughgarden (1985), using computer simulation, have extended this model to compare the effects of coevolution on community structure with the effects of invasion and competitive exclusion without evolution (see also Case 1981). In their model, coevolution between asymmetric competitors leads to a community of few species with large niche separations. As in Figure 5, most of the species shift to a lower part of the resource spectrum, leaving resources available for more large species to invade. The coevolutionary shift toward the lower end of the spectrum is then repeated, in a theoretically unending TAXON CYCLE (Chapter 13). In contrast, a community in which coevolution does not occur, but which continually receives invaders of various sizes, builds up a higher diversity of species across the full range of the resource spectrum. Excessively similar species are unable to become established, but the community reaches equilibrium with a complement of species with smaller niche separations than in the coevolving community. Case's (1981) model also predicts that com-

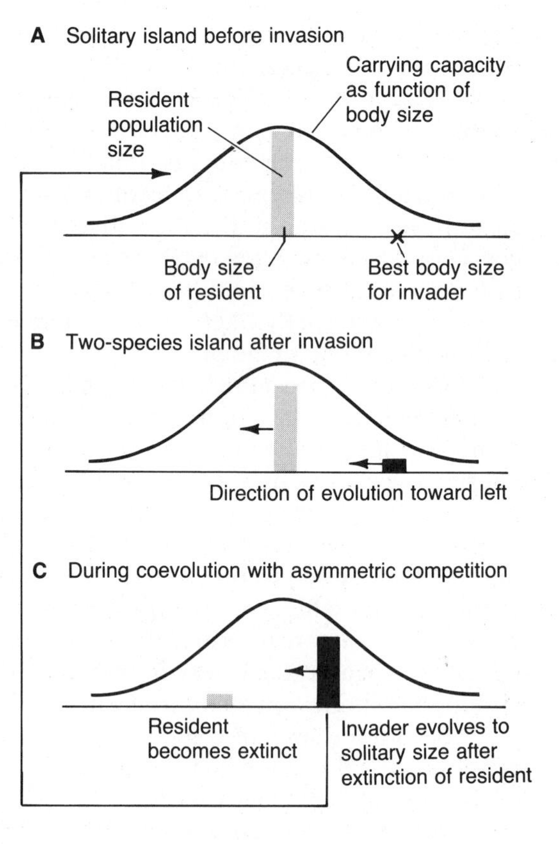


FIGURE 5

**A model of species turnover on islands, caused by coevolution between competing species. The model has been proposed to account for the distribution of *Anolis* lizards of various body sizes in the West Indies. The horizontal axis represents body size and the vertical axis population density. Larger species are superior competitors, so it is assumed that an island with one species (a "solitary island," A) can be invaded only by a larger species. The invader evolves smaller body size because of the distribution of resources (the carrying capacity function), and the resident evolves smaller body size to reduce competition (B). With sufficently strong and asymmetrical competition, the resident is driven to extinction (C). (From Roughgarden 1983b)**

peting species will be more similar in resource use in invasion-structured than in coevolution-structured communities. These predictions have not yet been tested.

The above models deal only with competition via exploitation of resources. When individuals of two species interfere with each other by aggression, mutual predation, allelopathy, or other mechanisms, the competitive effect of species *j* on species *i* can evolve either if *i* becomes more resistant to *j*'s onslaughts or if *j* improves its ability to interfere with *i*. Gill (1972) has termed the latter effect "α-selection."

### Examples of the evolution of competitive interactions

Character displacement among potentially competing species, i.e., greater divergence in sympatry than in allopatry, has been documented in rather few cases (summarized by Taper and Case 1985). Two species of mud snails, *Hydrobia ventrosa* and *H. ulvae*, have diverged in size (and in the size distribution of the particles on which they feed) in a Danish fjord that they colonized in the nineteenth century; allopatric populations of the species are similar in body size, but so are sympatric populations outside the fjord (Fenchel and Christiansen 1977).

The geospizine finches of the Galápagos Islands are a classical case of adaptive radiation, made famous by Darwin's comments on them and by David Lack's subsequent work (1947). Lack interpreted the great variation in size and shape of the beak among species as the consequence of adaptation to different foods under the pressure of interspecific competition. Peter Grant and his colleagues have analyzed many of the finches in detail (e.g., Abbott et al. 1977, Grant and Grant 1982, Schluter and Grant 1982, 1984). They have found that the differences in beak size among species of ground finches (*Geospiza*) that coexist on an island are generally greater than if the species were drawn at random from the entire pool of *Geospiza* species that occupy the archipelago (see also Hendrickson 1981, Case and Sidell 1983). There is evidence that the beak size of a population is related to the size and hardness of the seeds the birds eat, and that food is at least sometimes a limiting resource. Some pairs of species show competitive exclusion; for example, low-elevation islands have either *G. fuliginosa* or *G. difficilis*, whereas on islands with more topographic relief *G. fuliginosa* occurs at low and *G. difficilis* at high elevations. Character displacement in beak size and feeding habits is evident in several instances. For example, *G. fortis* and *G. fuliginosa* have the same beak length on small islands where each species occurs without the other, whereas their beak lengths differ on islands on which they are sympatric (Figure 9 in Chapter 2). Perhaps the best analyzed case is the relationship between *G. conirostris* and *G. magnirostris*. *G. conirostris* occurs on only two islands, Isla Española and Isla Genovese, from which two smaller species that otherwise are widely distributed throughout the archipelago are absent. On Española, *G. conirostris* has a beak of intermediate size, and takes a wide variety of foods, including those typically taken by both the smaller species and a larger congener, *G. magnirostris*, which does not occur on this island. *G. magnirostris* occurs on Isla Genovese, however, and here *G. conirostris* has a smaller beak, does not feed on the large seeds taken by *G. magnirostris*, and is a specialized cactus feeder, much like one of the missing smaller species (Grant and Grant 1982).

Several authors have described evolutionary responses to competition that, although dramatic, seem not to have had morphological consequences. For example, the altitudinal distributions of the salamanders *Plethodon jordani* and *P. glutinosus* overlap broadly in the Balsam Mountains, but only narrowly in the Great Smoky Mountains in the southern Appalachians. Hairston (1980) demonstrated that the species compete by showing that the population of each increases if the other species is removed. By replacing *P. jordani* in each overlap zone with *P. jordani* from the other zone of overlap, Hairston provided evidence that the Balsam populations of each species do not suppress each other's population growth very much, whereas the Great Smoky populations have a strong competitive impact. It is not known whether the direction of evolution has been from strong to weak competitive interactions or vice versa. Evidence of evolution of competitive relationships has also been found in plants. For example, seed production of mixtures of *Erodium obtusiplicatum* and *E. cicutarium* (Geraniaceae) from sympatric populations was greater than in mixtures from allopatric populations (Martin and Harding 1981). Evidence of very localized natural selection by interspecific competition was described by Turkington and Harper (1979), who found that different genotypes of white clover (*Trifolium repens*) from a single field produced more biomass when grown in competition with species with which they were associated in the field than with species that the other genotypes were associated with.

## EVOLUTION OF PREDATOR–PREY RELATIONSHIPS

In general, individual selection should cause prey species to evolve protective characteristics and predators to become more proficient at catching and subduing their victims, even if this results in the diminution or even the extinction of the prey population. Selection for proficiency in predation is most intense, in fact, if the predator population is limited by scarcity of prey.

Most models of predator–prey coevolution have considered one species of predator and one species of prey, which might be expected to evolve in a never-ending "arms race" that would eventually result in the extinction of one or both species unless each evolutionary advance of one species were precisely counteracted by an advance of the other. The existing models of exploiter–victim relationships are not explicit genetic models, and they differ in their predictions about the ubiquity and consequences of such arms races. Stenseth and Maynard Smith (1984) have developed a very general model of coevolution in which species either coevolve indefinitely, some becoming extinct in the process (the "Red Queen effect," Chapter 12), or else arrive at a static, nonevolving equilibrium. This model, however, does not take into account the extent to which genetic variation may limit the rate of evolution. Schaffer and Rosenzweig (1978), in contrast, postulate that coevolution of predator and prey continues indefinitely but seldom leads to extinction (Figure 6). In this model, the fixation rate of mutations that improve a predator's proficiency or a victim's resistance depends on how proficient or resistant the species already is. Proficiency evolves faster in an inefficient predator than resistance in an already resistant prey, and resistance evolves faster in a highly vulnerable victim than proficiency in an efficient predator. Thus evolutionary rates are balanced, and neither species is likely to win the race. However, such an evolutionary race might grind to a halt if selection for greater resistance or proficiency were counteracted by their greater "cost" (e.g.,

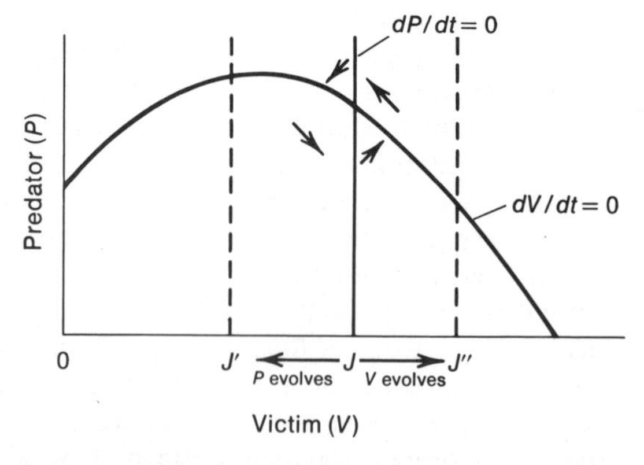


FIGURE 6
**Evolution of the predator isocline, as visualized by Rosenzweig. A point in the space represents the number of predator individuals *P* and prey individuals *V*. If *V* is low, the predator population declines; if high, it increases. Therefore there is a value $V = J$ at which the predator population does not change ($dP/dt = 0$). Similarly, the prey population remains unchanged at those combinations of *V* and *P* values that fall on the curve $dV/dt = 0$. The arrows in the diagram are vectors showing changes in *P* and *V* when population densities are not on the zero isoclines. The system is stable if the predator zero isocline $dP/dt = 0$ falls to the right of the peak of the prey zero isocline $dV/dt = 0$; it is unstable if the predator zero isocline is to the left. *J* will move toward *J'*, and the system becomes destabilized, if the predator evolves greater proficiency (can support itself at a lower value of *V*). The system becomes stabilized (*J* moves toward *J''*) if the prey evolves better defenses (so that the predator needs a larger prey population in order to support itself). (After Rosenzweig 1973)**

Levin 1972). One likely reason for such cost is that the features of a prey species that provide resistance to one predator may make it more vulnerable to others (Futuyma 1983b). Cucurbitacins in cucumber plants (*Cucumis sativus*), for example, enhance resistance to mites, but attract certain herbivorous beetles (Dacosta and Jones 1971). Similarly, we might expect that the proficiency of a predator in handling one prey species comes at the expense of handling others. Therefore the more predators a prey species must withstand, and the more prey species a predator relies upon, the more sluggish coevolution is likely to be (Slobodkin 1974, Futuyma and Slatkin 1983c).

Perhaps for this reason, there is little evidence of long-continued coevolution between predators and their prey. During the Tertiary several groups of ungulates evolved to be more fleet, but the lineages of carnivores that presumably preyed on them did not (Bakker 1983; Figure 7). Predator–prey coevolution, when it occurs, appears to be diffuse, and often includes the extinction of one group and its replacement by another. For example, the evolution in the Mesozoic of more efficient marine predators such as crabs was accompanied by the evolution of more heavily defended molluscs, but there is no evidence of gradual, coupled change between particular lineages of predators and prey (Vermeij 1983). When more effective grazers appeared in the Mesozoic, solenopore algae were largely replaced by the better-defended coralline algae, but the solenopores did not evolve better defenses (Stanley et al. 1983).

Models of single prey and single predators, then, are probably less realistic than models of multiple species interactions, but few such models have been developed. Coevolution is likely to be slower and more sporadic if each species is engaged in a multiplicity of predator–prey interactions than if a species is the sole prey of a specialized predator. In the latter case each species might evolve quite refined adaptations to the other. When several species are subject to a common predator (or predators), it is likely that selection will favor divergence in mechanisms of escape or defense, in the same way that predation may maintain polymorphism by frequency-dependent selection within a population (Levin 1983; see Chapter 6). If, for example, cryptic color patterns in moths are a defense against birds, and if birds learn by experience to discern common cryptic moths, divergence from the common pattern is likely to be selectively advantageous. Ricklefs and O'Rourke (1975) found that the "aspect diversity"—the variety of cryptic patterns—is greater among tropical moths than in less diverse temperate zone communities, and hypothesized that species may so diverge as to maintain some minimal difference in this respect. This hypothesis has not yet been rigorously tested.

### Herbivores and plants

Occasionally a prey species may evolve a "broad spectrum" defense that releases it from attack by most predators. This species and its descendants may later become the resource base for particular lineages of predators that acquire adaptations for overcoming the defense. But such predators may evolve long after the prey lineage diversifies, and they may be unrelated to the predators that originally fed on the prey (Figure 8).

A scenario rather like this appears to describe the evolution of relationships among herbivorous insects, such as butterfly larvae, and their host plants (Ehrlich

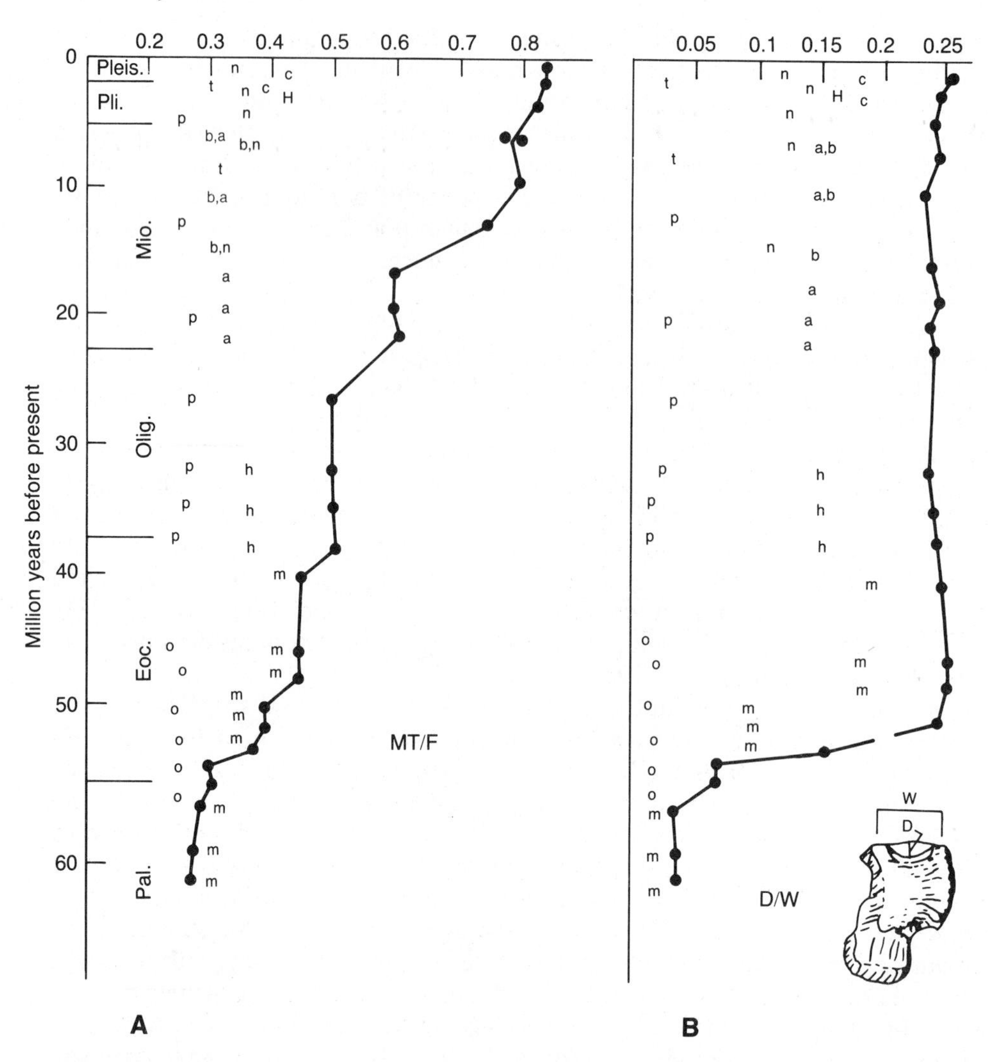


FIGURE 7

**Indices of running speeds of North American carnivores and ungulates throughout the Cenozoic. The horizontal axis in A is the ratio of the length of the metatarsus (MT) to the femur (F); in B it is the ratio of the depth (D) to the width (W) of the groove on the astragalus. Both indices are greater in rapidly running than in slow mammals. In each graph, the solid line connects the mean values for all species of ungulates in paleontological faunas: both indices increased, indicating that later ungulates were swifter than earlier forms. Each letter represents the mean value for a particular group of carnivorous mammals at that time. Within most carnivorous taxa (e.g., paleofelids, *p*), there was little if any evolution of greater speed. The data suggest that coevolution between predators and their prey did not occur, at least with respect to speed. (Abbreviations for the predators are: *m*, mesonychid; *o*, oxyaenid; *p*, paleofelid; *h*, hyaenodontid; *a*, amphicyonid; *n*, neofelid; *b*, borophagine; *c*, canine; *H*, hyaenid; and *t*, *Thylacinus*, the Tasmanian wolf, a marsupial given for comparison with North American fauna.) (From Bakker 1983)**

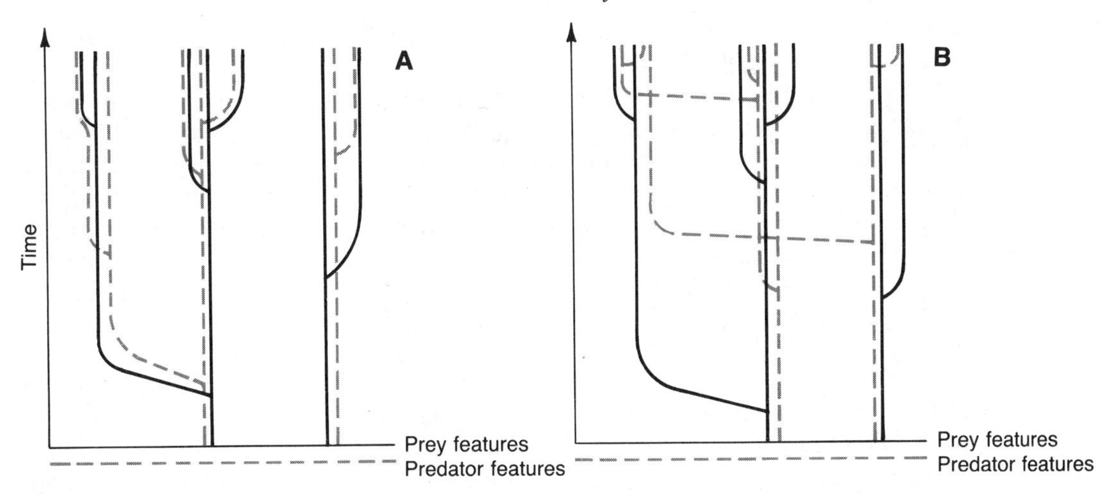


FIGURE 8

**(A) Coevolution of lineages of prey (e.g., plants), shown by solid lines, and of specialized predators (e.g., herbivorous insects), shown by broken lines. In this case, divergence of prey in, for example, defensive characteristics is accompanied by counteradaptation (and, usually, by speciation) of the predator. (B) The usual pattern for plants and herbivorous insects. Divergence of the prey (plants) is not necessarily accompanied by divergence of the predators (insects). Insects may evolve association with unrelated, phenotypically distinct plants, and adapt to them, at various times subsequent to diversification of the plants.**

and Raven 1964, Mitter and Brooks 1983). Closely related insects often feed on related plants, such as mustards and other Cruciferae in the case of pierine butterflies. The cues used to select the host plant are often "secondary compounds" such as mustard oil glycosides, which are often repellent or toxic to nonadapted insects and which may have evolved originally as defenses against herbivory. (But not all authors agree that these compounds have evolved because of their adaptive function; see Jermy 1984.) Ehrlich and Raven proposed that different groups of plants evolved different protective compounds, and that various groups of butterflies (and other insects) subsequently evolved the ability to overcome one or another kind of chemical defense, and indeed to use as a cue for host recognition the very compounds that confer protection against other insects (reviewed by Futuyma 1983b). In some cases, several steps of defense and counterdefense can be seen; for example, certain umbelliferous plants are protected by linear furanocoumarins against all insects except a few that are adapted to these compounds (Berenbaum 1983); but these insects are not adapted to the angular furanocoumarins that certain other umbellifers possess, and which may be a later evolutionary acquisition. Plants with angular furanocoumarins are attacked by other, quite specialized, species of insects. These, however, are not closely related to the species that can tolerate linear furanocoumarins, so there is no direct correspondence between the phylogeny of the umbellifers with different compounds, and the phylogeny of their associated insects. Nor is the phylogeny of the Lepidoptera as a whole closely correlated with that of the plants on which they feed (Mitter and Brooks 1983). It is likely, then, that a plant group with distinctive protective compounds can become the host of many unrelated insects at various points in evolutionary time.

Although the phylogenetic relationships among herbivorous insects seldom correspond to those among their host plants (Futuyma 1983b, Mitter and Brooks 1983), closely related insects are often similar in their host affiliation, as in other characteristics. For example, members of the neotropical butterfly genus *Heliconius* all feed as larvae on species of passion vines (*Passiflora*) and related genera in the Passifloraceae (Benson et al. 1976, Turner 1981). They are among the few insects that feed on *Passiflora,* having apparently evolved adaptations to the alkaloids and other secondary compounds of these plants. Perhaps because the plants contend with few other herbivores, some species of *Passiflora* have evolved special defenses against *Heliconius*: hooked hairs that immobilize newly-hatched larvae (Gilbert 1971) and structures that mimic *Heliconius* eggs. (Presumably to reduce the competition to which their offspring would be subjected, the females of some species of *Heliconius* tend not to lay eggs on plants that already have eggs—or egg-mimicking structures; Williams and Gilbert 1981.) It is in cases such as this, in which ecological interaction is so specialized that each species exerts a primary selective pressure on the other, that we may expect to see reciprocal coevolutionary responses of individual species to each other.

### Parasites and hosts

Coevolutionary relationships among parasites and their hosts can be considerably more complicated than among predators and their prey (Ewald 1983, May and Anderson 1983). We still expect hosts to evolve more effective defenses; and under many conditions parasites can evolve to become more virulent. The extent to which a parasite weakens or kills its host is often correlated with the reproductive rate of the parasite, which should be maximized by individual selection. There is no limit to the evolution of virulence in parasites such as the nuclear polyhedrosis viruses of insect larvae; the progeny of these viruses are released in vast numbers into the environment when the host larva dies, and are ingested by other larvae. Such parasites are not expected to evolve so as to prolong the life of their hosts. If, however, the parasite must be transmitted from one live host to another, a lower degree of virulence may evolve because the early death of a host may result in the death of the parasites before they can be transmitted. For example, the myxoma virus that was introduced to control rabbits (*Oryctolagus cuniculus*) in Australia evolved a lower degree of virulence over the course of a decade (Table I), apparently because the virus is transmitted by mosquitoes that feed only on live rabbits. Very avirulent strains of virus, however, do not have a high reproductive rate, so selection favored an intermediate level of virulence (Fenner 1965, May and Anderson 1983). During this period, the rabbits also evolved greater resistance to the virus, so this case illustrates coevolution in the most restricted sense.

Even if effective transmission depends on a low level of virulence, individual selection on the parasite favors high reproductive rates, and hence high virulence, if different genotypes of the virus colonize the same individual host. The parasites in each host may be considered a temporary deme, or trait-group, however. According to D.S. Wilson's (1980) model of group selection (Chapter 9), trait-groups with high average virulence may contribute less to the parasite gene pool

TABLE I

**Frequency of field-collected strains of myxoma virus of different grades of virulence, collected from rabbits in Australia**

| | Virulence grade[a] | | | | | |
|---|---|---|---|---|---|---|
| | I | II | IIIA | IIIB | IV | V |
| 1950-1951[b] | 100 | — | — | — | — | — |
| 1958-1959 | 0 | 25.0 | 29.0 | 27.0 | 14.0 | 5.0 |
| 1963-1964 | 0 | 0.3 | 26.0 | 34.0 | 31.3 | 8.3 |

From May and Anderson (1983)
[a]Grades I to V are a descending series of degrees of virulence.
[b]The virus was introduced in 1950.

than those with lower virulence, so it is possible that more benign parasites can evolve by group selection.

Among parasites that disperse actively from host to host, there is opportunity for transfer among unrelated host species. It is perhaps not surprising then that related species in such groups of parasites frequently occupy unrelated hosts, so that the phylogenies of hosts and parasites are not strongly congruent. For example, species of the trematode genus *Acanthostomum* are parasites of teleost fishes and of crocodiles, and one species has been found in both turtles and snakes (Mitter and Brooks 1983). In contrast, parasites that are directly transferred by contact between individual host organisms may have little opportunity for dispersal across taxonomic lines. The phylogenetic relationships among pinworms of the genus *Enterobius*, for example, appear to reflect those of their primate hosts (Figure 9), as if divergence in the hosts were accompanied by divergence in their associated parasites.

## MUTUALISM

Mutualistically interacting species do not contribute to each other's success out of good will; rather, they reciprocally exploit each other as resources. The distinction between mutualism and parasitism can be a fine one, as with mycorrhizal fungi that benefit plants by enhancing the uptake of mineral nutrients at the cost of carbohydrates that they extract from the plant. Roughgarden (1975, 1983a), in a cost–benefit model, has concluded that a "guest" should forego excessive exploitation of its "host" if this will enhance the host's, and thereby the guest's, survival.

Some mutualistic interactions are highly specific; for example, almost every species of fig is pollinated by a single species of host-specific fig wasp. However, most mutualisms, such as the majority of pollination systems, involve groups of interacting species. As for predators and prey, the rate and precision of coevolution is likely to be lower the greater the number of interacting species (Wheelwright and Orians 1982, Schemske 1983b, Howe 1984). Conversely, species that engage in highly specific mutualistic interactions often have highly refined adaptations to each other. Thus the consequences of coevolution, strictly defined,

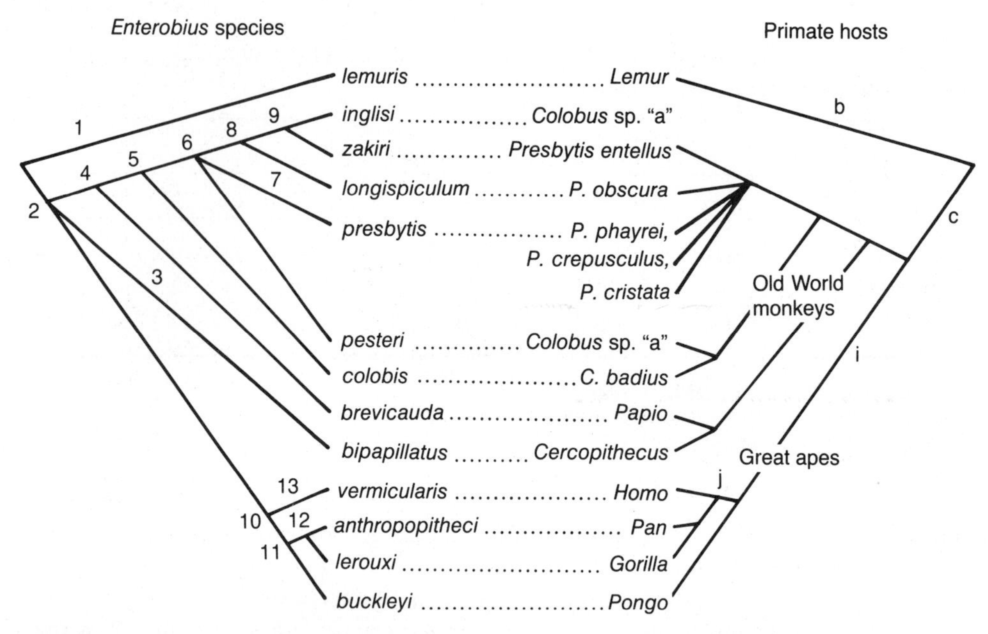


FIGURE 9

**Phylogenies of a group of parasitic nematodes (pinworms of the genus *Enterobius*) and of their primate hosts. In many respects, the phylogenies are congruent, as expected if parasites diverged in concert with their hosts and were not transferred between host species. Note, for example, that the parasite of *Lemur*, the sister group of the other primates, is the sister group of the other *Enterobius* species. The phylogeny of the parasites of the great apes, however, is not strictly congruent with that of their hosts. (From Mitter and Brooks 1983)**

are most evident in pairs of strongly interacting species such as *Ensifera ensifera*, a hummingbird with an extraordinarily long bill (more than 80 mm) that feeds almost exclusively from passion flowers in which the corolla is up to 114 mm long. Often it must be advantageous for one species to be the exclusive resource of its partners, but advantageous for each of the partners to be more generalized. A plant may avoid cross-pollination from other species if it has a specialized pollinator, but each pollinator may find it advantageous to garner nectar and pollen from many plants. The outcome of such a conflict has not been modeled in detail, but the subject has been discussed for some specific mutualisms (e.g., Janzen 1983, Feinsinger 1983).

Some mutualisms probably arise from commensal relationships. For example, arboreal ants may generally provide some incidental protection against herbivores to the plants in which they nest and on which they forage. This association has been exploited by many plants that have evolved extrafloral nectaries (nectar glands outside the flowers). These nectaries often attract many species of pugnacious ants that defend the plants (Bentley 1977). The relationship has been accentuated in some species of *Acacia*, which provide not only extrafloral nectar, but proteinaceous food bodies and hollow thorns that are the exclusive nesting site of some ants of the genus *Pseudomyrmex*. The ants are unusually hostile to

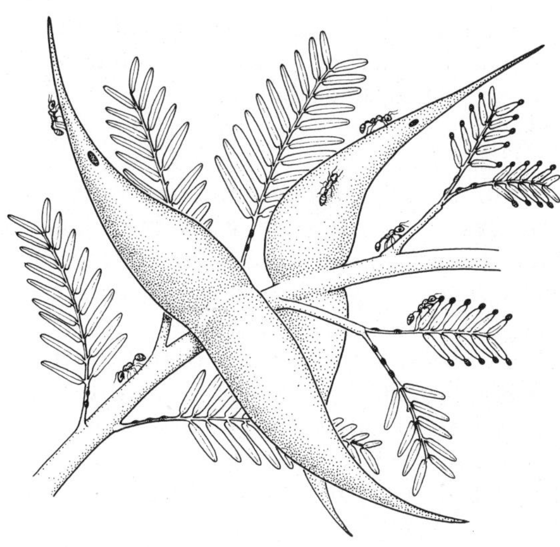

FIGURE 10

**The mutualistic interaction between the tree *Acacia* and the ant *Pseudomyrmex*. The ants inhabit the thorns and feed at the nectaries on the petioles and on the proteinaceous Beltian bodies at the tips of the young leaflets. The tree is defended by the pugnacious ants against herbivores and competing vegetation.**

intruders, and protect their hosts not only against herbivores but also against vines and competing plants (Janzen 1966; Figure 10).

Other mutualisms appear to arise from parasite–host relationships, although not all parasites evolve to become mutualists. For example, yuccas (*Yucca*) are pollinated exclusively by moths (*Tegeticula*) whose larvae feed on some of the developing seeds that result from their parents' pollination activity. Likewise, figs are pollinated in a very complex way by agaonid wasps whose larvae feed on some of the developing ovules (Figure 11). Both the moths and the wasps are derived from ancestors that are "seed parasites" that pollinate their hosts only incidentally if at all (Feinsinger 1983).

Each of most of the hundreds of species of figs is pollinated by only a single host-specific species of wasp. It is likely that divergence in a local population of either the fig or the wasp, whether by genetic drift or selection, will induce a coevolutionary change in the other. Host specificity provides reproductive isolation among both the wasps and the figs, so coevolutionary divergence among

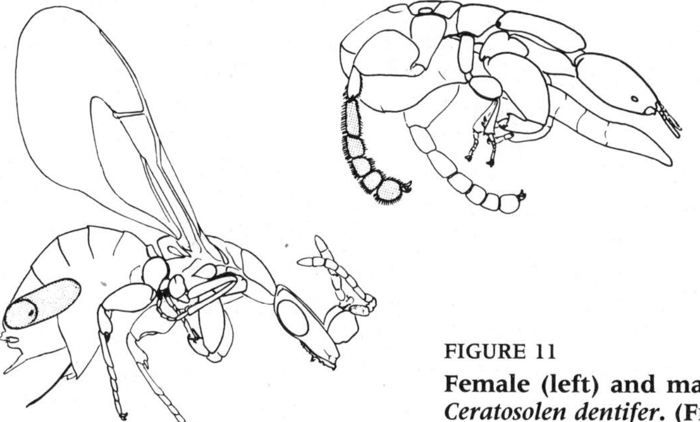

FIGURE 11

**Female (left) and male (right) of a fig wasp, *Ceratosolen dentifer*. (From Wiebes 1979)**

populations is likely to lead to speciation (Kiester et al. 1984; see Chapter 8). Because of their highly specific obligate mutualism, figs and their pollinators are likely to exemplify coevolution in the strictest sense of the term.

## GENETIC STUDIES OF COEVOLUTION

Much of the study of the evolution of interspecific interactions has focused on the results rather than the process of coevolution. In only a few cases has the genetic basis of interspecific interactions been explored. One of the most intriguing results has been the description of "gene-for-gene" systems governing the interaction between certain parasites and their hosts. In several crop plants, dominant alleles at a number of loci have been described that confer resistance to a pathogenic fungus; for each such gene, the fungus appears to have a recessive allele for "virulence" that enables the fungus to attack the otherwise resistant host. It is possible, however, that such gene-for-gene systems are an artifact of the screening techniques used by plant breeders in their search for resistant germ plasm (Barrett 1983).

Studies of geographic variation in ecologically interacting species should shed light on the dynamics of coevolution. Cases of character displacement among competing species (see above) are among the best evidence that interspecific interactions can result in genetic change. Perhaps the most striking examples of coevolution are among mimetic butterflies (Turner 1981, Gilbert 1983). For example, geographic populations of the distasteful neotropical butterfly *Heliconius erato* differ in the pattern of red, yellow, and white markings on the wings. Crosses among populations have shown that the presence or absence of each mark is governed by one of about eight single loci. The Müllerian mimic *H. melpomene* has an almost perfectly concordant pattern of geographic variation (Figure 12), and the wing markings are governed by a number of major loci as in *H. erato*.

Assuming that parasites and their hosts coevolve in an "arms race," we might deduce that the parasite is "ahead" if local populations are more capable of attacking the host population with which they are associated than other populations, whereas the host may be "ahead" if local populations are more resistant to the local parasite than to other populations of the parasite. Parker (1985) found that each of several local populations of a pathogenic fungus (*Synchytrium decipiens*) was more capable of growth and reproduction on its native population of the sole host, the wild hog peanut (*Amphicarpaea bracteata*), than on plants from other populations of the same species. Both the plant and the fungus varied genetically over short distances. In this instance it appears that the parasite is capable of more rapid adaptation to its host than vice versa.

Several authors have used laboratory selection experiments as models of coevolution. Interactions between competing species or strains of *Drosophila*, muscoid flies, and *Tribolium* beetles have evolved rather rapidly in some cases (see Pimentel et al. 1965, Seaton and Antonovics 1967) but did not do so in other cases (Park and Lloyd 1955, Futuyma 1970, Sulzbach 1980). Pimentel et al. (1965) suggested that continued evolution of competitive ability might stabilize competitive interactions, if the rarer of the two species is selected to improve its ability to compete with the more common one, while the more common one is selected to improve its intraspecific competitive ability. As the interspecific competitive ability of the rarer species increases, it becomes numerically predominant,

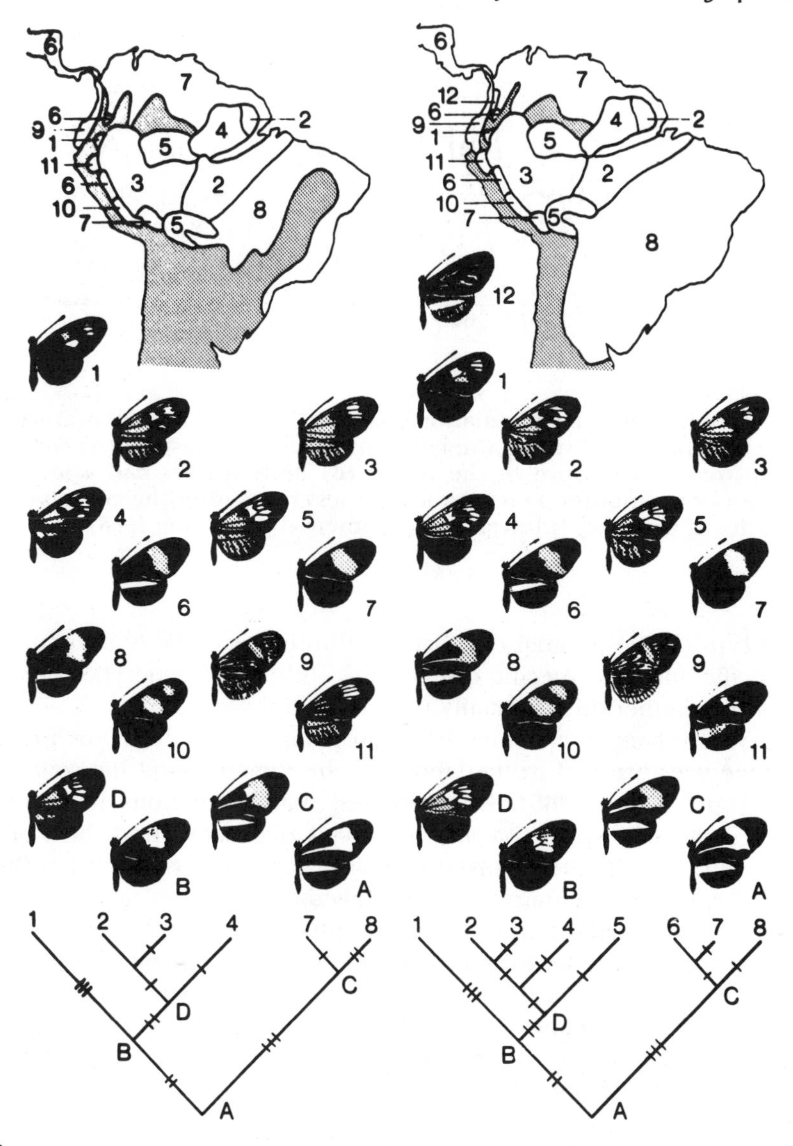


FIGURE 12

**Parallel geographic variation in two Müllerian mimics, the butterflies *Heliconius melpomene* (left) and *H. erato* (right). Similar color patterns in the two species bear the same numbers, and have similar geographic distributions in tropical America, as shown on the maps. The inferred phylogenetic relationships among several of the races of each species are displayed below (letters represent hypothetical ancestors). (From Turner 1981)**

and the nature of selection in the two species is reversed. Thus the two species might oscillate indefinitely in abundance and in genetic composition. Indeed, after 25 generations of competition between houseflies and blowflies, the initially inferior blowflies became genetically superior competitors and eliminated the houseflies; however, the hypothesized stabilization of the interaction did not

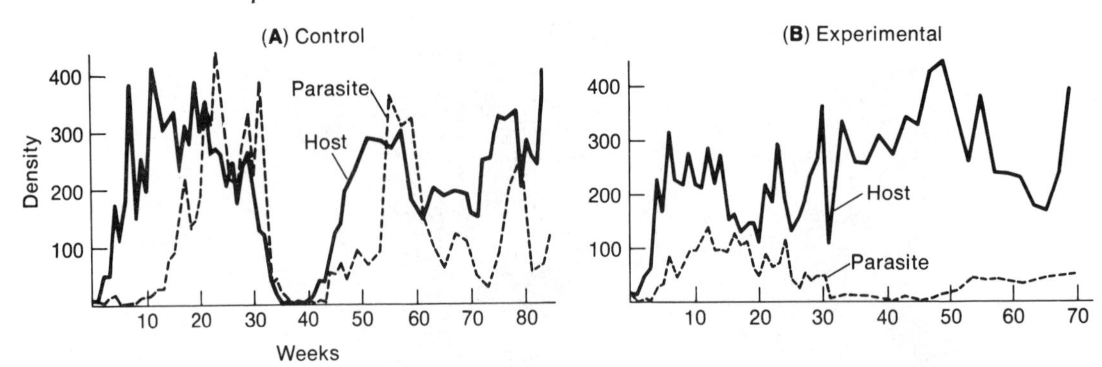


FIGURE 13

**Effects of coevolution on population sizes and stability in a laboratory system of houseflies as hosts and wasps (*Nasonia vitripennis*) as parasites. (A) Neither species had a history of exposure to the other. (B) Both species had a long history of exposure to one another. Fluctuations are less marked in the coevolved case, and the density of parasites is lower. (After Pimentel and Stone 1968)**

occur. Levin (1971) pointed out that stabilization is unlikely to occur unless the intraspecific and interspecific competitive abilities of genotypes are negatively correlated; whether this is usually the case is unknown.

Coevolutionary interactions between parasites and hosts, or predators and prey, have been studied in the laboratory by Pimentel and his coworkers (e.g., Pimentel and Stone 1968), who reported that fluctuations in populations of houseflies and of a wasp (*Nasonia vitripennis*) that parasitized them became less pronounced after the flies evolved increased resistance to the wasp (Figure 13). The most extensive laboratory studies have been done on rapidly evolving systems of bacteria (*Escherichia coli*) and the phage that attack them. For example, improved resistance in bacteria and greater virulence of phage have been observed to coevolve (Chao et al. 1977; see also Levin and Lenski 1983); the system apparently remained stable because the "cost" of both greater bacterial resistance and enhanced capacity of the phage to attack are great enough to prevent either species from gaining the upper hand.

## EVOLUTION AND THE STRUCTURE OF COMMUNITIES

Although far too little is known to justify any firm conclusions (Järvinen et al. 1986), we may venture some speculations on the overall effect of evolution on community structure. It seems likely that evolution can affect the stability, species diversity, and connectance of communities (Futuyma 1986).

As we have seen, evolution can either enhance stability, as when competing species undergo character displacement, or reduce stability, as when competitors evolve greater competitive ability or predators become more efficient. After temporary destabilization, a community becomes more stable when species that do not fit in become extinct, leaving a residue of species that by definition form stable associations. Because changes in population size occur more rapidly than genetic changes, it is likely that ecological processes (e.g., competitive exclusion) contribute more to the formation of stable associations than evolutionary processes do.

Over long periods of time, the diversity of species in a community tends to increase and apparently reaches an equilibrium (Chapter 12). Species arise with different specializations; interactions among species provide niches for yet other species such as parasites and mimics. Nevertheless, there appear to be upper limits on the diversity of competing species, and some communities have retained rather similar structure and diversity over long periods of evolutionary time (Boucot 1978, Bambach 1983). Evolutionary rates in such communities are likely to be low, perhaps because of conflicting selection pressures imposed by the various species with which each species interacts. Rapid evolutionary change, including adaptive radiation, is most pronounced when competition is relieved by the extinction of prevailing species. Occasionally, however, perhaps even in "saturated" communities, a species may escape from the web of interactions by evolving new characteristics, as when a plant escapes its complex of herbivores by evolving a new chemical defense. Such a species becomes a qualitatively new resource, to which some other species eventually adapt. A chemically defended plant and its specialized herbivores become, then, an isolated "component community" within the larger community.

CONNECTANCE is a term that describes the number of trophic links between any one species and others in the community. In a community with high connectance, each prey species might be fed on by many species of predators; in contrast, connectance would be lowest if each prey species is fed on by only one species-specific predator. From the viewpoint of ecological theory, low connectance may enhance community stability (May 1981; but see Pimm 1982). But an isolated subcommunity of, say, one prey and one predator is precisely where we should expect coevolution to proceed most rapidly, since each species is free from conflicting adaptations necessary to cope with other species. Such coevolution may take the form of an "arms race" that may lead ultimately to extinction. Moreover, highly specialized species are susceptible to extinction if extrinsic factors lower the abundance of their prey. Both evolutionary and ecological processes, therefore, may set a lower limit to the connectance of communities.

Ecological theory predicts that specialized ecological relationships between mutualists or between predators and prey should be more prevalent in relatively constant environments such as those that are sometimes thought to characterize coral reefs and tropical rainforests (Futuyma 1973, May 1981). However, the evidence for this view is incomplete (Boucher et al. 1982), and it is possible that specialized relationships are more prevalent in the tropics because of their greater age (Chapter 13), not because of a more constant environment. It is possible that specialized component communities (e.g., pairs of mutualists) arise continually by coevolution, but have a high extinction rate. No one really knows whether tropical rainforests and coral reefs experience more constant environments than less diverse communities, but ecologists have come to suspect that these are fragile ecosystems composed of intricately interacting, often specialized species (Farnworth and Golley 1974). Such communities may well be irreparably changed by even slight alterations of the environment, much less the massive destruction that they now suffer from human activity.

## SUMMARY

The species composition of communities and the ecological relationships among their members are determined by ecological processes of immigration and extinction and by evolutionary responses of species to each other. Although coevolution has been demonstrated in various kinds of interspecific interactions, its prevalence and its effects on community structure cannot be assessed at present. Interactions among species can be either stabilized or destabilized by their evolution.

## FOR DISCUSSION AND THOUGHT

1. How might we determine whether the characteristics of coexisting species (e.g., niche differences among competitors) are attributable to coevolution or to differential persistence and extinction?
2. Is there any evidence from the literature on introduced species (e.g., biological control agents) that they undergo rapid genetic changes in their interaction with native species?
3. The generation time of insects and pathogens is commonly far shorter than that of their hosts, so their potential rate of evolution may be greater. How is it, then, that they have not evolved such proficiency that they extinguish their hosts?
4. Some members of ancient groups of plants, such as cycads and the ginkgo, are host to very few species of insects, whereas some of the more recently evolved groups such as composites harbor a diverse insect fauna. What are the possible explanations of this pattern?
5. Discuss the likely similarities and differences in the course of coevolution between plants and insects, hosts and parasites, and prey and their vertebrate predators.
6. During the Tertiary, high-crowned teeth evolved in horses and several other groups of ungulates, apparently as an adaptation to the spread of grasslands. Discuss the evidence that would be required to show that there had been true coevolutionary responses between grasses and grazers (see Stebbins 1981, Coughenour 1985).
7. How can we measure the degree to which there has been convergence in the structure of communities that have developed independently in similar physical environments? What factors will influence the degree to which full convergence of community structure is attained?
8. It is often supposed that parasites will have a more benign effect on hosts with which they have long been associated than on new hosts. Discuss reasons for and against this view.
9. Discuss the conditions that would favor the evolution of specific (one-on-one) versus nonspecific (many-on-many) associations among species of plants and pollinators, plants and seed-dispersing animals, and hosts and parasites.

## MAJOR REFERENCES

Futuyma, D.J. and M. Slatkin (eds.). 1983. *Coevolution*. Sinauer Associates, Sunderland, MA. 555 pages. Essays by numerous authors on processes of coevolution and their operation in various interspecific interactions.

Thompson, J.N. 1982. *Interaction and coevolution*. Wiley, New York. 179 pages. A stimulating presentation of ideas and evidence on the evolution of interspecific interactions.

Nitecki, M.H. (ed.). 1983. *Coevolution*. University of Chicago Press, Chicago. 392 pages. Essays on a number of special topics in coevolution.

# Human Evolution and Social Issues

# Chapter Seventeen

*Homo sum: humani nil a me alienum puto.*
(I am human: nothing human is foreign to me).

Terence, about 1 B.C.

The mechanisms of coevolution, the nature of linkage disequilibrium, or the history of species diversity may be of absorbing interest to relatively few people, but the topic of human evolution evokes almost universal interest. This is the topic that lies at the heart of creationists' fulminations against evolution, the topic that for others may offer clues to the mystery of "the paragon of animals" and insights into the potentialities and limitations of "human nature." But we must bear in mind that while evolutionary biology may have something to say about the human condition, this is equally the province of anthropology and sociology, of psychology and history, and of philosophy, religion, and the arts. Humans are too complex to be understood from the narrow perspective of biology or of any other single way of knowing.

## THE PROBLEM OF OBJECTIVITY

The process of science, we may hope, approaches ever nearer to an objective description and understanding of its subject matter—objective both in the sense that one scientist's description or conclusion can be achieved independently by others, and in the sense that its conclusions are not dictated by emotion, desire, or *a priori* expectations. But individual scientists are often far from objective. To the extent that objectivity is achieved in science, it is commonly brought about by the discovery of error and the demolition of ideas that are defended passionately, even at times in the face of overwhelming evidence, by those who have held them.

Because of their social implications, human genetics and evolution are subjects highly charged with emotion, and much of the literature on these topics suffers from statements unsupported by evidence and from unspoken and often untested assumptions. This is equally true of other areas of science, but in human biology it can have especially dangerous consequences. Much of the history of scientists' ideas on human races, for example, has served to legitimize the racist beliefs of the societies in which those ideas held sway (see, for example, Gould 1981 for a partial history). H.H. Goddard (1920), a pioneer in mental testing, "discovered" by administering IQ tests to immigrants that 79 percent of the Italians, 83 percent of the Jews, and 87 percent of the Russians were "feeble-minded," and warned against the social consequences of immigration, because of "the fixed character of mental levels," a fixity for which there was not, and could not be, a shred of evidence. To the scientists of the nineteenth and early twentieth centuries, the record of history made it self-evident that the white race (to which they themselves belonged) is genetically superior to others. Freud and Jung merely voiced the conventional wisdom of their times when they developed psychological theories that assumed, without evidence, intrinsic differences between the sexes in aggressiveness, nurturance, and emotionality.

Unconscious assumptions affect the interpretation of data, but sometimes a scientist's predilections seem to affect the data themselves and the way in which they are obtained. Some psychologists who believed that homosexual behavior is "unnatural" and pathological "proved" that homosexuals were neurotic and maladjusted—by studying biased samples of homosexuals who underwent psychiatric treatment (see Churchill 1967, Tripp 1975). More controlled studies revealed no differences in mental health between homosexuals and heterosexuals (Hooker 1957, Saghir and Robbins 1973). The most comprehensive study claiming a strong genetic basis for variation in IQ, a study that was largely responsible for the British examination that irrevocably determined whether an 11-year-old child would prepare for a university or a technical school, appears to have included fraudulent data (Kamin 1974, Hearnshaw 1979).

Scientists who hold that there are biological differences between races or sexes are not necessarily, or even usually, racist or sexist. But scientific pronouncements about human genetics and behavior find a ready audience, some of which is ready to use these statements to justify oppressive political and social policies. The "scientific racism" of the Nazis is the most egregious and evil example, but there are dangers closer to home. A long paper by University of California psychologist Arthur Jensen (1969), claiming that most variation in intelligence is heritable, was enthusiastically read into the *Congressional Record.* Because of the high stakes, then, scientists should demand at least as much rigorous evidence for conclusions about human genetics and evolution as they do of research on ducks or *Drosophila.*

## THE PHYLOGENETIC POSITION OF THE HUMAN SPECIES

The order Primates is among the earliest eutherian (placental) orders of mammals to appear in the fossil record, in the late Cretaceous and Paleocene. The earliest primates were very similar to the other generalized eutherians, and most primates have retained a great many primitive eutherian traits, neither limbs nor teeth having undergone as much modification for specialized ways of life as in most other mammalian orders. The major distinguishing features of primates are those associated with a primarily arboreal way of life, including flexible digits and the development of vision as the primary sensory faculty. Especially in the "higher" primates, the optic lobes are large, and the eyes are large and directed forward. The separation of the eye sockets from the temporal fossa by a postorbital bar distinguishes most primates from most other mammals. Most of the "advanced" primates or Anthropoidea (Table I; Figure 1) have relatively large brains and foreshortened faces. They include the New World monkeys (Platyrrhini) and the Old World monkeys and apes (Catarrhini), which differ in nose structure, tooth formula, and a number of other characters, and apparently have been distinct since the Miocene. Within the Catarrhini, the Hominoidea (apes and humans) have a distinctive pattern of cusps on the molars; most are large in size and have some corresponding skeletal modifications. Modern humans differ from the other living species of Hominoidea in the structure of the vertebral column, pelvis, leg bones, and foot associated with bipedality; the more fully opposable thumb and other structural differences in the hand and arm; the short face and jaw and the position of the foramen magnum under, rather than at the rear of, the skull; differences in the teeth and in the parabolic shape of the tooth row, which in apes has parallel sides; and of course, the greatly enlarged brain and the capacity

for language and enormously more complicated behavior. Nevertheless, humans and apes have the same bones and muscles; the distinctions are differences in form.

Although our close relationship to the great apes (the orangutan, *Pongo pygmaeus*, of Borneo and Sumatra, and the chimpanzees—*Pan troglodytes* and *Pan*

FIGURE 1

**Representative primates. (A) A prosimian, the black-and-white lemur *Lemur variegatus* (Lemuridae). (B) A New World platyrrhine monkey, the mantled howler monkey *Alouatta villosa* (Cebidae). (C) An Old World catarrhine monkey, the colobus monkey *Colobus polykomos* (Cercopithecidae). (D) A member of the Hominoidea, the lar gibbon *Hylobates lar* (Hylobatidae). (E) The orangutan *Pongo pygmaeus* (Pongidae). (F) The gorilla *Gorilla gorilla* (Pongidae). (G) The common chimpanzee *Pan troglodytes* (Pongidae). (New York Zoological Society photos)**

*paniscus*—and gorilla, *Gorilla gorilla*, of Africa) has long been evident, the degree of our relationship has only been recently revealed. Humans and chimpanzees have identical amino acid sequences in a number of proteins; about half the alleles found at 44 proteins are electrophoretically indistinguishable between the species; and the thermal stability of heteroduplexes between chimpanzee and

TABLE I
**Classification of the living primates**

Order Primates
  Suborder Prosimii
    Infraorder Lemuriformes
      Families Lemuridae, Indriidae, Daubentoniidae (lemurs, indris, aye-aye)
    Infraorder Lorisiformes
      Family Lorisidae (lorises, galagos)
    Infraorder Tarsiiformes
      Family Tarsiidae (tarsiers)
  Suborder Anthropoidea
    Infraorder Platyrrhini
      Superfamily Ceboidea
        Families Callitrichidae, Cebidae (marmosets, New World monkeys)
    Infraorder Catarrhini
      Superfamily Cercopithecoidea
        Family Cercopithecidae (Old World monkeys)
      Superfamily Hominoidea
        Family Hylobatidae (gibbons, siamang)
        Family Pongidae (genera *Pongo*, orangutan; *Pan*, chimpanzees; *Gorilla*, gorilla)
        Family Hominidae (genus *Homo*)

Modified from Napier, J. R., and P. H. Napier. 1967. *A handbook of living primates*. Academic Press, London and New York. Tupaiids deleted as they are generally no longer included in Primates.

human DNA indicates that only about 1.1 percent of the base pairs differ between their genomes (King and Wilson 1975). The genetic distance (Chapter 4) between human and chimpanzee estimated from electrophoretic data is substantially less than that between some sibling species of *Drosophila* and rodents. A detailed analysis of the banding patterns of the chromosomes revealed that humans and chimpanzees differ in only nine pericentric inversions and in the amount and distribution of heterochromatin, which consists primarily of nontranscribed highly repeated sequences (Yunis et al. 1980). Assuming that molecular divergence occurs at a constant rate (Chapter 10), Wilson and Sarich (1969) estimated from the immunological distance between human and chimpanzee serum albumin that these species diverged from a common ancestor only four to five Myr ago, an estimate that is now generally accepted among anthropologists.

The most extensive data bearing on relationships among the living Hominoidea are macromolecular data, including amino acid sequences of hemoglobins and several other proteins (Goodman et al. 1983), the nucleotide sequence of part of the mitochondrial genome (Brown et al. 1982), and thermal stability of heteroduplexes of single-copy nuclear DNA of pairs of species (Sibley and Ahlquist 1984). (Some of this evidence is discussed in detail in Chapter 10.) All these lines of evidence indicate that the gibbons (Hylobatidae) are the sister group of the African apes plus humans (Figure 2). The relationships among chimpanzee, gorilla, and human are so close that many analyses cannot resolve the trichotomy (Figure 2A). For example, the $\alpha$ and $\beta$ hemoglobins of *Homo* and *Pan* are identical in amino acid sequence, and *Gorilla* differs from them by only a single amino acid in each chain. From a parsimony analysis of restriction enzyme data, Tem-

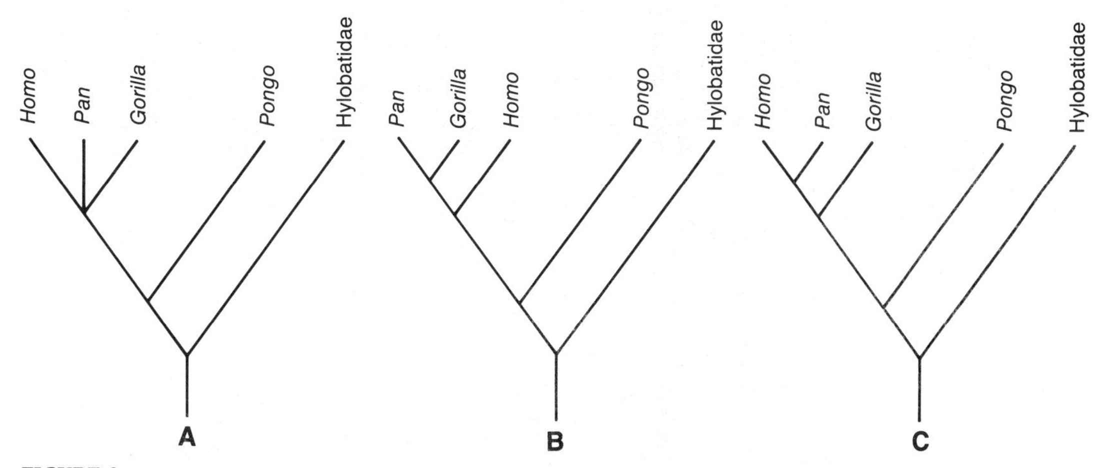


FIGURE 2

**Phylogenetic relationships in the Hominoidea. (A) Most studies at the molecular level have been unable to resolve the relationships among *Homo, Pan,* and *Gorilla.* (B) Several authors have concluded that *Homo* diverged before *Pan* and *Gorilla.* (C) Other analyses have concluded that *Gorilla* diverged before *Homo* and *Pan* diverged from each other.**

pleton (1983) concluded that *Homo* diverged before *Pan* and *Gorilla* diverged from each other (Figure 2B). Nei and Tajima (1985) criticized some aspects of Templeton's analysis on statistical grounds, and concluded from the same data that *Homo* and *Pan* are more closely related to each other than either is to *Gorilla* (Figure 2C). A detailed comparison of chromosome banding patterns also supports the close relationship of *Homo* and *Pan* (Yunis and Prakash 1982), as does the thermal stability analysis of "hybrid" DNA performed by Sibley and Ahlquist (1984; see Figure 3 and Chapter 10). If, as now seems likely, humans and chimpanzees are more closely related to each other than to any other living primate, it is probable that humans descended from an ancestor that, like the chimpanzee and gorilla, walked on its knuckles. Also, many morphological characteristics have evolved more rapidly in the human lineage than in either of the African apes. It is also clear that from a cladistic point of view, humans should be classified within the Pongidae rather than as a separate family.

## THE HOMINOID FOSSIL RECORD

The earliest fossils with dental features characteristic of the superfamily Hominoidea are *Propliopithecus* and *Aegyptopithecus,* monkey-sized Oligocene primates dated at 35–30 Myr B.P. These may be ancestral to the widespread Miocene dryopithecines (20–14 Myr B.P.), which were clearly apes, some of which attained large size. An early fossil that was once thought to bear on human origins is *Ramapithecus* (15 Myr B.P.), known from jaw fragments that were interpreted to have hominid rather than pongid characters. If *Ramapithecus* truly had apomorphic hominid characters, the divergence between modern apes and humans occurred more than ten million years earlier than the molecular similarity among the species implies. However, the characters of *Ramapithecus* appear to be plesio-

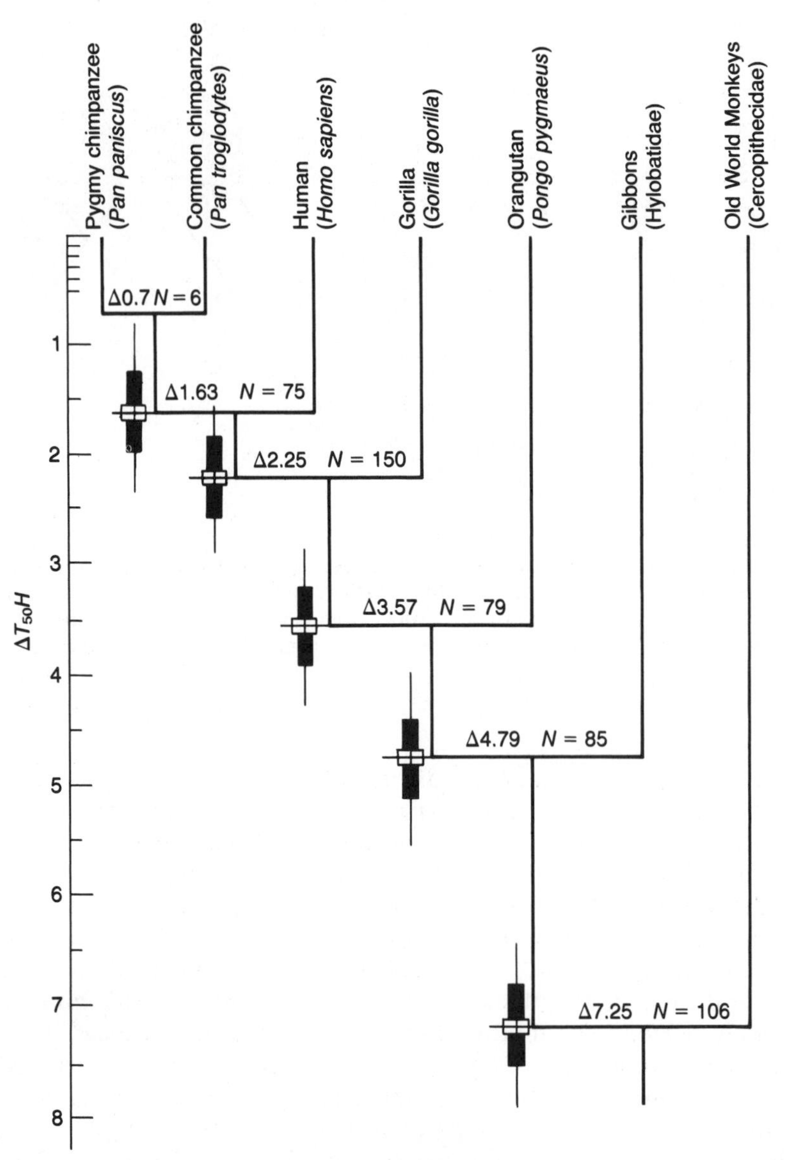


FIGURE 3

**The similarity between single-copy DNA of pairs of primates, based on the temperature ($\Delta T_{50}H$) required for separation of "hybrid" DNA duplexes formed by mixing DNA from both primates. At each level, the horizontal bar is the mean melting temperature, the vertical bar is the range of values observed, and the white and black rectangles represent standard errors and standard deviations. *N* is the number of comparisons. (Diagram, based on manuscript in preparation, courtesy of C.G. Sibley)**

morphic (primitive) within the Hominoidea, so that it almost surely does not mark the divergence between humans and apes.

Indubitably hominid fossils are few, and have been arranged by different authors into numerous phylogenetic trees (Figure 4). It is hard to interpret the literature on the subject, partly because almost every individual fossil has been

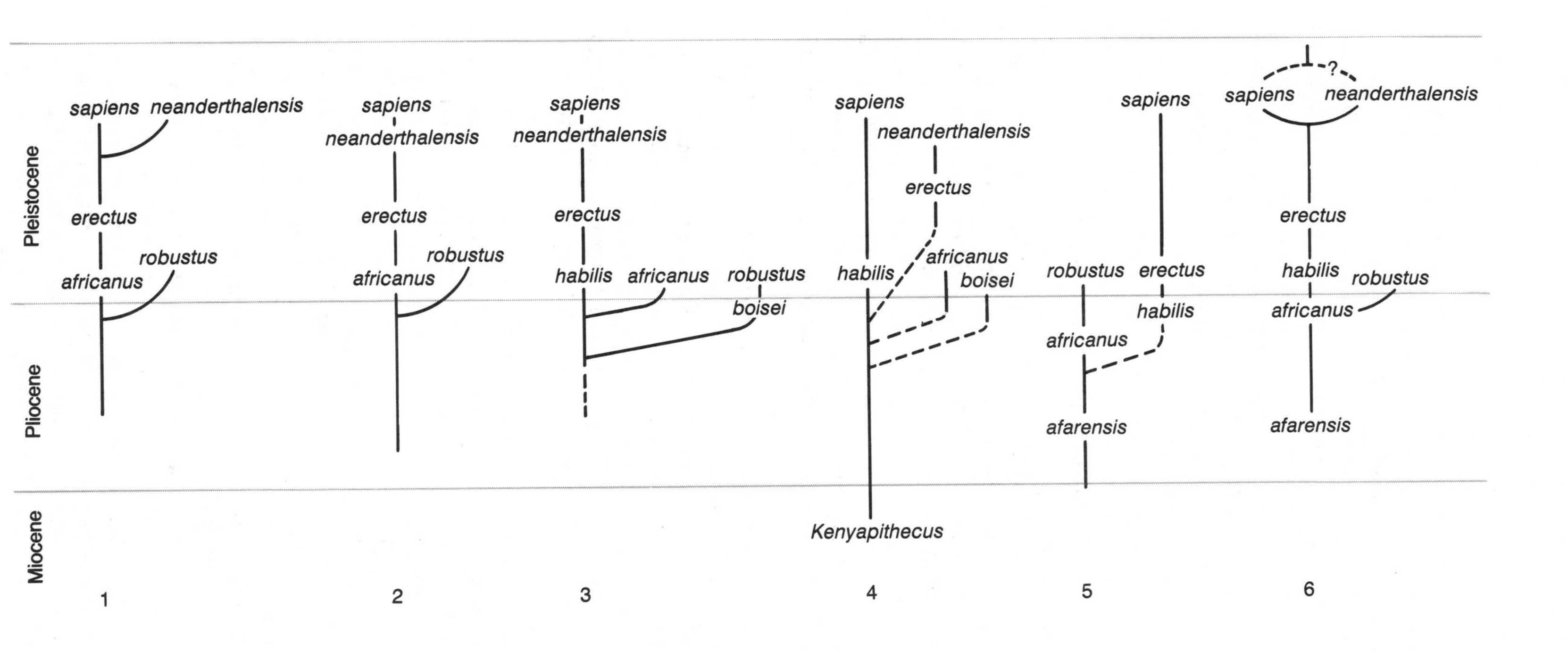


FIGURE 4

**Some of the many postulated phylogenetic relationships among hominids described from fossils. Diagrams 1–4 were published before the discovery of *afarensis*. Diagram 5, among the most recently published, differs from most treatments by recognizing *robustus* as a direct descendant of *africanus* rather than as a contemporaneous species. A representation that appears compatible with existing literature and which minimizes the number of biological species for which there is insufficient evidence, is offered as diagram 6, in which *sapiens* and *neanderthalensis* are to be interpreted as conspecific geographic populations that may have hybridized subsequent to divergence. Authors of the diagrams are: (1) LeGros Clark 1955; (2) Brace 1967; (3) Tobias 1967; (4) Louis Leakey 1971; (5) Johanson 1980. (After Campbell 1982)**

given a different specific or even generic name; fortunately, many of these names have been discarded as an appreciation of individual variation has developed. Another problem is that "species" has often been used typologically to indicate degree of difference, so different species names do not indicate whether the fossils are thought to represent contemporaneous reproductively isolated populations, successively modified members of the same lineage (chronospecies), or different geographic populations. More serious, substantive problems arise from the inability to date many of the fossils precisely, from their often fragmentary nature, and especially from the very small sample sizes. Sample sizes are usually insufficient to tell whether two fossils are individual variants of the same population or members of different biological populations, whether a feature changed gradually or intermittently, or even whether later fossils are direct descendants of an earlier population. Nevertheless, the broad outlines of human evolution are fairly clear, even if few of the details are.

The earliest hominid fossils, assigned to the genus *Australopithecus*, are in East African deposits estimated at 3.8–3.6 Myr B.P. These have some apelike features, such as curved finger bones, slightly projecting canines, and small skulls. Their most striking feature is that they were bipedal, as indicated by the structure of the pelvis, leg, and foot and by the remarkable discovery of 3.6 million-year-old footprints. Johanson and White (1979) named this form *Australopithecus afarensis*. Australopithecines from more recent African deposits are more or less distinguishable as "gracile" and "robust" forms; *afarensis* clearly falls into the gracile category.

Most authors argue that *afarensis* developed into the later gracile australopithecines of Africa (from about 2.75 Myr B.P. onward), which are generally named *A. africanus* (Figure 5). These have smaller canines and probably greater cranial capacity (mean about 494 cc). From about 2.0–1.0 Myr B.P., a larger form with a larger brain (mean 500 cc), generally named *A. robustus*, also existed; most authors agree that this was a separate biological species that became extinct without issue, because fossils assigned to *robustus* and to *erectus* (see below) have been found in the same stratum.

Gracile specimens with still larger crania (mean 656 cc), from about 1.8 to about 1.0 Myr B.P., have been called *Homo habilis*. Whether these are direct descendants of *africanus* or were contemporaneous with late *africanus* and hence a distinct biological species appears to be a matter of some controversy. A Kenyan skull (ER 1470; cranial capacity about 775 cc), the dating of which is uncertain but is probably 2.4-1.8 Myr B.P., may be transitional between *africanus* and *habilis*, but there is enough temporal overlap between *africanus*-like and *habilis*-like specimens for many anthropologists to consider them separate biological species.

Hominid fossils from about 1.5 Myr B.P. to about 200,000 B.P. are called *Homo erectus*, and have been found in Africa, southeastern Asia, and (from about 300,000 B.P. forward) in southern Europe. *Homo erectus* displays rapid evolutionary change (Wolpoff 1984) in the jaw, which became reduced, and in cranial capacity, which progressed from about 850 cc at 1.5 Myr to 900 cc at 0.8 Myr and 1000–1200 cc at 0.5 to 0.3 Myr (Lasker and Tyzzer 1982). Early specimens, especially, differ from modern humans in the shape of the skull and in certain other features such as the heavy brow ridges and lack of a chin, but many of the later specimens (from about 0.4–0.2 Myr) are so modern in aspect that they can be called *sapiens*.

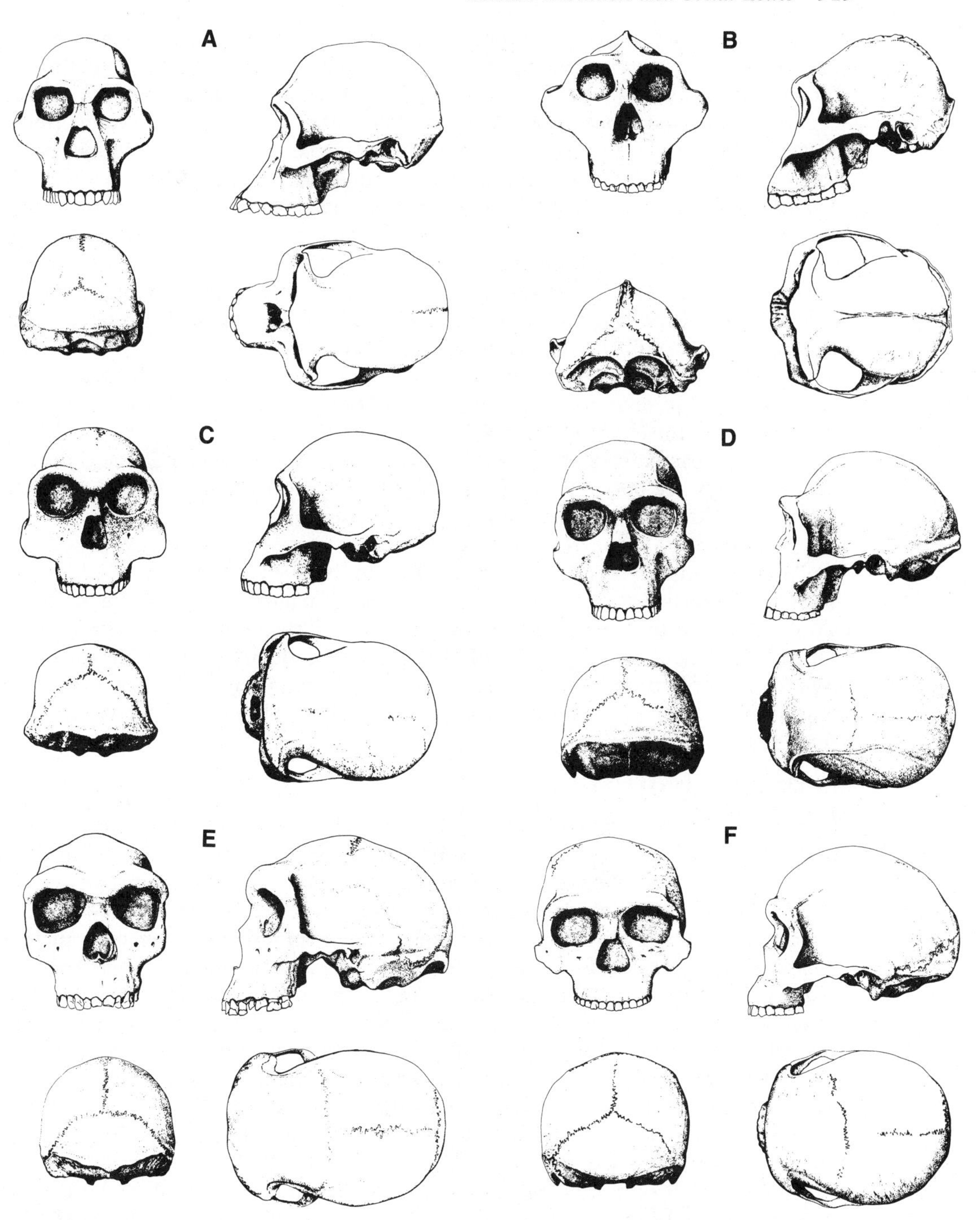


FIGURE 5

**Frontal, lateral, posterior, and superior views of the skulls of each of several fossil hominids. (A) *Australopithecus africanus*. (B) *Australopithecus robustus* (*A. boisei*). (C) *Homo habilis*. (D) *Homo erectus*. (E) *Homo sapiens* (specimen referred to subspecies *rhodensiensis* from upper Pleistocene of Zambia, about 110,000 years old). (F) *Homo sapiens neanderthalensis* (specimen from upper Pleistocene of Morocco, very approximately 47,000 years old). (From Howell 1978)**

From about 0.2 Myr on, hominid fossils are modern in character, and are referred to *Homo sapiens*. During this period, the brow ridges became smaller, the foramen magnum moved forward, the chin became evident, and the cranial capacity increased from about 1175 cc at 0.2 Myr to its modern value (mean 1400 cc). In western Europe from about 70,000–30,000 B.P., *H. sapiens* had heavy brow ridges and a stocky form, and is usually called *H. s. neanderthalensis*. The Neanderthals, however, had fully modern posture and large brains—perhaps even larger than the modern average. During the same period (from 40,000 B.P. and perhaps as far back as 60,000 B.P.), populations referred to as *H. s. sapiens* existed in the Near East. The disappearance of the Neanderthals is the subject of considerable speculation: they may have been extinguished by Pleistocene climatic changes or by conflict with invading populations from the east, or may simply have interbred with eastern populations so that their distinctive characters have been lost. It is quite plausible to suppose that the gene pool of modern Europeans includes contributions from the Neanderthal gene pool.

Human fossils from about 20,000 to 10,000 B.P. are widespread and numerous; by about 12,000 B.P., and possibly earlier, humans had spread over the Bering land bridge from Asia into the New World.

In summary, then, the fossil evidence indicates that the lineage leading to modern humans diverged from that leading to the apes at least 3.5 or 4 Myr ago, with the evolution of a bipedal posture preceding major changes in the hand and skull. Although the specimens are too few to provide a detailed picture of hominid phylogeny, a broad pattern of change is evident: there were almost surely two (or possibly more) contemporary species of australopithecines in Africa, the gracile form of which evolved fairly gradually, although not at a steady rate, into *Homo sapiens*.

## CULTURAL EVOLUTION

Cultural evolution consists of changes in behavior based not on changes in gene frequencies, but on learning. It may be either vertical (transmission from older to younger generations) or horizontal (as when we imitate the practices of our peers). Mathematical models of cultural evolution (e.g., Cavalli-Sforza and Feldman 1981) are only in the earliest stages of development. They may include genetic considerations, for example by postulating that certain genotypes are more likely to learn or adopt a cultural trait than others, or they may treat cultural traits as entities subject to their own rules of (nongenetic) inheritance and nongenetic selection, migration, and drift. Culture and genetics can clearly interact. For example, many Asian people are genetically incapable of digesting milk products as adults, and these products are traditionally not part of their diet—but whether the cultural practice of not drinking milk has influenced the gene frequency or vice versa is not known.

It is easy to draw analogies between biological and cultural evolution, but they should not be taken too far (Harris 1975). Cultural innovations, the analogues of mutations, are acted on by selective factors, in the sense that some become entrenched in the culture and others do not; they may become entrenched because of their perceived utility, because they are popularly associated with high status, or because they are forcibly imposed. At least in the evolution of languages a kind of drift seems to operate; seemingly random changes that lead, for example, to the formation of dialects. Cultural traits and languages diverge if geographically separated, and there exist many examples of cultural divergence, convergence, and parallelism among human societies. It is conceivable that group selection of cultural traits occurs. Many cultural peculiarities, for example, seem to maintain ecological balance between a tribe and its environment without conferring any special advantage on particular members of the group (Harris 1974).

Cultural evolution differs from biological evolution in important ways. Perhaps most important is that it is Lamarckian: behavior, language, or property that an individual acquires during his or her lifetime is transmitted to descendants or to other individuals. Consequently, cultural change can occur at far greater rates than biological evolution, and sweeping changes can occur within a single generation. Cultural evolution is far more reticulate than biological evolution, for societies adopt each others' habits, and it even entails a kind of blending inheritance: it is very hard to identify units of culture that are passed on without change, as genes are (consider the pronunciation of the Spanish name "Los Angeles" in English).

The selection guiding cultural change is usually selection of the traits themselves, not of the individuals practicing them; the automobile replaces the horse because of its perceived advantage, not because car drivers have more children than horse riders. The advantage may be illusory, of course, rather than real: many cultural traits, ranging from cigarette smoking to nuclear arms policies, act to the long-term, and often to the short-term, disadvantage of both individuals and groups.

Some cultural traits increase in frequency not merely because of cultural transmission (e.g., learning), but because they affect the growth and spread of populations. The adoption of agriculture, for example, enabled pastoral societies to achieve much higher densities than hunter-gatherer societies, leading to increase both in total population size and in the proportion of farmers (Ammerman and Cavalli-Sforza 1984). As a consequence, other cultural traits typical of farmers became prevalent, as did the frequency of alleles that the agricultural tribes happened to bear.

In genetic evolution a trait changes simply as the relative numbers of individuals with one or another genotype are altered, and the properties of a population are often the summation of the properties of the individuals that make it up; a herd of fleet deer is a fleet herd. But cultural events do not follow simply from numerical changes, and the behavior of a society cannot be understood merely as the consequence of its members' desires or actions. Wars do not occur because most of a society's members are belligerent, but because of economic and political forces that arise from the society's structure and govern its members' behavior. Social and economic forces interact with unique historical events and social attitudes to shape history and culture.

A rudimentary capacity for culture is found in many species of social animals; for example, great tits (*Parus major*) in England have acquired the habit, presumably by imitation of other tits, of pecking through the paper lids of milk bottles on people's doorsteps. Among the primates, cultural "traditions" may vary from troop to troop. Japanese macaques (*Macaca fuscata*), for instance, have developed a variety of cultural traditions that are spread by learning, such as separating wheat from sand by floating it on water. Chimpanzees likewise learn from their elders how to use twigs to extract termites from their nests. Several captive chimpanzees have demonstrated the ability to learn and use sign language, a use of symbolic language that had been considered uniquely human.

The evolution of human mental faculties from such rudiments was almost certainly gradual, as was the evolution of the size of the brain. Stone tools show a sporadic but consistent increase in variety and sophistication of design and manufacture from their first appearance, 2.5 Myr ago, onward. Possible evidence of fire 700,000 years ago has been found, and fire was certainly used extensively by 500,000 years ago. Perhaps giving evidence of religious or mythical belief, statuettes that may represent fertility symbols date back about 27,000 years; the exquisite Cro-Magnon cave paintings date from about 28,000 to 10,000 years ago; and ritualized burials were practiced at least 23,000 and perhaps 60,000 years ago. Agriculture, which began the human transformation of the face of the earth, is about 11,000 years old. There is, at least at present, no way of knowing which of these cultural advances were associated with genetic changes in the capacity for reason, imagination, and awareness; for all we know, hominid brains half a million years ago were the equal of ours, and all of cultural evolution since then has been the revelation of unending potentialities.

## THE PHYSICAL AND MENTAL EVOLUTION OF THE HUMAN SPECIES

Many authors have speculated on the forces of natural selection that led to the distinctive characteristics of humans. It is often difficult to test such adaptive hypotheses for any organism (Chapter 9), and is especially difficult when a species differs greatly from even its nearest relatives. Nor can we assume that all features are adaptive. Many human characteristics, for example, appear neotenic (Figure 6), and may be consequences of the delay in maturation that distinguishes humans from other primates. The delay in maturation may well be a consequence of selection for the great capacity for learning (selection for "intelligence") that is our overridingly important trait. A long prereproductive period of growth may have been advantageous if it was necessary to acquire mental skills. As mental skills improved in evolution, there may well have been selection for changes in morphological features such as the structure of the hand, as manual dexterity evolved to fashion the cultural implements that mental acuity made possible. As the capacity for culture increased, humans fashioned for themselves an ever more complex and variable social environment that surely imposed major new selection pressures. We may speculate, as many authors have, that cultural pressures may have been responsible for such features as the large penis (compared to apes), perhaps as a consequence of sexual selection, and the continuous sexual receptivity of the female (rather than being seasonal as in most other species), a possible adaptation for maintaining the pair bond. But there is little evidence for or against such hypotheses, and it is difficult to imagine ways of testing them. We do not even know why humans are the most nearly hairless of apes—although speculations abound (Morris 1967).

Why did intelligence evolve to such an extraordinary degree? Almost surely selection for intelligence is positively frequency dependent. That is, the more frequent the trait, the more advantageous it is. In such cases, evolution is a "runaway" process, as in the evolution of male secondary sexual characters by

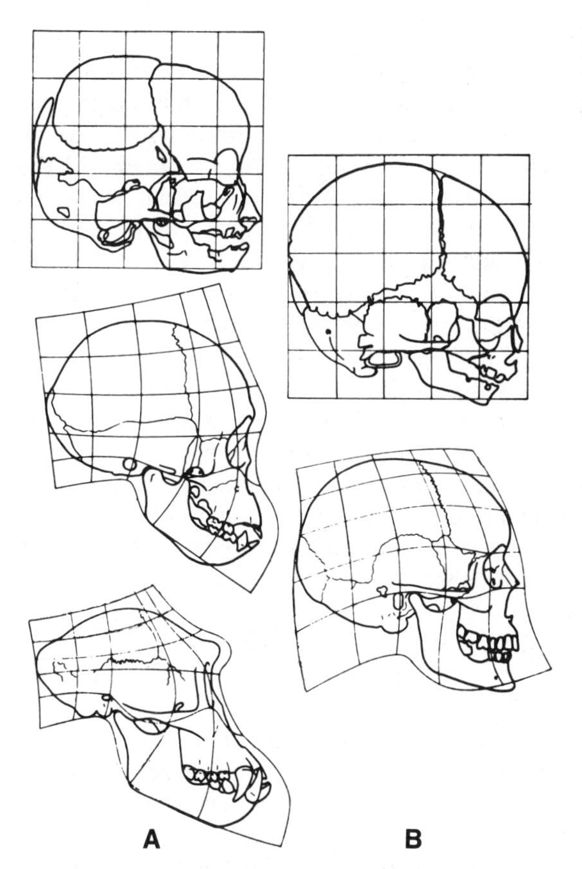


FIGURE 6
**Development of the skull of chimpanzee (A) and human (B), interpreted as human neoteny. The fetal skulls of the two species are very similar; as growth progresses, the chimpanzee departs further from the juvenile condition than the human does. Transformed coordinates mark corresponding points during ontogeny, as in Figure 20 of Chapter 14. (From Starck and Kummer 1962)**

female choice (Chapter 9). It has frequently been suggested that competition and aggression between early hominid groups was the motive force of selection for greater mental abilities. This may well have been so—there is no evidence one way or the other. But we need not invoke aggression, for it is equally easy to imagine more peaceful scenarios in which less intelligent individuals would have a disadvantage in survival or reproduction. Social status, and with it greater access to mates and resources, could well have been correlated with mental prowess in early hominids.

## GENETIC VARIATION WITHIN POPULATIONS

In most respects, *Homo sapiens* is an unexceptional species from the viewpoint of population genetics; its patterns of genetic variation are much like those of many other species. Our self-created cultural environment, however, has affected some aspects of genetic variation.

Electrophoretic surveys of polymorphism on proteins and other enzymes have revealed levels of heterozygosity of about the same magnitude as in many other outcrossing species; one survey of about 62 proteins gave an average within-population heterozygosity of about 0.135 (Nei and Roychoudhury 1982). By restriction enzyme analysis of mitochondrial DNA, 21 individual humans of diverse origin each proved to have a unique distribution of cleavage sites (Brown

1980). Human populations, like those of *Drosophila* and other outbreeding organisms, carry numerous deleterious recessive alleles, each of which typically has a very low frequency. Some deleterious alleles, however, are much more common. For example, color blindness, caused by alleles at two sex-linked loci, has an incidence of about 7 percent in most industrial societies. The frequency of the allele for sickle-cell hemoglobin, which causes severe anemia and is usually lethal in homozygous condition, is as high as 0.16 in some African populations, and about 0.05 among American blacks (Cavalli-Sforza and Bodmer 1971).

Because controlled genetic crosses cannot be performed between humans, it is often difficult to determine the mode of inheritance of a trait, or even if the variation is genetically based. Analysis of the transmission of a trait within families is often complex, and is most successful at identifying single-locus variants such as phenylketonuria (PKU), a recessive trait with several manifestations including mental retardation that is caused by deficiency of the enzyme that converts dietary phenylalanine to tyrosine. Determining the genetic basis of morphological and behavioral variations is usually more difficult than for biochemical traits. From studies of twins, various studies have calculated heritability values of 0.79–0.93 for height, 0.80–0.87 for arm length, 0.19–0.66 for hip circumference, and so on (Cavalli-Sforza and Bodmer 1971). As we shall see, however, the difficulties of estimating heritability for human traits are considerable.

### Genetic variation among populations

*Homo sapiens* is a single, cosmopolitan, biological species. There exist no biological isolating mechanisms among any human populations, although there frequently are cultural barriers to interbreeding. For example, there is assortative mating in Australia between people of Scottish and Irish origin. Assortative mating based on "racial" physical features is common, and in some societies has been enforced by law. Even in the face of social taboos against interbreeding, though, there is considerable gene flow among "races." During the period in which slavery was practiced in the United States, there was extensive gene flow from the white to the black population. A blood-group allele *Fy* that is moderately frequent in European populations but virtually absent in Africa was found to have a frequency of 0.11 in the black population of Detroit, Michigan, from which it was calculated that the admixture of genes from the white population has been 26 percent (Cavalli-Sforza and Bodmer 1971).

Physical characteristics such as skin color, hair texture, shape of the incisors, head shape, and stature vary geographically in humans, just as every other widespread species varies in some characters. These features have been used to define "races," the equivalent of the subspecies in other organisms. As for other species, the number of races recognized is arbitrary, depending only on the number of characters studied and the degree of difference used to make distinctions. Three major racial groups have long been recognized, often termed Caucasoid (Europe and western Asia), Negroid (Africa south of the Sahara), and Mongoloid (eastern Asian and native American populations). Many anthropologists recognize native Americans and the peoples of Australia and the Polynesian region as distinct races; moreover, each of these racial groups can be subdivided into an indefinite number of distinct populations. In Africa, for example, Congo pygmies are the shortest of humans, and Masai among the tallest. Many geographically variable

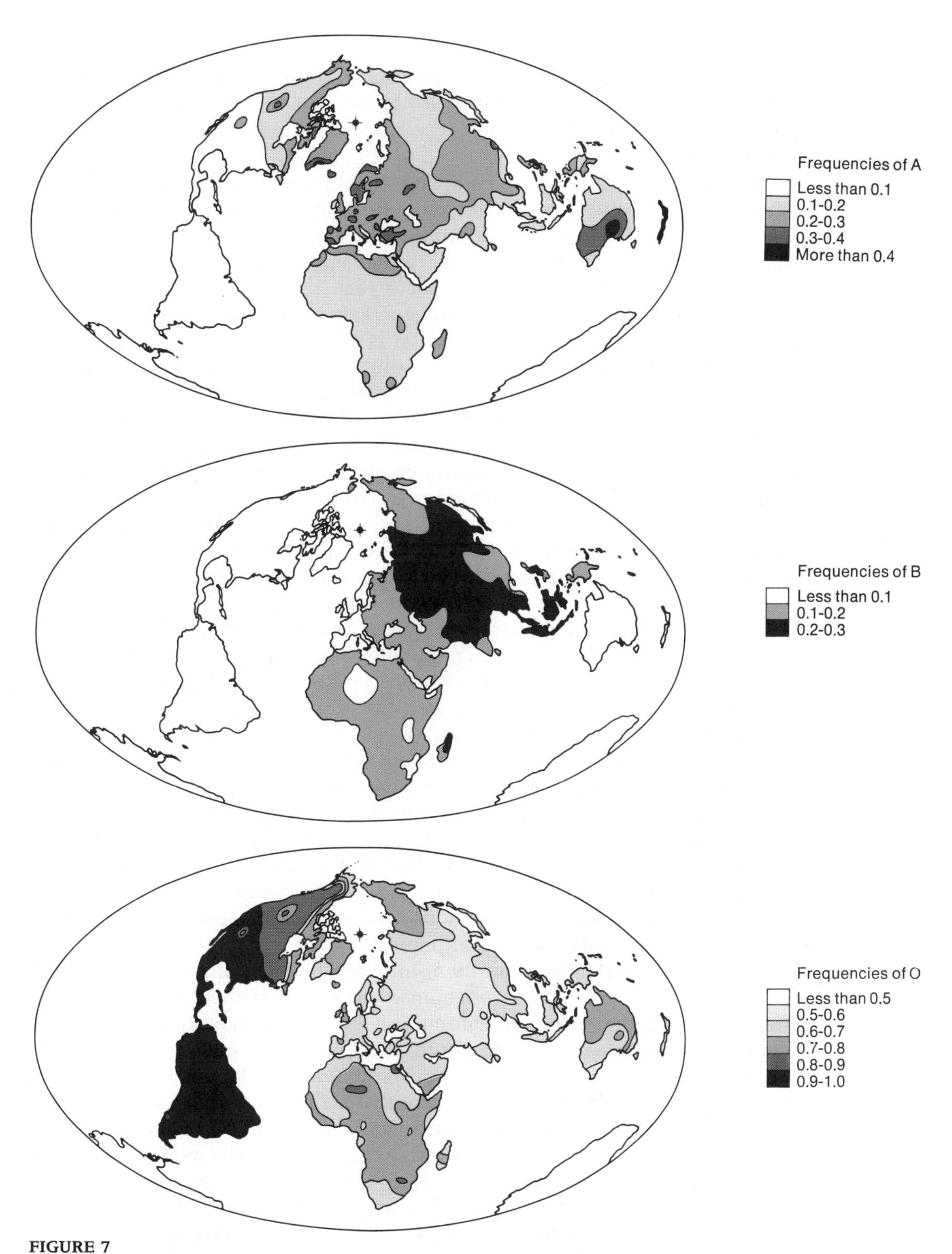


FIGURE 7

**Distribution of allele frequencies of the ABO blood group system. If races were defined by this character, they would have different distributions from those presently recognized. (From Harris 1975)**

characters do not coincide with the traditional racial divisions. For example, a classification based on gene frequencies at the ABO blood group locus would divide the world's populations into very different races from the traditional ones (Figure 7).

Obviously populations can be distinguished from each other, but overall, the degree of genetic divergence is very slight. Of the more than 150 protein and blood group loci that have been examined by electrophoresis and related techniques, 75 percent are monomorphic throughout the world's population, and among the variable loci, there is no locus that is fixed for different alleles in different "races." Partitioning the total genetic variation among and within populations, about 85 percent of the variation is among individuals within populations, about 8 percent is among tribes within "races," and only 7 to 9 percent is among the major "races" (Lewontin 1972, Nei and Roychoudhury 1982). The average "genetic distance" (Chapter 4) among major racial groups, based on 62 protein loci, ranges from 0.011 (Caucasoids versus Mongoloids) to only 0.029 (Mongoloids versus Negroids)—considerably less than the distance typically found between subspecies of other animals, which usually exceeds 0.05 (Nei and Roychoudhury 1982). As Lewontin et al. (1984) say, "If everyone on earth became extinct except for the Kikuyu of East Africa, about 85 percent of all human variability would still be present in the reconstituted species."

Based on variable protein and blood group loci, Nei and Roychoudhury (1982) have provided a phenogram (Figure 8) that describes the degree of similarity among various of the world's populations. Assuming that gene frequencies have diverged by genetic drift and that one unit of genetic distance accrues per $3.75 \times 10^6$ years (see Chapter 10), Nei and Roychoudhury estimate that the Negroid group diverged from Caucasoids and Mongoloids about 110,000 years ago, and that the Caucasoid and Mongoloid groups diverged about 41,000 years ago. These are very imprecise estimates; moreover, there has clearly been gene flow among groups, which will reduce the degree of divergence. In general, the genetic similarity of populations is related to their geographic propinquity, as expected if there has been gene flow. Morphologically similar peoples are not necessarily genetically most similar; for example, Philippine and Malay Negritos share many morphological features with Africans, but are genetically as distinct as many morphologically less similar populations.

Studies of genetic variation, then, indicate that even the most different of peoples diverged very recently in human history; that there is overall slight difference among the "races" except in those few obvious morphological features by which they are usually recognized; that many geographically variable features do not correspond to racial boundaries; and that unless there is evidence to the contrary, there is little reason to assume that traits other than skin color, hair form, and the like will vary substantially among "races." For example, there is no reason to expect mental abilities to vary among "races," any more than protein-encoding loci or the structure of the hand—even though white Europeans have assumed such differences, and for 200 years sought evidence of the moral and intellectual inferiority of the races they had typologically defined.

### Population structure

Until the advent of agriculture, humans subsisted by hunting and gathering, as some populations do still. From the study of such contemporary populations, it

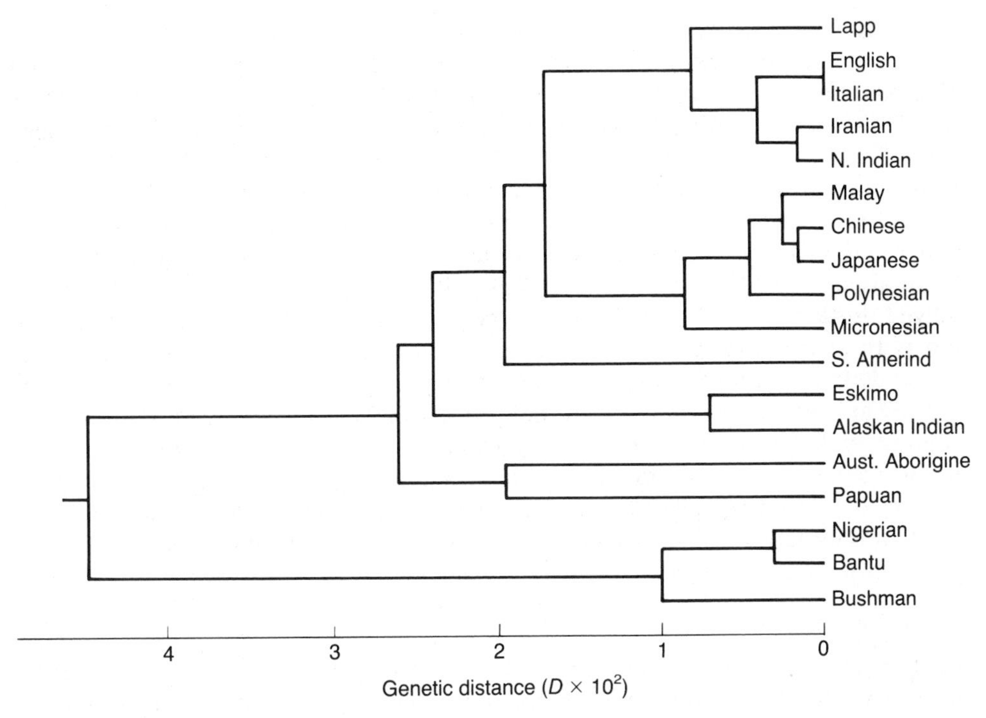


FIGURE 8

**A phenogram of some human populations based on genetic distances calculated for allele frequencies at loci coding for enzymes and blood groups. Although the branching sequence may describe in part the history of divergence, this diagram describes the degree of difference, not necessarily phylogenetic or genealogical relationships (see Chapter 10). (From Nei and Roychoudhury 1982)**

is likely that the population density was low, and that most populations consisted of rather small bands of nomads whose travels were quite localized. The total population of the world in 10,000 B.C. has been estimated at about ten million—not much more than the present population of New York City (Ammerman and Cavalli-Sforza 1984). A population structure of scattered small demes, confined by topographic barriers that we today can surmount in hours, is highly favorable to divergence by genetic drift, which doubtless accounts for much of the geographic variation of the human species. Surely some geographic variations, however, are a consequence of selection; for example, the length of arms and legs relative to body size is less in high-latitude than in tropical populations, as is expected if selection favored reduced loss of heat. Geographic variation in skin color is probably adaptive, but its advantage has not yet been identified with certainty.

Agricultural populations are larger than hunter-gatherer populations, chiefly because of a higher birth rate rather than a lower death rate (Ammerman and Cavalli-Sforza 1984). They are far more sedentary, however, so gene flow among such populations is often low, even today. Extreme inbreeding is generally prevented by marriage rules that often foster a certain amount of exchange among neighboring clans, but the breeding structure of human populations has been one of division into rather highly localized demes for most of human history. For example, gene frequency differences were discerned even among Indian villages along the shores of Lake Atitlán in Guatemala (Cavalli-Sforza and Bodmer 1971).

Throughout history, however, this pattern of local differentiation has been altered by migrations and conflicts. The genetic structure of North American Indian populations was altered forever when they were slaughtered and moved onto reservations by whites; the Atitlán villages studied in the 1960s have since suffered displacement and massacre under a series of brutal government regimes. From about 8000 B.C. to 5500 B.C., agricultural populations from the Near East moved northwestward over Europe at the rate of about 1 km per year, interbreeding with at least some of the indigenous hunting and gathering tribes as they proceeded. This movement has left its stamp on gene frequencies, which vary clinally at several loci in a pattern that reflects the historical spread of agriculture (Figure 9; Sokal and Menozzi 1982, Ammerman and Cavalli-Sforza 1984). In industrial societies of the modern world, the rate of gene flow among population centers is now probably higher than ever before in history, but much of the world's population still lives in traditional communities among which there is only limited exchange.

### Mutation and selection

Numerous deleterious alleles, of which many have been biochemically characterized, have been identified by students of human genetics. The average mutation rate per locus is about $10^{-5}$ per gamete, as in other organisms. Of every 1000 live births, about 12 to 13 have genetic disorders attributable to known loci (Crow and Denniston 1985). For many of these disorders, it is possible to determine if a parent is heterozygous, and so to estimate the likelihood that a prospective parent will have a defective child. GENETIC COUNSELING of prospective parents, to provide them with this information, is one of the chief applications of human genetics (see Cavalli-Sforza and Bodmer 1971).

The frequency of dominant and recessive deleterious alleles, many of which prevent reproduction or cause death before reproductive age, is reduced by natural selection. A few loci in human populations display heterozygous advantage. The best known of these is the polymorphism for sickle-cell hemoglobin, which, like thalassemia and several other hemoglobin variants, confers some resistance to malaria in heterozygous condition. Homozygotes, however, are anemic and seldom reach reproductive age. In populations that carry such alleles but are no longer exposed to malaria, the frequency of the alleles may be expected to decline, although at the unfortunate cost of the death of afflicted homozygous children.

The intensity of selection on a phenotypic trait depends on the age at which it is expressed. For example, the ABO blood group polymorphism is associated with several organic disorders: persons with type O blood, for instance, have a higher incidence of stomach ulcers than persons with other blood types, and those with type A blood have a higher incidence of stomach cancer. But because these disorders generally do not occur until after reproduction has ceased, they exert very little selection, and cannot explain the blood group polymorphism, the causes of which are unknown.

Very little is known about the operation of natural selection on physical characteristics other than rare, highly deleterious birth defects. Probably the best

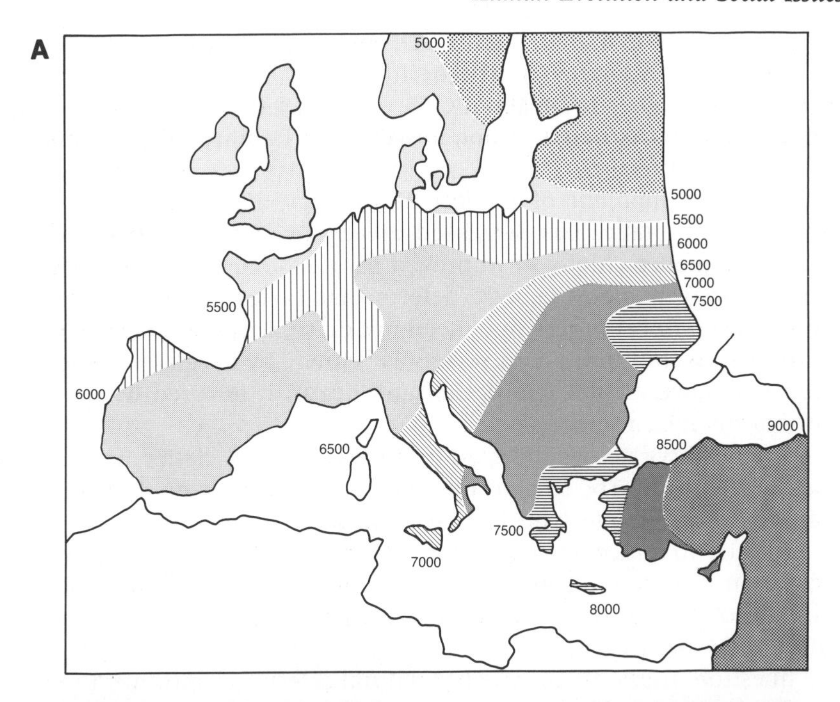


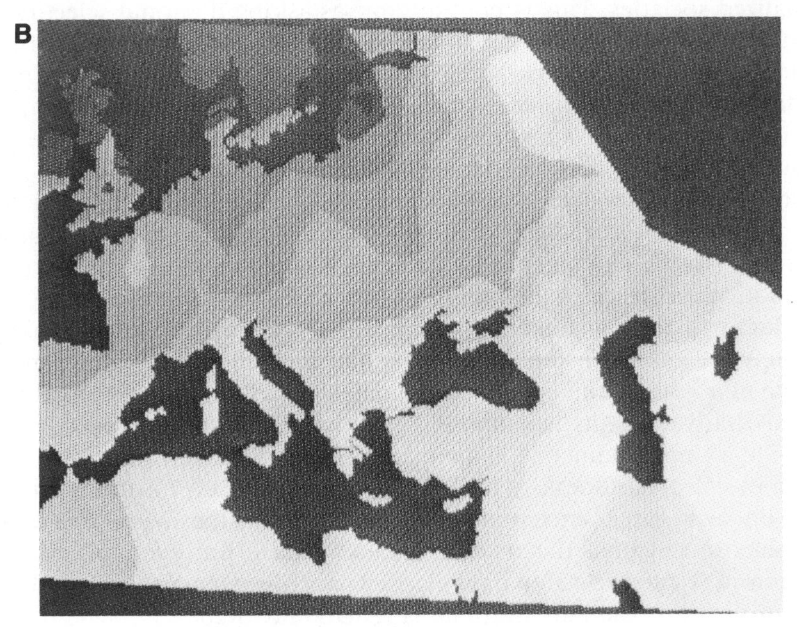


FIGURE 9

**The movement of agricultural peoples into Europe from the Near East has affected the distribution of gene frequencies. (A) The distribution of agriculture in Europe at 500-year intervals, based on radiocarbon ages of artifacts. Time is expressed in years before present. (B) A map of clinal variation in the genetic composition of modern European populations, based on the distribution of 39 alleles at 9 loci. The composite of the allele frequencies is represented as the first principal component of a multivariate statistical analysis. (After Ammerman and Cavalli-Sforza 1984)**

example of selection on a human metric character is stabilizing selection on birth weight (Figure 5 in Chapter 7). The heritability of birth weight has been estimated from twin studies to be about 0.63, so the selection may indeed be acting on genetic variation. As noted earlier, however, heritability values in humans are generally not very precise.

With the development of modern technology and education, the rate of prereproductive mortality has decreased greatly in most industrialized nations. This is a consequence chiefly of improved sanitation and, to a lesser degree, of medicine. Moreover, many genetic defects that in less technological societies reduced (or still reduce) fitness now need not do so. Myopia in a tribe of hunters may be disastrous, but in modern society is "cured" by eyeglasses. Phenylketonuria need not cause mental retardation and death; it is cured by prescribing a diet without phenylalanine.

As the mortality rate has dropped in industrialized societies, so has the birth rate, usually after a lag. The reasons for this are complex and not fully understood, but it appears that people who feel economically secure, and do not need numerous children to tend their farms and to care for them in old age, opt to have fewer children. Throughout much of the world, however, the birth rate is still very high, and the rapid growth of populations is perhaps the greatest crisis confronting our species.

The question arises, then, whether natural selection continues to operate in industrialized societies. This is not the same as asking if natural selection is still operating on the human species: much of the world is not industrialized, and continues to suffer from famine and infectious disease. If there is genetic variation in resistance to infectious diseases, it is certainly still subject to selection in such populations. Aside from a few genes that confer resistance to malaria, though, we know little about genetic variation in disease resistance.

If selection against deleterious alleles such as PKU is relaxed, as by medical cure, the alleles will increase in frequency. Recall (from Chapter 6) that a deleterious recessive allele changes in frequency at a rate of $\Delta q = u(1 - q) - s(1 - q)q^2$ per generation, where $u$ is the mutation rate and $s$ the selection coefficient. If $s$ were reduced to zero by technology, the allele frequency would increase at approximately the mutation rate. With an initial incidence of $1 \times 10^{-5}$ and a mutation rate of $u = 10^{-5}$, the incidence of a defect that was formerly lethal but now suffers no disadvantage in fitness would be only $1.73 \times 10^{-5}$ after 100 generations (Cavalli-Sforza and Bodmer 1971). An interesting possible example of such increase is in color blindness, which has an incidence of 7 percent in modern societies but less than 2 percent in some hunting and gathering societies. Perhaps natural selection against the trait has been relaxed, but even so, not enough generations of relaxed selection have elapsed to account for such a great difference in frequency, unless the mutation rate is remarkably high.

Should we be alarmed that selection against deleterious mutations has been relaxed? Note, first, that their rate of increase is very low. More importantly, these alleles are not deleterious in the new environment in which we now live. Unaided vision is an adaptation to an environment that the readers of this book no longer know; myopia and astigmatism are no longer detrimental. The same is true of curable genetic disorders like PKU. There is a social cost of an increased incidence of disorders that require medical treatment or other care; but the alternative is either the human cost of the afflictions of individual human beings, or the socially intolerable policy of regulating who may or may not reproduce on the basis of their genes.

Selection requires that there be variation in survival or reproduction. Crow (1958) introduced an index of the opportunity for selection, $I = V/\overline{w}^2$, where $V$ is the variance in fitness and $\overline{w}$ is mean fitness. This may be divided into two components corresponding to the variance in prereproductive survival ($I_m$) and the variance in fertility ($I_f$). These variances indicate only the opportunity for selection; whether or not any natural selection is actually occurring depends on whether or not either of the variances has a genetic component.

Modern changes in demography have altered the opportunity for selection strikingly. For example, the prereproductive mortality rate was higher in an Andean tribe of Chilean nomads than in an agricultural village, and higher in the village than in an industrial town; the opportunity for selection by mortality varied accordingly (Table II). However, even though the average number of children per family was highest in the nomads, their variance in fertility, and consequently the index $I_f$, was lowest. Overall, however, the least "modern" of the three populations had the highest index of total selection. In the United States, the index of mortality selection ($I_m$) has declined greatly since 1840. Even though mean family size has declined fairly steadily since then, the variance, and hence the index $I_f$, has fluctuated, and has been higher in the twentieth century than in the nineteenth (Crow 1966, Kirk 1968, Cavalli-Sforza and Bodmer 1971). Variance in fertility now accounts for over 90 percent of the total opportunity for selection. Consequently, most of the selection that might be occurring in the United States population is on genetically variable traits that are correlated with whether, and to what extent, people reproduce. It is not clear whether or not there are any such traits.

TABLE II

**Demographic differences and indices of opportunity for selection in three contemporary Chilean populations**

| | Industrial coastal town | Upland pastoral village | Highland nomadic shepherds |
|---|---|---|---|
| Mean number of children | 4.3 | 5.9 | 6.1 |
| Variance in number of children | 8.5 | 7.5 | 6.4 |
| Proportion surviving to adulthood | 0.87 | 0.75 | 0.42 |
| Selection index (mortality), $I_m$ | 0.15 | 0.33 | 1.38 |
| Selection index (fertility), $I_f$ | 0.45 | 0.22 | 0.17 |
| Index of total opportunity for selection, $I$ | 0.67 | 0.62 | 1.78 |

(From Crow 1966)

## EVOLUTION AND HUMAN BEHAVIOR

The cultural conditions of human societies have influenced and continue to influence our biological evolution. They affect population size, gene flow, and many of the agents of natural selection. We turn now to issues of the evolution of human behavior itself. Few questions have evoked as much interest and controversy as the problem of how our behavior may be constrained by our genes.

The arguments concern two distinct but related problems. In some cases the question is whether *differences* among individuals and groups in intelligence or other behavioral traits are based on genetic differences among them. In other cases the argument is whether a seemingly *invariant, universal* human trait is encoded in the genome; claims have been made, for example, that humans are aggressive—or altruistic—by nature. These arguments raise the problem of what "human nature" may be.

The concept that something is "natural" is a typological view; it is closely related to the notion that beings have "essences" to which they conform in greater or lesser degree. For example, it is commonly supposed that it is "natural" for humans to be aggressive or heterosexual, and "unnatural" to be pacific or homosexual. This is quite analogous to saying that it is "natural" for white oak trees (*Quercus alba*) to have a high, short-branched growth form; most white oaks, after all, have this shape. But that is because most oaks grow in forests; open-grown white oaks have a broadly spreading growth form. Is this unnatural? Clearly not. The genotype of a white oak determines a NORM OF REACTION: a variety of phenotypes that develop as environmental circumstances differ. No one of these phenotypes is any more "natural" than another. Similarly, unless we suppose that individual humans differ in aggressiveness or in sexual orientation because of genetic differences among them, the phenotypic variations in behavior are manifestations of the norm of reaction of the species-typical human genotype, and none can be termed more "natural" than another. "Human nature" includes whatever human beings do: it includes everything from coprophilia to the composition of symphonies. These behaviors are indeed highly "abnormal," if by that we simply mean that they are statistically rare. But there is no clear criterion for determining that they are "unnatural."

The behavioral norm of reaction of the human genotype is extraordinarily broad: because of our unparalleled capacity for learning and thought, and the unique history of experience that each of us has had, every human being experiences a different environment and the variation in our behavior is correspondingly diverse. Of course it is true that our behavior is genetically determined, in the trivial sense that the size and complexity of our nervous system is encoded in DNA, and in the equally trivial sense that our behavior would be very different if we were six inches tall, or lacked opposable thumbs, or were aquatic rather than terrestrial. The behavioral differences between humans and orangutans, including the differences in mental abilities, are a consequence of genetic differences. But it is equally true that our behavior is a consequence of environment: every phenotypic trait is determined by the response of the genotype to environmental conditions.

It is very difficult to define what is meant by those who say that humans are "naturally" or "genetically" aggressive (or peaceful, or whatever). Moreover, it is hard to imagine a way of testing this claim. The argument that our behavior, as a species, is genetically determined can only mean that it has limits—that the norm of reaction is so constrained by our genes that aggressiveness (or whatever) develops irrespective of environmental conditions, or at least develops under the majority of environmental conditions to which we might be exposed. But since there is almost no imaginable limit to the variety of cultural and social environments in which humans do or could develop, it is almost impossible to imagine how we could ever conclude that some conceivable form of human behavior (e.g., pacifism) does not lie within the behavioral norm of reaction. In fact, of course, human variation does embrace almost every imaginable behavior.

But is it possible that the variations among human beings are the consequence of genetic variation? Is the difference between Hitler and the Buddha a reflection not of different phenotypic manifestations of the same genetically determined norm of reaction, but of different genetic reaction norms? Indeed, this is conceivable, and this hypothesis is testable in principle, for the methodology of genetic analysis can be brought to bear on the question of whether *variation* is genetic or environmental in origin. Thus claims about the biological basis of "human nature" and about variations in human behavior are very different in meaning and in the methodology needed to evaluate them.

## TWO VIEWS OF HUMAN NATURE

The claim that the human species is "naturally" inclined toward certain kinds of behavior is usually supported in one of two ways. The traditional way (e.g., Ardrey 1966, Lorenz 1966, Tiger and Fox 1971) is to recall that humans developed rather recently from common ancestors with other primates, and to point out similarities between human behavior and that of other primates or more distantly related species. Territoriality is common among mammals, and so it is cited as the evolutionary foundation of our aggressive inclinations; heterosexual pair bonding and sexual reproduction are the norm among other species, so heterosexual behavior is said to be natural to humans, and homosexual behavior is not. That is, the traits are presumed to be genetically homologous between humans and other species.

There are several problems with this argument. First, almost any such behavior is so variable among humans and among nonhuman species that almost any correspondence between the two can be discerned. Some human tribes are highly warlike, and others are not; gibbons are highly territorial, but troops of chimpanzees are not, and exchange members readily. It is certainly possible to infer homology from the distributions of a trait among related species (Chapter 10), but most behavioral traits of humans have not yet been analyzed by this method. It is often difficult even with morphology, and is very much more difficult with behavior, to distinguish homologous from homoplasious characters (Chapter 10). It may be possible to determine if territorial defense in gibbons is genetically homologous to the behavior of urban street gangs or the conflict between nations, but the criteria have not yet been identified in this case. In contrast, although homosexual and heterosexual pair-bonding are both known in nonhuman species, they appear in different behavioral contexts from, and so are probably not homologous to, their manifestation in humans (Ford and Beach 1951). Moreover, we know that features that are ordinarily conservative during the evolution of a group often change very rapidly when one lineage in the group enters a new adaptive zone and is subject to very different selective pressures. Possession of two aortic arches is conservative in the reptiles—except for those that became birds and mammals. With the acquisition of reason, culture, and the most complex brain in the animal kingdom, humans have clearly entered a unique adaptive zone, and it is specifically in the realm of behavior that we should expect them to be least bound by the genetic constraints of their ancestors.

The second form of argument for genetic determination of "human nature" is the argument from adaptation. It has been the chief style of argument by some sociobiologists (e.g., E.O. Wilson 1975, Symons 1979, Barash 1982) and others who do not identify themselves by that name (e.g., Alexander 1979). Their arguments have drawn a great deal of criticism (e.g., Sahlins 1976, Sociobiology Study Group 1977, Lewontin et al. 1984, Kitcher 1985). Sociobiological argument consists chiefly of hypothesizing what behaviors should be adaptive, and finding congruence between predicted and observed behaviors. In an extreme example (which is not, however, typical of all sociobiologists) Barash (1982) speculated that the behavioral differences between men and women are due in large part to a genetically programmed difference in the phenotypic expression of the human genome. Barash draws on the argument that underlies sexual selection theory (see Chapter 9), that mating entails greater investment for the females than for males, so the cost (in Darwinian fitness) of a reproductive mistake is greater. Moreover, the female of a mated pair is certain that the offspring are hers, but the male is not; so the coefficient of relationship is greater, on average, between offspring and female than between offspring and male, and confers greater opportunity for kin selection to shape the evolution of parental care.

From these principles, Barash argues (see also Symons 1979) that women are likely to be more discriminating in mate choice than men, and will choose men, often older than themselves, who are likely to have more resources to provide for their offspring. Men are expected to compete with each other for the resources that make them sexually attractive, will feel threatened by successful, competent women, and will be more promiscuous than women and have a less highly developed urge to parental care. They will, instead, devote their energies to increasing their social status and sexual attractiveness. Thus, suggests Barash, "women have almost universally found themselves relegated to the nursery whereas men derive their greatest satisfaction from their jobs." The predictions that emerge from this evolutionary logic are similar to the sex roles with which we are too familiar, although Barash rightly points out that what may be biological is not necessarily ethical or immutable. The supposition that the sex role differences observed in modern society are biologically "natural" has long been used, however, as an argument for keeping women in the nursery and out of economic and political affairs.

An adaptive scenario of this kind may sound plausible, but it certainly does not have the force of evidence. It rests on numerous untested assumptions, and it is easy to make up adaptive scenarios that could predict entirely different sex roles. There is no evidence that men's reproductive success is enhanced by promiscuity, aggressiveness, or a negligent approach to child care, nor that women's fitness is enhanced by coyness, submissiveness, or monogamy. Males might maximize their fitness by being promiscuous and not maintaining a pair bond—or, if the survival of offspring depends on parental care, males might maximize fitness by being faithful to their mates and deriving from the pair bond the twofold advantage of helping their offspring and preventing insemination by other males. Choosing between these diametrically opposed sociobiological hypotheses would require knowledge of a whole complex of interacting factors that impinged on our ancestors, such as population density, risk of infant mortality, degree of kinship among competing males, and even the frequency of promiscuous versus nonpromiscuous males–none of which we know. In such a vacuum, it is easy to construct a sociobiological scenario that would "predict" any conceivable observation.

It is equally easy to provide an evolutionary rationale for the "environmentalist" view, which holds that except for a few behavioral traits such as suckling by infants, our behavior is canalized by genes only within wide, largely unknown limits. One might argue, for example, that once an elaborate social organization evolved, it became highly selectively advantageous to recognize other individuals, to respond to each of them differently in different social contexts, and to be able to imitate behavioral innovations that provided greater immunity to the vagaries of the environment. The more elaborate the social behavior becomes, the more there is to learn, and the less fit a behaviorally canalized genotype might be. It would be naive to imagine that the brain evolved piecemeal, with parts of the circuitry evolving while other parts, such as those affecting sexual behavior and parental behavior, remained unchanged. This would be as naive as to suppose that there is a separate gene for each of a fly's bristles, a notion that geneticists abandoned long ago. Thus the remodeling of the nervous system to provide flexibility may have extended to the centers that had controlled "instinctive" behaviors in other species. As Dobzhansky (in Dobzhansky et al. 1977) wrote, "Natural selection for educability and plasticity of behavior, rather than for genetically fixed egoism or altruism, has been the dominant directive factor in human evolution."

As evidence, environmentalists point to the vast amount of variation that humans display in every aspect of behavior, including those that biological determinists hold to be "natural." Aggression in its various forms may be common, but many individuals, and some whole tribes, do not fit this description, and there is no reason to believe that the differences are genetic—so how can "aggressiveness" be any more "natural" than its absence? The urge to hold and defend private property strikes us as "natural" because it is so familiar—but historically, there have been and continue to be many societies in which this is a foreign concept. Patriarchal societies, in which men are socially and economically dominant over women, are almost (but not quite) universal, but we do not know whether this owes to genetically encoded sex differences in mental and emotional qualities, or to cultural and economic factors, reinforced simply by the greater physical strength of men. Because in every society the social environment of males and females is drastically different from the moment of birth, the differences between men and women could be a matter of genes or entirely of culture—we cannot tell. Among newborns, slight sexual differences are evident in precision of digital movement, but almost nothing else (Fairweather 1976, cited in Lewontin et al. 1984); numerous other differences have been claimed, but the role of early conditioning cannot be ruled out (Birns 1976).

But the behavior of infants is almost irrelevant in any case. Strong cultural differences in the experience of men and women during development are a universal fact of life, and have manifest effects. To ask if boys and girls would develop identically if treated identically is to pose a question that has no bearing on reality. The only question is whether it is impossible, under some cultural conditions, for men to take on "feminine" roles such as parental care, or for women to take on "masculine" roles such as social and economic leadership. The empirical evidence clearly indicates that they can. Similarly, if there are genetic limits to our other social behaviors, they are very wide limits indeed.

## VARIATION IN BEHAVIORAL TRAITS

The existence of variation in human behavioral traits would not support the environmentalist view if the variation were largely genetic—possibly each of us is constrained to a particular part of the behavioral spectrum by our particular genotype. This raises two questions: Is the behavioral variation among individuals due to genetic differences to a substantial degree? And, if so, does this imply that our individual behavior is fixed by our genes?

Let us quickly dispose of the second question by its obvious answer: no. Each genotype has a norm of reaction, a variety of phenotypes expressed under the influence of different environments. Thus two genotypes that manifest different phenotypes in one environment may have the same phenotype or reversed phenotypes in another environment, as we know from genetic studies of numerous organisms. Suppose, moreover, we found that the heritability of a trait in some population was 0.90. This does not mean at all that the individuals in this population necessarily have different phenotypes fixed by their genes. It may simply mean that their environment is very homogeneous; place the same population in a more variable environment, and the trait's heritability may approach zero (Chapter 7). Moreover, the heritability could be very high, yet the mean phenotype could be changed enormously by altering the environment, as is clear in the simple experiment of fertilizing crop plants. For example, twin studies have suggested that the heritability of human height is 0.8 or more. In many industrial nations, mean height has increased rapidly in this century—at a rate of 3 cm/generation in Italy, for example. But this increase is almost entirely, if not entirely, a consequence of an altered environment. In a single generation, Italian-Swiss immigrants to the United States gained an average of 4 cm in height over nonimmigrants (Cavalli-Sforza and Bodmer 1971).

What is the evidence for substantial genetic variation in human behavioral traits? For the vast majority of personality traits, little evidence bears on the issue one way or the other. The problem is that even if a trait "runs in families," the variation can as well be attributable to environment as to genes, because members of a family share a common environment. One attempt to assess genetic factors is to compare monozygotic (MZ), so-called "identical" twins with dizygotic (DZ), or "fraternal" twins, on the theory that the concordance should be greater among MZ twins because of their identical genotype. However, MZ twins reared together are commonly treated more similarly than DZ twins, so a stronger correlation among MZ than DZ twins could be a consequence of this treatment, to an unknown degree. For this reason, attempts have been made to partition genetic from environmental variation in human behavioral traits by comparing MZ twins that have been reared apart in uncorrelated environments. Even this is not foolproof, because the similarity of MZ twins can be influenced by whether or not they shared a single placenta (Melnick et al. 1978), and especially because children are often placed by adoption agencies in homes like those from which they are taken, or are reared by relatives of their parents (Kamin 1974).

For these reasons, many estimates of the heritability of human behavioral traits are highly inconclusive. For example, Wilson (1978) remarks that "nowhere has the sanctification of premature biological hypotheses inflicted more pain than in the treatment of homosexuals," treatment that has been justified by the argument that heterosexual behavior is biologically "natural" and homosexual behavior "unnatural." As somewhere from 8 to 15 percent of the population is predominantly homosexual, the magnitude of the injustice is rather considerable in western society, although many other cultures have been tolerant or, as in the case of classical Hellenistic culture, actively approving of homosexual behavior. Wilson argues that homosexuality is largely genetically based and may have evolved by kin selection, whereby homosexuals helped to rear their relatives' children. So, he suggests, there is "a strong possibility that homosexuality is normal in a biological sense, that it is a distinct beneficent behavior that evolved as an important element of early human social organization." This may be true, but the data on which Wilson draws to support his view that differences in sexual orientation are genetically determined are inadequate for any conclusion; they constitute small samples of twins reared together rather than apart, and are flawed in several other ways as well (Futuyma and Risch 1984). Whether variation in sexual orientation has a genetic basis or not, anyone who holds humanitarian ideals will agree with Wilson that "it would be tragic to continue to discriminate against homosexuals on the basis of religious dogma supported by the unlikely assumption that they are biologically unnatural"—or on any other basis.

## VARIATION IN INTELLIGENCE

To a greater extent than in almost any other context, the "nature versus nurture" debate has centered around variations in intelligence, or, more properly, IQ ("intelligence quotient") score, since no one knows what intelligence is or how to define it other than by an IQ score. IQ tests are supposed to be "culture-free," but they have been strongly criticized as favoring white, middle-class individuals (see, for example, Kamin 1974, Gould 1981).

Few would deny that the components of intelligence (whatever it is) have a physical and biochemical basis in the brain, nor that certain genetic lesions (such as phenylketonuria) can impair intellectual development. The debate, rather, is whether the continuous variation in IQ also represents genetic variation. People with severe genetic lesions cannot be used as evidence of a genetic basis of IQ variation any more than a mutant plant without chlorophyll is evidence that other naturally occurring variations in chlorophyll content are genetically based. The wild plants may simply have been exposed to different environments (light intensity, trace elements such as magnesium) that affect chlorophyll synthesis.

The most recent major round in the argument about IQ was touched off by Jensen (1969, 1973), who concluded from other workers' data that the heritability ($V_G/V_P$; see Chapter 7) of IQ scores within European and American Caucasian populations is about 0.80. Jensen (1969) asked in his title, "How much can we boost IQ and scholastic achievement?" and concluded that if IQ is so highly heritable, it cannot easily be improved by education, so that programs of compensatory education are doomed to failure.

Even if the heritability of IQ were as high as 0.80, this conclusion would not be justified: we know of many genetic errors of metabolism that have high heritability yet can be treated by dietary or medical means (Lewontin 1975). In any case, the facts contradict the conclusion. In one study, adopted children had a mean IQ score of 117 while that of their biological mothers was only 86 (Skodak and Skeels 1949); in another, children who remained in their homes had an average IQ score of 107, those adopted into different homes an average of 116, and those who were returned to their biological mothers after a period of adoption, only 101 (Tizard 1973). The mean IQ of black children adopted by white families in Minnesota is equal to the national average, about 15 points higher than the mean for the black population (Scarr and Weinberg 1976).

Moreover, there are good reasons to doubt that the heritability of IQ is as great as Jensen claims (Kamin 1974, Gould 1981, Lewontin et al. 1984). The data most crucial to Jensen's argument were several studies of twins reared apart; but as Kamin (1974) showed, all but one of these studies entailed very small samples and in all the studies, there is abundant reason to believe that the twins were reared in correlated environments—often one twin was reared by relatives, and the twins commonly knew each other and sat together at school. The only study that reported a large sample, that of Cyril Burt, has become a famous scientific scandal, entailing fraudulent data and "coauthorship" with nonexistent coauthors (Kamin 1974, Hearnshaw 1979).

The best data on the heritability of IQ have come from recent studies that correlate the IQ of adoptive parents with their biological children and with adopted children in the same household, the theory being that the correlation between parent and biological child should be due to both genes and environment, while that between parent and adopted child should be due to environment alone. In two such studies (Scarr and Weinberg 1976, Horn et al. 1979), the IQ of the mother was equally correlated with that of her adopted and her biological child. The correlation of the fathers' IQ with the biological children's was higher, although not statistically significantly so, than with the adopted children's. Within these families, the correlation in IQ between pairs of biological siblings was not significantly higher than between biological and adopted children of the same parents. Thus even though it would be surprising if human populations contained no genetic variation in characteristics that affect performance on IQ tests, the evidence that genetic differences account for much of the phenotypic variation is inconclusive at best. There is certainly no evidence that genetic factors could prevent society from boosting IQ and scholastic achievement; the evidence on this point is clearly to the contrary.

The most controversial of Jensen's conclusions was that the mean differences among groups of different socioeconomic standing and among different "races" are largely genetic in origin. The mean IQ score of American blacks is about 15 points (one standard deviation) below that of whites, but there is great overlap between the groups, so a substantial fraction of blacks has higher IQ scores than the white mean. Jensen, having concluded that IQ within the white population is highly heritable, argued that the racial difference was also likely to be genetically based in large part.

This conclusion, however, has two major weaknesses aside from the ambiguity about the heritability within the white population. Many psychologists hold that IQ tests are biased in favor of whites (see also Gould 1981). The other fault, which any geneticist will recognize, is that even if a trait is 100 percent heritable within a population, the difference between two populations may be entirely due to differences in their environment—and the environment of most blacks in the United States is clearly very different from that of most whites. In point of fact, whites and blacks do not differ in IQ when they are carefully matched for such variables as family size, medical care, and other social variables (Sanday 1972). Tizard (1973) found no differences in IQ between black and white children who had spent six months in the enriching environment of a residential nursery. As noted earlier, the IQ of black children adopted by white families in Minnesota did not differ from that of the families' biological children (Scarr and Weinberg 1976). Finally, studies of individuals of mixed ancestry, using blood group profiles as an indicator of genetic background, provide no evidence of a correlation between IQ and racial background (Loehlin et al. 1975, Lewontin 1976). As far as genetic differences among socioeconomic classes are concerned, adoption studies such as those cited previously invariably show that IQ is best predicted by the socioeconomic status of the family in which children are raised from an early age, rather than that into which they are born (Lewontin et al. 1984).

Occasionally the alarm has been raised that intelligence is undergoing a genetic decline. People of low socioeconomic status, it is said, have lower IQ, and have more children than do people with higher IQ. Consequently, genes for low IQ are on the increase. But the few data that exist do not support this argument. There is no evidence that IQ differences among social classes is genetically determined; rather, deprived environments are reflected in low IQ scores. Moreover, there is no clear correlation between IQ and reproductive rate. Bajema (1963) calculated the rate of natural increase $r$ (Chapter 2), which defines fitness, for a sample of Michigan residents that was divided into five IQ classes (Table III). Fertility was bimodally distributed, being highest for the group in the range 80–94 and in the group with IQ greater than 120. The group with the lowest IQ ($< 80$) had a very low fertility, because many apparently did not reproduce.

## EVOLUTION AND SOCIETY

Evolution, one of the fundamental discoveries and concepts in modern thought, is central to modern biology and to the use of biology in modern society. Without it, genetics, physiology, ecology, and every other aspect of biology would lack coherence; numerous practical applications of biology would be purely empirical, and would have only a weak theoretical foundation, if any. From a philosophical point of view, surely little can be more satisfying than to have attained an understanding of our origin and that of other living beings, and we may well agree with Darwin that "there is grandeur in this view of life," in which "from so simple a beginning endless forms most beautiful and most wonderful have been, and are being evolved."

Knowledge, it is often said, is power, and power can be used to the good or to the detriment of society. It surely ought to be good, for example, to learn how to identify genes that make certain people susceptible to industrial toxins, so that

TABLE III
**Reproduction and relative fitness in relation to IQ in Michigan**[a]

| IQ range | Sample size | Percentage leaving no offspring | Average number of offspring per individual | Per capita rate of increase ($r$) | Average generation time ($T$) in years | Relative fitness calculated by number of offspring | Relative fitness calculated as $e^{rT}$ |
|---|---|---|---|---|---|---|---|
| ≥120 | 82 | 13.41 | 2.598 | +0.008885 | 29.42 | 1.0000 | 1.0000 |
| 105–119 | 282 | 17.02 | 2.238 | +0.003890 | 28.86 | 0.8614 | 0.8674 |
| 95–104 | 318 | 22.01 | 2.019 | +0.000332 | 28.41 | 0.7771 | 0.7838 |
| 80–94 | 267 | 22.47 | 2.464 | +0.007454 | 28.01 | 0.9484 | 0.9600 |
| 69–79 | 30 | 30.00 | 1.500 | −0.010001 | 28.76 | 0.5774 | 0.5839 |

(From Bajema 1963)
[a]Individuals (non-immigrant whites) were scored for IQ at mean age 11.6 years in 1916–1917 in Kalamazoo, Michigan; survival and reproduction through age 45 were recorded. Per capita rate of increase $r$ calculated from the expression $\sum_x l_x m_x e^{-rx} = 1$ (see Chapter 2); average generation length is $T = [\ln(\Sigma l_x m_x)]/r$. The relative fitness calculated from $e^{rT}$ expresses the rate of population growth of a subclass relative to that of the ≥120 class, taking into account the generation time of the subclass compared to that of the population as a whole.

they might be advised of the risks of some occupations. But this knowledge, which is growing rapidly, bears with it the risk of discrimination, if employers use such genetic screening to bar members of high-risk groups from employment. Scientific knowledge, then, bears implications for social policy. Scientists do not set social policy, but they can and perhaps should bear responsibility for alerting the public to the abuses to which their discoveries might be put.

It is even more clearly the duty of scientists to protest the unwarranted conclusions that others may draw from the ideas and data of science. Just as in prescientific Europe there were those who cited the Bible to justify crusades, inquisitions, and witch hunts, so the ideas of science have been pressed into the service of social inequity. From its earliest days (Hofstadter 1955), evolution was misappropriated by social Darwinists to justify racism and imperialist domination, to exclude women from political and economic power on the grounds of their supposed genetic inferiority, and to ascribe poverty, illiteracy, and crime not to the social conditions that exclude large parts of society from access to wealth and learning, but to genetic inferiority in intellectual and moral qualities. At no time has there been evidence of a genetic basis for the qualities that supposedly justified discrimination. Moreover, at no time has there existed a scientific justification for the notion that heredity is destiny. Nor has there ever been a scientific justification for the naturalistic fallacy that has pervaded the social Darwinist view: the theme that because natural selection, the "survival of the fittest," is the law of nature, so it is right and proper that it be the law of society.

Darwin's great defender Thomas Henry Huxley (1893) said, "Cosmic evolution may teach us how the good and the evil tendencies of man may have come about but, in itself, it is incompetent to furnish any better reason why what we call good is preferable to what we call evil than what we had before." Moreover, "the influence of the cosmic process [natural selection] on the evolution of society is the greater the more rudimentary its civilization. Social progress means a checking of the cosmic process at every step and the substitution for it of another, which may be called the ethical process; the end of which is the survival not of those who may happen to be the fittest, in respect of the whole of the conditions which obtain, but of those who are ethically the best." That is, evolution provides no philosophical basis for esthetics or ethics; in itself, it contains neither morality nor immorality; it bears no moral force or obligation, and should not be used to rationalize violations of the ethical codes that we, as sentient, empathic beings who can act as if we possess free will, decide on. Evolutionary biology should not constrain our ethics, demean the vision of our poets, philosophers, and spiritual leaders, or restrain us from the ideals to which we aspire.

To the contrary, evolutionary biology, like other knowledge, can serve the cause of human freedom and dignity. Together with genetics and ecology, its applications in medicine, food production, and environmental management can help to liberate us from disease and hunger. As we learn human genetics, we come to appreciate still more the unity of humankind. As we extend scientific explanation into the realms of human biology, we gain confidence—or are terrified—by the realization that our fate as a species lies in our own knowledge and compassion, not in the whims of the unknowable supernatural. As we think with humility of our place in biological history, as we reflect on our common origin with other living things, we may even come to feel at one with, and to care for, those endless forms most beautiful and most wonderful.

## SUMMARY

Anatomical, macromolecular, and chromosomal evidence indicates that humans are most closely related to the African apes, having diverged from a common ancestor probably 4–8 Myr ago. Although the detailed relationships among fossil hominids are in dispute, the fossil record indicates that hominids arose in Africa, that the evolution of bipedality preceded major changes in the hand or skull, and that cranial capacity increased monotonically, if not steadily, over the last 3 Myr. A considerable number of fossils cannot be assigned clearly to one or another of the major named forms, because of the gradual nature of hominid evolution.

Genetic variation within and among human populations is much like that in other geographically widespread species. Geographic variation in some physical characteristics is probably the consequence of natural selection, but the adaptive significance of most geographically variable traits is not understood. Some geographic variation is almost surely a consequence of genetic drift, operating in the rather small, localized populations into which the species has been divided for much of its history. Overall, the degree of genetic differentiation among groups that have been termed races is less than is observed among subspecies of many other species. There is no evidence that human populations differ genetically in intelligence or other behavioral traits. Instances of balanced polymorphism and selection against deleterious alleles are known. In industrialized societies, most of the opportunity for selection is provided by variation in reproduction rather than mortality, but whether or not selection is actually acting on any traits via differential reproduction is not known.

Whether or not evolutionary biology is important for understanding human behavior is highly controversial. Although many human behavioral traits can be seen in rudimentary form in other primates, the dominant factor in recent human evolution has been the evolution of enormous behavioral flexibility, and the ability to learn and transmit culture. Although some biologists and anthropologists view supposedly universal human traits as genetically canalized adaptations inherited from early hominid ancestors, others see these same features as highly variable cultural and learned responses, and point to rapid historical changes and cultural variation among peoples to support their view. There is little unequivocal evidence that much of the variation in human behavioral qualities is genetically based, and no evidence that behavioral qualities are genetically constrained in any meaningful way.

## FOR DISCUSSION AND THOUGHT

1. It is often claimed that if we could arrive at a deep understanding of the evolution of human behavior, we would be in a better position to shape our social institutions to our benefit. Take an example of socially important behavior and discuss exactly how social institutions might profit from such understanding.
2. Suppose it should turn out that men are indeed biologically more prone to aggressive, dominant behavior than women, or that sexual orientation is in part genetically based. Should any social policies be affected by such discoveries? Exactly what would they be, and why?
3. Many excellent treatments of human genetics (e.g., Dobzhansky 1962, Cavalli-Sforza and Bodmer 1971, Lerner 1968) accept the view that the heritability of IQ is fairly high. After reading one or more of these treatments, discuss the differences in the conclusions reached by those authors and this book.
4. To deny that human behavior is genetically controlled seems to many people to imply that our behavior is determined entirely by conditional reflexes, as argued by the behavioral psychologist B.F. Skinner. Is there a middle course between the Scylla of biological determinism and the Charybdis of environmental determinism?
5. In arguing against sociobiological interpretations of human behavior, I have made little reference to the innumerable parallels that may be drawn between the behavior of humans and that of many other species, nor to the limbic system, a part of the brain that is structurally homologous throughout the vertebrates and which appears to be the seat of many instincts in other vertebrates. How can I cavalierly ignore such evidence?
6. Much of the sociobiological theory applied to humans is based on the theory of what form of behavior would be adaptive and perhaps optimal. Evaluate this theoretical approach from the standpoint of evolutionary theory in general.
7. Lewontin et al. (1984) argue that most biological determinism is deeply flawed because it takes a reductionist approach, explaining the structure of societies in terms of genetically determined behavior of their constituent individuals. Discuss the causal relationships between individual behavior and social structure, and evaluate their complaint.
8. For each of the following observations about Western culture, think of an evolutionary and of a nonevolutionary (cultural) explanation, and judge their relative merits. (a) Young people are often more adventurous and less conservative than older people. (b) Most people would rather rear their own children than adopt. (c) Incest is considered immoral. (d) Many people accept religious or political doctrines with little question. (e) Children (reputedly) don't like spinach.
9. A population exercise in science fiction is to imagine the physical and mental changes in the human species that might evolve in the next few thousand years. Assuming

that the technology of our society continues to increase, and is not obliterated by a nuclear catastrophe, what do you predict on the basis of information in this chapter?

10. Read Shelley's poem *Ozymandias*, reflect on what has happened in the last few million years of evolutionary history and the last few thousand years of human history, and discuss the likely fate over the next 5000 years of such institutions as democracy, technology, the United States, and the Soviet Union.

## MAJOR REFERENCES

For physical anthropology and the hominid fossil record, recent textbooks should be consulted, for the interpretation of the fossil record changes rapidly. Among the more recent treatments are:

Lasker, G.W., and R.N. Tyzzer. 1982. *Physical Anthropology*. Third edition. Holt, Rinehart and Winston, New York.

Campbell, B.G. 1982. *Humankind emerging*. Third edition. Little, Brown and Company, Boston.

The anthropological aspects of variation within and among races are treated by:

Molnar, S. 1983. *Human variation: Races, types, and ethnic groups*. Prentice-Hall, Englewood Cliffs, N.J.

The most comprehensive book on human population genetics is:

Cavalli-Sforza, L.L., and W.F. Bodmer. 1971. *The genetics of human populations*. Freeman, San Francisco.

Major treatments of sociobiology and the role of genetics in human behavior include:

Wilson, E.O. 1978. *On human nature*. Harvard University Press, Cambridge, MA. A non-technical argument in favor of the application of sociobiological theory to humans.

Lewontin, R.C., S. Rose, and L.J. Kamin. 1984. *Not in our genes: Biology, ideology, and human nature*. Pantheon, New York. Although the style and rhetoric are distracting, the content of this attack on sociobiology and biological determinism is generally excellent.

Kitcher, P. 1985. *Vaulting ambition: The quest for human nature*. MIT Press, Cambridge, MA. A very detailed, carefully reasoned, and usually gentle critique of sociobiology by a philosopher of science.

Gould, S.J. 1981. *The mismeasure of man*. W.W. Norton, New York. A history of modern mental testing and its antecedents.

# MEANS, VARIANCES, AND CORRELATIONS

This appendix introduces the statistical concepts and notation used in the text for readers to whom such material is unfamiliar. More extensive treatments are found in introductory books on statistics; *Quantitative Zoology* by Simpson, Roe, and Lewontin (1960) is an excellent introductory treatment that stresses biological applications.

## THE ARITHMETIC MEAN

Let $X$ be the value of some measured variable, for example the length of a snake's tail. $X_i$ is the value for the $i$th observation (snake 3 might have the value $X_3 = 10$ cm). If there are $n$ observations, the sum of the $n$ tail lengths is denoted $\sum_{i=1}^{n} X_i$ (or simply $\Sigma X_i$). The ARITHMETIC MEAN (commonly known as the AVERAGE) is then

$$\bar{x} = \frac{\Sigma X_i}{n}$$

(Other kinds of means can be calculated and are used in evolutionary theory, but I have used them very little in this book.) If $X$ is a discrete variable such as the number of scales, there may be $n_1$ individuals with $X_1$ scales, $n_2$ with $X_2$, and so on, where $n_1 + n_2 + \ldots + n_k = n$. Then the arithmetic mean is

$$\bar{x} = \frac{n_1X_1 + n_2X_2 + \ldots + n_kX_k}{n_1 + n_2 + \ldots + n_k}$$

$$= \frac{n_1X_1 + n_2X_2 + \ldots + n_kX_k}{n}$$

$$= \frac{n_1}{n}X_1 + \frac{n_2}{n}X_2 + \ldots + \frac{n_k}{n}X_k$$

If we set $n_i/n = f_i$, the *frequency* of individuals with value $X_i$, this becomes

$$\bar{x} = \sum_{i=1}^{k} (f_iX_i)$$

A special case that is most important in genetics is the binomial distribution, in which the probability that an event of the $i$th type will occur is $p_i$. Let there be two possible events, 0 (*e.g.*, heads if a coin is tossed) and 1 (tails) with probabilities $q$ and $p$ respectively. Since there are only two possible events, $q = 1 - p$. The weighted sum of the values in a series of $n$ trials (coin tosses) is then

$n(1 - p)(0) + np(1) = np$. Dividing by $n$, we find the mean of the probability distribution of the two events, $\bar{x} = p$.

## VARIATION

We are also interested in the variation represented by a series of measurements. The most useful measure of variation is the VARIANCE, defined as the mean value of the square of an observation's deviation from the arithmetic mean of the population or sample:

$$V = \frac{(X_1 - \bar{x})^2 + (X_2 - \bar{x})^2 + \ldots + (X_n - \bar{x})^2}{n} = \frac{1}{n}\sum_{i=1}^{n}(X_i - \bar{x})^2$$

Each $X_i$ value might occur several times, with frequency $f_i$; if there are $k$ different values (classes) of $X_i$, the variance can be written

$$V = \frac{n_1(X_1 - \bar{x})^2 + n_2(X_2 - \bar{x})^2 + \ldots + n_k(X_k - \bar{x})^2}{n} = \sum_{i=1}^{k} f_i(X_i - \bar{x})^2$$

or

$$V = \left(\sum_{i=1}^{k} f_i X_i^2\right) - \bar{x}^2$$

Note that a variance is always positive; that the farther an observation is from the mean, the more it contributes to the variance; and that the more observations there are that deviate greatly from the mean, the greater the variance is. $V$ is thus more sensitive to variation than the simple range between the smallest and largest observations. If $X_i - \bar{x}$ is written as the deviation $d_i$, the variance is the same if the mean is set to 0 by subtracting $\bar{x}$ from each $X_i$, in which case the value of each observation is $X_i - \bar{x} = d_i$.

For the binomial distribution, the probability, or expected frequency $f_i$, of 0 is $1 - p$, and that of 1 is $p$. Then because $\bar{x} = p$,

$$V = \Sigma f_i(X_i - \bar{x}) = (1 - p)(0 - p)^2 + p(1 - p)^2 = p(1 - p)$$

This is the variance of the probability of heads and tails, for example. If we toss the coin $n$ times, we expect the proportion of tails to be $p$ (the mean of the probability distribution), but in practice it may not be exactly $p$. In repeated sets of $n$ tosses, $p$ will vary from one set to another; and the smaller $n$ is, the larger the variation in $p$ will be. The variance of $p$, in repeated sets of $n$ tosses, is

$$V = \frac{p(1 - p)}{n}$$

This is important in genetics. If the proportion of $A$ alleles in a population is $p$, repeated samples of $n$ individuals ($2n$ genes) will vary in allele frequency, with variance $p(1 - p)/2n$.

Because it is expressed in squared units ($cm^2$, for example), the variance is not as easily visualized as a related measure of variation, the STANDARD DEVIATION $S$ (often denoted $s$ or $\sigma$, just as the variance is often denoted $s^2$ or $\sigma^2$). This is the square root of the variance: $S = \sqrt{V}$. It is most easily visualized if the frequency distribution of $X$ values forms a bell-shaped, or normal, curve (Figure 1). Because of the mathematical form of the distribution, $S$ constitutes a fixed fraction of the

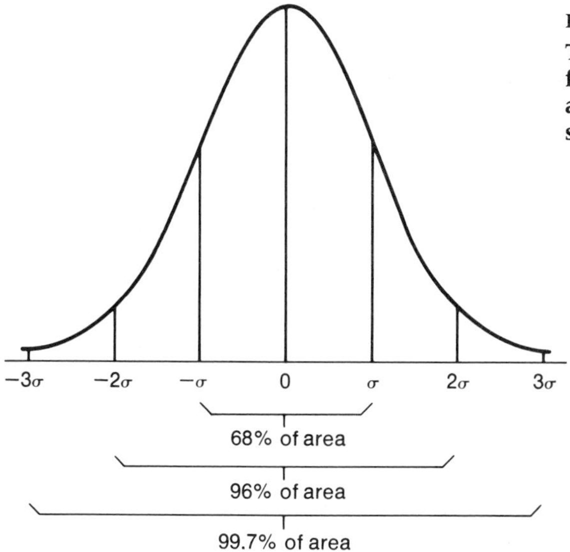


FIGURE 1
**The normal distribution curve, showing the fraction of the area embraced by one, two, and three standard deviations (σ) on either side of the mean.**

area under the curve; for example, 68 percent of the observations fall within one standard deviation on either side of the mean. If $S$ is large, this fraction of observations must spread out farther from the mean than if $S$ is small; thus a large $S$ implies a broad, variable distribution.

An important property of variances is their additivity. If a set of observations falls into several groups, the variance of the entire set is the sum of the within-group and among-group variances. A simple example is provided in Table I.

There may be many components to the total variance. Table II offers some hypothetical weights of eight individual plants grown under four combinations of low and high levels of nitrogen and phosphorus. In part *a*, the total variance

TABLE I
**Additivity of variances**

| | Group 1 | Group 2 |
|---|---|---|
| $X$ values | 4, 4, 5, 6, 6 | 1, 2, 3, 4, 5 |
| Group means | 5 | 3 |
| Sum of squared deviations from group means | 4 | 10 |
| Within-group variances | 0.8 | 2.0 |

The mean within-group variance is $(0.8 + 2.0)/2 = 1.4$. The mean of the group means is $(5 + 3)/2 = 4$, so the variance between group means is $[(5 - 4)^2 + (3 - 4)^2]/2 = 1.0$. The sum of within-group and between-group variances is thus $1.4 + 1.0 = 2.4$.

When sample sizes are equal, this is identical to the variance calculated for the data as a whole, for which $\bar{x} = 4$. $V = [(4 - 4)^2 + (4 - 4)^2 + \ldots + (5 - 4)^2]/10 = 2.4$.

TABLE II
**Hypothetical effects of fertilizers on plant weights**

(*a*) No interaction

| | | PHOSPHORUS | |
|---|---|---|---|
| | | low | high |
| NITROGEN | low | 9,11 | 14,16 |
| | high | 19,21 | 24,26 |

Increased phosphorus adds 5, on average

Increased nitrogen adds 10, on average

(High N, high P) − (low N, low P) = 15, on average

(*b*) Interaction

| | | PHOSPHORUS | |
|---|---|---|---|
| | | low | high |
| NITROGEN | low | 9,11 | 14,16 |
| | high | 19,21 | 29,31 |

High phosphorus adds 5 if nitrogen is low, 10 if nitrogen is high

High nitrogen adds 10 if phosphorus is low, 15 if phosphorus is high

is the sum of (1) the average variance within each treatment, (2) the variance among rows (nitrogen levels), and (3) the variance among columns (phosphorus treatments). In part *b*, however, there is a fourth component of the variance, that due to the interaction between the two fertilizers ("the whole is greater than the sum of its parts"). In statistical terms the fertilizers do not interact to influence weight in case *a*; they are said to have additive effects.

## CORRELATIONS AMONG VARIABLES

Table III presents hypothetical data on the body length $X$ and tail length $Y$ of each of five individuals. The mean total length $\bar{z}$ is the sum of the separate means $\bar{x}$ and $\bar{y}$. The variance of the total length $V_Z$ would be the sum of the separate variances $V_X$ and $V_Y$, except that there is a clear correlation between $X$ and $Y$. Rather, $V_Z$ is found to be

$$V_Z = \frac{1}{n}\sum_{i=1}^{n}[(X_i + Y_i) - (\bar{x} + \bar{y})]^2$$

$$= \frac{1}{n}\Sigma\,(X_i - \bar{x} + Y_i - \bar{y})^2$$

$$= \frac{1}{n}[\Sigma(X_i - \bar{x})^2 + \Sigma\,(Y_i - \bar{y})^2 + 2\,\Sigma\,(X_i - \bar{x})(Y_i - \bar{y})]$$

$$= V_X + V_Y + 2\,\mathrm{Cov}(X, Y)$$

That is, the variance is inflated by twice the term $(1/n)\,\Sigma\,(X_i - \bar{x})\,(Y_i - \bar{y})$, the COVARIANCE, which measures the dispersion of values around the joint mean $\bar{x}, \bar{y}$.

The covariance is closely related to the CORRELATION COEFFICIENT between the variables $X$ and $Y$, which may be written

$$r_{XY} = \frac{1}{n}\,\frac{\Sigma_i\,(X_i - \bar{x})\,(Y_i - \bar{y})}{S_X S_Y}$$

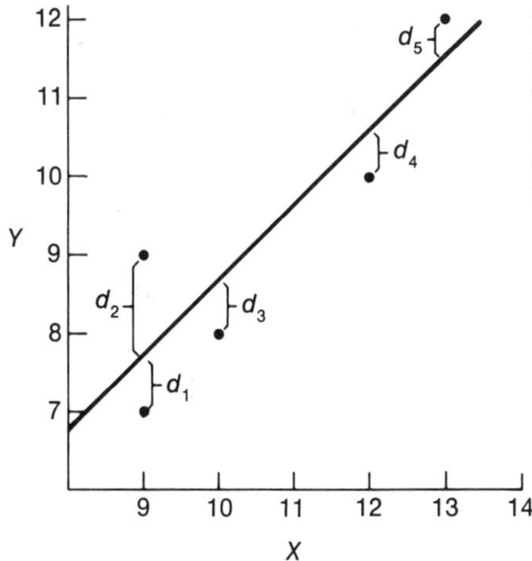


FIGURE 2

**Example of linear regression of a dependent variable *Y* on an independent variable *X*. The regression line has a slope *b* and *Y*-intercept *a* such that the sum of squared deviations of the points from the line ($\Sigma_i d_i^2$) is minimized.**

where $S_x$ and $S_y$ are the standard deviations of *X* and *Y* respectively. The correlation coefficient $r_{XY}$ ranges from $+1$ for variables that are perfectly positively correlated to $-1$ for those that are perfectly negatively correlated. It measures the degree of association of the two variables.

Figure 2 plots the data from Table III and introduces another measure of association between *Y* and *X*. If *Y* depends on or is caused by *X*, it is a dependent variable, and *X* is the independent variable; for example, *Y* might be the phenotype of offspring, *X* that of parents. In this case it is possible not only to specify that they are correlated, but to predict *Y* from *X*. If the points approximate a straight line, the predictive equation is $Y = a + b_{Y \cdot X}X$, where $b_{Y \cdot X}$ is the COEFFICIENT OF REGRESSION of *Y* on *X*, the slope of the line illustrated in the figure. This

TABLE III

**Correlation and regression**

| Specimen | Tail length (Y) | Body length (X) | Total length (Z) |
|---|---|---|---|
| 1 | 7 | 9 | 16 |
| 2 | 9 | 9 | 18 |
| 3 | 8 | 10 | 18 |
| 4 | 10 | 12 | 22 |
| 5 | 12 | 13 | 25 |
| $\Sigma$ | 46 | 53 | 99 |
| Mean | $\bar{y} = 9.2$ | $\bar{x} = 10.6$ | $\bar{z} = 19.8$ |
| Variance | $s_Y^2 = 2.96$ | $s_X^2 = 2.64$ | $s_Z^2 = 10.56$ |

| | |
|---|---|
| Covariance | $\text{cov}(X, Y) = 4.96$ |
| Correlation coefficient | $r_{X \cdot Y} = 0.89$ |
| Regression coefficient, *Y* on *X* | $b_{Y \cdot X} = 0.939$ |
| *Y*-intercept | $a = -0.753$ |
| Regression equation | $Y = -0.753 + 0.939X$ |

is the line from which the sum of squared deviations of the $Y$ values is minimal; consequently, a regression analysis of this kind is called a least squares regression. The regression coefficient $b_{Y \cdot X}$ is related to the correlation coefficient:

$$b_{Y \cdot X} = r_{XY} \frac{S_Y}{S_X}$$

but it does not equal the correlation coefficient unless $X$ and $Y$ have equal standard deviations. The regression coefficient describing the relation between the phenotypes of parents ($X$) and their offspring ($Y$) is used to calculate the heritability of a trait (Chapter 7). Because the phenotypes of parents and offspring are usually not perfectly correlated ($r_{XY} < 1$), $b_{Y \cdot X}$, and therefore the heritability, are usually less than 1.

The regression coefficient can be calculated as

$$b_{Y \cdot X} = \frac{\sum_{i=1}^{n} [(X_i - \bar{x})(Y_i - \bar{y})]^2}{\sum_{i=1}^{n} (X_i - \bar{x})^2}$$

# LIST OF SYMBOLS

- $A$ An allele (also $A'$, $a$) or locus; area; used as a subscript in $V_A$, additive genetic variance
- $a$ An allele; allometric coefficient; the phenotypic effect of an allele in homozygous form
- $\alpha_{ij}$ Competitive effect of species $i$ on species $j$
- $b$ Per capita birth rate; constant in the allometric equation; an allele (likewise $B$, $B'$)
- $b_{Y \cdot X}$ Regression coefficient
- bp Base pairs
- $c$ A constant; conversion advantage in biased gene conversion
- $D$ Coefficient of linkage disequilibrium; $D_N$, genetic distance between populations; frequency of dominant homozygote; population density
- $d$ Derivative (as in $dq/dt$); phenotypic value of a heterozygote; distance between niches
- $\Delta$ Difference, or change in (as in $\Delta q/\Delta t$)
- $\delta$ Per capita death rate; deletion rate of multiple genes; measure of bias in gene conversion
- $E$ Extinction rate; energy content of food
- $e$ Base of natural logarithms, 2.718
- $\epsilon$ Coefficient of epistasis
- $F$ Inbreeding coefficient of a population; $F_t$, $F$ at time $t$
- $f$ Inbreeding coefficient of an individual; $f(\ )$, function of; $f_i$, frequency of event or item $i$; probability of identity of gene copies
- $g$ Gamete frequency; probability of identity of gene copies
- $H$ Frequency of heterozygotes; $H_E$, $H_F$, frequencies of heterozygotes in randomly mating and inbred populations, respectively; $H_t$, frequency of heterozygotes at time $t$
- $h$ Degree of dominance, with respect to fitness, of an allele with fitness $1 - s$ in homozygous condition; probability of identity of gene copies; handling time
- $h^2$ Heritability
- $I$ Immigration rate; $I_N$ (or $I$), genetic similarity of two populations
- $i, j$ Counters; for example, $A_i$ and $A_j$, the $i$th and $j$th alleles at locus $A$
- $K$ Equilibrium density of a population, or "carrying capacity"
- $k$ Growth rate of a morphological feature
- kb Kilobases (thousand base pairs)
- $L$ Genetic load; maximum age of survival; map length of a genome or chromosome
- $l_x$ Probability of survival from birth to age $x$
- ln Natural log ($\log_e$)
- $\lambda$ Encounter rate during search; rate of gene conversion
- Myr Million years
- $m$ Fraction of reproducing individuals in a population that are immigrants, in a given generation; i.e., rate of gene flow; number of chromosome sites occupiable by a transposable element
- $m_x$ Average fecundity of a female of age $x$
- $N$ Usually population size ($N_0$ and $N_t$, population sizes at times zero and $t$, respectively); also number of chromosome complements in the genome (e.g., $2N$ = diploid)
- $N_e$ Effective population size
- $n$ Any number (of generations, individuals, and so on)
- $\nu$ Origination rate of pseudogene

$O$ Origination rate of taxa

$P$ Density of predators; probability of survival

$p, q$ Allele frequencies; $\hat{p}, \hat{q}$, allele frequencies at equilibrium; $p$ also fraction of transposon insertions with lethal or sterile effects

$q_m$ Allele frequency among immigrants into a population

$R$ Frequency of crossing-over; the replacement rate, or ratio of population sizes in successive generations (hence also the fitness of a genotype); the phenotypic response to selection in one generation; rate of increase of species diversity; frequency of recessive homozygote

$r$ The instantaneous rate of increase of a population; occasionally an allele frequency; fitness of an allele or genotype; correlation coefficient; coefficient of relatedness

$r_m$ The intrinsic rate of increase of a population, i.e., the instantaneous rate of increase ($r$) when a population has a stable age distribution and is at low density

$S$ Number of species; selection differential; rate of speciation; $S_R$, genetic similarity of populations

$s$ Selection coefficient; rarely a standard deviation or, as $s^2$, a variance

$\Sigma$ Addition symbol; for example, $\Sigma_{i=0}^{k} X_i$, the sum of the values $X_i$, from $X_0$ to $X_k$

$\sigma$ Standard deviation: the square root of $\sigma^2$, the variance; also, breadth of niche

$T$ Generation time; time to equilibrium

$t$ Time; a selection coefficient; the name of an allele

$u$ Mutation rate; transposition rate

$V$ Variance ($= \sigma^2$); $V_G, V_E$, components of total variance; density of prey

$v$ Rate of back mutation; excision rate of transposable elements

$W$ Fitness of a genotype

$\overline{w}$ Average fitness of an individual in a population

$x$ A variable; often, age

$\bar{x}$ The arithmetic mean of some variable

$z$ Slope of the relation between species number and area on a log–log plot; value of a metric character

# CHAPTER-OPENING FIGURES

CHAPTER ONE Charles Darwin. (Down House, Kent)

CHAPTER TWO Grevy zebra (*Equus grevyi*) and reticulated giraffe (*Giraffa camelopardalis*) feeding in an African savannah. Interactions among species, as among plants and herbivores, are important features of the environment of every species. (Photograph courtesy of J. R. Ginsberg)

CHAPTER THREE The polytene chromosomes of the salivary gland of a larval *Drosophila melanogaster.* The banding pattern of *Drosophila* chromosomes has been useful for genetic and evolutionary studies. (P. J. Bryant, University of California/BPS)

CHAPTER FOUR Geographic variants of the salamander *Ensatina eschscholtzi;* two geographic races that meet in the mountains of southern California. (Photograph courtesy of S. J. Arnold)

CHAPTER FIVE *Stylidia biarticulata,* a fly in the family Nycteribiidae. A parasite of bats, its population structure is affected by its ability to disperse among bats. These wingless, eyeless flies hold the head reflexed over the thorax. [From O. Theodor, *An illustrated catalogue of the Rothschild Collection of Nycteribiidae (Diptera) in the British Museum (Natural History),* 1967. Courtesy of the Trustees of the British Museum (Natural History)]

CHAPTER SIX Among the consequences of natural selection are exquisitely refined morphological adaptations, such as those illustrated by the skeleton of a rattlesnake (*Crotalus*): the loosely articulated jaws, the movable, replaceable fangs, the flexible backbone of more than 150 vertebrae, and the rattle formed by terminal scales that are retained at each molt.

CHAPTER SEVEN Several domestic breeds of pigeons descended from the rock dove, *Columba livia.* From *The variation of animals and plants under domestication* (1868) by Charles Darwin, whose investigations of artificial selection, as illustrated by domestic breeds, provided part of his evidence for evolution.

CHAPTER EIGHT The drawings depict 8 of the 71 species of the beetle genus *Cryptocephalus* (Chrysomelidae) known from America north of Mexico. [From R. E. White, 1968, A review of the genus *Cryptocephalus* in America north of Mexico (Chrysomelidae: Coleoptera), U.S. Nat. Mus. Bull. 290. Reproduced courtesy of R. E. White]

CHAPTER NINE A dense population of *Senecio brassica* (whitish leaves) grows with some taller individuals of *S. keniodendron* and a *Lobelia keniensis* (right center foreground) on the slopes of Mt. Kenya. Their adaptations to abrupt temperature changes and the freezing temperatures that occur on most clear nights all year include heavily pubescent leaves that undergo a daily cycle of leaf movement, opening during the day and closing at night. (Photograph courtesy of G. H. Orians)

CHAPTER TEN The phylogeny of living things as conceived by Ernst Haeckel, who introduced the representation of phylogeny by a tree-like diagram. (From E. Haeckel, 1866, *Generelle Morphologie der Organismen,* Georg Reimer, Berlin)

CHAPTER ELEVEN *Protomimosoidea,* the earliest known mimosoid legume, from the Paleocene/Eocene of Tennessee. The fossil reveals that this subfamily of the legumes had become distinct by this time. (Photograph courtesy of W. L. Crepet)

CHAPTER TWELVE Some of the numerous forms of ammonites, chiefly of the Jurassic and Triassic. Details of the pattern of sutures are also included for several of the species. (From W. Keferstein, 1862–1866, Malacozoa, in H. G. Bronn's *Die Klassen und Ordnungen des Thier-Reichs* 3(2). C. F. Winter'sche Verlagshandlung, Leipzig und Heidelberg)

CHAPTER THIRTEEN A landscape in Baja California, Mexico. In the foreground the cactus *Pachycereus* (Cactaceae) appears; in the rear is *Idria columnaris* (Fouquieriaceae). Except for a genus in West Africa, and one in Ceylon which has probably been introduced by humans, the Cactaceae are restricted to the Americas. The Fouquieriaceae are restricted to deserts of Mexico and the United States, and *Idria* is endemic to Baja California. (Photograph by the author)

CHAPTER FOURTEEN Embryos of six mammals in three developmental stages. Left to right, a monotreme (spiny anteater), a marsupial (koala), and four placental mammals (deer, cat, macaque, human). From Ernst Haeckel, *Natürliche Schöpfungs-Geschichte,* 1898, Georg Reimer, Berlin. Haeckel's explorations of the relationship between ontogeny and phylogeny were extremely influential in the study of this subject.

CHAPTER FIFTEEN The DNA of a small portion of a human metaphase chromosome, from which the histone proteins have been removed. (Courtesy of Professor U. K. Laemmli and The Cold Spring Harbor Laboratory)

CHAPTER SIXTEEN The larva of a Central American saturniid moth bears the cocoons of braconid wasp parasitoids that have developed within the larva's body. (Photograph by the author)

CHAPTER SEVENTEEN *Diskobolos,* ca. 450 B.C., by Myron. A Roman copy of a Greek bronze original. (Museo delle Terme, Rome)

# GLOSSARY

The list that follows includes frequently used terms or terms that in the text have been defined implicitly by context, rather than explicitly. A few terms are included that may not appear in the text, but are frequent enough in the evolutionary literature to warrant listing. This glossary is not exhaustive; refer to the index for text entries of terms not defined here.

ADAPTATION A process of genetic change of a population, owing to natural selection, whereby the average state of a character becomes improved with reference to a specific function, or whereby a population is thought to have become better suited to some feature of its environment. Also, *an* adaptation: a feature that has become prevalent in a population because of a selective advantage owing to its provision of an improvement in some function. A complex and poorly defined concept; see Chapter 9.

ADAPTIVE PEAK That allele frequency, or combination of allele frequencies at two or more loci, at which the mean fitness of a population has a (local) maximum.

ADAPTIVE RADIATION Evolutionary divergence of members of a single phylogenetic line into a variety of different adaptive forms; usually with reference to diversification in the use of resources or habitats.

ADAPTIVE ZONE A set of similar ECOLOGICAL NICHES occupied by a group of (usually) related species, often constituting a higher taxon.

ADDITIVE EFFECT Of an allele, its effect on the phenotype, averaged over the variety of genetic combinations in which it occurs. It is said to act additively if its effect is independent of the other alleles it is combined with.

ALLELE One of the several forms of the same gene, presumably differing by mutation of the DNA sequence, and capable of segregating as a unit Mendelian factor. Alleles are usually recognized by their phenotypic effects; DNA sequence variants recognized by direct sequencing are usually called *haplotypes.*

ALLELE FREQUENCY See GENE FREQUENCY.

ALLOMETRIC GROWTH Growth of a feature during ontogeny at a rate different from that of another feature with which it is compared.

ALLOPATRIC Of a population or species, occupying a geographic region different from that of another population or species. See PARAPATRIC, SYMPATRIC.

ALLOPOLYPLOID A POLYPLOID in which the several chromosome sets are derived from more than one species.

ALLOZYME One of several forms of an enzyme coded for by different alleles at a locus. See ISOZYME.

ANAGENESIS Introduced by Rensch to describe progressive evolution toward higher taxonomic levels; now more generally used to describe directional evolution of a feature over an arbitrarily short segment of a lineage.

ANEUPLOID Of a cell or organism, possessing an unbalanced chromosome complement by virtue of an excess or deficiency in number of one or more of the chromosomes compared to the others.

APOMIXIS Parthenogenetic reproduction in which an individual develops from an unfertilized egg or somatic cell.

APOSEMATIC Coloration or other features that advertise noxious properties; warning coloration.

ARTIFICIAL SELECTION Selection by humans of a consciously chosen trait or combination of traits in a (usually captive) population; differing from natural selection in that the criterion of survival and reproduction is the trait chosen, rather than fitness as determined by the entire genotype.

ASSORTATIVE MATING Nonrandom mating on the basis of phenotype; usually used for positive assortative mating, the propensity to mate with others of like phenotypes.

AUTOPOLYPLOID A POLYPLOID in which the several chromosome sets are derived from the same species.

AUTOSOME A chromosome other than a sex chromosome.

AUTOTROPH An organism that synthesizes the compounds it requires for energy.

CANALIZATION The operation of internal factors during development that reduce the effect of perturbing influences, thereby constraining variation in the phenotype around one or more modes.

CARRYING CAPACITY The population density that can be sustained by limiting resources.

CATEGORY In taxonomy, one of the ranks (e.g., genus, family) of classification. See TAXON.

CHARACTER A feature, or trait. CHARACTER STATE, one of the variant conditions of a character.

CHRONOSPECIES A segment of an evolving lineage preserved in the fossil record, that differs enough from earlier or later members of the lineage to be given a different binomial (name). Not equivalent to biological species (q.v.).

CIS Of two genetic elements, located on the same individual chromosome rather than on homologous chromosomes; of gene action, influence of one genetic element on the function or expression of another genetic element on the same individual chromosome. See TRANS.

CISTRON A gene that differs in location on the chromosome, and usually in function, from other genes. Usually a length of DNA that codes for a polypeptide.

CLADE Properly, the set of species descended from a particular ancestral species. Sometimes used more loosely, as a set of related species, from which some descendants are excluded.

CLADISTIC Pertaining to branching patterns; a cladistic classification classifies organisms on the basis of the historical sequences by which they have diverged from common ancestors.

CLADOGENESIS Branching of lineages during phylogeny.

CLEISTOGAMY Self-pollination within a flower that does not open.

CLINE A gradual change in an allele frequency or in the mean of a character over a geographic transect.

CLONE A lineage of individuals reproduced asexually by parthenogenesis or vegetative reproduction, by mitotic division.

COEFFICIENT OF VARIATION (C.V.) The standard deviation divided by the mean, multiplied by 100. C.V. = $100 \times (s/\bar{x})$.

COHORT Those members of a population that are of the same age.

COMMENSALISM An ecological relationship between species in which one is benefited but the other is little affected.

CONCERTED EVOLUTION Maintenance of a homogeneous nucleotide sequence among the members of a gene family (*q.v.*), which sequence evolves over time.

CONVERGENT EVOLUTION Evolution of similar features independently in unrelated taxa, usually from different antecedent features or by different developmental pathways.

CORRELATION A statistical relationship that quantifies the degree to which two variables are associated (see *correlation coefficient* in Appendix I). For *phenotypic correlation, genetic correlation, environmental correlation* as applied to the relationship between two traits, see Chapter 7.

COUPLING Of a gamete or chromosome, bearing at two or more loci alleles that have been designated to be alike in some way (e.g., both wild type rather than mutant).

COVARIANCE A statistical term for a component of the correlation coefficient; see Appendix I.

DEME A local population, usually a small, panmictic population.

DENSITY-DEPENDENT Affected by population density.

DETERMINISTIC Causing certain outcomes to be more probable than others. See STOCHASTIC.

DIPLOID A cell or organism possessing two chromosome complements; *ploidy* thus refers to the number of chromosome complements (see HAPLOID, POLYPLOID).

DIRECTIONAL SELECTION Selection for a higher or lower value of a character than its current mean value.

DISRUPTIVE SELECTION Selection in favor of two or more modal phenotypes and against those intermediate between them. Equals *diversifying selection*.

DOMINANCE Of an allele, the extent to which it produces when heterozygous the same phenotype as when homozygous. Of a species, the extent to which it is numerically (or otherwise) predominant in a community.

ECOLOGICAL NICHE The range of combinations of all relevant environmental variables under which a species or population can persist; often more loosely used to describe the "role" of a species, or the resources it utilizes.

ECOTYPE A genetically determined phenotype of a species that is found as a local variant associated with certain ecological conditions.

EFFICIENT CAUSE Aristotle's term for the mechanical reason for an event.

ENDEMIC Of a species, restricted to a specified region or locality.

ENVIRONMENT Usually, the complex of external physical, chemical, and biotic factors that may affect a population, an organism, or the expression of an organism's genes; more generally, anything external to the object of interest (e.g., a gene, an organism, a population) that may influence its function or activity. Thus other genes within an organism may be part of a gene's environment, or other individuals in a population may be part of an organism's environment.

EPIGENETIC Developmental; pertaining especially to interactions among developmental processes above the level of primary gene action.

EPISTASIS A synergistic effect, on the phenotype or fitness, of two or more gene loci, whereby their joint effect differs from the sum of the loci taken separately.

EQUILIBRIUM A condition of stasis, as of population size or genetic composition. Also the value (of population size, gene frequency) at which stasis occurs. See also STABILITY, UNSTABLE EQUILIBRIUM.

ETHOLOGICAL Behavioral.

EVOLUTION In a broad sense, the origin of entities possessing different states of one or more characteristics, and changes in their proportions over time. *Organic evolution*, or *biological evolution*, is a change over time of the proportions of individual organisms differing genetically in one or more traits; such changes transpire by the origin and subsequent alteration of the frequencies of alleles or genotypes from generation to generation within populations, by the alterations of the proportions of genetically differentiated populations of a species, or by changes in the numbers of species with different characteristics, thereby altering the frequency of one or more traits within a higher taxon.

EURYTOPIC Of a species or population, capable of persistence in a wide variety of conditions or habitats; relative to a STENOTOPIC species or population (q.v.).

EXON A part of an interrupted gene that is translated into a polypeptide.

FECUNDITY The quantity of gametes, usually eggs, produced.

FINAL CAUSE Aristotle's term for a goal, attainment of which is the reason for being of an event or process.

FITNESS The average contribution of one allele or genotype to the next generation or to succeeding generations, compared with that of other alleles or genotypes.

FIXATION Attainment of a frequency of 1 (i.e., 100 percent) by an allele in a population, which thereby becomes MONOMORPHIC for the allele.

FOUNDER EFFECT The principle that the founders of a new colony carry only a fraction of the total genetic variation in the source population.

FUGITIVE SPECIES One that occupies temporary environments or habitats and so does not persist for many generations at any one site.

GENE The functional unit of heredity; a cistron. A complex concept; see Chapter 3.

GENE FAMILY Two or more loci with similar nucleotide sequences, that have been derived from a common ancestral sequence.

GENE FLOW The incorporation of genes into the gene pool of one population from one or more other populations.

GENE FREQUENCY The proportion of gene copies in a population that an allele accounts for; i.e., the probability of finding this allele when a gene is taken randomly from the population. Equals ALLELE FREQUENCY.

GENE POOL The totality of the genes of a given sexual population.

GENETIC DEATH A death due to genotype, hence contributing to natural selection. Deaths that are random with respect to genotype are not genetic deaths.

GENETIC DISTANCE Any of several measures of the degree of genetic difference between populations, based on differences in allele frequencies.

GENETIC DRIFT Random changes in the frequencies of two or more alleles or genotypes within a population.

GENETIC LOAD Any reduction of the mean fitness of a population owing to the existence of genotypes with lower fitness than that of the most fit genotype.

GENIC SELECTION The differential propagation of different alleles within a population because of properties of the alleles rather than of the genotypes; i.e., a form of natural selection in which the frequency of an allele is determined by fitness averaged over the variety of genotypes in which it occurs. See INDIVIDUAL SELECTION, KIN SELECTION, NATURAL SELECTION.

GENOTYPE The set of genes possessed by an individual organism; often, its genetic composition at a specific locus or set of loci singled out for discussion.

GEOMETRIC MEAN The *n*th root of the product of *n* values. $G = (x_1 \cdot x_2 \cdot \ldots \cdot x_n)^{1/n}$.

GRADE A level of phenotypic organization attained by one or more species during evolution.

GROUP SELECTION The differential rate of origination or extinction of whole populations (or species, if the term is used broadly) on the basis of differences among them in one or more characteristics. See INTERDEMIC SELECTION, SPECIES SELECTION.

HABITAT SELECTION The capacity of an organism (usually an animal) to choose a habitat in which to perform activities.

HAPLOID A cell or organism possessing a single chromosome complement, hence a single gene copy at each locus.

HERITABILITY The proportion of the variance (q.v.) among individuals in a trait, that is attributable to differences in genotype. For heritability in the narrow and broad senses, see Chapter 7.

HETEROCHRONY An evolutionary change in phenotype based on an alteration of timing of development.

HETEROKARYOTYPE A genome or individual that is heterozygous for a chromosomal rearrangement such as an inversion. A *homokaryote* is homozygous in this respect.

HETEROSIS Equivalent to *hybrid vigor*: the superiority in one or more characteristics (e.g., size, yield) of crossbred organisms compared with correspondingly inbred organisms, as a result of differences in the genetic constitutions of the uniting parental gametes. Sometimes used to describe the higher fitness of heterozygous than homozygous genotypes, which is better termed *euheterosis*, and which is distinguished from *luxuriance*, a superiority in size, etc., that does not increase fitness. See HETEROZYGOUS ADVANTAGE, OVERDOMINANCE.

HETEROTROPH An organism that does not synthesize the compounds it uses for energy.

HETEROZYGOTE An individual organism that possesses different alleles at a locus.

HETEROZYGOUS ADVANTAGE The manifestation of higher fitness by heterozygotes at a specific locus than by homozygotes.

HISTONE One of a class of proteins that are constituents of the chromosomes in eukaryotes.

HOMEOSTASIS Maintenance of an equilibrium state by some self-regulating capacity.

HOMEOTIC MUTATION A mutational change of one structure into another of the organism's structures.

HOMOLOGY Possession of two or more species of a trait derived, with or without modification, from their common ancestor. Also, HOMOLOGOUS CHROMOSOMES, the members of a chromosome complement that bear the same genes.

HOMOPLASY Possession by two or more species of a similar or identical trait that has not been derived by both species from their common ancestor; embraces convergence, parallel evolution, and evolutionary reversal.

HOMOZYGOTE An individual organism that has the same allele at each of its copies of a gene locus.

HOST RACE An ill-defined term in entomology, denoting a differentiated form, which may or may not interbreed with other host races, that feeds on a specific host plant.

HYBRID An individual formed by mating between unlike forms, usually genetically differentiated populations or species; occasionally in genetics, the offspring of a mating between phenotypically distinguishable genotypes of any kind.

HYPERMORPHOSIS Exaggeration of the features of a descendant form compared to those of its ancestor due to an increase during evolution of the duration of ontogenetic development.

IDENTICAL BY DESCENT Describes copies of an allele that may be traced back through an arbitrary number of generations without mutation to a common ancestor of the organisms that carry the copies.

INBREEDING DEPRESSION Reduction, as a consequence of inbreeding, of the mean value of a character.

INCLUSIVE FITNESS The fitness of a gene or genotype measured by its effect on survival or reproduction both of the organism bearing it, and of the genes, identical by descent, borne by the organism's relatives.

INDIVIDUAL SELECTION A form of natural selection consisting of non-random differences among different genotypes within a population in their contribution to subsequent generations. See GENIC SELECTION, NATURAL SELECTION.

INTERACTION In a statistical sense, a joint effect of two independent variables ("causes") on a dependent variable, whereby the effect differs from the sum of the two causal effects taken separately: synergism. *Genotype × environment interaction* (Chapter 7) is consequently variation in phenotype arising from the difference in the effect of environment on the expression of different genotypes.

INTERDEMIC SELECTION Group selection of populations within a species.

INTRINSIC RATE OF NATURAL INCREASE The potential rate of increase of a population with a stable age distribution, whose growth is not depressed by the negative effects of density.

INTRON A part of an interrupted gene that is not translated into a polypeptide.

INVERSION A 180° reversal of the orientation of a part of a chromosome, relative to some standard chromosome.

ISOLATING MECHANISM A genetically determined difference between populations that restricts or prevents gene flow between them. The term does not include spatial segregation by extrinsic geographic or topographic barriers.

ISOZYME (ISOENZYME) One of several forms of an enzyme, produced by different, nonallelic, loci in an individual organism's genome. Often misused in place of ALLOZYME.

ITERATIVE EVOLUTION The repeated evolution of similar phenotype characteristics at different times during the history of a clade.

KARYOTYPE The chromosome complement of an individual.

KIN SELECTION A form of genic selection whereby alleles differ in their rate of propagation by influencing the survival of individuals (kin) who carry the same alleles by common descent.

LINKAGE Occurrence of two loci on the same chromosome: they are functionally linked only if they are so close together that they do not segregate independently in meiosis.

LINKAGE EQUILIBRIUM and LINKAGE DISEQUILIBRIUM If two alleles at two or more loci are associated more or less frequently than predicted by their individual frequencies, they are in linkage disequilibrium; if not, they are in linkage equilibrium.

LOCUS A site on a chromosome occupied by a specific gene; more loosely, the gene itself, in all its allelic states.

LOGISTIC EQUATION A specific equation describing the idealized growth of a population subject to a density-dependent limiting factor.

MACROEVOLUTION A vague term for the evolution of great phenotypic changes, usually great enough to allocate the changed lineage and its descendants to a distinct genus or higher taxon.

MATERNAL EFFECT A nongenetic effect of the mother on the phenotype of the offspring, owing to factors such as cytoplasmic inheritance, transmission of disease from mother to offspring, or nutritional conditions.

MEAN Usually the arithmetic mean or average; the sum of $n$ values, divided by $n$. The mean $\bar{x} = (x_1 + x_2 + \ldots + x_n)/n$.

MEIOTIC DRIVE Used broadly to denote a preponderance (>50 percent) of one allele among the gametes produced by a heterozygote; results in genic selection.

MERISTIC TRAIT A discretely varying, countable trait; e.g., number of digits.

MICROEVOLUTION A vague term for slight evolutionary changes within species.

MIGRATION Used in theoretical population genetics as a synonym for gene flow among populations; in other contexts, directed large-scale movement of organisms that does not necessarily result in gene flow.

MODIFIER GENE A gene that is recognized by its alteration of the phenotypic expression of genes at one or more other loci.

MONOMORPHIC A population in which virtually all individuals have the same genotype at a locus. Cf. POLYMORPHISM.

MONOPHYLETIC Of a taxon, consisting of species all of which are derived from a common ancestral taxon. In cladistic taxonomy, the term describes a taxon consisting of all the known species descended from a single ancestral species.

MORPHOCLINE An ordered array of character states such that each state could plausibly be derived only from neighboring states in the array.

MOSAIC EVOLUTION Evolution of different characters within a lineage or clade at different rates, hence more or less independently of one another.

MUTATION An error in replication of a nucleotide sequence, or any other alteration of the genome that is not manifested as reciprocal recombination. A complex, poorly defined concept; see Chapter 3.

MUTUALISM A symbiotic relation in which each of two species benefits by their interaction.

NATURAL SELECTION The differential survival and/or reproduction of classes of entities that differ in one or more hereditary characteristics; the difference in survival and/or reproduction is not due to chance, and it must have the potential consequence of altering the proportions of the different entities, to constitute natural selection. Thus natural selection is also definable as a partly or wholly deterministic difference in the contribution of hereditarily different classes of entities to subsequent generations. The entities may be alleles, genotypes or subsets of genotypes, populations, or in the broadest sense, species. A complex concept; see Chapter 6. See also GENIC SELECTION, INDIVIDUAL SELECTION, KIN SELECTION, GROUP SELECTION.

NEGATIVE FEEDBACK A dynamic relation whereby the product of a process inhibits the process that produces it, usually enhancing stability.

NEOTENY Heterochronic evolution whereby development of some or all somatic features is retarded relative to sexual maturation, resulting in sexually mature individuals with juvenile features. See PAEDOMORPHOSIS, PROGENESIS.

NEUTRAL ALLELES Alleles that do not differ measurably in their effect on fitness.

NORM OF REACTION The set of phenotypic expressions of a genotype under different environmental conditions.

ONTOGENY The development of an individual organism, from fertilized zygote until death.

OUTGROUP A taxon that diverged from a group of other taxa before they diverged from each other.

OVERDOMINANCE The expression by two alleles in heterozygous condition of a phenotypic value for some characteristic that lies outside the range of the two corresponding homozygotes; a possible basis for HETEROSIS, but not the only one. Higher fitness of a heterozygote than of homozygotes at that locus (HETEROZYGOUS ADVANTAGE) is often termed *overdominance for fitness*.

PAEDOMORPHOSIS Possession in the adult stage of features typical of the juvenile stage of the organism's ancestor.

PANMIXIA Random mating among members of a population.

PARALLEL EVOLUTION The evolution of similar or identical features independently in related lineages, thought usually to be based on similar modifications of the same developmental pathways.

PARALOGOUS Two or more gene loci, or their polypeptide products, derived by duplication of an ancestral locus, and occurring together in a haploid chromosome complement.

PARAPATRIC Populations that have contiguous but nonoverlapping geographic distributions.

PARSIMONY Economy in the use of means to an end (*Webster's New Collegiate Dictionary*); the principle of accounting for observations by that hypothesis requiring the fewest or simplest assumptions that lack evidence; in systematics, the principle of invoking the minimal number of evolutionary changes to infer phylogenetic relationships.

PARTHENOGENESIS Virgin birth; development from an egg to which there has been no paternal contribution of genes.

PERIPATRIC Of populations, situated peripheral to most of the populations of a species; PERIPATRIC SPECIATION, speciation by evolution of isolating mechanisms in such populations.

PHENETIC Pertaining to phenotypic similarity, as in a phenetic classification.

PHENOCOPY A phenotype, developed in response to an environmental stimulus, that resembles one known to be produced by a gene mutation.

PHENOTYPE The morphological, physiological, biochemical, behavioral, and other properties of an organism, manifested throughout its life, that develop through action of genes and environment; or any subset of such properties, especially those affected by a particular allele or other portion of the GENOTYPE.

PHYLOGENY The genealogy of a group of taxa such as species; also applied to the genealogy of genes derived from a common ancestral gene.

PLEIOTROPY The phenotypic effect of a gene on more than one characteristic.

POLYGENIC CHARACTER A trait, the variation in which is based wholly or in part on allelic variation at several or many loci.

POLYMORPHISM The existence within a population of two or more genotypes for a given trait, the rarest of which exceeds some arbitrarily low frequency (say, 1 percent); more rarely, the existence of phenotypic variation within a population, whether or not genetically based.

POLYPHAGOUS Feeding on many kinds of food; usually used to describe insects that feed on many plants.

POLYPHYLETIC Of a taxon, comprised of members derived by evolution from ancestors in more than one ancestral taxon.

POLYPLOID Possessing more than two entire chromosome complements.

POLYTOPY Geographic variation in which each of one or more distinctive forms is found in each of several separate localities, between which other forms are distributed.

POLYTYPY The existence of named geographic races or subspecies within a species.

POPULATION A group of conspecific organisms that occupy a more or less well defined geographic region and exhibit reproductive continuity from generation to generation; it is generally presumed that ecological and reproductive interactions are more frequent among these individuals than between them and members of other populations of the same species.

POSITION EFFECT A difference in the phenotypic expression of a gene associated with a change in its location on the chromosomes.

PREADAPTATION Possession of the necessary properties to permit a shift into a new niche or habitat. A structure is preadapted if it can assume a new function before it itself becomes modified.

PROGENESIS A decrease during evolution of the duration of ontogenetic development, resulting in retention of juvenile features in the sexually mature adult. See NEOTENY, PAEDOMORPHOSIS.

PSEUDOGENE A nonfunctional member of a gene family.

QUANTUM EVOLUTION A rapid evolutionary shift in a lineage to a phenotypic state distinctly unlike the ancestral condition. Applied often to the rapid evolution of features that define higher taxa.

RACE A poorly defined term for a set of populations occupying a particular region that differ in one or more characteristics from populations elsewhere; equivalent to SUBSPECIES. In some writings, a distinctive phenotype, whether or not allopatric from others (see HOST RACE).

RECAPITULATION The ontogenetic passage of an organism's features through stages that resemble the adult features of its phylogenetic ancestors.

REFUGIA Locations in which species have persisted while becoming extinct elsewhere.

RELICT A species that has been "left behind," for example, the last survivor of an otherwise extinct group. Sometimes, a species or population left in a locality after extinction throughout most of the region.

RELICTUAL The geographic distribution of a species or group that persists in localities that it occupied at an earlier time, but which has been extinguished over much of its former range.

REPRODUCTIVE VALUE Of an individual of a specific age, its likely contribution to the growth of the population.

REPULSION Of a gamete or chromosome, bearing at two or more loci alleles that have been designated to be unlike in some way. Cf. COUPLING.

RESTRICTION ENZYME An enzyme that cuts double-stranded DNA at specific short nucleotide sequences.

RETICULATE EVOLUTION Union of different lineages of a clade by hybridization.

SALTATION A jump; a discontinuous mutational change in one or more phenotypic traits, usually of considerable magnitude.

SELECTION Nonrandom differential survival or reproduction of classes of phenotypically different entities. See NATURAL SELECTION, ARTIFICIAL SELECTION.

SELECTION COEFFICIENT The difference between the mean relative fitness of individuals of a given genotype and those of a reference genotype.

SEMISPECIES One of several groups of populations that are partially but not entirely isolated from each other by biological factors (isolating mechanisms).

SERIAL HOMOLOGY A relationship among repeated, often differentiated, structures of a single organism, defined by their similarity of developmental origin; for example, the several legs and other appendages of an arthropod.

SEX-LINKED A gene carried by one of the sex chromosomes; it may be expressed phenotypically in both sexes.

SEXUAL REPRODUCTION Production of offspring whose genetic constitution is a mixture of that of two potentially genetically different gametes.

SIBLING SPECIES Species that are difficult or impossible to distinguish by morphological characters.

SPECIES In the sense of biological species, the members in aggregate of a group of populations that interbreed or potentially interbreed with each other under natural conditions; a complex concept (see Chapter 4). Also, a basic taxonomic category to which individual specimens are assigned, which often but not always corresponds to the biological species.

SPECIES SELECTION A form of GROUP SELECTION in which species with different characteristics increase (by speciation) or decrease (by extinction) in number at different rates, because of a difference in their characteristics.

STABILITY Often used to mean constancy; more often in this book, the propensity to return to a condition (a stable equilibrium) after displacement from that condition.

STABILIZING SELECTION Selection against phenotypes that deviate in either direction from an optimal value of a character.

STANDARD DEVIATION The square root of the VARIANCE.

STENOTOPIC Of a species or population, restricted to or capable of persistence in a narrow range of conditions or habitats; see EURYTOPIC.

STOCHASTIC Random.

SUBSPECIES A named geographic race; a set of populations of a species that share one or more distinctive features and occupy a different geographic area from other subspecies.

SUBSTITUTION The complete replacement of one allele by another within a population or species; in the term NUCLEOTIDE SUBSTITUTION, the complete replacement of one nucleotide pair by another within a lineage over evolutionary time. Cf. FIXATION.

SUPERGENE A group of two or more loci between which recombination is so reduced that they are usually inherited together as a single entity.

SUPERSPECIES A group of semispecies.

SYMBIOSIS An association between two or more species that benefits at least one of them.

SYMPATRIC Of two species or populations, occupying the same geographic locality so that the opportunity to interbreed is presented.

SYSTEMATICS In a restricted sense, the study of the historical evolutionary and genetic relationships among organisms, and of their phenotypic similarities and differences.

TAXON (pl. TAXA) The named taxonomic unit (e.g., *Homo sapiens*, Hominidae, or Mammalia) to which individuals, or sets of species, are assigned. HIGHER TAXA are those above the species level.

TAXON CYCLE Replacement of species of a given taxon within a community over evolutionary time, such that predominant species vary cyclically in character.

TAXONOMY The naming and assignment of organisms to taxa.

TERRITORY An area or volume of habitat defended by an organism or a group of organisms against other individuals, usually of the same species; TERRITORIAL BEHAVIOR, the behavior by which the territory is defended.

TRAIT GROUP A highly localized aggregation of interacting conspecific organisms that may differ in the mean of some trait from other such groups, and which remains aggregated only temporarily (often for less than a generation).

TRANS Of two genetic elements, located on different homologous chromosomes within an individual genome; of gene action, influence of one genetic element on the function or expression of another genetic element located on another chromosome.

TRANSLOCATION The transfer of a segment of a chromosome to another, nonhomologous, chromosome; or the chromosome formed by the addition of such a segment.

UNSTABLE EQUILIBRIUM An unchanging state, to which a system (e.g., a population density or gene frequency) does not return if disturbed.

VARIANCE ($\sigma^2$, $s^2$, $V$) The average squared deviation of an observation from the arithmetic mean; hence, a measure of variation. $s^2 = [\Sigma(x_i - \bar{x})^2]/(n - 1)$, where $\bar{x}$ is the mean and $n$ is the number of observations. See Appendix I.

WILD TYPE The allele, genotype, or phenotype that is most prevalent (if there is one) in wild populations; with reference to the wild type allele, other alleles are often termed mutations.

# LITERATURE CITED

THE NUMBERS IN BRACKETS IDENTIFY THE CHAPTER(S) IN WHICH THE REFERENCE IS CITED


Abbott, I., L. K. Abbott, and P. R. Grant. 1977. Comparative ecology of Galápagos ground finches (*Geospiza* Gould): evaluation of the importance of floristic diversity and interspecific competition. *Ecol. Monogr.* 47: 151–184. [16]

Abrams, P. 1983. The theory of limiting similarity. *Annu. Rev. Ecol. Syst.* 14: 359–376. [2, 16]

Ajioka, J.W., and W. F. Eanes. In preparation. Rapid accumulation of parasitic DNA: *de novo* versus naturally occurring insertions of P-elements in *Drosophila melanogaster*. [15]

Alberch, P. 1982. Developmental constraints in evolutionary processes. In J. T. Bonner (ed.), *Evolution and development*, pp. 313–332. Dahlem Conference Report No. 20, Springer-Verlag, Berlin. [9]

Alberch, P. 1983. Morphological variation in the neotropical salamander genus *Bolitoglossa*. *Evolution* 37: 906–919. [9]

Alberch, P., and J. Alberch. 1981. Heterochronic mechanisms of morphological diversification and evolutionary change in the neotropical salamander *Bolitoglossa occidentalis* (Amphibia: Plethodontidae). *J. Morphol.* 167: 249–264. [14]

Alberch, P., and E. A. Gale. 1985. A developmental analysis of an evolutionary trend: digital reduction in amphibians. *Evolution* 39: 8–23. [14]

Alberch, P., S. J. Gould, G. F. Oster, and D. B. Wake. 1979. Size and shape in ontogeny and phylogeny. *Paleobiology* 5: 296–317. [14]

Alexander, R. D. 1979. *Darwinism and human affairs*. University of Washington Press, Seattle. [17]

Alexander, R. D., and R. S. Bigelow. 1960. Allochronic speciation in field crickets, and a new species, *Acheta veletis*. *Evolution* 14: 334–346. [8]

Allard, R. W. 1960. *Principles of plant breeding*. Wiley, New York. [3]

Allard, R. W., and C. Wehrhahn. 1964. A theory which predicts stable equilibrium for inversion polymorphisms in the grasshopper *Moraba scurra*. *Evolution* 18: 129–130. [7]

Allen, M. K., and C. Yanofsky. 1963. A biochemical and genetic study of reversion with the A-gene A-protein system of *Escherichia coli* tryptophan synthetase. *Genetics* 48: 1065–1083. [3]

Allendorf, F. W.. 1979. Rapid loss of duplicate gene expression by natural selection. *Heredity* 43: 247–258. [15]

Allendorf, F. W., and G. Thorgaard. 1984. Polyploidy and the evolution of salmonid fishes. In B. J. Turner (ed.), *The evolutionary genetics of fishes*, pp. 1–53. Plenum, New York. [15]

Allison, A. C. 1961. Genetic factors in resistance to malaria. *Ann. N. Y. Acad. Sci.* 91: 710–729. [6]

Alvarez, L. W., W. Alvarez, F. Asaro, and H.V. Michel. 1980. Extraterrestrial cause for the Cretaceous-Tertiary extinction. *Science* 208: 1095–1108. [12]

Ammerman, A. J., and L. L. Cavalli-Sforza. 1984. *The neolithic transition and the genetics of populations in Europe*. Princeton University Press, Princeton, New Jersey. [17]

Anderson, E. 1949. *Introgressive hybridization*. Wiley, New York. [4]

Anderson, W. W. 1966. Genetic divergence in M. Vetukhiv's experimental populations of *Drosophila pseudoobscura*. 3. Divergence in body size. *Genet. Res.* 7: 255–266. [7]

Anderson, W. W. 1971. Genetic equilibrium and population growth under density-regulated selection. *Amer. Natur.* 105:489–498. [9]

Anderson, W. W., Th. Dobzhansky, O. Pavlovsky, J. Powell, and D. Yardley. 1975. Genetics of natural populations. XLII. Three decades of genetic change in *Drosophila pseudoobscura*. *Evolution* 29: 24–36. [10]

Andersson, M. 1982a. Female choice selects for extreme tail length in a widowbird. *Nature* 199: 818–820. [5, 9]

Andersson, M. 1982b. Sexual selection: natural selection and quality advertisement. *Biol. J. Linn. Soc.* 17: 375–393. [9]

Andrewartha, H. B., and L. C. Birch. 1954. *The distribution and abundance of animals*. University of Chicago Press, Chicago. [2]

Anstey, R. L. 1978. Taxonomic survivorship and morphological complexity in Paleozoic bryozoan genera. *Paleobiology* 4: 407–418. [12]

Antonovics, J., A. D. Bradshaw, and R. G. Turner. 1971. Heavy metal tolerance in plants. *Adv. Ecol. Res.* 7: 1–85. [4, 6]

Antonovics, J. and N. C. Ellstrand. 1984. Experimental studies of the evolutionary significance of sexual reproduction. I. A test of the frequency-dependent selection hypothesis. *Evolution* 38: 103–115. [9]

Aoki, K. 1982. A condition for group selection to prevail over counteracting individual selection. *Evolution* 36: 832–842. [9]

Ardrey, R. 1966. *The territorial imperative*. Dell, New York. [17]

Armstrong, R. A. and R. McGehee. 1980. Competitive exclusion. *Amer. Natur.* 115: 157–170. [2, 16]

Arnheim, N. 1983. Concerted evolution of multi-gene families. In M. Nei and R. K. Koehn (eds.), *Evolution of genes and proteins*, pp. 38–61. Sinauer Associates, Sunderland, Massachusetts. [15]

Arnold, A. J., and K. Fristrup. 1982. The theory of evolution by natural selection: a hierarchical expansion. *Paleobiology* 8: 113–129. [14]

Arnold, S. J. 1983. Sexual selection: the interface of theory and empiricism. In P. Bateson (ed.), *Mate choice*, pp. 67–107. Cambridge University Press, Cambridge. [9]

Askew, R. R. 1968. Considerations on speciation in Chalcidoidea (Hymenoptera). *Evolution* 22: 642–645. [5, 8]

Atchley, W. R., and J. J. Rutledge. 1980. Genetic components of size and shape. I. Dynamics of components of phenotypic variability and covariability during ontogeny in the laboratory rat. *Evolution* 34: 1161–1173. [14]

Atkins, M. D. 1978. *Insects in perspective*. Macmillan, New York. [10]

Avise, J. C. 1983. Protein variation and phylogenetic reconstruction. In G. S. Oxford and D. Rollinson (eds.), *Protein polymorphism: adaptive and taxonomic significance*, pp. 103–130. Academic Press, New York. [10]

Avise, J. C. and C. F. Aquadro. 1982. A comparative summary of genetic distances in the vertebrates. *Evol. Biol.* 15: 151–185. [8, 10]

Avise, J. C. and R. A. Lansman. 1983. Polymorphism of mitochondrial DNA in populations of higher animals. In M. Nei and R. K. Koehn (eds.), *Evolution of genes and proteins*, pp. 147–164. Sinauer Associates, Sunderland, Massachusetts. [15]

Avise, J. C., R. A. Lansman, and R. O. Shade. 1979. The use of restriction endonucleases to measure mitochondrial DNA sequence relatedness in natural populations. I. Population structure and evolution in the genus *Peromyscus*. *Genetics* 92: 279–295. [4]

Avise, J. C., and R. K. Selander. 1972. Evolutionary genetics of cave-dwelling fish of the genus *Astyanax*. *Evolution* 26: 1–19. [6]

Avise, J. C., and M. H. Smith. 1974. Biochemical genetics of sunfish. I. Geographic variation and subspecific intergradation in the bluegill, *Lepomis macrochirus*. *Evolution* 28: 42–56. [4]

Ayala, F. J. 1966. Evolution of fitness. I. Improvement in the productivity and size of irradiated populations of *Drosophila serrata* and *Drosophila birchii*. *Genetics* 53: 883–895. [7]

Ayala, F. J. 1975. Genetic differentiation during the speciation process. *Evol. Biol.* 8: 1–78. [8]

Ayala, F. J. (ed.) 1976. *Molecular evolution*. Sinauer Associates, Sunderland, Massachusetts. [3]

Ayala, F. J., and C. A. Campbell. 1974. Frequency-dependent selection. *Annu. Rev. Ecol. Syst.* 5: 115–138. [6]

Ayala, F. J., and Th. Dobzhansky. (eds.) 1974. *Studies in the philosophy of biology. Reduction and related problems*. University of California Press, Berkeley and Los Angeles. [15]

Ayala, F. J., M. L. Tracy, D. Hedgecock, and R. C. Richmond. 1974. Genetic differentiation during the speciation process in *Drosophila*. *Evolution* 28: 576–592. [8]

Bajema, C. J. 1963. Estimation of the direction and intensity of natural selection in relation to human intelligence by means of the intrinsic rate of natural increase. *Eugenics Quarterly* 10: 175–187. [17]

Bakker, R. T. 1977. Tetrapod mass extinctions—a model of the regulation of speciation rates and immigration by cycles of topographic diversity. In A. Hallam (ed.), *Patterns of evolution, as illustrated by the fossil record*, pp. 439–468. Elsevier, Amsterdam. [12]

Bakker, R. T. 1983. The deer flees, the wolf pursues: incongruencies in predator–prey coevolution. In D. J. Futuyma and M. Slatkin (eds.), *Coevolution*, pp. 350–382. Sinauer Associates, Sunderland, Massachusetts. [12, 14, 16]

Balkau, B. J., and M. W. Feldman. 1973. Selection for migration modification. *Genetics* 74: 171–174. [9]

Baltimore, D. 1985. Retroviruses and retrotransposons: the role of reverse transcription in shaping the eukaryotic genome. *Cell* 40: 481–482. [3, 15]

Bambach, R. K. 1977. Species richness in marine benthic habitats throughout the Phanerozoic. *Paleobiology* 3: 152–167. [12]

Bambach, R. K. 1983. Ecospace utilization and guilds in marine communities through the Phanerozoic. In M. J. S. Tevesz and P. L. McCall (eds.), *Biotic interactions in recent and fossil benthic communities*, pp. 719–746. Plenum, New York. [12, 16]

Bambach, R. K., C. R. Scotese, and A. M. Ziegler. 1980. Before Pangaea: the geographies of the Paleozoic world. *Amer. Sci.* 68: 26–38. [11]

Bannister, J. V., and M. W. Parker. 1985. The presence of a copper/zinc superoxide dismutase in the bacterium *Photobacterium leiognathi*: a likely case of gene transfer from eukaryotes to prokaryotes. *Proc. Natl. Acad. Sci. U.S.A.* 82: 149–152. [15]

Barash, D. P. 1982. *Sociobiology and behavior*, Second edition. Elsevier, New York. [17]

Barker, W. C. and M. O. Dayhoff. 1980. Evolutionary and functional relationships of homologous physiological mechanisms. *BioScience*. 30: 593–599. [15]

Barrett, J. A. 1983. Plant–fungus symbioses. In D. J. Futuyma and M. Slatkin (eds.), *Coevolution*, pp. 139–160. Sinauer Associates, Sunderland, Massachusetts. [16]

Barrowclough, G. F. 1980. Gene flow, effective population sizes, and genetic variance components in birds. *Evolution* 34: 789–798. [5]

Barton, N. H. 1979. Gene flow past a cline. *Heredity* 43: 333–339. [8]

Barton, N. H. 1980. The fitness of hybrids between two chromosomal races of the grasshopper *Podisma pedestris*. *Heredity* 45: 49–61. [8]

Barton, N. H. 1983. Multilocus clines. *Evolution* 37: 454–471. [8]

Barton, N. H., and B. Charlesworth. 1984. Genetic revolutions, founder effects, and speciation. *Ann. Rev. Ecol. Syst.* 15: 133–164. [8, 14]

Barton, N. H., and G. M. Hewitt. 1981. Hybrid zones and speciation. In W. R. Atchley and D. S. Woodruff (eds.), *Evolution and speciation: Essays in honor of M. J. D. White*, pp. 109–145. Cambridge University Press, Cambridge. [6, 8]

Bateman, A. J. 1947a. Contamination of seed crops. I. Insect pollination. *J. Genet.* 48: 257–275. [5]

Bateman, A. J. 1947b. Contamination of seed crops. II. Wind pollination. *Heredity* 1: 235–246. [5]

Bateson, P. (ed.) 1983. *Mate choice*. Cambridge University Press, Cambridge. [9]

Battaglia, B. 1958. Balanced polymorphism in *Tisbe reticulata*, a marine copepod. *Evolution* 12: 358–364. [4]

Bellairs, A. 1970. *The life of reptiles*. Universe Books, New York. [13]

Bender, E. A., T. J. Case, and M. E. Gilpin. 1984. Perturbation experiments in community ecology: theory and practice. *Ecology* 65: 1–13. [2]

Bengtsson, B. O., and W. F. Bodmer. 1976. On the increase of chromosomal mutations under random mating. *Theoret. Pop. Biol.* 9: 260–281. [8]

Bennett, J. 1960. A comparison of selective methods and a test of the pre-adaptation hypothesis. *Heredity* 15: 65–77. [3]

Bennett, M. D. 1982. Nucleotypic basis of the spatial ordering of chromosomes in eukaryotes and the implication of the order for genome evolution and phenotypic variation. In G. A. Dover and R. B. Flavell (eds.), *Genome Evolution*, pp. 239–261. Academic Press, New York. [15]

Benson, W. W., K. S. Brown, Jr., and L. E. Gilbert. 1976. Coevolution of plants and herbivores: passion flower butterflies. *Evolution* 28: 659–680. [16]

Bentley, B. L. 1977. Extrafloral nectaries and protection by pugnacious bodyguards. *Annu. Rev. Ecol. Syst.* 8: 407–427. [16]

Benton, M. J. 1983. Dinosaur success in the Triassic: a noncompetitive ecological model. *Quart. Rev. Biol.* 58: 29–55. [10]

Benton, M. J. 1985. Classification and phylogeny of the diapsid reptiles. *Zool. J. Linn. Soc.* 84: 97–164. [10]

Berenbaum, M. 1983. Coumarins and caterpillars: a case for coevolution. *Evolution* 39: 163–179. [15, 16]

Berger, E. 1976. Heterosis and the maintenance of enzyme polymorphisms. *Amer. Natur.* 110: 823–839. [6]

Bertness, M. D. 1984. Habitat and community modifications by an introduced herbivorous snail. *Ecology* 65: 370–381. [2]

Berven, K. A., D. E. Gill, and S. J. Smith-Gill. 1979. Countergradient selection in the green frog, *Rana clamitans*. *Evolution* 33: 609–623. [4]

Birch, L. C., Th. Dobzhansky, P. D. Elliott, and R. C. Lewontin. 1963. Relative fitness of geographic races of *Drosophila serrata*. *Evolution* 17: 72–83. [2]

Birns, B. 1976. The emergence and socialization of sex differences in the earliest years. *Merrill-Palmer Quarterly* 22: 229–254. [17]

Bishop, J. A. 1981. A neoDarwinian approach to resistance: examples from mammals. In J. A. Bishop and L. M. Cook (eds.), *Genetic consequences of man-made change*, pp. 37–51. Academic Press, London. [6]

Bishop, J. A., and L. M. Cook. 1980. Industrial melanism and the urban environment. *Adv. Ecol. Res.* 11: 373–404. [4]

Bishop, J. A., and L. M. Cook (eds.) 1981. *Genetic consequences of man-made change*. Academic Press, London. [6]

Björkman, O., and P. Holmgren. 1963. Adaptibility of the photosynthetic apparatus to light intensity in ecotypes from exposed and shaded habitats. *Physiol. Plant.* 16: 889–914. [4]

Blum, M. S., and N. A. Blum (eds.). 1979. *Sexual selection and reproductive competition in insects*. Academic Press, New York. [9]

Boag, P. T. 1983. The heritability of external morphology in Darwin's ground finches (*Geospiza*) on Isla Daphne Major, Galápagos. *Evolution* 37: 877–894. [4]

Boag, P. T. and P. R. Grant. 1981. Intense natural selection in a population of Darwin's finches (Geospizinae) in the Galápagos. *Science* 214: 82–85. [7]

Bock, W. J. 1959. Preadaptation and multiple evolutionary pathways. *Evolution* 13: 194–211. [14]

Bock, W. J. 1970. Microevolutionary sequences as a fundamental concept in macroevolutionary models. *Evolution* 24: 704–722. [2]

Bodmer, M. and M. Ashburner. 1984. Conservation and change in the DNA sequences coding for alcohol dehydrogenase in sibling species of *Drosophila*. *Nature* 309: 425–430. [15]

Bodmer, W. F. 1965. Differential fertility in population genetic models. *Genetics* 51: 411–424. [6]

Bodmer, W. F. and P. A. Parsons. 1962. Linkage and recombination in evolution. *Adv. Genet.* 11: 1–100. [7]

Bonnell, M. L. and R. K. Selander. 1974. Elephant seals: genetic variation and near extinction. *Science* 184: 908–909. [5]

Bonner, J. R. (ed.). 1982. *Evolution and development*. Springer-Verlag, Berlin. [14]

Bookstein, F. L, P. D. Gingerich, and A. G. Kluge. 1978. Hierarchical linear modeling of the tempo and mode of evolution. *Paleobiology* 4: 120–134. [14]

Borror, D. J., D. M. Delong, and C. A. Triplehorn. 1981. *An introduction to the study of insects*, Fifth edition. Saunders College Publishing Co., Philadelphia. [10]

Boucher, D. H., S. James, and K. H. Keeler. 1982. The ecology of mutualisms. *Annu. Rev. Ecol. Syst.* 13: 315–347. [2, 16]

Boucot, A. J. 1978. Community evolution and rates of cladogenesis. *Evol. Biol.* 11: 545–655. [16]

Boudreaux, H. B. 1979. *Arthropod phylogeny, with special reference to insects*. Wiley, New York. [11]

Bowler, P. 1984. *Evolution: The history of an idea*. University of California Press, Berkeley. [1]

Briggs, B. G. 1962. Hybridization in *Ranunculus*. *Evolution* 16: 372–390. [4]

Britten, R. J., and E. H. Davidson. 1971. Repetitive and nonrepetitive DNA and a speculation on the origin of evolutionary novelty. *Quart. Rev. Biol.* 46: 111–133. [3]

Brockmann, H. J. 1984. The evolution of social behavior in insects. In J. R. Krebs and N. B. Davies (eds.), *Behavioral ecology: an evolutionary approach*, Second edition, pp. 340–361. Sinauer Associates, Sunderland, Massachusetts. [9]

Brookfield, J. F. Y., E. Montgomery, and C. H. Langley. 1984. Apparent absence of transposable elements related to the P elements of *D. melanogaster* in other species of *Drosophila*. *Nature* 310: 330–332. [15]

Brown, A. H. D. 1979. Enzyme polymorphisms in plant populations. *Theoret. Pop. Biol.* 15: 1–42. [5]

Brown, D. D., P. C. Wensink, and E. Jordan. 1972. A comparison of the ribosomal DNAs of *Xenopus laevis* and *Xenopus mulleri*: the evolution of tandem genes. *J. Molec. Biol.* 63: 59–73. [15]

Brown, J. H. 1971a. Mammals on mountaintops: nonequilibrium insular biogeography. *Amer. Natur.* 105: 467–478. [13]

Brown, J. H. 1971b. Mechanisms of competitive exclusion between two species of chipmunks. *Ecology* 52: 305–319. [2]

Brown, J. H. 1975. Geographical ecology of desert rodents. In M. L. Cody and J. M. Diamond (eds.), *Ecology and evolution of communities*, pp. 315–341. Harvard University Press, Cambridge, Massachusetts. [2, 13]

Brown, J. H. 1981. Two decades of homage to Santa Rosalia: toward a general theory of diversity. *Amer. Zool.* 21: 877–888. [2, 13]

Brown, J. H. 1984. On the relationship between abundance and distribution of species. *Amer. Natur.* 124: 255–279. [13]

Brown, J. H., D. W. Davidson, and O. J. Reichman. 1979. An experimental study of competition between seed-eating desert rodents and ants. *Amer. Zool.* 19: 1129–1143. [2]

Brown, J. H., and A. C. Gibson. 1983. *Biogeography*. Mosby, St. Louis. [2, 13]

Brown, W. L., Jr. 1959. General adaptation and evolution. *Syst. Zool.* 7: 157–168. [10, 14]

Brown, W. L., Jr., and E. O. Wilson. 1956. Character displacement. *Syst. Zool.* 5: 49–64. [2, 16]

Brown, W. M. 1980. Polymorphism in mitochondrial DNA of humans as revealed by restriction endonuclease analysis. *Proc. Natl. Acad. Sci. U.S.A.* 77: 3605–3609. [4, 17]

Brown, W. M. 1983. Evolution of animal mitochondrial DNA. In M. Nei and R. K. Koehn (eds.), *Evolution of genes and proteins*, pp. 62–88. Sinauer Associates, Sunderland, Massachusetts. [3, 15]

Brown, W. M., M. George, Jr., and A. C. Wilson. 1979. Rapid evolution of animal mitochondrial DNA. *Proc. Natl. Acad. Sci. U.S.A.* 76: 1967–1971. [15]

Brown, W. M., E. M. Prager, A. Wang, and A. C. Wilson. 1982. Mitochondrial DNA sequences of primates: tempo and mode of evolution. *J. Molec. Evol.* 18: 225–239. [17]

Brundin, L. 1965. On the real nature of transantarctic relationships. *Evolution*. 19: 496–505. [13]

Brussard, P. F., P. R. Ehrlich, and M. C. Singer. 1974. Adult movements and population structure in *Euphydras editha*. *Evolution* 28: 408–415. [2]

Bryant, E. H. 1974. On the adaptive significance of enzyme polymorphisms in relation to environmental variability. *Amer. Natur.* 108: 1–19. [6]

Bulmer, M. G. 1973. Inbreeding in the great tit. *Heredity* 30: 313–325. [5]

Bumpus, H. C. 1899. The elimination of the unfit as illustrated by the introduced sparrow. *Biol. Lec. Mar. Biol. Woods Hole* 11: 209–226. [7]

Buri, P. 1956. Gene frequency drift in small population of mutant *Drosophila*. *Evolution* 10: 367–402. [5]

Bush, G. L. 1969. Sympatric host race formation and speciation in frugivorous flies of the genus *Rhagoletis* (Diptera, Tephritidae). *Evolution* 23: 237–251. [8]

Bush, G. L. 1975a. Modes of animal speciation. *Annu. Rev. Ecol. Syst.* 6: 334–364. [8]

Bush, G. L. 1975b. Sympatric speciation in phytophagous parasitic insects. In P. W. Price (ed.), *Evolutionary strategies of parasitic insects and mites*, pp. 187–206. Plenum, New York. [8]

Bush, G. L., S. M. Case, A. C. Wilson, and J. L. Patton. 1977. Rapid speciation and chromosomal evolution in mammals. *Proc. Natl. Acad. Sci. U.S.A.* 74: 3942–3946. [8]

Buth, D. G. 1984. The application of electrophoretic data in systematic studies. *Annu. Rev. Ecol. Syst.* 15: 501–522. [10]

Cahn, P. H. 1959. Comparative optic development in *Astyanax mexicanus* and two of its blind cave derivatives. *Bull. Amer. Mus. Nat. Hist.* 115: 72–112. [14]

Cain, A. J., and P. M. Sheppard. 1954. Natural selection in *Cepaea*. *Genetics* 39: 89–116. [4]

Caisse, M., and J. Antonovics. 1978. Evolution in closely adjacent plant populations. IX. Evolution of reproductive isolation in clinal populations. *Heredity* 40: 371–384. [8]

Calos, M. P., and J. H. Miller. 1980. Transposable elements. *Cell* 20: 579–595. [15]

Campbell, A. 1983. Transposons and their evolutionary significance. In M. Nei and R. K. Koehn (eds.), *Evolution of genes and proteins*, pp. 258–279. Sinauer Associates, Sunderland, Massachusetts. [15]

Campbell, B. G. 1982. *Humankind emerging*, Third edition. Little, Brown, Boston [17]

Campbell, J. H. 1982. Autonomy in evolution. In R. Milkman (ed.), *Perspectives on evolution*, pp. 190–201. Sinauer, Sunderland, Massachusetts. [3]

Caple, G. R., R. T. Balda, and W. R. Willis. 1983. The physics of leaping animals and the evolution of preflight. *Amer. Natur.* 121: 455–467. [14]

Caraco, T. 1980. On foraging time allocation in a stochastic environment. *Ecology* 61: 119–128. [16]

Carpenter, F. M. 1976. Geological history and evolution

of the insects. *Proc. XV Internat. Congr. Entomol.*, 63–70. [11]

Carpenter, F. M., and L. Burnham. 1985. The geological record of insects. *Annu. Rev. Earth Planet. Sci.* 13: 297–314. [11]

Carr, G. D., and D. W. Kyhos. 1981. Adaptive radiation in the Hawaiian silversword alliance (Compositae-Madiinae). I. Cytogenetics of spontaneous hybrids. *Evolution* 35: 543–556. [8]

Carroll, R. L. 1982. Early evolution of reptiles. *Annu. Rev. Ecol. Syst.* 13: 87–109. [10]

Carson, H. L. 1959. Genetic conditions which promote or retard the formation of species. *Cold Spring Harbor Symp. Quant. Biol.* 24: 87–105. [9]

Carson, H. L. 1970. Chromosome tracers of the origin of species. *Science* 168: 1414–1418. [8]

Carson, H. L.. 1975. The genetics of speciation at the diploid level. *Amer. Natur.* 109: 83–92. [8]

Carson, H. L. 1982. Speciation as a major reorganization of polygenic balances. In C. Barigozzi (ed.), *Mechanisms of speciation*, pp. 411–433. Alan R. Liss, New York. [8]

Carson, H. 1983. Chromosomal sequences and interisland colonizations in Hawaiian *Drosophila*. *Genetics* 103: 465–482. [10]

Carson, H. L., L. S. Chang, and T. W. Lyttle. 1982. Decay of female sexual behavior under parthenogenesis. *Science* 218: 68–70. [9]

Carson, H. L., and K. Y. Kaneshiro. 1976. *Drosophila* of Hawaii: systematics and evolutionary genetics. *Annu. Rev. Ecol. Syst.* 7: 311–346. [8, 10]

Carson, H. L., and A. R. Templeton. 1984. Genetic revolutions in relation to speciation phenomena: the founding of new populations. *Annu. Rev. Ecol. Syst.* 15: 97–131. [8]

Case, T. J. 1981. Niche packing and coevolution in competition communities. *Proc. Natl. Acad. Sci. U.S.A.* 78: 5021–5025. [16]

Case, T. J. 1982. Coevolution in resource-limited competition communities. *Theor. Pop. Biol.* 21: 69–91. [16]

Case, T. J., J. Faaborg, and R. Sidell. 1983. The role of body size in the assembly of West Indian bird communities. *Evolution* 37: 1062–1074. [2]

Case, T. J., and R. Sidell. 1983. Pattern and chance in the structure of model and natural communities. *Evolution* 37: 832–849. [16]

Caugant, D. A., B. R. Levin, and R. K. Selander. 1981. Genetic diversity and temporal variation in the *E. coli* population of a human host. *Genetics* 98: 467–490. [4, 5, 6]

Cavalli-Sforza, L. L., and W. F. Bodmer. 1971. *The genetics of human populations.* Freeman, San Francisco. [4, 5, 6, 7, 9, 17]

Cavalli-Sforza, L. L., and M. W. Feldman, 1981. *Cultural transmission and evolution.* Princeton University Press, Princeton, New Jersey. [17]

Cavener, D. 1979. Preference for ethanol in *Drosophila melanogaster* associated with the alcohol dehydrogenase polymorphism. *Behav. Genet.* 9: 349–365. [7]

Cavener, D. R., and M. T. Clegg. 1981. Multigenic response to ethanol in *Drosophila melanogaster*. *Evolution* 35: 1–10. [7]

Chao, L., and E. C. Cox. 1983. Competition between high and low mutating strains of *Escherichia coli*. *Evolution* 37: 125–134. [9]

Chao, L., B. R. Levin, and F. M. Stewart. 1977. A complex community in a simple habitat: an experimental study with bacteria and phage. *Ecology* 58: 369–378. [16]

Chao, L., C. Vargas, B. B. Spear, and E. C. Cox. 1983. Transposable elements as mutator genes in evolution. *Nature* 303: 633–635. [15]

Charlesworth, B. 1980. *Evolution in age-structured populations.* Cambridge University Press, Cambridge. [9]

Charlesworth, B. 1982. Hopeful monsters cannot fly. *Paleobiology* 8: 469–474. [14]

Charlesworth, B. 1984a. Some quantitative methods for studying evolutionary patterns in single characters. *Paleobiology* 10: 308–318. [14]

Charlesworth, B. 1984b. The cost of phenotypic evolution. *Paleobiology* 10: 319–327. [14]

Charlesworth, B. 1986. The population genetics of transposable elements. In T. Ohta and K.-I. Aoki (eds.), *Population genetics and molecular evolution.* Springer-Verlag, Berlin. [15]

Charlesworth, B., and D. Charlesworth. 1983. The population dynamics of transposable elements. *Genet. Res.* 42: 1–27. [15]

Charlesworth, B., D. Charlesworth, M. Loukas, and K. Morgan. 1979. A study of linkage disequilibrium in British populations of *Drosophila subobscura*. *Genetics* 92: 983–994. [7]

Charlesworth, B., R. Lande, and M. Slatkin. 1982. A neo-Darwinian commentary on macroevolution. *Evolution* 36: 474–498. [8, 14]

Charlesworth, B., and C. H. Langley. 1986. The evolution of self-regulated transposition of transposable elements. *Genetics* 112: 359–383. [15]

Charlesworth, B., C. H. Langley, and W. Stephan. 1986. The evolution of restricted recombination and the accumulation of repeated DNA sequences. *Genetics* 112: 947–962. [15]

Charnov, E. 1976. Optimal foraging: attack strategy of a mantid. *Amer. Natur.* 110: 141–151. [9]

Charnov, E. L. 1982. *The theory of sex allocation.* Princeton University Press, Princeton, New Jersey. [9]

Charnov, E. L., and W. M. Schaffer. 1973. Life-history consequences of natural selection: Cole's result revisited. *Amer. Natur.* 107: 791–793. [9]

Chesson, P. L., and T. J. Case. 1986. Overview: nonequilibrium community theories: chance, variability, history, and coexistence. In J. Diamond and T. J. Case (eds.), *Community ecology.* Harper & Row, New York. [2]

Chesson, P. L., and R. R. Warner. 1981. Environmental variability promotes coexistence in lottery competitive systems. *Amer. Natur.* 117: 923–943. [16]

Cheverud, J. 1982. Phenotypic, genetic, and environmental morphological integration in the cranium. *Evolution* 36: 499–516. [14]

Cheverud, J. M. 1984. Quantitative genetics and developmental constraints on evolution by selection. *J. Theor. Biol.* 101: 155–171. [14]

Cheverud, J. M., M. M. Dow, and W. W. Leutenegger. 1985. The quantitative assessment of phylogenetic constraints in comparative analyses: sexual dimorphism in body weight among primates. *Evolution* 39: 1335–1351. [9]

Chinnici, J. P. 1971. Modification of recombination frequency in *Drosophila.* I. Selection for increased and decreased crossing over. II. The polygenic control of crossing over. *Genetics* 69: 71–83; 85–96. [3, 9]

Churchill, W. 1967. *Homosexual behavior among males. A cross-cultural and cross-species investigation.* Prentice-Hall, Englewood Cliffs, New Jersey. [17]

Cifelli, R. 1969. Radiation of Cenozoic planktonic Foraminifera. *Syst. Zool.* 18: 154–168. [12]

Clark, B. C. 1966. The evolution of morph-ratio clines. *Amer. Natur.* 100: 389–402. [8]

Clarke, B. 1962. Balanced polymorphism and the diversity of sympatric species. *Syst. Assoc. Publ.* 4: 47–70. [6]

Clarke, B. 1976. The ecological genetics of host-parasite relationships. In A. E. R. Taylor and R. Muller (eds.), *Genetic aspects of host-parasite relationships. Symp. Brit. Soc. Parasitol.* 14: 87–103. [6]

Clarke, C. A., C. G. C. Dickson, and P. M. Sheppard. 1963. Larval color pattern in *Papilio demodocus. Evolution* 17: 130–137. [6]

Clarke, C. A., and P. M. Sheppard. 1960. Supergenes and mimicry. *Heredity* 14: 175–185. [7]

Clarke, P. H. 1974. The evolution of enzymes for the utilisation of novel substrates. In M. J. Carlile and J. J. Skehel (eds.), *Evolution in the microbial world,* pp. 183–217. Cambridge University Press, Cambridge. [3, 15]

Clausen, J., D. D. Keck, and W. M. Hiesey. 1947. Heredity of geographically and ecologically isolated races. *Amer. Natur.* 81: 114 -133. [4]

Clay, K. 1982. Environmental and genetic determinants of cleistogamy in a natural population of the grass *Danthonia spicata. Evolution* 36: 734–741. [5]

Clayton, G. A., and A. Robertson. 1955. Mutation and quantitative variation. *Amer. Natur.* 89: 151–158. [3, 4]

Clayton, G. A., and A. Robertson. 1957. An experimental check on quantitative genetical theory. II. The long-term effects of selection. *J. Genet.* 55: 152–170. [7]

Clayton, G. A., G. R. Knight, J. A. Morris, and A. Robertson. 1957. An experimental check on quantitative genetic theory. III. Correlated responses. *J. Genet.* 55: 171–180. [7]

Clemens, W. A., J. D. Archibald, and L. J. Hickey. 1981. Out with a whimper not a bang. *Paleobiology* 7: 293–297. [12]

Cloud, P. 1976. Beginnings of biospheric evolution and their biochemical consequences. *Paleobiology* 2: 351–387. [11]

Cloud, P. 1978. *Cosmos, earth and man.* Yale University Press, New Haven, Connecticut. [11]

Cloud, P., and M. F. Glaessner. 1982. The Ediacaran period and system: Metazoa inherit the earth. *Science* 217: 783–792. [11]

Clutton-Brock, T. H., and P. H. Harvey. 1984. Comparative approaches to investigating adaptation. In J. R. Krebs and N. B. Davies (eds.), *Behavioral ecology: an evolutionary approach,* Second edition, pp. 7–29. Sinauer Associates, Sunderland, Massachusetts. [9]

Cock, A. G. 1966. Genetical aspects of metrical growth and form in animals. *Quart. Rev. Biol.* 41: 131–190. [14]

Cody, M. L., and J. Diamond (eds.). 1975. *Ecology and evolution of communities.* Harvard University Press, Cambridge, Massachusetts. [2, 16]

Cohan, F. M. 1984. Can uniform selection retard random genetic divergence between isolated conspecific populations? *Evolution* 38: 495–504. [4, 7]

Colbert, E. H. 1949. Progressive adaptations as seen in the fossil record. In G. L. Jepsen, G. G. Simpson, and E. Mayr (eds.), *Genetics, paleontology, and evolution,* pp. 390–402. Princeton University Press, Princeton, New Jersey. [12]

Colbert, E. H. 1980. *Evolution of the vertebrates.* Wiley, New York. [11]

Cold Spring Harbor Symposia on Quantitative Biology. 1957. *Population studies: animal ecology and demography.* The Biological Laboratory, Cold Spring Harbor, New York. [2]

Colgan, D. J., and J. Cheney. 1980. The inversion polymorphism of *Keyacris scurra* and the adaptive topography. *Evolution* 34: 181–192. [7]

Collins, J. 1959. Darwin's impact on philosophy. *Thought.* 34: 185–248. [1]

Colwell, R. K. 1981. Group selection is implicated in the evolution of female-biased sex ratios. *Nature* 290: 401–404. [9]

Colwell, R. K. 1986. Community biology and sexual selection: lessons from hummingbird flower mites. In J. Diamond and T. Case (eds.), *Community ecology,* pp. 406–424. Harper & Row, New York. [16]

Conant, R. 1958. *A field guide to reptiles and amphibians.* Houghton Mifflin, Boston. [4, 13, 14]

Connell, J. H. 1983. On the prevalence and relative importance of interspecific competition: evidence from field experiments. *Amer. Natur.* 122: 661–696. [2, 16]

Connor, E. F., and D. Simberloff. 1984. Neutral models of species co-occurrence patterns. In D. R. Strong Jr., D. Simberloff, L. G. Abele, and A. B. Thistle (eds.), *Ecological communities: conceptual issues and the evidence,* pp. 316–331. Princeton University Press, Princeton, New Jersey. [13]

Conway Morris, S., and H. B. Whittington. 1979. The animals of the Burgess Shale. *Sci. Amer.* 241 (1): 122–133. [11]

Cooke, F., and F. Cooch. 1968. The genetics of polymorphism in the goose *Anser coerulescens. Evolution* 22: 289–300. [4]

Coope, G. R.. 1979. Late Cenozoic fossil Coleoptera: evolution, biogeography, and ecology. *Annu. Rev. Ecol. Syst.* 10: 249–267. [11]

Corbin, K. W., C. G. Sibley, and A. Ferguson. 1979. Genic changes associated with the establishment of sympatry in orioles of the genus *Icterus*. *Evolution* 33: 624–633. [4]

Coughenour, M. B. 1985. Graminoid responses to grazing by large herbivores—adaptations, exaptations, and interacting processes. *Ann. Missouri Bot. Gard.* 72: 852–863. [16]

Cox, C. R., and B. J. Le Boeuf. 1977. Female incitation of male competition: a mechanism in sexual selection. *Amer. Natur.* 111: 317–335. [9]

Cox, G. W., and D. G. Cox. 1974. Substrate color matching in the grasshopper *Circotettix rabula* (Orthoptera: Acrididae). *Great Basin Nat.* 34: 60–70. [6]

Coyne, J. A. 1974. The evolutionary origin of hybrid inviability. *Evolution* 28: 505–506. [8]

Coyne, J. 1976. Lack of genic similarity between two sibling species of Drosophila as revealed by varied techniques. *Genetics* 84: 593–607. [4]

Coyne, J. A. 1983a. Genetic basis of differences in genital morphology among three sibling species of *Drosophila*. *Evolution* 37: 1101–1118. [8]

Coyne, J. A. 1983b. Correlation between heterogygosity and rate of chromosome evolution in animals. *Amer. Natur.* 123: 725–729. [8]

Coyne, J. A., A. A. Felton, and R. C. Lewontin. 1978. Extent of genetic variation at a highly polymorphic esterase locus in *Drosophila pseudoobscura*. *Proc. Natl. Acad. Sci. U.S.A.* 75: 5090–5093. [4]

Cracraft, J. 1974. Phylogeny and evolution of the ratite birds. *Ibis* 116: 494–521. [13]

Cracraft, J. 1983. The significance of phylogenetic classifications for systematic and evolutionary biology. In J. Felsenstein (ed.), *Numerical Taxonomy*, pp. 1–17. Springer-Verlag, Berlin. [10]

Crawford, I. P. 1982. Nucleotide sequences and bacterial evolution. In R. Milkman (ed.), *Perspectives on evolution*, pp. 148–163. Sinauer Associates, Sunderland, Massachusetts. [15]

Crow, J. F. 1958. Some possibilities for measuring selection intensities in man. *Human Biol.* 30: 1- 13. [17]

Crow, J. F. 1966. The quality of people: human evolutionary changes. *BioScience* 16: 863–867. [17]

Crow, J. F., and C. Denniston. 1985. Mutation in human populations. *Adv. Hum. Genet.* 14: 59–123. [17]

Crow, J. F., and M. Kimura. 1965. Evolution in sexual and asexual populations. *Amer. Natur.* 99: 439–450. [9]

Crow, J. F. and M. Kimura. 1970. *An introduction to population genetics theory*. Harper & Row, New York. [4, 5, 6, 15]

Crow, J. F., and T. Nagylaki. 1976. The rate of change of a character correlated with fitness. *Amer. Natur.* 110: 207–213. [7]

Crumpacker, D. W. 1967. Genetic loads in maize (*Zea mays* L.) and other cross-fertilized plants and animals. *Evol. Biol.* 1: 306–324. [4]

Cullis, C. A. 1983. Environmentally induced DNA changes in plants. *CRC Critical Reviews in Plant Science* 1: 117–131. [3]

Culver, D. C. 1982. *Cave life*. Harvard University Press, Cambridge, Massachusetts. [9, 14]

Dacosta, C. P., and C. M. Jones. 1971. Cucumber beetle resistance and mite susceptibility controlled by the bitter gene in *Cucumis sativus*. *Science* 172: 1145–1146. [16]

Daday, H. 1954. Gene frequencies in wild populations of *Trifolium repens* L. I. Distribution by latitude. *Heredity* 8: 61–78. [4]

Danforth, C. H. 1950. Evolution and plumage traits in pheasant hybrids, *Phasianus* × *Chrysolophus*. *Evolution* 4: 301–315. [8]

Daniels, S. B., L. D. Strasbaugh, L. Ehrman, and R. Armstrong. 1984. Sequences homologous to *P* elements occur in *Drosophila paulistorum*. *Proc. Natl. Acad. Sci. U.S.A.* 81: 6794–6797. [15]

Darlington, P. J. Jr. 1957. *Zoogeography: the geographical distribution of animals*. Wiley, New York. [13]

Darnell, J. E. 1982. Variety in the level of gene control in eukaryotic cells. *Nature* 297: 365–371. [3]

Davidson, E. H., and R. J. Britten. 1973. Organization, transcription and regulation in the animal genome. *Quart. Rev. Biol.* 48: 565–613. [3, 14, 15]

Davies, R. W. 1971. The genetic relationship of two quantitative characters in *Drosophila melanogaster*. II. Location of the effects. *Genetics* 69: 363–375. [7]

Davies, R. W., and P. L. Workman. 1971. The genetic relationship of two quantitative characters in *Drosophila melanogaster*. I. Responses to selection and whole chromosome analysis. *Genetics* 69: 353–361. [7]

Davis, M. B. 1976. Pleistocene biogeography of temperate deciduous forests. *Geoscience and Man* 13: 13–26. [11, 13]

Dawkins, R. 1976. *The selfish gene*. Oxford University Press, New York. [9, 15]

Dawid, I., et al. 1982. Genomic change and morphological evolution: group report. In J. Bonner (ed.), *Evolution and development*, pp. 19–39. Springer-Verlag, Berlin. [14]

Dawson, P. S. 1970. Linkage and the elimination of deleterious mutant genes from experimental populations. *Genetica* 41: 147–169. [6]

Dawson, W. R., G. A. Bartholomew, and A. F. Bennett. 1977. A reappraisal of the aquatic specialization of the Galápagos marine iguana (*Amblyrhynchus cristatus*). *Evolution* 31: 891–897. [14]

DeBach, P. 1966. The competitive displacement and coexistence principles. *Annu. Rev. Entomol.* 11: 183–212. [2]

Delevoryas, T. 1962. *Morphology and evolution of fossil plants*. Holt, Rinehart, Winston, New York. [11]

del Solar, E. 1966. Sexual isolation by selection for positive and negative phototaxis and geotaxis in *Drosophila pseudoobscura*. *Proc. Natl. Acad. Sci. U.S.A.* 56: 484–487. [8]

Den Boer, P. J., and G. R. Gradwell (eds.). 1970. *Dynamics of populations*. Proc. Adv. Study Inst. Dynamics Numb. Pop., Osterbeek, Wageningen, Netherlands. [2]

Dent, J. N. 1968. Survey of amphibian metamorphosis. In W. Etkin and L. I. Gilbert (eds.), *Metamorphosis*, pp. 271–311. Appleton-Century-Crofts, New York. [14]

de Oliveira, A. K., and A. R. Cordeiro. 1980. Adaptation of *Drosophila willistoni* experimental populations to extreme pH medium. II. Development of incipient reproductive isolation. *Heredity* 44: 123–130. [8]

de Queiroz, K. 1985. The ontogenetic method for determining character polarity and its relevance to phylogenetic systematics. *Syst. Zool.* 34: 280–299. [10]

Dewey, J. 1910. *The influence of Darwin on philosophy, and other essays in contemporary thought.* P. Smith, New York. [1]

Dhouailly, D. 1973. Dermo-epidermal interactions between birds and mammals: differentiation of cutaneous appendages. *J. Embryol. Exp. Morphol.* 30: 587–603. [14]

Diamond, J. M. 1975. Assembly of species communities. In M. L. Cody and J. M. Diamond (eds.), *Ecology and evolution of communities*, pp. 342–445. Harvard University Press, Cambridge, Massachusetts. [2, 13]

Diamond, J. M. 1984. Historic extinctions: a Rosetta Stone for understanding prehistoric extinctions. In P. S. Martin and R. G. Klein (eds.), *Quaternary extinctions*, pp. 824–862. University of Arizona Press, Tucson, Arizona. [2]

Dickinson, H., and J. Antonovics. 1973. Theoretical considerations of sympatric divergence. *Amer. Natur.* 107: 256–274. [8]

Dickerson, R. E. 1978. Chemical evolution and the origin of life. *Sci. Amer.* 239(3): 70–86. [11]

Diver, C. 1929. Fossil records of Mendelian mutants. *Nature* 124: 183. [4]

Dobzhansky, Th. 1936. Studies on hybrid sterility. II. Localization of sterility factors in *Drosophila pseudoobscura* hybrids. *Genetics* 21: 113–135. [4]

Dobzhansky, Th. 1937. *Genetics and the origin of species*, First edition. Columbia University Press, New York. [1, 8]

Dobzhansky, Th. 1948. Genetics of natural populations. XVIII. Experiments on chromosomes of *Drosophila pseudoobscura* from different geographic regions. *Genetics* 33: 588–602. [6]

Dobzhansky, Th. 1955. A review of some fundamental concepts and problems of population genetics. *Cold Spring Harbor Group Symp. Quant. Biol.* 20: 1–15. [6, 7]

Dobzhansky, Th. 1962. *Mankind evolving*. Yale University Press, New Haven, Connecticut. [17]

Dobzhansky, Th. 1970. *Genetics of the evolutionary process*. Columbia University Press, New York. [3, 4, 6, 8]

Dobzhansky, Th., F. J. Ayala, G. L. Stebbins, and J. W. Valentine. 1977. *Evolution*. Freeman, San Francisco. [15, 17]

Dobzhansky, Th., C. Krimbas, and M. G. Krimbas. 1960. Genetics of natural populations. XXX. Is the genetic load in *Drosophila pseudoobscura* a mutational or a balanced one? *Genetics* 45: 741–753. [4, 6]

Dobzhansky, Th., and H. Levene 1955. Genetics of natural populations. XXIV. Developmental homeostasis in natural populations of *Drosophila pseudoobscura*. *Genetics* 40: 797–808. [6]

Dobzhansky, Th., and O. Pavlovsky. 1953. Indeterminate outcome of certain experiments on *Drosophila* populations. *Evolution* 7: 198–210. [4, 7]

Dobzhansky, Th., and O. Pavlovsky. 1957. An experimental study of interaction between genetic drift and natural selection. *Evolution* 11: 311–319. [8]

Dobzhansky, Th., and O. Pavlovsky. 1971. An experimentally created incipient species of *Drosophila*. *Nature* 23: 289–292. [8]

Dobzhansky, Th., and S. Wright. 1943. Genetics of natural populations. X. Dispersal rates in *Drosophila pseudoobscura*. *Genetics* 28: 304–340. [5]

Dominey, W. J. 1984. Effects of sexual selection and life history on speciation: species flocks in African cichlids and Hawaiian *Drosophila*. In A. A. Echelle and I. Kornfield (eds.), *Evolution of fish species flocks*, pp. 231–249. University of Maine Press, Orono. [8]

Doolittle, R. F. 1985. The genealogy of some recently evolved vertebrate proteins. *Trends Biochem. Sciences* 10: 233–237. [15]

Doolittle, W. F., and C. Sapienza. 1980. Selfish genes, the phenotypic paradigm and genomic evolution. *Nature* 284: 601–603. [15]

Douglas, M. E., and J. C. Avise. 1982. Speciation rates and morphological divergence in fishes: tests of gradual versus rectangular modes of evolutionary change. *Evolution* 36: 224–232. [14]

Dover, G. A. 1980. Ignorant DNA? *Nature* 285: 618–620. [15]

Dover, G. 1982. Molecular drive: a cohesive mode of species evolution. *Nature* 299: 111–117. [15]

Dover, G., S. Brown, E. Coen, J. Dallas, T. Strachan, and M. Trick. 1982. The dynamics of genome evolution and species differentiation. In G. A. Dover and R. B. Flavell (eds.), *Genome evolution*, pp. 343–372. Academic Press, New York. [15]

Dover, G. A., and R. Flavell (eds.). 1982. *Genome evolution*. Academic Press, New York. [8]

Doyle, J. A. 1978. Origin of angiosperms. *Annu. Rev. Ecol. Syst.* 9: 365–392. [11]

Drake, J. W. 1974. The role of mutation in microbial evolution. In M. J. Carlile and J. J. Skehel (eds.), *Evolution in the microbial world*, pp. 41–58. Cambridge University Press, Cambridge. [3]

Drummond, M. 1979. Crown gall disease. *Nature* 281: 343–347. [15]

Durham, J. W. 1971. The fossil record and the origin of Deuterostomata. *N. Am. Paleontol. Conv., Chicago, 1969, Proc.* Part H, pp. 1104–1131. [11]

Durrant, A. 1962. The environmental induction of heritable change in *Linum*. *Heredity* 17: 27–61. [3]

Dykhuizen, D. 1978. Selection for tryptophan auxotrophs of *Escherichia coli* in glucose limited chemostats as a test of the energy conservation hypothesis of evolution. *Evolution* 32: 125–150. [3]

Dykhuizen, D. E., and D. L. Hartl. 1980. Selective neutrality of 6PGD allozymes in *E. coli* and the effects of genetic background. *Genetics* 96: 801–817. [3, 6]

Dykhuizen, D. E., and D. L. Hartl. 1983a. Selection in chemostats. *Microbiology Reviews* 47: 150–168. [3]

Dykhuizen, D., and D. L. Hartl. 1983b. Functional effects of PGI allozymes in *Escherichia coli*. *Genetics* 105: 1–18. [6]

Eakin, R. M. 1968. Evolution of photoreceptors. *Evol. Biol.* 2: 194–242. [14]

Eanes, W. F. 1984. Viability interactions, *in vivo* activity and the G6PD polymorphism in *Drosophila melanogaster*. *Genetics* 106: 95–107. [6]

Eanes, W. F., and R. K. Koehn. 1978. An analysis of genetic structure in the monarch butterfly, *Danaus plexippus* L. *Evolution* 32: 784–797. [5]

Eberhard, W. G. 1986. *Sexual selection and animal genitalia.* Harvard University Press, Cambridge, Massachusetts. [8]

Echelle, A. A., and E. Kornfield (eds.). 1984. *Evolution of fish species flocks.* University of Maine Press, Orono. [8]

Edwards, A. W. F., and L. L. Cavalli-Sforza. 1964. Reconstruction of evolutionary trees. In V. H. Heywood and J. McNeill (eds.), *Phenetic and phylogenetic classification*, pp. 67–76. Systematics Association Publication No. 6, London. [10]

Ehrlich, P. R., D. E. Breedlove, P. F. Brussard, and M. A. Sharp. 1972. Weather and the "regulation" of subalpine populations. *Ecology* 53: 243–247. [12]

Ehrlich, P. R., and A. Ehrlich. 1981. *Extinction: the causes and consequences of the disappearance of species.* Random House, New York. [11]

Ehrlich, P. R., and P. H. Raven. 1964. Butterflies and plants: a study in coevolution. *Evolution* 18: 586–608. [16]

Ehrlich, P. R., and P. H. Raven. 1969. Differentiation of populations. *Science* 165: 1228–1232. [5]

Ehrman, L. 1962. Hybrid sterility as an isolating mechanism in the genus *Drosophila*. *Quart. Rev. Biol.* 37: 279–302. [4]

Ehrman, L. 1964. Genetic divergence in M. Vetukhiv's experimental populations of *Drosophila pseudoobscura*. 1. Rudiments of sexual isolation. *Genet. Res.* 5: 150–157. [8]

Ehrman, L. 1965. Direct observation of sexual isolation between allopatric and between sympatric strains of the different *Drosophila paulistorum* races. *Evolution* 19: 459–464. [4]

Ehrman, L. 1967. Further studies in genotype frequency and mating success in *Drosophila*. *Amer. Natur.* 101: 415–424. [5, 6]

Ehrman, L., and D. L. Williamson. 1969. On the etiology of the sterility of hybrids between certain strains of *Drosophila paulistorum*. *Genetics* 62: 193–199. [8]

Eicher, D. L. 1976. *Geologic time.* Prentice-Hall, Englewood Cliffs, New Jersey. [11]

Eigen, M., W. Gardiner, P. Schuster, and R. Winkler. 1981. The origin of genetic information. *Sci. Amer.* 244(4): 88–118. [11]

Eisen, E. J. 1975. Population size and selection intensity effects on long-term selection response in mice. *Genetics* 79: 305–323. [5]

Eisen, E. J., J. P. Hanrahan, and J. E. Legates. 1973. Effects of population size and selection intensity on correlated resonses to selection for post-weaning gain in mice. *Genetics* 74: 157–170. [7]

Eldredge, N. 1971. The allopatric model and phylogeny in Paleozoic invertebrates. *Evolution* 25: 156–167. [8, 14]

Eldredge, N., and S. J. Gould. 1972. Punctuated equilibria: an alternative to phyletic gradualism. In T. J. M. Schopf (ed.), *Models in paleobiology*, pp. 82–115. Freeman, Cooper and Company, San Francisco. [8, 14]

Eldredge, N., and J. Cracraft. 1980. *Phylogenetic patterns and the evolutionary process.* Columbia University Press, New York. [10]

Elliott, D. K. (ed.). 1986. *Dynamics of extinction.* Wiley, New York. [12]

Elton, C. S. 1958. *The ecology of invasions by animals and plants.* Methuen, London.

Emerson, S. 1939. A preliminary survey of the *Oenothera organensis* population. *Genetics* 24: 524–537. [5]

Emlen, J. M. 1966. The role of time and energy in food preference. *Amer. Natur.* 100: 611–617. [16]

Emlen, S. T. 1984. Cooperative breeding in birds and mammals. In J. R. Krebs and N. B. Davies (eds.), *Behavioural ecology: an evolutionary approach*, Second edition, pp. 305–339. Sinauer Associates, Sunderland, Massachusetts. [9]

Endler, J. A. 1973. Gene flow and population differentiation. *Science* 179: 243–250. [6, 8]

Endler, J. A. 1977. *Geographic variation, speciation, and clines.* Princeton University Press, Princeton, New Jersey. [4, 6, 8]

Enfield, F. D. 1980. Long term effects of selection: the limits to reponse. *Proc. Symp. on Selection Exp. in Lab. and Domestic Anim.*, Commonwealth Ag. Bureaux, Slough, U.K., pp. 69–86. [7]

Engels, W. R. 1983. The P family of transposable elements in *Drosophila*. *Annu. Rev. Genet.* 17: 319–344. [3, 8, 15]

Eshel, J., and M. W. Feldman. 1970. On the evolutionary effect of recombination. *Theoret. Pop. Biol.* 1: 88–100. [3]

Estabrook, G. F. 1978. Some concepts for the estimation of evolutionary relationships in systematic botany. *Syst. Bot.* 3: 146–158. [10]

Ewald, P. W. 1983. Host-parasite relations, vectors, and the evolution of disease severity. *Annu. Rev. Ecol. Syst.* 14: 465–485. [16]

Ewens, W. J. 1977. Population genetics theory in relation to the neutralist-selectionist controversy. *Adv. Hum. Genet.* 8: 67–134. [6]

Faegri, K., and L. van der Pijl. 1971. *The principles of pollination ecology*, Second edition. Pergamon, New York. [5]

Fairweather, H. 1976. Sex differences in cognition. *Cognition* 4: 31–280. [17]

Falconer, D. S. 1981. *Introduction to quantitative genetics*, Second edition. Longman, London. [7]

Farnworth, E. G., and F. B. Golley (eds.). 1974. *Fragile ecosystems: Evaluation of research and applications in the neotropics.* Springer-Verlag, New York. [16]

Farris, J. S. 1970. Methods for computing Wagner trees. *Syst. Zool.* 19: 83–92. [10]

Farris, J. S. 1983. The logical basis of phylogenetic analysis. *Advances in Cladistics* 2: 7–36. Columbia University Press, New York. [10]

Feinsinger, P. 1983. Coevolution and pollination. In D. J. Futuyma and M. Slatkin (eds.), *Coevolution*, pp. 282–310. Sinauer Associates, Sunderland, Massachusetts. [16]

Felsenstein, J. 1965. The effect of linkage on directional selection. *Genetics* 52: 349–363. [7]

Felsenstein, J. 1971. Inbreeding and variance effective number in populations with overlapping generations. *Genetics* 68: 581–597. [5]

Felsenstein, J. 1974. The evolutionary advantage of recombination. *Genetics* 78: 737–756. [3]

Felsenstein, J. 1976. The theoretical population genetics of variable selection and migration. *Annu. Rev. Genet.* 10: 253–280. [6]

Felsenstein, J. 1979. Excursions along the interface between disruptive and stabilizing selection. *Genetics* 93: 773–795. [7]

Felsenstein, J. 1981. Skepticism towards Santa Rosalia, or why are there so few kinds of animals? *Evolution* 35: 124–138. [8]

Felsenstein, J. 1982. Numerical methods for inferring evolutionary trees. *Quart. Rev. Biol.* 57: 379–404. [10]

Felsenstein, J. 1985a. Phylogenies and the comparative method. *Amer. Natur.* 125: 1–15. [9]

Felsenstein, J. 1985b. Phylogenies from gene frequencies: a statistical problem. *Syst. Zool.* 34: 300–311. [10]

Fenchel, T., and F. Christiansen. 1977. Selection and interspecific competition. In F. Christiansen and T. Fenchel (eds.), *Measuring selection in natural populations*, pp. 477–498. Springer-Verlag, New York. [16]

Fenner, F. 1965. Myxoma virus and *Oryctolagus cuniculus*: two colonizing species. In H. G. Baker and G. L. Stebbins (eds.), *The genetics of colonizing species*, pp. 485–501. Academic Press, New York. [9, 16]

Ferris, S. D., R. D. Sage, C.-M. Huang, J. T. Nielsen, U. Ritte, and A. C. Wilson. 1983. Flow of mitochondrial DNA across a species boundary. *Proc. Natl. Acad. Sci. U.S.A.* 80: 2290–2294. [15]

Ferris, S. D., and G. S. Whitt. 1979. Evolution of the differential regulation of duplicate genes after polyploidization. *J. Molec. Evol.* 12: 267–317. [15]

Ferris, S. D., A. C. Wilson, and W. M. Brown. 1981. Evolutionary tree for apes and humans based on cleavage maps of mitochondrial DNA. *Proc. Natl. Acad. Sci. U.S.A.* 78: 2432–2436. [17]

Findlay, C. S., and F. Cooke. 1983. Genetic and environmental components of clutch size variation in a wild population of lesser snow geese (*Anser caerulescens caerulescens*). *Evolution* 37: 724–734. [7]

Finnegan, D. J., B. H. Will, A. A. Bayev, A. M. Bowcock, and L. Brown. 1982. Transposable DNA sequences in eukaryotes. In G. A. Dover and R. B. Flavell (eds.), *Genome evolution*, pp. 29–40. Academic Press, New York. [15]

Fischer, A. G. 1984. Biological innovations and the sedimentary record. In H. D. Holland and A. F. Trendall (eds.), *Patterns of change in earth evolution*, pp. 145–157. Springer-Verlag, Berlin. [11]

Fisher, R. A. 1928. The possible modification of the response of wild type to recurrent mutations. *Amer. Natur.* 62: 115–126. [7]

Fisher, R. A. 1930. *The genetical theory of natural selection.* Clarendon Press, Oxford, England. [1, 2, 6, 8, 9, 14]

Fisher, R. A. 1941. Average excess and average effect of a gene substitution. *Ann. Eugenics* 11: 53–63. [9]

Fitch, W. M. 1977a. On the problem of discovering the most parsimonious tree. *Amer. Natur.* 111: 223–257. [10]

Fitch, W. M. 1977b. The phyletic interpretation of macromolecular sequence information: simple methods. In M. K. Hecht, P. C. Goody, and B. M. Hecht (eds.), *Major patterns in vertebrate evolution*, pp. 169–204. Plenum, New York. [10]

Fitch, W. M. 1982. The challenges to Darwinism since the last centennial and the impact of molecular studies. *Evolution* 36: 1133–1143. [1, 3]

Fitch, W. M., and E. Margoliash. 1970. The usefulness of amino acid and nucleotide sequences in evolutionary studies. *Evol. Biol.* 4: 67–109. [10]

Flavell, R. B. 1982. Sequence amplification, deletion, and rearrangement: major sources of variation during species divergence. In G. A. Dover and R. B. Flavell (eds.), *Genome evolution*, pp. 301–323. Academic Press, New York. [15]

Flessa, K. W., and D. Jablonski. 1985. Declining Phanerozoic background extinction rates: effects of taxonomic structure? *Nature* 313: 216–218. [12]

Ford, C. S., and F. A. Beach. 1951. *Patterns of sexual behavior.* Harper & Row, New York. [17]

Ford, E. B. 1971. *Ecological genetics*, Third edition. Chapman and Hall, London. [4]

Ford, E. B. 1975. *Ecological genetics*, Fourth edition. Chapman and Hall, London. [6, 7]

Fox, W. 1951. Relationships among the garter snakes of the *Thamnophis elegans Rassenkreis. Univ. California Publ. Zool.* 50: 485–530. [8]

Frakes, L. A. 1979. *Climates through geologic time.* Elsevier, Amsterdam. [11]

Frazzetta, T. H. 1975. *Complex adaptations in evolving populations.* Sinauer Associates, Sunderland, Massachusetts. [14]

Fryer, G. 1959. Some aspects of evolution in Lake Nyasa. *Evolution* 13: 440–451. [8]

Fryer, G., and T. D. Iles. 1972. *The cichlid fishes of the great lakes of Africa*. T. F. H. Publications, Neptune City, New Jersey. [8]

Fuerst, P. A., R. Chakraborty, and M. Nei. 1977. Statistical studies on protein polymorphism in natural populations. I. Distribution of single locus heterozygosity. *Genetics* 86: 455–483. [6]

Futuyma, D. J. 1970. Variation in genetic response to interspecific competition in laboratory populations of *Drosophila*. *Amer. Natur.* 104: 239–252. [16]

Futuyma, D. J. 1973. Community structure and stability in constant environments. *Amer. Natur.* 107: 443–446. [16]

Futuyma, D. J. 1983a. *Science on trial: the case for evolution*. Pantheon, New York. [1]

Futuyma, D. J. 1983b. Evolutionary interactions among herbivorous insects and plants. In D. J. Futuyma and M. Slatkin (eds.), *Coevolution*, pp. 207–231. Sinauer Associates, Sunderland, Massachusetts. [16]

Futuyma, D. J. 1986. Evolution and coevolution in communities. In D. Raup and D. Jablonski (eds.), *Patterns and processes in the evolution of life*. Springer-Verlag, Berlin and New York. [8, 13, 14, 16]

Futuyma, D. J., and F. Gould. 1979. Associations of plants and insects in a deciduous forest. *Ecol. Monogr.* 49: 33–50. [13]

Futuyma, D. J., and G. C. Mayer. 1980. Non-allopatric speciation in animals. *Syst. Zool.* 29: 254–271. [8]

Futuyma, D. J., and S. J. Risch. 1984. Sexual orientation, sociobiology, and evolution. *J. Homosexuality* 9: 157–168. [17]

Futuyma, D. J., and S. C. Peterson. 1985. Genetic variation in the use of resources by insects. *Annu. Rev. Entomol.* 30: 217–238. [6, 8]

Futuyma, D. J., and M. Slatkin (eds.). 1983a. *Coevolution*. Sinauer Associates, Sunderland, Massachusetts. [16]

Futuyma, D. J., and M. Slatkin. 1983b. Introduction. In D. J. Futuyma and M. Slatkin (eds.), *Coevolution*, pp. 1–13. Sinauer Associates, Sunderland, Massachusetts. [16]

Futuyma, D. J., and M. Slatkin. 1983c. The study of coevolution. In D. J. Futuyma and M. Slatkin (eds.), *Coevolution*, pp. 459–464. Sinauer Associates, Sunderland, Massachusetts. [16]

Gadgil, M., and W. H. Bossert. 1970. Life history consequences of natural selection. *Amer. Natur.* 102: 52–64. [9]

Gans, C. 1974. *Biomechanics: an approach to vertebrate biology*. Lippincott, Philadelphia. [9]

Garstang, W. 1922. The theory of recapitulation: a critical restatement of the biogenetic law. *J. Linn. Soc. Zool.* 35: 81–101. [14]

Gehring, W. J. 1985. The homeo box: a key to the understanding of development? *Cell* 40: 3- 5. [15]

Geist, V. 1971. *Mountain sheep*. University of Chicago Press, Chicago. [9]

Gerhart, J. C., et al. 1982. The cellular basis of morphogenetic change: group report. In J. T. Bonner (ed.), *Evolution and development*, pp. 87–114. Springer-Verlag, Berlin. [3, 14]

Ghiselin, M. T. 1969. *The triumph of the Darwinian method*. University of California Press, Berkeley. [1]

Gibson, T. C., M. L. Scheppe, and E. C. Cox. 1970. Fitness of an *Escherichia coli* mutator gene. *Science* 169: 686–688. [9]

Giesel, J. T. 1971. The relations between population structure and rate of inbreeding. *Evolution* 25: 491–496. [5]

Gilbert, J. J., and J. K. Waage. 1967. *Asplancha, Asplancha*-substance, and posterolateral spine length variation of the rotifer *Brachionus calyciflorus* in a natural environment. *Ecology* 48: 1027–1031. [2]

Gilbert, L. E. 1971. Butterfly–plant coevolution: has *Passiflora adenopoda* won the selectional race with heliconiine butterflies? *Science* 172: 585–586. [16]

Gilbert, L. E. 1983. Coevolution and mimicry. In D. J. Futuyma and M. Slatkin (eds.), *Coevolution*, pp. 263–281. Sinauer Associates, Sunderland, Massachusetts. [16]

Gilbert, S. F. 1985. *Developmental biology*. Sinauer Associates, Sunderland, Massachusetts. [3]

Gill, D. E. 1972. Intrinsic rates of increase, saturation densities, and competitive ability. I. An experiment with *Paramecium*. *Amer. Natur.* 106: 465–471. [16]

Gillespie, J. H., and C. H. Langley. 1974. A general model to account for enzyme variation in natural populations. *Genetics* 76: 837–884. [6]

Gilpin, M. E., and J. M. Diamond. 1984. Are species co-occurrences on islands non-random, and are null hypotheses useful in community ecology? In D. R. Strong, Jr., D. Simberloff, L. G. Abele, and A. B. Thistle (eds.), *Ecological communities: conceptual issues and the evidence*, pp. 297–315. Princeton University Press, Princeton, New Jersey. [13]

Gingerich, P. D. 1976. Paleontology and phylogeny: patterns of evolution at the species level in early Tertiary mammals. *Am. J. Sci.* 276: 1–28. [14]

Gingerich, P. D. 1977. Patterns of evolution in the mammalian fossil record. In A. Hallam (ed.), *Patterns of evolution, as illustrated by the fossil record*, pp. 469–500. Elsevier, Amsterdam. [12]

Gingerich, P. D. 1983. Rates of evolution: effects of time and temporal scaling. *Science* 222: 159–161. [14]

Glesener, R. R., and D. Tilman. 1978. Sexuality and the components of environmental uncertainty: clues from geographic parthenogenesis in terrestrial animals. *Amer. Natur.* 112: 659–673. [9]

Goddard, H. H. 1920. *Human efficiency and levels of intelligence*. Princeton University Press, Princeton, New Jersey. [17]

Godfrey, L. R. (ed.). 1983. *Scientists confront creationism*. Norton, New York. [1]

Gojobori, T., W.-H. Li, and D. Grauer. 1982. Patterns of nucleotide substitution in pseudogenes and functional genes. *J. Mol. Evol.* 18: 360–369. [3]

Golding, G. B., and C. Strobeck. 1983. Increased number of alleles found in hybrid populations due to intragenic recombination. *Evolution* 17: 17–19. [3, 4]

Goldschmidt, R. B. 1938. *Physiological genetics*. McGraw-Hill, New York. [14]

Goldschmidt, R. 1940. *The material basis of evolution*. Yale University Press, New Haven, Connecticut. [1, 4, 14]

Gooch, J. M., and T. J. M. Schopf. 1973. Genetic variability in the deep sea: relation to environmental variability. *Evolution* 26: 545–562. [6]

Goodman, M., G. Braunitzer, A. Stangl, and B. Shrank. 1983. Evidence on human origins from haemoglobins of African apes. *Nature* 303: 546–548. [17]

Goodman, M., M. L. Weiss, and J. Czelusniak. 1982. Molecular evolution above the species level: branching patterns, rates, and mechanisms. *Syst. Zool.* 31: 376–399. [5, 10, 15]

Gorczynski, R. M., and E. J. Steele. 1981. Simultaneous yet independent inheritance of somatically acquired inheritance of two distinct H-2 antigenic haplotype determinants in mice. *Nature* 289: 678–681. [3]

Gottlieb, L. D. 1974. Genetic confirmation of the origin of *Clarkia lingulata*. *Evolution* 28: 244–250. [8]

Gottlieb, L. D. 1984. Genetics and morphological evolution in plants. *Amer. Natur.* 123: 681–709. [8]

Gould, S. J. 1966. Allometry and size in ontogeny and phylogeny. *Biol. Rev.* 41: 587–680. [14]

Gould, S. J. 1972. Allometric fallacies and the evolution of *Gryphaea*: a new interpretation based on White's criterion of geometric similarity. *Evol. Biol.* 6: 91–119. [12]

Gould, S. J. 1974. The origin and funcion of "bizarre" structures: antler size and skull size in the "Irish elk," *Megaloceros giganteus*. *Evolution* 28: 191–220. [14]

Gould, S. J. 1976. Palaeontology plus ecology as palaeobiology. In R. M. May (ed.), *Theoretical ecology: principles and applications*, pp. 218–236. Saunders, Philadelphia. [11]

Gould, S. J. 1977. *Ontogeny and phylogeny*. Harvard University Press, Cambridge, Massachusetts. [10, 14]

Gould, S. J. 1981. *The mismeasure of man*. Norton, New York. [17]

Gould, S. J. 1982. Darwinism and the expansion of evolutionary theory. *Science* 216: 380–387. [14]

Gould, S. J., and C. B. Calloway. 1980. Clams and brachiopods—ships that pass in the night. *Paleobiology* 6: 383–396. [12]

Gould, S. J., and N. Eldredge. 1977. Punctuated equilibria: the tempo and mode of evolution reconsidered. *Paleobiology* 3: 115–151. [12, 14]

Gould, S. J., and R. F. Johnston. 1971. Geographic variation. *Annu. Rev. Ecol. Syst.* 3: 457–498. [4]

Gould, S. J., and R. C. Lewontin. 1979. The spandrels of San Marco and the Panglossian paradigm: a critique of the adaptationist programme. *Proc. Roy. Soc. Lond. B* 205: 581–598. [9]

Gould, S. J., and E. S. Vrba. 1982. Exaptation—a missing term in the science of form. *Paleobiology* 8: 4–15. [9]

Grant, P. R. 1972. Convergent and divergent character displacement. *Biol. J. Linn. Soc.* 4: 39–68. [2]

Grant, B. R., and P. R. Grant. 1982. Niche shifts and competition in Darwin's finches: *Geospiza conirostris* and congeners. *Evolution* 36: 637–657. [16]

Grant, V. 1966. The selective origin of incompatibility barriers in the plant genus *Gilia*. *Amer. Natur.* 100: 99–118. [8]

Grant, V. 1981. *Plant speciation*, Second edition. Columbia University Press, New York. [4, 8]

Grant, K. A., and V. Grant. 1964. Mechanical isolation of *Salvia apiana* and *Salvia mellifera* (Labiatae). *Evolution* 18: 196–212. [4]

Graves, J., R. H. Rosenblatt, and G. N. Somero. 1983. Kinetic and electrophoretic differentiation of lactate dehydrogenases of teleost species-pairs from the Atlantic and Pacific coasts of Panama. *Evolution* 37: 30–37. [6]

Gray, J., and A. J. Boucot (eds.). 1979. *Historical biogeography, plate tectonics, and the changing environment*. Oregon State University Press, Corvallis. [11]

Greene, J. C. 1959. *The death of Adam: Evolution and its impact on Western thought*. Iowa State University Press, Ames. [1]

Greenslade, P. J. M. 1968. Island patterns in the Solomon Islands bird fauna. *Evolution* 22: 751–761. [13]

Greenwood, P. J., and P. H. Harvey. 1982. The natal and breeding dispersal of birds. *Annu. Rev. Ecol. Syst.* 13: 1–21. [5]

Greenwood, P. J., P. H. Harvey, and C. M. Perrins. 1978. Inbreeding and dispersal in the great tit. *Nature* 271: 52–54. [5]

Gregory, W. K. 1951. *Evolution emerging*. Macmillan, New York. [11]

Griffiths, I. 1963. The phylogeny of the Salientia. *Biol. Rev.* 38: 241–292. [10]

Grimaldi, D., and J. Jaenike. 1984. Competition in natural populations of mycophagous *Drosophila*. *Ecology* 65: 1113–1120. [2]

Grula, J. W., and O. R. Taylor, Jr. 1980. The effect of X-chromosome inheritance on mate-selection behavior in the sulfur butterflies, *Colias eurytheme* and *C. philodice*. *Evolution* 34: 688–695. [8]

Grun, P. 1976. *Cytoplasmic genetics and evolution*. Columbia University Press, New York. [3]

Guerrant, E. O., Jr. 1982. Neotenic evolution of *Delphinium nudicaule* (Ranunculaceae): a hummingbird-pollinated larkspur. *Evolution* 36: 699–712. [14]

Gupta, A. P., and R. C. Lewontin. 1982. A study of reaction norms in natural populations of *Drosophila pseudoobscura*. *Evolution* 36: 934–948. [7]

Haeckel, E. 1866. *Generelle Morphologie der Organismen: Allgemeine Grundzüge der organischen Formen-Wissenschaft, mechanisch begründet durch die von Charles Darwin reformirte Descendenz-Theorie*. Georg Riemer, Berlin. [10]

Hairston, N. G. 1951. Interspecies competition and its probable influence upon the vertical distribution of Applachian salamanders of the genus *Plethodon*. *Ecology* 32: 266–274. [2]

Hairston, N. G. 1980. Evolution under interspecific competition: field experiments in terrestrial salamanders. *Evolution* 34: 409–420. [16]

Hairston, N. G., F. E. Smith, and L. B. Slobodkin. 1960. Community structure, population control, and competition. *Amer. Natur.* 94: 421–425. [2]

Haldane, J. B. S. 1932. *The causes of evolution*. Longmans, Green, New York. [1]

Haldane, J. B. S. 1949. Suggestions as to the quantitative measurement of rates of evolution. *Evolution* 3: 51–56. [14]

Haldane, J. B. S. 1956. The relation between density regulation and natural selection. *Proc. Roy. Soc. Lond. B* 145: 306–308. [2]

Haldane, J. B. S. 1957. The cost of natural selection. *J. Genet.* 55: 511–524. [6]

Haldane, J. B. S., and S. D. Jayakar. 1963. Polymorphism due to selection of varying direction. *J. Genet.* 58: 318–323. [6]

Hall, B. G. 1982. Evolution on a petri dish. The evolved β-galactosidase system as a model for studying acquisitive evolution in the laboratory. *Evol. Biol.* 15: 85–150. [15]

Hall, B. G. 1983. Evolution of new metabolic functions in laboratory organisms. In M. Nei and R. K. Koehn (eds.), *Evolution of genes and proteins*, pp. 234–257. Sinauer Associates, Sunderland, Massachusetts. [3, 15]

Hall, B. K. 1984. Developmental mechanisms underlying the formation of atavisms. *Biol. Rev.* 59: 89–124. [14]

Hallam, A. 1974. Changing patterns of provinciality and diversity of fossil animals in relation to plate tectonics. *J. Biogeogr.* 1: 213–225. [11]

Hallam, A. (ed.). 1977. *Patterns of evolution, as illustrated by the fossil record*. Elsevier, Amsterdam. [11]

Hallam, A. 1982. Patterns of speciation in Jurassic *Gryphaea*. *Paleobiology* 8: 354–366. [14]

Hallam, A. 1983. Plate tectonics and evolution. In D. S. Bendall (ed.), *Evolution from molecules to men*, pp. 367–386. Cambridge University Press, Cambridge. [11]

Hallam, A. 1984. Pre-Quaternary sea-level changes. *Annu. Rev. Earth Planet. Sci.* 12: 205–243. [11, 12]

Hamilton, W. D. 1964. The genetical evolution of social behavior, I and II. *J. Theor. Biol.* 7: 1–52. [9]

Hampé, A. 1960. La compétition entre les éléments osseux du zeugopode de poulet. *J. Embryol. Exp. Morphol.* 8: 241–245. [14]

Hamrick, J. L., and R. W. Allard. 1972. Microgeographical variation in allozyme frequencies in *Avena barbata*. *Proc. Natl. Acad. Sci. U.S.A.* 69: 2100–2104. [4, 5]

Hamrick, J. L., Y. B. Linhart, and J. B. Mitton. 1979. Relationships between life history characteristics and electrophoretically detectable genetic variation in plants. *Annu. Rev. Ecol. Syst.* 10: 173–200. [4, 5]

Hanken, J. 1984. Miniaturization and its effect on cranial morphology in plethodontid salamanders, genus *Thorius* (Amphibia: Plethodontidae). I. Osteological variation. *Biol. J. Linn. Soc.* 23: 55–75. [14]

Hansche, P. E. 1975. Gene duplication as a mechanism of genetic adaptation in *Saccharomyces cerevisiae*. *Genetics* 79: 661–674. [3]

Hansen, T. A. 1980. Influence of larval dispersal and geographic distribution on species longevity in neogastropods. *Paleobiology* 6: 193–207. [12]

Hansen, T. A. 1982. Modes of larval development in Early Tertiary neogastropods. *Paleobiology* 8: 367–377. [12]

Harding, K., C. Wedeen, W. McGinnis, and M. Levine. 1985. Spatially regulated expression of homeotic genes in *Drosophila*. *Science* 229: 1236–1242. [14]

Harlan, J. R., and J. M. J. de Wet. 1953. The compilospecies concept. *Evolution* 17: 497–501. [3]

Harland, W. B., A. V. Cox, P. G. Llewellyn, C. A. G. Pickton, A. G. Smith, and R. Walters. 1982. *A geologic time scale*. Cambridge University Press, Cambridge. [11]

Harper, J. L., and J. White. 1974. The demography of plants. *Annu. Rev. Ecol. Syst.* 5: 419–463. [9]

Harris, H. 1966. Enzyme polymorphisms in man. *Proc. Roy. Soc. Lond. B* 164: 298–310. [4]

Harris, M. 1974. *Cows, pigs, wars and witches. The riddles of culture*. Random House, New York. [17]

Harris, M. 1975. *Culture, people, nature. An introduction to general anthropology*. Thomas Y. Crowell, New York. [17]

Harrison, R. G. 1977. Parallel variation at an enzyme locus in sibling species of field crickets. *Nature* 266: 168–170. [6]

Harrison, R. G. 1980. Dispersal polymorphisms in insects. *Annu. Rev. Ecol. Syst.* 11: 95–118. [9]

Harshman, L. G., and D. J. Futuyma. 1985. The origin and distribution of clonal diversity in *Alsophila pometaria* (Lepidoptera: Geometridae). *Evolution* 39: 315–324. [4]

Hartl, D. L. 1981. *A primer of population genetics*. Sinauer Associates, Sunderland, Massachusetts. [4]

Hartl, D. L., and D. E. Dykhuizen. 1984. The population genetics of *Escherichia coli*. *Annu. Rev. Genet.* 18: 31–68. [15]

Harvey, P. H., R. K. Colwell, J. W. Silvertown, and R. M. May. 1983. Null models in ecology. *Annu. Rev. Ecol. Syst.* 14: 189–211. [2]

Haverschmidt, F. 1968. *Birds of Surinam*. Oliver & Boyd, London. [13]

Hearnshaw, L. S. 1979. *Cyril Burt, psychologist*. Hodder and Stoughton, London. [17]

Hebert, P. 1974. Enzyme variability in natural populations of *Daphnia magna*. II. Genotypic frequencies in permanent populations. *Genetics* 77: 323–334. [6]

Hecht, M. K. 1952. Natural selection in the lizard genus *Aristelliger*. *Evolution* 6: 112–124. [7]

Hecht, M. K. 1965. The role of natural selection and evolutionary rates in the origin of higher levels of organization. *Syst. Zool.* 14: 301–317. [10]

Hecht, M. K., and J. L. Edwards. 1977. The methodology of phylogenetic inference above the species level. In M. K. Hecht, P. C. Goody, and B. M. Hecht (eds.), *Major patterns in vertebrate evolution*, pp. 3–51. Plenum, New York. [10]

Hedrick, P. W. 1981. The establishment of chromosomal variants. *Evolution* 35: 322–332. [6]

Hedrick, P. W. 1983. *Genetics of populations*. Science Books International, Boston. [4, 5, 6]

Hedrick, P. W., M. E. Ginevan, and E. P. Ewing. 1976. Genetic polymorphism in heterogeneous environments. *Annu. Rev. Ecol. Syst.* 7: 1–32. [6]

Hedrick, P. W., S. Jain, and L. Holden. 1978. Multilocus systems in evolution. *Evol. Biol.* 11: 101–182. [4, 7]

Henry, G. M. 1971. *Birds of Ceylon*. Oxford University Press, Oxford. [6]

Henry, C. S. 1985. Sibling species, call differences, and speciation in green lacewings (Neuroptera: Chrysopidae: *Chrysoperla*). *Evolution* 39: 965–984. [8]

Hendrickson, J. A., Jr. 1981. Community-wide character displacement reexamined. *Evolution* 35: 795–809. [16]

Hennig, W. 1979. *Phylogenetic systematics*. University of Illinois Press, Urbana. [10]

Hersh, A. H. 1930. The facet-temperature relation in the Bar series in *Drosophila*. *J. Exp. Zool.* 57: 283–306. [3]

Hickey, D. A. 1982. Selfish DNA: a sexually transmitted nuclear parasite. *Genetics* 106: 519–531. [15]

Hickey, L. J., and J. A. Doyle. 1977. Early Cretaceous fossil evidence for angiosperm evolution. *Bot. Rev.* 43: 3–104. [11]

Highton, R., and A. Larson. 1979. The genetic relationships of the salamanders of the genus *Plethodon*. *Syst. Zool.* 28: 579–599. [10]

Highton, R., and T. P. Webster. 1976. Geographic protein variation and divergence in populations of the salamander *Plethodon cinereus*. *Evolution* 30: 33–45. [5]

Hill, J. 1967. The environmental induction of heritable changes in *Nicotiana rustica* parental and selection lines. *Genetics* 55: 735–754. [3]

Hill, W. G. 1982. Predictions of response to artificial selection from new mutations. *Genet. Res.* 40: 255–278. [7]

Hill, W. G., and A. Robertson. 1968. Linkage disequilibrium in finite populations. *Theoret. Appl. Genet.* 38: 226–231. [7]

Hilu, K. W. 1983. The role of single-gene mutations in the evolution of flowering plants. *Evol. Biol.* 16: 97–128. [8]

Himmelfarb, G. 1959. *Darwin and the Darwinian revolution*. Doubleday, Garden City, New York. [1]

Hinchliffe, J. R., and P. J. Griffiths. 1983. The prechondrogenic patterns in tetrapod limb development and their phylogenetic significance. In B. C. Goodwin, N. Holden, and C. C. Wylie (eds.), *Development and evolution*, pp. 99–121. Cambridge University Press, Cambridge. [14]

Hirschberg, J., and L. McIntosh. 1983. Molecular basis of herbicide resistance in *Amaranthus hybridus*. *Science* 222: 1346–1348. [6]

Ho, M.-W., and P. T. Saunders. 1984. *Beyond neo-Darwinism. An introduction to the new evolutionary paradigm*. Academic Press, London. [14]

Hochachka, P. W., and G. N. Somero. 1973. *Strategies of biochemical adaptation*. Saunders, Philadelphia. [14]

Hoffman, A. 1982. Punctuated versus gradual mode of evolution—a reconsideration. *Evol. Biol.* 15: 411–436. [14]

Hofstader, R. 1955. *Social Darwinism in American thought*. Beacon Press, Boston. [1, 17]

Holmes, J. C. 1983. Evolutionary relationships between parasitic helminths and their hosts. In D. J. Futuyma and M. Slatkin (eds.), *Coevolution*, pp. 161–185. Sinauer Associates, Sunderland, Massachusetts. [2]

Holt, R. D. 1977. Predation, apparent competition, and the structure of prey communities. *Theoret. Pop. Biol.* 12: 197–229. [2]

Hooker, E. 1957. The adjustment of the male overt homosexual. *J. of Projective Techniques* 21: 18–31. [17]

Hopson, J. A. 1975. The evolution of cranial display structures in hadrosaurian dinosaurs. *Paleobiology* 1: 21–43. [11]

Hopson, J. A., and A. W. Crompton. 1969. Origin of mammals. *Evol. Biol.* 3: 15–72. [10]

Horn, H. S., and R. H. MacArthur. 1972. Competition among fugitive species in a harlequin environment. *Ecology* 53: 749–752. [2]

Horn, J. M., J. L. Loehlin, and L. Willerman. 1979. Intellectual resemblances among adoptive and biological relatives: the Texas adoption project. *Behav. Genet.* 9: 177–207. [17]

House, M. R. (ed.). 1979. *The origin of major invertebrate groups*. Academic Press, New York. [11]

Howard, D. J., and R. G. Harrison. 1984. Habitat segregation in ground crickets: the role of interspecific competition and habitat selection. *Ecology* 65: 69–76. [2]

Howard, W. E. 1949. Dispersal, amount of inbreeding, and longevity in a local population of prairie deermice on the George Reserve, southern Michigan. *Contr. Lab. Vert. Biol. Univ. Mich.* 43: 1–50. [5]

Howe, H. F. 1984. Constraints on the evolution of mutualisms. *Amer. Natur.* 123: 764–777. [16]

Howell, F. C. 1978. Hominidae. In V. J. Maglio and H. B. S. Cooke (eds.), *Evolution of African mammals*, pp. 154–248. Harvard University Press, Cambridge, Massachusetts. [17]

Hoy, R. R., J. Hahn, and R. C. Paul. 1977. Hybrid cricket auditory behavior: evidence for genetic coupling in animal communication. *Science* 195: 82–83. [8]

Hsiao, T. H. 1978. Host plant adaptation among geographic populations of the Colorado potato beetle. *Ent. Exp. Appl.* 24: 237–247. [16]

Hudson, R. R. 1983. Testing the constant-rate neutral allele model with protein sequence data. *Evolution* 37: 203–217. [5]

Hudson, R. R., and J. L. Kaplan. 1985. Statistical properties of the number of recombination events in the history of a sample of DNA sequences. *Genetics* 111: 147–164. [15]

Hull, D. L. 1973. *Darwin and his critics*. Harvard University Press, Cambridge, Massachusetts. [1]

Hunkapiller, T., H. Huang, L. Hood, and J. H. Campbell. 1982. The impact of modern genetics on evolutionary theory. In R. Milkman (ed.), *Perspectives on evolution*, pp. 164–189. Sinauer Associates, Sunderland, Massachusetts. [15]

Hurd, L. E., and R. M. Eisenberg. 1975. Divergent selection for geotactic response and evolution of reproductive isolation in sympatric and allopatric populations of houseflies. *Amer. Natur.* 109: 353–358. [8]

Hutchinson, G. E. 1957. Concluding remarks. *Cold Spring Harbor Symp. Quant. Biol.* 22: 415–427. [2]

Hutchinson, G. E. 1968. When are species necessary? In R. C. Lewontin (ed.), *Population biology and evolution*, pp. 177–186. Syracuse University Press, Syracuse, New York. [8]

Hutchinson, J. 1969. *Evolution and phylogeny of flowering plants*. Academic Press, New York. [14]

Huxley, J. S. 1932. *Problems of relative growth*. MacVeagh, London. (Second edition 1972, Dover, New York). [14]

Huxley, J. S. 1942. *Evolution, the modern synthesis*. Allen and Unwin, London. [1]

Huxley, T. H. 1893. Evolution and ethics (The Romanes Lecture, 1893). In T. H. Huxley, *Evolution and ethics and other essays*, D. Appleton and Company, New York (1898). [1, 17]

Imai, H. T. 1983. Quantitative analysis of karyotypic alteration and species differentiation in mammals. *Evolution* 37: 1154–1161. [8]

Inger, R. F. 1967. The development of a phylogeny of frogs. *Evolution* 21: 369–384. [10]

Istock, C. A. 1983. The extent and consequences of heritable variation for fitness characters. In C. R. King and P. W. Dawson (eds.), *Population biology: retrospect and prospect*, pp. 61–96. Columbia University Press, New York. [7]

Ives, P. T. 1950. The importance of mutation rate genes in evolution. *Evolution* 4: 236–252. [3]

Ivie, G. W., D. L. Bull, R. C. Beier, N. W. Pryor, and E. H. Vertli. 1983. Metabolic detoxification: mechanisms of insect resistance to plant psoralens. *Science* 221: 374–376. [14, 15]

Jablonski, D. 1984. Keeping time with mass extinctions. *Paleobiology* 10: 139–145. [12]

Jablonski, D. 1986a. Evolutionary consequences of mass extinctions. In D. M. Raup and D. Jablonski (eds.), *Patterns and processes in the history of life*. Springer-Verlag, Berlin. [11, 12]

Jablonski, D. 1986b. Causes and consequences of mass extinctions: a comparative approach. In D. K. Elliott (ed.), *Dynamics of extinctions*. Wiley, New York. [12]

Jablonski, D., S. J. Gould, and D. M. Raup. 1986. The nature of the fossil record: a biological perspective. In D. M. Raup and D. Jablonski (eds.), *Patterns and processes in the history of life*. Springer-Verlag, Berlin. [11]

Jablonski, D., and R. A. Lutz. 1983. Larval ecology of marine benthic invertebrates: paleobiological implications. *Biol. Rev.* 58: 21–89. [12]

Jablonski, D., J. J. Sepkoski, Jr., D. J. Bottjer, and P. M. Sheehan. 1983. Onshore-offshore patterns in the evolution of Phanerozoic shelf communities. *Science* 222: 1123–1125. [13]

Jackson, J. B. C. 1974. Biogeographic consequences of eurytopy and stenotopy among marine bivalves and their evolutionary significance. *Amer. Natur.* 108: 541–560. [12]

Jaeger, R. G. 1980. Density-dependent and density-independent causes of extinction of a salamander population. *Evolution* 34: 617–621. [5, 12]

Jaenike, J. 1982. Environmental modification of oviposition behavior in *Drosophila*. *Amer. Natur.* 119: 784–802. [6]

Jaenike, J. and D. Grimaldi. 1983. Genetic variation for host preference within and among populations of *Drosophila tripunctata*. *Evolution* 37: 1023–1033. [4]

Jaenike, J., E. D. Parker, Jr., and R. K. Selander. 1980. Clonal niche structure in the parthenogenic earthworm *Octolasion tyrtaeum*. *Amer. Natur.* 116: 196–205. [7]

Jain, S. K. 1976. Evolution of inbreeding in plants. *Annu. Rev. Ecol. Syst.* 7: 468–495. [9]

Jain, S. K., and D. R. Marshall. 1967. Population studies on predominantly self-pollinating species. X. Variation in natural populations of *Avena fatua* and *A. barbata*. *Amer. Natur.* 101: 19–33. [5]

Janzen, D. H., 1966. Coevolution of mutualism between ants and acacias in Central America. *Evolution* 20: 249–275. [16]

Janzen, D. H. 1970. Herbivores and the number of tree species in tropical forests. *Amer. Natur.* 104: 501–528. [2]

Janzen, D. H. 1980. When is it coevolution? *Evolution* 34: 611–612. [16]

Janzen, D. H. 1983. Dispersal of seeds by vertebrate guts. In D. J. Futuyma and M. Slatkin (eds.), *Coevolution*, pp. 222–262. Sinauer Associates, Sunderland, Massachusetts. [16]

Janzen, D. H., and P. S. Martin. 1982. Neotropical anachronisms: the fruits the gomphotheres ate. *Science* 215: 19–27. [9, 11]

Järvinen, O., et al. 1986. The neontologico-paleontological interface of community evolution: how do the pieces in the kaleidoscopic biosphere move? In D. M. Raup and D. Jablonski (eds.), *Patterns and processes in the history of life*. Springer-Verlag, Berlin. [16]

Jeffreys, A. J. 1982. Evolution of globin genes. In G. A. Dover and R. B. Flavell (eds.), *Genome evolution*, pp. 157–176. Academic Press, New York. [15]

Jensen, A. R. 1969. How much can we boost IQ and scholastic achievement? *Harvard Educational Review* 33: 1–123. [17]

Jensen, A. R. 1973. *Educability and group differences*. Harper & Row, New York. [17]

Jensen, R. A. 1976. Enzyme recruitment in evolution of new function. *Annu. Rev. Microbiol.* 30: 409–425. [15]

Jerison, H. J. 1973. *Evolution of the brain and intelligence*. Academic Press, New York. [14]

Jermy, T. 1984. Evolution of insect/host plant relationships. *Amer. Natur.* 124: 609–630. [9, 16]

Jinks, J. L. 1964. *Extrachromosomal inheritance.* Prentice-Hall, Englewood Cliffs, New Jersey. [3]

Jinks, J. L., and K. Mather. 1955. Stability in development of heterozygotes and homozygotes. *Proc. Roy. Soc. Lond. B* 143: 561–578. [9]

Jinks, J. L., J. M. Perkins, and H. S. Pooni. 1973. The incidence of epistasis in normal and extreme environments. *Heredity* 31: 263–269. [7]

Johanson, D. C., and T. D. White. 1979. A systematic assessment of early African hominids. *Science* 203: 321–330. [17]

John, B. 1981. Chromosome change and evolutionary change: a critique. In W. R. Atchley and D. S. Woodruff (eds.), *Evolution and speciation: Essays in honor of M. J. D. White,* pp. 23–51. Cambridge University Press, Cambridge. [3]

Johnsgard, P. A. 1983. *The grouse of the world.* University of Nebraska Press, Lincoln. [9]

Johnson, C. 1976. *Introduction to natural selection.* University Park Press, Baltimore, Maryland. [4]

Johnson, L. K. 1982. Sexual selection in a brentid weevil. *Evolution* 36: 251–262. [9]

Johnston, R. F. 1969. Taxonomy of house sparrows and their allies in the Mediterranean basin. *Condor* 71: 129–139. [8]

Johnston, R. F., D. M. Niles, and S. A. Rohwer. 1972. Hermon Bumpus and natural selection in the house sparrow *Passer domesticus. Evolution* 26: 20–31. [7]

Johnston, R. F., and R. K. Selander. 1964. House sparrows: rapid evolution of races in North America. *Science* 144: 548–550. [14]

Jones, D. A. 1973. Co-evolution and cyanogenesis. In V. H. Heywood (ed.), *Taxonomy and ecology,* pp. 213–242. Academic Press, New York. [4]

Jones, D. F. 1924. The attainment of homozygosity in inbred strains of maize. *Genetics* 9: 405–418. [5]

Jones, J. S., B. H. Leith, and P. Rawlings. 1977. Polymorphism in *Cepaea*: a problem with too many solutions? *Annu. Rev. Ecol. Syst.* 8: 109–143. [4, 7]

Jukes, T. H. 1983. Evolution of the amino acid code. In M. Nei and R. K. Koehn (eds.), *Evolution of genes and proteins,* pp. 191–207. Sinauer Associates, Sunderland, Massachusetts. [3]

Kacser, H., and J. A. Burns. 1981. The molecular basis of dominance. *Genetics* 97: 639–666. [7]

Kaestner, A. 1970. *Invertebrate zoology. Volume III.* (Translated by H. Levi and L. R. Levi). Wiley, New York. [14]

Kamin, L. J. 1974. *The science and politics of IQ.* Wiley, New York. [17]

Kaneshiro, K. Y. 1983. Sexual selection and direction of evolution in the biosystematics of Hawaiian Drosophilidae. *Annu. Rev. Entomol.* 28: 161–178. [8]

Karlin, S. 1975. General two-locus selection models: some objectives, results, and interpretations. *Theoret. Pop. Biol.* 7: 364–398. [7]

Karn, M. N., and L. S. Penrose. 1951. Birth weight and gestation time in relation to maternal age, parity, and infant survival. *Ann. Eugenics.* 16: 147–164. [7]

Kaufman, P. K., F. D. Enfield, and R. E. Comstock. 1977. Stabilizing selection for pupa weight in *Tribolium castaneum. Genetics* 87: 327–341. [7]

Kearsey, M. J., and B. W. Barnes. 1970. Variation for metrical characters in *Drosophila* populations. II. Natural selection. *Heredity* 25: 11–21. [7]

Kearsey, M. J., and K. Kojima. 1967. The genetic architecture of body weight and egg hatchability in *Drosophila melanogaster. Genetics* 56: 23–37. [7]

Kellogg, D. E. 1973. The role of phyletic change in the evolution of *Pseudocubus vema* (Radiolaria). *Paleobiology* 1: 359–370. [10]

Kemp, T. S. 1982. *Mammal-like reptiles and the origin of mammals.* Academic Press, New York. [10, 11]

Kerr, W. E., and S. Wright. 1954. Experimental studies of the distribution of gene frequencies in very small populations of *Drosophila melanogaster* I. Forked. *Evolution* 8: 172–177. [5]

Kerster, H. W. 1964. Neighborhood size in the rusty lizard, *Sceloporus olivaceus. Evolution* 18: 445–457. [5]

Kettlewell, H. B. D. 1955. Selection experiments on industrial melanism in the Lepidoptera. *Heredity* 10: 287–301. [6]

Kettlewell, H. B. D. 1973. *The evolution of melanism.* Clarendon, Oxford. [4]

Kettlewell, H. B. D., and D. L. T. Conn. 1977. Further background-choice experiments on cryptic Lepidoptera. *J. Zool., Lond.* 181: 371–376. [6]

Key, K. H. L. 1968. The concept of stasipatric speciation. *Syst. Zool.* 17: 14–22. [8]

Kidd, K. K., and L. L. Cavalli-Sforza. 1974. The role of genetic drift in the differentiation of Icelandic and Norwegian cattle. *Evolution* 28: 381–395. [5]

Kidwell, M. G., J. F. Kidwell, and J. A. Sved. 1977. Hybrid dysgenesis in *Drosophila melanogaster*: a syndrome of aberrant traits including mutation, sterility, and male recombination. *Genetics* 86: 813–833. [3, 8]

Kiester, A. R., R. Lande, and D. W. Schemske. 1984. Models of coevolution and speciation in plants and their pollinators. *Amer. Natur.* 124: 220–243. [8, 16]

Kimura, M. 1955. Solution of a process of random genetic drift with a continuous model. *Proc. Natl. Acad. Sci. U.S.A.* 41: 144–150. [5]

Kimura, M. 1981. Possibility of extensive neutral evolution under stabilizing selection with special reference to non-random usage of synonymous codons. *Proc. Natl. Acad. Sci. U.S.A.* 78: 5773-5777. [7]

Kimura, M. 1982. The neutral theory as a basis for understanding the mechanism of evolution and variation at the molecular level. In M. Kimura (ed.), *Molecular evolution, protein polymorphism, and the neutral theory,* pp. 3–56. Japan Scientific Societies Press, Tokyo, and Springer-Verlag, Berlin. [5]

Kimura, M. 1983a. *The neutral theory of molecular evolution.* Cambridge University Press, Cambridge. [5, 6, 10]

Kimura, M. 1983b. The neutral theory of molecular evolution. In M. Nei and R. K. Koehn (eds.), *Evolution of genes and proteins*, pp. 208–233. Sinauer Associates, Sunderland, Massachusetts. [5]

Kimura, M., and T. Ohta. 1971. *Theoretical aspects of population genetics.* Princeton University Press, Princeton, New Jersey. [5]

King, J. L. 1967. Continuously distributed factors affecting fitness. *Genetics* 55: 483–492. [6]

King, M.-C., and A. C. Wilson. 1975. Evolution at two levels: molecular similarities and biological differences between humans and chimpanzees. *Science* 188: 107–116. [10, 17]

Kirk, D. 1968. Patterns of survival and reproduction in the United States: implications for selection. *Proc. Natl. Acad. Sci. U.S.A.* 59: 662–670. [17]

Kirkpatrick, M. 1982. Sexual selection and the evolution of female choice. *Evolution* 36: 1–12. [8, 9]

Kitcher, P. 1982. *Abusing science: the case against creationism.* MIT Press, Cambridge, Massachusetts. [1]

Kitcher, P. 1985. *Vaulting ambition: Sociobiology and the quest for human nature.* MIT Press, Cambridge, Massachusetts. [17]

Kitchell, J. A., and T. R. Carr. 1985. Nonequilibrium model of diversification: faunal turnover dynamics. In J. W. Valentine (ed.), *Phanerozoic diversity patterns: profiles in macroevolution*, pp. 277–309. Princeton University Press, Princeton, New Jersey. [12]

Klauber, L. M. 1972. *Rattlesnakes: their habits, life histories, and influence on mankind*, Second edition. University of California Press, Berkeley. [13]

Kleckner, N. 1981. Transposable elements in prokaryotes. *Annu. Rev. Genet.* 15: 341–404. [15]

Klug, W. S., and M. R. Cummings. 1983. *Concepts of genetics.* Charles E. Merrill, Columbus, Ohio. [3]

Kluge, A. G. 1983. Cladistics and the classification of the great apes. In R. L. Ciochon and R. S. Corrucini (eds.), *New interpretations of ape and human ancestry*, pp. 151–177. Plenum, New York. [10]

Kluge, A. G., and J. S. Farris. 1969. Quantitative phyletics and the evolution of anurans. *Syst. Zool.* 18: 1–32. [10]

Kluge, A. G., and R. E. Strauss. 1985. Ontogeny and systematics. *Annu. Rev. Ecol. Syst.* 16: 247–268. [10]

Knoll, A. H., and G. W. Rothwell. 1980. Paleobotany: perspectives in 1980. *Paleobiology* 7: 7–35. [11]

Kodric-Brown, A., and J. H. Brown. 1984. Truth in advertising: the kinds of traits favored by sexual selection. *Amer. Natur.* 124: 309–323. [9]

Koehn, R. K., and W. F. Eanes. 1978. Molecular structure and protein variation within and among populations. *Evol. Biol.* 11: 39–100. [6]

Koehn, R. K., R. I. E. Newell, and F. Immermann. 1980. Maintenance of an aminopeptidase allele frequency cline by natural selection. *Proc. Natl. Acad. Sci. U.S.A.* 77: 5385–5389. [6]

Koehn, R. K., A. J. Zera, and J. G. Hall. 1983. Enzyme polymorphism and natural selection. In M. Nei and R. K. Koehn (eds.), *Evolution of genes and proteins*, pp. 115–136. Sinauer Associates, Sunderland, Massachusetts. [6]

Kollar, E. J., and C. Fischer. 1980. Tooth induction in chick epithelium: expression of quiescent genes for enamel synthesis. *Science* 207: 993–995. [14]

Koopman, K. F. 1950. Natural selection for reproductive isolation between *Drosophila pseudoobscura* and *Drosophila persimilis*. *Evolution* 4: 135–145. [8]

Krebs, C. J. 1978. *Ecology: the experimental analysis of distribution and abundance*, Second edition. Harper & Row, New York. [2]

Krebs, J. R., and N. B. Davies (eds.). 1984. *Behavioural ecology: an evolutionary approach*, Second edition. Sinauer Associates, Sunderland, Massachusetts. [9]

Kreitman, M. 1983. Nucleotide polymorphism at the alcohol dehydrogenase locus of *D. melanogaster*. *Nature* 304: 412–417. [6, 15]

Kropotkin, P. 1902. *Mutual aid. A factor of evolution.* Reprinted 1955 by Extending Horizons, Boston. [1]

Kruckeberg, A. R. 1957. Variation in fertility of hybrids between isolated populations of the serpentine species, *Streptanthus glandulosus* Cook. *Evolution* 11: 185–211. [4, 8]

Kurtén, B. 1959. Rates of evolution in fossil mammals. *Cold Spring Harbor Symp. Quant. Biol.* 24: 205–215. [14]

Kurtén, B. 1963. Return of a lost structure in the evolution of the felid dentition. *Soc. Scient. Fenn., Comment. Biol.* 26: 3–11. [10]

Kurtén, B. 1968. *Pleistocene mammals of Europe.* Aldine, Chicago. [11]

Lack, D. 1947. *Darwin's finches.* Cambridge University Press, Cambridge. [2, 16]

Lack, D. 1954. *The natural regulation of animal numbers.* Oxford University Press, Oxford. [2, 9]

Lack, D. 1969. Tit niches in two worlds, or homage to Evelyn Hutchinson. *Amer. Natur.* 103: 43–50. [4]

Lacy, R. C., and P. W. Sherman. 1983. Kin recognition by phenotypic matching. *Amer. Natur.* 121: 489–512. [9]

Lagler, K. F., J. E. Bardach, and R. R. Miller. 1962. *Ichthyology.* Wiley, New York. [9]

Lamotte, M. 1959. Polymorphism of natural populations of *Cepaea nemoralis*. *Cold Spring Harbor Symp. Quant. Biol.* 24: 65–86. [4]

Lande, R. 1976a. Natural selection and random genetic drift in phenotypic evolution. *Evolution* 30: 314–334. [7]

Lande, R. 1976b. The maintenance of genetic variability by mutation in a polygenic character with linked loci. *Genet. Res.* 26: 221–235. [3, 7, 14]

Lande, R. 1978. Evolutionary mechanisms of limb loss in tetrapods. *Evolution* 32: 73–92. [7]

Lande, R. 1979. Effective deme sizes during long-term evolution estimated from rates of chromosome rearrangement. *Evolution* 33: 234–251. [3, 6, 8]

Lande, R. 1980a. The genetic covariance between characters maintained by pleiotropic mutations. *Genetics* 94: 203–215. [7]

Lande, R. 1980b. Sexual dimorphism, sexual selection, and adaptation in polygenic characters. *Evolution* 34: 292–307. [9]

Lande, R. 1980c. Genetic variation and phenotypic evolution during allopatric speciation. *Amer. Natur.* 116: 463–479. [14]

Lande, R. 1981a. The minimum number of genes contributing to quantitative variation between and within populations. *Genetics* 99: 541–553. [7, 8]

Lande, R. 1981b. Models of speciation by sexual selection on polygenic traits. *Proc. Natl. Acad. Sci. U.S.A.* 78: 3721–3725. [8, 9]

Lande, R. 1982a. A quantitative genetic theory of life history evolution. *Ecology* 63: 607–615. [7]

Lande, R. 1982b. Rapid origin of sexual isolation and character divergence in a cline. *Evolution* 36: 213–223. [8]

Lande, R. 1984. The expected fixation rate of chromosomal inversions. *Evolution* 38: 743–752. [8]

Langley, C. H. 1977. Nonrandom associations between allozymes in natural populations of *Drosophila melanogaster*. In F. B. Christiansen and T. M. Fenchel (eds.), *Measuring selection in natural populations*, pp. 265–273. Springer-Verlag, Berlin. [4]

Langley, C. H., J. F. Y. Brookfield, and N. L. Kaplan. 1983. Transposable elements in Mendelian populations. I. A theory. *Genetics* 104: 457–472. [15]

Langley, C. H., and W. M. Fitch. 1974. An examination of the constancy of the rate of molecular evolution. *J. Mol. Evol.* 3: 161–177. [5]

Langley, C. H., E. Montgomery, and W. F. Quattlebaum. 1982. Restriction map variation in the *Adh* region of *Drosophila*. *Proc. Natl. Acad. Sci. U.S.A.* 79: 5631–5635. [15]

Larson, A., D. B. Wake, L. R. Maxson, and R. Highton. 1981. A molecular phylogenetic perspective on the origins of morphological novelties in the salamanders of the tribe Plethodontini (Amphibia, Plethodontidae). *Evolution* 35: 405–422. [10]

Lasker, G. W., and R. N. Tyzzer. 1982. *Physical anthropology*, Third edition. Holt, Rinehart and Winston, New York. [17]

Laurie-Ahlberg, C. C. 1985. Genetic variation affecting the expression of enzyme-coding genes in *Drosophila*: an evolutionary perspective. *Isozymes: Current topics in biological and medical research* 12: 33–88. [7]

Laven, H. 1958. Speciation by cytoplasmic isolation in the *Culex pipiens*-complex. *Cold Spring Harbor Symp. Quant. Biol.* 24: 166–173. [8]

Lawton, J. H., and D. R. Strong. 1981. Community patterns and competition in folivorous insects. *Amer. Natur.* 118: 317–338. [2, 13]

Lederberg, J., and E. M. Lederberg. 1952. Replica plating and indirect selection of bacterial mutants. *J. Bacteriol.* 63: 399–406. [3]

Lees, D. R. 1981. Industrial melanism: genetic adaptation of animals to air pollution. In J. A. Bishop and L. M. Cook (eds.), *Genetic consequences of man made change*, pp. 129–176. Academic Press, London. [6]

Leigh, E. G., Jr. 1973. The evolution of mutation rates. *Genetics Suppl.* 73: 1–18. [9]

Leipold, M., and J. Schmidtke. 1982. Gene expression in phylogenetically polyploid organisms. In G. A. Dover and R. B. Flavell (eds.), *Genome evolution*, pp. 219–236. Academic Press, New York. [15]

Lerner, I. M. 1954. *Genetic homeostasis*. Oliver & Boyd, Edinburgh. [6, 7]

Lerner, I. M. 1958. *The genetic basis of selection*. Wiley, New York. [7]

Lerner, I. M. 1968. *Heredity, evolution and society*. Freeman, San Francisco. [17]

Lerner, I. M., and C. A. Gunns. 1952. Egg size and reproductive fitness. *Poultry Sci.* 31: 537–544. [7]

Lessios, H. A. 1981. Divergence in allopatry: molecular and morphological divergence between sea urchins separated by the Isthmus of Panama. *Evolution* 35: 618–634. [10]

Levene, H. 1953. Genetic equilibrium when more than one ecological niche is available. *Amer. Natur.* 87: 331–333. [6]

Levin, B. 1971. The operation of selection in situations of interspecific competition. *Evolution* 25: 249–264. [16]

Levin, B. R., and W. L. Kilmer. 1974. Interdemic selection and the evolution of altruism: a computer simulation study. *Evolution* 28: 527–545. [9]

Levin, B., and R. E. Lenski. 1983. Coevolution in bacteria and their viruses and plasmids. In D. J. Futuyma and M. Slatkin, *Coevolution*, pp. 99–127. Sinauer Associates, Sunderland, Massachusetts. [16]

Levin, D. A. 1970. Developmental instability in species and hybrids of *Liatris*. *Evolution* 24: 613–624. [8]

Levin, D. A. 1975. Pest pressure and recombination systems in plants. *Amer. Natur.* 109: 437–451. [9]

Levin, D. A. 1978. The origin of isolating mechanisms in flowering plants. *Evol. Biol.* 11: 185–317. [4, 8]

Levin, D. A. 1979. The nature of plant species. *Science* 204: 381–384. [4]

Levin, D. A. 1981. Dispersal versus gene flow in plants. *Ann. Missouri. Bot. Gard.* 68: 233–253. [5]

Levin, D. A. 1983. Polyploidy and novelty in flowering plants. *Amer. Natur.* 122: 1–25. [3, 14]

Levin, D. A., and H. W. Kerster. 1967. Natural selection for reproductive isolation in *Phlox*. *Evolution* 21: 679–687. [4]

Levin, D. A., and H. W. Kerster. 1974. Gene flow in seed plants. *Evol. Biol.* 17: 139–220. [5]

Levin, S. A. 1972. A mathematical analysis of the genetic feedback mechanism. *Amer. Natur.* 106: 145–164. [16]

Levin, S. A. 1983. Some approaches to the modelling of coevolutionary interactions. In M. H. Nitecki (ed.), *Coevolution*, pp. 21–65. University of Chicago Press, Chicago. [16]

Levin, S. A., and C. A. Segel. 1982. Models of the influence of predation on aspect diversity in prey populations. *J. Math. Biol.* 14: 253–284. [2]

Levins, R. 1968. *Evolution in changing environments.* Princeton University Press, Princeton, New Jersey. [16]

Levins, R., and R. Lewontin. 1985. *The dialectical biologist.* Harvard University Press, Cambridge, Massachusetts. [15]

Levinton, J. S., and R. K. Bambach. 1975. A comparative study of Silurian and Recent deposit-feeding bivalve communities. *Paleobiology* 1: 97–124. [13]

Lewin, B. 1985. *Genes II.* Wiley, New York. [3]

Lewis, E. B. 1978. A gene complex controlling segmentation in *Drosophila. Nature* 276: 565–570. [14]

Lewis, H. 1973. The origin of diploid neospecies in *Clarkia. Amer. Natur.* 107: 161–170. [8]

Lewis, H., and M. Lewis. 1955. The genus *Clarkia. Univ. Calif. Publ. Bot.* 20: 241–392. [8]

Lewontin, R. C. 1962. Interdeme selection controlling a polymorphism in the house mouse. *Amer. Natur.* 96: 65–78. [9]

Lewontin, R. C. 1964. The interaction of selection and linkage. II. Optimum models. *Genetics* 50: 757–782. [7]

Lewontin, R. C. 1965. Selection for colonizing ability. In H. G. Baker and G. L. Stebbins (eds.), *The genetics of colonizing species,* pp. 77–94. Academic Press, New York. [2]

Lewontin, R. C. 1966. Is nature probable or capricious? *BioScience* 16: 25–27. [2]

Lewontin, R. C. 1970. The units of selection. *Annu. Rev. Ecol. Syst.* 1: 1–18. [9]

Lewontin, R. C. 1972. The apportionment of human diversity. *Evol. Biol.* 6: 381–398. [4, 17]

Lewontin, R. C. 1974a. *The genetic basis of evolutionary change.* Columbia University Press, New York. [4, 6, 7, 10]

Lewontin, R. C. 1974b. The analysis of variance and the analysis of causes. *Amer. J. Hum. Genet.* 26: 400–411. [7]

Lewontin, R. C. 1975. Genetic aspects of intelligence. *Annu. Rev. Genet.* 9: 387–405. [17]

Lewontin, R. C. 1976. Review of *Race differences in intelligence* by J. C. Loehlin, G. Lindzey, and J. N. Spuhler. *Amer. J. Hum. Genet.* 28: 92–97. [17]

Lewontin, R. C. 1977. Sociobiology—a caricature of Darwinism. PSA 2, Philosophy of Science Association. [9]

Lewontin, R. C. 1983. The organism as the subject and object of evolution. *Scientia* 118: 65–82. [2, 9, 14]

Lewontin, R. C. 1985. Population genetics. In P. J. Greenwood, P. H. Harvey, and M. Slatkin (eds.), *Evolution: Essays in honour of John Maynard Smith,* pp. 3–18. Cambridge University Press, Cambridge. [1]

Lewontin, R. C., and L. C. Birch. 1966. Hybridization as a source of variation for adaptation to new environments. *Evolution* 20: 315–336. [3]

Lewontin, R. C., L. R. Ginzburg, and S. D. Tuljapurkar. 1978. Heterosis as an explanation for large amounts of genetic polymorphism. *Genetics* 88: 149–169. [6]

Lewontin, R. C., and J. L. Hubby. 1966. A molecular approach to the study of genic heterozygosity in natural populations. II. Amount of variation and degree of heterozygosity in natural populations of *Drosophila pseudoobscura. Genetics* 54: 595–609. [4, 6]

Lewontin, R. C., and K. Kojima. 1960. The evolutionary dynamics of complex polymorphisms. *Evolution* 14: 458–472. [7]

Lewontin, R. C., S. Rose, and L. J. Kamin. 1984. *Not in our genes: Biology, ideology, and human nature.* Pantheon, New York. [17]

Lewontin, R. C., and M. J. D. White. 1960. Interaction between inversion polymorphisms of the two chromosome pairs in the grasshopper, *Moraba scurra. Evolution* 14: 116–129. [7]

Li, C. C. 1976. *First course in population genetics.* Boxwood Press, Pacific Grove, California. [6]

Li, W.-H. 1980. Rate of gene silencing at duplicate loci: a theoretical study and interpretation of data from tetraploid fishes. *Genetics* 95: 237–258. [15]

Li, W.-H. 1983. Evolution of duplicate genes and pseudogenes. In M. Nei and R. K. Koehn (eds.), *Evolution of genes and proteins,* pp. 14–37. Sinauer Associates, Sunderland, Massachusetts. [15]

Li, W.-H., and T. Gojobori. 1983. Rapid evolution of goat and sheep globin genes following gene duplication. *Mol. Biol. Evol.* 1: 94–108. [15]

Li, W.-H., T. Gojobori, and M. Nei. 1981. Pseudogenes as a paradigm of neutral evolution. *Nature* 292: 237–239. [15]

Liem, K. F. 1973. Evolutionary strategies and morphological innovations: cichlid pharyngeal jaws. *Syst. Zool.* 22: 425–441. [14]

Lillegraven, J. A., Z. Kielan-Jaworoska, and W. A. Clemens (eds.). 1979. *Mesozoic mammals: The first two-thirds of mammalian history.* University of California Press, Berkeley. [10, 11]

Littlejohn, M. J., and J. J. Loftus-Hills. 1968. An experimental evaluation of premating isolation in the *Hyla ewingi* complex (Anura: Hylidae). *Evolution* 22: 659–663. [4]

Livingstone, F. B. 1964. The distributions of the abnormal hemoglobin genes and their significance for human evolution. *Evolution* 18: 685–699. [3]

Lloyd, D. G. 1980. Benefits and handicaps of sexual reproduction. *Evol. Biol.* 13: 69–111. [9]

Lloyd, J. E. 1966. Studies on the flash communication system in *Photinus* fireflies. *Misc. Publ. Mus. Zool. Univ. Michigan* 130: 1–195. [4]

Loehlin, J. C., G. Lindzey, and J. N. Spuhler. 1975. *Race differences in intelligence.* Freeman, San Francisco. [17]

Lorenz, K. 1966. *On aggression.* Harcourt, Brace and World, New York. [17]

Lovejoy, A. D. 1936. *The great chain of being: A study of the history of an idea.* Harvard University Press, Cambridge, Massachusetts. [1]

Lovejoy, A. D. 1959. The argument for organic evolution before the *Origin of Species,* 1830–1858. In B. Glass, O. Temkin, and W. Strauss, Jr. (eds.), *Forerunners of Darwin, 1745–1859.* Johns Hopkins University Press, Baltimore, Maryland. [1]

Luckinbill, L. S., R. Arking, M. G. Clare, W. C. Cirocco, and S. A. Buck. 1984. Selection for delayed senescence in *Drosophila melanogaster*. *Evolution* 38: 996–1003. [9]

Ludwin, I. 1951. Natural selection in *Drosophila melanogaster* under laboratory conditions. *Evolution* 5: 231–242. [3]

MacArthur, R. H. 1972. *Geographical ecology. Patterns in the distribution of species.* Harper & Row, New York. [2]

MacArthur, R. H., and R. Levins. 1964. Competition, habitat selection and character displacement in a patchy environment. *Proc. Natl. Acad. Sci. U.S.A.* 51: 1207–1210. [2]

MacArthur, R. H., and R. Levins. 1967. The limiting similarity, convergence, and divergence of coexisting species. *Amer. Natur.* 101: 377–387. [16]

MacArthur, R. H., and E. R. Pianka. 1966. On optimal use of a patchy environment. *Amer. Natur.* 100: 603–609. [9, 16]

MacArthur, R. H., and E. O. Wilson. 1967. *The theory of island biogeography.* Princeton University Press, Princeton, New Jersey. [2, 9]

MacFadden, B. J. 1985. Patterns of phylogeny and rates of evolution in fossil horses: hipparions from the Miocene and Pliocene of North America. *Paleobiology* 11: 245–257. [14]

MacGregor, H. C. 1982. Big chromosomes and speciation amongst Amphibia. In G. A. Dover and R. B. Flavell (eds.), *Genome evolution*, pp. 325–341. Academic Press, New York. [15]

MacIntyre, R. J. 1982. Regulatory genes and adaptation—past, present, and future. *Evol. Biol.* 15: 247–286. [14]

MacIntyre, R. J. (ed.). 1986. *Molecular evolutionary genetics.* Plenum, New York. [15]

Mackay, T. F. C. 1984. Jumping genes meet abdominal bristles: hybrid dysgenesis-induced quantitative variation in *Drosophila melanogaster*. *Genet. Res.* 44: 231–237. [3]

Macnair, M. R. 1981. Tolerance of higher plants to toxic materials. In J. A. Bishop and L. M. Cook (eds.), *Genetic consequences of man made change*, pp. 177–207. Academic Press, New York. [6, 8]

Maglio, V. J. 1972. Evolution of mastication in the Elephantidae. *Evolution* 26: 638–658. [10, 12]

Maglio, V. J. 1973. Origin and evolution of the Elephantidae. *Amer. Phil. Soc. Trans.* 63: 1–149. [14]

Margulis, L. 1970. *The origin of eukaryotic cells.* Yale University Press, New Haven, Connecticut. [11]

Margulis, L. 1981. *Symbiosis in cell evolution: life and its environment on the early earth.* Freeman, San Francisco. [15]

Mark, G. A., and K. W. Flessa. 1977. A test for evolutionary equilibria: Phanerozoic brachiopods and Cenozoic mammals. *Paleobiology* 3: 17–22. [12]

Markow, T. A. 1981. Mating preferences are not predictive of the direction of evolution in experimental populations of *Drosophila*, *Science* 213: 1405–1407. [8]

Martin, M. M., and J. Harding. 1981. Evidence for the evolution of competition between two species of annual plants. *Evolution* 35: 975–987. [16]

Martin, P. S., and R. G. Klein (eds.). 1984. *Quaternary extinctions: a prehistoric revolution.* University of Arizona Press, Tucson. [11]

Maruyama, T., and M. Kimura. 1980. Genetic variability and effective population size when local extinction and recolonization of subpopulations are frequent. *Proc. Natl. Acad. Sci. U.S.A.* 77: 6710–6714. [5]

Mason, L. G. 1964. Stabilizing selection for mating fitness in natural populations of *Tetraopes*. *Evolution* 18: 492–497. [7]

Mather, K. 1941. Variation and selection of polygenic characters. *J. Genet.* 41: 159–193. [7]

Mather, K. 1949. *Biometrical genetics: the study of continuous variation.* Methuen, London. [4]

Mather, K. 1979. Historical overview: quantitative variation and polygenic systems. In J. N. Thompson, Jr. and J. M. Thoday (eds.), *Quantitative genetic variation*, pp. 5–34. Academic Press, New York. [7]

Mather, K., and B. J. Harrison. 1949. The manifold effect of selection. *Heredity* 3: 1–52; 131–162. [6, 7]

Matthew, W. D. 1915. Climate and evolution. *Ann. N. Y. Acad. Sci.* 24: 171–318. [13]

May, R. M. 1973. *Stability and complexity in model ecosystems.* Princeton University Press, Princeton, New Jersey. [2]

May, R. M. 1981. Patterns in multi-species communities. In R. M. May (ed.), *Theoretical ecology: principles and applications*, Second edition, pp. 197–227. Sinauer Associates, Sunderland, Massachusetts. [16]

May, R. M., and R. M. Anderson. 1983. Parasite-host coevolution. In D. J. Futuyma and M. Slatkin (eds.), *Coevolution*, pp. 186–206. Sinauer Associates, Sunderland, Massachusetts. [2, 6, 9, 16]

May, R. M., and G. F. Oster. 1976. Bifurcations and dynamic complexity in simple ecological models. *Amer. Natur.* 110: 573–599. [2]

Maynard Smith, J. 1964. Group selection and kin selection. *Nature* 201: 1145–1147. [9]

Maynard Smith, J. 1966. Sympatric speciation. *Amer. Natur.* 100: 637–650. [8]

Maynard Smith, J. 1976a. Group selection. *Quart. Rev. Biol.* 51: 277–283. [9]

Maynard Smith, J. 1976b. Sexual selection and the handicap principle. *J. Theor. Biol.* 57: 239–242. [9]

Maynard Smith, J. 1976c. A comment on the Red Queen. *Amer. Natur.* 110: 325–330. [12]

Maynard Smith, J. 1978a. Optimization theory in evolution. *Annu. Rev. Ecol. Syst.* 9: 31–56. [9]

Maynard Smith, J. 1978b. *The evolution of sex.* Cambridge University Press, Cambridge. [9]

Maynard Smith, J. 1982. *Evolution and the theory of games.* Cambridge University Press, Cambridge. [9]

Maynard Smith, J. 1983. The genetics of stasis and punctuation. *Annu. Rev. Genet.* 17: 11–25. [14]

Maynard Smith, J., et al. 1985. Developmental constraints and evolution. *Quart. Rev. Biol.* 60: 265–287. [14]

Maynard Smith, J., and J. Haigh. 1974. The hitchhiking effect of a favourable gene. *Genet. Res.* 23: 23–35. [6, 7]

Maynard Smith, J., and R. Hoekstra. 1980. Polymorphism in a varied environment: how robust are the models? *Genet. Res.* 35: 45–57. [6]

Maynard Smith, J., and G. R. Price. 1973. The logic of animal conflict. *Nature* 246: 15–18. [9]

Maynard Smith, J., and K. Sondhi. 1960. The genetics of a pattern. *Genetics* 45: 1039–1050. [4]

Mayr, E. 1942. *Systematics and the origin of species.* Columbia University Press, New York. [1, 4, 8]

Mayr, E. 1954. Change of genetic environment and evolution. In J. Huxley, A. C. Hardy, and E. B. Ford (eds.), *Evolution as a process,* pp. 157–180. Macmillan, New York. [5, 8, 14]

Mayr, E. 1960. The emergence of evolutionary novelties. In S. Tax (ed.), *The evolution of life,* pp. 349–380. University of Chicago Press, Chicago. [14]

Mayr, E. 1963. *Animal species and evolution.* Belknap Press of Harvard University Press, Cambridge, Massachusetts. [4, 5, 8, 14]

Mayr, E. 1976. *Evolution and the diversity of life.* Harvard University Press, Cambridge, Massachusetts. [1]

Mayr, E. 1981. Biological classification: toward a synthesis of opposing methodologies. *Science* 214: 510–516. [10]

Mayr, E. 1982a. *The growth of biological thought. Diversity, evolution and inheritance.* Harvard University Press, Cambridge, Massachusetts. [1, 14]

Mayr, E. 1982b. Processes of speciation in animals. In C. Barigozzi (ed.), *Mechanisms of speciation,* pp. 1–19. Alan R. Liss, New York. [8]

Mayr, E., and W. B. Provine (eds.). 1980. *The evolutionary synthesis: perspectives on the unification of biology.* Harvard University Press, Cambridge, Massachusetts. [1]

Mayr, E., and C. Vaurie. 1948. Evolution in the family Dicruridae (birds). *Evolution* 3: 238–265. [4]

McNeilly, T. 1968. Evolution in closely adjacent plant populations. III. *Agrostis tenuis* on a small copper mine. *Heredity* 23: 99–108. [6]

McPhail, J. D. 1969. Predation and the evolution of a stickleback (*Gasterosteus*). *J. Fish Res. Bd. Can.* 26: 3183–3208. [8]

Meacham, C. A., and G. F. Estabrook. 1985. Compatibility methods in systematics. *Annu. Rev. Ecol. Syst.* 16: 431–446. [10]

Melnick, M., N. C. Myrianthopoulos, and J. C. Christian. 1978. The effects of chorion type on variation in IQ in the NCPP twin population. *Amer. J. Hum. Genet.* 30: 425–433. [17]

Merrell, D. J. 1968. A comparison of the estimated size and the "effective size" of breeding populations of the leopard frog, *Rana pipiens. Evolution* 22: 274–283. [5]

Merritt, R. B. 1972. Geographic distribution and enzymatic properties of lactate dehydrogenase allozymes in the fathead minnow, *Pimephales promelas. Amer. Natur.* 196: 173–184. [6]

Michod, R. E. 1982. The theory of kin selection. *Annu. Rev. Ecol. Syst.* 13: 23–55. [9]

Milkman, R. D. 1967. Heterosis as a major cause of heterozygosity in nature. *Genetics.* 55: 493–495. [6]

Milkman, R. 1973. Electrophoretic variation in *Escherichia coli* from natural sources. *Science* 1982: 1024–1026. [4]

Miller, S. L., and H. Urey. 1959. Organic compound synthesis on the primitive earth. *Science* 130: 245–251. [11]

Mitter, C., and D. R. Brooks. 1983. Phylogenetic aspects of coevolution. In D. J. Futuyma and M. Slatkin (eds.), *Coevolution,* pp. 65–98. Sinauer Associates, Sunderland, Massachusetts. [16]

Mitter, C., and D. J. Futuyma. 1979. Population genetic consequences of feeding habits in some forest Lepidoptera. *Genetics* 92: 1005–1021. [6]

Molnar, S. 1983. *Human variation: Races, types, and ethnic groups.* Prentice-Hall, Englewood Cliffs, New Jersey. [17]

Monod, J. 1971. *Chance and necessity.* Knopf, New York. [15]

Montgomery, E. A., and C. H. Langley. 1983. Transposable elements in Mendelian populations. I. Distribution of three *copia*-like elements in a natural population. *Genetics* 104: 473–483. [15]

Moore, J. A. 1957. An embryologist's view of the species concept. In E. Mayr (ed.), *The species problem,* pp. 325–338. Amer. Assoc. Advancement Sci., Washington. [8]

Moore, J. A. 1961. A cellular basis for genetic isolation. In W. F. Blair (ed.), *Vertebrate speciation,* pp. 62–68. University of Texas Press, Austin. [4]

Morris, D. 1967. *The naked ape: a zoologist's study of the human animal.* McGraw-Hill, New York. [17]

Mortlock, R. P. 1982. Regulatory mutations and the development of new metabolic pathways by bacteria. *Evol. Biol.* 14: 205–268. [15]

Morton, N. E., J. F. Crow, and H. J. Muller. 1956. An estimate of the mutational damage in man from data on consanguineous marriages. *Proc. Natl. Acad. Sci. U.S.A.* 42: 855–863. [4, 5]

Mosquin, T. 1967. Evidence for autopolyploidy in *Epilobium angustifolium* (Onagraceae). *Evolution* 21: 713–719. [3]

Moulton, M. P., and S. L. Pimm. 1983. The introduced Hawaiian avifauna: biogeographic evidence for competition. *Amer. Natur.* 121: 669–690. [2]

Mukai, T. 1969. The genetic structure of natural populations of *Drosophila melanogaster.* VII. Synergistic interactions of mutant polygenes controlling viability. *Genetics* 61: 749–761. [15]

Mukai, T., S. I. Chigusa, and S.-I. Kusakaba. 1982. The genetic structure of natural populations of *Drosophila melanogaster.* XV. Nature of developmental homeostasis for viability. *Genetics* 101: 279–300. [7]

Mukai, T., S. I. Chigusa, L. E. Mettler, and J. F. Crow. 1972. Mutation rate and dominance of genes affecting viability in *Drosophila melanogaster. Genetics* 72: 335–355. [3]

Mukai, T., and C. C. Cockerham. 1977. Spontaneous mutation rates at enzyme loci in *Drosophila melanogaster. Proc. Natl. Acad. Sci. U.S.A.* 74: 2514–2517. [3]

Mukai, T., and R. A. Voelker. 1977. The genetic structure of natural populations of *Drosophila melanogaster.* XIII. Further studies on linkage disequilibrium. *Genetics* 86: 175–185. [7]

Mukai, T., and T. Yamazaki. 1980. Test for selection in polymorphic isozyme genes using the population cage method. *Genetics* 96: 537–542. [6]

Muller, H. J. 1950. Our load of mutations. *Amer. J. Hum. Genet.* 2: 111–176. [6]

Murdoch, W. W., and A. Oaten. 1975. Predation and population stability. *Adv. Ecol. Res.* 9: 2–131. [2]

Murray, J. D. 1981. A pre-pattern formation mechanism for animal coat markings. *J. Theor. Biol.* 88: 161–199. [14]

Nagylaki, T. 1975. Conditions for the existence of clines. *Genetics* 80: 595–615. [6]

Nagylaki, T. 1984a. The evolution of multigene families under intrachromosomal gene conversion. *Genetics* 106: 529–548. [15]

Nagylaki, T. 1984b. The evolution of multigene families under interchromosomal gene conversion. *Proc. Natl. Acad. Sci. U.S.A.* 81: 3796–3800. [15]

Nagylaki, T., and T. D. Petes. 1982. Intrachromosomal gene conversion and the maintenance of sequence homogeneity among repeated genes. *Genetics* 100: 315–337. [15]

Napier, J. R., and P. H. Napier. 1967. *A handbook of living primates.* Academic Press, New York. [17]

Neel, J. V. 1983. Frequency of spontaneous and induced "point" mutations in higher eukaryotes. *J. Hered.* 74: 2-15. [3]

Nei, M. 1971. Interspecific differences and evolutionary time estimated from electrophoretic data on protein identity. *Amer. Natur.* 105: 385–398. [8, 10]

Nei, M. 1972. Genetic distance between populations. *Amer. Natur.* 106: 283–292. [4]

Nei, M. 1975. *Molecular population genetics and evolution.* American Elsevier, New York. [4, 6, 8]

Nei, M. 1983. Genetic polymorphism and the role of mutation in evolution. In M. Nei and R. K. Koehn (eds.), *Evolution of genes and proteins,* pp. 165–190. Sinauer Associates, Sunderland, Massachusetts. [5, 6, 15]

Nei, M., P. A. Fuerst, and R. Chakraborty. 1976. Testing the neutral mutation hypothesis by distribution of single locus heterozygosity. *Nature* 262: 491–493. [6]

Nei, M., and R. K. Koehn (eds.). 1983. *Evolution of genes and proteins.* Sinauer Associates, Sunderland, Massachusetts. [15]

Nei, M., T. Maruyama, and R. Chakraborty. 1975. The bottleneck effect and genetic variability in populations. *Evolution* 29: 1–10. [5, 8]

Nei, M., T. Maruyama, and C.-I. Wu. 1983. Models of evolution of reproductive isolation. *Genetics* 103: 557–579. [8]

Nei, M., and A. K. Roychoudhury. 1972. Gene differences between Caucasian, Negro, and Japanese populations. *Science* 177: 434–436. [4]

Nei, M., and A. K. Roychoudhury. 1982. Genetic relationship and evolution of human races. *Evol. Biol.* 14: 1–59. [17]

Nei, M., and F. Tajima. 1985. Evolutionary change of restriction cleavage sites and phylogenetic inference for man and apes. *Mol. Biol. Evol.* 2: 189–205. [10, 17]

Neill, W. E. 1974. The community matrix and interdependence of the competition coefficients. *Amer. Natur.* 108: 399–408. [2]

Nelson, G., and N. Platnick. 1981. *Systematics and biogeography: cladistics and vicariance.* Columbia University Press, New York. [13]

Nelson, G., and N. I. Platnick. 1984. Systematics and evolution. In M.-W. Ho and P. J. Saunders (eds.)., *Beyond neo-Darwinism,* pp. 143–158. Academic Press, New York. [10]

Nestmann, E. R., and R. F. Hill. 1973. Population genetics in continuously growing mutator cultures of *Escherichia coli. Genetics* 73 (Suppl): 41–44. [9]

Nevo, E., and H. Bar-El. 1976. Hybridization and speciation in fossorial mole rats. *Evolution* 30: 831–840. [8]

Newell, N. D. 1967. Revolutions in the history of life. *Geol. Soc. Amer. Special Papers.* 89: 63–91. [12]

Newell, N. D. 1982. *Creation and evolution: myth or reality.* Columbia University Press, New York. [1]

Nicholson, A. J. 1958. The self-adjustment of populations to change. *Cold Spring Harbor Symp. Quant. Biol.* 22: 153–173. [2]

Nielsen, C. 1985. Animal phylogeny in the light of the trochaea theory. *Biol. J. Linn. Soc.* 25: 243–299. [11]

Nielsen, J. T. 1977. Variation in the number of genes coding for salivary amylase in the bank vole, *Clethrionomys glareola. Genetics* 85: 155–169. [7]

Niklas, K. J. 1983. The influence of Palaeozoic ovule and cupule morphologies on wind pollination. *Evolution* 37: 968–986. [9]

Niklas, K. J. 1986. Large-scale changes in animal and plant terrestrial communities. In D. M. Raup and D. Jablonski (eds.), *Patterns and processes in the history of life.* Springer-Verlag, Berlin. [12]

Niklas, K. J., B. H. Tiffney, and A. H. Knoll. 1980. Apparent changes in the diversity of fossil plants. *Evol. Biol.* 12: 1–89. [11, 12]

Nitecki, M. H. (ed.). 1983. *Coevolution.* University of Chicago Press, Chicago. [16]

Noble, G. K. 1931. *The biology of the Amphibia.* McGraw-Hill, New York. [10, 14]

North, G. 1984. How to make a fruitfly. *Nature* 311: 214–216. [14]

Novacek, M. J. 1982. Information for macromolecular studies from anatomical and fossil evidence on higher eutherian phylogeny. In M. Goodman (ed.), *Macromolecular sequences in systematic and evolutionary biology*, pp. 2–41. Plenum, New York. [10]

O'Brien, R. D. 1967. *Insecticides: action and metabolism.* Academic Press, New York. [14]

O'Donald, P. 1980. *Genetic models of sexual selection.* Cambridge University Press, New York. [9]

Officer, C. B., and C. L. Drake. 1983. The Cretaceous-Tertiary transition. *Science* 219: 1383–1390. [12]

Ohno, S. 1970. *Evolution by gene duplication.* Springer-Verlag, New York. [3, 15]

Ohno, S., C. Stenius, L. Christian, and G. Schipmann. 1969. *De novo* mutation-like events observed at the 6PGD locus of the Japanese quail, and the principle of polymorphism breeding more polymorphisms. *Biochem. Genet.* 3: 417–428. [3]

Ohta, T. 1974. Mutational pressure as the main cause of molecular evolution and polymorphism. *Nature* 252: 351–354. [6]

Ohta, T. 1980. *Evolution and variation of multigene families.* Springer-Verlag, Berlin. [15]

Ohta, T. 1985. A model of duplicative transposition and gene conversion for repetitive DNA families. *Genetics* 110: 513–524. [15]

Ohta, T., and G. A. Dover. 1984. The cohesive population genetics of molecular drive. *Genetics* 108: 501–521. [15]

Oliver, C. G. 1972. Genetic and phenotypic differentiation and geographic distance in four species of Lepidoptera. *Evolution* 26: 221–241. [8]

Olson, E. C. 1971. *Vertebrate paleozoology.* Wiley-Interscience, New York. [11]

Olson, E. C. 1981. The problem of missing links: today and yesterday. *Quart. Rev. Biol.* 56: 405–482. [14]

Olson, E. C., and R. L. Miller. 1958. *Morphological integration.* University of Chicago Press, Chicago. [14]

Oparin, A. I. 1953. *The origin of life.* Dover, New York. [11]

Orgel, L. E. 1973. *The origins of life: molecular and natural selection.* Wiley, New York. [11]

Orgel, L. E., and F. H. C. Crick. 1980. Selfish DNA: the ultimate parasite. *Nature* 284: 604–606. [15]

Orians, G. H., and R. T. Paine. 1983. Convergent evolution at the community level. In D. J. Futuyma and M. Slatkin (eds.), *Coevolution*, pp. 431–458. Sinauer Associates, Sunderland, Massachusetts. [2, 13]

Ostrom, J. H. 1976. *Archaeopteryx* and the origin of birds. *Biol. J. Linn. Soc.* 8: 91–182. [11, 14]

Padian, K. 1985. The origins and aerodynamics of flight in extinct vertebrates. *Paleontology* 28: 413–433. [14]

Palmer, E. J. 1948. Hybrid oaks of North America. *J. Arnold Arboretum* 29: 1–48. [4]

Park, T., and M. Lloyd. 1955. Natural selection and the outcome of competition. *Amer. Natur.* 89: 235–240. [16]

Parker, E. D., Jr. 1979. Ecological implications of clonal diversity in parthenogenetic morphospecies. *Amer. Zool.* 19: 753–762. [4]

Parker, G. A. 1974. The reproductive behavior and the nature of sexual selection in *Scatophaga stercoraria* L. (Diptera: Scatophagidae). IX. Spatial distribution of fertilization rates and evolution of male search strategy within the reproductive area. *Evolution* 28: 93–108. [9]

Parker, G. A., R. R. Baker, and V. G. F. Smith. 1972. The origin and evolution of gamete dimorphism and the male-female phenomenon. *J. Theor. Biol.* 36: 529–553. [9]

Parker, M. A. 1985. Local population differentiation for compatibility in an annual legume and its host-specific pathogen. *Evolution* 39: 713–723. [16]

Paterniani, E. 1969. Selection for reproductive isolation between two populations of maize, *Zea mays* L. *Evolution* 23: 534–547. [8]

Paterson, H. E. H. 1982. Perspective on speciation by reinforcement. *S. Afr. J. Sci.* 78: 53–57. [8]

Patterson, B. D. 1983. Grasshopper mandibles and the niche-variation hypothesis. *Evolution* 37: 375–388. [6]

Patterson, C. 1982. Morphological characters and homology. In K. A. Joysey and A. E. Friday (eds.), *Problems of phylogenetic reconstruction*, pp. 21–74. Academic Press, New York. [10]

Patthy, L. 1985. Evolution of the proteases of blood coagulation and fibrinolysis by assembly from modules. *Cell* 41: 657–663. [15]

Patton, J. L. 1969. Chromosome evolution in the pocket mouse, *Perognathus goldmani* Osgood. *Evolution* 23: 645–662. [8]

Patton, J. L. 1972. Patterns of geographic variation in karotype in the pocket gophers, *Thomomys bottae* (Eydoux and Gervais). *Evolution* 26: 574–586. [8]

Paul, C. R. C. 1977. Evolution of primitive echinoderms. In A. Hallam (ed.), *Patterns of evolution as illustrated by the fossil record*, pp. 123–158. Elsevier, Amsterdam. [12]

Peterson, R. T. 1961. *A field guide to western birds.* Houghton-Mifflin, Boston. [9]

Pianka, E. R. 1983. *Evolutionary ecology*, Third edition. Harper & Row, New York. [13]

Pimentel, D., and A. C. Bellotti. 1976. Parasite-host population systems and genetic stability. *Amer. Natur.* 110: 877–888. [7]

Pimentel, D., E. H. Feinberg, D. W. Wood, and J. T. Hayes. 1965. Selection, spatial distribution, and the coexistence of competing fly species. *Amer. Natur.* 99: 97–108. [16]

Pimentel, D., and F. A. Stone. 1968. Evolution and population ecology of parasite-host systems. *Canad. Entomol.* 100: 655–662. [16]

Pimm, S. L. 1982. *Food Webs.* Chapman and Hall, London. [16]

Place, A. R., and D. A. Powers. 1979. Genetic variation and relative catalytic efficiencies: lactate dehydrogenase B allozymes of *Fundulus heteroclitus*. *Proc. Natl. Acad. Sci. U.S.A.* 76: 2354–2358. [6]

Pollard, J. W. 1984. Is Weissman's barrier absolute? In M.-W. Ho and P. T. Saunders (eds.), *Beyond neo-Darwinism: an introduction to the new evolutionary paradigm*, pp. 291–314. Academic Press, London. [1]

Popham, E. J. 1942. Further experimental studies on the selective action of predators. *Proc. Zool. Soc. Lond. (A)* 112: 105–117. [6]

Popper, K. R. 1968. *Conjectures and refutations: the growth of scientific knowledge*. Harper Torchbooks, Harper & Row, New York. [1]

Porter, K. R. 1972. *Herpetology*. Saunders, Philadelphia. [10]

Pough, R. H. 1951. *Audubon water bird guide*. Doubleday, New York. [4]

Powell, J. R. 1978. The founder-flush speciation theory: an experimental approach. *Evolution* 32: 465–474. [8]

Powell, J. R., and R. C. Richmond. 1974. Founder effects and linkage disequilibrium in experimental populations of *Drosophila*. *Proc. Natl. Acad. Sci. U.S.A.* 71: 1663–1665. [7]

Powell, J. R., and H. Wistrand. 1978. The effect of heterogeneous environments and a competitor on genetic variation in *Drosophila*. *Amer. Natur.* 112: 935–947. [6]

Prakash, S., and R. C. Lewontin. 1968. A molecular approach to the study of genic heterozygosity in natural populations. III. Direct evidence of coadaptation in gene arrangements of *Drosophila*. *Proc. Natl. Acad. Sci. U.S.A.* 59: 398–405. [4, 7]

Prance, G. T. (ed.). 1982. *Biological diversification in the tropics*. Columbia University Press, New York. [11]

Price, M. V., and N. M. Waser. 1982. Population structure, frequency-dependent selection, and the maintenance of sexual reproduction. *Evolution* 36: 35–43. [9]

Price, P. W. 1980. *Evolutionary biology of parasites*. Princeton University Press, Princeton, New Jersey. [2]

Price, P. W., C. E. Bouton, P. Gross, B. A. McPheron, J. N. Thompson, and A. E. Weis. 1980. Interactions among three trophic levels: influence of plants on interactions between insect herbivores and natural enemies. *Annu. Rev. Ecol. Syst.* 11: 41–65. [2]

Prosser, C. L., and F. A. Brown, Jr. 1961. *Comparative animal physiology*. Saunders, Philadelphia. [2]

Prout, T. 1964. Observations on structural reduction in evolution. *Amer. Natur.* 97: 239–249. [9]

Provine, W. B. 1971. *The origins of theoretical population genetics*. University of Chicago Press, Chicago. [1]

Pulliam, H. R. 1985. Foraging efficiency, resource partitioning, and the coexistence of sparrow species. *Ecology* 66: 1829–1836. [16]

Pyke, G. H. 1984. Optimal foraging theory: a critical review. *Annu. Rev. Ecol. Syst.* 15: 523–575. [9]

Pyke, G. H., H. R. Pulliam, and E. L. Charnov. 1977. Optimal foraging: a selective review of theory and tests. *Quart. Rev. Biol.* 52: 137–154. [16]

Quinn, J. F. 1983. Mass extinctions in the fossil record. *Science* 219: 1239–1240. [12]

Radinsky, L. 1978a. Do albumin clocks run on time? *Science* 200: 1182–1183. [10]

Radinsky, L. B. 1978b. Evolution of brain size in carnivores and ungulates. *Amer. Natur.* 11: 815–831. [12]

Radinsky, L. B. 1982. Evolution of skull shape in carnivores. 3. The origin and early radiation of the modern carnivore families. *Paleobiology* 8: 177–195. [12]

Radinsky, L. 1984. Ontogeny and phylogeny in horse skull evolution. *Evolution* 38: 1–15. [14]

Raff, R. A., and T. C. Kaufman. 1983. *Embryos, genes, and evolution: the developmental-genetic basis of evolutionary change*. Macmillan, New York. [3, 14]

Raff, R. A., and H. R. Mahler. 1972. The non-symbiotic origin of mitochondria. *Science* 177: 575–582. [11]

Rao, S. V., and P. DeBach. 1969. Experimental studies in hybridization and sexual isolation between some *Aphytis* species (Hymenoptera: Aphelinidae). III. The significance of reproductive isolation between interspecific hybrids and parental species. *Evolution* 23: 525–533. [4]

Raup, D. W. 1962. Computer as aid in describing form in gastropod shells. *Science* 138: 150–152. [14]

Raup, D. M. 1966. Geometric analysis of shell coiling: general problems. *J. Paleontol.* 40: 1178–1190. [14]

Raup, D. M. 1972. Taxonomic diversity during the Phanerozoic. *Science* 177: 1065–1071. [12]

Raup, D. M. 1978. Cohort analysis of generic survivorship. *Paleobiology* 4: 1–15. [12]

Raup, D. M. 1979. Biases in the fossil record of species and genera. *Bull. Carnegie Mus. Nat. Hist.* 13: 85–91. [12]

Raup, D. M. 1984. Evolutionary radiations and extinctions. In H. D. Holland and A. F. Trendall (eds.), *Patterns of change in earth evolution*, pp. 5–14. Springer-Verlag, Berlin. [12]

Raup, D. M., and S. J. Gould. 1974. Stochastic simulation and evolution of morphology—towards a nomothetic paleontology. *Syst. Zool.* 23: 305–322. [14]

Raup, D. M., S. J. Gould, T. J. M. Schopf, and D. S. Simberloff. 1973. Stochastic models of phylogeny and the evolution of diversity. *J. Geol.* 81: 525–542. [12]

Raup, D. M., and D. Jablonski (eds.). 1986. *Patterns and processes in the history of life*. Springer-Verlag, Berlin. [12]

Raup, D. M., and J. J. Sepkoski, Jr. 1982. Mass extinctions in the fosil record. *Science* 215: 1501–1503. [12]

Raup, D. M., and J. J. Sepkoski, Jr., 1984. Periodicities of extinctions in the geologic past. *Proc. Natl. Acad. Sci. U.S.A.* 81: 801–805. [12]

Raven, P. H. 1963. Amphitropical relationships in the floras of North and South America. *Quart. Rev. Biol.* 38: 151–177. [13]

Raven, P. H. 1980. Hybridization and the nature of species in higher plants. *Canad. Bot. Assoc. Bull.* Suppl., 13: 3–10. [4]

Real, L., J. Ott, and E. Silverfine. 1982. On the tradeoff between the mean and the variance in foraging: effect of spatial distribution and color preference. *Ecology* 63: 1617–1623. [16]

Reanny, D. 1976. Extrachromosomal elements as possible agents of adaptation and development. *Bact. Rev.* 40: 552–590. [3]

Rees, H., G. Jenkins, A. G. Seal, and J. Hutchinson. 1982. Assays of the phenotypic effects of changes in DNA amounts. In G. A. Dover and R. Flavell (eds.), *Genome evolution*, pp. 287–297. Academic Press, London. [8, 15]

Reichle, D. E. 1966. Some pselaphid beetles with boreal affinities and their distribution along the post-glacial fringe. *Syst. Zool.* 15: 330–334. [13]

Reig, O. A. 1981. A refreshed orthodox view of paleobiogeography of South American mammals. *Evolution* 35: 1032–1035. [13]

Rendel, J. M. 1951. Mating of ebony vestigial and wild type *Drosophila melanogaster* in light and dark. *Evolution* 5: 226–230. [3]

Rendel, J. M. 1953. Variations in the weights of hatched and unhatched ducks' eggs. *Biometrika* 33: 48–58. [7]

Rendel, J. M. 1967. *Canalisation and gene control.* Logos Press, London. [4, 7]

Rensch, B. 1947. *Neuere Probleme der Abstammungslehre.* Enke, Stuttgart. [1]

Rensch, B. 1959. *Evolution above the species level.* Columbia University Press, New York. [1, 10, 14]

Repetski, J. E. 1978. A fish from the Upper Cambrian of North America. *Science* 200: 529–531. [11]

Reyment, R. A. 1982. Analysis of trans-specific evolution in Cretaceous ostracods. *Paleobiology* 8: 293–306. [14]

Reynoldson, T. B. 1966. The distribution and abundance of lake-dwelling triclads—towards a hypothesis. *Adv. Ecol. Res.* 3: 1–71. [2]

Rice, W. R. 1985. Disruptive selection on habitat preference and the evolution of reproductive isolation: an exploratory experiment. *Evolution* 39: 645–656. [8]

Ricklefs, R. E. 1979. *Ecology*, Second edition. Chiron Press, New York. [2]

Ricklefs, R. E., and G. W. Cox. 1972. Taxon cycles in the West Indian avifauna. *Amer. Natur.* 106: 195–219. [13]

Ricklefs, R., and K. O'Rourke. 1975. Aspect diversity in moths: a temperate-tropical comparison. *Evolution* 29: 313–324. [16]

Riedl, R. 1977. A systems-analytical approach to macroevolutionary phenomena. *Quart. Rev. Biol.* 52: 351–370. [14]

Riedl, R. 1978. *Order in living organisms: A systems analysis of evolution.* (English edition 1978, trans. R. P. S. Jeffries). Wiley, New York. [14]

Riley, H. P. 1938. A character analysis of colonies of *Iris fulva* and *I. hexagona* var. *giganticaerulea*. *Amer. J. Bot.* 29: 323–331. [4]

Riley, R. 1982. Cytogenetic evidence on the nature of speciation in wheat and its relatives. In C. Barigozzi (ed.), *Mechanisms of speciation*, pp. 471–478. Alan R. Liss, New York. [8]

Riska, B., and W. R. Atchley. 1985. Genetics of growth predicts patterns of brain-size evolution. *Science* 229: 668–671. [14]

Robertson, A. 1955. Selection in animals: synthesis. *Cold Spring Harbor Symp. Quant. Biol.* 20: 225–229. [7]

Robertson, A. 1962. Selection for heterozygotes in small populations. *Genetics* 47: 1291–1300. [6]

Robertson, F. W., and E. C. R. Reeve. 1952. Heterozygosity, environmental variation and heterosis. *Nature* 170: 296. [7]

Rogers, J. 1985. Origins of repeated DNA. *Nature* 317: 765–766. [3]

Rogers, J. S. 1972. Measures of genetic similarity and genetic distance. *Univ. Tex. Publ.* 7213: 145–153. [4]

Rohlf, F. J., and G. D. Schnell. 1971. An investigation of the isolation-by-distance model. *Amer. Natur.* 105: 295–324. [5]

Romer, A. S. 1949. Time series and trends in animal evolution. In G. L. Jepsen, E. Mayr, and G. G. Simpson (eds.), *Genetics, paleontology, and evolution*, pp. 103–120. Princeton University Press, Princeton, New Jersey. [12]

Romer, A. S. 1956. *Osteology of the reptiles.* University of Chicago Press, Chicago. [14]

Romer, A. S. 1966. *Vertebrate paleontology*, Third edition. University of Chicago Press, Chicago. [10, 11, 12]

Rose, M. R., and B. Charlesworth. 1981. Genetics of life history in *Drosophila melanogaster*. I. Sib analysis of adult females. *Genetics* 97: 173–186. II. Exploratory selection experiments. *Genetics* 97: 187–196. [7, 9]

Rose, M. R., and W. F. Doolittle. 1983. Molecular biological mechanisms of speciation. *Science* 220: 157–162. [8]

Rosenzweig, M. L. 1973. Evolution of the predator isocline. *Evolution* 27: 84–94. [16]

Ross, H. H., 1958. Evidence suggesting a hybrid origin for certain leafhopper species. *Evolution* 12: 337–446. [8]

Ross, H. H., G. C. Decker, and H. B. Cunningham. 1964. Adaptation and differentiation of temperate phylogenetic lines from tropical ancestors in *Empoasca*. *Evolution* 18: 639–651. [5]

Roughgarden, J. 1971. Density-dependent natural selection. *Ecology* 52: 453–468. [9]

Roughgarden, J. 1972. Evolution of niche width. *Amer. Natur.* 106: 683–718. [8]

Roughgarden, J. 1975. Evolution of marine symbioses—a simple cost-benefit model. *Ecology* 56: 1201–1208. [16]

Roughgarden, J. 1976. Resource partitioning among competing species—a coevolutionary approach. *Theoret. Pop. Biol.* 9: 388–424. [16]

Roughgarden, J. 1979. *Theory of population genetics and evolutionary ecology: an introduction.* Macmillan, New York. [6]

Roughgarden, J. 1983a. The theory of coevolution. In D. J. Futuyma and M. Slatkin (eds.), *Coevolution*, pp. 33–64. Sinauer Associates, Sunderland, Massachusetts. [2, 16]

Roughgarden, J. 1983b. Coevolution between competitors. In D. J. Futuyma and M. Slatkin (eds.), *Coevolution*, pp. 383–403. Sinauer Associates, Sunderland, Massachusetts. [16]

Rubin, G. M. 1983. Dispersed repetitive DNAs in *Drosophila*. In J. A. Shapiro (ed.), *Mobile genetic elements*, pp. 329–361. Academic Press, New York. [15]

Rummel, J. D., and J. Roughgarden. 1985. A theory of faunal buildup for competition communities. *Evolution* 39: 1009–1033. [16]

Rutgers, A. 1969. *Birds of Asia*. Methuen, London. [8]

Sage, R. D., and R. K. Selander. 1979. Hybridization between species of the *Rana pipiens* complex in central Texas. *Evolution* 33: 1069–1088. [3, 4]

Saghir, M. T., and E. Robbins. 1973. *Male and female homosexuality*. Williams and Wilkins, Baltimore, Maryland. [17]

Sahlins, M. 1976. *The use and abuse of biology: an anthropological critique of sociobiology*. University of Michigan Press, Ann Arbor. [17]

Salzburg, M. A. 1984. *Anolis sagrei* and *Anolis cristatellus* in southern Florida: a case study in interspecific competition. *Ecology* 65: 14–19. [2]

Sanday, P. R. 1972. On the causes of IQ differences between groups and implications for social policy. *Human Organization* 31: 411–424. Reprinted in A. Montagu (ed.), *Race and IQ*, Oxford University Press, London (1975). [17]

Sargent, T. D. 1969. Background selections of the pale and melanic forms of the cryptic moth *Phigalia titea* (Cramer). *Nature* 222: 585–586. [6]

Sarich, V. M. 1977. Rates, sample sizes, and the neutrality hypothesis for electrophoresis in evolutionary studies. *Nature* 265: 24–28. [10]

Savage, J. M. 1973. The geographic distribution of frogs: patterns and predictions. In J. L. Vial (ed.), *Evolutionary biology of the anurans*, pp. 351–445. University of Missouri Press, Columbia. [10]

Savage, J. M. 1982. The enigma of the Central American herpetofauna: dispersals or vicariance? *Ann. Missouri Bot. Gard.* 69: 464–547. [13]

Scarr, S., and R. A. Weinberg. 1976. IQ test performance of black children adopted by white families. *Amer. Psychol.* 31: 726–739. [17]

Schaal, B. A., and W. G. Smith. 1980. The apportionment of genetic variation within and among populations of *Desmodium nudiflorum*. *Evolution* 34: 214–221. [4]

Schaeffer, B. 1956. Evolution in the subholostean fishes. *Evolution* 10: 201–212. [10]

Schaeffer, B., M. K. Hecht, and N. Eldredge. 1972. Phylogeny and paleontology. *Evol. Biol.* 6: 31–46. [10]

Schaffer, W. M., and M. L. Rosenzweig. 1978. Homage to the Red Queen. I. Coevolution of predators and their victims. *Theoret. Pop. Biol.* 14: 135–157. [16]

Schemske, D. W. 1983a. Breeding system and habitat effects on fitness components in three neotropical *Costus* (Zingiberaceae). *Evolution* 37: 523–539. [5]

Schemske, D. W. 1983b. Limits to specialization and coevolution in plant-animal mutualisms. In M. Nitecki (ed.), *Coevolution*, pp. 67–109. University of Chicago Press, Chicago. [16]

Schindewolf, O. H. 1936. *Palaeontologie, Entwicklungslehre, und Genetik*. Bornträger, Berlin. [14]

Schluter, D., and P. R. Grant. 1984. Determinants of morphological patterns in communities of Darwin's finches. *Amer. Natur.* 123: 175–196. [2, 16]

Schmalhausen, I. I. 1949. *Factors of evolution*. Blakiston, Philadelphia. [2]

Schoen, D. J. 1982. The breeding system of *Gilia achilleifolia*: variation in floral characteristics and outcrossing rate. *Evolution* 36: 352–360. [5]

Schoener, T. W. 1968. The *Anolis* lizards of Bimini: resource partitioning in a complex fauna. *Ecology* 49: 704–726. [2]

Schoener, T. W. 1971. Theory of feeding strategies. *Annu. Rev. Ecol. Syst.* 22: 103–114. [16]

Schoener, T. W. 1983. Field experiments on interspecific competition. *Amer. Natur.* 122: 240–285. [2, 16]

Schoener, T. W. 1984. Size differences among sympatric, bird-eating hawks: a worldwide survey. In D. R. Strong, Jr., D. Simberloff, L. G. Abele, and A. B. Thistle (eds.), *Ecological communities: conceptual issues and the evidence*, pp. 254–281. Princeton University Press, Princeton, New Jersey. [16]

Schopf, J. W. (ed.). 1983. *Earth's earliest biosphere*. Princeton University Press, Princeton, New Jersey. [11]

Schopf, T. J. M. 1974. Permo-Triassic extinctions: relation to sea-floor spreading. *J. Geol.* 82: 129–143. [12]

Schopf, T. J. M., D. M. Raup, S. J. Gould, and D. S. Simberloff. 1975. Genomic versus morphologic rates of evolution: influence of morphologic complexity. *Paleobiology* 1: 63–70. [12]

Schull, W. J., and J. V. Neel. 1965. *The effects of inbreeding on Japanese children*. Harper & Row, New York. [5]

Schwartz, D., and W. J. Laughner. 1969. A molecular basis for heterosis. *Science* 166: 626–627. [6]

Scossiroli, R. E. 1954. Artificial selection of a quantitative trait in *Drosophila melanogaster* under increased mutation rate. *Atti IX Congr. Intern. Genet. Caryologia* 4 (Suppl.): 861–864. [7]

Seaton, A. P. C., and J. Antonovics. 1967. Population inter-relationships. I. Evolution in mixtures of *Drosophila* mutants. *Heredity* 22: 19–33. [16]

Seilacher, A. 1984. Late Precambrian and early Cambrian Metazoa: preservational or real extinctions? In H. D. Holland and A. F. Trendall (eds.), *Patterns of change in earth evolution*, pp. 159–168. Springer-Verlag, Berlin. [11]

Selander, R. K. 1966. Sexual dimorphism and differential niche utilization in birds. *Condor* 68: 113–151. [2]

Selander, R. K. 1970. Behavior and genetic variation in natural populations. *Amer. Zool.* 10: 53–66. [5]

Selander, R. K. 1976. Genic variation in natural populations. In F. J. Ayala (ed.), *Molecular evolution*, pp. 21–45. Sinauer Associates, Sunderland, Massachusetts. [4]

Selander, R. K. 1982. Phylogeny. In R. Milkman (ed.), *Perspectives on evolution*, pp. 32–59. Sinauer Associates, Sunderland, Massachusetts. [5]

Selander, R. K., and B. R. Levin. 1980. Genetic diversity and structure in populations of *Escherichia coli*. *Science* 210: 545–547. [5]

Selander, R. K., and T. S. Whittam. 1983. Protein polymorphism and the genetic structure of populations. In M. Nei and R. K. Koehn (eds.), *Evolution of genes and proteins*, pp. 89–114. Sinauer Associates, Sunderland, Massachusetts. [6]

Sepkoski, J. J., Jr. 1976. Species diversity in the Phanerozoic: species-area effects. *Paleobiology* 2: 298–303. [12]

Sepkoski, J. J. 1978. A kinetic model of Phanerozoic taxonomic diversity. I. Analysis of marine orders. *Paleobiology* 4: 223–251. [11, 12]

Sepkoski, J. J., Jr. 1979. A kinetic model of Phanerozoic taxonomic diversity. II. Early Phanerozoic families and multiple equilibria. *Paleobiology* 5: 222–251. [12]

Sepkoski, J. J., Jr. 1981. A factor analytic description of the Phanerozoic marine fossil record. *Paleobiology* 7: 36–53. [12]

Sepkoski, J. J., Jr. 1984. A kinetic model of Phanerozoic taxonomic diversity. III. Post-Paleozoic families and mass extinctions. *Paleobiology* 10: 246–267. [12]

Sepkoski, J. J., R. K. Bambach, D. M. Raup, and J. W. Valentine. 1981. Phanerozoic marine diversity and the fossil record. *Nature* 293: 435–437. [12]

Shapiro, J. A. (ed.). 1983. *Mobile genetic elements*. Academic Press, New York. [3, 15]

Sharp, P. A. 1985. On the origin of RNA splicing and introns. *Cell* 42: 397–400. [3]

Shaw, D. D. 1981. Chromosomal hybrid zones in orthopteroid insects. In W. R. Atchley and D. S. Woodruff (eds.), *Evolution and speciation: essays in honor of M. J. D. White*, pp. 146–170. Cambridge University Press, Cambridge. [4, 8]

Sheldon, B. L., and M. K. Milton. 1972. Studies on the scutellar bristles of *Drosophila melanogaster*. II. Long-term selection for high bristle number in the Oregon RC strain and correlated responses in abdominal chaetae. *Genetics* 71: 567–595. [7]

Shepherd, J. C. W., W. McGinnis, A. E. Carrasco, E. M. DeRobertis, and W. J. Gehring. 1984. Fly and frog homoeo domains show homologies with yeast mating type structural proteins. *Nature* 310: 70–71. [15]

Shin, H.-S., T. A. Bargiello, B. T. Clark, F. R. Jackson, and M. W. Young. 1985. An unusual coding sequence from a *Drosophila* clock gene is conserved in vertebrates. *Nature* 317: 445–448. [15]

Sibley, C. G., and J. E. Ahlquist. 1981. The phylogeny and relationships of the ratite birds as indicated by DNA-DNA hybridization. In G. G. E. Scudder and J. L. Reveal (eds.), *Evolution today*, pp. 301–335. Hunt Inst. Botan. Document., Pittsburgh, Pennsylvania. [13]

Sibley, C. G., and J. E. Ahlquist. 1983. Phylogeny and classification of birds based on the data of DNA-DNA hybridization. *Current Ornithology* 1: 245–292. [10]

Sibley, C. G., and J. E. Ahlquist. 1984. The phylogeny of the hominoid primates as indicated by DNA-DNA hybridization. *J. Mol. Evol.* 20: 2–15. [10, 17]

Simberloff, D. 1974. Permo-Triassic extinctions: effects of area on biotic equilibrium. *J. Geol.* 82: 267–274. [12]

Simberloff, D. 1983. Sizes of coexisting species. In D. J. Futuyma and M. Slatkin (eds.), *Coevolution*, pp. 404–430. Sinauer Associates, Sunderland, Massachusetts. [2, 16]

Simberloff, D. 1986. The proximate causes of extinction. In D. M. Raup and D. Jablonski (eds.), *Patterns and processes in the history of life*. Springer-Verlag, Berlin. [11, 12]

Simberloff, D., and W. Boecklen. 1981. Santa Rosalia reconsidered: size ratios and competition. *Evolution* 35: 1206–1228. [16]

Simmons, M. J., and J. F. Crow. 1977. Mutations affecting fitness in *Drosophila* populations. *Annu. Rev. Genet.* 11: 49–78. [4, 6]

Simpson, B. B., and J. Haffer. 1978. Speciation patterns in the Amazonian forest biota. *Annu. Rev. Ecol. Syst.* 9: 497–518. [11]

Simpson, G. G. 1944. *Tempo and mode in evolution*. Columbia University Press, New York. [1, 8, 12, 14]

Simpson, G. G. 1952. How many species? *Evolution* 6: 342. [12]

Simpson, G. G. 1953. *The major features of evolution*. Columbia University Press, New York. [1, 10, 12, 14]

Simpson, G. G. 1959. The nature and origin of supraspecific taxa. *Cold Spring Harbor Symp. Quant. Biol.* 24: 255–271. [10]

Simpson, G. G. 1961. *Principles of animal taxonomy*. Columbia University Press, New York. [10]

Simpson, G. G. 1964. *This view of life: the world of an evolutionist*. Harcourt, Brace and World, New York. [15]

Simpson, G. G. 1980. *Splendid isolation: the curious history of South American mammals*. Yale University Press, New Haven, Connecticut. [13]

Sinclair, D. 1969. *Human growth after birth*. Oxford University Press, Oxford. [14]

Skodak, M., and H. M. Skeels. 1949. A final follow-up study of one hundred adopted children. *J. Genet. Psychol.* 75: 83–125. [17]

Skutch, A. F. 1973. *The life of the hummingbird*. Crown Publishers, New York. [8]

Slack, J. 1984. A Rosetta stone for pattern formation in animals? *Nature* 310: 364–365. [14]

Slatkin, M. 1973. Gene flow and selection in a cline. *Genetics* 75: 735–756. [6, 8]

Slatkin, M. 1975. Gene flow and selection in a two-locus system. *Genetics* 81: 209–222. [8]

Slatkin, M. 1977. Gene flow and genetic drift in a species subject to frequent local extinctions. *Theoret. Pop. Biol.* 12: 253–262. [5]

Slatkin, M. 1978. On the equilibration of fitnesses by natural selection. *Amer. Natur.* 112: 845–859. [9]

Slatkin, M. 1979. Frequency- and density-dependent selection on a quantitative character. *Genetics* 93: 755–771. [7]

Slatkin, M. 1980. Ecological character displacement. *Ecology* 61: 163–177. [16]

Slatkin, M. 1981. Estimating levels of gene flow in natural populations. *Genetics* 99: 323–335. [5]

Slatkin, M. 1985a. Gene flow in natural populations. *Annu. Rev. Ecol. Syst.* 16: 393–430. [5]

Slatkin, M. 1985b. Genetic differentiation of tranposable elements under mutation and unbiased gene conversion. *Genetics* 110: 145–158. [15]

Slobodkin, L. B. 1961. *Growth and regulation of animal populations.* Holt, Rinehart and Winston, New York. [2]

Slobodkin, L. B. 1968. Toward a predictive theory of evolution. In R. C. Lewontin (ed.), *Population biology and evolution,* pp. 187–205. Syracuse University Press, Syracuse, New York. [2]

Slobodkin, L. B. 1974. Prudent predation does not require group selection. *Amer. Natur.* 108: 665–678. [2]

Slobodkin, L. B., and H. L. Sanders. 1969. On the contribution of environmental predictability to species diversity. In G. M. Woodwell and H. H. Smith (eds.), *Diversity and stability in ecological systems,* pp. 82–95. Brookhaven National Laboratory, Upton, New York. [13]

Slobodkin, L. B., F. E. Smith, and N. G. Hairston. 1967. Regulation in terrestrial ecosystems, and the implied balance of nature. *Amer. Natur.* 101: 104–124. [2]

Smith, J. N. M., and A. A. Dhondt. 1980. Experimental confirmation of heritable morphological variation in a natural population of song sparrows. *Evolution* 34: 1155–1158. [4]

Smith, N. G. 1966. Evolution of some arctic gulls (*Larus*): an experimental study of isolating mechanisms. *Amer. Ornithologist's Union Onithol. Monogr. No. 4.* [4]

Smith, N. G. 1968. The advantage of being parasitized. *Nature* 219: 690–694. [2]

Sneath, P. H. A., and R. R. Sokal. 1973. *Numerical taxonomy: the principles and practice of numerical classification.* Freeman, San Francisco. [10]

Snell, T. W., and C. E. King. 1977. Lifespan and fecundity patterns in rotifers: the cost of reproduction. *Evolution* 31: 882–890. [9]

Snodgrass, R. E. 1935. *Principles of insect morphology.* McGraw-Hill, New York. [14]

Soans, A. B., D. Pimentel, and J. S. Soans. 1974. Evolution of reproductive isolation in allopatric and sympatric populations. *Amer. Natur.* 108: 116–124. [8]

Sober, E. 1984. *The nature of selection: evolutionary theory in philosophical focus.* MIT Press, Cambridge, Massachusetts. [9]

Sober, E., and R. C. Lewontin. 1982. Artifact, cause and genic selection. *Philosophy of Science* 49: 157–180. [9]

Sociobiology Study Group. 1977. Sociobiology—a new biological determinism. In Ann Arbor Science for the People Collective (ed.), *Biology as a social weapon.* Burgess, Minneapolis. [17]

Sokal, R. R., and T. J. Crovello. 1970. The biological species concept: a critical evaluation. *Amer. Natur.* 104: 127–154. [4]

Sokal, R. R., and P. Menozzi. 1982. Spatial autocorrelations of HLA frequencies in Europe support demic diffusion of early farmers. *Amer. Natur.* 119: 1–17. [17]

Somero, G. N. 1978. Temperature adaptation of enzymes: biological optimization through structure-function compromises. *Annu. Rev. Ecol. Syst.* 9: 1–29. [6]

Sondhi, K. C. 1962. The evolution of a pattern. *Evolution* 16: 186–191. [14]

Sondhi, K. C. 1963. The biological foundations of animal patterns. *Quart. Rev. Biol.* 38: 289–327. [14]

Sorenson, G. 1969. Embryonic genetic load in coastal Douglas fir, *Pseudotsuga menziesii* var. *menziesii. Amer. Natur.* 103: 389–398. [4]

Soulé, M. 1966. Trends in the insular radiation of a lizard. *Amer. Natur.* 100: 47–64. [8]

Soulé, M. 1967. Phenetics of natural populations. II. Asymmetry and evolution in a lizard. *Amer. Natur.* 101: 141–160. [7]

Southwood, T. R. E. 1961. The number of species of insects associated with various trees. *J. Anim. Ecol.* 30: 1–8. [13, 16]

Spassky, B., Th. Dobzhansky, and W. W. Anderson. 1965. Genetics of natural populations. XXXVI. Epistatic interactions of the components of the genetic load in *Drosophila pseudoobscura. Genetics* 52: 653–664. [7]

Spencer, W. P. 1957. Genetic studies on *Drosophila mulleri.* I. Genetic analysis of a population. *Tex. Univ. Publ.* 5721: 186–205. [4]

Spiess, E. B. 1977. *Genes in populations.* Wiley, New York. [6]

Spradling, A. C., and G. M. Rubin. 1981. Drosophila genome organization: conserved and dynamic aspects. *Annu. Rev. Genet.* 15: 219–264. [3]

Sprinkle, J., and B. M. Bell. 1978. Paedomorphosis in edrioasteroid echinoderms. *Paleobiology* 4: 82–88. [14]

Spuhler, J. N. 1968. Assortative mating with respect to physical characteristics. *Eugen. Quart.* 15: 128–140. [5]

Srb, A. M., R. D. Owen, and R. S. Edgar. 1965. *General genetics.* Freeman, San Francisco. [3, 4]

Stanley, S. M. 1975. A theory of evolution above the species level. *Proc. Natl. Acad. Sci. U.S.A.* 72: 646–650. [12, 14]

Stanley, S. M. 1979. *Macroevolution: pattern and process.* Freeman, San Francisco. [8, 11, 12, 13, 14]

Stanley, S. M. 1985. Rates of evolution. *Paleobiology* 11: 13–26. [12]

Stanley, S. M., B. Van Valkenburgh, and R. S. Steneck. 1983. Coevolution and the fossil record. In D. J. Futuyma and M. Slatkin (eds.), *Coevolution,* pp. 328–349. Sinauer Associates, Sunderland, Massachusetts. [16]

Starck, D., and B. Kummer. 1962. Zur Ontogenese des Schimpansenschädels. *Anthrop. Anz.* 25: 204–215. [17]

Stearns, S. C. 1976. Life history tactics: a review of the ideas. *Quart. Rev. Biol.* 51: 3–47. [9]

Stearns, S. C. 1977. The evolution of life history traits. *Annu. Rev. Ecol. Syst.* 8: 145–171. [9]

Stearns, S. C. 1980. A new view of life-history evolution. *Oikos* 35: 266–281. [9]

Stebbins, G. L. 1950. *Variation and evolution in plants.* Columbia University Press, New York. [1, 3, 4, 8]

Stebbins, G. L. 1970. Variation and evolution in plants: progress during the past twenty years. In M. K. Hecht and W. C. Steere (eds.), *Essays in evolution and genetics in honor of Theodosius Dobzhansky*, pp. 173–208. Appleton-Century-Crofts, New York. [3]

Stebbins, G. L. 1971. *Processes of organic evolution*, Second edition. Prentice-Hall, Englewood Cliffs, New Jersey. [3]

Stebbins, G. L. 1974. *Flowering plants: evolution above the species level.* Belknap Press of Harvard University Press, Cambridge, Massachusetts. [4, 13, 14]

Stebbins, G. L. 1981. Coevolution of grasses and herbivores. *Ann. Missouri Bot. Gard.* 68: 75–86. [16]

Stebbins, G. L. 1982. Plant speciation. In C. Barigozzi (ed.), *Mechanisms of speciation*, pp. 21–39. Alan R. Liss, New York. [4, 8]

Stebbins, G. L., and F. J. Ayala. 1981. Is a new evolutionary synthesis necessary? *Science* 213: 967–971. [14]

Stebbins, G. L., and K. Daly. 1961. Changes in the variation pattern of a hybrid population of *Helianthus* over an eight-year period. *Evolution* 15: 60–71. [8]

Stebbins, G. L., and A. Day. 1967. Cytogenetic evidence for long continued stability in the genus *Plantago. Evolution* 21: 409–428. [8]

Stebbins, R. C. 1954. *Amphibians and reptiles of western North America.* McGraw-Hill, New York. [4]

Stehli, F. G., R. G. Douglas, and N. D. Newell. 1969. Generation and maintenance of gradients in taxonomic diversity. *Science* 164: 947–949. [13]

Steneck, R. S. 1983. Escalating herbivory and resulting adaptive trends in calcareous algal crusts. *Paleobiology* 9: 44–61. [12]

Stenseth, N. C., and J. Maynard Smith. 1984. Coevolution in ecosystems: Red Queen evolution or stasis? *Evolution* 38: 870–880. [12, 16]

Sterba, G. 1962. *Freshwater fishes of the world.* Longacre Press, London. [2]

Stern, C. 1968. *Genetic mosaics and other essays.* Harvard University Press, Cambridge, Massachusetts. [14]

Stern, C. 1973. *Principles of human genetics*, Third edition. Freeman, San Francisco. [5]

Stevens, P. F. 1980. Evolutionary polarity of character states. *Annu. Rev. Ecol. Syst.* 11: 333–358. [10]

Stewart, W. N. 1983. *Paleobiology and the evolution of plants.* Cambridge University Press, Cambridge. [11]

Strathmann, R. R. 1978. Progressive vacating of adaptive types during the Phanerozoic. *Evolution* 32: 907–914. [12]

Strathmann, R. R., and M. Slatkin. 1983. The improbability of animal phyla with few species. *Paleobiology* 9: 97–106. [12]

Straw, R. M. 1955. Hybridization, homogamy, and sympatric speciation. *Evolution* 9: 441–444. [8]

Strickberger, M. W. 1968. *Genetics.* Macmillan, New York. [3, 5]

Strong, D. S., Jr., D. Simberloff, L. G. Abele, and A. B. Thistle (eds.). 1984. *Ecological communities: conceptual issues and the evidence.* Princeton University Press, Princeton, New Jersey. [16]

Sturtevant, A. H. 1920-21. Genetic studies on *Drosophila simulans. Genetics* 5: 488–500; 6: 179–207. [8]

Sulloway, F. J. 1979. Geographic isolation in Darwin's thinking: the vicissitudes of a crucial idea. *Stud. Hist. Biol.* 3: 23–65. [1]

Sulzbach, D. S. 1980. Selection for competitive ability: negative results in *Drosophila. Evolution* 34: 431–436. [16]

Surlyk, F., and M. B. Johansen. 1984. End-Cretaceous brachiopod extinctions in the chalk of Denmark. *Science* 223: 1174–1177. [12]

Sved, J. A. 1981. A two-sex polygenic model for the evolution of premating isolation. I. Deterministic theory for natural populations. *Genetics* 97: 197–215. [8]

Sved, J. A., and O. Mayo. 1970. The evolution of dominance. In K. Kojima (ed.), *Mathematical topics in population genetics*, pp. 289–316. Springer-Verlag, New York. [7]

Sved, J. A., T. E. Reed, and W. F. Bodmer. 1967. The number of balanced polymorphisms that can be maintained in a natural population. *Genetics* 55: 469–481. [6]

Swanson, C. P., T. Merz, and W. J. Young. 1981. *Cytogenetics*, Second edition. Prentice-Hall, Englewood Cliffs, New Jersey. [3]

Symons, D. 1979. *The evolution of human sexuality.* Oxford University Press, Oxford. [17]

Szostak, J. W., T. L. Orr-Weaver, R. J. Rothstein, and F. W. Stahl. 1983. The double-strand break–repair model for recombination. *Cell* 33: 25–35. [3]

Tabachnik, W. J., L. E. Munstermann, and J. R. Powell. 1979. Genetic distinctness of sympatric forms of *Aedes aegypti* in East Africa. *Evolution* 33: 287–295. [6]

Takahata, N., and M. Slatkin. 1984. Mitochondrial gene flow. *Proc. Natl. Acad. Sci. U.S.A.* 81: 1764–1767. [15]

Taper, M. L., and T. J. Case. 1985. Quantitative genetic models for the coevolution of character displacement. *Ecology* 66: 355–371. [16]

Tauber, C. A., and M. J. Tauber. 1982. Maynard Smith's model and corroborating evidence: no reason for misinterpretation. *Ann. Ent. Soc. Amer.* 75: 5–6. [8]

Temin, H. M. 1985. Reverse transcription in the eukaryotic genome: retroviruses, pararetroviruses, retrotransposons, and retrotranscripts. *Mol. Biol. Evol.* 2: 455–468. [3]

Templeton, A. R. 1980. The theory of speciation via the founder principle. *Genetics* 94: 1011–1038. [8]

Templeton, A. R. 1981. Mechanisms of speciation—a population genetic approach. *Annu. Rev. Ecol. Syst.* 12: 23–48. [8]

Templeton, A. R. 1982a. Genetic architectures of speciation. In C. Barigozzi (ed.), *Mechanisms of speciation*, pp. 105–121. Alan R. Liss, New York. [8]

Templeton, A. R. 1982b. Why read Goldschmidt? *Paleobiology* 8: 474–481. [14]

Templeton, A. R. 1983. Phylogenetic inference from restriction endonuclease cleavage site maps with particular reference to the evolution of humans and the apes. *Evolution* 37: 221–244. [10, 17]

Thoday, J. M. 1953. Components of fitness. *Symp. Soc. Exp. Biol.* 7: 96–113. [2]

Thoday, J. M. 1955. Balance, heterozygosity and developmental stability. *Cold Spring Harbor Symp. Quant. Biol.* 20: 318–326. [7]

Thoday, J. M., and J. B. Gibson. 1962. Isolation by disruptive selection. *Nature* 193: 1164–1166. [8]

Thoday, J. M., and J. B. Gibson. 1970. The probability of isolation by disruptive selection. *Amer. Natur.* 104: 219–230. [8]

Thomas, P. A. 1968. Geographic variation of the rabbit tick, *Haemaphysalis leporispalustris*, in North America. *Univ. Kansas Sci. Bull.* 47: 787–828. [4]

Thompson, D. W. 1917. *On growth and form*. Cambridge University Press, Cambridge. [14]

Thompson, J. N. 1982. *Interaction and coevolution*. Wiley, New York. [16]

Thomson, G. 1977. The effect of a selected locus on linked neutral loci. *Genetics* 85: 753–788. [7]

Thornhill, R. 1980. Mate choice in *Hylobittacus apicalis* (Insecta: Mecoptera) and its relation to some models of female choice. *Evolution* 34: 519–538. [9]

Thornhill, R., and R. Alcock. 1983. *The evolution of insect mating systems*. Harvard University Press, Cambridge, Massachusetts. [9]

Thorpe, J. P. 1982. The molecular clock hypothesis: biochemical evolution, genetic differentiation and systematics. *Annu. Rev. Ecol. Syst.* 13: 139–168. [10]

Thorpe, W. H. 1956. *Learning and instinct in animals*. Methuen, London. [6]

Throckmorton, L. H. 1965. Similarity *versus* relationship in *Drosophila*. *Syst. Zool.* 14: 221–236. [10]

Thulborn, R. A., and T. L. Hamley. 1982. The reptilian relationships of *Archaeopteryx*. *Aust. J. Zool.* 30: 611–634. [11]

Tiger, L, and R. Fox. 1971. *The imperial animal*. Holt, Rinehart and Winston, New York. [17]

Tinkle, D. W. 1969. The concept of reproductive effort and its relation to the evolution of life histories of lizards. *Amer. Natur.* 103: 501–516. [9]

Tizard, B. 1973. IQ and race. *Nature* 247: 316. [17]

Tompkins, R. 1978. Genic control of axolotl metamorphosis. *Amer. Zool.* 18: 313–319. [14]

Tripp, C. A. 1975. *The homosexual matrix*. McGraw-Hill, New York. [17]

Trivers, R. L. 1972. Parental investment and sexual selection. In B. Campbell (ed.), *Sexual selection and the descent of man, 1871–1971*, pp. 136–179. Aldine, Chicago. [9]

Turelli, M. 1984. Heritable genetic variation via mutation-selection balance: Lerch's zeta meets the abdominal bristle. *Theoret. Pop. Biol.* 25: 138–193. [3, 7]

Turesson, G. 1922. The genotypical response of the plant species to the habitat. *Hereditas* 3: 211–350. [4]

Turing, A. M. 1952. The chemical basis of morphogenesis. *Phil. Trans. Roy. Soc. Lond.* (B) 237: 37–72. [14]

Turkington, R., and J. L. Harper. 1979. The growth, distribution, and neighbor relationships of *Trifolium repens* in a permanent pasture. *J. Ecol.* 67: 245–254. [16]

Turner, J. R. G. 1971. Two thousand generations of hybridisation in a *Heliconius* butterfly. *Evolution* 25: 471–482. [4]

Turner, J. R. G. 1981. Adaptation and evolution in *Heliconius*: a defense of NeoDarwinism. *Annu. Rev. Ecol. Syst.* 12: 99–121. [16]

Turner, J. R. G. 1986. The genetics of adaptive radiation: a neo-Darwinian theory of punctuational evolution. In D. M. Raup and D. Jablonski (eds.), *Pattern and process in the history of life*, Springer-Verlag, Berlin. [14]

Uyenoyama, M., and M. W. Feldman. 1980. Theories of kin and group selection: a population genetics perspective. *Theoret. Pop. Biol.* 17: 380–414. [9]

Val, F. C. 1977. Genetic analysis of the morphological differences between two interfertile species of Hawaiian *Drosophila*. *Evolution* 31: 611–629. [8]

Valentine, J. W. 1973. *Evolutionary paleoecology of the marine biosphere*. Prentice-Hall, Englewood Cliffs, New Jersey. [11, 12]

Valentine, J. W. 1977. General patterns of metazoan evolution. In A. Hallam (ed.), *Patterns of evolution as illustrated by the fossil record*, pp. 27–57. Elsevier, Amsterdam. [11]

Valentine, J. W. (ed.). 1985. *Phanerozoic diversity patterns: profiles in macroevolution*. Princeton University Press, Princeton, New Jersey. [12]

Valentine, J. W., T. C. Foin, and D. Peart. 1978. A provincial model of Phanerozoic marine diversity. *Paleobiology* 4: 55–66. [11, 12]

van Delden, W. 1982. The alcohol dehydrogenase polymorphism in *Drosophila melanogaster*: selection at an enzyme locus. *Evol. Biol.* 15: 187–222. [6]

Vandermeer, J. H. 1969. The competitive structure of communities: an experimental approach with Protozoa. *Ecology* 50: 362–371. [2]

Vandermeer, J. H. 1980. Indirect mutualism: variations on a theme by Stephen Levine. *Amer. Natur.* 116: 441–448. [2]

van Noordwijk, A. J., J. H. Van Balen, and W. Scharloo. 1980. Heritability of ecologically important traits in the great tit (*Parus major*). *Ardea* 68: 193–203. [7]

van Noordwijk, A. J., and W. Scharloo. 1981. Inbreeding in an island population of the great tit. *Evolution* 35: 674–688. [5]

Van Tyne, J., and A. J. Berger. 1959. *Fundamentals of ornithology*. Wiley, New York. [13]

Van Valen, L. 1971. Group selection and the evolution of dispersal. *Evolution* 25: 591–598. [9, 12]

Van Valen, L. 1973. A new evolutionary law. *Evolutionary Theory* 1: 1–30. [12]

Van Valen, L. M. 1984. Catastrophes, expectations, and the evidence. *Paleobiology* 10: 121–127. [12]

Vasek, F. C. 1964. The evolution of *Clarkia unguiculata* derivatives adapted to relatively xeric environments. *Evolution* 18: 26–42. [4]

Vawter, A. T., R. Rosenblatt, and G. C. Gorman. 1980. Genetic divergence among fishes of the eastern Pacific and Caribbean: support for the molecular clock. *Evolution* 34: 705–711. [10]

Vermeij, G. J. 1983. Intimate associations and coevolution in the sea. In D. J. Futuyma and M. Slatkin (eds.), *Coevolution*, pp. 311–327. Sinauer Associates, Sunderland, Massachusetts. [16]

von Borstel, R. C., S.-K. Quah, C. M. Steinberg, F. Flury, and D. J. C. Gottlieb. 1973. Mutants of yeast with enhanced spontaneous mutation rates. *Genetics* 73 (suppl.): 141–151. [3]

Vrba, E. S. 1980. Evolution, species and fossils: How does life evolve? *S. Afr. J. Sci.* 76: 61–84. [12]

Waage, J. K. 1979. Reproductive character displacement in *Calopteryx* (Odonata: Calopterygidae). *Evolution* 33: 104–116. [4, 8]

Waddington, C. H. 1953. Genetic assimilation of an acquired character. *Evolution* 7: 118–126. [7]

Waddington, C. H. 1956a. *Principles of embryology*. Allen and Unwin, London. [7, 14]

Waddington, C. G. 1956b. Genetic assimilation of the *bithorax* phenotype. *Evolution* 10: 1–13. [14]

Wade, M. J. 1977. An experimental study of group selection. *Evolution* 31: 134–153. [9]

Wade, M. J. 1978. A critical review of the models of group selection. *Quart. Rev. Biol.* 53: 101–114. [9]

Wade, M. J. 1979. The evolution of social interactions by family selection. *Amer. Natur.* 113: 399–417. [9]

Wade, M. J. 1980a. Kin selection: its components. *Science* 210: 665–667. [9]

Wade, M. J. 1980b. An experimental study of kin selection. *Evolution* 34: 844–855. [9]

Wade, M. J., and S. J. Arnold. 1980. The intensity of sexual selection in relation to male sexual behavior, female choice, and sperm precedence. *Anim. Behav.* 28: 446–461. [9]

Wagner, G. P. 1986. The systems approach: an interface between developmental and population genetics aspects of evolution. In D. M. Raup and D. Jablonski (eds.), *Patterns and processes in the history of life*. Springer-Verlag, Berlin. [14]

Wagner, R. P., B. H. Judd, B. G. Saunders, and R. H. Richardson. 1980. *Introduction to modern genetics*. Wiley, New York. [3]

Wake, D. B. 1966. Comparative osteology and evolution of the lungless salamanders, family Plethodontidae. *Mem. So. California Acad. Sci.* 4: 1–111. [10]

Wake, D. B., G. Roth, and M. H. Wake. 1983. On the problem of stasis in organismal evolution. *J. Theor. Biol.* 101: 211–224. [14]

Walker, E. P. 1975. *Mammals of the world*, Third edition. Johns Hopkins University Press, Baltimore, Maryland. [13]

Walker, J. W., G. J. Brenner, and A. G. Walker. 1983. Winteraceous pollen in the lower Cretaceous of Israel: early evidence of a magnolialean angiosperm family. *Science* 220: 1273–1275. [11]

Walker, T. D., and J. W. Valentine. 1984. Equilibrium models of evolutionary species diversity and the number of empty niches. *Amer. Natur.* 124: 887–899. [12]

Wallace, B. 1966. On the dispersal of Drosophila. *Amer. Natur.* 100: 551–564. [5]

Wallace, B. 1968a. *Topics in population genetics*. Norton, New York. [6]

Wallace, B. 1968b. Polymorphism, population size, and genetic load. In R. C. Lewontin (ed.), *Population biology and evolution*, pp. 87–108. Syracuse University Press, Syracuse, New York. [6]

Wallace, B., and Th. Dobzhansky. 1962. Experimental proof of balanced genetic loads in Drosophila. *Genetics* 47: 1027–1042. [4]

Wallace, B., and M. Vetukhiv. 1955. Adaptive organization of the gene pools of *Drosophila* populations. *Cold Spring Harbor Symp. Quant. Biol.* 20: 303–310. [7]

Walsh, J. B. 1982. Rate of accumulation of reproductive isolation by chromosome arrangements. *Amer. Natur.* 120: 510–532. [8]

Walsh, J. B. 1983. Role of biased gene conversion in one-locus neutral theory and genome evolution. *Genetics* 105: 461–468. [15]

Walsh, J. B. 1985a. How many processed pseudogenes are accumulated in a gene family? *Genetics* 110: 345–364. [3, 15]

Walsh, J. B. 1985b. Interaction of selection and biased gene conversion in a multigene family. *Proc. Natl. Acad. Sci. U.S.A.* 82: 153–157. [15]

Walther, F. R. 1984. *Communication and expression in hoofed mammals*. Indiana University Press, Bloomington. [9]

Ward, P. D., and P. W. Signor III. 1983. Evolutionary tempo in Jurassic and Cretaceous ammonites. *Paleobiology* 9: 183–198. [12]

Wasserman, A. O. 1957. Factors affecting interbreeding in sympatric species of spadefoots (genus *Scaphiopus*). *Evolution* 11: 320–338. [4]

Watt, W. B. 1972. Intragenic recombination as a source of population genetic variability. *Amer. Natur.* 106: 737–753. [3]

Watterson, G. A. 1978. The homozygosity test of neutrality. *Genetics* 88: 405–417. [6]

Webb, S. D. 1969. Extinction-origination equilbria in late Cenozoic land mammals of North America. *Evolution* 23: 688–702. [12]

Weichert, C. K. 1958. *Anatomy of the chordates.* McGraw-Hill, New York. [9]

Wessels, N. K. 1982. A catalogue of processes responsible for metazoan morphogenesis. In J. T. Bonner (ed.), *Evolution and development,* pp. 115–154. Springer-Verlag, Berlin. [14]

West-Eberhard, M. J. 1975. The evolution of social behavior by kin-selection. *Quart. Rev. Biol.* 50: 1–33. [9]

West-Eberhard, M. J. 1983. Sexual selection, social competition, and speciation. *Quart. Rev. Biol.* 58: 155–183. [8, 9]

Wheelwright, N. T., and G. H. Orians. 1982. Seed dispersal by animals: contrasts with pollen dispersal, problems of terminology, and constraints on coevolution. *Amer. Natur.* 119: 402–413. [16]

White, M. J. D. 1968. Models of speciation. *Science* 159: 1065–1070. [8]

White, M. J. D. 1973. *Animal cytology and evolution,* Third edition. Cambridge University Press, London. [3]

White, M. J. D. 1978. *Modes of speciation.* Freeman, San Francisco. [6, 8, 14]

Whitt, G. S., D. P. Phillip, and W. F. Childers. 1977. Aberrant gene expression during the development of hybrid sunfishes (Perciformes, Teleostei). *Differentiation* 9: 97–109. [8]

Whittaker, R. H. 1975. *Communities and ecosystems.* Macmillan, New York. [2]

Wickler, W. 1968. *Mimicry in plants and animals.* McGraw-Hill, New York. [2]

Wiebes, J. T. 1979. Co-evolution of figs and their insect pollinators. *Annu. Rev. Ecol. Syst.* 10: 1–12. [16]

Wiley, E. O. 1981. *Phylogenetics: the theory and practice of phylogenetic systematics.* Wiley, New York. [10]

Wilkens, H. 1971. Genetic interpretation of regressive evolutionary processes: studies on hybrid eyes of two *Astyanax* cave populations (Characidae, Pisces). *Evolution* 25: 530–544. [14]

Williams, A., and J. M. Hurst. 1977. Brachiopod evolution. In A. Hallam (ed.), *Patterns of evolution as illustrated by the fossil record,* pp. 79–121. Elsevier, Amsterdam. [12]

Williams, G. C. 1957. Pleiotropy, natural selection and the evolution of senescence. *Evolution* 11: 398–411. [9]

Williams, G. C. 1966. *Adaptation and natural selection.* Princeton University Press, Princeton, New Jersey. [9]

Williams, G. C. 1975. *Sex and evolution.* Princeton University Press, Princeton, New Jersey. [3, 9, 12]

Williams, G. C., and D. C. Williams. 1957. Natural selection of individually harmful social adaptations among sibs with special reference to social insects. *Evolution* 11: 32–39. [9]

Williams, K. S., and L. E. Gilbert. 1981. Insects as selective agents in plant vegetative morphology: egg mimicry reduces egg laying by butterflies. *Science* 212: 467–469. [16]

Williams, N. E. 1984. An apparent disjunction between the evolution of form and substance in the genus *Tetrahymena. Evolution* 38: 25–33. [14]

Willis, J. C. 1922. *Age and area. A study in geographical distribution and origin of species.* Cambridge University Press, Cambridge. [13]

Wills, C. 1978. Rank-order selection is capable of maintaining all genetic polymorphisms. *Genetics* 89: 403–417. [6]

Willson, M. F., and N. Burley. 1983. *Mate choice in plants.* Princeton University Press, Princeton, New Jersey. [9]

Wilson, A. C., S. S. Carlson, and T. J. White. 1977. Biochemical evolution. *Annu. Rev. Biochem.* 46: 573–639. [5, 10]

Wilson, A. C., and V. M. Sarich. 1969. A molecular time scale for human evolution. *Proc. Natl. Acad. Sci. U.S.A.* 63: 1088–1093. [17]

Wilson, D. S. 1975. The adequacy of body size as a niche difference. *Amer. Natur.* 109: 769–784. [16]

Wilson, D. S. 1980. *The natural selection of populations and communities.* Benjamin Cummings, Menlo Park, California. [2, 9, 16]

Wilson, D. S. 1983. The group selection controversy: history and current status. *Annu. Rev. Ecol. Syst.* 14: 159–187. [9]

Wilson, D. S., and R. K. Colwell. 1981. Evolution of sex ratio in structured demes. *Evolution* 35: 882–897. [9]

Wilson, E. O. 1961. The nature of the taxon cycle in the Melanesian ant fauna. *Amer. Natur.* 95: 169–193. [13]

Wilson, E. O. 1965. The challenge from related species. In H. G. Baker and G. L. Stebbins (eds.), *The genetics of colonizing species,* pp. 7–24. Academic Press, New York. [13]

Wilson, E. O. 1969. The species equilibrium. In G. M. Woodwell and H. H. Smith (eds.), *Diversity and stability in ecological systems,* pp. 38–47. Brookhaven National Laboratory, Upton, New York. [13]

Wilson, E. O. 1971. *The insect societies.* Harvard University Press, Cambridge, Massachusetts. [11]

Wilson, E. O. 1975. *Sociobiology: the new synthesis.* Harvard University Press, Cambridge, Massachusetts. [2, 17]

Wilson, E. O. 1978. *On human nature.* Harvard University Press, Cambridge, Massachusetts. [17]

Wilson, E. O., and W. H. Bossert. 1971. *A primer of population biology.* Sinauer Associates, Sunderland, Massachusetts. [2]

Wilson, E. O., and W. L. Brown. 1953. The subspecies concept and its taxonomic applications. *Syst. Zool.* 2: 97–111. [4]

Woese, C. R. 1981. Archaebacteria. *Sci. Amer.* 244 (6): 98–122. [11]

Wolpert, L. 1982. Pattern formation and change. In J. T. Bonner (ed.), *Evolution and development,* pp. 169–188. Springer-Verlag, Berlin. [14]

Wolpoff, M. H. 1984. Evolution in *Homo erectus*: the question of stasis. *Paleobiology* 10: 389–406. [17]

Wood, A. E. 1959. Eocene radiation and phylogeny of the rodents. *Evolution* 13: 354–360. [10]

Wood, R. J. 1981. Insecticide resistance: genes and mechanisms. In J. A. Bishop and L. M. Cook (eds.), *Genetic consequences of man made change*, pp. 53–96. Academic Press, London. [6]

Wood, R. J., and J. A. Bishop. 1981. Insecticide resistance: populations and evolution. In J. A. Bishop and L. M. Cook (eds.), *Genetic consequences of man made change*, pp. 97–127. Academic Press, London. [6]

Wood, T. K., and S. I. Guttman. 1983. *Enchenopa binotata* complex: sympatric speciation? *Science* 220: 310–312. [8]

Woodburn, M. O., and B. J. MacFadden. 1982. A reappraisal of the systematics, biogeography, and evolution of fossil horses. *Paleobiology* 8: 315–327. [12]

Woodruff, D. 1981. Toward a genodynamics of hybrid zones: studies of Australian frogs and West Indian land snails. In W. R. Atchley and D. S. Woodruff (eds.), *Evolution and speciation: essays in honor of M. J. D. White*, pp. 171–197. Cambridge University Press, Cambridge. [3]

Wright, S. 1921. Systems of mating. *Genetics* 6: 111–178. [5]

Wright, S. 1929. Fisher's theory of dominance. *Amer. Natur.* 63: 274–279. [7]

Wright, S. 1931. Evolution in Mendelian populations. *Genetics* 16: 97–159. [1, 5, 6, 8]

Wright, S. 1932. The roles of mutation, inbreeding, crossbreeding, and selection in evolution. *Proc. XI Internat. Congr. Genetics* 1: 356–366. [1, 6, 8]

Wright, S. 1935. The analysis of variance and the correlations between relatives with respect to deviations from an optimum. *J. Genet.* 30: 243–256. [7]

Wright, S. 1937. The distribution of gene frequencies in populations. *Proc. Natl. Acad. Sci. U.S.A.* 23: 307–320. [6]

Wright, S. 1941. On the probability of fixation of reciprocal translocations. *Amer. Natur.* 75: 513–522. [6]

Wright, S. *Evolution and the genetics of populations*. Vol. 1, 1968, *Genetic and biometric foundations*; Vol. 2, 1969, *The theory of gene frequencies*; Vol. 3, 1977, *Experimental results and evolutionary deductions*; Vol 4, 1978, *Variability within and among natural populations*. University of Chicago Press, Chicago. [4, 5, 6, 7]

Wright, S. 1982. Character change, speciation, and the higher taxa. *Evolution* 36: 427–443. [14]

Wright, S., Th. Dobzhansky, and W. Hovanitz. 1942. Genetics of natural populations. VII. The allelism of lethals in the third chromosome of *Drosophila pseudoobscura*. *Genetics* 27: 363–394. [4]

Wu, C.-I., and W.-H. Li. 1985. Evidence for higher rates of nucleotide substitution in rodents than in man. *Proc. Natl. Acad. Sci. U.S.A.* 82: 1741–1745. [10]

Wyles, J., and G. C. Gorman. 1980. The albumin immunological and Nei electrophoretic distance correlation: a calibration for the saurian genus *Anolis* (Iguanidae). *Copeia* 1980: 66–71. [10]

Wynne-Edwards, V. C. 1962. *Animal dispersion in relation to social behaviour*. Oliver & Boyd, Edinburgh. [9]

Yamazaki, T. 1971. Measurement of fitness at the esterase-5 locus in *Drosophila pseudoobscura*. *Genetics* 67: 579–603. [3, 6]

Yoo, B. H. 1980. Long-term selection for a quantitative character in large replicate populations of *Drosophila melanogaster*. I. Response to selection. *Genet. Res.* 35: 1–17. II. Lethals and visible mutants with large effects. *Genet. Res.* 35: 19–31. [7]

Yunis, J. J., and O. Prakash. 1982. The origin of man: a chromosomal pictorial legacy. *Science* 212: 1525–1530. [17]

Yunis, J. Y., J. R. Sawyer, and K. Dunham. 1980. The striking resemblance of high-resolution G-banded chromosomes of man and chimpanzee. *Science* 208: 1145–1148. [17]

Zachar, Z., and P. M. Bingham. 1982. Regulation of *white* locus expression: the structure of mutant alleles of the *white* locus of *Drosophila melanogaster*. *Cell* 30: 529–541. [3, 15]

Zahavi, A. 1975. Mate selection—a selection for a handicap. *J. Theor. Biol.* 53: 205–214. [9]

Zamenhof, S., and H. H. Eichhorn. 1967. Study of microbial evolution through loss of biochemical function: establishment of "defective" mutants. *Nature* 216: 456–458. [3]

Zaret, T. M., and R. T. Paine. 1973. Species introduction in a tropical lake. *Science* 182: 449–455. [2]

Zera, A. J. 1981. Genetic structure of two species of waterstriders (Gerridae: Hemiptera) with differing degrees of winglessness. *Evolution* 35: 218–225. [5]

Zera, A. J. 1984. Differences in survivorship, development rate and fertility between the longwinged and wingless morphs of the waterstrider, *Limnoporus canaliculatus*. *Evolution* 38: 1023–1032. [9]

Zimmer, E. A., S. L. Martin, S. M. Beverly, Y. W. Kan, and A. C. Wilson. 1980. Rapid duplication and loss of genes coding for the α chains of hemoglobin. *Proc. Natl. Acad. Sci. U.S.A.* 77: 2158–2162. [15]

Zouros, E. 1981. The chromosomal basis of sexual isolation in two sibling species of *Drosophila*: *D. arizonensis* and *D. mojavensis*. *Genetics* 97: 703–718. [8]

Zouros, E., S. M. Singh, and H. E. Miles. 1980. Growth rate in oysters: an overdominant phenotype and its possible explanations. *Evolution* 34: 856–867. [6]

# INDEX


ABO blood group, 521–522, 524
*Acacia*, 498
*Acanthiza*, 225
*Accipiter*, 487–488
Acquired characters, 4, 9, 43–44
Acrocentric chromosomes, 64–65
Actinopterygii, 331
Adaptation
 evolutionary success and, 358
 general, 293–294
 human nature and, 530
 illusory, 259
 molecular perspective and, 472
 phylogenetic analysis and, 305–306
 recognition of, 251–253
 special, 294
 as term, 251
 theoretical approaches to, 266–271
Adaptationist program, 254–258
Adaptive constraints, 437
Adaptive landscape, 172–173, 191, 255
Adaptive peak, 191, 255
Adaptive radiation
 competition and, 32–34
 as evolutionary trend, 368–369
Adaptive zone, 355–358. *See also* Ecological niche
Additive effects, 87
Additive genetic covariance, 200–201
Additive genetic variance, 198, 200
*Adh*. *See* Alcohol dehydrogenase
African apes, and human phylogeny, 311–315, 508–512. *See also Gorilla gorilla; Pan*
Agaonid wasp. *See Ceratosolen dentifer*
Age of Fishes, 331
Age of Reptiles, 334
Agnathans, 328–329
Agriculture
 human population structure and, 522–524, 525
 mass extinction and, 343
*Agrostis tenuis*, 161, 163
Alcohol dehydrogenase (*Adh*), 190–191, 195, 447
Allele frequency, 82. *See also* Gene flow; Genetic drift
 distribution of, 178
 equilibrium, 133, 155
 gene flow and, 137
 genetic death and, 176
 inbreeding and, 121–122, 127
 mutation and, 133
 neighborhood size and, 138
 probability of, 172–174
 selection and, 137, 155–159
 sex-linked loci and, 84–85
 variation in, 137
Alleles, 65. *See also* Allele frequency; Gene
 additive effects and, 54–55
 advantageous, 155–157
 deleterious, 124, 155–156, 520, 526
 dominance and, 55
 fixation of, 130, 155, 157
 identity by descent and, 121
 lethal, 94–96
 penetrance and, 53
 rare, 140, 155–157
 relative abundance and, 82–83
 selectively neutral, 143
Allochthonous groups, 385. *See also* Dispersal
Allometry, 255, 412–419
Allopolyploidy, 61–62, 78
Allozygous individuals, 122
Allozyme frequency, 140–141, 307
Allozymes, 97–99, 177–181
Altruistic trait, 254. *See also* Kin selection
*Ambystoma mexicanum*, 421, 427, 437–438
Amino acids
 cladistic analysis of, 452
 in genetic code, 46
 phylogenetic inference and, 307, 309–310
 species differences and, 409–410
 substitution of, 56
Ammonites, 335, 351–352
Ammonoidea, 363
Amphibians, 331–333,335–336
*Amphicarpaea bracteata*, 500
Amphitropical distribution, 378
Ampullae of Lorenzini, 251–252
Anachronistic feature, 257
Anagenesis, 286, 407
Andrenidae, 294–295
*Aneides*, 311–312
Aneuploidy, 114
Angiosperms, 113, 338, 340
*Anguilla rostrata*, 120
Animals. *See also* specific species
 in fossil record, 349–353
 heritability in, 193
 inbreeding in, 124–125
 phylogenetic relationships and, 325–328
*Anopheles*, 182
Antagonism, 29
Antbirds. *See* Formicariidae
Antennapedia complex (ANT-C), 434–435, 438
Antireductionism, 478
*Apatosaurus*, 336
Aphroteniinae, 386–387
*Archaeopteryx*, 336–338, 424, 435
Archosaurs, 334
Arithmetic mean, 541–542
*Artenkreis*, 103
Arthropoda, 430–435
Artificial selection, 90–91, 207–210
Asexual reproduction, 194
Asexual species, 247
Aspect diversity, 493
Assortative mating, 146, 520
Atavistic alteration, 434–436
Australian grasshopper. *See Caledia captiva; Keyacris scurra*
*Australopithecus*, 514–515
Autochthonous groups, 385.
Autotetraploidy, 60–61
Autozygosity, 139
*Avena*, 100–101, 127
Average. *See* Arithmetic mean

Backcross hybrid, 220–222
Background extinction, 360–362
Back mutation, 66
Bacteria. *See also* Cyanobacteria; *Escherichia coli*
 gene regulation in, 57
Balancing selection. *See* Stabilizing selection
*Baluchitherium*, 341
Base pair, 45, 65–67
Batesian mimicry, 28–29
Beneficial interactions, 29
Bergmann's rule, 105
Binomial distribution, 541–542
Biochemical change, 409–410
Biochemical pathways, 409–410, 474, 476–477
Biogenetic law, 303, 416
Biogeography, 374–376. *See also* Historical biogeography; Is-

land biogeography; Vicariance biogeography
Biological evolution, defined, 7
Biological species concept, 111, 114–115. *See also* Isolating mechanisms; Reproductive isolation
Biotic environment, 26–29
Birth rate, and population growth, 21–22, 24
Birth weight, stabilizing selection for, 202
*Biston betularia*, 158
Bithorax complex (BX-C), 434–435, 438
Bivalves, 366
Blending inheritance, 9, 43
*Bothriochloa intermedia*, 78
Bottleneck effect, 132, 135
Brachiopoda, 325–326, 358–359
*Brachiosaurus*, 336
*Brentus anchorago*, 277
Britten-Davidson model, 58
*Brontosaurus*, 336
Buffon, Count of, 3–4
*Bubulcus ibis*, 356
Burgess Shale, 325

*Caledia captiva*, 115
Cambrian period, 319–320, 325–328
Canalization, 53, 212–215, 294
Carboniferous period, 319–322, 332–333
Carnivora, 341–342
Catarrhini, 507, 510
Catostomidae, 456
Cattle egret. *See Bubulcus ibis*
Causes, purposive vs. material, 2
Cell differentiation, 427–428
Cells, morphogenesis and, 58–59, 428
Center of origin, 378
Cenozoic era, 340–343
*Cepaea nemoralis*, 93–94, 101, 180, 193–194
*Ceratosolen dentifer*, 499–500
Chambers, R., 5
Character displacement, 109–110, 488, 490, 500
Characteristic. *See also* Character displacement; Character states; Metric characteristic
  ancestral, 286–287
  canalization and, 53 (*See also* Canalization)
  conservative, 293–294
  covariance and, 205
  derived, 286–287, 307–308
  developmentally buffered, 53
  evolutionary rate and, 398–400
  genetic versus environmental factors and, 53–54
  meristic, 214–215
  polygenic, 55
  spurious, 255, 306
  variation in, 87–90
Character states, 286, 299, 300–306
Chasmogamous flowers, 126
Chemistry
  biology and, 479
  constraints on phenotype and, 430–431, 437
*Chen caerulescens*, 92–93
Chetverikov, S., 10
Chimpanzee. *See Pan*
China, prehistoric continent of, 319
Chironomidae, 385–387
Choanichthyes, 331
Chondrichthyes, 251–252, 330–331
Christian theology, 3
Chromosomes
  arrangement in *Drosophila*, 302
  crossing technique and, 95–96
  differentiation of, 236–238
  fusion and, 64–65
  heterochromatic regions of, 50
  inversions of, 62–63, 101
  mutational changes in, 60–65, 170–172
  reciprocal translation in, 63–64
  transposable elements and, 456–457, 458–459
Chronospecies, 346
*Chrysopa*, 230–231
Cichlidae, 26–27, 32, 245–246
*Cinclus mexicanus*, 257
Circular overlap, 224
Clade, 288, 289, 365, 368
Cladistic school, 291–292
Cladogenesis, 219, 286
Cladogram, 287, 291–292
*Clarkia*, 229, 238
Classification. *See* Cladistic school; Phenetic school; Systematics; Taxonomic classification
Cleistogamous flowers, 126
Climate, prehistoric, 321–322
Cline, 104-107, 180
Clover. *See Trifolium repens*
Coadaptation, 194–195
Codon, 45
Coelacanth, 331, 351
Coevolution. *See also* Coexisting species
  competing species and, 370, 485–491
  diffuse, 483
  genetic studies of, 500–502
  models of, 488–490
  parasite–host relationship and, 496–497, 499
  predator–prey relationship and, 28, 493, 494
  selection experiments and, 500–502
  as term, 483–484
Coexisting species, 31, 388–391. *See also* Coevolution
Cohort, 23
Coloration, 28. *See also* Pattern formulation
*Columba*, 336
Commensalism, 29, 498–499
Communities
  diversity of, 35–37
  equilibrium and, 388–391
  stability of, 35–37
  structure in, 502–503
Comparative method, 14, 252–253
Competitive exclusion principle, 7, 30–31, 488
Competitive interactions, 30–34, 358–359, 485–491
Competition coefficient, 30
*Compsognathus*, 337
Concentration gradient, 428–429
Concerted evolution, 464–472
  identity coefficients for, 469–470
  natural selection and, 471–472
  by transposition, 464–465
  by unequal exchange, 465–469
Concordant cline, 105–107
Condylarthra, 339–341
Connectance, 503
Consanguinity, 122
Consensus sequence, 450
Continental drift, 374–375, 379–380
Continent–island model, 136, 138
Continuous variation, 43–44
Convergence, 292–293, 295–299, 306, 337
Conversion bias, 465. *See also* Gene conversion
Copepod. *See Tisbe reticulata*
Cope's rule, 368
Copy number, 460–461
Corixid bug. *See Sigara distincta*
Corn. *See Zea mays*
Correlation coefficient, 544–546
Cost–benefit mutualism model, 497
Cost of selection, 176
Countergradient variation, 109
Coupling gametes, 84
Covariance, 544
Cowbirds. See *Scaphidura oryzivora*
Crab, 37–38
Creationism, 15
Cretaceous period, 321–322, 337–339
Crick, F.H.C., 13
Crocodilia, 358–359

Crossopterygian, 331–332
Crossover. *See* Gene conversion; Recombination; Unequal crossing over
*Crotalus*, 387–388
Crypsis, 28
Cryptic variation, 94–96
Cultural evolution, 516–518
Cultural inheritance, 254
Culture, in nonhuman animals, 517–518
Cuvier, G., 4
Cyanobacteria, 323
Cytodifferentiation, 58
Cytoplasmic inheritance, 45

Darwin, C. R.
- on adaptation, 257
- biogeography and, 374
- career of, 5–6
- and heredity, 43
- and materialist philosophy, 2
- on reproductive excess, 24–25
- sexual selection and, 232–233

*darwin* (unit measure), 398–399
Death rate, population growth and, 21–22, 24
Defense mechanisms, 27–28
*Deinonychus*, 336, 424
*Delphinium*, 419
Demes, 129–130, 265
Demography, and selection, 527
Density, 24–26
Deoxyribonucleic acid (DNA). *See also* Amino acids; Deoxyribonucleic acid sequences; DNA viruses; Macromolecules; Molecular biology; Nucleotide sequences
- as basic genetic material, 45
- in eukaryotes, 48
- highly repetitive, 50
- hybridization of, 314–315
- moderately repetitive fraction of, 48
- nonrepeated, 48–50
- repeated, 48–50
- replication of, 50–53

Deoxyribonucleic acid sequences
- divergence in, 145
- evolution of, 444, 447–451
- variation in, 98–99, 446–448

*Desmodium nudiflorum*, 102
Detoxification, 409
Development. *See also* Developmental constraints; Developmental homeostasis; Developmental integration; Ontogeny
- as epigenetic, 426
- mechanisms of, 425, 435–436
- processes of, 58–59

Developmental constraints, 91, 255–257, 369, 437–438
Developmental homeostasis, 210–215
Developmental integration, 439–440
Devonian period, 320, 329–332
*Dicrurus paradiseus*, 103–104
Dinosaurs, 334–338
*Diplodocus*, 336
Diptera, 77, 295–296
Direct flanking repeats, 70–71
Directional selection, 154–159
- artificial, 207–208
- constant fitnesses and, 155–159
- in *Drosophila*, 204
- evolutionary trends and, 369
- examples of, 157–159
- genetic variance and, 203
- trait evolution and, 201, 203–204
- at two loci, 188

Discordant cline, 105–107
Disjunct distribution, 377–378
Dispersal, 378–380, 382–385
Disruptive selection, 155, 201–202
Divergence
- constancy of, 308–309
- predator–prey coevolution and, 493–496
- in protein sequences, 145
- relative rate of, 308–309
- speciation and, 231–234

Diversifying selection, 155, 201–202
Diversity. *See* Species diversity
Diversity-dependent controlling factors, 347
DNA. *See* Deoxyribonucleic acid
DNA viruses, 69
Dobzhansky, Th., 10, 11–12
Dollo's law, 297
Dominance, 198, 200, 211–212, 392–394. *See also* Competitive exclusion principle; Competitive interactions; Sexual selection
Dosage compensation, 460
Double invasion, 224–225
Drepanididae, 32–33
Drongo. *See Dicrurus paradiseus*
*Drosophila*. *See also* specific species
- chromosomal inversions in, 63
- cryptic variation in, 94
- Hawaiian species of, 241–242, 303–305
- mutation in, 75, 433–434
- sterility and, 114, 222
- transposable elements and, 70–71

*Drosophila heteroneura*, 221
*Drosophila melanogaster*, 72–74, 95, 101, 142, 180–182, 190–191, 204, 206, 208–209, 213–214, 235, 420–421, 430–435, 446–447, 456–457, 470–471
*Drosophila mulleri*, 179
*Drosophila paulistorum*, 187, 246
*Drosophila persimilis*, 302
*Drosophila pseudoobscura*, 96–98, 140, 165, 177, 195–197, 212, 241
*Drosophila repleta*, 298
*Drosophila serrata*, 23
*Drosophila subobscura*, 429–430
*Drosophila willistoni*, 220
Dung fly. *See Scatophaga stercoraria*

Echinoidea, 363
Ecological factors, and variation, 178–180
Ecological niche, 19–20, 356
Ecological release, 32
Ecological replacement, 358–359
Ecotypes, 102
Eel. *See Anguilla rostrata*
Effective population size, 131–132, 136–142
*Elaphe obsoleta*, 104, 106
Electromorph, 98
Electrophoresis, 46, 97–99, 102–103, 115
Elephant seal. *See Mirounga angustirostris*
Embryology and phylogenetic analysis, 303
Endemic species, 375–376
*Ensatina*, 311–312
*Enterobius*, 497–498
Environment
- adaptation to, 406
- carrying capacity of, 24
- changes in, 158–159, 369-370
- development of dominance and, 392–394
- ecological, 19
- extinction and, 360
- heritability and, 532, 535
- human effects on, 158–159 (*See also* Agriculture)
- mutation and, 75–76
- variation in, 37–40, 167–170

Environmental correlation, 205
Environmental factors, 26, 36–40, 168–169. *See also* Temporal variation
Environmental variance, 89–90, 195–198
Enzymes. *See also* Molecular data; Nucleotide sequences; Proteins
- heterozygosity and, 97–99, 165
- regulation of, 474–478

substrate specific, 474
Epicontinental seas, 321–322
Epigenetic landscape, 213
Epistasis, 53, 185–187
Epistatic interaction, 185–195
in laboratory, 185–192
of loci, 194
in natural populations, 192–195
Epistemological reductionism, 479
Equidae, 292–293, 303, 407–409. *See also Equus; Hyracotherium*
Equilibria, multiple, 188–192
Equilibrium density, 24
*Equus,* 368–369, 416–417
*Escherichia coli,* 72–73, 141, 181–182, 259, 477, 502
ESS. *See* Evolutionarily stable strategy
Essentialism, 3, 108
Eukaryotes, 45, 47, 57–58, 76–78, 323
Eurytopic species, 20
*Eusthenopteron,* 332
*Eutamias,* 31
Evolution
biological versus cultural, 517–518
since Darwin, 8–10 (*see also* Modern Synthesis; Neo-Darwinian theory)
as fact, 15
factors in, 8–9, 13, 172–175 (*see also* Gene flow; Genetic drift; Mutation; Selection)
iterative, 365–367
method of study, 13–15
misconceptions of, 7–8
opposition to, 15 (*see also* Creationism)
origins of idea, 2–7
rate of, 292–294, 397–400
trends in, 366–373, 402–403 (*see also* Parallel evolution)
Evolutionarily stable strategy (ESS), 268–271
Evolutionary history, 8–9. *See also* Fossil record
Evolutionary mechanisms. *See* Evolution, factors in
Evolutionary reversal, 295–299
Evolutionary synthesis. *See* Modern Synthesis; Neo-Darwinian theory
Exon, 47–48, 474
Exploitation, defined, 30
Expressivity, 53
Extinction. *See also* Extinction rates; Mass extinction
evolutionary rate and, 397
gene flow and, 141
interspecific interactions and, 389
in marine animal families, 361
overspecialization and, 362–363
patterns of, 359–360
probability of, 362
Extinction rates, 346–347
distribution of, 360–364
origination rate and, 352–353, 355

$F_2$ breakdown, 195
Fecundity, 22–23, 272–273
Ferns, 51
Fig, 499–500
Fig wasp. *See Ceratosolen dentifer*
Finch, 5, 225, 486–487. *See also Geospiza*
Finite population. *See* Population size
Firefly. *See* Lampyridae
Fisher, R. A., 10–11
Fitness
calculation of, 151, 251
contributions to differences in, 152
environment and, 152–153
epistasis for, 185–187
inclusive, 261–262
natural selection and, 158
resource abundance and, 489
as term, 151
transposable elements and, 462
Flax, 44–45
Flight, in birds, 423
Flour beetle. *See Tribolium*
Flycatcher. *See Petroica multicolor*
Foraminifera, 363, 365, 367
Formicariidae, 378–379
Fossil record, 318–343
classification and, 299
gradual versus punctuated change and, 404–405
Hominoidea and, 511–516
phylogenetic analysis and, 303
species distribution and, 381–382
Fossils, living, 407
Founder effect, 133–135, 238–242
Frameshift mutation, 66
Frequency (statistic), 541
Frequency-dependent optimal models. *See* Evolutionarily stable strategy
Frequency-dependent selection, 155, 166–167
Frequency-independent optimal models. *See* Optimal models
Freud, S., 506
Function, and evolutionary innovation, 423–424
Fusion, morphological, 410–411
Fusion, versus speciation, 243
Galápagos finches, 31, 34
Gause's axiom. *See* Competitive exclusion principle
β-Galactosidase system, operons and, 57
Gamete frequency, 84
Gametic excess. *See* Linkage disequilibrium
*Gasterosteus aculeatus,* 244
Garter snake. *See Thamnophis*
Gastropods, 366
Gel electrophoresis. *See* Electrophoresis
Gene. *See also* Alleles; Gene flow; Genetics; Horizontal gene transfer
duplication and, 67–68, 451–456
eukaryotic, 47
evolution of, 451–456, 472–478
discovery of, 43
mutational change in, 65–71
regulation of expression of, 56–58
structure of, 47–48
as term, 84
transfer of, 69, 78–79
Gene conversion, 67–68. *See also* Recombination
concerted evolution and, 465–472
crossover frequency and, 77
interchromosomal, 469–470
intrachromosomal, 465–470
Gene family, 50, 451, 461, 464–472
Gene flow, 136–139
effective population size and, 139–142
in evolutionary synthesis, 12
fixation by selection and, 159, 160–162
genetic barriers to (*see* Isolating mechanisms)
in local breeding populations, 120
selection and, 160–162
Gene-for-gene system, 500
Gene frequency. *See* Allele frequency
Gene pool and coadaptation, 194–195
Gene linkage relationships, 451. *See also* Chromosomes; Linkage disequilibrium; Polyploidy
Generalized species, 20
Generation time and molecular clock, 309
Genetic assimilation, 196–197, 212
Genetic code, 45–46
Genetic correlation, 205–207, 306, 414
Genetic counseling, 524

Genetic death, 176
Genetic difference. *See* Genetic distance
Genetic distance. *See also* Molecular clock; Phylogenetic inference
  allozyme frequency and, 307
  closely related species and, 220
  between human and chimpanzee, 509–510
  human population groups and, 522, 523
  immunological distance and, 307
  measures of, 102
  Nei's index of, 102
  time of divergence and, 308
Genetic drift, 143–145, 255, 469. *See also* Divergence
  fixation of mutation and, 134
  human variation and, 523
  linkage disequilibrium and, 187
  in natural populations, 142–143
  population size and, 129–132
  speciation and, 234
  variation patterns and, 178–180
Genetic homeostasis, 210–215
Genetic load, 175–177
Genetic revolution, 231, 239
Genetics
  and Darwinian theory, 10 (*see also* Modern Synthesis; Neo-Darwinian theory)
  methods in, 46–47
  principles of, 43–45
  quantitative theory in, 399–400
Genetic similarity, 102. *See also* Genetic distance
Genetic variance, 195–198
Genetic variation, 159–170
  adaptation and, 12
  at allozyme loci, 99
  assay of (*see* Electrophoresis)
  environment and, 167–170
  gene flow and, 78
  in *Homo sapiens,* 519–527
  mutation and, 74, 133
  organization of, 99–101
  among populations, 102–103
  selection and, 177–181, 202–204
Genic selection, 153, 260–261
Genome, 48–49, 420, 444, 459–464. *See also* Molecular data
Genotype. *See also* Phenotype
  defined, 43
  fixation of, and selection, 484–485
  interaction with environment and, 196–198
  and phenotype, 53–56
  phenotypic versus fitness scales and, 192
Genotype frequency, 82
Geographic area and species diversity, 353
Geographic distribution, 21, 120, 375–380
Geographic isolation and speciation, 223
Geographic races. *See* Subspecies
Geographic variation, 103–111. *See also* Spatial variation; Temporal variation
  coevolution and, 500
  in ecological characteristics, 109–111
  patterns of, 103–107
  in reproductive characteristics, 109–111
Geological history, 319–322
Geological time scale, 318–319, 320
*Geospiza,* 31, 34, 91, 490–491
*Gilia achilleifolia,* 127
*Ginkgo biloba,* 334, 351
Glaciation, 322, 342
Globin family, phylogeny of genes in, 452–455
α-Glycerol phosphate dehydrogenase (α-Gpdh), 190–191
Goat, 454
Gondwanaland, 319, 321–322, 381–382
*Gorilla gorilla,* 509–511. *See also* African apes
Grade, 288–289
Gradual evolution, 402–403. *See also* Neo-Darwinian theory; Phyletic gradualism; Punctuated equilibrium; Speciation, sympatric
Grass. *See Agrostis tenuis*
Grasshopper. *See Caledia captiva; Keyacris scurra*
Great Chain of Being, 3, 16
Green frog. *See Rana clamitans*
Ground finch. *See Geospiza*
Group selection, 153, 258–259, 264–266
Grouse. *See* Tetraoninae
Growth limitation, 25–26
Growth rate. *See* Population growth
*Gryphaea,* 401
Guilds, 351, 390
Gymnosperms, 334
Gypsy moth. *See Porthetria dispar*

Haeckel's law, 303, 416
*Haemaphysalis leporispalustris,* 106–107
Haldane, J.B.S., 398–399
Halictidae, 294–295
Hamilton's rule, 262
Haplotype, 65
Hard selection, 175, 208
Hardy-Weinberg theorem, 82–87
  assumptions of, 85–87
  derivation of, 83
  deviation from equilibrium and, 99–100
  extensions of, 83–85
  and multiple loci, 84–85, 86
Hawaiian honeycreepers. *See* Drepanididae
*Heliconius,* 496, 500–501
*Hemidactylium scutatum,* 438
Hemiptera, 295–296
Herbicides, resistance to, 159
Herbivore–plant relationship, 493–496
Heredity, and neo-Darwinian theory, 13
Heritability, 198–200. *See also* Characteristic
  broad sense ($h^2_B$), 198
  components of variance and, 193
  environment and, 532, 535
  human traits and, 520
  narrow sense ($h^2_N$), 90, 198–200
  response to selection and, 200–205
Heterochrony, 412, 415–419
Heterosis, 124, 128
Heterostyly, 100–101, 126, 193
Heterozygosity
  as advantage, 162–166
  decrease in, 123
  developmental homeostasis and, 210–211
  effective population size and, 135
  genetic drift and, 134
  inbreeding and, 129
  invariant enzymes and, 97–99
  mutation and, 133
  selection against, 170–72
  specific loci and, 164–165
Higher taxa. *See also* Taxonomic classification
  diversity profiles and, 350–351
  origin of, 419–423
Historical biogeography. *See also* Dispersal; Vicariance
  ecology and, 387–388
  paleontology and, 381–382
  systematics and, 382–385
Hitchhiking, 179–180, 188
Homeotic mutation, 75, 430–435
Hominoidea
  origins of, 342
  phylogenetic relationships in, 311–315, 509–511, 512–513
*Homo erectus,* 514, 515
*Homo habilis,* 514, 515
Homology, 295–299, 306

Homoplasy, 295–299
*Homo sapiens. See also* Human behavior; Intelligence, human
  allometric growth in, 414
  cultural evolution and, 516–518
  evolution of mental faculties in, 518–519
  genetic variation in, 519–527
  globin gene families of, 51
  globin phylogeny in, 452–453
  mutation in, 73, 524–527
  natural selection in, 524–527
  phylogeny of, 311–315, 507–511, 514–516
  physical evolution in, 518
  population structure and, 522–524
  racial groups and, 520–522
*Homo sapiens neanderthalensis,* 515, 516
Horizontal gene transfer, 478
Horse. *See Equidae; Equus: Hyracotherium*
Housefly, 502
House mouse. *See Mus musculus*
Human behavior. *See also* Agriculture; Industrialization
  environmental impact of, 158–159
  environmentalist view of, 531–532
  evolution of, 528–533
  genetic determination of, 528, 529–533
  sociobiological argument and, 530–531
  variation in, and genetic differences, 532–533
Human species. *See Homo sapiens*
Hummingbirds, 114, 233
Huntington's chorea, 273
Hutton, J., 4
Huxley, J., 12
Huxley, T. H., 536–537
Hybrid dysgenesis, 235
Hybridization, 46–47, 78–79, 114–115
Hybrid sterility, 111, 114, 222–223, 235, 471. *See also* Isolating mechanisms; Reproductive isolation
Hybrid zone, 104, 115, 226–227
Hypermorphosis, 415–417
Hypothetico-deductive method, 6–7
*Hyracotherium,* 341, 368–369, 400, 408

*Icterus galbula,* 104
*Icthyostega,* 331–332
Identity-by-descent, 129, 134
Identity coefficients, 469–470
Immunological distance, 307
Impact hypothesis, 364–365
Inbred populations, 122, 127–128. *See also* Inbreeding depression
Inbreeding, 120–127, 129–131, 282–283. *See also* Selfing
Inbreeding coefficient, 121–122, 129
Inbreeding depression, 124, 127–128, 282
Independent assortment, 51
Individual organisms, ontogeny of, and evolution, 7
Individual selection, 150, 154–159, 259–260, 370
Induction, and morphogenesis, 59
Industrialization, 526
Insectivores, 340–342
Insects, in fossil record, 333, 340
Instantaneous speciation. *See* Speciation, sympatric
Intelligence, human, 506–507, 533–535
Interaction variance, 198, 200
Interbreeding, 111, 520
Interdemic selection. *See* Group selection
Interference, 30–31
Intersexual selection, 277
Interspecific allometry, 413–414
Interspecific interactions, 389, 483–503. *See also* Competitive interactions
Intragenic recombination, 67
Intrasexual selection, 277
Intraspecific allometry, 413
Intraspecific competition, 30
Intrinsic rate of natural increase, 24
Introduced species, 390–391
Introgressive hybridization, 115
Introns, 47–48
Inverse frequency dependence, 166
Invertebrates, fossil record of, 334, 348–349, 351, 393
IQ. *See* Intelligence, human
Island biogeography, 35–36, 388–391
Island model, 120, 136–138
Isolating mechanisms, 112–114, 219. *See also* Reproductive isolation
  ethological, 112
  and *Homo sapiens,* 520
  postmating, 112–114
  premating, 112–113, 242
Isolation-by-distance model, 120, 136, 138
Iteroparity, 272, 274
Jumping genes. *See* Transposable elements
Jung, C., 506
Jurassic period, 321–322, 335–337

Karyotype. *See also* Chromosomes
  changes in, and mutation, 60–65
  geographic variation in, 236–238
*Keyacris scurra,* 188–189
Key innovation, 356
King snake. *See Lampropeltis getulus*
Kin selection, 254, 261–263, 265
K-selected population, 276

Lacewing flies. *See Chrysopa*
Lamarck, J.—B. de, 4, 9–10
*Lampropeltis getulus,* 92–93
Lampyridae, 112–113
Land masses, prehistoric, 319–322
Land snail. *See Cepaea nemoralis*
Languages, evolution of, 516–517
*Latimeria chalumnae,* 331, 351
Latitude, and origination rate, 391–394
Laurasia, 321
Laurussia, 319
*Layia glandulosa,* 116
Lepidoptera, 296–297, 302, 420–421
Leucine aminopeptidase, 97
Life
  origin of, 322–323
  Precambrian, 323–324
Life history characteristics, 272–276
Linkage disequilibrium. *See also* Linkage equilibrium
  in artificial selection, 209–210
  coefficient of, 85
  in *Drosophila,* 101
  directional selection and, 188, 207
  epistasis for fitness and, 185–187
  genetic correlation and, 205–206
  genetic drift and, 187
  inbreeding and, 127
  selection at loci and, 178–180
Linkage equilibrium, 85, 99, 100–101
Linnaean hierarchy, 288. *See also* Taxonomic classification
Linnaeus, C., 3, 107–108
*Littorina littorea,* 35
Lotka-Volterra interspecific competition model, 485
Lungfish, 407
Lyell, C., 4
*Lymantria. See Porthetria dispar*

*Macaca mulatta,* 438–439
Macroevolution, 397, 409, 439–440

Macromolecules, phylogenetic inference and, 307–315, 510–511. *See also* Deoxyribonucleic acid; Molecular biology; Proteins; Ribonucleic acid
Maize. *See Zea mays*
Malthus, T. R., 5–6
Mammals. *See also* specific species
  in the Americas, 382–385
  in Cenozoic era, 348
  In Cretaceous period, 338
  in Jurassic period, 336–337
  multituberculate, 334, 337
  placental, 330–331, 339–340
  in Tertiary period, 340–341
  in Triassic period, 334–335
*Mammuthus,* 341
Marginal overdominance, 165
Mass extinction. *See also* Extinction
  of ammonite cephalopods, 352
  background extinction and, 360–362
  causes of, 364–366
  in Cretaceous period, 338–339, 364–365
  human agriculture and, 343
  in Jurassic period, 335
  in Ordovician period, 328–329
  in Phanerozoic era, 365
  proliferation after, 347–351, 359
  survivors of, 365
*Mastodon,* 341
Maternal effects, 44
Mathematical models, role of, 14–15
Mayr, E. 11
Megafauna, 343
*Megaloceros giganteus,* 416–417
Meiotic drive. *See* Segregation distortion
Mendel, G., 9–10, 43
Mesozoic era, 320, 334–339
Messenger RNA (mRNA), 46–47
Metacentric chromosomes, 64–65
Metapopulation, 137
Methanogenic bacteria, 323
Methodological reductionism, 479. *See also* Reductionism
Metric characteristic, 87–90, 125
Microevolution, 397, 409
Midparent value, 199
Mimicry, 28–29
*Mirounga angustirostris,* 135
Mitochondrial DNA, 448, 450
Mobile genetic elements. *See* Transposable elements
Models, role of, 14–15
Modern Synthesis, 10–12, 397. *See also* Neo-Darwinian theory
Molecular biology, 444–447, 478–480
Molecular clock, 308–309
Molecular drive, 471
Molecular evolution. *See also* Molecular clock
  rate of, 143–144
  speciation and, 234
Monophyletic group, 288, 299–306. *See also* Strictly monophyletic group
*Moraba scurra. See Keyacris scurra*
Morphocline, 301–302
Morphogenesis, 58–59, 427–428
Morphological features
  evolutionary change in, 399–400, 410–412, 425–430
  species differences in, 219
  variation in, 91–92
Mosaic evolution, 293
Moth. *See Biston betularia*; Lepidoptera
Mouse. *See Mus musculus; Perognathus goldmani*
mRNA. *See* Messenger RNA
Müllerian mimicry, 28–29
Multiple invasion, 225
*Muntiacus,* 425–426
*Mus musculus,* 73, 124, 260
Mussel. *See Mytilus edulis*
Mutation. *See also* Frameshift mutation; Homeotic mutation; Neutral mutation; Point mutation; Recurrent mutation; Transversion mutation
  chromosomal changes and, 60–65
  concerted evolution and, 469
  deleterious, 75
  in *Drosophila,* 72–74, 430–435
  fixation by selection and, 159–160
  genetic, 65–71
  in *Homo sapiens,* 524–527
  phenotypic effects of, 75–76
  pleiotropic effects and, 421
  in polygenic character, 204–205
  randomness of, 76
  rates of, 72–74, 259
  transposable elements and, 458–459
  viability and, 74
Mutator allele, 259–260
Mutualism, 29, 497–503
*Mytilus edulis,* 97, 161–162
Myxoma virus, 496–497

*Nasonia vitripennis,* 502
Naturalistic fallacy, 8
Naturalness, as concept, 528. *See also* Human behavior
Natural science, 3
Natural selection, 6–7. *See also* Selection
  concerted evolution and, 471–472
  and density, 24–25
  in evolutionary synthesis, 12–13
  Fisher's fundamental theorem of, 200
  fitness and, 151–159
  in *Homo sapiens,* 524–527
  misconceptions and, 150
  and mutation, 10
  as statistical measure, 150
  strength of, 181–182
  and technology, 526–527
Neanderthals. *See Homo sapiens neanderthalensis*
Neighborhood, 131, 140
Neo-Darwinian theory, 10–13
  explanatory power of, 440–441
  higher taxa and, 420–423
  microevolution and, 409
  morphological change and, 402–403
Neoteny, 415–418, 518–519
Neutralist–selectionist controversy, 177–181, 448, 180–182
Neutral mutation, 75
*Neurospora crassa,* 72–73
Niche overlap, 21. *See also* Ecological niche
*Nicotiana longiflora,* 88
Nomenclature, 288
Nonrandom mating, phenotype and, 146
Normal distribution curve, 543
Normal extinction. *See* Background extinction
Norm of reaction, 53–54, 528
Nucleotide sequences
  in genetic code, 46
  hybridization and, 46–47
  phylogenetic inference and, 307–315
  polarity of, 45
  substitution in, 144–146
  variation in, 446–447
Numerical taxonomy. *See* Phenetic school

Objectivity, 506–507
Oligocene epoch, 322
Ontogenetic method, 303
Ontogeny. *See* Biogenetic Law; Development
Ontological reductionism, 479. *See also* Reductionism
Onychophora, 431–432
Operon, 57
Optimal diet, 267–268
Optimal model, 266–268
Orangutan. *See Pongo pygmaeus*
Ordovician period, 320, 328–329

Organic evolution, 7
Origination patterns, 354–359. *See also* Speciation
Origination rate, 346–347, 352–353, 355
Oriole. *See Icterus galbula*
Ornithischia, 336
*Ornitholestes,* 337
Oropendola, 34–35
*Ortalis,* 310
Orthogenesis, 369
Orthologous gene, 451–453
*Oryctolagus cuniculus,* 496–497
*Oryx beisa,* 269, 270
Osteichthyes, 331, 357
Ostracoderms, 328
Outbreeding, 124–127, 282–283. *See also* Sexual reproduction
Outcrossing, 126–127
Outgroup, 302–303
Overdominance, 55, 124, 165

Paedomorphosis, 306, 415–418
Paleontology
  biogeography and, 381–382
  taxonomy and, 346
Paleozoic era, 320, 324–334
*Pan,* 508–511.
Pangaea, 319–321
Panmictic species, 120, 265–266
*Papilio dardanus,* 202, 212
*Papilio demodocus,* 169
*Papilio polyxenes,* 409, 475–477
Parallel evolution, 288–289, 295–299, 366
Paralogous gene, 451–453
*Paramys,* 337
Parapatric forms, 103
Paraphyletic group, 289–290
Parasite–host relationship, 496–497, 499
Parasitism, 28–29, 78–79
Parsimony, 300–302
Parthenogenesis, 279–282
*Passiflora,* 496
Pattern formation, 428–430
Peak shift, 231, 235–236, 238–242. *See also* Founder effect
Penetrance, 53
Peramorphosis, 415–416
Periodic selection, 260
Peripatric speciation, 402. *See also* Speciation
Periwinkle. *See Littorina littorea*
Permian period, 319–322, 333–334
*Perognathus goldmani,* 237–238
*Petroica multicolor,* 238, 239
Phanerozoic time, 324–325, 361, 365
Phenetic school, 289
Phenocopies, 53, 212
Phenogram, 289, 291
Phenotype
  and amount of gene product, 55
  defined, 43
  and environmental conditions, 54
  and genotype, 53–56
  mating and, 146
  mean, 88
  and ontogenic stage
  variation in, and inbreeding, 124
Phenotypic correlation, 205
Phenotypic covariance, 205
Phenotypic evolution, 409–412
Phenotypic gaps, 436–439
Phenotypic variance, 195–198
Phyletic gradualism, 402–403
Phylogenetic analysis
  biogeography and, 374–375
  comparative method and, 253
  compatibility methods, 300
  dispersal and, 382–385
  inference difficulties and, 292–299
  molecular data and, 307–315, 444
  morphological data and, 299–306
  outgroup criterion, 302–303
  parsimony methods and, 300
  phenogram and, 291
  phenotypic evolution and, 409
  vicariance and, 385–387
Phylogeny. *See also* Biogenetic law; Phylogenetic analysis
  of genes, 452–453
  of organism groups, 324. *See also* specific group
Phytosaurs,334
Placoderms, 329–330
Plants. *See also* specific species
  effective population size in, 140
  extinct, 333, 338, 340
  gene flow in, 140
  inbreeding in, 125–126
  species concept and, 115
Plasmids, 68–69
Plate tectonics, 319, 379–382
Plato, philosophy of, 2–3
Platyrrhini, 507, 510
Pleiotropy
  antagonistic, 206
  in artificial selection, 209
  direct, 55
  genetic variation and, 206–207
  relational, 55–56
Pleistocene epoch, 320, 322, 342–343
Plenitude, principle of, 3, 8
Plesiomorphic character state, 286
*Plethodon,* 48, 141, 311–312, 405, 418, 459, 491
Pocket mouse. *See Perognathus goldmani*
*Podocnemis expansa,* 378
Podonominae, 386–387
Point mutation, 65–67
Polygenic inheritance, 43, 195–200
  evolution and, 200–201
  mutation and, 204–205
Polymorphism, 91–99. *See also* Genetic variation
  balanced, 93–94
  in *Drosophila,* 97–98, 177, 194–195
  environmental heterogeneity and, 168
  factors influencing, 170
  frequency dependent selection and, 166
  multiple niche, 169–170
  transitional, 93
Polyploidy
  genetic evolution and, 456
  genome size and, 48–49
  Hardy-Weinberg theorem and, 84
  interbreeding and, 228–229
  meiosis and, 61
  mutation and, 60–62
Polytypic species. *See Rassenkreis*
Pongidae, 510–511
*Pongo pygmaeus,* 311–315, 508–509
Population genetics, 10–12, 399, 444
Population growth, 21–24
  density and, 24–26, 275–276
  fecundity and, 22
  idealized curve for, 22
  individual selection and, 151–152
  in limited environment, 24
  mathematical formulation of, 21–22, 24
  potential versus actual rates of, 23–24
  predator–prey interactions and, 26–28
Populations. *See also* Effective population size; Population genetics; Population growth; Small populations
  average fitness and, 175–177
  geographic variation and, 103–111
  natural, 90–96, 192–194
  variation in, 12, 90–96, 102–103
Population size, 129–131, 133–134. *See also* Effective population size; Population growth
Porcupines, 384
*Porthetria dispar,* 110
Position effects, 65

Pre-adaptation, 424
Precambrian era, 320, 322–324
Predation, 27–28, 167
Predator–prey interactions, 26–28, 492–497, 503
Predator switching, 27
Prepattern, 59, 429
Primates, 341–342, 507–511. *See also* specific species
Primitive characteristics. *See* Characteristics, ancestral
Primrose. *See Primula vulgaris*
*Primula vulgaris,* 100–101, 193–194, 206
Probability, 541–542
Progenesis, 415–416, 418–419
Progress, as concept, 8, 16
Prokaryotes, 45, 323
Promoter, 48
Protandry, 127
Proteases, evolution of, 475
Protein electrophoresis. *See* Electrophoresis
Proteins. *See also* Molecular biology
  evolution of, 472–478
  phylogenetic inference and, 307–315
  rate of divergence in, 145
  variation in, 96–99
Provincialization, 334, 349
*Pseudocubus vema,* 293–294
Pseudoextinction, 346
Pseudogene, 50, 453–455
  processed, 69, 458
*Pseudomonas aeruginosa,* 75
*Pseudomyrmex,* 498–499
Pseudo-overdominance, 164–165
Pseudospeciation, 346
Pteridosperms, 333
Pull of the Recent, 346
Punctuated equilibrium, 239, 401–409

Quantitative characteristic. *See* Metric characteristic
Quantum evolution, 246. *See also* Divergence
Quaternary period, 342–343

Rabbit tick. *See Haemaphysalis leporispalustris*
Race
  as concept, 107–109
  IQ scores and, 534–535
  and physical characteristics, 520–521
Racism, scientific, 507
Radioactive dating, 318–319
*Ramapithecus,* 511–512
*Rana clamitans,* 109
Random genetic drift, 11–13. *See also* Genetic drift
*Rassenkreis,* 103
Rat, 128. *See also* Rodents
Ratite birds, 374–375
Rat snake. *See Elaphe obsoleta*
Rattlesnake. *See Crotalus*
Recapitulation. *See* Biogenetic law
Recombination, 50–53
  concerted evolution and, 470
  evolution of, 279–283
  gene duplication and, 67–68
  model of, 52
  and mutation, 60
  suppression of, 194
  and variation, 76–78
Recombination fraction, 51
Recurrent mutation, 160
Red-backed salamander. *See Plethodon cinereus*
Red Queen hypothesis, 362, 492
Reduction, morphological, 410–411
Reductionism, 478–480
Reductionist fallacy, 16
Refugia, 380–381
Regression coefficient, 545–546
Regressions, of sea level, 321–322
Relative fitness, 152
Relative rate test, 308–309, 314–315
Relict species, 381
Rensch, B., 12
Replica plating, 76–77
Replicons, 50
Reproduction, 22, 272–274. *See also* Asexual reproduction; Fecundity; Inbreeding; Reproductive isolation; Selfing; Sexual reproduction
Reproductive excess, 24
Reproductive isolation, 111–114, 219–223. *See also* Inbreeding; Isolating mechanisms; Selfing
  genetics of, 222, 238
  selection for, 242–244
Reproductive value, 22
Reptiles, 290, 330–331, 334
Resource utilization, 484–485, 487
Restriction enzymes, 47, 307, 313, 445, 447, 510–512
Reticulate evolution, 286
Retrotransposons, 457
Retrovirus, 69. *See also* Reverse transcription
Reverse recapitulation, 416
Reverse transcription, 69–70, 457
Ribonucleic acid, 323, 444, 450, 458. *See also* Genetic code; Macromolecules: Messenger RNA; Molecular biology; Ribosomal RNA; Transfer RNA
Ribosomal RNA (rRNA), 45. *See also* Ribonucleic acid
Ritualized aggressive behavior, 269–271
RNA. *See* Ribonucleic acid
Robertsonian fusion, 65
Rodents, 35, 37, 342. *See also* Rat
Rogers' index of genetic similarity, 102
*Roxira serpentina. See Layia glandulosa*
rRNA. *See* Ribosomal RNA
*r*-selected population, 276

*Sagittaria sagittifolia,* 39
Salamander. *See Ambystoma mexicanum; Aneides; Hemidactylium scutatum*; Neoteny; *Plethodon; Thorius*
*Salmonella typhimurium,* 72–73
Saltation, 397, 420–423
*Salvia nodosa,* 126
Satellite DNA, 50
Saurischia, 334–335
Sauropods, 335–336
*Scala Naturae,* 3, 9
*Scaphidura oryzivora,* 34–35
*Scatophaga stercoraria,* 269–271
Search image, 27, 167
Secondary contact, 104
Sedimentary rock, radioactive dating of, 319
Segregation, 50–53
Segretation distortion, 51–52, 153, 260
α-Selection, 490
Selection. *See also* Artificial selection; Directional selection; Diversifying selection; Frequency-dependent selection; Group selection; Individual selection; Natural selection; α-selection; Sexual selection
  gene flow and, 160–162
  at gene level, 457
  genetic variation and, 202–204
  heritability and, 200–205
  levels of, 153, 258–263
  recurrent mutation and, 160
  opportunity for, 527
  responses to, 200–205
Selection coefficient, 152
Selection differential, 200
Selection plateau, 208, 210
Self-fertilization. *See* Selfing
Selfing, 121, 123–124, 126—127, 279–280, 282. *See also* Inbreeding
Selfish DNA, 457, 461–464
Semelparity, 272, 274
Semispecies, 115
Senescence, 273
Serially homologous organs, 410–412

Sex races, 110
Sex ratio, 261, 265–266
Sexual isolation. *See* Reproductive isolation
Sexual orientation, genetic basis for, 533
Sexual reproduction, 279–283
Sexual selection, 231–234, 276–279. *See also* Divergence
 Fisher's model of, 211–212, 261, 265, 277
 runaway, 278–279, 518–519
Shark. *See* Chondrichthyes
Shifting balance theory, 174–175, 403–404
Sibling species, 111
Sickle-cell hemoglobin, 56, 163, 524
*Sigara distincta,* 167
Silent substitution, 45, 66, 75
Silurian period, 320, 329
Simpson, G. G., 11–12
Singular events, causes of, 13–14
Small populations, 21. *See also* Geographic distribution
Snail shell, 422–423
Snow goose. *See Chen caerulescens*
Social Darwinism, 8, 536
Social policy, 535–537
Soft selection, 175
Spatial distribution. *See* Geographic distribution
Spatial variation, 36–37, 39. *See also* Geographic variation
Specialized species, 20–21
Speciation. *See also* Founder effect; Origination rate
 allopatric, 103, 223–227
 chromosomal evolution and, 246
 in evolutionary synthesis, 12
 evolution of characters and, 403–404
 versus fusion, 243
 by genetic revolution, 231
 genetic theories of, 231–238
 modes of, 223–224
 molecular evolution and, 234–235
 parapatric, 227–228
 peripatric, 238–242
 sympatric, 228–231, 487–488
 time required for, 244–246
Species, 111–115. *See also* Eurytopic species; Geographic distribution; Speciation; Stenotopic species; Taxonomic classification
 as biological concept, 219
 complex interactions among, 34–35 (*see also* Competitive interactions)
 diversity of, and equilibrium, 503
 significance of, 246–247
 as taxonomic category, 219
Species distributions, causes of, 386–388
Species diversity, 35–36, 219–223, 254–258, 346–354, 391–393
Species equilibrium, 389, 503
Species selection, 153, 370–371, 406–409
Specific mate-recognition system, 244
*Sphecomyrma,* 340–341
*Sphenodon punctatus,* 377
Stabilizing selection, 155, 201–204, 405–406
Standard sequence, 301–302
Stasipatric speciation model, 227–228
Stasis, 2, 404–406
Stebbins, G. L., 12
*Stegosaurus,* 336
Stenotopic species, 20–21
Step cline, 105
Stepping stone model, 136
Stickleback. *See Gasterosteus aculeatus*
Strictly monophyletic group, 292
Subspecies, 103, 107–109
Supergene, 100, 193–194
Superiority, competitive. *See* Competitive interactions
Superspecies. *See Artenkreis*
Survival
 age-specific reproduction and, 272–273
 clutch size and, 274–275
 genetic variation in, 23
 and population growth, 22
Swallowtail butterfly. *See Papilio dardanus; Papilio polyxenes*
Symplesiomorphic character, 287
Synapomorphic character, 287
*Synchytrium decipiens,* 500
Systematics, 287–292. *See also* Taxonomic classification
 schools of, 288–292
 historical biogeography and, 382–385
 and evolutionary history, 9

*Taeniolabis,* 337
*Tapirus,* 378–379
Taxon, 288
 characteristic extinction rate in, 362–364
 disjunct distribution and, 377–378
 polyphyletic, 288
 as strictly monophyletic, 292
Taxon cycle, 393, 489
Taxonomic classification, 12, 111, 116, 287–288, 346. *See also* Higher taxa; Species; Systematics; Taxon
Taxonomic survivorship curves, 363
Technology, natural selection and, 526–527
Temporal variation, 37–40
Tertiary period, 340–342
Tetraoninae, 255–256
*Thamnophilis doliatus,* 379
*Thamnophis,* 224–225
Thecodontia, 334, 336
Theory, in evolution, 15
*Thorius,* 424
Threshold character, 214–215
Tick trefoil, *See Desmodium nudiflorum*
*Tisbe reticulata,* 94
Tissue differentiation, 427
Tissue interactions, 58–59
Titanotheres, 341, 366, 368
Trait. *See* Characteristic
Trait group, 264
Transcription, 45
Transfer RNA, 45
Transformation series, 301–302
Transgressions, Paleozoic, 321
Transient polymorphism, 159
Transilience. *See* Peak shift
Transition mutation, 65–66
Translation, 45
Transposable elements, 70–71, 457–459
 fitness and, 462
 multiplication of, 458–459
 mutation and, 68–71
Transposition, 462–465
Transposons. *See* Transposable elements
Transversion mutation, 65–66
Treffer's mutator, 259
Trend, as term, 366. *See also* Evolutionary trends
Triassic period, 320–321, 334
*Tribolium,* 159, 203–204, 263–264, 266–267
*Triceratops,* 336
*Trifolium repens,* 491
Trilobites, 325–326, 333, 351
*Triops cancriformis,* 398
tRNA. *See* Transfer RNA
Typological thinking. *See* Essentialism
Typology, 107–109. *See also* Systematics; Taxonomic classification

Unequal crossing over, 67, 69, 451–456, 465–469

Ungulates, 340–341
Uniformitarianism, 4
Utilization function, 485–486

Variance, 542–544
Vertebrates. *See also* specific species
  chronological distribution of, 330–331
  extinct species of, 328
  interspecific allometry and, 414
  origination rates and, 357
Viability, 74
  variation in, 190–191, 206
Vicariance, 378–381, 385–387
Vicariance biogeography, 385–387
Virus, gene transfer and, 69, 78–79
von Baer's law, 303–305

Wagner method, 300–301
Wahlund effect, 138–139
Wallace, A. R., 6, 374, 376
Water ouzel. *See Cinclus mexicanus*
Watson, J. D., 13
Weismann, A., 9
Wild oat. *See Avena fatua*
Wild types, 92, 94–96, 211–212, 420–421
Wright, S., 10–12

*Xenopus,* 464

*Zarhynchus wagleri. See* Oropendola
*Zea mays,* 73, 141, 230
Zones of intergradation, 104

ABOUT THE BOOK

This book was set in Linotron 212 Meridien at DEKR Corporation, and manufactured at the Murray Printing Company. The book was edited by Andrew N. Davis. Joseph Vesely created the format and supervised production.